ASTM STANDARDS ON GROUND WATER AND VADOSE ZONE INVESTIGATIONS

Sponsored by ASTM Committee D-18
on Soil and Rock

Second Edition
1994

ASTM Publication Code Number (PCN): 03-418094-38

ASTM
1916 Race St., Philadelphia, PA 19103

Library of Congress Cataloging-in-Publication Data

ASTM standards on ground water and vadose zone investigations /
sponsored by ASTM Committee D-18 on Soil and Rock.—2nd ed.
Includes index.
ISBN 0-8031-1786-8
1. Groundwater flow—Measurements—Standards. 2. Aquifers—
Measurement—Standards. I. ASTM Committee D-18 on Soil and Rock.
GB1197.7.A78 1994
551.49′0287—dc20 93-39811
CIP

Library of Congress Catalog Card Number: 93-39811

ISBN 0-8031-1786-8

Printed in Ann Arbor, MI
January 1994

FOREWORD

This compilation is sponsored by ASTM Subcommittee D18.21 on Ground Water and Vadose Zone Investigations. It contains 46 ASTM standards and a listing of related standards and draft documents under development by Committee D18. This second edition contains 28 more standards than the first edition.

Suggestions and comments to enhance the compilation's usefulness are welcome and should be addressed to Product Manager, Publications, ASTM, 1916 Race Street, Philadelphia, PA 19103.

ASTM

ASTM, founded in 1898, is a developer and publisher of technical information designed to promote the understanding and development of technology and to ensure the quality of commodities and services and safety of products.

ASTM's primary mission is to develop voluntary full consensus standards for materials, products, systems, and services. It provides a forum for producers, users, ultimate consumers, and those having a general interest (representatives of government and academia) to meet on common ground to write standards that best meet their needs.

COMMITTEE D-18

Committee D-18 on Soil and Rock is one of 131 technical committees within ASTM. Its subcommittee D18.21 on Ground Water and Vadose Zone Investigations comprises sections on vadose zone monitoring; well drilling and soil sampling procedures; determination of hydrogeological parameters; monitoring well design and construction; well maintenance, rehabilitation and abandonment; ground water sample collection and handling; network design and data interpretation; special problems of monitoring in karst terrains; and ground water modeling. Its subcommittee D18.01 on Surface and Subsurface Characterization comprises a section on surface and borehole geophysics.

For additional information on ASTM technical committee activities and membership, see the application at the end of this book, or contact: ASTM, Member Services, 1916 Race Street, Philadelphia, PA 19103.

RELATED PUBLICATIONS

ASTM has issued several publications that may be of interest to users of this compilation of standards. These are listed as follows.

Special Technical Publications (STPs):

STP 1118 Current Practices in Ground Water and Vadose Zone Investigations (1992)
STP 1053 Ground Water and Vadose Zone Monitoring (1992)
STP 963 Ground Water Contamination: Field Methods (1988)

Compilations of ASTM Standards:

ASTM Standards on Soil Compaction, Second Edition (1993)
ASTM Standards on Soil Stabilization with Admixtures, Second Edition (1992)

For further information on these and other ASTM publications, contact ASTM Customer Service, 1916 Race Street, Philadelphia, PA 19103 or call (215) 299-5585. A publications catalogue is available upon request.

CONTENTS

In the serial designations prefixed to the following titles, the number following the dash indicates the year of original issue or, in the case of revision, the year of last revision. Thus, standards adopted or revised during the year 1994 have as their final number 94. A letter following this number indicates more than one revision during that year, that is 94a indicates the second revision in 1994, 94b the third revision, etc. Standards that have been reapproved without change are indicated by the year of last reapproval in parentheses as part of the designation number, for example, (1994). A superscript epsilon (ϵ) indicates an editorial change since the last revision or reapproval—$\epsilon 1$ for the first change, $\epsilon 2$ for the second change, etc.

D 5269 – 92	Test Method for Determining Transmissivity of Nonleaky Confined Aquifers by the Theis Recovery Method
D 5270 – 92	Test Method for Determining Transmissivity and Storage Coefficient of Bounded, Nonleaky, Confined Aquifers
D 5299 – 92	Guide for Decommissioning of Ground Water Wells, Vadose Zone Monitoring Devices, Boreholes, and Other Devices for Environmental Activities
D 5314 – 93	Guide for Soil Gas Monitoring in the Vadose Zone
D 5408 – 93	Guide for Set of Data Elements to Describe a Ground-Water Site; Part One—Additional Identification Descriptors
D 5409 – 93	Guide for Set of Data Elements to Describe a Ground-Water Site; Part Two—Physical Descriptors
D 5410 – 93	Guide for Set of Data Elements to Describe a Ground-Water Site; Part Three—Usage Descriptors
F 480 – 93	Specification for Thermoplastic Well Casing Pipe and Couplings Made in Standard Dimension Ratios (SDR), SCH 40 and SCH 80

RELATED MATERIAL

STANDARD ACTIVITIES RELATED TO GROUND WATER AND VADOSE ZONE INVESTIGATIONS OF QUANTITY, QUALITY, AND CONTAMINATION

All standards are available from ASTM individually or may be found in Volumes 4.08 and 4.09 of the *Annual Book of ASTM Standards*. Draft documents will be published as they become ASTM standards.

Subcommittee D18.01 on Surface and Subsurface Characterization

D 420	Guide to Site Characterization for Engineering, Design and Construction Purposes
Draft	Guide for Preparing Hydrologic and Geologic Reports
Draft	Guide for Planning Ground Water Projects
Draft	Guide for Site Characterization for Environmental Purposes
Draft	Guide for Planning and Conducting Borehole Geophysical Investigations
Draft	Guide for Site Characterization for Construction Purposes
Draft	Guide for Using the Seismic Refraction Method for Subsurface Investigation
Draft	Guide for Reporting Geostatistical Site Investigations
Draft	Guide for Preparing Hydrologic and Geologic Investigation Reports
Draft	Guide for Selection of Geostatistical Methods for Ground Water Investigations
Draft	Guide for Reporting of Geostatistical Studies
Draft	Specification for Metadata Support in Geographic Information Systems and Spatial Data Exchange
Draft	Guidelines for Surface Resistivity
Draft	Guidelines for Electromagnetic Induction
Draft	Guidelines for Ground Penetrating Radar
Draft	Guide for the Graphical Display of the Chemistry of Natural Ground Water—Water-Analysis Diagram
Draft	Guide to the Selection of Purging and Sampling Devices for Ground-Water Monitoring Wells

Subcommittee D18.02 on Sampling and Related Field Testing for Soil Investigations

D 1452	Practice for Soil Investigation and Sampling by Auger Borings
D 1586	Test Method for Penetration Test and Split-Barrel Sampling of Soils
D 1587	Practice for Thin-Walled Tube Sampling of Soils
D 2113	Practice for Diamond Core Drilling for Site Investigation
D 3550	Practice for Ring-Lined Barrel Sampling of Soils
D 4220	Practices for Preserving and Transporting Soil Samples

Subcommittee D18.03 on Texture, Plasticity and Density Characteristics of Soils

D 421	Practice for Dry Preparation of Soil Samples for Particle-Size Analysis and Determination of Soil Constants
D 422	Method for Particle-Size Analysis of Soils
D 425	Test Method for Centrifuge Moisture Equivalent of Soils
D 854	Test Method for Specific Gravity of Soils
D 2216	Test Method for Laboratory Determination of Water (Moisture) Content of Soil and Rock
D 4718	Practice for Correction of Unit Weight and Water Content for Soils Containing Oversize Particles

Subcommittee D18.04 on Hydrologic Properties of Soil and Rocks

D 2325	Test Method for Capillary-Moisture Relationships for Coarse- and Medium-Textured Soils by Porous-Plate Apparatus
D 2434	Test Method for Permeability of Granular Soils (Constant Head)
D 3152	Test Method for Capillary-Moisture Relationships for Fine-Textured Soils by Pressure-Membrane Apparatus
D 3385	Test Method for Infiltration Rate of Soils in Field Using Double-Ring Infiltrometers
D 5084	Test Method for Hydraulic Conductivity of Saturated Porous Materials Using a Flexible Wall Permeameter
D 5093	Test Method for Field Measurement of Infiltration Rate Using a Double-Ring Infiltrometer with a Sealed-Inner Ring
D 5298	Test Method for Measurement of Soil Potential (Suction) Using Filter Paper
Draft	Test Method for Measurement of Hydraulic Conductivity of Porous Material Using a Rigid-Wall, Compaction-Mold Permeameter
Draft	Test Method for Field Measurement of Hydraulic Conductivity Limits of Porous Materials Using the Two-Stage Borehole Procedure
Draft	Test Method for Determination of Matric Potential (Suction)-Water Content Relationship by Pressure Plate and Pressure Membrane Extraction

Subcommittee D18.06 on Physico-Chemical Properties of Soils and Rocks

D 4404	Test Method for Determination of Pore Volume and Pore Volume Distribution of Soil and Rock by Mercury Intrusion Porosimetry
D 4972	Test Method for pH of Soils

Subcommittee D18.07 on Identification and Classification of Soils

D 2488	Practice for Description and Identification of Soils (Visual-Manual Procedure)
D 5434	Guide for Field Logging of Subsurface Explorations of Soil and Rock

Subcommittee D18.08 on Special and Construction Control Tests

D 2937	Test Method for Density of Soil in Place by the Drive-Cylinder Method
D 4564	Test Method for Density of Soil in Place by the Sleeve Method
D 5195	Test Method for Density of Soil and Rock In-Place at Depths Below the Surface by Nuclear Methods
D 5220	Test Method for Water Content of Soil and Rock In-Place by the Neutron Depth Probe Method
Draft	Specification for Nuclear Soil Density Gage Calibration Standards

Subcommittee D18.12 on Rock Mechanics

D 4525	Test Method for Permeability of Rocks by Flowing Air

Subcommittee D18.13 on Marine and Coastal Geotechnics

D 4542	Test Method for Pore Water Extraction and Determination of the Soluble Salt Content of Soils by Refractomer

Subcommittee D18.21 on Ground Water and Vadose Zone Investigations

D 3404	Guide to Measuring Matric Potential in the Vadose Zone Using Tensiometers
D 4043	Guide for Selection of Aquifer-Test Field and Analytical Procedures in Determination of Hydraulic Properties by Well Techniques
D 4044	Test Method (Field Procedure) for Instantaneous Change in Head (Slug Tests) for Determining Hydraulic Properties of Aquifers
D 4050	Test Method (Field Procedure) for Withdrawal and Injection Well Tests for Determining Hydraulic Properties of Aquifer Systems
D 4104	Test Method (Analytical Procedure) for Determining Transmissivity of Nonleaky Confined Aquifers by Overdamped Well Response to Instantaneous Change in Head (Slug Test)
D 4105	Test Method (Analytical Procedure) for Determining Transmissivity and Storage Coefficient of Nonleaky Confined Aquifers by the Modified Theis Nonequilibrium Method
D 4106	Test Method (Analytical Procedure) for Determining Transmissivity and Storage Coefficient of Nonleaky Confined Aquifers by the Theis Nonequilibrium Method

D 4630 Test Method for Determining Transmissivity and Storativity of Low-Permeability Rocks by In Situ Measurements Using the Constant Head Injection Test

D 4631 Test Method for Determining Transmissivity and Storativity of Low Permeability Rocks by In Situ Measurements Using the Pressure Pulse Technique

D 4696 Guide for Pore-Liquid Sampling from the Vadose Zone

D 4700 Guide for Soil Sampling from the Vadose Zone

D 4750 Test Method for Determining Subsurface Liquid Levels in a Borehole or Monitoring Well (Observation Well)

D 5092 Practice for Design and Installation of Ground Water Monitoring Wells in Aquifers

D 5126 Guide for Comparison of Field Methods for Determining Hydraulic Conductivity in the Vadose Zone

D 5254 Practice for the Minimum Set of Data Elements to Identify a Ground-Water Site

D 5269 Test Method for Determining Transmissivity of Nonleaky Confined Aquifers by the Theis Recovery Method

D 5270 Test Method for Determining Transmissivity and Storage Coefficient of Bounded, Nonleaky, Confined Aquifers

D 5299 Guide for Decommissioning of Ground Water Wells, Vadose Zone Monitoring Devices, Boreholes, and Other Devices for Environmental Activities

D 5314 Guide for Soil Gas Monitoring in the Vadose Zone

D 5408 Guide for the Set of Data Elements to Describe a Ground-Water Site; Part 1—Additional Identification Descriptors

D 5409 Guide for the Set of Data Elements to Describe a Ground-Water Site; Part 2—Physical Descriptors

D 5410 Guide for the Set of Data Elements to Describe a Ground-Water Site; Part 3—Usage Descriptors

D 5447 Guide for Application of a Ground-Water Flow Model to a Site-Specific Problem

D 5472 Test Method for Determining Specific Capacity and Estimating Transmissivity at the Control Well

D 5473 Test Method (Analytical Procedure) for Analyzing the Effects of Partial Penetration of Control Well and Determining the Horizontal and Vertical Hydraulic Conductivity in a Nonleaky Confined Aquifer

D 5474 Guide for Selection of Data Elements for Ground-Water Investigations

Draft Guide for Review of an Application of Statistics to Environmental Data

Draft Guide for the Use of Dual Wall Reverse Circulation Drilling for Geoenvironmental Exploration and the Installation of Subsurface Water Quality Monitoring Devices

Draft Guide for the Use of Direct Air Rotary Drilling for Geoenvironmental Exploration and the Installation of Subsurface Water Quality Monitoring Devices

Draft Guide for the Use of Direct Rotary Drilling with Water-Based Drilling Fluid for Geoenvironmental Exploration and the Installation of Subsurface Water Quality Monitoring Devices

Draft Guide for the Use of Hollow-Stem Augers for Geoenvironmental Exploration and the Installation of Subsurface Water Quality Monitoring Devices

Draft Test Method (Field Procedure) for Constant Drawdown Tests in Flowing Wells for Determining Hydraulic Properties of Aquifer Systems

Draft Test Method (Analytical Procedures) for Determining Transmissivity and Storage Coefficient of Aquifers Using the Constant Head Method

Draft Guide for Measuring the Rate of Discharge of a Control Well

Draft Test Method to Measure the Rate of Well Discharge by Circular Orifice Wire (Orifice Tube)

Draft Development of Ground-Water Monitoring Wells in Aquifers

Draft Design of Filter Pack Grain Size Monitoring Well Screen Slot Sizes in Aquifers

Draft Guide for Maintenance and Rehabilitation of Ground Water Monitoring Wells

Draft Designing and Constructing Screened Ground-Water Production Wells

Draft Development of Ground Water Monitoring Wells in Aquifers

Draft Guide for Review of an Application of Statistics to Environmental Data

Draft Guide for the Location of Abandoned Wells

Draft Guide for the Design of Ground-Water Monitoring Systems in Karst and Fractured-Rock Aquifers

Draft Guide for Review of Statistical Applications

Draft Practice for Use of Aquifer Identification Code

Draft Guide for Presentation of Ground Water Level Data

Draft Test Method (Analytical Procedure) for Tests of Anisotropic Unconfined Aquifers by the Neuman Method

Draft Test Method (Analytical Procedure) for Determining Transmissivity and Storage Coefficient of a Leaky Confined Aquifer with No Storage in the Confining Bed

Draft Guide for Defining Boundary Conditions in Ground-Water Modeling

Draft Guide for Defining Initial Conditions in Ground-Water Modeling

Draft Guide for Application of the Principle of Superposition in Ground-Water Modeling

Draft	Guide for Describing a Ground-Water Model Code
Draft	Guide for Developing a Ground-Water Model Code
Draft	Classification of Ground-Water Model Applications
Draft	Guide for Developing a Conceptual Ground-Water Flow Model
Draft	Guide for Documenting a Ground-Water Model
Draft	Guide for Graphical Representation of Water Quality Data; Single Sample Graphics

Subcommittee D18.93 on Terminology Relating to Soil, Rock, and Contained Fluids

D 653	Terminology Relating to Soil, Rock, and Contained Fluids
D 3584	Practice for Indexing Papers and Reports on Soil and Rock for Engineering Purposes

Designation: C 136 – 92

Standard Test Method for Sieve Analysis of Fine and Coarse Aggregates[1]

This standard is issued under the fixed designation C 136; the number immediately following the designation indicates the year of original adoption or, in the case of revision, the year of last revision. A number in parentheses indicates the year of last reapproval. A superscript epsilon (ε) indicates an editorial change since the last revision or reapproval.

This standard has been approved for use by agencies of the Department of Defense. Consult the DoD Index of Specifications and Standards for the specific year of issue which has been adopted by the Department of Defense.

1. Scope

1.1 This test method covers the determination of the particle size distribution of fine and coarse aggregates by sieving.

1.2 Some specifications for aggregates which reference this method contain grading requirements including both coarse and fine fractions. Instructions are included for sieve analysis of such aggregates.

1.3 The values stated in acceptable metric units (SI units and units specifically approved in ASTM E 380 for use with SI units) are to be regarded as the standard. The values in parentheses are provided for information purposes only.

1.4 *This standard does not purport to address all of the safety problems, if any, associated with its use. It is the responsibility of the user of this standard to establish appropriate safety and health practices and determine the applicability of regulatory limitations prior to use.*

2. Referenced Documents

2.1 *ASTM Standards:*

C 117 Test Method for Materials Finer Than 75-µm (No. 200) Sieve in Mineral Aggregates by Washing[2]

C 125 Terminology Relating to Concrete and Concrete Aggregates[3]

C 670 Practice for Preparing Precision and Bias Statements for Test Methods for Construction Materials[3]

C 702 Practice for Reducing Field Samples of Aggregate to Testing Size[3]

D 75 Practice for Sampling Aggregates[2]

E 11 Specification for Wire-Cloth Sieves for Testing Purposes[4]

E 380 Practice For Use of the International System of Units (SI) (the Modernized Metric System)[5]

2.2 *AASHTO Standard:*

AASHTO No. T 27 Sieve Analysis of Fine and Coarse Aggregates[6]

3. Terminology

3.1 *Definitions*—For definitions of terms used in this standard, refer to Terminology C 125.

4. Summary of Test Method

4.1 A weighed sample of dry aggregate is separated through a series of sieves of progressively smaller openings for determination of particle size distribution.

5. Significance and Use

5.1 This test method is used primarily to determine the grading of materials proposed for use as aggregates or being used as aggregates. The results are used to determine compliance of the particle size distribution with applicable specification requirements and to provide necessary data for control of the production of various aggregate products and mixtures containing aggregates. The data may also be useful in developing relationships concerning porosity and packing.

5.2 Accurate determination of material finer than the 75-µm (No. 200) sieve cannot be achieved by use of this method alone. Test Method C 117 for material finer than 75-µm sieve by washing should be employed.

6. Apparatus

6.1 *Balances*—Balances or scales used in testing fine and coarse aggregate shall have readability and accuracy as follows:

6.1.1 For fine aggregate, readable to 0.1 g and accurate to 0.1 g or 0.1 % of the test load, whichever is greater, at any point within the range of use.

6.1.2 For coarse aggregate, or mixtures of fine and coarse aggregate, readable and accurate to 0.5 g or 0.1 % of the test load, whichever is greater, at any point within the range of use.

6.2 *Sieves*—The sieves shall be mounted on substantial frames constructed in a manner that will prevent loss of material during sieving. The sieves shall conform to Specification E 11. Sieves with openings larger than 125 mm (5 in.) shall have a permissible variation in average opening of ±2 % and shall have a nominal wire diameter of 8.0 mm (5/16 in.) or larger.

NOTE 1—It is recommended that sieves mounted in frames larger than standard 203-mm (8 in.) diameter frames be used for testing coarse aggregate.

6.3 *Mechanical Sieve Shaker*—A mechanical sieve shaker, if used, shall impart a vertical, or lateral and vertical, motion to the sieve, causing the particles thereon to bounce and turn so as to present different orientations to the sieving

---

[1] This test method is under the jurisdiction of ASTM Committee C-9 on Concrete and Concrete Aggregates and is the direct responsibility of Subcommittee C09.20 on Normal Weight Aggregates.

Current edition approved Nov. 15, 1992. Published January 1993. Originally published as C 136 – 38 T. Last previous edition C 136 – 84a.

[2] *Annual Book of ASTM Standards*, Vols 04.02 and 04.03.

[3] *Annual Book of ASTM Standards*, Vol 04.02.

[4] *Annual Book of ASTM Standards*, Vol 14.02.

[5] *Annual Book of ASTM Standards*, Vol 14.02. Excerpts in all volumes.

[6] Available from American Association of State Highway and Transportation Officials, 444 North Capitol St. N.W., Suite 225, Washington, DC 20001.

surface. The sieving action shall be such that the criterion for adequacy of sieving described in 8.4 is met in a reasonable time period.

NOTE 2—Use of a mechanical sieve shaker is recommended when the size of the sample is 20 kg or greater, and may be used for smaller samples, including fine aggregate. Excessive time (more than approximately 10 min) to achieve adequate sieving may result in degradation of the sample. The same mechanical sieve shaker may not be practical for all sizes of samples, since the large sieving area needed for practical sieving of a large nominal size coarse aggregate very likely could result in loss of a portion of the sample if used for a small sample of coarse aggregate or fine aggregate.

6.4 *Oven*—An oven of appropriate size capable of maintaining a uniform temperature of 110 ± 5°C (230 ± 9°F).

7. Sampling

7.1 Sample the aggregate in accordance with Practice D 75. The weight of the field sample shall be the weight shown in Practice D 75 or four times the weight required in 7.4 and 7.5 (except as modified in 7.6), whichever is greater.

7.2 Thoroughly mix the sample and reduce it to an amount suitable for testing using the applicable procedures described in Practice C 702. The sample for test shall be approximately of the weight desired when dry and shall be the end result of the reduction. Reduction to an exact predetermined weight shall not be permitted.

NOTE 3—Where sieve analysis, including determination of material finer than the 75-µm sieve, is the only testing proposed, the size of the sample may be reduced in the field to avoid shipping excessive quantities of extra material to the laboratory.

7.3 *Fine Aggregate*—The test sample of fine aggregate shall weigh, after drying, approximately the following amount:

Aggregate with at least 95 % passing a 2.36-mm (No. 8) sieve	100 g
Aggregate with at least 85 % passing a 4.75-mm (No. 4) sieve and more than 5 % retained on a 2.36-mm (No. 8) sieve	500 g

7.4 *Coarse Aggregate*—The weight of the test sample of coarse aggregate shall conform with the following:

Nominal Maximum Size, Square Openings, mm (in.)	Minimum Weight of Test Sample, kg (lb)
9.5 (3/8)	1 (2)
12.5 (1/2)	2 (4)
19.0 (3/4)	5 (11)
25.0 (1)	10 (22)
37.5 (1 1/2)	15 (33)
50 (2)	20 (44)
63 (2 1/2)	35 (77)
75 (3)	60 (130)
90 (3 1/2)	100 (220)
100 (4)	150 (330)
112 (4 1/2)	200 (440)
125 (5)	300 (660)
150 (6)	500 (1100)

7.5 *Coarse and Fine Aggregate Mixtures*—The weight of the test sample of coarse and fine aggregate mixtures shall be the same as for coarse aggregate in 7.4.

7.6 The size of sample required for aggregates with large nominal maximum size is such as to preclude testing except with large mechanical sieve shakers. However, the intent of this method will be satisfied for samples of aggregate larger than 50 mm nominal maximum size if a smaller weight of sample is used, provided that the criterion for acceptance or rejection of the material is based on the average of results of several samples, such that the sample size used times the number of samples averaged equals the minimum weight of sample shown in 7.4.

7.7 In the event that the amount of material finer than the 75-µm (No. 200) sieve is to be determined by Test Method C 117, proceed as follows:

7.7.1 For aggregates with a nominal maximum size of 12.5 mm (1/2 in.) or less, use the same test sample for testing by Test Method C 117 and this method. First test the sample in accordance with Test Method C 117 through the final drying operation, then dry sieve the sample as stipulated in 8.2 through 8.7 of this method.

7.7.2 For aggregates with a nominal maximum size greater than 12.5 mm (1/2 in.), a single test sample may be used as described in 7.7.1, or separate test samples may be used for Test Method C 117 and this method.

7.7.3 Where the specifications require determination of the total amount of material finer than the 75-µm sieve by washing and dry sieving, use the procedure described in 7.7.1.

8. Procedure

8.1 Dry the sample to constant weight at a temperature of 110 ± 5°C (230 ± 9°F).

NOTE 4—For control purposes, particularly where rapid results are desired, it is generally not necessary to dry coarse aggregate for the sieve analysis test. The results are little affected by the moisture content unless: (*1*) the nominal maximum size is smaller than about 12.5 mm (1/2 in.); (*2*) the coarse aggregate contains appreciable material finer than 4.75 mm (No. 4); or (*3*) the coarse aggregate is highly absorptive (a lightweight aggregate, for example). Also, samples may be dried at the higher temperatures associated with the use of hot plates without affecting results, provided steam escapes without generating pressures sufficient to fracture the particles, and temperatures are not so great as to cause chemical breakdown of the aggregate.

8.2 Suitable sieve sizes shall be selected to furnish the information required by the specifications covering the material to be tested. The use of additional sieves may be desirable to provide other information, such as fineness modulus, or to regulate the amount of material on a sieve. Nest the sieves in order of decreasing size of opening from top to bottom and place the sample on the top sieve. Agitate the sieves by hand or by mechanical apparatus for a sufficient period, established by trial or checked by measurement on the actual test sample, to meet the criterion for adequacy or sieving described in 8.4.

8.3 Limit the quantity of material on a given sieve so that all particles have opportunity to reach sieve openings a number of times during the sieving operation. For sieves with openings smaller than 4.75-mm (No. 4), the weight retained on any sieve at the completion of the sieving operation shall not exceed 6 kg/m^2 (4 g/in.^{2}) of sieving surface. For sieves with openings 4.75 mm (No. 4) and larger, the weight in kg/m^2 of sieving surface shall not exceed the product of 2.5 × (sieve opening in mm). In no case shall the weight be so great as to cause permanent deformation of the sieve cloth.

NOTE 5—The 6 kg/m^2 amounts to 194 g for the usual 203-mm (8 in.) diameter sieve. The amount of material retained on a sieve may be regulated by (*1*) the introduction of a sieve with larger openings immediately above the given sieve or (*2*) testing the sample in a number of increments.

8.4 Continue sieving for a sufficient period and in such

manner that, after completion, not more than 1 weight % of the residue on any individual sieve will pass that sieve during 1 min of continuous hand sieving performed as follows: Hold the individual sieve, provided with a snug-fitting pan and cover, in a slightly inclined position in one hand. Strike the side of the sieve sharply and with an upward motion against the heel of the other hand at the rate of about 150 times per minute, turn the sieve about one sixth of a revolution at intervals of about 25 strokes. In determining sufficiency of sieving for sizes larger than the 4.75-mm (No. 4) sieve, limit the material on the sieve to a single layer of particles. If the size of the mounted testing sieves makes the described sieving motion impractical, use 203-mm (8 in.) diameter sieves to verify the sufficiency of sieving.

8.5 In the case of coarse and fine aggregate mixtures, the portion of the sample finer than the 4.75-mm (No. 4) sieve may be distributed among two or more sets of sieves to prevent overloading of individual sieves.

8.5.1 Alternatively, the portion finer than the 4.75-mm (No. 4) sieve may be reduced in size using a mechanical splitter according to Practice C 702. If this procedure is followed, compute the weight of each size increment of the original sample as follows:

$$A = \frac{W_1}{W_2} \times B$$

where:

A = weight of size increment on total sample basis,
W_1 = weight of fraction finer than 4.75-mm (No. 4) sieve in total sample,
W_2 = weight of reduced portion of material finer than 4.75-mm (No. 4) sieve actually sieved, and
B = weight of size increment in reduced portion sieved.

8.6 Unless a mechanical sieve shaker is used, hand sieve particles larger than 75 mm (3 in.) by determining the smallest sieve opening through which each particle will pass. Start the test on the smallest sieve to be used. Rotate the particles, if necessary, in order to determine whether they will pass through a particular opening; however, do not force particles to pass through an opening.

8.7 Determine the weight of each size increment by weighing on a scale or balance conforming to the requirements specified in 5.1 to the nearest 0.1 % of the total original dry sample weight. The total weight of the material after sieving should check closely with original weight of sample placed on the sieves. If the amounts differ by more than 0.3 %, based on the original dry sample weight, the results should not be used for acceptance purposes.

8.8 If the sample has previously been tested by Test Method C 117, add the weight finer than the 75-µm (No. 200) sieve determined by that method to the weight passing the 75-µm (No. 200) sieve by dry sieving of the same sample in this method.

9. Calculation

9.1 Calculate percentages passing, total percentages retained, or percentages in various size fractions to the nearest 0.1 % on the basis of the total weight of the initial dry sample. If the same test sample was first tested by Test Method C 117, include the weight of material finer than the 75-µm (No. 200) size by washing in the sieve analysis calculation; and use the total dry sample weight prior to washing in Test Method C 117 as the basis for calculating all the percentages.

9.2 Calculate the fineness modulus, when required, by adding the total percentages of material in the sample that is coarser than each of the following sieves (cumulative percentages retained), and dividing the sum by 100: 150-µm (No. 100), 300-µm (No. 50), 600-µm (No. 30), 1.18-mm (No. 16), 2.36-mm (No. 8), 4.75-mm (No. 4), 9.5-mm (3/8-in.), 19.0-mm (3/4-in.), 37.5-mm (1½-in.), and larger, increasing in the ratio of 2 to 1.

10. Report

10.1 Depending upon the form of the specifications for use of the material under test, the report shall include the following:

10.1.1 Total percentage of material passing each sieve, or

10.1.2 Total percentage of material retained on each sieve, or

10.1.3 Percentage of material retained between consecutive sieves.

10.2 Report percentages to the nearest whole number, except if the percentage passing the 75-µm (No. 200) sieve is less than 10 %, it shall be reported to the nearest 0.1 %.

10.3 Report the fineness modulus, when required, to the nearest 0.01.

11. Precision

11.1 The estimates of precision of this method listed in Table 1 are based on results from the AASHTO Materials Reference Laboratory Reference Sample Program, with testing conducted by this method and AASHTO Method T 27. While there are differences in the minimum weight of the test sample required for other nominal maximum sizes of aggregate, no differences entered into the testing to affect the determination of these precision indices. The data are based on the analyses of more than 100 paired test results from 40 to 100 laboratories. The values in the table are given for different ranges of percentage of aggregate passing one sieve and retained on the next finer sieve.

12. Keywords

12.1 aggregate; coarse aggregate; fine aggregate; gradation; grading; sieve analysis; size analysis

TABLE 1 Precision

	% of Size Fraction Between Consecutive Sieves	Coefficient of Variation (1S %), %[B]	Standard Deviation (1S), %[A]	Acceptable Range of Test Results	
				(D2S %)[B] % of Avg.	(D2S),[A] %
Coarse Aggregates:[C]					
Single-Operator Precision	0 to 3	30[D]	...	85[D]	...
	3 to 10		1.4[D]		4.0[D]
	10 to 20		0.95		2.7
	20 to 50		1.38		3.9
Multilaboratory Precision	0 to 3	35[D]	...	99[D]	...
	3 to 10		1.06		3.0
	10 to 20		1.66		4.7
	20 to 30		2.01		5.7
	30 to 40		2.44		6.9
	40 to 50		3.18		9.0
Fine Aggregates:					
Single-Operator Precision	0 to 3		0.14		0.4
	3 to 10		0.43		1.2
	10 to 20		0.60		1.7
	20 to 30		0.64		1.8
	30 to 40		0.71		2.0
	40 to 50		...		...
Multilaboratory Precision	0 to 3		0.21		0.6
	3 to 10		0.57		1.6
	10 to 20		0.95		2.7
	20 to 30		1.24		3.5
	30 to 40		1.41		4.0
	40 to 50		...		...

[A] These numbers represent, respectively, the (1s) and (d2s) limits as described in Practice C 670.
[B] These numbers represent, respectively, the (1s %) and (d2s %) limits as described in Practice C 670.
[C] The precision estimates are based on coarse aggregates with nominal maximum size of 19.0 mm (¾ in.).
[D] These values are from precision indices first included in Method C 136 – 77. Other indices were developed in 1982 from more recent AASHTO Materials Reference Laboratory sample data, which did not provide sufficient information to revise the values so noted.

The American Society for Testing and Materials takes no position respecting the validity of any patent rights asserted in connection with any item mentioned in this standard. Users of this standard are expressly advised that determination of the validity of any such patent rights, and the risk of infringement of such rights, are entirely their own responsibility.

This standard is subject to revision at any time by the responsible technical committee and must be reviewed every five years and if not revised, either reapproved or withdrawn. Your comments are invited either for revision of this standard or for additional standards and should be addressed to ASTM Headquarters. Your comments will receive careful consideration at a meeting of the responsible technical committee, which you may attend. If you feel that your comments have not received a fair hearing you should make your views known to the ASTM Committee on Standards, 1916 Race St., Philadelphia, PA 19103.

Designation: C 150 – 92

Standard Specification for Portland Cement[1]

This standard is issued under the fixed designation C 150; the number immediately following the designation indicates the year of original adoption or, in the case of revision, the year of last revision. A number in parentheses indicates the year of last reapproval. A superscript epsilon (ε) indicates an editorial change since the last revision or reapproval.

This standard has been approved for use by agencies of the Department of Defense. Consult the DoD Index of Specifications and Standards for the specific year of issue which has been adopted by the Department of Defense.

1. Scope

1.1 This specification covers eight types of portland cement, as follows (see Note):

1.1.1 *Type I*—For use when the special properties specified for any other type are not required.

1.1.2 *Type IA*—Air-entraining cement for the same uses as Type I, where air-entrainment is desired.

1.1.3 *Type II*—For general use, more especially when moderate sulfate resistance or moderate heat of hydration is desired.

1.1.4 *Type IIA*—Air-entraining cement for the same uses as Type II, where air-entrainment is desired.

1.1.5 *Type III*—For use when high early strength is desired.

1.1.6 *Type IIIA*—Air-entraining cement for the same use as Type III, where air-entrainment is desired.

1.1.7 *Type IV*—For use when a low heat of hydration is desired.

1.1.8 *Type V*—For use when high sulfate resistance is desired.

1.2 The values stated in inch-pound units are to be regarded as the standard.

2. Referenced Documents

2.1 *ASTM Standards:*

C 33 Specification for Concrete Aggregates[2]

C 109 Test Method for Compressive Strength of Hydraulic Cement Mortars (Using 2-in. or 50-mm Cube Specimens)[3]

C 114 Test Methods for Chemical Analysis of Hydraulic Cement[3]

C 115 Test Method for Fineness of Portland Cement by the Turbidimeter[3]

C 151 Test Method for Autoclave Expansion of Portland Cement[3]

C 183 Practice for Sampling and the Amount of Testing of Hydraulic Cement[3]

C 185 Test Method for Air Content of Hydraulic Cement Mortar[3]

C 186 Test Method for Heat of Hydration of Hydraulic Cement[3]

C 191 Test Method for Time of Setting of Hydraulic Cement by Vicat Needle[3]

C 204 Test Method for Fineness of Portland Cement by Air Permeability Apparatus[3]

C 226 Specification for Air-Entraining Additions for Use in the Manufacture of Air-Entraining Portland Cement[3]

C 266 Test Method for Time of Setting of Hydraulic Cement Paste by Gillmore Needles[3]

C 451 Test Method for Early Stiffening of Portland Cement (Paste Method)[3]

C 452 Test Method for Potential Expansion of Portland Cement Mortars Exposed to Sulfate[3]

C 465 Specification for Processing Additions for Use in the Manufacture of Hydraulic Cements[3]

C 563 Test Method for Optimum SO_3 in Portland Cement[3]

C 1038 Test Method for Expansion of Portland Cement Mortar Bars Stored in Water[3]

3. Terminology

3.1 *Definitions:*

3.1.1 *portland cement*—a hydraulic cement produced by pulverizing clinker consisting essentially of hydraulic calcium silicates, usually containing one or more of the forms of calcium sulfate as an interground addition.

3.1.2 *air-entraining portland cement*—a hydraulic cement produced by pulverizing clinker consisting essentially of hydraulic calcium silicates, usually containing one or more of the forms of calcium sulfate as an interground addition, and with which there has been interground an air-entraining addition.

4. Ordering Information

4.1 Orders for material under this specification shall include the following:

4.1.1 This specification number and date,

4.1.2 Type or types allowable. If no type is specified, Type I shall be supplied,

4.1.3 Any optional chemical requirements from Table 2, if desired,

4.1.4 Type of setting-time test required, Vicat or Gillmore. If not specified, the Vicat shall be used,

4.1.5 Any optional physical requirements from Table 4, if desired.

NOTE—Attention is called to the fact that cements conforming to the requirements for all types may not be carried in stock in some areas. In advance of specifying the use of other than Type I cement, it should be determined whether the proposed type of cement is or can be made available.

---

[1] This specification is under the jurisdiction of ASTM Committee C-1 on Cement and is the direct responsibility of Subcommittee C01.10 on Portland Cement.

Current edition approved March 15, 1992. Published May 1992. Originally published as C 150 – 40 T. Last previous edition C 150 – 89.

[2] *Annual Book of ASTM Standards*, Vol 04.02.

[3] *Annual Book of ASTM Standards*, Vol 04.01.

TABLE 1 Standard Chemical Requirements

Cement Type[A]	I and IA	II and IIA	III and IIIA	IV	V
Silicon dioxide (SiO_2), min, %	...	20.0	...	...	...
Aluminum oxide (Al_2O_3), max, %	...	6.0	...	...	...
Ferric oxide (Fe_2O_3), max, %	...	6.0	...	6.5	...
Magnesium oxide (MgO), max, %	6.0	6.0	6.0	6.0	6.0
Sulfur trioxide (SO_3),[B] max, %					
When (C_3A)[C] is 8 % or less	3.0	3.0	3.5	2.3	2.3
When (C_3A)[C] is more than 8 %	3.5	[D]	4.5	[D]	[D]
Loss on ignition, max, %	3.0	3.0	3.0	2.5	3.0
Insoluble residue, max, %	0.75	0.75	0.75	0.75	0.75
Tricalcium silicate (C_3S)[C] max, %	...	...	...	35[E]	...
Dicalcium silicate (C_2S)[C] min, %	...	...	...	40[E]	...
Tricalcium aluminate (C_3A)[C] max, %	...	8	15	7[E]	5[F]
Tetracalcium aluminoferrite plus twice the tricalcium aluminate[C] ($C_4AF + 2(C_3A)$), or solid solution ($C_4AF + C_2F$), as applicable, max, %	...	...	...	...	25[F]

[A] See Note.

[B] There are cases where optimum SO_3 (using Test Method C 563) for a particular cement is close to or in excess of the limit in this specification. In such cases where properties of a cement can be improved by exceeding the SO_3 limits stated in this table, it is permissible to exceed the values in the table, provided it has been demonstrated by Test Method C 1038 that the cement with the increased SO_3 will not develop expansion in water exceeding 0.020 % at 14 days. When the manufacturer supplies cement under this provision, he shall, upon request, supply supporting data to the purchaser.

[C] The expressing of chemical limitations by means of calculated assumed compounds does not necessarily mean that the oxides are actually or entirely present as such compounds.

When expressing compounds, C = CaO, S = SiO_2, A = Al_2O_3, F = Fe_2O_3. For example, $C_3A = 3CaO \cdot Al_2O_3$.

Titanium dioxide and phosphorus pentoxide (TiO_2 and P_2O_5) shall be included with the Al_2O_3 content. The value historically and traditionally used for Al_2O_3 in calculating potential compounds for specification purposes is the ammonium hydroxide group minus ferric oxide ($R_2O_3 - Fe_2O_3$) as obtained by classical wet chemical methods. This procedure includes as Al_2O_3 the TiO_2, P_2O_5 and other trace oxides which precipitate with the ammonium hydroxide group in the classical wet chemical methods. Many modern instrumental methods of cement analysis determine aluminum or aluminum oxide directly without the minor and trace oxides included by the classical method. Consequently, for consistency and to provide comparability with historic data and among various analytical methods, when calculating potential compounds for specification purposes, those using methods which determine Al or Al_2O_3 directly should add to the determined Al_2O_3 weight quantities of P_2O_5, TiO_2 and any other oxide except Fe_2O_3 which would precipitate with the ammonium hydroxide group when analyzed by the classical method and which is present in an amount of 0.05 weight % or greater. The weight percent of minor or trace oxides to be added to Al_2O_3 by those using direct methods may be obtained by actual analysis of those oxides in the sample being tested or estimated from historical data on those oxides on cements from the same source, provided that the estimated values are identified as such.

When the ratio of percentages of aluminum oxide to ferric oxide is 0.64 or more, the percentages of tricalcium silicate, dicalcium silicate, tricalcium aluminate, and tetracalcium aluminoferrite shall be calculated from the chemical analysis as follows:

Tricalcium silicate = $(4.071 \times \%\ CaO) - (7.600 \times \%\ SiO_2) - (6.718 \times \%\ Al_2O_3) - (1.430 \times \%\ Fe_2O_3) - (2.852 \times \%\ SO_3)$

Dicalcium silicate = $(2.867 \times \%\ SiO_2) - (0.7544 \times \%\ C_3S)$

Tricalcium aluminate = $(2.650 \times \%\ Al_2O_3) - (1.692 \times \%\ Fe_2O_3)$

Tetracalcium aluminoferrite = $3.043 \times \%\ Fe_2O_3$

When the alumina-ferric oxide ratio is less than 0.64, a calcium aluminoferrite solid solution (expressed as $ss(C_4AF + C_2F)$) is formed. Contents of this solid solution and of tricalcium silicate shall be calculated by the following formulas:

$ss(C_4AF + C_2F) = (2.100 \times \%\ Al_2O_3) + (1.702 \times \%\ Fe_2O_3)$

Tricalcium silicate = $(4.071 \times \%\ CaO) - (7.600 \times \%\ SiO_2) - (4.479 \times \%\ Al_2O_3) - (2.859 \times \%\ Fe_2O_3) - (2.852 \times \%\ SO_3)$.

No tricalcium aluminate will be present in cements of this composition. Dicalcium silicate shall be calculated as previously shown.

In the calculation of all compounds the oxides determined to the nearest 0.1 % shall be used.

All values calculated as described in this note shall be reported to the nearest 1 %.

[D] Not applicable.

[E] Does not apply when the heat of hydration limit in Table 4 is specified.

[F] Does not apply when the sulfate resistance limit in Table 4 is specified.

5. Additions

5.1 The cement covered by this specification shall contain no addition except as follows:

5.1.1 Water or calcium sulfate, or both, may be added in amounts such that the limits shown in Table 1 for sulfur trioxide and loss-on-ignition shall not be exceeded.

5.1.2 At the option of the manufacturer, processing additions may be used in the manufacture of the cement, provided such materials in the amounts used have been shown to meet the requirements of Specification C 465.

5.1.3 Air-entraining portland cement shall contain an interground addition conforming to the requirements of Specification C 226.

6. Chemical Composition

6.1 Portland cement of each of the eight types shown in Section 1 shall conform to the respective standard chemical requirements prescribed in Table 1. In addition, optional chemical requirements are shown in Table 2.

7. Physical Properties

7.1 Portland cement of each of the eight types shown in Section 1 shall conform to the respective standard physical requirements prescribed in Table 3. In addition, optional physical requirements are shown in Table 4.

8. Sampling

8.1 When the purchaser desires that the cement be sampled and tested to verify compliance with this specification, sampling and testing should be performed in accordance with Practice C 183.

8.2 Practice C 183 is not designed for manufacturing quality control and are not required for manufacturer's certification.

9. Test Methods

9.1 Determine the applicable properties enumerated in this specification in accordance with the following test methods:

9.1.1 *Air Content of Mortar*—Test Method C 185.

TABLE 2 Optional Chemical Requirements[A]

Cement Type	I and IA	II and IIA	III and IIIA	IV	V	Remarks
Tricalcium aluminate (C_3A),[B] max, %	...	...	8	...	...	for moderate sulfate resistance
Tricalcium aluminate (C_3A),[B] max, %	...	...	5	...	...	for high sulfate resistance
Sum of tricalcium silicate and tricalcium aluminate,[B] max, %	...	58[C]	...	...	...	for moderate heat of hydration
Alkalies (Na_2O + $0.658K_2O$), max, %	0.60[D]	0.60[D]	0.60[D]	0.60[D]	0.60[D]	low-alkali cement

[A] These optional requirements apply only if specifically requested. Availability should be verified. See note in Section 4.

[B] The expressing of chemical limitations by means of calculated assumed compounds does not necessarily mean that the oxides are actually or entirely present as such compounds.

When expressing compounds, C = CaO, S = SiO_2, A = Al_2O_3, F = Fe_2O_3. For example, $C_3A = 3CaO \cdot Al_2O_3$.

Titanium dioxide and phosphorus pentoxide (TiO_2 and P_2O_5) shall be included with the Al_2O_3 content. The value historically and traditionally used for Al_2O_3 in calculating potential compounds for specification purposes is the ammonium hydroxide group minus ferric oxide ($R_2O_3 - Fe_2O_3$) as obtained by classical wet chemical methods. This procedure includes as Al_2O_3 the TiO_2, P_2O_5 and other trace oxides which precipitate with the ammonium hydroxide group in the classical wet chemical methods. Many modern instrumental methods of cement analysis determine aluminum or aluminum oxide directly without the minor and trace oxides included by the classical method. Consequently, for consistency and to provide comparability with historic data and among various analytical methods, when calculating potential compounds for specification purposes, those using methods which determine Al or Al_2O_3 directly should add to the determined Al_2O_3 weight quantities of P_2O_5, TiO_2 and any other oxide except Fe_2O_3 which would precipitate with the ammonium hydroxide group when analyzed by the classical method and which is present in an amount of 0.05 weight % or greater. The weight percent of minor or trace oxides to be added to Al_2O_3 by those using direct methods may be obtained by actual analysis of those oxides in the sample being tested or estimated from historical data on those oxides on cements from the same source, provided that the estimated values are identified as such.

When the ratio of percentages of aluminum oxide to ferric oxide is 0.64 or more, the percentages of tricalcium silicate, dicalcium silicate, tricalcium aluminate and tetracalcium aluminoferrite shall be calculated from the chemical analysis as follows:

Tricalcium silicate = $(4.071 \times \% CaO) - (7.600 \times \% SiO_2) - (6.718 \times \% Al_2O_3) - (1.430 \times \% Fe_2O_3) - (2.852 \times \% SO_3)$

Dicalcium silicate = $(2.867 \times \% SiO_2) - (0.7544 \times \% C_3S)$

Tricalcium aluminate = $(2.650 \times \% Al_2O_3) - (1.692 \times \% Fe_2O_3)$

Tetracalcium aluminoferrite = $3.043 \times \% Fe_2O_3$

When the alumina-ferric oxide ratio is less than 0.64, a calcium aluminoferrite solid solution (expressed as ss ($C_4AF + C_2F$)) is formed. Contents of this solid solution and of tricalcium silicate shall be calculated by the following formulas:

$ss(C_4AF + C_2F) = (2.100 \times \% Al_2O_3) + (1.702 \times \% Fe_2O_3)$

Tricalcium silicate = $(4.071 \times \% CaO) - (7.600 \times \% SiO_2) - (4.479 \times \% Al_2O_3) - (2.859 \times \% Fe_2O_3) - (2.852 \times \% SO_3)$.

No tricalcium aluminate will be present in cements of this composition. Dicalcium silicate shall be calculated as previously shown.

In the calculation of all compounds the oxides determined to the nearest 0.1 % shall be used.

All values calculated as described in this note shall be reported to the nearest 1 %.

[C] The optional limit for heat of hydration in Table 4 shall not be requested when this optional limit is requested.

[D] This limit may be specified when the cement is to be used in concrete with aggregates that may be deleteriously reactive. Reference should be made to Specification C 33 for suitable criteria of deleterious reactivity.

9.1.2 *Chemical Analysis*—Test Methods C 114.

9.1.3 *Strength*—Test Method C 109.

9.1.4 *False Set*—Test Method C 451.

9.1.5 *Fineness by Air Permeability*—Test Method C 204.

9.1.6 *Fineness by Turbidimeter*—Test Method C 115.

9.1.7 *Heat of Hydration*—Test Method C 186.

9.1.8 *Autoclave Expansion*—Test Method C 151.

9.1.9 *Time of Setting by Gillmore Needles*—Test Method C 266.

9.1.10 *Time of Setting by Vicat Needles*—Test Method C 191.

9.1.11 *Sulfate Resistance*—Test Method C 452 (sulfate expansion).

9.1.12 *Calcium Sulfate (expansion of) Mortar*—Test Method C 1038.

9.1.13 *Optimum SO_3*—Test Method C 563.

10. Inspection

10.1 Inspection of the material shall be made as agreed upon by the purchaser and the seller as part of the purchase contract.

11. Rejection

11.1 The cement may be rejected if it fails to meet any of the requirements of this specification.

11.2 Cement remaining in bulk storage at the mill, prior to shipment, for more than 6 months, or cement in bags in local storage in the hands of a vendor for more than 3 months, after completion of tests, may be retested before use and may be rejected if it fails to conform to any of the requirements of this specification.

11.3 Packages shall identify the weight as net weight. Packages more than 2 % below the weight marked thereon may be rejected; and if the average weight of packages in any shipment, as shown by weighing 50 packages taken at random, is less than that marked on the packages, the entire shipment may be rejected.

12. Manufacturer's Statement

12.1 At the request of the purchaser, the manufacturer shall state in writing the nature, amount, and identity of the air-entraining agent used, and of any processing addition that may have been used, and also, if requested, shall supply test data showing compliance of such air-entraining addition with the provisions of Specification C 226, and of any such processing addition with Specification C 465.

13. Packaging and Package Marking

13.1 When the cement is delivered in packages, the words "Portland Cement," the type of cement, the name and brand of the manufacturer, and the weight of the cement contained therein shall be plainly marked on each package. When the cement is an air-entraining type, the words "air-entraining" shall be plainly marked on each package. Similar information shall be provided in the shipping documents accompanying the shipment of packaged or bulk cement. All packages shall be in good condition at the time of inspection.

TABLE 3 Standard Physical Requirements

Cement Type[A]	I	IA	II	IIA	III	IIIA	IV	V
Air content of mortar,[B] volume %:								
max	12	22	12	22	12	22	12	12
min	...	16	...	16	...	16	...	...
Fineness,[C] specific surface, m^2/kg (alternative methods):								
Turbidimeter test, min	160	160	160	160	...	...	160	160
Air permeability test, min	280	280	280	280	...	...	280	280
Autoclave expansion, max, %	0.80	0.80	0.80	0.80	0.80	0.80	0.80	0.80
Strength, not less than the values shown for the ages indicated below:[D]								
Compressive strength, psi (MPa):								
1 day	...	...	...	...	1800 (12.4)	1450 (10.0)	...	...
3 days	1800 (12.4)	1450 (10.0)	1500 (10.3) 1000[F] (6.9)[F]	1200 (8.3) 800[F] (5.5)[F]	3500 (24.1)	2800 (19.3)	...	1200 (8.3)
7 days	2800 (19.3)	2250 (15.5)	2500 (17.2) 1700[F] (11.7)[F]	2000 (13.8) 1350[F] (9.3)[F]	...	...	1000 (6.9)	2200 (15.2)
28 days	...	...	...	...	...	...	2500 (17.2)	3000 (20.7)
Time of setting (alternative methods):[E]								
Gillmore test:								
Initial set, min, not less than	60	60	60	60	60	60	60	60
Final set, min, not more than	600	600	600	600	600	600	600	600
Vicat test:[G]								
Time of setting, min, not less than	45	45	45	45	45	45	45	45
Time of setting, min, not more than	375	375	375	375	375	375	375	375

[A] See Note.

[B] Compliance with the requirements of this specification does not necessarily ensure that the desired air content will be obtained in concrete.

[C] Either of the two alternative fineness methods may be used at the option of the testing laboratory. However, when the sample fails to meet the requirements of the air-permeability test, the turbidimeter test shall be used, and the requirements in this table for the turbidimetric method shall govern.

[D] The strength at any specified test age shall be not less than that attained at any previous specified test age.

[E] The purchaser should specify the type of setting-time test required. In case he does not so specify, the requirements of the Vicat test only shall govern.

[F] When the optional heat of hydration or the chemical limit on the sum of the tricalcium silicate and tricalcium aluminate is specified.

[G] The time of setting is that described as initial setting time in Test Method C 191.

TABLE 4 Optional Physical Requirements[A]

Cement Type	I	IA	II	IIA	III	IIIA	IV	V
False set, final penetration, min, %	50	50	50	50	50	50	50	50
Heat of hydration:								
7 days, max, cal/g (kJ/kg)	...	...	70 (290)[B]	70 (290)[B]	...	...	60[C] (250)	...
28 days, max, cal/g (kJ/kg)	...	...	...	...	...	...	70[C] (290)	...
Strength, not less than the values shown:								
Compressive strength, psi (MPa)								
28 days	4000 (27.6)	3200 (22.1)	4000 (27.6) 3200[B] (22.1)[B]	3200 (22.1) 2560[B] (17.7)[B]	...	...	...	...
Sulfate resistance,[D] 14 days, max, % expansion	...	...	...	...	...	...	...	0.040

[A] These optional requirements apply only if specifically requested. Availability should be verified. See Note in Section 4.

[B] The optional limit for the sum of the tricalcium silicate and tricalcium aluminate in Table 2 shall not be requested when this optional limit is requested. These strength requirements apply when either heat of hydration or the sum of tricalcium silicate and tricalcium aluminate requirements are requested.

[C] When the heat of hydration limit is specified, it shall be instead of the limits of C3S, C2S, and C3A listed in Table 1.

[D] When the sulfate resistance is specified, it shall be instead of the limits of C_3A and $C_4AF + 2\ C_3A$ listed in Table 1.

14. Storage

14.1 The cement shall be stored in such a manner as to permit easy access for proper inspection and identification of each shipment, and in a suitable weather-tight building that will protect the cement from dampness and minimize warehouse set.

15. Manufacturer's Certification

15.1 Upon request of the purchaser in the contract or order, a manufacturer's report shall be furnished at the time of shipment stating the results of tests made on samples of the material taken during production or transfer and certifying that the cement conforms to applicable requirements of this specification.

Designation: D 420 – 93

Standard Guide to Site Characterization for Engineering, Design, and Construction Purposes[1]

This standard is issued under the fixed designation D 420; the number immediately following the designation indicates the year of original adoption or, in the case of revision, the year of last revision. A number in parentheses indicates the year of last reapproval. A superscript epsilon (ε) indicates an editorial change since the last revision or reapproval.

INTRODUCTION

Investigation and identification of subsurface materials involves both simple and complex techniques that may be accomplished by many different procedures and may be variously interpreted. These studies are frequently site specific and are influenced by geological and geographical settings, by the purpose of the investigation, by design requirements for the project proposed, and by the background, training, and experience of the investigator. This guide has been extensively rewritten and enlarged since the version approved in 1987. Material has been added for clarification and for expansion of concepts. Many new ASTM standards are referenced and a bibliography of non-ASTM references is appended.

This document is a guide to the selection of the various ASTM standards that are available for the investigation of soil, rock, and ground water for projects that involve surface or subsurface construction, or both. It is intended to improve consistency of practice and to encourage rational planning of a site characterization program. Since the subsurface conditions at a particular site are usually the result of a combination of natural, geologic, topographic, and climatic factors, and of historical modifications both natural and manmade, an adequate and internally consistent exploration program will allow evaluation of the results of these influences.

1. Scope

1.1 This guide refers to ASTM methods by which soil, rock, and ground water conditions may be determined. The objective of the investigation should be to identify and locate, both horizontally and vertically, significant soil and rock types and ground water conditions present within a given site area and to establish the characteristics of the subsurface materials by sampling or in situ testing, or both.

1.2 Laboratory testing of soil, rock, and ground water samples is specified by other ASTM standards not listed herein. Subsurface exploration for environmental purposes will be the subject of a separate ASTM document.

1.3 Prior to commencement of any intrusive exploration the site should be checked for underground utilities. Should evidence of potentially hazardous or otherwise contaminated materials or conditions be encountered in the course of the investigation, work should be interrupted until the circumstances have been evaluated and revised instructions issued before resumption.

1.4 The values stated in (SI) inch-pound units are to be regarded as the standard.

1.5 *This standard does not purport to address all of the safety problems, if any, associated with its use. It is the responsibility of the user of this standard to establish appropriate safety and health practices and determine the applicability of regulatory limitations prior to use.*

2. Referenced Documents

2.1 *ASTM Standards:*

C 119 Terminology Relating to Dimension Stone[2]

C 294 Descriptive Nomenclature for Constituents of Natural Mineral Aggregates[3]

C 851 Practice for Estimating Scratch Hardness of Coarse Aggregate Particles[3]

D 75 Practice for Sampling Aggregates[4]

D 653 Terminology Relating to Soil, Rock, and Contained Fluids[2]

D 1194 Test Method for Bearing Capacity of Soil for Static Load and Spread Footings[2]

D 1195 Test Method for Repetitive Static Plate Load Tests of Soils and Flexible Pavement Components, for Use in Evaluation and Design of Airport and Highway Pavements[2]

D 1196 Test Method for Nonrepetitive Static Plate Load Tests of Soils and Flexible Pavement Components, for Use in Evaluation and Design of Airport and Highway Pavements[2]

D 1452 Practice for Soil Investigation and Sampling by Auger Borings[2]

D 1586 Test Method for Penetration Test and Split-Barrel Sampling of Soils[2]

---

[1] This guide is under the jurisdiction of ASTM Committee D-18 on Soil and Rock and is the direct responsibility of Subcommittee D18.01 on Surface and Subsurface Characterization.

Current edition approved June 15, 1993. Published August 1993. Originally published as D 425 – 65 T. Last previous edition D 420 – 87.

[2] *Annual Book of ASTM Standards*, Vol 04.08.

[3] *Annual Book of ASTM Standards*, Vol 04.02.

[4] *Annual Book of ASTM Standards*, Vol 04.03.

D 1587 Practice for Thin-Walled Tube Sampling of Soils[2]
D 2113 Practice for Diamond Core Drilling for Site Investigation[2]
D 2487 Classification of Soils for Engineering Purposes (Unified Soil Classification System)[2]
D 2488 Practice for Description and Identification of Soils (Visual-Manual Procedure)[2]
D 2573 Test Method for Field Vane Shear Test in Cohesive Soil[2]
D 2607 Classification for Peats, Mosses, Humus, and Related Products[2]
D 3017 Test Method for Water Content of Soil and Rock in Place by Nuclear Methods (Shallow Depth)[2]
D 3213 Practices for Handling, Storing, and Preparing Soft Undisturbed Marine Soil[2]
D 3282 Classification of Soils and Soil-Aggregate Mixtures for Highway Construction Purposes[2]
D 3385 Test Method for Infiltration Rate of Soils in Field Using Double-Ring Infiltrometers[2]
D 3404 Guide to Measuring Matric Potential in the Vadose Zone Using Tensiometers[2]
D 3441 Test Method for Deep, Quasi-Static, Cone and Friction-Cone Penetration Tests of Soil[2]
D 3550 Practice for Ring-lined Barrel Sampling of Soils[2]
D 3584 Practice for Indexing Papers and Reports on Soil and Rock for Engineering Purposes[2]
D 4083 Practice for Description of Frozen Soils (Visual-Manual Procedure)[2]
D 4220 Practices for Preserving and Transporting Soil Samples[2]
D 4394 Test Method for Determining the In Situ Modulus of Deformation of Rock Mass Using the Rigid Plate Loading Method[2]
D 4395 Test Method for Determining the In Situ Modulus of Deformation of Rock Mass Using the Flexible Plate Loading Method[2]
D 4403 Practice for Extensometers Used in Rock[2]
D 4428 Test Methods for Crosshole Seismic Testing[2]
D 4429 Test Method for Bearing Ratio of Soils in Place[2]
D 4452 Methods for X-Ray Radiography of Soil Samples[2]
D 4506 Test Method for Determining the In Situ Modulus of Deformation of Rock Mass Using a Radial Jacking Test[2]
D 4544 Practice for Estimating Peat Deposit Thickness[2]
D 4553 Test Method for Determining the In Situ Creep Characteristics of Rock[2]
D 4554 Test Method for In Situ Determination of Direct Shear Strength of Rock Discontinuities[2]
D 4555 Test Method for Determining Deformability and Strength of Weak Rock by an In Situ Uniaxial Compressive Test[2]
D 4622 Test Method for Rock Mass Monitoring Using Inclinometers[2]
D 4623 Test Method for Determination of In Situ Stress in Rock Mass by Overcoring Method—USBM Borehole Deformation Gage[2]
D 4630 Test Method for Determining Transmissivity and Storativity of Low Permeability Rocks by In Situ Measurements Using the Constant Head Injection Test[2]
D 4631 Test Method for Determining Transmissivity and Storativity of Low Permeability Rocks by In Situ Measurements Using the Pressure Pulse Technique[2]
D 4633 Test Method for Stress Wave Energy Measurement for Dynamic Penetrometer Testing Systems[2]
D 4645 Test Method for Determination of the In Situ Stress in Rock Using the Hydraulic Fracturing Method[2]
D 4700 Guide for Soil Sampling from the Vadose Zone[2]
D 4719 Test Method for Pressuremeter Testing in Soils[2]
D 4729 Test Method for In Situ Stress and Modulus of Deformation Using the Flatjack Method[2]
D 4750 Test Method for Determining Subsurface Liquid Levels in a Borehole or Monitoring Well (Observation Well)[2]
D 4879 Guide for Geotechnical Mapping of Large Underground Openings in Rock[2]
D 4971 Test Method for Determining the In Situ Modulus of Deformation of Rock Using the Diametrically Loaded 76-mm (3-in.) Borehole Jack[2]
D 5079 Practices for Preserving and Transporting Rock Core Samples[2]
D 5088 Practice for Decontamination of Field Equipment Used at Nonradioactive Waste Sites[2]
D 5092 Practice for Design and Installation of Ground Water Monitoring Wells in Aquifers[2]
D 5093 Test Method for Field Measurement of Infiltration Rate Using a Double-Ring Infiltrometer with a Sealed-Inner Ring[2]
D 5126 Guide for Comparison of Field Methods for Determining Hydraulic Conductivity in the Vadose Zone[2]
D 5195 Test Method for Density of Soil and Rock In-Place at Depths Below the Surface by Nuclear Methods[2]
E 177 Practice for the Use of the Terms Precision and Bias in ASTM Test Methods[5]
E 380 Practice for the Use of the International System of Units (SI) (the Modernized Metric System)[5]
G 51 Test Method for pH of Soil for Use in Corrosion Testing[6]
G 57 Method for Field Measurement of Soil Resistivity Using the Wenner Four-Electrode Method[6]

3. Significance and Use

3.1 An adequate soil, rock, and ground water investigation will provide pertinent information for decision making on one or more of the following subjects:

3.1.1 Optimum location of the structure, both vertically and horizontally, within the area of the proposed construction.

3.1.2 Location and preliminary evaluation of suitable borrow and other local sources of construction aggregates.

3.1.3 Need for special excavating and dewatering techniques with the corresponding need for information, even if only approximate, on the distribution of soil water content or pore pressure, or both, and on the piezometric heads and apparent permeability (hydraulic conductivity) of the various subsurface strata.

---

[5] *Annual Book of ASTM Standards*, Vol 14.02.
[6] *Annual Book of ASTM Standards*, Vol 03.02.
[7] The boldface numbers in parentheses refer to the list of references at the end of this standard.

3.1.4 Investigation of slope stability in natural slopes, cuts, and embankments.

3.1.5 Conceptual selection of embankment types and hydraulic barrier requirements.

3.1.6 Conceptual selection of alternate foundation types and elevations of the corresponding suitable bearing strata.

3.1.7 Development of additional detailed subsurface investigations for specific structures or facilities.

3.2 The investigation may require the collection of sufficiently large soil and rock samples of such quality as to allow adequate testing to determine the soil or rock classification or mineralogic type, or both, and the engineering properties pertinent to the proposed design.

3.3 This guide is not meant to be an inflexible description of investigation requirements; methods defined by other ASTM standards or non-ASTM techniques may be appropriate in some circumstances. The intent is to provide a checklist to assist in the design of an exploration/investigation plan.

4. Reconnaissance of Project Area

4.1 Available technical data from the literature or from personal communication should be reviewed before any field program is started. These include, but are not limited to, topographic maps, aerial photography, satellite imagery, geologic maps, statewide or county soil surveys and mineral resource surveys, and engineering soil maps covering the proposed project area. Reports of subsurface investigations of nearby or adjacent projects should be studied.

NOTE 1—While certain of the older maps and reports may be obsolete and of limited value in the light of current knowledge, a comparison of the old with the new will often reveal valuable information.

4.1.1 The United States Geological Survey and the geological surveys of the various states are the principal sources of geologic maps and reports on mineral resources and ground water.

4.1.2 United States Department of Agriculture Soil Conservation Service soil surveys, where available and of recent date, should enable the investigator to estimate the range in soil profile characteristics to depths of 5 or 6 ft (1.5 or 2 m) for each soil mapped.

NOTE 2—Each soil type has a distinctive soil profile due to age, parent material, relief, climatic condition, and biological activity. Consideration of these factors can assist in identifying the various soil types, each requiring special engineering considerations and treatment. Similar engineering soil properties are often found where similar soil profiles characteristics exist. Changes in soil properties in adjacent areas often indicate changes in parent material or relief.

4.2 In areas where descriptive data are limited by insufficient geologic or soil maps, the soil and rock in open cuts in the vicinity of the proposed project should be studied and various soil and rock profiles noted. Field notes of such studies should include data outlined in 11.6.

4.3 Where a preliminary map covering the area of the project is desired, it can be prepared on maps compiled from aerial photography that show the ground conditions. The distribution of the predominant soil and rock deposits likely to be encountered during the investigation may be shown using data obtained from geologic maps, landform analysis and limited ground reconnaissance. Experienced photo-interpreters can deduce much subsurface data from a study of black and white, color, and infrared photographs because similar soil or rock conditions, or both, usually have similar patterns of appearance in regions of similar climate or vegetation.

NOTE 3—This preliminary map may be expanded into a detailed engineering map by locating all test holes, pits, and sampling stations and by revising boundaries as determined from the detailed subsurface survey.

4.4 In areas where documentary information is insufficient, some knowledge of subsurface conditions may be obtained from land owners, local well drillers, and representatives of the local construction industry.

5. Exploration Plan

5.1 Available project design and performance requirements must be reviewed prior to final development of the exploration plan. Preliminary exploration should be planned to indicate the areas of conditions needing further investigation. A complete soil, rock, and ground water investigation should encompass the following activities:

5.1.1 Review of available information, both regional and local, on the geologic history, rock, soil, and ground water conditions occurring at the proposed location and in the immediate vicinity of the site.

5.1.2 Interpretation of aerial photography and other remote sensing data.

5.1.3 Field reconnaissance for identification of surficial geologic conditions, mapping of stratigraphic exposures and outcrops, and examination of the performance of existing structures.

5.1.4 On site investigation of the surface and subsurface materials by geophysical surveys, borings, or test pits.

5.1.5 Recovery of representative disturbed samples for laboratory classification tests of soil, rock, and local construction material. These should be supplemented by undisturbed specimens suitable for the determination of those engineering properties pertinent to the investigation.

5.1.6 Identification of the position of the ground water table, or water tables, if there is perched ground water, or of the piezometric surfaces if there is artesian ground water. The variability of these positions in both short and long time frames should be considered. Color mottling of the soil strata may be indicative of long-term seasonal high ground water positions.

5.1.7 Identification and assessment of the location of suitable foundation material, either bedrock or satisfactory load-bearing soils.

5.1.8 Field identification of soil sediments, and rock, with particular reference to type and degree of decomposition (for example, saprolite, karst, decomposing or slaking shales), the depths of their occurrence and the types and locations of their structural discontinuities.

5.1.9 Evaluation of the performance of existing installations, relative to their structure foundation material and environment in the immediate vicinity of the proposed site.

6. Equipment and Procedures for Use in Exploration

6.1 *Pertinent ASTM Standards*—Practices D 1452, D 2113, D 4544, D 5088, D 5092; Method D 1586; and Test Methods D 4622, D 4633, D 4750.

6.2 The type of equipment required for a subsurface investigation depends upon various factors, including the type of subsurface material, depth of exploration, the nature of the terrain, and the intended use of the data.

6.2.1 *Hand Augers, Hole Diggers, Shovels, and Push Tube Samplers* are suitable for exploration of surficial soils to depths of 3 to 15 ft (1 to 5 m).

6.2.2 *Earth Excavation Equipment*, such as backhoes, draglines, and drilled pier augers (screw or bucket) can allow in situ examination of soil deposits and sampling of materials containing very large particles. The investigator should be aware of the possiblity of permanent disturbance of potential bearing strata by unbalanced pore pressure in test excavations.

6.2.3 Soil and rock boring and drilling machines and proofing devices may be used to depths of 200 to 300 ft in soil and to a much greater depth in rock.

6.2.4 Well drilling equipment may be suitable for deep geologic exploration. Normally samples are in the form of sand-sized cuttings captured from the return flow, but coring devices are available.

7. Geophysical Exploration

7.1 *Pertinent ASTM Standards*—Test Methods D 4428, and Method G 57.

7.2 Remote sensing techniques may assist in mapping the geological formations and for evaluating variations in soil and rock properties. Satellite and aircraft spectral mapping tools, such as LANDSAT, may be used to find and map the areal extent of subsurface materials and geologic structure. Interpretation of aircraft photographs and satellite imagery can locate and identify significant geologic features that may be indicative of faults and fractures. Some ground control is generally required to verify information derived from remote sensing data.

7.3 Geophysical survey methods may be used to supplement borehole and outcrop data and to interpolate between holes. Seismic, ground penetrating radar, and electrical resistivity methods can be particularly valuable when distinct differences in the properties of contiguous subsurface materials are indicated.

7.4 Shallow seismic refraction/reflection and ground penetrating radar techniques can be used to map soil horizons and depth profiles, water tables, and depth to bedrock in many situations, but depth penetration and resolution vary with local conditions. Electromagnetic induction, electrical resistivity, and induced polarization (or complex resistivity) techniques may be used to map variations in water content, clay horizons, stratification, and depth to aquifer/bedrock. Other geophysical techniques such as gravity, magnetic, and shallow ground temperature methods may be useful under certain specific conditions. Deep seismic and electrical methods are routinely used for mapping stratigraphy and structure of rock in conjunction with logs. Crosshole shear wave velocity measurements can provide soil and rock parameters for dynamic analyses.

7.4.1 The seismic refraction method may be especially useful in determining depth to, or rippability of, rock in locations where successively denser strata are encountered.

7.4.2 The seismic reflection method may be useful in delineating geological units at depths below 10 ft (3 m). It is not constrained by layers of low seismic velocity and is especially useful in areas of rapid stratigraphic change.

7.4.3 The electrical resistivity method, Method G 57, may be similarly useful in determining depth to rock and anomalies in the stratigraphic profile, in evaluating stratified formations where a denser stratum overlies a less dense stratum, and in location of prospective sand-gravel or other sources of borrow material. Resistivity parameters also are required for the design of grounding systems and cathodic protection for buried structures.

7.4.4 The ground penetrating radar method may be useful in defining soil and rock layers and manmade structures in the depth range of 1 to 30 ft (⅓ to 10 m).

NOTE 4—Surface geophysical investigations can be a useful guide in determining boring or test hole locations. If at all possible, the interpretation of geophysical studies should be verified by borings or test excavations.

8. Sampling

8.1 *Pertinent ASTM Standards*—Practices D 75, D 1452, D 1587, D 2113, D 3213, D 3550, D 4220, D 5079; Test Method D 1586; Methods D 4452; and Guide D 4700.

8.2 Obtain samples that adequately represent each subsurface material that is significant to the project design and construction. The size and type of sample required is dependent upon the tests to be performed, the relative amount of coarse particles present, and the limitations of the test equipment to be used.

NOTE 5—The size of disturbed or bulk samples for routine tests may vary at the discretion of the geotechnical investigator, but the following quantities are suggested as suitable for most materials: (*a*) Visual classification—50 to 500 g (2 oz to 1 lb); (*b*) Soil constants and particle size analysis of non-gravelly soil—500 g to 2.5 kg (1 to 5 lb); (*c*) Soil compaction tests and sieve analysis of gravelly soils—20 to 40 kg (40 to 80 lb); (*d*) Aggregate manufacture or aggregate properties tests—50 to 200 kg (100 to 400 lb).

8.3 Accurately identify each sample with the boring, test hole, or testpit number and depth below reference ground surface from which it was taken. Place a waterproof identification tag inside the container, securely close the container, protect it to withstand rough handling, and mark it with proper identification on the outside. Keep samples for natural water content determination in sealed containers to prevent moisture loss. When drying of samples may affect classification or engineering properties test results, protect them to minimize moisture loss. Practices D 4220 and D 5079 address the transportation of samples from field to laboratory. Most of the titles of the referenced standards are self-explanatory, but some need elaboration for the benefit of the users of this guide.

8.3.1 Practice D 75 describes the sampling of coarse and fine aggregates for the preliminary investigation of a potential source of supply.

8.3.2 Practice D 1452 describes the use of augers in soil investigations and sampling where disturbed soil samples can be used. Depths of auger investigations are limited by ground water conditions, soil characteristics, and equipment used.

8.3.3 Test Method D 1586 describes a procedure to obtain representative soil samples for identification and classification laboratory tests.

8.3.4 Practice D 1587 describes a procedure to recover

relatively undisturbed soil samples suitable for laboratory testing.

8.3.5 Practice D 2113 describes a procedure to recover intact samples of rock and certain soils too hard to sample by Test Method D 1586 or Practice D 1587.

8.3.6 Practice D 3550 describes a procedure for the recovery of moderately disturbed, representative samples of soil for classification testing and, in some cases, shear or consolidation testing.

9. Classification of Earth Materials

9.1 *Pertinent ASTM Standards*—Terminology C 119; Descriptive Nomenclature C 294; Classifications D 2487, D 2607, D 3282; Practices D 2488, D 4083.

9.2 Additional description of samples of soil and rock may be added after submission to the laboratory for identification and classification tests in accordance with one or more ASTM laboratory standards or other applicable references, or both. Section 10.6.3 discusses the use, for identification and for classification purposes, of some of the standards listed in 9.1.

10. Determination of Subsurface Conditions

10.1 Subsurface conditions are positively defined only at the individual test pit, hole, boring, or open cut examined. Conditions between observation points may be significantly different from those encountered in the exploration. A stratigraphic profile can be developed by detailed investigations only where determinations of a continuous relationship of the depths and locations of various types of soil and rock can be inferred. This phase of the investigation may be implemented by plotting logs of soil and rock exposures in walls of excavations or cut areas and by plotting logs of the test borings. Then one may interpolate between, and extrapolate a reasonable distance beyond, these logs. The spacing of these investigations should depend on the geologic complexity of the project area and on the importance of soil and rock continuity to the project design. Exploration should be deep enough to identify all strata that might be significantly affected by the proposed use of the site and to develop the engineering data required to allow analysis of the items listed in Section 4 for each project.

NOTE 6—Plans for a program of intrusive subsurface investigation should consider possible requirements for permits for installation and proper closure of bore holes and wells at the completion of the investigation.

10.2 The depth of exploratory borings or test pits for roadbeds, airport paving, or vehicle parking areas should be to at least 5 ft (1.5 m) below the proposed subgrade elevation. Special circumstances may increase this depth. Borings for structures, excavations, or embankments should extend below the level of significant stress or ground water influence from the proposed load as determined by subsurface stress analysis.

10.3 When project construction or performance of the facility may be affected by either previous water-bearing materials or impervious materials that can block internal drainage, borings should extend sufficiently to determine those engineering and hydrogeologic properties that are relevant to the project design.

10.4 In all borrow areas the borings or test pits should be sufficient in number and depth to outline the required quantities of material meeting the specified quality requirements.

10.5 Where frost penetration or seasonal desiccation may be significant in the behavior of soil and rock, borings should extend well below the depth from finished grade of the anticipated active zone.

10.6 Exploration records shall be kept in a systematic manner for each project. Such records shall include:

10.6.1 Description of each site or area investigated. Each test hole, boring, test pit, or geophysical test site shall be clearly located (horizontally and vertically) with reference to some established coordinate system, datum, or permanent monument.

10.6.2 Logs of each test hole, boring, test pit, or cut surface exposure shall show clearly the field description and location of each material and any water encountered, either by symbol or word description. Reference to a Munsell color chart designation is a substantial aid to an accurate description of soil and rock materials.

NOTE 7—Color photographs of rock cores, soil samples, and exposed strata may be of considerable value. Each photograph should include an identifying number or symbol, a date, and reference scale.

10.6.3 Identification of all soils based on Classification D 2487, Practice D 2488, Classification D 2607, or Practice D 4083. Identification of rock materials based on Terminology C 119, Descriptive Nomenclature C 294, or Practice C 851. Classification of soil and rock is discussed in Section 9.

10.6.4 Location and description of seepage and water-bearing zones and records of piezometric elevations found in each hole, boring, piezometer, or test pit.

10.6.5 The results and precise locations of in situ test results such as the penetration resistance or vane shear discussed in 8.3, plate load tests, or other in situ test-engineering properties of soils or rock.

10.6.6 Percentage of core recovery and rock quality designation in core drilling as outlined in 8.3.5.

10.6.7 Graphical presentation of field and laboratory and its interpretation facilitates comprehensive understanding subsurface conditions.

11. In Situ Testing

11.1 *Pertinent ASTM Standards*—Test Methods D 1194, D 1195, D 1196, D 1586, D 2573, D 3017, D 3441, D 3885, D 4394, D 4395, D 4429, D 4506, D 4553, D 4554, D 4555, D 4623, D 4630, D 4631, D 4645, D 4719, D 4729, D 4971, D 5093, D 5195, G 51; Guides D 3404, D 5126; and Practice D 4403.

11.2 In situ testing is useful for: (*a*) measurement of soil parameters in their undisturbed condition with all of the restraining or loading effects, or both, of the surrounding soil or rock mass active, and (*b*) for rapid or closely spaced measurements, or both, of earth properties without the necessity of sampling. Most of the titles of the various referenced standards are self-explanatory, but some need elaboration for the users of this guide.

11.2.1 Test Method D 1586 describes a penetration test that has been correlated by many authors with various strength properties of soils.

11.2.2 Test Method D 2573 describes a procedure to

measure the in situ unit shear resistance of cohesive soils by rotation of a four-bladed vane in a horizontal plane.

11.2.3 Test Method D 3441 describes the determination of the end bearing and side friction components of the resistance to penetration of a conical penetrometer into a soil mass.

11.2.4 Practice D 4403 describes the application of various types of extensometers used in the field of rock mechanics.

11.2.5 Test Method D 4429 describes the field determination of the California Bearing Ratio for soil surfaces in situ to be used in the design of pavement systems.

11.2.6 Test Method D 4719 describes an in situ stress-strain test performed on the walls of a bore hole in soil.

NOTE 8—Other standards for in situ test procedures and automated data collection are being prepared by ASTM Committee D-18 for publication at a later date.

12. Interpretation of Results

12.1 Interpret the results of an investigation in terms of actual findings and make every effort to collect and include all field and laboratory data from previous investigations in the same area. Extrapolation of data into local areas not surveyed and tested should be made only for conceptual studies. Such extrapolation can be done only where geologically uniform stratigraphic and structural relationships are known to exist on the basis of other data. Cross sections may be developed as part of the site characterization if required to demonstrate the site conditions.

12.1.1 Cross sections included with the presentation of basic data from the investigation should be limited to the ground surface profile and the factual subsurface data obtained at specific exploration locations. Stratigraphic units between the locations of intrusive explorations should only be indicated if supported by continuous geophysical profiles.

12.1.2 Cross sections showing interpretations of stratigraphic units and other conditions between intrusive explorations but without support of continuous geophysical profiles should be presented in an interpretative report appendix or in a separate interpretative report. The interpretive cross sections must be accompanied by notes describing anomalies or otherwise significant variations in the site conditions that should be anticipated for the intended design or construction activities.

NOTE 9—Additional exploration should be considered if there is not sufficient knowledge to develop interpretative cross sections, with realistic descriptions of anticipated variations in subsurface conditions, to meet project requirements.

12.2 Subject to the restrictions imposed by state licensing law, recommendations for design parameters can be made only by professional engineers and geologists specializing in the field of geotechnical engineering and familiar with purpose, conditions, and requirements of the study. Soil mechanics, rock mechanics, and geomorphological concepts must be combined with a knowledge of geotechnical engineering or hydrogeology to make a complete application of the soil, rock, and ground water investigation. Complete design recommendations may require a more detailed study than that discussed in this guide.

12.3 Delineate subsurface profiles only from actual geophysical, test-hole, test-pit, or cut-surface data. Interpolation between locations should be made on the basis of available geologic knowledge of the area and should be clearly identified. The use of geophysical techniques as discussed in 7.2 is a valuable aid in such interpolation. Geophysical survey data should be identified separately from sample data or in situ test data.

13. Report

13.1 *Pertinent ASTM Standards*—Terminology D 653; Practices D 3584, E 177, E 380; and Guide D 4879.

13.2 The report of a subsurface investigation shall include:

13.2.1 The location of the area investigated in terms pertinent to the project. This may include sketch maps or aerial photos on which the test pits, bore holes, and sample areas are located, as well as geomorphological data relevant to the determination of the various soil and rock types. Such data includes elevation contours, streambeds, sink holes, cliffs, and the like. Where feasible, include in the report a geologic map or an agronomic soils map, or both, of the area investigated.

13.2.2 A description of the investigation procedures, including all borings and testhole logs, graphic presentation of all compaction, consolidation, or load test data tabulation of all laboratory test results, and graphical interpretations of geophysical measurements.

13.2.3 A summary of the findings obtained under Sections 4, 10, and 12, using subhead titles for the respective sections and appropriate recommendations and disclaimers for the use of the report.

14. Precision and Bias

14.1 This guide provides qualitative data only; therefore, a precision and bias statement is not applicable.

15. Keywords

15.1 explorations; feasibility studies; field investigations; foundation investigations; geological investigations; geophysical investigation; ground water; hydrologic investigations; maps; preliminary investigations; reconnaissance surveys; sampling; site investigations (see Practice D 3584); soil surveys; subsurface investigations

REFERENCES

(1) *Engineering Geology Field Manual*, U.S. Bureau of Reclamation, 1989.

(2) Dietrich, R. V., Dutro, J. V., Jr., and Foose, R. M., (Compilers), "AGI Data Sheets for Geology in Field, Laboratory, and Office," Second Edition, American Geological Institute, 1982.

(3) Pelsner, A. (Ed.), "Manual on Subsurface Investigations," American Association of State Highway and Transportation Officials, Washington, DC.

(4) Shuter, E., and Teasdale, W. E., "Applications of Drilling, Coring and Sampling Techniques to Test Holes and Wells," Techniques of

Water-Resources Investigation, Book 2, U.S. Geological Survey, Washington, DC, 1989.
(5) Keys, W. S., "Borehole Geophysics Applied to Ground Water Investigations," U.S Geological Survey Open-File Report R87-539, Denver, CO, 1988.
(6) Dowding, C. H. (Ed.), "Site Characterization Exploration," American Society of Civil Engineers, Proceedings of Specialty Workshop, New York, NY, 1978.
(7) "Earth Manual," U.S. Bureau of Reclamation, Denver, CO.
(8) "Engineering and Design—Geotechnical Investigation Engineer Manual," EM 1110-1-1804, Headquarters, Department of the Army, Washington, DC, 1984.
(9) "Agricultural Handbook, No. 436, Soil Taxonomy," Soil Conservation Service, U.S. Dept. of Agriculture, U.S. Printing Office, Washington, DC, December, 1975.

Designation: D 421 – 85

Standard Practice for Dry Preparation of Soil Samples for Particle-Size Analysis and Determination of Soil Constants[1]

This standard is issued under the fixed designation D 421; the number immediately following the designation indicates the year of original adoption or, in the case of revision, the year of last revision. A number in parentheses indicates the year of last reapproval. A superscript epsilon (ε) indicates an editorial change since the last revision or reapproval.

1. Scope

1.1 This practice covers the dry preparation of soil samples as received from the field for particle-size analysis and the determination of the soil constants.

1.2 *This standard may involve hazardous materials, operations, and equipment. This standard does not purport to address all of the safety problems associated with its use. It is the responsibility of whoever uses this standard to consult and establish appropriate safety and health practices and determine the applicability of regulatory limitations prior to use.*

2. Referenced Documents

2.1 *ASTM Standards:*

D 2217 Practice for Wet Preparation of Soil Samples for Particle-Size Analysis and Determination of Soil Constants[2]

E 11 Specification for Wire-Cloth Sieves for Testing Purposes[3]

3. Significance and Use

3.1 This practice can be used to prepare samples for particle-size and plasticity tests where it is desired to determine test values on air-dried samples, or where it is known that air drying does not have an effect on test results relative to samples prepared in accordance with Practice D 2217.

4. Apparatus

4.1 *Balance*, sensitive to 0.1 g.

4.2 *Mortar and Rubber-Covered Pestle*, suitable for breaking up the aggregations of soil particles.

4.3 *Sieves*—A series of sieves, of square mesh woven wire cloth, conforming to Specification E 11. The sieves required are as follows:

No. 4 (4.75-mm)
No. 10 (2.00-mm)
No. 40 (425-μm)

4.4 *Sampler*—A riffle sampler or sample splitter, for quartering the samples.

5. Sampling

5.1 Expose the soil sample as received from the field to the air at room temperature until dried thoroughly. Break up the aggregations thoroughly in the mortar with a rubber-covered pestle. Select a representative sample of the amount required to perform the desired tests by the method of quartering or by the use of a sampler. The amounts of material required to perform the individual tests are as follows:

5.1.1 *Particle-Size Analysis*—For the particle-size analysis, material passing a No. 10 (2.00-mm) sieve is required in amounts equal to 115 g of sandy soils and 65 g of either silt or clay soils.

5.1.2 *Tests for Soil Constants*—For the tests for soil constants, material passing the No. 40 (425-μm) sieve is required in total amount of 220 g, allocated as follows:

Test	Grams
Liquid limit	100
Plastic limit	15
Centrifuge moisture equivalent	10
Volumetric shrinkage	30
Check tests	65

6. Preparation of Test Sample

6.1 Select that portion of the air-dried sample selected for purpose of tests and record the mass as the mass of the total test sample uncorrected for hygroscopic moisture. Separate the test sample by sieving with a No. 10 (2.00-mm) sieve. Grind that fraction retained on the No. 10 sieve in a mortar with a rubber-covered pestle until the aggregations of soil particles are broken up into the separate grains. Then separate the ground soil into two fractions by sieving with a No. 10 sieve.

6.2 Wash that fraction retained after the second sieving free of all fine material, dry, and weigh. Record this mass as the mass of coarse material. Sieve the coarse material, after being washed and dried, on the No. 4 (4.75-mm) sieve and record the mass retained on the No. 4 sieve.

7. Test Sample for Particle-Size Analysis

7.1 Thoroughly mix together the fractions passing the No. 10 (2.00-mm) sieve in both sieving operations, and by the method of quartering or the use of a sampler, select a portion weighing approximately 115 g for sandy soils and approximately 65 g for silt and clay soil for particle-size analysis.

8. Test Sample for Soil Constants

8.1 Separate the remaining portion of the material passing the No. 10 (2.00-mm) sieve into two parts by means of a No. 40 (425-μm) sieve. Discard the fraction retained on the No. 40 sieve. Use the fraction passing the No. 40 sieve for the determination of the soil constants.

---

[1] This practice is under the jurisdiction of ASTM Committee D-18 on Soil and Rock and is the direct responsibility of Subcommittee D18.03 on Texture, Plasticity, and Density Characteristics of Soils.

Current edition approved July 26, 1985. Published September 1985. Originally published as D 421 – 35 T. Last previous edition D 421 – 58 (1978)$^{\epsilon 1}$.

[2] *Annual Book of ASTM Standards*, Vol 04.08.

[3] *Annual Book of ASTM Standards*, Vol 14.02.

Designation: D 422 – 63 (Reapproved 1990)$^{\epsilon 1}$

Standard Test Method for Particle-Size Analysis of Soils[1]

This standard is issued under the fixed designation D 422; the number immediately following the designation indicates the year of original adoption or, in the case of revision, the year of last revision. A number in parentheses indicates the year of last reapproval. A superscript epsilon (ε) indicates an editorial change since the last revision or reapproval.

ε1 NOTE—Section 19 was added editorially in September 1990.

1. Scope

1.1 This test method covers the quantitative determination of the distribution of particle sizes in soils. The distribution of particle sizes larger than 75 µm (retained on the No. 200 sieve) is determined by sieving, while the distribution of particle sizes smaller than 75 µm is determined by a sedimentation process, using a hydrometer to secure the necessary data (Notes 1 and 2).

NOTE 1—Separation may be made on the No. 4 (4.75-mm), No. 40 (425-µm), or No. 200 (75-µm) sieve instead of the No. 10. For whatever sieve used, the size shall be indicated in the report.

NOTE 2—Two types of dispersion devices are provided: (*1*) a high-speed mechanical stirrer, and (*2*) air dispersion. Extensive investigations indicate that air-dispersion devices produce a more positive dispersion of plastic soils below the 20-µm size and appreciably less degradation on all sizes when used with sandy soils. Because of the definite advantages favoring air dispersion, its use is recommended. The results from the two types of devices differ in magnitude, depending upon soil type, leading to marked differences in particle size distribution, especially for sizes finer than 20 µm.

2. Referenced Documents

2.1 *ASTM Standards:*

D 421 Practice for Dry Preparation of Soil Samples for Particle-Size Analysis and Determination of Soil Constants[2]

E 11 Specification for Wire-Cloth Sieves for Testing Purposes[3]

E 100 Specification for ASTM Hydrometers[4]

3. Apparatus

3.1 *Balances*—A balance sensitive to 0.01 g for weighing the material passing a No. 10 (2.00-mm) sieve, and a balance sensitive to 0.1 % of the mass of the sample to be weighed for weighing the material retained on a No. 10 sieve.

3.2 *Stirring Apparatus*—Either apparatus A or B may be used.

3.2.1 Apparatus A shall consist of a mechanically operated stirring device in which a suitably mounted electric motor turns a vertical shaft at a speed of not less than 10 000 rpm without load. The shaft shall be equipped with a replaceable stirring paddle made of metal, plastic, or hard rubber, as shown in Fig. 1. The shaft shall be of such length that the stirring paddle will operate not less than ¾ in. (19.0 mm) nor more than 1½ in. (38.1 mm) above the bottom of the dispersion cup. A special dispersion cup conforming to either of the designs shown in Fig. 2 shall be provided to hold the sample while it is being dispersed.

3.2.2 Apparatus B shall consist of an air-jet dispersion cup[5] (Note 3) conforming to the general details shown in Fig. 3 (Notes 4 and 5).

NOTE 3—The amount of air required by an air-jet dispersion cup is of the order of 2 ft³/min; some small air compressors are not capable of supplying sufficient air to operate a cup.

NOTE 4—Another air-type dispersion device, known as a dispersion tube, developed by Chu and Davidson at Iowa State College, has been shown to give results equivalent to those secured by the air-jet dispersion cups. When it is used, soaking of the sample can be done in the sedimentation cylinder, thus eliminating the need for transferring the slurry. When the air-dispersion tube is used, it shall be so indicated in the report.

NOTE 5—Water may condense in air lines when not in use. This water must be removed, either by using a water trap on the air line, or by blowing the water out of the line before using any of the air for dispersion purposes.

3.3 *Hydrometer*—An ASTM hydrometer, graduated to read in either specific gravity of the suspension or grams per litre of suspension, and conforming to the requirements for hydrometers 151H or 152H in Specifications E 100. Dimensions of both hydrometers are the same, the scale being the only item of difference.

3.4 *Sedimentation Cylinder*—A glass cylinder essentially 18 in. (457 mm) in height and 2½ in. (63.5 mm) in diameter, and marked for a volume of 1000 mL. The inside diameter shall be such that the 1000-mL mark is 36 ± 2 cm from the bottom on the inside.

3.5 *Thermometer*—A thermometer accurate to 1°F (0.5°C).

3.6 *Sieves*—A series of sieves, of square-mesh woven-wire cloth, conforming to the requirements of Specification E 11. A full set of sieves includes the following (Note 6):

[1] This test method is under the jurisdiction of ASTM Committee D-18 on Soil and Rock and is the direct responsibility of Subcommittee D18.03 on Texture, Plasticity, and Density Characteristics of Soils.

Current edition approved Nov. 21, 1963. Originally published 1935. Replaces D 422 – 62.

[2] *Annual Book of ASTM Standards*, Vol 04.08.

[3] *Annual Book of ASTM Standards*, Vol 14.02.

[4] *Annual Book of ASTM Standards*, Vol 14.03.

[5] Detailed working drawings for this cup are available at a nominal cost from the American Society for Testing and Materials, 1916 Race St., Philadelphia, PA 19103. Order Adjunct No. 12-404220-00.

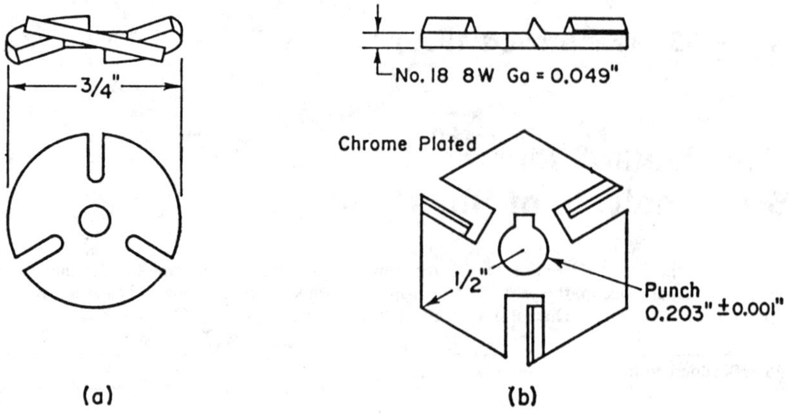

Metric Equivalents

in.	0.001	0.049	0.203	½	¾
mm	0.03	1.24	5.16	12.7	19.0

FIG. 1 Detail of Stirring Paddles

3-in. (75-mm)
2-in. (50-mm)
1½-in. (37.5-mm)
1-in. (25.0-mm)
¾-in. (19.0-mm)
⅜-in. (9.5-mm)
No. 4 (4.75-mm)

No. 10 (2.00-mm)
No. 20 (850-μm)
No. 40 (425-μm)
No. 60 (250-μm)
No. 140 (106-μm)
No. 200 (75-μm)

NOTE 6—A set of sieves giving uniform spacing of points for the graph, as required in Section 17, may be used if desired. This set consists of the following sieves:

3-in. (75-mm)
1½-in. (37.5-mm)
¾-in. (19.0-mm)
⅜-in. (9.5-mm)
No. 4 (4.75-mm)
No. 8 (2.36-mm)

No. 16 (1.18-mm)
No. 30 (600-μm)
No. 50 (300-μm)
No. 100 (150-μm)
No. 200 (75-μm)

3.7 *Water Bath or Constant-Temperature Room*—A water bath or constant-temperature room for maintaining the soil suspension at a constant temperature during the hydrometer analysis. A satisfactory water tank is an insulated tank that maintains the temperature of the suspension at a convenient constant temperature at or near 68°F (20°C). Such a device is illustrated in Fig. 4. In cases where the work is performed in a room at an automatically controlled constant temperature, the water bath is not necessary.

3.8 *Beaker*—A beaker of 250-mL capacity.

3.9 *Timing Device*—A watch or clock with a second hand.

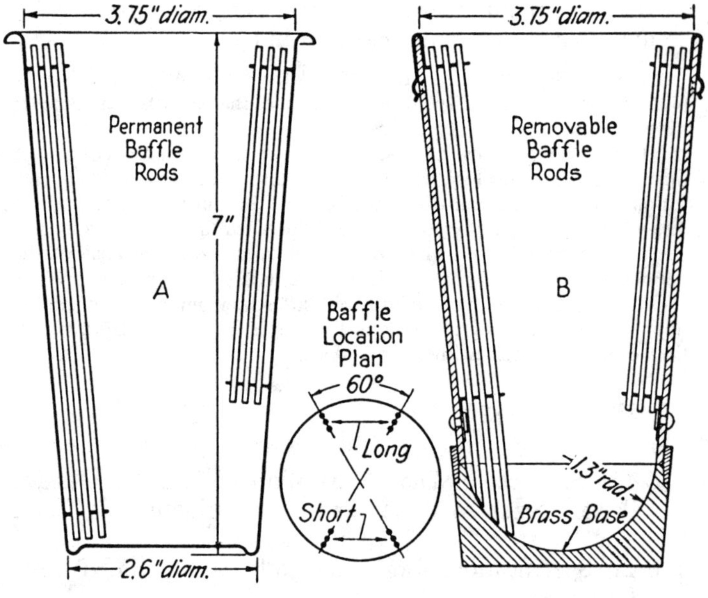

Metric Equivalents

in.	1.3	2.6	3.75
mm	33	66	95.2

FIG. 2 Dispersion Cups of Apparatus

4. Dispersing Agent

4.1 A solution of sodium hexametaphosphate (sometimes called sodium metaphosphate) shall be used in distilled or demineralized water, at the rate of 40 g of sodium hexametaphosphate/litre of solution (Note 7).

NOTE 7—Solutions of this salt, if acidic, slowly revert or hydrolyze back to the orthophosphate form with a resultant decrease in dispersive action. Solutions should be prepared frequently (at least once a month) or adjusted to pH of 8 or 9 by means of sodium carbonate. Bottles containing solutions should have the date of preparation marked on them.

4.2 All water used shall be either distilled or demineralized water. The water for a hydrometer test shall be brought to the temperature that is expected to prevail during the hydrometer test. For example, if the sedimentation cylinder is to be placed in the water bath, the distilled or demineralized water to be used shall be brought to the temperature of the controlled water bath; or, if the sedimentation cylinder is used in a room with controlled temperature, the water for the test shall be at the temperature of the room. The basic temperature for the hydrometer test is 68°F (20°C). Small variations of temperature do not introduce differences that are of practical significance and do not prevent the use of corrections derived as prescribed.

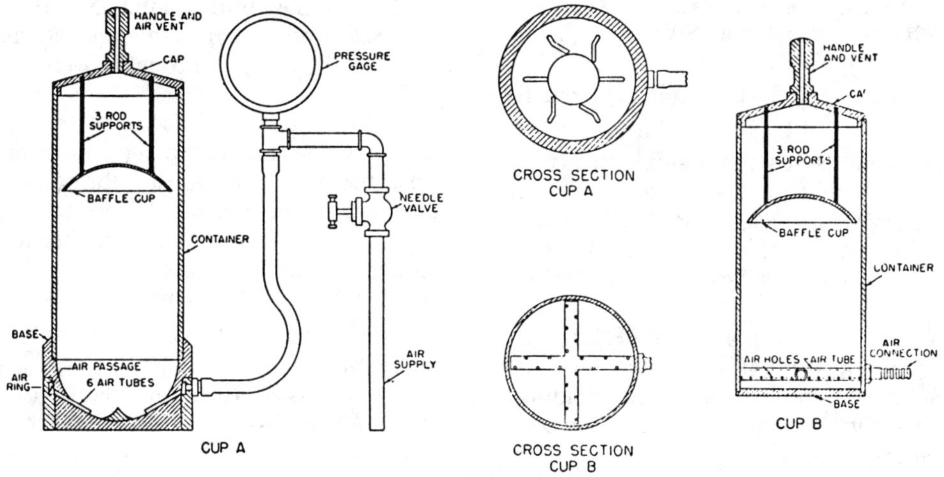

FIG. 3 Air-Jet Dispersion Cups of Apparatus B

5. Test Sample

5.1 Prepare the test sample for mechanical analysis as outlined in Practice D 421. During the preparation procedure the sample is divided into two portions. One portion contains only particles retained on the No. 10 (2.00-mm) sieve while the other portion contains only particles passing the No. 10 sieve. The mass of air-dried soil selected for purpose of tests, as prescribed in Practice D 421, shall be sufficient to yield quantities for mechanical analysis as follows:

5.1.1 The size of the portion retained on the No. 10 sieve shall depend on the maximum size of particle, according to the following schedule:

Nominal Diameter of Largest Particles, in. (mm)	Approximate Minimum Mass of Portion, g
3/8 (9.5)	500
3/4 (19.0)	1000
1 (25.4)	2000
1 1/2 (38.1)	3000
2 (50.8)	4000
3 (76.2)	5000

5.1.2 The size of the portion passing the No. 10 sieve shall be approximately 115 g for sandy soils and approximately 65 g for silt and clay soils.

5.2 Provision is made in Section 5 of Practice D 421 for weighing of the air-dry soil selected for purpose of tests, the separation of the soil on the No. 10 sieve by dry-sieving and washing, and the weighing of the washed and dried fraction retained on the No. 10 sieve. From these two masses the percentages retained and passing the No. 10 sieve can be calculated in accordance with 12.1.

NOTE 8—A check on the mass values and the thoroughness of pulverization of the clods may be secured by weighing the portion passing the No. 10 sieve and adding this value to the mass of the washed and oven-dried portion retained on the No. 10 sieve.

SIEVE ANALYSIS OF PORTION RETAINED ON NO. 10 (2.00-mm) SIEVE

6. Procedure

6.1 Separate the portion retained on the No. 10 (2.00-mm) sieve into a series of fractions using the 3-in. (75-mm), 2-in. (50-mm), 1½-in. (37.5-mm), 1-in. (25.0-mm), ¾-in. (19.0-mm), ⅜-in. (9.5-mm), No. 4 (4.75-mm), and No. 10 sieves, or as many as may be needed depending on the sample, or upon the specifications for the material under test.

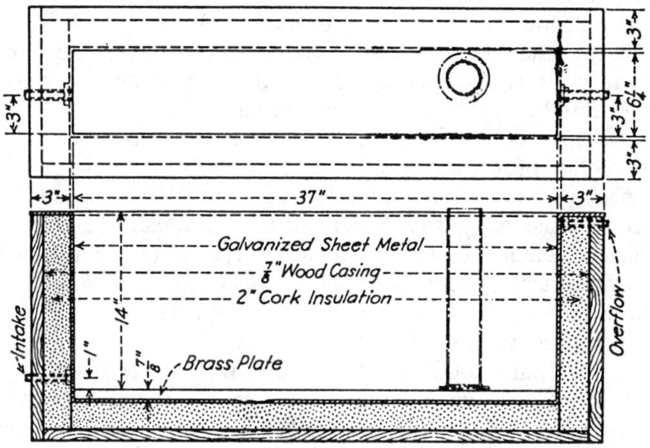

Metric Equivalents

in.	7/8	1	3	6¼	14	37
mm	22.2	25.4	76.2	158.2	356	940

FIG. 4 Insulated Water Bath

6.2 Conduct the sieving operation by means of a lateral and vertical motion of the sieve, accompanied by a jarring action in order to keep the sample moving continuously over the surface of the sieve. In no case turn or manipulate fragments in the sample through the sieve by hand. Continue sieving until not more than 1 mass % of the residue on a sieve passes that sieve during 1 min of sieving. When mechanical sieving is used, test the thoroughness of sieving by using the hand method of sieving as described above.

6.3 Determine the mass of each fraction on a balance conforming to the requirements of 3.1. At the end of weighing, the sum of the masses retained on all the sieves used should equal closely the original mass of the quantity sieved.

HYDROMETER AND SIEVE ANALYSIS OF PORTION PASSING THE NO. 10 (2.00-mm) SIEVE

7. Determination of Composite Correction for Hydrometer Reading

7.1 Equations for percentages of soil remaining in suspension, as given in 14.3, are based on the use of distilled or demineralized water. A dispersing agent is used in the water, however, and the specific gravity of the resulting liquid is appreciably greater than that of distilled or demineralized water.

7.1.1 Both soil hydrometers are calibrated at 68°F (20°C), and variations in temperature from this standard temperature produce inaccuracies in the actual hydrometer readings. The amount of the inaccuracy increases as the variation from the standard temperature increases.

7.1.2 Hydrometers are graduated by the manufacturer to be read at the bottom of the meniscus formed by the liquid on the stem. Since it is not possible to secure readings of soil suspensions at the bottom of the meniscus, readings must be taken at the top and a correction applied.

7.1.3 The net amount of the corrections for the three items enumerated is designated as the composite correction, and may be determined experimentally.

7.2 For convenience, a graph or table of composite corrections for a series of 1° temperature differences for the range of expected test temperatures may be prepared and used as needed. Measurement of the composite corrections may be made at two temperatures spanning the range of expected test temperatures, and corrections for the intermediate temperatures calculated assuming a straight-line relationship between the two observed values.

7.3 Prepare 1000 mL of liquid composed of distilled or demineralized water and dispersing agent in the same proportion as will prevail in the sedimentation (hydrometer) test. Place the liquid in a sedimentation cyclinder and the cylinder in the constant-temperature water bath, set for one of the two temperatures to be used. When the temperature of the liquid becomes constant, insert the hydrometer, and, after a short interval to permit the hydrometer to come to the temperature of the liquid, read the hydrometer at the top of the meniscus formed on the stem. For hydrometer 151H the composite correction is the difference between this reading and one; for hydrometer 152H it is the difference between the reading and zero. Bring the liquid and the hydrometer to the other temperature to be used, and secure the composite correction as before.

8. Hygroscopic Moisture

8.1 When the sample is weighed for the hydrometer test, weigh out an auxiliary portion of from 10 to 15 g in a small metal or glass container, dry the sample to a constant mass in an oven at 230 ± 9°F (110 ± 5°C), and weigh again. Record the masses.

9. Dispersion of Soil Sample

9.1 When the soil is mostly of the clay and silt sizes, weigh out a sample of air-dry soil of approximately 50 g. When the soil is mostly sand the sample should be approximately 100 g.

9.2 Place the sample in the 250-mL beaker and cover with 125 mL of sodium hexametaphosphate solution (40 g/L). Stir until the soil is thoroughly wetted. Allow to soak for at least 16 h.

9.3 At the end of the soaking period, disperse the sample further, using either stirring apparatus A or B. If stirring apparatus A is used, transfer the soil - water slurry from the beaker into the special dispersion cup shown in Fig. 2, washing any residue from the beaker into the cup with distilled or demineralized water (Note 9). Add distilled or demineralized water, if necessary, so that the cup is more than half full. Stir for a period of 1 min.

NOTE 9—A large size syringe is a convenient device for handling the water in the washing operation. Other devices include the wash-water bottle and a hose with nozzle connected to a pressurized distilled water tank.

9.4 If stirring apparatus B (Fig. 3) is used, remove the cover cap and connect the cup to a compressed air supply by means of a rubber hose. A air gage must be on the line between the cup and the control valve. Open the control valve so that the gage indicates 1 psi (7 kPa) pressure (Note 10). Transfer the soil - water slurry from the beaker to the air-jet dispersion cup by washing with distilled or demineralized water. Add distilled or demineralized water, if necessary, so that the total volume in the cup is 250 mL, but no more.

NOTE 10—The initial air pressure of 1 psi is required to prevent the soil - water mixture from entering the air-jet chamber when the mixture is transferred to the dispersion cup.

9.5 Place the cover cap on the cup and open the air control valve until the gage pressure is 20 psi (140 kPa). Disperse the soil according to the following schedule:

Plasticity Index	Dispersion Period, min
Under 5	5
6 to 20	10
Over 20	15

Soils containing large percentages of mica need be dispersed for only 1 min. After the dispersion period, reduce the gage pressure to 1 psi preparatory to transfer of soil - water slurry to the sedimentation cylinder.

10. Hydrometer Test

10.1 Immediately after dispersion, transfer the soil - water slurry to the glass sedimentation cylinder, and add distilled or demineralized water until the total volume is 1000 mL.

10.2 Using the palm of the hand over the open end of the cylinder (or a rubber stopper in the open end), turn the cylinder upside down and back for a period of 1 min to complete the agitation of the slurry (Note 11). At the end of 1 min set the cylinder in a convenient location and take hydrometer readings at the following intervals of time (measured from the beginning of sedimentation), or as many as may be needed, depending on the sample or the specification for the material under test: 2, 5, 15, 30, 60, 250, and 1440 min. If the controlled water bath is used, the sedimentation cylinder should be placed in the bath between the 2- and 5-min readings.

NOTE 11—The number of turns during this minute should be approximately 60, counting the turn upside down and back as two turns.

Any soil remaining in the bottom of the cylinder during the first few turns should be loosened by vigorous shaking of the cylinder while it is in the inverted position.

10.3 When it is desired to take a hydrometer reading, carefully insert the hydrometer about 20 to 25 s before the reading is due to approximately the depth it will have when the reading is taken. As soon as the reading is taken, carefully remove the hydrometer and place it with a spinning motion in a graduate of clean distilled or demineralized water.

NOTE 12—It is important to remove the hydrometer immediately after each reading. Readings shall be taken at the top of the meniscus formed by the suspension around the stem, since it is not possible to secure readings at the bottom of the meniscus.

10.4 After each reading, take the temperature of the suspension by inserting the thermometer into the suspension.

11. Sieve Analysis

11.1 After taking the final hydrometer reading, transfer the suspension to a No. 200 (75-μm) sieve and wash with tap water until the wash water is clear. Transfer the material on the No. 200 sieve to a suitable container, dry in an oven at 230 ± 9°F (110 ± 5°C) and make a sieve analysis of the portion retained, using as many sieves as desired, or required for the material, or upon the specification of the material under test.

CALCULATIONS AND REPORT

12. Sieve Analysis Values for the Portion Coarser than the No. 10 (2.00-mm) Sieve

12.1 Calculate the percentage passing the No. 10 sieve by dividing the mass passing the No. 10 sieve by the mass of soil originally split on the No. 10 sieve, and multiplying the result by 100. To obtain the mass passing the No. 10 sieve, subtract the mass retained on the No. 10 sieve from the original mass.

12.2 To secure the total mass of soil passing the No. 4 (4.75-mm) sieve, add to the mass of the material passing the No. 10 sieve the mass of the fraction passing the No. 4 sieve and retained on the No. 10 sieve. To secure the total mass of soil passing the ⅜-in. (9.5-mm) sieve, add to the total mass of soil passing the No. 4 sieve, the mass of the fraction passing the ⅜-in. sieve and retained on the No. 4 sieve. For the remaining sieves, continue the calculations in the same manner.

12.3 To determine the total percentage passing for each sieve, divide the total mass passing (see 12.2) by the total mass of sample and multiply the result by 100.

13. Hygroscopic Moisture Correction Factor

13.1 The hydroscopic moisture correction factor is the ratio between the mass of the oven-dried sample and the air-dry mass before drying. It is a number less than one, except when there is no hygroscopic moisture.

14. Percentages of Soil in Suspension

14.1 Calculate the oven-dry mass of soil used in the hydrometer analysis by multiplying the air-dry mass by the hygroscopic moisture correction factor.

14.2 Calculate the mass of a total sample represented by the mass of soil used in the hydrometer test, by dividing the oven-dry mass used by the percentage passing the No. 10 (2.00-mm) sieve, and multiplying the result by 100. This value is the weight W in the equation for percentage remaining in suspension.

TABLE 1 Values of Correction Factor, α, for Different Specific Gravities of Soil Particles[A]

Specific Gravity	Correction Factor[A]
2.95	0.94
2.90	0.95
2.85	0.96
2.80	0.97
2.75	0.98
2.70	0.99
2.65	1.00
2.60	1.01
2.55	1.02
2.50	1.03
2.45	1.05

[A] For use in equation for percentage of soil remaining in suspension when using Hydrometer 152H.

14.3 The percentage of soil remaining in suspension at the level at which the hydrometer is measuring the density of the suspension may be calculated as follows (Note 13): For hydrometer 151H:

$$P = [(100\,000/W) \times G/(G - G_1)](R - G_1)$$

NOTE 13—The bracketed portion of the equation for hydrometer 151H is constant for a series of readings and may be calculated first and then multiplied by the portion in the parentheses.

For hydrometer 152H:

$$P = (Ra/W) \times 100$$

where:

a = correction faction to be applied to the reading of hydrometer 152H. (Values shown on the scale are computed using a specific gravity of 2.65. Correction factors are given in Table 1),

P = percentage of soil remaining in suspension at the level at which the hydrometer measures the density of the suspension,

R = hydrometer reading with composite correction applied (Section 7),

W = oven-dry mass of soil in a total test sample represented by mass of soil dispersed (see 14.2), g,

G = specific gravity of the soil particles, and

G_1 = specific gravity of the liquid in which soil particles are suspended. Use numerical value of one in both instances in the equation. In the first instance any possible variation produces no significant effect, and in the second instance, the composite correction for R is based on a value of one for G_1.

15. Diameter of Soil Particles

15.1 The diameter of a particle corresponding to the percentage indicated by a given hydrometer reading shall be calculated according to Stokes' law (Note 14), on the basis that a particle of this diameter was at the surface of the suspension at the beginning of sedimentation and had settled to the level at which the hydrometer is measuring the density of the suspension. According to Stokes' law:

$$D = \sqrt{[30n/980(G - G_1)] \times L/T}$$

where:

D = diameter of particle, mm,

TABLE 1 Conversion Factors Between Units in the Specific Test Method and Other Units in Common Use

To Convert	Into	Multiply By
mL (or cm^3) of dissolved oxygen/L	mg/L	1.429
grains/US gal	mg/L	17.12
grains/Imperial gal	mg/L	14.25
grams/L	mg/L	1000
normality	meq/L	1000
mg/L as $CaCO_3$	meq/L	0.0200
grains/US gal as $CaCO_3$	meq/L	0.342
grains/Imperial gal as $CaCO_3$	meq/L	0.285
mg of dissolved oxygen per L	mL (or cm^3) of dissolved oxygen per L	0.700
mg/L	grains/US gal	0.0584
mg/L	grains/Imperial gal	0.0702
mg/L	grams/L	0.001
meq/L	normality	0.001
meq/L	percent of normal	0.1
meq/L	mg/L as $CaCO_3$	50.0
meq/L	grains/US gal as $CaCO_3$	2.92
meq/L	grains/Imperial gal as $CaCO_3$	3.51

TABLE 2 Factors for Interconversion Between Milligrams Per Litre and Milliequivalents Per Litre[A,B]

Ion	Multiplier	
	mg/L to meq/L	meq/L to mg/L
Al^{+3}	0.1112	8.994
Ba^{+2}	0.01456	68.67
Br^{-}	0.01252	79.90
Ca^{+2}	0.04990	20.04
Cl^{-}	0.02821	35.45
CN^{-}	0.03844	26.02
CO_3^{-2}	0.03333	30.00
Cr^{+3}	0.05770	17.33
Cr^{+6}	0.1154	8.666
CrO_4^{-2}	0.01724	58.00
Cu^{+2}	0.03147	31.77
F^{-}	0.05264	19.00
Fe^{+2}	0.03581	27.92
Fe^{+3}	0.05372	18.62
H^{+}	0.9921	1.008
HCO_3^{-}	0.01639	61.02
HPO_4^{-2}	0.02084	47.99
$H_2PO_4^{-}$	0.01031	96.99
HS^{-}	0.03024	33.07
HSO_3^{-}	0.01233	81.07
HSO_4^{-}	0.01030	97.07
I^{-}	0.007880	126.9
K^{+}	0.02558	39.10
Mg^{+2}	0.08229	12.15
Mn^{+2}	0.03640	27.47
Mn^{+4}	0.07281	13.73
Na^{+}	0.04350	22.99
NH_4^{+}	0.05544	18.04
Ni^{+2}	0.03407	29.35
NO_2^{-}	0.02174	46.01
NO_3^{-}	0.01613	62.00
OH^{-}	0.05880	17.01
Pb^{+2}	0.009653	103.6
PO_4^{-3}	0.03159	31.66
S^{-2}	0.06238	16.03
SiO_3^{-2}	0.02629	38.04
SO_3^{-2}	0.02498	40.03
SO_4^{-2}	0.02082	48.03
Sr^{+2}	0.02283	43.81
Zn^{+2}	0.03059	32.69

[A] Based on ^{12}C = 12 amu (atomic mass units).
[B] It is assumed that reactions proceed to the zero oxidation state.

Designation: D 653 – 90

Standard Terminology Relating to Soil, Rock, and Contained Fluids[1]

These definitions were prepared jointly by the American Society of Civil Engineers and the American Society for Testing and Materials.

INTRODUCTION

A number of the definitions include symbols and indicate the units of measurement. The symbols appear in italics immediately after the name of the term, followed by the unit in parentheses. No significance should be placed on the order in which the symbols are presented where two or more are given for an individual term. The applicable units are indicated by capital letters, as follows:

F—Force, such as pound-force, ton-force, newton
L—Length, such as inch, foot, centimetre
T—Time, such as second, minute
D—Dimensionless

Positive exponents designate multiples in the numerator. Negative exponents designate multiples in the denominator. Degrees of angle are indicated as "degrees."

Expressing the units either in SI or the inch-pound system has been purposely omitted in order to leave the choice of the system and specific unit to the engineer and the particular application, for example:

FL^{-2}—may be expressed in pounds-force per square inch, kilopascals, tons per square foot, etc.
LT^{-1}—may be expressed in feet per minute, centimetres per second, etc.

Where synonymous terms are cross-referenced, the definition is usually included with the earlier term alphabetically. Where this is not the case, the later term is the more significant.

Definitions marked with (ISRM) are taken directly from the publication in Ref 42 and are included for the convenience of the user.

For a list of ISRM symbols relating to soil and rock mechanics, refer to Appendix X1.

A list of references used in the preparation of these definitions appears at the end.

AASHTO compaction—see **compaction test**.

"A" Horizon—see **horizon**.

abrasion—a rubbing and wearing away. (ISRM)

abrasion—the mechanical wearing, grinding, scraping or rubbing away (or down) of rock surfaces by friction or impact, or both.

abrasive—any rock, mineral, or other substance that, owing to its superior hardness, toughness, consistency, or other properties, is suitable for grinding, cutting, polishing, scouring, or similar use.

abrasiveness—the property of a material to remove matter when scratching and grinding another material. (ISRM)

absorbed water—water held mechanically in a soil or rock mass and having physical properties not substantially different from ordinary water at the same temperature and pressure.

absorption—the assimilation of fluids into interstices.

absorption loss—that part of transmitted energy (mechanical) lost due to dissipation or conversion into other forms (heat, etc.).

accelerator—a material that increases the rate at which chemical reactions would otherwise occur.

activator—a material that causes a catalyst to begin its function.

active earth pressure—see **earth pressure**.

active state of plastic equilibrium—see **plastic equilibrium**.

additive—any material other than the basic components of a grout system.

adhesion—shearing resistance between soil and another material under zero externally applied pressure.

	Symbol	Unit
Unit Adhesion	c_a	FL^{-2}
Total Adhesion	C_a	F or FL^{-1}

adhesion—shearing resistance between two unlike materials under zero externally applied pressure.

admixture—a material other than water, aggregates, or cementitious material, used as a grout ingredient for cement-based grouts.

adsorbed water—water in a soil or rock mass attracted to the particle surfaces by physiochemical forces, having properties that may differ from those of pore water at the same temperature and pressure due to altered molecular ar-

[1] This terminology is under the jurisdiction of ASTM Committee D-18 on Soil and Rock and is the direct responsibility of Subcommittee D18.93 on Terminology for Soil, Rock, and Contained Fluids.

Current edition approved Oct. 27, 1989, and Feb. 1, 1990. Published March 1990. Originally published as D 653 – 42 T. Last previous edition D 653 – 89.

This extensive list of definitions represents the joint efforts of Subcommittee D18.93 on Terminology for Soil, Rock, and Contained Fluids of ASTM Committee D-18 on Soil and Rock, and the Committee on Definitions and Standards of the Geotechnical Engineering Division of the American Society of Civil Engineers. These two groups function together as the Joint ASCE/ASTM Committee on Nomenclature in Soil and Rock Mechanics. This list incorporates some terms from ASTM Definitions D 1707, Terms Relating to Soil Dynamics, which were discontinued in 1967.

Designation: D 596 – 91

Standard Practice for Reporting Results of Analysis of Water[1]

This standard is issued under the fixed designation D 596 ; the number immediately following the designation indicates the year of original adoption or, in the case of revision, the year of last revision. A number in parentheses indicates the year of last reapproval. A superscript epsilon (ε) indicates an editorial change since the last revision or reapproval.

1. Scope

1.1 This practice provides guidelines for the reporting of results of water analyses, including drinking water, wastewater, process water, ground water, and surface water, to laboratory clients in a complete and systematic fashion. Adequate documentation must be provided on the sample analyzed, the methods of analysis used, the results obtained, the precision and bias of the measurements, and related quality assurance information.

1.2 Results of chemical analysis of water shall be reported as a weight/volume ratio, such as milligrams per litre (mg/L), milliequivalents per litre (meq/L), etc., when concentration is being determined.

1.3 Results of other tests, such as pH, radioactivity, or turbidity, shall be reported as specified in the individual test methods.

1.4 The values stated in SI units are to be regarded as the standard.

1.5 *This standard does not purport to address all of the safety problems, if any, associated with its use. It is the responsibility of the user of this standard to establish appropriate safety and health practices and determine the applicability of regulatory limitations prior to use.*

2. Referenced Documents

2.1 *ASTM Standards:*

D 933 Practice for Reporting Results of Examination and Analysis of Water-Formed Deposits[2]

D 1129 Terminology Relating to Water[3]

D 2777 Practice for Determination of Precision and Bias of Applicable Methods of Committee D-19 on Water[3]

D 3856 Guide for Good Laboratory Practices in Laboratories Engaged in Sampling and Analysis of Water[3]

D 4210 Practice for Intralaboratory Quality Control Procedures and a Discussion on Reporting Low-Level Data[3]

D 4460 Practice for Calculating Precision Limits Where Values are Calculated from Other Test Methods[4]

E 29 Practice for Using Significant Digits in Test Data to Determine Conformance With Specifications[5]

ES 16 Practice for Generation of Environmental Data Related to Waste Management Activities[6]

3. Terminology

3.1 *Definitions*—For definition of terms used in this practice, refer to Terminology D 1129 and Practice ES 16.

3.2 *Descriptions of Terms Specific to This Standard:*

3.2.1 *milliequivalent per litre (meq/L)*—a weight-volume measurement obtained by multiplying the concentration expressed in moles per litre by the ionic charge or by the change in oxidation number of the substance in a defined reaction.

3.2.2 *milligrams per litre (mg/L)*—a weight-volume measurement that expresses the concentration of a solute in milligrams (10^{-3} g) in a litre of solution.

3.2.3 *micrograms per litre (μg/L)*—a weight-volume measurement that expresses the concentration of a solute in micrograms (10^{-6} g) in a litre of solution.

3.2.4 *surrogates*—compounds that are similar to analytes of interest in chemical composition and behavior, separation, and measurement, but that are not normally found in environmental samples. These compounds are added to blanks, standards, samples, or spiked samples prior to analysis to confirm the proper operation of the analytical system.

4. Significance and Use

4.1 The proper use of analytical data requires adequate documentation of the source and history of the sample, laboratory performing the analysis, method of analysis, date of analysis, precision and bias of the measurements, and related quality assurance information.

4.2 Tables are included for interconversion of data between units in common use.

4.3 Other information on reporting results may be included in individual test methods for the analysis of water.

4.4 For corresponding information regarding the reporting of results for water-formed deposits, see Practice D 933.

5. Sample Documentation

5.1 Information regarding the source and history of the sample to be included in the analytical report should define the sample and include the following, as appropriate:

5.1.1 Laboratory performing analysis,

5.1.2 Name and address of organization or person requesting analysis,

5.1.3 Specific location of sampling and complete identification of sample,

5.1.4 Date and time of sampling,

5.1.5 Sample identification number, and

5.1.6 Sampling method, treatment, and preservation.

5.2 In addition to the information in 5.1, the following information should be included as appropriate:

---

[1] This practice is under the jurisdiction of ASTM Committee D-19 on water and is the direct responsibility of Subcommittee D19.02 on General Specifications, Technical Resources, and Statistical Methods.

Current edition approved Sept. 15, 1991. Published February 1992. Originally published as D 596 – 40. Last previous edition D 596 – 83.

[2] *Annual Book of ASTM Standards*, Vol 11.02.

[3] *Annual Book of ASTM Standards*, Vol 11.01.

[4] *Annual Book of ASTM Standards*, Vol 04.03.

[5] *Annual Book of ASTM Standards*, Vol 14.02.

[6] *Annual Book of ASTM Standards*, Vol 11.04.

5.2.1 Identification of sampling organization and individual sampler,

5.2.2 Pressure and temperature of system sampled,

5.2.3 Flow rate of water in a stream,

5.2.4 Copies of sampling logs with signatures,

5.2.5 Chain of custody forms with signatures,

5.2.6 Results of field measurements, and

5.2.7 Description information (color, odor, etc.) clearly presented.

6. Analysis Documentation

6.1 As described in Guide D 3856, the test method of analysis should be specified in the analytical report for each determination performed on a sample. A reference of sufficient definition or a copy of the test method should be provided to the requestor of the analytical services.

6.2 The precision, bias, and detection limit of each analytical test method should be disclosed as part of either the test method or the analytical report. Consult Guide D 3856 for the quality control system from which estimates of precision and bias could be made, or review the procedure for determining single-operator precision of a test method as provided in Practice D 2777 for guidance. The procedure used to derive the detection limit should be identified along with any specific definitions associated with the derivation. Practice D 4210 is one of many sources for computing detection limits.

6.3 The date on which each determination is performed should be reported, as should other time-critical processes such as extractions. In some cases, such as in microbiology tests, it is critical to record the time that the test was started.

6.4 The analytical reports should clearly specify the form in which multi-atomic analytes, such as nitrate and orthophosphate, are reported.

6.5 If a sample is diluted prior to analysis, the sample dilution ratio(s) for the sample or involved constituents, and the reason for the dilution(s) should be documented on the analytical report.

7. Significant Figures

7.1 When recording direct measurements, test results should be reported by recording all digits that are known plus one that may be subject to change on repeated analysis. When calculating results from test data, rounding should be performed only on the final result, not upon the intermediate values employed in the calculation. For a discussion of the principles and practices for determining significant figures, refer to Practice E 29.

7.2 When a value is computed from two or more other test results, refer to Practice D 4460 for techniques of determining precision limits of the calculated value.

8. Documentation of Quality

8.1 Each sample analysis may have different quality needs based on the use of the data. This information should be determined before sampling and analysis. Based on this information, an analytical report may include the following information, as appropriate:

8.1.1 Recovery of any surrogate compounds with laboratory control limits,

8.1.2 Results of corresponding check samples or blank spikes with laboratory control limits,

8.1.3 Results of analysis of duplicate samples or duplicate matrix spiked samples and the percent difference with laboratory control limits,

8.1.4 Recoveries of any matrix spikes (and matrix spike duplicates) with laboratory control limits,

8.1.5 Results of all blanks,

8.1.6 Results of any reference samples run during sample analysis with laboratory control limits,

8.1.7 Calibration and tuning data, and

8.1.8 Chromatograms.

9. Review of Analytical Results

9.1 The deviation from a perfect balance between cations and anions determined in water samples may be appraised by totalling separately the determined concentrations in milliequivalents per litre (meq/L) of anions and cations. This can only be done if all major ions have been determined. The cation-anion difference, either positive or negative, may be calculated from the following empirical formula in which cations and anions are expressed in meq/L:[7]

$$\textit{percent cation-anion difference} = \frac{\Sigma\ \textit{cations} - \Sigma\ \textit{anions}}{\Sigma\ \textit{cations} + \Sigma\ \textit{anions}} \times 100$$

NOTE—With careful work, the difference will not generally exceed 2 % of the total cations or anions in waters of moderate concentrations (250 for 1000 mg/L). A somewhat larger percentage can be tolerated if the sum of cations and anions is less than about 5.00 meq/L.[8]

9.2 A comparison of results from the dissolved solids determination with dissolved solids calculated from individual major ion determinations is a helpful review procedure.

9.3 The procedure used to report analytical values when constituents of interest are found in a blank analysis should be described by the laboratory.

9.4 A quality assurance narrative should be used to explain any discrepancies in the data or unusual conditions that resulted in data of questionable quality (that is, matrix interferences, elevated detection limits).

9.5 The report should include the signature of the laboratory manager or a designee attesting to the review of the results reported and that the results meet the specified quality of data.

10. Conversion Factors

10.1 Table 1 lists factors for interconversion between units in common use.

10.2 Table 2 lists factors for interconversion of milligrams per litre and milliequivalents per litre of common ions.

11. Keywords

11.1 analysis; blank; cation-anion balance; results; surrogate; water

[7] Friedman, L. C., and Erdmann, D. E., "Quality Assurance Practices for the Chemical and Biological Analyses of Water and Fluvial Sediments," *Techniques of Water-Resources Investigations of the U.S. Geological Survey*, Book 5, Chapter A6, U.S. Government Printing Office, 1982.

[8] Hem, J. D., "Study and Interpretation of the Chemical Characteristics of Natural Water," *U.S. Geological Survey Water-Supply Paper 2254*, 1985.

n = coefficient of viscosity of the suspending medium (in this case water) in poises (varies with changes in temperature of the suspending medium),

L = distance from the surface of the suspension to the level at which the density of the suspension is being measured, cm. (For a given hydrometer and sedimentation cylinder, values vary according to the hydrometer readings. This distance is known as effective depth (Table 2)),

T = interval of time from beginning of sedimentation to the taking of the reading, min,

G = specific gravity of soil particles, and

G_l = specific gravity (relative density) of suspending medium (value may be used as 1.000 for all practical purposes).

NOTE 14—Since Stokes' law considers the terminal velocity of a single sphere falling in an infinity of liquid, the sizes calculated represent the diameter of spheres that would fall at the same rate as the soil particles.

15.2 For convenience in calculations the above equation may be written as follows:

$$D = K\sqrt{L/T}$$

where:

K = constant depending on the temperature of the suspension and the specific gravity of the soil particles. Values of K for a range of temperatures and specific gravities are given in Table 3. The value of K does not change for a series of readings constituting a test, while values of L and T do vary.

15.3 Values of D may be computed with sufficient accuracy, using an ordinary 10-in. slide rule.

NOTE 15—The value of L is divided by T using the A- and B-scales, the square root being indicated on the D-scale. Without ascertaining the value of the square root it may be multiplied by K, using either the C- or CI-scale.

16. Sieve Analysis Values for Portion Finer than No. 10 (2.00-mm) Sieve

16.1 Calculation of percentages passing the various sieves used in sieving the portion of the sample from the hydrometer test involves several steps. The first step is to calculate the mass of the fraction that would have been retained on the No. 10 sieve had it not been removed. This mass is equal to the total percentage retained on the No. 10 sieve (100 minus total percentage passing) times the mass of the total sample represented by the mass of soil used (as calculated in 14.2), and the result divided by 100.

16.2 Calculate next the total mass passing the No. 200 sieve. Add together the fractional masses retained on all the sieves, including the No. 10 sieve, and subtract this sum from the mass of the total sample (as calculated in 14.2).

16.3 Calculate next the total masses passing each of the other sieves, in a manner similar to that given in 12.2.

16.4 Calculate last the total percentages passing by dividing the total mass passing (as calculated in 16.3) by the total mass of sample (as calculated in 14.2), and multiply the result by 100.

17. Graph

17.1 When the hydrometer analysis is performed, a graph of the test results shall be made, plotting the diameters of the particles on a logarithmic scale as the abscissa and the percentages smaller than the corresponding diameters to an

TABLE 2 Values of Effective Depth Based on Hydrometer and Sedimentation Cylinder of Specified Sizes[A]

Hydrometer 151H		Hydrometer 152H			
Actual Hydrometer Reading	Effective Depth, L, cm	Actual Hydrometer Reading	Effective Depth, L, cm	Actual Hydrometer Reading	Effective Depth, L, cm
1.000	16.3	0	16.3	31	11.2
1.001	16.0	1	16.1	32	11.1
1.002	15.8	2	16.0	33	10.9
1.003	15.5	3	15.8	34	10.7
1.004	15.2	4	15.6	35	10.6
1.005	15.0	5	15.5		
1.006	14.7	6	15.3	36	10.4
1.007	14.4	7	15.2	37	10.2
1.008	14.2	8	15.0	38	10.1
1.009	13.9	9	14.8	39	9.9
1.010	13.7	10	14.7	40	9.7
1.011	13.4	11	14.5	41	9,6
1.012	13.1	12	14.3	42	9.4
1.013	12.9	13	14.2	43	9.2
1.014	12.6	14	14.0	44	9.1
1.015	12.3	15	13.8	45	8.9
1.016	12.1	16	13.7	46	8.8
1.017	11.8	17	13.5	47	8.6
1.018	11.5	18	13.3	48	8.4
1.019	11.3	19	13.2	49	8.3
1.020	11.0	20	13.0	50	8.1
1.021	10.7	21	12.9	51	7.9
1.022	10.5	22	12.7	52	7.8
1.023	10.2	23	12.5	53	7.6
1.024	10.0	24	12.4	54	7.4
1.025	9.7	25	12.2	55	7.3
1.026	9.4	26	12.0	56	7.1
1.027	9.2	27	11.9	57	7.0
1.028	8.9	28	11.7	58	6.8
1.029	8.6	29	11.5	59	6.6
1.030	8.4	30	11.4	60	6.5
1.031	8.1				
1.032	7.8				
1.033	7.6				
1.034	7.3				
1.035	7.0				
1.036	6.8				
1.037	6.5				
1.038	6.2				

[A] Values of effective depth are calculated from the equation:

$$L = L_1 + \tfrac{1}{2}[L_2 - (V_B/A)]$$

where:

L = effective depth, cm,

L_1 = distance along the stem of the hydrometer from the top of the bulb to the mark for a hydrometer reading, cm,

L_2 = overall length of the hydrometer bulb, cm,

V_B = volume of hydrometer bulb, cm^3, and

A = cross-sectional area of sedimentation cylinder, cm^2

Values used in calculating the values in Table 2 are as follows:

For both hydrometers, 151H and 152H:

L_2 = 14.0 cm

V_B = 67.0 cm^3

A = 27.8 cm^2

For hydrometer 151H:

L_1 = 10.5 cm for a reading of 1.000

= 2.3 cm for a reading of 1.031

For hydrometer 152H:

L_1 = 10.5 cm for a reading of 0 g/litre

= 2.3 cm for a reading of 50 g/litre

TABLE 3 Values of K for Use in Equation for Computing Diameter of Particle in Hydrometer Analysis

Temperature, °C	Specific Gravity of Soil Particles								
	2.45	2.50	2.55	2.60	2.65	2.70	2.75	2.80	2.85
16	0.01510	0.01505	0.01481	0.01457	0.01435	0.01414	0.01394	0.01374	0.01356
17	0.01511	0.01486	0.01462	0.01439	0.01417	0.01396	0.01376	0.01356	0.01338
18	0.01492	0.01467	0.01443	0.01421	0.01399	0.01378	0.01359	0.01339	0.01321
19	0.01474	0.01449	0.01425	0.01403	0.01382	0.01361	0.01342	0.1323	0.01305
20	0.01456	0.01431	0.01408	0.01386	0.01365	0.01344	0.01325	0.01307	0.01289
21	0.01438	0.01414	0.01391	0.01369	0.01348	0.01328	0.01309	0.01291	0.01273
22	0.01421	0.01397	0.01374	0.01353	0.01332	0.01312	0.01294	0.01276	0.01258
23	0.01404	0.01381	0.01358	0.01337	0.01317	0.01297	0.01279	0.01261	0.01243
24	0.01388	0.01365	0.01342	0.01321	0.01301	0.01282	0.01264	0.01246	0.01229
25	0.01372	0.01349	0.01327	0.01306	0.01286	0.01267	0.01249	0.01232	0.01215
26	0.01357	0.01334	0.01312	0.01291	0.01272	0.01253	0.01235	0.01218	0.01201
27	0.01342	0.01319	0.01297	0.01277	0.01258	0.01239	0.01221	0.01204	0.01188
28	0.01327	0.01304	0.01283	0.01264	0.01244	0.01255	0.01208	0.01191	0.01175
29	0.01312	0.01290	0.01269	0.01249	0.01230	0.01212	0.01195	0.01178	0.01162
30	0.01298	0.01276	0.01256	0.01236	0.01217	0.01199	0.01182	0.01165	0.01149

arithmetic scale as the ordinate. When the hydrometer analysis is not made on a portion of the soil, the preparation of the graph is optional, since values may be secured directly from tabulated data.

18. Report

18.1 The report shall include the following:

18.1.1 Maximum size of particles,

18.1.2 Percentage passing (or retained on) each sieve, which may be tabulated or presented by plotting on a graph (Note 16),

18.1.3 Description of sand and gravel particles:

18.1.3.1 Shape—rounded or angular,

18.1.3.2 Hardness—hard and durable, soft, or weathered and friable,

18.1.4 Specific gravity, if unusually high or low,

18.1.5 Any difficulty in dispersing the fraction passing the No. 10 (2.00-mm) sieve, indicating any change in type and amount of dispersing agent, and

18.1.6 The dispersion device used and the length of the dispersion period.

NOTE 16—This tabulation of graph represents the gradation of the sample tested. If particles larger than those contained in the sample were removed before testing, the report shall so state giving the amount and maximum size.

18.2 For materials tested for compliance with definite specifications, the fractions called for in such specifications shall be reported. The fractions smaller than the No. 10 sieve shall be read from the graph.

18.3 For materials for which compliance with definite specifications is not indicated and when the soil is composed almost entirely of particles passing the No. 4 (4.75-mm) sieve, the results read from the graph may be reported as follows:

(*1*) Gravel, passing 3-in. and retained on No. 4 sieve .... %
(*2*) Sand, passing No. 4 sieve and retained on No. 200 sieve .... %
(*a*) Coarse sand, passing No. 4 sieve and retained on No. 10 sieve .... %
(*b*) Medium sand, passing No. 10 sieve and retained on No. 40 sieve .... %
(*c*) Fine sand, passing No. 40 sieve and retained on No. 200 sieve .... %
(*3*) Silt size, 0.074 to 0.005 mm .... %
(*4*) Clay size, smaller than 0.005 mm .... %
Colloids, smaller than 0.001 mm .... %

18.4 For materials for which compliance with definite specifications is not indicated and when the soil contains material retained on the No. 4 sieve sufficient to require a sieve analysis on that portion, the results may be reported as follows (Note 17):

SIEVE ANALYSIS

Sieve Size	Percentage Passing
3-in.	....
2-in.	....
1½-in.	....
1-in.	....
¾-in.	....
⅜-in.	....
No. 4 (4.75-mm)	....
No. 10 (2.00-mm)	....
No. 40 (425-μm)	....
No. 200 (75-μm)	....
HYDROMETER ANALYSIS	
0.074 mm	....
0.005 mm	....
0.001 mm	....

NOTE 17—No. 8 (2.36-mm) and No. 50 (300-μm) sieves may be substituted for No. 10 and No. 40 sieves.

19. Keywords

19.1 grain-size; hydrometer analysis; hygroscopic moisture; particle-size; sieve analysis rangement; adsorbed water does not include water that is chemically combined within the clay minerals.

adsorption—the attachment of water molecules or ions to the surfaces of soil particles.

advancing slope grouting—a method of grouting by which the front of a mass of grout is caused to move horizontally by use of a suitable grout injection sequence.

aeolian deposits—wind-deposited material such as dune sands and loess deposits.

aggregate—*as a grouting material*, relatively inert granular mineral material, such as sand, gravel, slag, crushed stone, etc. "Fine aggregate" is material that will pass a No. 4 (6.4-mm) screen,

"Coarse aggregate" is material that will not pass a No. 4 (6.4-mm) screen. Aggregate is mixed with a cementing agent (such as Portland cement and water) to form a grout material.

agitator tank—a tank, usually vertical and with open top, with rotation paddles used to prevent segregation of grout after mixing.

air-space ratio, G_a (D)—ratio of: (*1*) volume of water that can be drained from a saturated soil or rock under the action of force of gravity, to (*2*) total volume of voids.

air-void ratio, G_v (D)—the ratio of: (*1*) the volume of air space, to (*2*) the total volume of voids in a soil or rock mass.

alkali aggregate reaction—a chemical reaction between Na_2O and K_2O in the cement and certain silicate minerals in the cement and certain silicate minerals in the aggregate, which causes expansion resulting in weakening and cracking of Portland cement grout. See **reactive aggregate**.

allowable bearing value (allowable soil pressure), q_a, p_a (FL^{-2})—the maximum pressure that can be permitted on foundation soil, giving consideration to all pertinent factors, with adequate safety against rupture of the soil mass or movement of the foundation of such magnitude that the structure is impaired.

allowable pile bearing load, Q_a, P_a (F)—the maximum load that can be permitted on a pile with adequate safety against movement of such magnitude that the structure is endangered.

alluvium—soil, the constituents of which have been transported in suspension by flowing water and subsequently deposited by sedimentation.

amplification factor—ratio of dynamic to static displacement.

amorphous peat—see **sapric peat**.

angle of external friction (angle of wall friction), δ (degrees)—angle between the abscissa and the tangent of the curve representing the relationship of shearing resistance to normal stress acting between soil and surface of another material.

angle of friction (angle of friction between solid bodies), ϕs (degrees)—angle whose tangent is the ratio between the maximum value of shear stress that resists slippage between two solid bodies at rest with respect to each other, and the normal stress across the contact surfaces.

angle of internal friction (angle of shear resistance), ϕ (degrees)—angle between the axis of normal stress and the tangent to the Mohr envelope at a point representing a given failure-stress condition for solid material.

angle of obliquity, α, β, ϕ, Ψ (degrees)—the angle between the direction of the resultant stress or force acting on a given plane and the normal to that plane.

angle of repose, α (degrees)—angle between the horizontal and the maximum slope that a soil assumes through natural processes. For dry granular soils the effect of the height of slope is negligible; for cohesive soils the effect of height of slope is so great that the angle of repose is meaningless.

angle of shear resistance—see **angle of internal friction**.

angle of wall friction—see **angle of external friction**.

angular aggregate—aggregate, the particles of which possess well-defined edges formed at the intersection of roughly planar faces.

anisotropic mass—a mass having different properties in different directions at any given point.

anisotropy—having different properties in different directions. (ISRM)

apparent cohesion—see **cohesion**.

aquiclude—a relatively impervious formation capable of absorbing water slowly but will not transmit it fast enough to furnish an appreciable supply for a well or spring.

aquifer—a water-bearing formation that provides a ground water reservoir.

aquitard—a confining bed that retards but does not prevent the flow of water to or from an adjacent aquifer; a leaky confining bed.

arching—the transfer of stress from a yielding part of a soil or rock mass to adjoining less-yielding or restrained parts of the mass.

area grouting—grouting a shallow zone in a particular area utilizing holes arranged in a pattern or grid.

Discussion—This type of grouting is sometimes referred to as blanket or consolidation grouting.

area of influence of a well, α (L^2)—area surrounding a well within which the piezometric surface has been lowered when pumping has produced the maximum steady rate of flow.

area ratio of a sampling spoon, sampler, or sampling tube, A_r (D)—the area ratio is an indication of the volume of soil displaced by the sampling spoon (tube), calculated as follows:

$$A_r = [(D_e^2 - D_i^2/D_i^2] \times 100$$

where:

D_e = maximum external diameter of the sampling spoon, and

D_i = minimum internal diameter of the sampling spoon at the cutting edge.

armor—the artificial surfacing of bed, banks, shore, or embankment to resist erosion or scour.

armor stone—(generally one ton to three tons in weight) stone resulting from blasting, cutting, or by other methods to obtain rock heavy enough to require handling two individual pieces by mechanical means.

ash content—the percentage by dry weight of material remaining after an oven dry organic soil or peat is burned by a prescribed method.

attenuation—reduction of amplitude with time or distance.

"B" horizon—see **horizon**.

average interstitial velocity—see **velocity, average interstitial**.

backpack grouting—the filling with grout of the annular space between a permanent tunnel lining and the surrounding formation.

Discussion—Same as crown grouting and backfill grouting.

back-packing—any material (usually granular) that is used to fill the empty space between the lagging and the rock surface. (ISRM)

baffle—a pier, weir, sill, fence, wall, or mound built on the bed of a stream to parry, deflect, check, or regulate the flow or to float on the surface to dampen the wave action.

base—*in grouting*, main component in a grout system.

base course (base)—a layer of specified or selected material of planned thickness constructed on the subgrade or subbase for the purpose of serving one or more functions such as distributing load, providing drainage, minimizing frost action, etc.

base exchange—the physicochemical process whereby one species of ions adsorbed on soil particles is replaced by another species.

batch—*in grouting*, quantity of grout mixed at one time.

batch method—*in grouting*, a quantity of grout materials are mixed or catalyzed at one time prior to injection.

batch mixer—*in grouting*, a machine that mixes batches of grout, in contrast to a continuous mixer.

bearing capacity—see **ultimate bearing capacity**.

bearing capacity (of a pile), Q_p, P_p (F)—the load per pile required to produce a condition of failure.

bedding—applies to rocks resulting from consolidation of sediments and exhibiting surfaces of separation (bedding planes) between layers of the same or different materials, that is, shale, siltstone, sandstone, limestone, etc. (ISRM)

bedding—collective term signifying the existence of layers of beds. Planes or other surfaces dividing sedimentary rocks of the same or different lithology.

bedrock—the more or less continuous body of rock which underlies the overburden soils. (ISRM)

bedrock (ledge)—rock of relatively great thickness and extent in its native location.

bench—(*1*) the unexcavated rock having a nearly horizontal surface which remains after a top heading has been excavated, or (*2*) step in a slope; formed by a horizontal surface and a surface inclined at a steeper angle than that of the entire slope. (ISRM)

bending—process of deformation normal to the axis of an elongated structural member when a moment is applied normal to its long axis. (ISRM)

bentonitic clay—a clay with a high content of the mineral montmorillonite, usually characterized by high swelling on wetting.

berm—a shelf that breaks the continuity of a slope.

biaxial compression—compression caused by the application of normal stresses in two perpendicular directions. (ISRM)

biaxial state of stress—state of stress in which one of the three principal stresses is zero. (ISRM)

binder (soil binder)—portion of soil passing No. 40 (425-μm) U.S. standard sieve,

binder—anything that causes cohesion in loosely assembled substances, such as clay or cement.

bit—any device that may be attached to or is an integral part of a drill string and is used as a cutting tool to bore into or penetrate rock or other materials.

blaine fineness—the fineness of powdered materials, such as cement and pozzolans, expressed as surface area usually in square centimetres per gram.

blanket grouting—a method in which relatively closely spaced shallow holes are drilled and grouted on a grid pattern over an area, for the purpose of making the upper portions of the bedrock stronger and less pervious.

blastibility—index value of the resistance of a rock formation to blasting. (ISRM)

blasting cap (detonator, initiator)—a small tube containing a flashing mixture for firing explosives. (ISRM)

bleeding—*in grouting*, the autogeneous flow of mixing water within, or its emergence from, newly placed grout caused by the settlement of the solid materials within the mass.

bleeding rate—*in grouting*, the rate at which water is released from grout by bleeding.

blocking—wood blocks placed between the excavated surface of a tunnel or shaft and the main bracing system. (ISRM)

body force—a force such as gravity whose effect is distributed throughout a material body by direct action on each elementary part of the body independent of the others. (ISRM)

bog—a peat covered area with a high water table and a surface dominated by a carpet of mosses, chiefly sphagnum. It is generally nutrient poor and acidic. It may be treed or treeless.

bond strength—*in grouting*, resistance to separation of set grout from other materials with which it is in contact; a collective expression for all forces such as adhesion, friction, and longitudinal shear.

bottom charge—concentrated explosive charge at the bottom of a blast hole. (ISRM)

boulder clay—a geological term used to designate glacial drift that has not been subjected to the sorting action of water and therefore contains particles from boulders to clay sizes.

boulders—a rock fragment, usually rounded by weathering or abrasion, with an average dimension of 12 in. (305 mm) or more.

breakwater stone—(generally three tons to twenty tons in weight) stone resulting from blasting, cutting, or other means to obtain rock heavy enough to require handling individual pieces by mechanical means.

buckling—a bulge, bend, bow, kink, or wavy condition produced in sheets, plates, columns, or beams by compressive stresses.

bulb of pressure—see **pressure bulb**.

bulkhead—a steep or vertical structure supporting natural or artificial embankment.

bulking—the increase in volume of a material due to manipulation. Rock bulks upon being excavated; damp sand bulks if loosely deposited, as by dumping, because the apparent cohesion prevents movement of the soil particles to form a reduced volume.

buoyant unit weight (*submerged unit weight*)—see **unit weight**.

dilatancy—property of volume increase under loading. (ISRM)

dilatancy—the expansion of cohesionless soils when subject to shearing deformation.

direct shear test—a shear test in which soil or rock under an applied normal load is stressed to failure by moving one section of the sample or sample container (shear box) relative to the other section.

discharge velocity, v, q (LT^{-1})—rate of discharge of water through a porous medium per unit of total area perpendicular to the direction of flow.

discontinuity surface—any surface across which some property of a rock mass is discontinuous. This includes fracture surfaces, weakness planes, and bedding planes, but the term should not be restricted only to mechanical continuity. (ISRM)

dispersing agent—*in grouting,* an addition or admixture that promotes dispersion of particulate grout ingredients by reduction of interparticle attraction.

dispersing agent—see **deflocculating agent**.

dispersion—the phenomenon of varying speed of transmission of waves, depending on their frequency. (ISRM)

displacement—a change in position of a material point. (ISRM)

displacement grouting—injection of grout into a formation in such a manner as to move the formation; it may be controlled or uncontrolled. See also **penetration grouting**.

distortion—a change in shape of a solid body. (ISRM)

divergence loss—that part of transmitted energy lost due to spreading of wave rays in accordance with the geometry of the system.

double amplitude—total or peak to peak excursion.

drag bit—a noncoring or full-hole boring bit, which scrapes its way through relatively soft strata. (ISRM)

drain—a means for intercepting, conveying, and removing water.

drainage curtain—*in grouting,* a row of open holes drilled parallel to and downstream from the grout curtain of a dam for the purpose of reducing uplift pressures.

Discussion—Depth is ordinarily approximately one-third to one-half that of the grout curtain.

drainage gallery—*in grouting,* an opening or passageway from which grout holes or drainage curtain holes, or both, are drilled. See also **grout gallery**.

drawdown (L)—vertical distance the free water elevation is lowered or the pressure head is reduced due to the removal of free water.

drill—a machine or piece of equipment designed to penetrate earth or rock formations, or both.

drillability—index value of the resistance of a rock to drilling. (ISRM)

drill carriage; jumbo—a movable platform, stage, or frame that incorporates several rock drills and usually travels on the tunnel track; used for heavy drilling work in large tunnels. (ISRM)

drilling pattern—the number, position, depth, and angle of the blastholes forming the complete round in the face of a tunnel or sinking pit. (ISRM)

drill mud—*in grouting,* a dense fluid or slurry used in rotary drilling; to prevent caving of the bore hole walls, as a circulation medium to carry cuttings away from the bit and out of the hole, and to seal fractures or permeable formations, or both, preventing loss of circulation fluid.

Discussion—The most common drill mud is a water-bentonite mixture, however, many other materials may be added or substituted to increase density or decrease viscosity.

dry pack—a cement-sand mix with minimal water content used to fill small openings or repair imperfections in concrete.

dry unit weight (*dry density*)—see **unit weight**.

ductility—condition in which material can sustain permanent deformation without losing its ability to resist load. (ISRM)

dye tracer—*in grouting,* an additive whose primary purpose is to change the color of the grout or water.

earth—see **soil**.

earth pressure—the pressure or force exerted by soil on any boundary.

	Symbol	Unit
Pressure	p	FL^{-2}
Force	P	F or FL^{-1}

active earth pressure, P_A, p_A—the minimum value of earth pressure. This condition exists when a soil mass is permitted to yield sufficiently to cause its internal shearing resistance along a potential failure surface to be completely mobilized.

earth pressure at rest, P_o, p_o—the value of the earth pressure when the soil mass is in its natural state without having been permitted to yield or without having been compressed.

passive earth pressure, P_p, p_p—the maximum value of earth pressure. This condition exists when a soil mass is compressed sufficiently to cause its internal shearing resistance along a potential failure surface to be completely mobilized.

effect diameter (effective size), D_{10}, D_e (L)—particle diameter corresponding to 10 % finer on the grain-size curve.

effective drainage porosity—see **effective porosity**.

effective force, $\bar{F}$ (F)—the force transmitted through a soil or rock mass by intergranular pressures.

effective porosity (effective drainage porosity), n_e (D)—the ratio of: (*1*) the volume of the voids of a soil or rock mass that can be drained by gravity, to (*2*) the total volume of the mass.

effective pressure—see **stress**.

effective size—see **effective diameter**.

effective stress—see **stress**.

effective unit weight—see **unit weight**.

efflux time—time required for all grout to flow from a flow cone.

elasticity—property of material that returns to its original form or condition after the applied force is removed. (ISRM)

elastic limit—point on stress strain curve at which transition from elastic to inelastic behavior takes place. (ISRM)

elastic state of equilibrium—state of stress within a soil mass when the internal resistance of the mass is not fully mobilized.

elastic strain energy—potential energy stored in a strained solid and equal to the work done in deforming the solid from its unstrained state less any energy dissipated by inelastic deformation. (ISRM)

electric log—a record or log of a borehole obtained by lowering electrodes into the hole and measuring any of the various electrical properties of the rock formations or materials traversed.

electrokinetics—involves the application of an electric field to soil for the purpose of dewatering materials of very low permeability to enhance stability. The electric field produces negative pore pressures near a grout pipe that facilitates grout injection.

emulsifier—a substance that modifies the surface tension of colloidal droplets, keeping them from coalescing, and keeping them suspended.

emulsion—a system containing dispersed colloidal droplets.

endothermic—pertaining to a reaction that occurs with the adsorption of heat.

envelope grouting—grouting of rock surrounding a hydraulic pressure tunnel for purpose of consolidation, and primarily, reduction of permeability.

epoxy—a multicomponent resin grout that usually provides very high, tensile, compressive, and bond strengths.

equipotential line—see **piezometric line**.

equivalent diameter (equivalent size), D (L)—the diameter of a hypothetical sphere composed of material having the same specific gravity as that of the actual soil particle and of such size that it will settle in a given liquid at the same terminal velocity as the actual soil particle.

equivalent fluid—a hypothetical fluid having a unit weight such that it will produce a pressure against a lateral support presumed to be equivalent to that produced by the actual soil. This simplified approach is valid only when deformation conditions are such that the pressure increases linearly with depth and the wall friction is neglected.

excess hydrostatic pressure—see **hydrostatic pressure**.

exchange capacity—the capacity to exchange ions as measured by the quantity of exchangeable ions in a soil or rock.

excitation (stimulus)—an external force (or other input) applied to a system that causes the system to respond in some way.

exothermic—pertaining to a reaction that occurs with the evolution of heat.

expansive cement—a cement that tends to increase in volume after it is mixed with water.

extender—an additive whose primary purpose is to increase total grout volume.

extension—linear strain associated with an increase in length. (ISRM)

external force—a force that acts across external surface elements of a material body. (ISRM)

extrados—the exterior curved surface of an arch, as opposed to intrados, which is the interior curved surface of an arch. (ISRM)

fabric—*for rock or soil*, the spatial configuration of all textural and structural features as manifested by every recognizable material unit from crystal lattices to large scale features requiring field studies.

fabric—the orientation in space of the elements composing the rock substance. (ISRM)

face (heading)—the advanced end of a tunnel, drift, or excavation at which work is progressing. (ISRM)

facing—the outer layer of revetment.

failure (in rocks)—exceeding the maximum strength of the rock or exceeding the stress or strain requirement of a specific design. (ISRM)

failure by rupture—see **shear failure**.

failure criterion—specification of the mechanical condition under which solid materials fail by fracturing or by deforming beyond some specified limit. This specification may be in terms of the stresses, strains, rate-of-change of stresses, rate-of-change of strains, or some combination of these quantities, in the materials.

failure criterion—theoretically or empirically derived stress or strain relationship characterizing the occurrence of failure in the rock. (ISRM)

false set—*in grouting*, the rapid development of rigidity in a freshly mixed grout without the evolution of much heat.

Discussion—Such rigidity can be dispelled and plasticity regained by further mixing without the addition of water; premature stiffening, hesitation set, early stiffening, and rubber set are other much used terms referring to the same phenomenon.

fatigue—the process of progressive localized permanent structural change occurring in a material subjected to conditions that produce fluctuating stresses and strains at some point or points and that may culminate in cracks or complete fracture after a sufficient number of fluctuations.

fatigue—decrease of strength by repetitive loading. (ISRM)

fatigue limit—point on stress-strain curve below which no fatigue can be obtained regardless of number of loading cycles. (ISRM)

fault—a fracture or fracture zone along which there has been displacement of the two sides relative to one another parallel to the fracture (this displacement may be a few centimetres or many kilometres). (See also **joint fault set** and **joint fault system**. (ISRM)

fault breccia—the assemblage of broken rock fragments frequently found along faults. The fragments may vary in size from inches to feet. (ISRM)

fault gouge—a clay-like material occurring between the walls of a fault as a result of the movement along the fault surfaces. (ISRM)

fiber—*for peats and organic soils*, a fragment or piece of plant tissue that retains a recognizable cellular structure and is large enough to be retained after wet sieving on a 100-mesh sieve (openings 0.15 mm).

fibric peat—peat in which the original plant fibers are slightly decomposed (greater than 67 % fibers).

fibrous peat—see **fibric peat**.

field moisture equivalent—see **moisture equivalent**.

fill—man-made deposits of natural soils or rock products and waste materials.

filling—generally, the material occupying the space between joint surfaces, faults, and other rock discontinuities. The filling material may be clay, gouge, various natural cementing agents, or alteration products of the adjacent rock. (ISRM)

filter bedding stone—(generally 6-in. minus material) stone placed under graded riprap stone or armor stone in a layer or combination of layers designed and installed in such a

compressibility—property of a soil or rock pertaining to its susceptibility to decrease in volume when subjected to load.

compression curve—see **pressure-void ratio curve.**

compression index, C_c (D)—the slope of the linear portion of the pressure-void ratio curve on a semi-log plot.

compression wave (irrotational)—wave in which element of medium changes volume without rotation.

compressive strength (unconfined or uniaxial compressive strength), p_c, q_u, C_o (FL^{-2})—the load per unit area at which an unconfined cylindrical specimen of soil or rock will fail in a simple compression test. Commonly the failure load is the maximum that the specimen can withstand in the test.

compressive stress—normal stress tending to shorten the body in the direction in which it acts. (ISRM)

concentration factor, n (D)—a parameter used in modifying the Boussinesq equations to describe various distributions of vertical stress.

cone of impression, *n*—a rise of the potentiometric surface in the approximate shape of a cone that develops around an injection well.

conjugate joints (faults)—two sets of joints (faults) that formed under the same stress conditions (usually shear pairs). (ISRM)

connate water, *n*-water entrapped in the voids of a sedimentary or extrusive igneous rock at the time of its deposition or emplacement.

consistency—the relative ease with which a soil can be deformed.

consistency—*in grouting,* the relative mobility or ability of freshly mixed mortar or grout to flow; the usual measurements are slump for stiff mixtures and flow for more fluid grouts.

consistency index—see **relative consistency**.

consolidated-drained test (slow test)—a soil test in which essentially complete consolidation under the confining pressure is followed by additional axial (or shearing) stress applied in such a manner that even a fully saturated soil of low permeability can adapt itself completely (fully consolidate) to the changes in stress due to the additional axial (or shearing) stress.

consolidated-undrained test (consolidated quick test)—a soil test in which essentially complete consolidation under the vertical load (in a direct shear test) or under the confining pressure (in a triaxial test) is followed by a shear at constant water content.

consolidation—the gradual reduction in volume of a soil mass resulting from an increase in compressive stress.

initial consolidation (*initial compression*)—a comparatively sudden reduction in volume of a soil mass under an applied load due principally to expulsion and compression of gas in the soil voids preceding primary consolidation.

primary consolidation (*primary compression*) (*primary time effect*)—the reduction in volume of a soil mass caused by the application of a sustained load to the mass and due principally to a squeezing out of water from the void spaces of the mass and accompanied by a transfer of the load from the soil water to the soil solids.

secondary consolidation (*secondary compression*) (*secondary time effect*)—the reduction in volume of a soil mass caused by the application of a sustained load to the mass and due principally to the adjustment of the internal structure of the soil mass after most of the load has been transferred from the soil water to the soil solids.

consolidation curve—see **consolidation time curve.**

consolidation grouting—injection of a fluid grout, usually sand and Portland cement, into a compressible soil mass in order to displace it and form a lenticular grout structure for support.

Discussion—In rock, grouting is performed for the purpose of strengthening the rock mass by filling open fractures and thus eliminating a source of settlement.

consolidation ratio, U_s (D)—the ratio of: (*1*) the amount of consolidation at a given distance from a drainage surface and at a given time, to (*2*) the total amount of consolidation obtainable at that point under a given stress increment.

consolidation test—a test in which the specimen is laterally confined in a ring and is compressed between porous plates.

consolidation-time curve (time curve) (consolidation curve) (theoretical time curve)—a curve that shows the relation between: (*1*) the degree of consolidation, and (*2*) the elapsed time after the application of a given increment of load.

constitutive equation—force deformation function for a particular material. (ISRM)

contact grouting—see **backpack grouting**.

contact pressure, p (FL^{-2})—the unit of pressure that acts at the surface of contact between a structure and the underlying soil or rock mass.

continuous mixer—a mixer into which the ingredients of the mixture are fed without stopping, and from which the mixed product is discharged in a continuous stream.

contraction—linear strain associated with a decrease in length. (ISRM)

controlled blasting—includes all forms of blasting designed to preserve the integrity of the remaining rocks, that is, smooth blasting or pre-splitting. (ISRM)

controlled-strain test—a test in which the load is so applied that a controlled rate of strain results.

controlled-stress test—a test in which the stress to which a specimen is subjected is applied at a controlled rate.

convergence—generally refers to a shortening of the distance between the floor and roof of an opening, for example, in the bedded sedimentary rocks of the coal measures where the roof sags and the floor heaves. Can also apply to the convergence of the walls toward each other. (ISRM)

core—a cylindrical sample of hardened grout, concrete, rock, or grouted deposits, usually obtained by means of a core drill.

core drilling; diamond drilling—a rotary drilling technique, using diamonds in the cutting bit, that cuts out cylindrical rock samples. (ISRM)

core recovery—ratio of the length of core recovered to the length of hole drilled, usually expressed as a percentage.

cover—the perpendicular distance from any point in the roof of an underground opening to the ground surface. (ISRM)

cover—*in grouting,* the thickness of rock and soil material overlying the stage of the hole being grouted.

crack—a small fracture, that is, small with respect to the scale of the feature in which it occurs. (ISRM)

crater—excavation (generally of conical shape) generated by an explosive charge. (ISRM)

creep—slow movement of rock debris or soil usually imperceptible except to observations of long duration. Time-dependent strain or deformation, for example, continuing strain with sustained stress.

critical circle (critical surface)—the sliding surface assumed in a theoretical analysis of a soil mass for which the factor of safety is a minimum.

critical damping—the minimum viscous damping that will allow a displaced system to return to its initial position without oscillation.

critical density—the unit weight of a saturated granular material below which it will lose strength and above which it will gain strength when subjected to rapid deformation. The critical density of a given material is dependent on many factors.

critical frequency, f_c—frequency at which maximum or minimum amplitudes of excited waves occur.

critical height, H_c (L)—the maximum height at which a vertical or sloped bank of soil or rock will stand unsupported under a given set of conditions.

critical hydraulic gradient—see **hydraulic gradient**.

critical slope—the maximum angle with the horizontal at which a sloped bank of soil or rock of given height will stand unsupported.

critical surface—see **critical circle**.

critical void ratio—see **void ratio**.

crown—also roof or back, that is, the highest point of the cross section. *In tunnel linings*, the term is used to designate either the arched roof above spring lines or all of the lining except the floor or invert. (ISRM)

cryology—the study of the properties of snow, ice, and frozen ground.

cure—*in grouting*, the change in properties of a grout with time.

cure time—*in grouting*, the interval between combining all grout ingredients or the formation of a gel and substantial development of its potential properties.

curtain grouting—injection of grout into a sub-surface formation in such a way as to create a barrier of grouted material transverse to the direction of the anticipated water flow.

cuttings—small-sized rock fragments produced by a rock drill. (ISRM)

damping—reduction in the amplitude of vibration of a body or system due to dissipation of energy internally or by radiation. (ISRM)

damping ratio—*for a system with viscous damping*, the ratio of actual damping coefficient to the critical damping coefficient.

decay time—the interval of time required for a pulse to decay from its maximum value to some specified fraction of that value. (ISRM)

decomposition—*for peats and organic soils*, see **humification**.

decoupling—the ratio of the radius of the blasthole to the radius of the charge. In general, a reducing of the strain wave amplitude by increasing the spacing between charge and blasthole wall. (ISRM)

deflocculating agent (deflocculant) (dispersing agent)—an agent that prevents fine soil particles in suspension from coalescing to form flocs.

deformability—*in grouting*, a measure of the elasticity of the grout to distort in the interstitial spaces as the sediments move.

deformation—change in shape or size.

deformation—a change in the shape or size of a solid body. (ISRM)

deformation resolution (deformation sensitivity), R_d (L)—ratio of the smallest subdivision of the indicating scale of a deformation-measuring device to the sensitivity of the device.

degree-days—the difference between the average temperature each day and 32°F (0°C). In common usage degree-days are positive for daily average temperatures above 32°F and negative for those below 32°F (see **freezing index**).

degree of consolidation (percent consolidation), U (D)—the ratio, expressed as a percentage, of: (*1*) the amount of consolidation at a given time within a soil mass, to (*2*) the total amount of consolidation obtainable under a given stress condition.

degrees-of-freedom—the minimum number of independent coordinates required in a mechanical system to define completely the positions of all parts of the system at any instant of time. In general, it is equal to the number of independent displacements that are possible.

degree of saturation—see **percent saturation**.

degree of saturation—the extent or degree to which the voids in rock contain fluid (water, gas, or oil). Usually expressed in percent related to total void or pore space. (ISRM)

degree of sensitivity (*sensitivity ratio*)—see **remolding index**.

delay—time interval (fraction of a second) between detonation of explosive charges. (ISRM)

density—the mass per unit, ρ (ML^{-3}) kg/m^3.

density of dry soil or rock, ρ_d (ML^{-3}) kg/m^3—the mass of solid particles per the total volume of soil or rock.

density of saturated soil or rock, ρ_{sat} (ML^{-3}) kg/m^3—the total mass per total volume of completely saturated soil or rock.

density of soil or rock (*bulk density*), ρ (ML^{-3}) kg/m^3—the total mass (solids plus water) per total volume.

density of solid particles, ρ_s (ML^{-3}) kg/m^3—the mass per volume of solid particles.

density of submerged soil or rock, ρ_{sub} (ML^{-3}) kg/m^3—the difference between the density of saturated soil or rock, and the density of water.

density of water, ρ_w (ML^{-3}) kg/m^3—the mass per volume of water.

detonation—an extremely rapid and violent chemical reaction causing the production of a large volume of gas. (ISRM)

deviator stress, Δ, σ (FL^{-2})—the difference between the major and minor principal stresses in a triaxial test.

deviator of stress (strain)—the stress (strain) tensor obtained by subtracting the mean of the normal stress (strain) components of a stress (strain) tensor from each normal stress (strain) component. (ISRM)

differential settlement—settlement that varies in rate or amount, or both, from place to place across a structure.

burden—*in an explosive blasting*, the distance between the charge and the free face of the material to be blasted.

burden—distance between charge and free surface in direction of throw. (ISRM)

"C" Horizon—see **horizon**.

California bearing ratio, *CBR* (D)—the ratio of: (*1*) the force per unit area required to penetrate a soil mass with a 3-in.^{2} (19-cm)^2 circular piston (approximately 2-in. (51-mm) diameter) at the rate of 0.05 in. (1.3 mm)/min, to (*2*) that required for corresponding penetration of a standard material. The ratio is usually determined at 0.1-in. (2.5-mm) penetration, although other penetrations are sometimes used. Original California procedures required determination of the ratio at 0.1-in. intervals to 0.5 in. (12.7 mm). Corps of Engineers' procedures require determination of the ratio at 0.1 in. and 0.2 in. (5.1 mm). Where the ratio at 0.2 in. is consistently higher than at 0.1 in., the ratio at 0.2 in. is used.

camouflet—the underground cavity created by a fully contained explosive. (ISRM)

capillary action (capillarity)—the rise or movement of water in the interstices of a soil or rock due to capillary forces.

capillary flow—see **capillary action.**

capillary fringe zone—the zone above the free water elevation in which water is held by capillary action.

capillary head, *h* (L)—the potential, expressed in head of water, that causes the water to flow by capillary action.

capillary migration—see **capillary action**.

capillary rise (height of capillary rise), h_c (L)—the height above a free water elevation to which water will rise by capillary action.

capillary water—water subject to the influence of capillary action.

catalyst—a material that causes chemical reactions to begin.

catalyst system—those materials that, in combination, cause chemical reactions to begin; catalyst systems normally consist of an initiator (catalyst) and an activator.

cation—an ion that moves, or would move toward a cathode; thus nearly always synonymous with positive ion.

cation exchange—see **base exchange**.

cavity—a natural underground opening that may be small or large.

cavity—underground opening created by a fully contained explosive. (ISRM)

cement factor—quantity of cement contained in a unit volume of concrete or grout, expressed as weight, or volume (specify which).

cement grout—a grout in which the primary cementing agent is Portland cement.

cementitious factor—quantity of cement and other cementitious materials contained in a unit volume of concrete or grout, expressed as weight or volume (specify which).

centrifuge moisture equivalent—see **moisture equivalent.**

chamber—a large room excavated underground, for example, for a powerhouse, pump station, or for storage. (ISRM)

chamber blasting (coyotehole blasting)—a method of quarry blasting in which large explosive charges are confined in small tunnel chambers inside the quarry face. (ISRM)

chemical grout—any grouting material characterized by being a true solution; no particles in suspension. See also **particulate grout**.

chemical grout system—any mixture of materials used for grouting purposes in which all elements of the system are true solutions (no particles in suspension).

chip—crushed angular rock fragment of a size smaller than a few centimetres. (ISRM)

chisel—the steel cutting tool used in percussion drilling. (ISRM)

circuit grouting—a grouting method by which grout is circulated through a pipe extending to the bottom of the hole and back up the hole via the annular space outside the pipe. Then the excess grout is diverted back over a screen to the agitator tank by means of a packing gland at the top of the hole. The method is used where holes tend to cave and sloughing material might otherwise clog openings to be grouted.

clay (clay soil)—fine-grained soil or the fine-grained portion of soil that can be made to exhibit plasticity (putty-like properties) within a range of water contents, and that exhibits considerable strength when air-dry. The term has been used to designate the percentage finer than 0.002 mm (0.005 mm in some cases), but it is strongly recommended that this usage be discontinued, since there is ample evidence from an engineering standpoint that the properties described in the above definition are many times more important.

clay size—that portion of the soil finer than 0.002 mm (0.005 mm in some cases) (see also **clay**).

clay soil—see **clay**.

cleavage—*in crystallography*, the splitting, or tendency to split, along planes determined by the crystal structure. *In petrology*, a tendency to cleave or split along definite, parallel, closely spaced planes. It is a secondary structure, commonly confined to bedded rocks.

cleavage—the tendency to cleave or split along definite parallel planes, which may be highly inclined to the bedding. It is a secondary structure and is ordinarily accompanied by at least some recrystallization of the rock. (ISRM)

cleavage planes—the parallel surfaces along which a rock or mineral cleaves or separates; the planes of least cohesion, usually parallel to a certain face of the mineral or crystal.

cleft water—water that exists in or circulates along the geological discontinuities in a rock mass.

closure—the opening is reduced in dimension to the extent that it cannot be used for its intended purpose. (ISRM)

closure—*in grouting*, closure refers to achieving the desired reduction in grout take by splitting the hole spacing. If closure is being achieved, there will be a progressive decrease in grout take as primary, secondary, tertiary, and quanternary holes are grouted.

cobble (cobblestone)—a rock fragment, usually rounded or semirounded, with an average dimension between 3 and 12 in. (75 and 305 mm).

coefficient of absolute viscosity—see **coefficient of viscosity.**

coefficient of active earth pressure—see **coefficient of earth pressure.**

coefficient of compressibility (coefficient of compression), α_v (L^2F^{-1})—the secant slope, for a given pressure increment, of the pressure-void ratio curve. Where a stress-strain curve is used, the slope of this curve is equal to $\alpha_v/(1 + e)$.

coefficient of consolidation, c_v (L^2T^{-1})—a coefficient utilized in the theory of consolidation, containing the physical constants of a soil affecting its rate of volume change.

$$c_v = k\,(1 + e)/\alpha_v\gamma_w$$

where:

k = coefficient of permeability, LT^{-1},
e = void ratio, D,
α_v = coefficient of compressibility, L^2F^{-1}, and
γ_w = unit weight of water, FL^{-3}.

Discussion—In the literature published prior to 1935, the coefficient of consolidation, usually designated c, was defined by the equation:

$$c = k/\alpha_v\gamma_w\,(1 + e)$$

This original definition of the coefficient of consolidation may be found in some more recent papers and care should be taken to avoid confusion.

coefficient of earth pressure, K (D)—the principal stress ratio at a point in a soil mass.

coefficient of earth pressure, active, K_A (D)—the minimum ratio of : (*1*) the minor principal stress, to (*2*) the major principal stress. This is applicable where the soil has yielded sufficiently to develop a lower limiting value of the minor principal stress.

coefficient of earth pressure, at rest, K_O (D)—the ratio of: (*1*) the minor principal stress, to (*2*) the major principal stress. This is applicable where the soil mass is in its natural state without having been permitted to yield or without having been compressed.

coefficient of earth pressure, passive, K_P (D)—the maximum ratio of: (*1*) the major principal stress, to (*2*) the minor principal stress. This is applicable where the soil has been compressed sufficiently to develop an upper limiting value of the major principal stress.

coefficient of friction (coefficient of friction between solid bodies), f (D)—the ratio between the maximum value of shear stress that resists slippage between two solid bodies with respect to each other, and the normal stress across the contact surfaces. The tangent of the angle of friction is ϕs.

coefficient of friction, f—a constant proportionality factor, μ, relating normal stress and the corresponding critical shear stress at which sliding starts between two surfaces: $T = \mu \cdot \sigma$. (ISRM)

coefficient of internal friction, μ (D)—the tangent of the angle of internal friction (angle of shear resistance) (see **internal friction**).

coefficient of permeability (permeability), k (LT^{-1})—the rate of discharge of water under laminar flow conditions through a unit cross-sectional area of a porous medium under a unit hydraulic gradient and standard temperature conditions (usually 20°C).

coefficient of shear resistance—see **coefficient of internal friction,** μ (D).

coefficient of subgrade reaction (modulus of subgrade reaction), k, k_s (FL^{-3})—ratio of: (*1*) load per unit area of horizontal surface of a mass of soil, to (*2*) corresponding settlement of the surface. It is determined as the slope of the secant, drawn between the point corresponding to zero settlement and the point of 0.05-in. (1.3-mm) settlement, of a load-settlement curve obtained from a plate load test on a soil using a 30-in. (762-mm) or greater diameter loading plate. It is used in the design of concrete pavements by the Westergaard method.

coefficient of transmissibility—the rate of flow of water in gallons per day through a vertical strip of the aquifer 1 ft (0.3 m) wide, under a unit hydraulic gradient.

coefficient of uniformity, C_u (D)—the ratio D_{60}/D_{10}, where D_{60} is the particle diameter corresponding to 60 % finer on the cumulative particle-size distribution curve, and D_{10} is the particle diameter corresponding to 10 % finer on the cumulative particle-size distribution curve.

coefficient of viscosity (coefficient of absolute viscosity), η (FTL^{-2})—the shearing force per unit area required to maintain a unit difference in velocity between two parallel layers of a fluid a unit distance apart.

coefficient of volume compressibility (modulus of volume change), m_v (L^2F^{-1})—the compression of a soil layer per unit of original thickness due to a given unit increase in pressure. It is numerically equal to the coefficient of compressibility divided by one plus the original void ratio, or $a_v/(1 + e)$.

cohesion—shear resistance at zero normal stress (an equivalent term in rock mechanics is intrinsic shear strength). (ISRM)

cohesion, c (FL^{-2})—the portion of the shear strength of a soil indicated by the term c, in Coulomb's equation, $s = c + p \tan \phi$. See **intrinsic shear strength**.

apparent cohesion—cohesion in granular soils due to capillary forces.

cohesionless soil—a soil that when unconfined has little or no strength when air-dried and that has little or no cohesion when submerged.

cohesive soil—a soil that when unconfined has considerable strength when air-dried and that has significant cohesion when submerged.

collar—*in grouting*, the surface opening of a borehole.

colloidal grout—*in grouting*, a grout in which the dispersed solid particles remain in suspension (colloids).

colloidal mixer—*in grouting*, a mixer designed to produce colloidal grout.

colloidal particles—particles that are so small that the surface activity has an appreciable influence on the properties of the aggregate.

communication—*in grouting*, subsurface movement of grout from an injection hole to another hole or opening.

compaction—the densification of a soil by means of mechanical manipulation.

compaction curve (Proctor curve) (moisture-density curve)—the curve showing the relationship between the dry unit weight (density) and the water content of a soil for a given compactive effort.

compaction test (moisture-density test)—a laboratory compacting procedure whereby a soil at a known water content is placed in a specified manner into a mold of given dimensions, subjected to a compactive effort of controlled magnitude, and the resulting unit weight determined. The procedure is repeated for various water contents sufficient to establish a relation between water content and unit weight.

manner as to prevent the loss of underlying soil or finer bedding materials due to moving water.

filter (protective filter)—a layer or combination of layers of pervious materials designed and installed in such a manner as to provide drainage, yet prevent the movement of soil particles due to flowing water.

final set—*in grouting*, a degree of stiffening of a grout mixture greater than initial set, generally stated as an empirical value indicating the time in hours and minutes that is required for cement paste to stiffen sufficiently to resist the penetration of a weighted test needle.

fineness—a measure of particle-size.

fineness modulus—an empirical factor obtained by adding the total percentages of an aggregate sample retained on each of a specified series of sieves, and dividing the sum by 100; in the United States, the U.S. Standard sieve sizes are: No. 100 (149 μm), No. 50 (297 μm), No. 30 (590 μm), No. 16 (1,190 μm), No. 8 (2,380 μm), and No. 4 (4,760 μm) and ⅜ in. (9.5 mm), ¾ in. (19 mm), 1½ in. (38 mm), 3 in. (76 mm), and 6 in. (150 mm).

fines—portion of a soil finer than a No. 200 (75-μm) U.S. standard sieve.

finite element—one of the regular geometrical shapes into which a figure is subdivided for the purpose of numerical stress analysis. (ISRM)

fishing tool—*in grouting*, a device used to retrieve drilling equipment lost or dropped in the hole.

fissure—a gapped fracture. (ISRM)

flash set—*in grouting*, the rapid development of rigidity in a freshly mixed grout, usually with the evolution of considerable heat; this rigidity cannot be dispelled nor can the plasticity be regained by further mixing without addition of water; also referred to as quick set or grab set.

floc—loose, open-structured mass formed in a suspension by the aggregation of minute particles.

flocculation—the process of forming flocs.

flocculent structure—see **soil structure**.

floor—bottom of near horizontal surface of an excavation, approximately parallel and opposite to the roof. (ISRM)

flow channel—the portion of a flow net bounded by two adjacent flow lines.

flow cone—*in grouting*, a device for measurement of grout consistency in which a predetermined volume of grout is permitted to escape through a precisely sized orifice, the time of efflux (flow factor) being used as the indication of consistency.

flow curve—the locus of points obtained from a standard liquid limit test and plotted on a graph representing water content as ordinate on an arithmetic scale and the number of blows as abscissa on a logarithmic scale.

flow failure—failure in which a soil mass moves over relatively long distances in a fluid-like manner.

flow index, F_w, I_f (D)—the slope of the flow curve obtained from a liquid limit test, expressed as the difference in water contents at 10 blows and at 100 blows.

flow line—the path that a particle of water follows in its course of seepage under laminar flow conditions.

flow net—a graphical representation of flow lines and equipotential (piezometric) lines used in the study of seepage phenomena.

flow slide—the failure of a sloped bank of soil in which the movement of the soil mass does not take place along a well-defined surface of sliding.

flow steady, *n*—a characteristic of a flow system where the specific discharge is constant in time at any point.

flow value, N_ϕ (degrees)—a quantity equal to tan [45 deg + (ϕ/2)].

flow velocity,—see **specific discharge**.

fluidifier—*in grouting*, an admixture employed in grout to increase flowability without changing water content.

fly ash—the finely divided residue resulting from the combustion of ground or powdered coal and which is transported from the firebox through the boiler by flue gases.

fold—a bend in the strata or other planar structure within the rock mass. (ISRM)

foliation—the somewhat laminated structure resulting from segregation of different minerals into layers parallel to the schistosity. (ISRM)

footing—portion of the foundation of a structure that transmits loads directly to the soil.

footwall—the mass of rock beneath a discontinuity surface. (ISRM)

forced vibration (forced oscillation)—vibration that occurs if the response is imposed by the excitation. If the excitation is periodic and continuing, the oscillation is steady-state.

forepoling—driving forepoles (pointed boards or steel rods) ahead of the excavation, usually over the last set erected, to furnish temporary overhead protection while installing the next set. (ISRM)

foundation—lower part of a structure that transmits the load to the soil or rock.

foundation soil—upper part of the earth mass carrying the load of the structure.

fracture—the general term for any mechanical discontinuity in the rock; it therefore is the collective term for joints, faults, cracks, etc. (ISRM)

fracture—a break in the mechanical continuity of a body of rock caused by stress exceeding the strength of the rock. Includes joints and faults.

fracture frequency—the number of natural discontinuities in a rock or soil mass per unit length, measured along a core or as exposed in a planar section such as the wall of a tunnel.

fracture pattern—spatial arrangement of a group of fracture surfaces. (ISRM)

fracturing—*in grouting*, intrusion of grout fingers, sheets, and lenses along joints, planes of weakness, or between the strata of a formation at sufficient pressure to cause the strata to move away from the grout.

fragmentation—the breaking of rock in such a way that the bulk of the material is of a convenient size for handling. (ISRM)

free water (gravitational water) (ground water) (phreatic water)—water that is free to move through a soil or rock mass under the influence of gravity.

free water elevation (water table) (ground water surface) (free water surface) (ground water elevation)—elevations at which the pressure in the water is zero with respect to the atmospheric pressure.

freezing index, *F* (degree-days)—the number of degree-days between the highest and lowest points on the cumulative degree-days—time curve for one freezing season. It is used as a measure of the combined duration and magnitude of below-freezing temperature occurring during any given freezing season. The index determined for air temperatures at 4.5 ft (1.4 m) above the ground is commonly designated as the air freezing index, while that determined for temperatures immediately below a surface is known as the surface freezing index.

free vibration—vibration that occurs in the absence of forced vibration.

frequency, f (T^{-1})—number of cycles occurring in unit time.

frost action—freezing and thawing of moisture in materials and the resultant effects on these materials and on structures of which they are a part or with which they are in contact.

frost boil—(*a*) softening of soil occurring during a thawing period due to the liberation of water form ice lenses or layers.

(*b*) the hole formed in flexible pavements by the extrusion of soft soil and melt waters under the action of wheel loads.

(*c*) breaking of a highway or airfield pavement under traffic and the ejection of subgrade soil in a soft and soupy condition caused by the melting of ice lenses formed by frost action.

frost heave—the raising of a surface due to the accumulation of ice in the underlying soil or rock.

fundamental frequency—lowest frequency of periodic variation.

gage length, L (L)—distance over which the deformation measurement is made.

gage protector—*in grouting,* a device used to transfer grout pressure to a gage without the grout coming in actual contact with the gage.

gage saver—see **gage protector**.

gel—*in grouting,* the condition where a liquid grout begins to exhibit measurable shear strength.

gel time—*in grouting,* the measured time interval between the mixing of a grout system and the formation of a gel.

general shear failure—see **shear failure**.

glacial till (till)—material deposited by glaciation, usually composed of a wide range of particle sizes, which has not been subjected to the sorting action of water.

gradation (grain-size distribution) (texture)—the proportions by mass of a soil or fragmented rock distributed in specified particle-size ranges.

grain-size analysis (mechanical analysis) (particle-size analysis)—the process of determining grain-size distribution.

gravel—rounded or semirounded particles of rock that will pass a 3-in. (76.2-mm) and be retained on a No. 4 (4.75-μm) U.S. standard sieve.

gravitational water—see **free water**.

gravity grouting—grouting under no applied pressure other than the height of fluid in the hole.

groin—bank or shore-protection structure in the form of a barrier placed oblique to the primary motion of water, designed to control movement of bed load.

ground arch—the theoretical stable rock arch that develops some distance back from the surface of the opening and supports the opening. (ISRM)

ground water—that part of the subsurface water that is in the saturated zone.

Discussion—Loosely, all subsurface water as distinct from surface water.

ground-water barrier—soil, rock, or artificial material which has a relatively low permeability and which occurs below the land surface where it impedes the movement of ground water and consequently causes a pronounced difference in the potentiometric level on opposite sides of the barrier.

ground-water basin—a ground-water system that has defined boundaries and may include more than one aquifer of permeable materials, which are capable of furnishing a significant water supply.

Discussion—A basin is normally considered to include the surface area and the permeable materials beneath it. The surface-water divide need not coincide with ground-water divide.

ground-water discharge—the water released from the zone of saturation; also the volume of water released.

ground-water divide—a ridge in the water table or other potentiometric surface from which ground water moves away in both directions normal to the ridge line.

ground-water elevation—see **free water elevation.**

ground-water flow—the movement of water in the zone of saturation.

ground-water level—the level below which the rock and subsoil, to unknown depths, are saturated. (ISRM)

ground-water, perched—see **perched ground-water.**

ground-water recharge—the process of water addition to the saturated zone; also the volume of water added by this process.

ground-water surface—see **free water elevation.**

grout—*in soil and rock grouting,* a material injected into a soil or rock formation to change the physical characteristics of the formation.

groutability—the ability of a formation to accept grout.

groutability ratio of granular formations—the ratio of the 15 % size of the formation particles to be grouted to the 85 % size of grout particles (suspension-type grout). This ratio should be greater than 24 if the grout is to successfully penetrate the formation.

groutable rock bolts—rock bolts with hollow cores or with tubes adapted to the periphery of the bolts and extending to the bottom of the bolts to facilitate filling the holes surrounding the bolts with grout.

grouted-aggregate concrete—concrete that is formed by injecting grout into previously placed coarse aggregate. See also **preplaced aggregate concrete**.

grout cap—a "cap" that is formed by placing concrete along the top of a grout curtain. A grout cap is often used in weak foundation rock to secure grout nipples, control leakage, and to form an impermeable barrier at the top of a grout curtain.

grout gallery—an opening or passageway within a dam utilized for grouting or drainage operations, or both.

grout header—a pipe assembly attached to a ground hole, and to which the grout lines are attached for injecting grout. Grout injector is monitored and controlled by means of valves and a pressure gate mounted on the header; sometimes called grout manifold.

oscillation—the variation, usually with time, of the magnitude of a quantity with respect to a specified reference when the magnitude is alternately greater and smaller than the reference.

outcrop—the exposure of the bedrock at the surface of the ground. (ISRM)

overbreak—the quantity of rock that is excavated or breaks out beyond the perimeter specified as the finished excavated tunnel outline. (ISRM)

overburden—the loose soil, sand, silt, or clay that overlies bedrock. In some usages it refers to all material overlying the point of interest (tunnel crown), that is, the total cover of soil and rock overlying an underground excavation. (ISRM)

overburden load—the load on a horizontal surface underground due to the column of material located vertically above it. (ISRM)

overconsolidated soil deposit—a soil deposit that has been subjected to an effective pressure greater than the present overburden pressure.

overconsolidation ratio, OCR—the ratio of preconsolidation vertical stress to the current effective overburden stress.

packer—*in grouting,* a device inserted into a hole in which grout or water is to be injected which acts to prevent return of the grout or water around the injection pipe; usually an expandable device actuated mechanically, hydraulically, or pneumatically.

paddle mixer—a mixer consisting essentially of a trough within which mixing paddles revolve about the horizontal axis, or a pan within which mixing blades revolve about the vertical axis.

pan mixer—a mixer comprised of a horizontal pan or drum in which mixing is accomplished by means of the rotating pan of fixed or rotating paddles, or both; rotation is about a vertical axis.

parent material—material from which a soil has been derived.

particle-size analysis—see **grain-size analysis**.

particle-size distribution—see **gradation, grain-size distribution**.

particulate grout—any grouting material characterized by undissolved (insoluble) particles in the mix. See also **chemical grout**.

passive earth pressure—see **earth pressure**.

passive state of plastic equilibrium—see **plastic equilibrium**.

path percolation (line of creep)—the path that water follows along the surface of contact between the foundation soil or rock and the base of a dam or other structure.

pavement pumping—ejection of soil and water mixtures from joints, cracks, and edges of rigid pavements, under the action of traffic.

peak shear strength—maximum shear strength along a failure surface. (ISRM)

peat—a naturally occurring highly organic substance derived primarily from plant materials.

Discussion—Peat is distinguished from other organic soil materials by its lower ash content (less than 25 % ash by dry weight) and from other phytogenic material of higher rank (that is, lignite coal) by its lower calorific value on a water saturated basis.

peatland—areas having peat-forming vegetation on which peak has accumulated or is accumulating.

penetrability—a grout property descriptive of its ability to fill a porous mass; primarily a function of lubricity and viscosity.

penetration—depth of hole cut in rock by a drill bit. (ISRM)

penetration grouting—filling joints or fractures in rock or pore spaces in soil with a grout without disturbing the formation; this grouting method does not modify the solid formation structure. See also **displacement grouting**.

penetration resistance (standard penetration resistance) (Proctor penetration resistance), p_R, N (FL^{-2} or Blows L^{-1})—(*a*) number of blows of a hammer of specified weight falling a given distance required to produce a given penetration into soil of a pile, casing, or sampling tube.

(*b*) unit load required to maintain constant rate of penetration into soil of a probe or instrument.

(*c*) unit load required to produce a specified penetration into soil at a specified rate of a probe or instrument. For a Proctor needle, the specified penetration is 2½ in. (63.5 mm) and the rate is ½ in. (12.7 mm)/s.

penetration resistance curve (Proctor penetration curve)—the curve showing the relationship between: (*1*) the penetration resistance, and (*2*) the water content.

percent compaction—the ratio, expressed as a percentage, of: (*1*) dry unit weight of a soil, to (*2*) maximum unit weight obtained in a laboratory compaction test.

percent consolidation—see **degree of consolidation**.

percent fines—amount, expressed as a percentage by weight, of a material in aggregate finer than a given sieve, usually the No. 200 (74 µm) sieve.

percent saturation (degree of saturation), S,S_r (D)—the ratio, expressed as a percentage, of: (*1*) the volume of water in a given soil or rock mass, to (*2*) the total volume of intergranular space (voids).

perched ground water—unconfined ground water separated from an underlying body of ground water by an unsaturated zone.

perched water table—a water table usually of limited area maintained above the normal free water elevation by the presence of an intervening relatively impervious confining stratum.

perched water table—groundwater separated from an underlying body of groundwater by unsaturated soil or rock. Usually located at a higher elevation than the groundwater table. (ISRM)

percolation—the movement of gravitational water through soil (see **seepage**).

percolation—movement, under hydrostatic pressure of water through the smaller interstices of rock or soil, excluding movement through large openings such as caves and solution channels. (ISRM)

percussion drilling—a drilling technique that uses solid or hollow rods for cutting and crushing the rock by repeated blows. (ISRM)

percussion drilling—a drilling process in which a hole is advanced by using a series of impacts to the drill steel and attached bit; the bit is normally rotated during drilling. See **rotary drilling**.

period—time interval occupied by one cycle.

permafrost—perennially frozen soil.

permanent strain—the strain remaining in a solid with respect to its initial condition after the application and removal of stress greater than the yield stress (commonly also called "residual" strain). (ISRM)

permeability—see **coefficient of permeability**.

permeability—the capacity of a rock to conduct liquid or gas. It is measured as the proportionality constant, k, between flow velocity, v, and hydraulic gradient, I; $v = k \cdot I$. (ISRM)

permeability intrinsic, n (L^{-3})—a measure of the ease with which a porous medium can transmit a fluid under a potential gradient.

Discussion—Intrinsic permeability is a property of the medium alone and is independent of the nature of the fluid and of the force field causing movement.

permeation grouting—filling joints or fractures in rock or pore spaces in soil with a grout, without disturbing the formation.

pH, *pH* (D)—an index of the acidity or alkalinity of a soil in terms of the logarithm of the reciprocal of the hydrogen ion concentration.

phase difference—difference between phase angles of two waves of same frequency.

phase of periodic quantity—fractional part of period through which independent variable has advanced, measured from an arbitrary origin.

phreatic line—the trace of the phreatic surface in any selected plane of reference.

phreatic line—see **line of seepage**.

phreatic surface—see **free water elevation**.

phreatic water—see **free water**.

piezometer—an instrument for measuring pressure head.

piezometric line (equipotential line)—line along which water will rise to the same elevation in piezometric tubes.

piezometric surface—the surface at which water will stand in a series of piezometers.

piezometric surface—an imaginary surface that everywhere coincides with the static level of the water in the aquifer. (ISRM)

pile—relatively slender structural element which is driven, or otherwise introduced, into the soil, usually for the purpose of providing vertical or lateral support.

pillar—in-situ rock between two or more underground openings: crown pillars; barrier pillars; rib pillars; sill pillars; chain pillars; etc. (ISRM)

pilot drift (pioneer tunnel)—a drift or tunnel first excavated as a smaller section than the dimensions of the main tunnel. A pilot drift or tunnel is usually used to investigate rock conditions in advance of the main tunnel, to permit installation of bracing before the principal mass of rock is removed, or to serve as a drainage tunnel. (ISRM)

piping—the progressive removal of soil particles from a mass by percolating water, leading to the development of channels.

pit—an excavation in the surface of the earth from which ore is obtained as in large open pit mining or as an excavation made for test purposes, that is, a testpit. (ISRM)

plane of weakness—surface or narrow zone with a (shear or tensile) strength lower than that of the surrounding material. (ISRM)

plane stress (strain)—a state of stress (strain) in a solid body in which all stress (strain) components normal to a certain plane are zero. (ISRM)

plane wave—wave in which fronts are parallel to plane normal to direction of propagation.

plastic deformation—see **plastic flow**.

plastic equilibrium—state of stress within a soil or rock mass or a portion thereof, which has been deformed to such an extent that its ultimate shearing resistance is mobilized.

active state of plastic equilibrium—plastic equilibrium obtained by an expansion of a mass.

passive state of plastic equilibrium—plastic equilibrium obtained by a compression of a mass.

plastic flow (plastic deformation)—the deformation of a plastic material beyond the point of recovery, accompanied by continuing deformation with no further increase in stress.

plasticity—the property of a soil or rock which allows it to be deformed beyond the point of recovery without cracking or appreciable volume change.

plasticity—property of a material to continue to deform indefinitely while sustaining a constant stress. (ISRM)

plasticity index, I_p, PI, I_w (D)—numerical difference between the liquid limit and the plastic limit.

plasticizer—*in grouting*, a material that increases the plasticity of a grout, cement paste, or mortar.

plastic limit, w_p, PL, P_w (D)—(*a*) the water content corresponding to an arbitrary limit between the plastic and the semisolid states of consistency of a soil. (*b*) water content at which a soil will just begin to crumble when rolled into a thread approximately ⅛ in. (3.2 mm) in diameter.

plastic soil—a soil that exhibits plasticity.

plastic state (plastic range)—the range of consistency within which a soil or rock exhibits plastic properties.

Poisson's ratio, (ν)—ratio between linear strain changes perpendicular to and in the direction of a given uniaxial stress change.

pore pressure (*pore water pressure*)—see **neutral stress** under stress.

pore water—water contained in the voids of the soil or rock.

porosity, n (D)—the ratio, usually expressed as a percentage, of: (*1*) the volume of voids of a given soil or rock mass, to (*2*) the total volume of the soil or rock mass.

porosity—the ratio of the aggregate volume of voids or interstices in a rock or soil to its total volume. (ISRM)

portal—the surface entrance to a tunnel. (ISRM)

positive displacement pump—a pump that will continue to build pressure until the power source is stalled if the pump outlet is blocked.

potential drop, Δh (L)—the difference in total head between two equipotential lines.

power spectral density—the limiting mean-square value (for example, of acceleration, velocity, displacement, stress, or other random variable) per unit bandwidth, that is the limit of the mean-square value in a given rectangular bandwidth divided by the bandwidth, as the bandwidth approaches zero.

pozzolan—a siliceous or siliceous and aluminous material, which in itself possesses little or no cementitious value but will, in finely divided form and in the presence of moisture, chemically react with calcium hydroxide at ordinary temperatures to form compounds possessing cementitious properties.

linear (normal) strain—the change in length per unit of length in a given direction. (ISRM)

line of creep (path of percolation)—the path that water follows along the surface of contact between the foundation soil and the base of a dam or other structure.

line of seepage (seepage line) (phreatic line)—the upper free water surface of the zone of seepage.

linear expansion, L_e (D)—the increase in one dimension of a soil mass, expressed as a percentage of that dimension at the shrinkage limit, when the water content is increased from the shrinkage limit to any given water content.

linear shrinkage, L_s (D)—decrease in one dimension of a soil mass, expressed as a percentage of the original dimension, when the water content is reduced from a given value to the shrinkage limit.

lineation—the parallel orientation of structural features that are lines rather than planes; some examples are parallel orientation of the long dimensions of minerals; long axes of pebbles; striae on slickensides; and cleavage-bedding plane intersections. (ISRM)

liquefaction—the process of transforming any soil from a solid state to a liquid state, usually as a result of increased pore pressure and reduced shearing resistance.

liquefaction potential—the capability of a soil to liquefy or develop cyclic mobility.

liquefaction (spontaneous liquefaction)—the sudden large decrease of the shearing resistance of a cohesionless soil. It is caused by a collapse of the structure by shock or other type of strain and is associated with a sudden but temporary increase of the prefluid pressure. It involves a temporary transformation of the material into a fluid mass.

liquid, limit, *LL*, L_w, w_L (D)—*(a)* the water content corresponding to the arbitrary limit between the liquid and plastic states of consistency of a soil.

(b) the water content at which a pat of soil, cut by a groove of standard dimensions, will flow together for a distance of ½ in. (12.7 mm) under the impact of 25 blows in a standard liquid limit apparatus.

liquidity index (water-plasticity ratio) (relative water content), *B*, R_w, I_L (D)—the ratio, expressed as a percentage, of: *(1)* the natural water content of a soil minus its plastic limit, to *(2)* its plasticity index.

liquid-volume measurement—*in grouting*, measurement of grout on the basis of the total volume of solid and liquid constituents.

lithology—the description of rocks, especially sedimentary clastics and especially in hand specimens and in outcrops, on the basis of such characteristics as color, structures, mineralogy, and particle size.

loam—a mixture of sand, silt, or clay, or a combination of any of these, with organic matter (see **humus**).

Discussion—It is sometimes called topsoil in contrast to the subsoils that contain little or no organic matter.

local shear failure—see **shear failure**.

loess—a uniform aeolian deposit of silty material having an open structure and relatively high cohesion due to cementation of clay or calcareous material at grain contacts.

Discussion—A characteristic of loess deposits is that they can stand with nearly vertical slopes.

logarithmic decrement—the natural logarithm of the ratio of any two successive amplitudes of like sign, in the decay of a single-frequency oscillation.

longitudinal rod wave—see **compression wave**.

longitudinal wave, v_l (LT^{-1})—wave in which direction of displacement at each point of medium is normal to wave front, with propagation velocity, calculated as follows:

$$y_l = \sqrt{(E/\rho)[(1 - \nu)/(1 + \nu)(1 - 2\nu)]} = \sqrt{(\lambda + 2\mu)/\rho}$$

where:

E = Young's modulus,
ρ = mass density,
λ and μ = Lamé's constants, and
ν = Poisson's ratio.

long wave (quer wave), *W* (LT^{-1})—dispersive surface wave with one horizontal component, generally normal to the direction of propagation, which decreases in propagation velocity with increase in frequency.

losing stream, *n*—a stream or reach of a stream in which water flows from the stream bed into the ground.

Discussion—synonymous with influent stream.

lubricity—*in grouting*, the physico-chemical characteristic of a grout material flow through a soil or rock that is the inverse of the inherent friction of that material to the soil or rock; comparable to "wetness."

lugeon—a measure of permeability defined by a pump-in test or pressure test, where one Lugeon unit is a water take of 1 L/min per metre of hole at a pressure of 10 bars.

major principal plane—see **principal plane**.

major principal stress—see **stress**.

manifold—see **grout header**.

marl—calcareous clay, usually containing from 35 to 65 % calcium carbonate ($CaCO_3$).

marsh—a wetland characterized by grassy surface mats which are frequently interspersed with open water or by a closed canopy of grasses, sedges, or other herbacious plants.

mass unit weight—see **unit weight**.

mathematical model—the representation of a physical system by mathematical expressions from which the behavior of the system can be deduced with known accuracy. (ISRM)

matrix—*in grouting*, a material in which particles are embedded, that is, the cement paste in which the fine aggregate particles of a grout are embedded.

maximum amplitude (L, LT^{-1}, LT^{-2})—deviation from mean or zero point.

maximum density (*maximum unit weight*)—see **unit weight**.

mechanical analysis—see **grain-size analysis**.

mesic peat—see **hemic peat**.

metering pump—a mechanical arrangement that permits pumping of the various components of a grout system in any desired proportions or in fixed proportions. (*Syn.* proportioning pump, variable proportion pump.)

microseism—seismic pulses of short duration and low amplitude, often occurring previous to failure of a material or structure. (ISRM)

minor principal plane—see **principal plane**.

minor principal stress—see **stress**.

mixed-in-place pile—a soil-cement pile, formed in place by forcing a grout mixture through a hollow shaft into the ground where it is mixed with the in-place soil with an auger-like head attached to the hollow shaft.

mixer—a machine employed for blending the constituents of grout, mortar, or other mixtures.

mixing cycle—the time taken for the loading, mixing, and unloading cycle.

mixing speed—the rotation rate of a mixer drum or of the paddles in an open-top, pan, or trough mixer, when mixing a batch; expressed in revolutions per minute.

modifier—*in grouting*, an additive used to change the normal chemical reaction or final physical properties of a grout system.

modulus of deformation—see **modulus of elasticity.**

modulus of elasticity (modulus of deformation), *E, M* (FL^{-2})—the ratio of stress to strain for a material under given loading conditions; numerically equal to the slope of the tangent or the secant of a stress-strain curve. The use of the term **modulus of elasticity** is recommended for materials that deform in accordance with Hooke's law; the term **modulus of deformation** for materials that deform otherwise.

modulus of subgrade reaction—see **coefficient of subgrade reaction.**

modulus of volume change—see **coefficient of volume compressibility.**

Mohr circle—a graphical representation of the stresses acting on the various planes at a given point.

Mohr circle of stress (strain)—a graphical representation of the components of stress (strain) acting across the various planes at a given point, drawn with reference to axes of normal stress (strain) and shear stress (strain). (ISRM)

Mohr envelope—the envelope of a sequence of Mohr circles representing stress conditions at failure for a given material. (ISRM)

Mohr envelope (rupture envelope) (rupture line)—the envelope of a series of Mohr circles representing stress conditions at failure for a given material.

Discussion—According to Mohr's rupture hypothesis, a rupture envelope is the locus of points the coordinates of which represent the combinations of normal and shearing stresses that will cause a given material to fail.

moisture content—see **water content.**

moisture-density curve—see **compaction curve.**

moisture-density test—see **compaction test.**

moisture equivalent:

centrifuge moisture equivalent, W_e, *CME* (D)—the water content of a soil after it has been saturated with water and then subjected for 1 h to a force equal to 1000 times that of gravity.

field moisture equivalent, FME—the minimum water content expressed as a percentage of the weight of the oven-dried soil, at which a drop of water placed on a smoothed surface of the soil will not immediately be absorbed by the soil but will spread out over the surface and give it a shiny appearance.

montmorillonite—a group of clay minerals characterized by a weakly bonded sheet-like internal molecular structure; consisting of extremely finely divided hydrous aluminum or magnesium silicates that swell on wetting, shrink on drying, and are subject to ion exchange.

muck—stone, dirt, debris, or useless material; or an organic soil of very soft consistency.

mud—a mixture of soil and water in a fluid or weakly solid state.

mudjacking—see **slab jacking.**

multibench blasting—the blasting of several benches (steps) in quarries and open pits, either simultaneously or with small delays. (ISRM)

multiple-row blasting—the drilling, charging, and firing of several rows of vertical holes along a quarry or opencast face. (ISRM)

muskeg—level, practically treeless areas supporting dense growth consisting primarily of grasses. The surface of the soil is covered with a layer of partially decayed grass and grass roots which is usually wet and soft when not frozen.

mylonite—a microscopic breccia with flow structure formed in fault zones. (ISRM)

natural frequency—the frequency at which a body or system vibrates when unconstrained by external forces. (ISRM)

natural frequency (displacement resonance), f_n—frequency for which phase angle is 90° between the direction of the excited force (or torque) vector and the direction of the excited excursion vector.

neat cement grout—a mixture of hydraulic cement and water without any added aggregate or filler materials.

Discussion—This may or may not contain admixture.

neutral stress—see **stress.**

newtonian fluid—a true fluid that tends to exhibit constant viscosity at all rates of shear.

node—point, line, or surface of standing wave system at which the amplitude is zero.

normal force—a force directed normal to the surface element across which it acts. (ISRM)

normal stress—see **stress.**

normally consolidated soil deposit—a soil deposit that has never been subjected to an effective pressure greater than the existing overburden pressure.

no-slump grout—grout with a slump of 1 in. (25 mm) or less according to the standard slump test (Test Method C 143).[2] See also **slump** and slump test.

open cut—an excavation through rock or soil made through a hill or other topographic feature to facilitate the passage of a highway, railroad, or waterway along an alignment that varies in topographic relief. An open cut can be comprised of single slopes or multiple slopes, or multiple slopes and horizontal benches, or both. (ISRM)

optimum moisture content (optimum water content), *OMC,* w_o (D)—the water content at which a soil can be compacted to a maximum dry unit weight by a given compactive effort.

organic clay—a clay with a high organic content.

organic silt—a silt with a high organic content.

organic soil—soil with a high organic content.

Discussion—In general, organic soils are very compressible and have poor load-sustaining properties.

organic terrain—see **peatland.**

[2] *Annual Book of ASTM Standards,* Vol 04.02.

grout mix—the proportions or amounts of the various materials used in the grout, expressed by weight or volume. (The words "by volume" or "by weight" should be used to specify the mix.)

grout nipple—*in grouting*, a short length of pipe, installed at the collar of the grout hole, through which drilling is done and to which the grout header is attached for the purpose of injecting grout.

grout slope—the natural slope of grout injected into preplaced-aggregate or other porous mass.

grout system—formulation of different materials used to form a grout.

grout take—the measured quantity of grout injected into a unit volume of formation, or a unit length of grout hole.

hanging wall—the mass of rock above a discontinuity surface. (ISRM)

hardener—*in grouting*, in a two component epoxy or resin, the chemical component that causes the base component to cure.

hardness—resistance of a material to indentation or scratching. (ISRM)

hardpan—a hard impervious layer, composed chiefly of clay, cemented by relatively insoluble materials, that does not become plastic when mixed with water and definitely limits the downward movement of water and roots.

head—pressure at a point in a liquid, expressed in terms of the vertical distance of the point below the surface of the liquid. (ISRM)

heat of hydration—heat evolved by chemical reactions with water, such as that evolved during the setting and hardening of Portland cement.

heave—upward movement of soil caused by expansion or displacement resulting from phenomena such as: moisture absorption, removal of overburden, driving of piles, frost action, and loading of an adjacent area.

height of capillary rise—see **capillary rise**.

hemic peat—peat in which the original plant fibers are moderately decomposed (between 33 and 67 % fibers).

heterogeneity—having different properties at different points. (ISRM)

homogeneity—having different properties at different points. (ISRM)

homogeneity—having the same properties at all points. (ISRM)

homogeneous mass—a mass that exhibits essentially the same physical properties at every point throughout the mass.

honeycomb structure—see **soil structure**.

horizon (soil horizon)—one of the layers of the soil profile, distinguished principally by its texture, color, structure, and chemical content.

"A" horizon—the uppermost layer of a soil profile from which inorganic colloids and other soluble materials have been leached. Usually contains remnants of organic life.

"B" horizon—the layer of a soil profile in which material leached from the overlying "A" horizon is accumulated.

"C" horizon—undisturbed parent material from which the overlying soil profile has been developed.

humic peat—see **sapric peat**.

humification—a process by which organic matter decomposes.

Discussion—The degree of humification for peats is indicated by the state of the fibers. In slightly decomposed material, most of the volume consists of fibers. In moderately decomposed material, the fibers may be preserved but may break down with disturbance, such as rubbing between the fingers. In highly decomposed materials, fibers will be virtually absent; see **von Post humification scale**.

humus—a brown or black material formed by the partial decomposition of vegetable or animal matter; the organic portion of soil.

hydration—formation of a compound by the combining of water with some other substance.

hydraulic conductivity—see **coefficient of permeability**.

hydraulic fracturing—the fracturing of an underground strata by pumping water or grout under a pressure in excess of the tensile strength and confining pressure; also called hydrofracturing.

hydraulic gradient, i, s (D)—the loss of hydraulic head per unit distance of flow, dh/dL.

critical hydraulic gradient, i_c (D)—hydraulic gradient at which the intergranular pressure in a mass of cohesionless soil is reduced to zero by the upward flow of water.

hydrostatic head—the fluid pressure of formation water produced by the height of water above a given point.

hydrostatic pressure, u_o (FL^{-2})—a state of stress in which all the principal stresses are equal (and there is no shear stress), as in a liquid at rest; the product of the unit weight of the liquid and the different in elevation between the given point and the free water elevation.

excess hydrostatic pressure (*hydrostatic excess pressure*), $\bar{u}$, u (FL^{-2})—the pressure that exists in pore water in excess of the hydrostatic pressure.

hydrostatic pressure—a state of stress in which all the principal stresses are equal (and there is no shear stress). (ISRM)

hygroscopic capacity (hygroscopic coefficient), w_e (D)—ratio of: (*1*) the weight of water absorbed by a dry soil or rock in a saturated atmosphere at a given temperature, to (*2*) the weight of the oven-dried soil or rock.

hygroscopic water content, w_H (D)—the water content of an air-dried soil or rock.

hysteresis—incomplete recovery of strain during unloading cycle due to energy consumption. (ISRM)

impedance, acoustic—the product of the density and sonic velocity of a material. The extent of wave energy transmission and reflection at the boundary of two media is determined by their acoustic impedances. (ISRM)

inelastic deformation—the portion of deformation under stress that is not annulled by removal of stress. (ISRM)

inert—not participating in any fashion in chemical reactions.

influence value, I (D)—the value of the portion of a mathematical expression that contains combinations of the independent variables arranged in dimensionless form.

influent stream, *n*—see preferred term **losing stream**.

inhibitor—a material that stops or slows a chemical reaction from occurring.

initial consolidation (*initial compression*)—see **consolidation**.

initial set—a degree of stiffening of a grout mixture generally stated as an empirical value indicating the time in hours and minutes that is required for a mixture to stiffen sufficiently to resist the penetration of a weighted test needle.

injectability—see **groutability**.

inorganic silt—see **silt**.

in situ—applied to a rock or soil when occurring in the situation in which it is naturally formed or deposited.

intergranular pressure—see **stress**.

intermediate principal plane—see **principal plane**.

intermediate principal stress—see **stress**.

internal friction (shear resistance), (FL^{-2})—the portion of the shearing strength of a soil or rock indicated by the terms $p \tan \phi$ in Coulomb's equation $s = c + p \tan \phi$. It is usually considered to be due to the interlocking of the soil or rock grains and the resistance to sliding between the grains.

interstice—see preferred term **void**.

interstitial—occurring between the grains or in the pores in rock or soil.

intrinsic shear strength, S_o (FL^{-2})—the shear strength of a rock indicated by Coulomb's equation when $p \tan \phi$ (shear resistance or internal friction) vanishes. Corresponds to cohesion, *c*, in soil mechanics.

invert—on the cross section, the lowest point of the underground excavation or the lowest section of the lining. (ISRM)

isochrome—a curve showing the distribution of the excess hydrostatic pressure at a given time during a process of consolidation.

isotropic mass—a mass having the same property (or properties) in all directions.

isotropic material—a material whose properties do not vary with direction.

isotropy—having the same properties in all directions. (ISRM)

jackhammer—an air driven percussion drill that imparts a rotary hammering motion to the bit and has a passageway to the bit for the injection of compressed air for cleaning the hole of cuttings.

Discussion—These two characteristics distinguish it from the pavement breaker which is similar in size and general appearance.

jack-leg—a portable percussion drill of the jack-hammer type, used in underground work; has a single pneumatically adjustable leg for support.

jet grouting—technique utilizing a special drill bit with horizontal and vertical high speed water jets to excavate alluvial soils and produce hard impervious columns by pumping grout through the horizontal nozzles that jets and mixes with foundation material as the drill bit is withdrawn.

jetty—an elongated artificial obstruction projecting into a body of water from a bank or shore to control shoaling and scour by deflection of the force of water currents and waves.

joint—a break of geological origin in the continuity of a body of rock occurring either singly, or more frequently in a set or system, but not attended by a visible movement parallel to the surface of discontinuity. (ISRM)

joint diagram—a diagram constructed by accurately plotting the strike and dip of joints to illustrate the geometrical relationship of the joints within a specified area of geologic investigation. (ISRM)

joint pattern—a group of joints that form a characteristic geometrical relationship, and which can vary considerably from one location to another within the same geologic formation. (ISRM)

joint (fault) set—a group of more or less parallel joints. (ISRM)

joint (fault) system—a system consisting of two or more joint sets or any group of joints with a characteristic pattern, that is, radiating, concentric, etc. (ISRM)

jumbo—a specially built mobile carrier used to provide a work platform for one or more tunneling operations, such as drilling and loading blast holes, setting tunnel supports, installing rock bolts, grouting, etc.

kaolin—a variety of clay containing a high percentage of kaolinite.

kaolinite—a common clay mineral having the general formula $Al_2(Si_2O_5)(OH_4)$; the primary constituent of kaolin.

karst—a geologic setting where cavities are developed in massive limestone beds by solution of flowing water. Caves and even underground river channels are produced into which surface runoff drains and often results in the land above being dry and relatively barren. (ISRM)

kelly—a heavy-wall tube or pipe, usually square or hexagonal in cross section, which works inside the matching center hole in the rotary table of a drill rig to impart rotary motion to the drill string.

lagging, *n*—*in mining or tunneling*, short lengths of timber, sheet steel, or concrete slabs used to secure the roof and sides of an opening behind the main timber or steel supports. The process of installation is also called lagging or lacing.

laminar flow (streamline flow) (viscous flow)—flow in which the head loss is proportional to the first power of the velocity.

landslide—the perceptible downward sliding or movement of a mass of earth or rock, or a mixture of both. (ISRM)

landslide (slide)—the failure of a sloped bank of soil or rock in which the movement of the mass takes place along a surface of sliding.

leaching—the removal in solution of the more soluble materials by percolating or moving waters. (ISRM)

leaching—the removal of soluble soil material and colloids by percolating water.

leakage, *n*—the flow of liquid from one hydrogeolic unit to another.

Discussion—the leakage may be natural, as through semi-impervious confining layer, or man-made, as through an uncased well.

leakance, *n*(T^{-1})—the ratio K′/b′, in which K′ and b′ are the vertical hydraulic conductivity and the thickness, respectively, of the confining beds.

leaky aquifier, *n*—aquifiers, whether artesian or unconfined, that lose or gain water through adjacent less permeable beds.

lime—specifically, calcium oxide (CaO_2); also loosely, a general term for the various chemical and physical forms of quicklime, hydrated lime, and hydraulic hydrated lime.

ledge—see **bedrock**.

preconsolidation pressure (prestress), p_e (FL^{-2})—the greatest effective pressure to which a soil has been subjected.

preplaced aggregate concrete—concrete produced by placing coarse aggregate in a form and later injecting a portland cement-sand or resin grout to fill the interstices.

pressure, p (FL^{-2})—the load divided by the area over which it acts.

pressure bulb—the zone in a loaded soil or rock mass bounded by an arbitrarily selected isobar of stress.

pressure testing—a method of permeability testing with water or grout pumped downhole under pressure.

pressure-void ratio curve (compression curve)—a curve representing the relationship between effective pressure and void ratio of a soil as obtained from a consolidation test. The curve has a characteristic shape when plotted on semilog paper with pressure on the log scale. The various parts of the curve and extensions to the parts of the curve and extensions to the parts have been designated as recompression, compression, virgin compression, expansion, rebound, and other descriptive names by various authorities.

pressure washing—the cleaning of soil or rock surfaces accomplished by injection of water, air, or other liquids, under pressure.

primary consolidation (*primary compression*) (*primary time effect*)—see **consolidation.**

primary hole—*in grouting,* the first series of holes to be drilled and grouted, usually at the maximum allowable spacing.

primary lining—the lining first placed inside a tunnel or shaft, usually used to support the excavation. The primary lining may be of wood or steel sets with steel or wood lagging or rock bolts and shot-crete. (ISRM)

primary permeability—internal permeability of intack rock; intergranular permeability (not permeability due to fracturing).

primary porosity—the porosity that developed during the final stages of sedimentation or that was present within sedimentary particles at the time of deposition.

primary state of stress—the stress in a geological formation before it is disturbed by man-made works. (ISRM)

principal plane—each of three mutually perpendicular planes through a point in a soil mass on which the shearing stress is zero.

intermediate principal plane—the plane normal to the direction of the intermediate principal stress.

major principal plane—the plane normal to the direction of the major principal stress.

minor principal plane—the plane normal to the direction of the minor principal stress.

principal stress—see **stress.**

principal stress (strain)—the stress (strain) normal to one of three mutually perpendicular planes on which the shear stresses (strains) at a point in a body are zero. (ISRM)

Proctor compaction curve—see **compaction curve.**

Proctor penetration curve—see **penetration resistance curve.**

Proctor penetration resistance—see **penetration resistance.**

profile—see **soil profile.**

progressive failure—failure in which the ultimate shearing resistance is progressively mobilized along the failure surface.

progressive failure—formation and development of localized fractures which, after additional stress increase, eventually form a continuous rupture surface and thus lead to failure after steady deterioration of the rock. (ISRM)

proportioning pump—see **metering pump.**

proprietary—made and marketed by one having the exclusive right to manufacture and sell; privately owned and managed.

protective filter—see **filter.**

pumpability—*in grouting,* a measure of the properties of a particular grout mix to be pumped as controlled by the equipment being used, the formation being injected, and the engineering objective limitations.

pumping of pavement (*pumping*)—see **pavement pumping.**

pumping test—a field procedure used to determine in situ permeability or the ability of a formation to accept grout.

pure shear—a state of strain resulting from that stress condition most easily described by a Mohr circle centered at the origin. (ISRM)

quarry—an excavation in the surface of the earth from which stone is obtained for crushed rock or building stone. (ISRM)

Quer-wave (love wave), W—dispersive surface wave with one horizontal component, generally normal to the direction of propagation, which decreases in propagation velocity with increase in frequency.

quick condition (quicksand)—condition in which water is flowing upwards with sufficient velocity to reduce significantly the bearing capacity of the soil through a decrease in intergranular pressure.

quick test—see **unconsolidated undrained test.**

radius of influence of a well—distance from the center of the well to the closest point at which the piezometric surface is not lowered when pumping has produced the maximum steady rate of flow.

raise—upwardly constructed shaft; that is, an opening, like a shaft, made in the roof of one level to reach a level above. (ISRM)

range (of a deformation-measuring instrument)—the amount between the maximum and minimum quantity an instrument can measure without resetting. In some instances provision can be made for incremental extension of the range.

Rayleigh wave, v_R (LT^{-1})—dispersive surface wave in which element has retrograding elliptic orbit with one major vertical and one minor horizontal component both in plane of propagation velocity:

$$v_R = \alpha v_t \text{ with } 0.910 < \alpha < 0.995 \text{ for } 0.25 < v < 0.5$$

reactant—*in grouting,* a material that reacts chemically with the base component of grout system.

reactive aggregate—an aggregate containing siliceous material (usually in amorphous or crypto-crystalline state) which can react chemically with free alkali in the cement.

Discussion—The reaction can result in expansion of the hardened material, frequently to a damaging extent.

reflected (or refracted) wave—components of wave incident upon second medium and reflected into first medium (or refracted) into second medium.

reflection and refraction loss—that part of transmitted energy lost due to nonuniformity of mediums.

refusal—*in grouting*, when the rate of grout take is low, or zero, at a given pressure.

relative consistency, I_c, C_r (D)—ratio of: (*1*) the liquid limit minus the natural water content, to (*2*) the plasticity index.

relative density, D_d, I_D (D)—the ratio of (*1*) the difference between the void ratio of a cohesionless soil in the loosest state and any given void ratio, to (*2*) the difference between the void ratios in the loosest and in the densest states.

relative water content—see **liquidity index.**

remodeled soil—soil that has had its natural structure modified by manipulation.

remolding index, I_R (D)—the ratio of: (*1*) the modulus of deformation of a soil in the undisturbed state, to (*2*) the modulus of deformation of the soil in the remolded state.

remodeling sensitivity (sensitivity ratio), S_t (D)—the ratio of: (*1*) the unconfined compressive strength of an undisturbed specimen of soil, to (*2*) the unconfined compressive strength of a specimen of the same soil after remolding at unaltered water content.

residual soil—soil derived in place by weathering of the underlying material.

residual strain—the strain in a solid associated with a state of residual stress. (ISRM)

residual stress—stress remaining in a solid under zero external stress after some process that causes the dimensions of the various parts of the solid to be incompatible under zero stress, for example, (*1*) deformation under the action of external stress when some parts of the body suffer permanent strain; or (*2*) heating or cooling of a body in which the thermal expansion coefficient is not uniform throughout the body. (ISRM)

resin—*in grouting*, a material that usually constitutes the base of an organic grout system.

resin grout—a grout system composed of essentially resinous materials such as epoxys, polyesters, and urethanes.

Discussion—In Europe, this refers to any chemical grout system regardless of chemical origin.

resolution (of a deformation-measuring instrument)—the ratio of the smallest divisional increment of the indicating scale to the sensitivity of the instrument. Interpolation within the increment may be possible, but is not recommended in specifying resolution.

resonance—the reinforced vibration of a body exposed to the vibration, at about the frequency, of another body.

resonant frequency—a frequency at which resonance exists.

response—the motion (or other output) in a device or system resulting from an excitation (stimulus) under specified conditions.

retard—bank-protection structure designed to reduce the riparian velocity and induce silting or accretion.

retardation—delay in deformation. (ISRM)

retarder—a material that slows the rate at which chemical reactions would otherwise occur.

reverse circulation—a drilling system in which the circulating medium flows down through the annulus and up through the drill rod, that is, in the reverse of the normal direction of flow.

revetment—bank protection by armor, that is, by facing of a bank or embankment with erosion-resistant material.

riprap stone—(generally less than 1 ton in weight) specially selected and graded quarried stone placed to prevent erosion through wave action, tidal forces, or strong currents and thereby preserve the shape of a surface, slope, or underlying structure.

rise time (pulse rise time)—the interval of time required for the leading edge of a pulse to rise from some specified small fraction to some specified larger fraction of the maximum value.

rock—natural solid mineral matter occurring in large masses or fragments.

rock—any naturally formed aggregate of mineral matter occurring in large masses or fragments. (ISRM)

rock anchor—a steel rod or cable installed in a hole in rock; in principle the same as rock bolt, but generally used for rods longer than about four metres. (ISRM)

rock bolt—a steel rod placed in a hole drilled in rock used to tie the rock together. One end of the rod is firmly anchored in the hole by means of a mechanical device or grout, or both, and the threaded projecting end is equipped with a nut and plate that bears against the rock surface. The rod can be pretensioned. (ISRM)

rock burst—a sudden and violent expulsion of rock from its surroundings that occurs when a volume of rock is strained beyond the elastic limit and the accompanying failure is of such a nature that accumulated energy is released instantaneously.

rock burst—sudden explosive-like release of energy due to the failure of a brittle rock of high strength. (ISRM)

rock flour—see **silt.**

rock mass—rock as it occurs in situ, including its structural discontinuities. (ISRM)

rock mechanics—the application of the knowledge of the mechanical behavior of rock to engineering problems dealing with rock. Rock mechanics overlaps with structural geology, geophysics, and soil mechanics.

rock mechanics—theoretical and applied science of the mechanical behaviour of rock. (ISRM)

roof—top of excavation or underground opening, particularly applicable in bedded rocks where the top surface of the opening is flat rather than arched. (ISRM)

rotary drilling—a drilling process in which a hole is advanced by rotation of a drill bit under constant pressure without impact. See **percussion drilling.**

round—a set of holes drilled and charged in a tunnel or quarry that are fired instantaneously or with short-delay detonators. (ISRM)

running ground—*in tunneling*, a granular material that tends to flow or "run" into the excavation.

rupture—that stage in the development of a fracture where instability occurs. It is not recommended that the term rupture be used in rock mechanics as a synonym for fracture. (ISRM)

rupture envelope (*rupture line*)—see **Mohr envelope.**

sagging—usually occurs in sedimentary rock formations as a separation and downward bending of sedimentary beds in the roof of an underground opening. (ISRM)

sand—particles of rock that will pass the No. 4 (4.75-mm) sieve and be retained on the No. 200 (75-μm) U.S. standard sieve.

texture—*of soil and rock*, geometrical aspects consisting of size, shape, arrangement, and crystallinity of the component particles and of the related characteristics of voids.

texture—the arrangement in space of the components of a rock body and of the boundaries between these components. (ISRM)

theoretical time curve—see **consolidation time curve**.

thermal spalling—the breaking of rock under stresses induced by extremely high temperature gradients. High-velocity jet flames are used for drilling blast holes with this effect. (ISRM)

thermo-osmosis—the process by which water is caused to flow in small openings of a soil mass due to differences in temperature within the mass.

thickness—the perpendicular distance between bounding surfaces such as bedding or foliation planes of a rock. (ISRM)

thixotropy—the property of a material that enables it to stiffen in a relatively short time on standing, but upon agitation or manipulation to change to a very soft consistency or to a fluid of high viscosity, the process being completely reversible.

throw—the projection of broken rock during blasting. (ISRM)

thrust—force applied to a drill in the direction of penetration. (ISRM)

tight—rock remaining within the minimum excavation lines after completion of a blasting record. (ISRM)

till—see **glacial till**.

time curve—see **consolidation time curve**.

time factor, T_v, T (D)—dimensionless factor, utilized in the theory of consolidation, containing the physical constants of a soil stratum influencing its time-rate of consolidation, expressed as follows:

$$T = k\,(1 + e)t/(a_v \gamma_w \cdot H^2) = (c_v \cdot t)/H^2$$

where:
k = coefficient of permeability (LT^{-1}),
e = void ratio (dimensionless),
t = elapsed time that the stratum has been consolidated (T),
a_v = coefficient of compressibility (L^2F^{-1}),
γ_w = unit weight of water (FL^{-3}),
H = thickness of stratum drained on one side only. If stratum is drained on both sides, its thickness equals $2H$ (L), and
c_v = coefficient of consolidation (L^2T^{-1}).

topsoil—surface soil, usually containing organic matter.

torsional shear test—a shear test in which a relatively thin test specimen of solid circular or annular cross-section, usually confined between rings, is subjected to an axial load and to shear in torsion. In-place torsion shear tests may be performed by pressing a dentated solid circular or annular plate against the soil and measuring its resistance to rotation under a given axial load.

total stress—see **stress**.

toughness index, I_T, T_w—the ratio of: (*1*) the plasticity index, to (*2*) the flow index.

traction, S_1, S_2, S_3 (FL^{-2})—applied stress.

transformed flow net—a flow net whose boundaries have been properly modified (transformed) so that a net consisting of curvilinear squares can be constructed to represent flow conditions in an anisotropic porous medium.

transported soil—soil transported from the place of its origin by wind, water, or ice.

transverse wave, v_t (LT^{-1})—wave in which direction of displacement of element of medium is parallel to wave front. The propagation velocity, v_t, is calculated as follows:

$$v_t = \sqrt{G/\rho} = \sqrt{\mu/\rho} = \sqrt{(E/\rho)[1/2(1+\nu)]}$$

where:
G = shear modulus,
ρ = mass density,
ν = Poisson's ratio, and
E = Young's modulus.

transverse wave (shear wave)—a wave in which the displacement at each point of the medium is parallel to the wave front. (ISRM)

tremie—material placed under water through a tremie pipe in such a manner that it rests on the bottom without mixing with the water.

trench—usually a long, narrow, near vertical sided cut in rock or soil such as is made for utility lines. (ISRM)

triaxial compression—compression caused by the application of normal stresses in three perpendicular directions. (ISRM)

triaxial shear test (triaxial compression test)—a test in which a cylindrical specimen of soil or rock encased in an impervious membrane is subjected to a confining pressure and then loaded axially to failure.

triaxial state of stress—state of stress in which none of the three principal stresses is zero. (ISRM)

true solution—one in which the components are 100 % dissolved in the base solvent.

tube A manchette—*in grouting*, a grout pipe perforated with rings of small holes at intervals of about 12 in. (305 mm).

Discussion—Each ring of perforations is enclosed by a short rubber sleeve fitting tightly around the pipe so as to act as a one-way valve when used with an inner pipe containing two packer elements that isolate a stage for injection of grout.

tunnel—a man-made underground passage constructed without removing the overlying rock or soil. Generally nearly horizontal as opposed to a shaft, which is nearly vertical. (ISRM)

turbulent flow—that type of flow in which any water particle may move in any direction with respect to any other particle, and in which the head loss is approximately proportional to the second power of the velocity.

ultimate bearing capacity, q_e, q_{ult} (FL^{-2})—the average load per unit of area required to produce failure by rupture of a supporting soil or rock mass.

unconfined compressive strength—the load per unit area at which an unconfined prismatic or cylindrical specimen of material will fail in a simple compression test without lateral support.

unconfined aquifer, *n*—an aquifer that has a water table.

unconfined compressive strength—see **compressive strength**.

unconsolidated-undrained test (quick test)—a soil test in which the water content of the test specimen remains practically unchanged during the application of the confining pressure and the additional axial (or shearing) force.

undamped natural frequency—*of a mechanical system*, the frequency of free vibration resulting from only elastic and inertial forces of the system.

underconsolidated soil deposit—a deposit that is not fully consolidated under the existing overburden pressure.

undisturbed sample—a soil sample that has been obtained by methods in which every precaution has been taken to minimize disturbance to the sample.

uniaxial (unconfined) compression—compression caused by the application of normal stress in a single direction. (ISRM)

uniaxial state of stress—state of stress in which two of the three principal stresses are zero. (ISRM)

unit weight, γ (FL^{-3})—weight per unit volume (with this, and all subsequent unit-weight definitions, the use of the term weight means force).

dry unit weight (*unit dry weight*), γ_d, γ_e (FL^{-3})—the weight of soil or rock solids per unit of total volume of soil or rock mass.

effective unit weight, γ_e (FL^{-3})—that unit weight of a soil or rock which, when multiplied by the height of the overlying column of soil or rock, yields the effective pressure due to the weight of the overburden.

maximum unit weight, γ_{max} (FL^{-3})—the dry unit weight defined by the peak of a compaction curve.

saturated unit weight, γ_G, γ_{sat} (FL^{-3})—the wet unit weight of a soil mass when saturated.

submerged unit weight (*buoyant unit weight*), γ_m, γ', γ_{sub} (FL^{-3})—the weight of the solids in air minus the weight of water displaced by the solids per unit of volume of soil or rock mass; the saturated unit weight minus the unit weight of water.

unit weight of water, γ_w (FL^{-3})—the weight per unit volume of water; nominally equal to 62.4 lb/ft^3 or 1 g/cm^3.

wet unit weight (*mass unit weight*), γ_m, γ_{wet} (FL^{-3})—the weight (solids plus water) per unit of total volume of soil or rock mass, irrespective of the degree of saturation.

zero air voids unit weight, γ_z, γ_s (FL^{-3})—the weight of solids per unit volume of a saturated soil or rock mass.

unloading modulus—slope of the tangent to the unloading stress-strain curve at a given stress value. (ISRM)

uplift—the upward water pressure on a structure.

	Symbol	Unit
unit symbol	u	FL^{-2}
total symbol	U	F or FL^{-1}

uplift—the hydrostatic force of water exerted on or underneath a structure, tending to cause a displacement of the structure. (ISRM)

uplift—*in grouting*, vertical displacement of a formation due to grout injection.

vane shear test—an in-place shear test in which a rod with thin radial vanes at the end is forced into the soil and the resistance to rotation of the rod is determined.

varved clay—alternating thin layers of silt (or fine sand) and clay formed by variations in sedimentation during the various seasons of the year, often exhibiting contrasting colors when partially dried.

vent hole—*in grouting*, a hole drilled to allow the escape of air and water and also used to monitor the flow of grout.

vent pipe—*in grouting*, a small-diameter pipe used to permit the escape of air, water, or diluted grout from a formation.

vibrated beam wall (injection beam wall)—barrier formed by driving an H-beam in an overlapping pattern of prints and filling the print of the beam with cement-bentonite slurry or other materials as it is withdrawn.

vibration—an oscillation wherein the quantity is a parameter that defines the motion of a mechanical system (see **oscillation**).

virgin compression curve—see **compression curve**.

viscoelasticity—property of materials that strain under stress partly elastically and partly viscously, that is, whose strain is partly dependent on time and magnitude of stress. (ISRM)

viscosity—the internal fluid resistance of a substance which makes it resist a tendency to flow.

viscous damping—the dissipation of energy that occurs when a particle in a vibrating system is resisted by a force that has a magnitude proportional to the magnitude of the velocity of the particle and direction opposite to the direction of the particle.

viscous flow—see **laminar flow**.

void—space in a soil or rock mass not occupied by solid mineral matter. This space may be occupied by air, water, or other gaseous or liquid material.

void ratio, e (D)—the ratio of: (*1*) the volume of void space, to (*2*) the volume of solid particles in a given soil mass.

critical void ratio, e_c (D)—the void ratio corresponding to the critical density.

volumetric shrinkage (volumetric change), V_s (D)—the decrease in volume, expressed as a percentage of the soil mass when dried, of a soil mass when the water content is reduced from a given percentage to the shrinkage limit.

von Post humification scale—a scale describing various stages of decomposition of peat ranging from H1, which is completely undecomposed, to H10, which is completely decomposed.

wall friction, f' (FL^{-2})—frictional resistance mobilized between a wall and the soil or rock in contact with the wall.

washing—*in grouting*, the physical act of cleaning the sides of a hole by circulating water, water and air, acid washes, or chemical substances through drill rods or tremie pipe in an open hole.

water-cement ratio—the ratio of the weight of water to the weights of Portland cement in a cement grout or concrete mix. See also **grout mix**.

water content, w (D)—the ratio of the mass of water contained in the pore spaces of soil or rock material, to the solid mass of particles in that material, expressed as a percentage.

water gain—see **bleeding**.

water-holding capacity (D)—the smallest value to which the water content of a soil or rock can be reduced by gravity drainage.

water-plasticity ratio (*relative water content*) (*liquidity index*)—see **liquidity index**.

water table—see **free water elevation**.

water table aquifer, *n*—see preferred term **unconfined aquifer**.

wave—disturbance propagated in medium in such a manner that at any point in medium the amplitude is a function of

soil profile (profile)—vertical section of a soil, showing the nature and sequence of the various layers, as developed by deposition or weathering, or both.

soil stabilization—chemical or mechanical treatment designed to increase or maintain the stability of a mass of soil or otherwise to improve its engineering properties.

soil structure—the arrangement and state of aggregation of soil particles in a soil mass.

flocculent structure—an arrangement composed of flocs of soil particles instead of individual soil particles.

honeycomb structure—an arrangement of soil particles having a comparatively loose, stable structure resembling a honeycomb.

single-grained structure—an arrangement composed of individual soil particles; characteristic structure of coarse-grained soils.

soil suspension—highly diffused mixture of soil and water.

soil texture—see **gradation.**

solution cavern—openings in rock masses formed by moving water carrying away soluble materials.

sounding well—*in grouting,* a vertical conduit in a mass of coarse aggregate for preplaced aggregate concrete which contains closely spaced openings to permit entrance of grout.

Discussion—The grout level is determined by means of a measuring line on a float within the sounding well.

spacing—the distance between adjacent blastholes in a direction parallel to the face. (ISRM)

spalling—(*1*) longitudinal splitting in uniaxial compression, or (*2*) breaking-off of plate-like pieces from a free rock surface. (ISRM)

specific discharge, n (LT^1)—the rate of flow of water through a porous medium per unit area measured at a right angle to the direction of flow.

specific gravity:

specific gravity of solids, G, G_s, S_s (D)—ratio of: (*1*) the weight in air of a given volume of solids at a stated temperature to (*2*) the weight in air of an equal volume of distilled water at a stated temperature.

apparent specific gravity, G_a, S_a (D)—ratio of: (*1*) the weight in air of a given volume of the impermeable portion of a permeable material (that is, the solid matter including its impermeable pores or voids) at a stated temperature to (*2*) the weight in air of an equal volume of distilled water at a stated temperature.

bulk specific gravity (*specific mass gravity*), G_m, S_m (D)—ratio of: (*1*) the weight in air of a given volume of a permeable material (including both permeable and impermeable voids normal to the material) at a stated temperature to (*2*) the weight in air of an equal volume of distilled water at a stated temperature.

specific surface (L^{-1})—the surface area per unit of volume of soil particles.

specific storage, n (L^{-1})—the volume of water released from or taken into storage per unit volume of the porous medium per unit change in head.

spherical wave—wave in which wave fronts are concentric spheres.

split spacing grouting—a grouting sequence in which initial (primary) grout holes are relatively widely spaced and subsequent grout holes are placed midway between previous grout holes to "split the spacing."; this process is continued until a specified hole spacing is achieved or a reduction in grout take to a specified value occurs, or both.

spring characteristics, c (FL^{-1})—ratio of increase in load to increase in deflection:

$$c = l/C$$

where:
C = compliance.

stability—the condition of a structure or a mass of material when it is able to support the applied stress for a long time without suffering any significant deformation or movement that is not reversed by the release of stress. (ISRM)

stability factor (stability number), N_s (D)—a pure number used in the analysis of the stability of a soil embankment, defined by the following equation:

$$N_s = H_c\gamma_e/c$$

where:
H_c = critical height of the sloped bank,
γ_e = effective unit of weight of the soil, and
c = cohesion of the soil

NOTE—Taylor's "stability number" is the reciprocal of Terzaghi's "stability factor."

stabilization—see **soil stabilization.**

stage—*in grouting,* the length of hole grouted at one time. See also **stage grouting.**

stage grouting—sequential grouting of a hole in separate steps or stages in lieu of grouting the entire length at once; holes may be grouted in ascending stages by using packers or in descending stages downward from the collar of the hole.

standard compaction—see **compaction test.**

standard penetration resistance—see **penetration resistance.**

standing wave—a wave produced by simultaneous transmission in opposite directions of two similar waves resulting in fixed points of zero amplitudes called nodes.

steady-state vibration—vibration in a system where the velocity of each particle is a continuing periodic quantity.

stemming—(*1*) the material (chippings, or sand and clay) used to fill a blasthole after the explosive charge has been inserted. Its purpose is to prevent the rapid escape of the explosion gases. (*2*) the act of pushing and tamping the material in the hole. (ISRM)

stick-slip—rapid fluctuations in shear force as one rock mass slides past another, characterized by a sudden slip between the rock masses, a period of no relative displacement between the two masses, a sudden slip, etc. The oscillations may be regular as in a direct shear test, or irregular as in a triaxial test.

sticky limit, T_w (D)—the lowest water content at which a soil will stick to a metal blade drawn across the surface of the soil mass.

stiffness—the ratio of change of force (or torque) to the corresponding change in translational (or rotational) deflection of an elastic element.

stiffness-force—displacement ratio. (ISRM)

stone—crushed or naturally angular particles of rock.

stop—*in grouting,* a packer setting at depth.

stop grouting—the grouting of a hole beginning at the lowest packer setting (stop) after the hole is drilled to total depth.

Discussion—Packers are placed at the top of the zone being grouted. Grouting proceeds from the bottom up. Also called upstage grouting.

strain, ϵ (D)—the change in length per unit of length in a given direction.

strain (linear or normal), ϵ (D)—the change in length per unit of length in a given direction.

strain ellipsoid—the representation of the strain in the form of an ellipsoid into which a sphere of unit radius deforms and whose axes are the principal axes of strain. (ISRM)

strain (stress) rate—rate of change of strain (stress) with time. (ISRM)

strain resolution (strain sensitivity), R_s (D)—the smallest subdivision of the indicating scale of a deformation-measuring device divided by the product of the sensitivity of the device and the gage length. The deformation resolution, R_d, divided by the gage length.

strain (stress) tensor—the second order tensor whose diagonal elements consist of the normal strain (stress) components with respect to a given set of coordinate axes and whose off-diagonal elements consist of the corresponding shear strain (stress) components. (ISRM)

streamline flow—see **laminar flow.**

strength—maximum stress which a material can resist without failing for any given type of loading. (ISRM)

stress, σ, p, f (FL^{-2})—the force per unit area acting within the soil mass.

effective stress (*effective pressure*) (*intergranular pressure*), $\bar{\sigma}$, f (FL^{-2})—the average normal force per unit area transmitted from grain to grain of a soil mass. It is the stress that is effective in mobilizing internal friction.

neutral stress (*pore pressure*) (*pore water pressure*), u, u_w (FL^{-2})—stress transmitted through the pore water (water filling the voids of the soil).

normal stress, σ, p (FL^{-2})—the stress component normal to a given plane.

principal stress, σ_1, σ_2, σ_3 (FL^{-2})—stresses acting normal to three mutually perpendicular planes intersecting at a point in a body, on which the shearing stress is zero.

major principal stress, σ_1 (FL^{-2})—the largest (with regard to sign) principal stress.

minor principal stress, σ_3 (FL^{-2})—the smallest (with regard to sign) principal stress.

intermediate principal stress, σ_2 (FL^{-2})—the principal stress whose value is neither the largest nor the smallest (with regard to sign) of the three.

shear stress (*shearing stress*) (*tangential stress*), τ, s FL^{-2})—the stress component tangential to a given plane.

total stress, σ, f (FL^{-2})—the total force per unit area acting within a mass of soil. It is the sum of the neutral and effective stresses.

stress ellipsoid—the representation of the state of stress in the form of an ellipsoid whose semi-axes are proportional to the magnitudes of the principal stresses and lie in the principal directions. The coordinates of a point P on this ellipse are proportional to the magnitudes of the respective components of the stress across the plane normal to the direction OP, where O is the center of the ellipsoid. (ISRM)

stress (strain) field—the ensemble of stress (strain) states defined at all points of an elastic solid. (ISRM)

stress relaxation—stress release due to creep. (ISRM)

strike—the direction or azimuth of a horizontal line in the plane of an inclined stratum, joint, fault, cleavage plane, or other planar feature within a rock mass. (ISRM)

structure—one of the larger features of a rock mass, like bedding, foliation, jointing, cleavage, or brecciation; also the sum total of such features as contrasted with texture. Also, in a broader sense, it refers to the structural features of an area such as anti-clines or synclines. (ISRM)

structure—see **soil structure.**

subbase—a layer used in a pavement system between the subgrade and base coarse, or between the subgrade and portland cement concrete pavement.

subgrade—the soil prepared and compacted to support a structure or a pavement system.

subgrade surface—the surface of the earth or rock prepared to support a structure or a pavement system.

submerged unit weight—see **unit weight.**

subsealing—*in grouting,* grouting under concrete slabs for the purpose of filling voids without raising the slabs.

subsidence—the downward displacement of the overburden (rock or soil, or both) lying above an underground excavation or adjoining a surface excavation. Also the sinking of a part of the earth's crust. (ISRM)

subsoil—(*a*) soil below a subgrade of fill. (*b*) that part of a soil profile occurring below the "A" horizon.

sulfate attack—*in grouting,* harmful or deleterious reactions between sulfates in soil or groundwater and the grout.

support—structure or structural feature built into an underground opening for maintaining its stability. (ISRM)

surface force—any force that acts across an internal or external surface element in a material body, not necessarily in a direction lying in the surface. (ISRM)

surface wave—a wave confined to a thin layer at the surface of a body. (ISRM)

suspension—a mixture of liquid and solid materials.

suspension agent—an additive that decreased the settlement rate of particles in liquid.

swamp—a forested or shrub covered wetland where standing or gently flowing water persists for long periods on the surface.

syneresis—*in grouting,* the exudation of liquid (generally water) from a set gel which is not stressed, due to the tightening of the grout material structure.

take—see **grout take.**

talus—rock fragments mixed with soil at the foot of a natural slope from which they have been separated.

tangential stress—see **stress.**

tangent modulus—slope of the tangent to the stress-strain curve at a given stress value (generally taken at a stress equal to half the compressive strength). (ISRM)

tensile strength (unconfined or uniaxial tensile strength), T_o (FL^{-2})—the load per unit area at which an unconfined cylindrical specimen will fail in a simple tension (pull) test.

tensile stress—normal stress tending to lengthen the body in the direction in which it acts. (ISRM)

tertiary hole—*in grouting,* the third series of holes to be drilled and grouted usually spaced midway between previously grouted primary and secondary holes.

sand boil—the ejection of sand and water resulting from piping.

sand equivalent—a measure of the amount of silt or clay contamination in fine aggregate as determined by test (Test Method D 2419).[3]

sanded grout—grout in which sand is incorporated into the mixture.

sapric peat—peat in which the original plant fibers are highly decomposed (less than 33 % fibers).

saturated unit weight—see **unit weight**.

saturation curve—see **zero air voids curve**.

scattering loss—that part of transmitted energy lost due to roughness of reflecting surface.

schistosity—the variety of foliation that occurs in the coarser-grained metamorphic rocks and is generally the result of the parallel arrangement of platy and ellipsoidal mineral grains within the rock substance. (ISRM)

secant modulus—slope of the line connecting the origin and a given point on the stress-strain curve. (ISRM)

secondary consolidation (*secondary compression*) (*secondary time effect*)—see **consolidation**.

secondary hole—*in grouting*, the second series of holes to be drilled and grouted usually spaced midway between primary holes.

secondary lining—the second-placed, or permanent, structural lining of a tunnel, which may be of concrete, steel, or masonry. (ISRM)

secondary state of stress—the resulting state of stress in the rock around man-made excavations or structures. (ISRM)

sediment basin—a structure created by construction of a barrier or small dam-like structure across a waterway or by excavating a basin or a combination of both to trap or restrain sediment.

seep—a small area where water oozes from the soil or rock.

seepage—the infiltration or percolation of water through rock or soil to or from the surface. The term seepage is usually restricted to the very slow movement of ground water. (ISRM)

seepage (percolation)—the slow movement of gravitational water through the soil or rock.

seepage face, *n*—a boundary between the saturated flow field and the atmosphere along which a subsurface liquid discharges, either by evaporation or movement "downhill" along the land surface or in a well as a thin film in response to the force of gravity.

seepage force—the frictional drag of water flowing through voids or interstices in rock, causing an increase in the intergranular pressure, that is, the hydraulic force per unit volume of rock or soil which results from the flow of water and which acts in the direction of flow. (ISRM)

seepage force, J (F)—the force transmitted to the soil or rock grains by seepage.

seepage line—see **line of seepage**.

seepage line, *n*—the uppermost level at which a flowing liquid emerges along a seepage face.

seepage velocity, V_a, V_1 (LT^{-1})—the rate of discharge of seepage water through a porous medium per unit area of void space perpendicular to the direction of flow.

segregation—*in grouting*, the differential concentration of the components of mixed grout, resulting in nonuniform proportions in the mass.

seismic support—mass (heavy) supported on springs (weak) so that mass remains almost at rest when free end of springs is subjected to sinusoidal motion at operating frequency.

seismic velocity—the velocity of seismic waves in geological formations. (ISRM)

seismometer—instrument to pick up linear (vertical, horizontal) or rotational displacement, velocity, or acceleration.

self-stressing grout—expansive-cement grout in which the expansion induces compressive stress in grout if the expansion movement is restrained.

sensitivity—the effect of remolding on the consistency of a cohesive soil.

sensitivity (of an instrument)—the differential quotient dQ_0/dQ_1, where Q_0 is the scale reading and Q_1 is the quantity to be measured.

sensitivity (of a transducer)—the differential quotient dQ_0/dQ_1, where Q_0 is the output and Q_1 is the input.

series grouting—similar to stage grouting, except each successively deeper zone is grouted by means of a newly drilled hole, eliminating the need for washing grout out before drilling the hole deeper.

set—*in grouting*, the condition reached by a cement paste, or grout, when it has lost plasticity to an arbitrary degree, usually measured in terms of resistance to penetration or deformation; initial set refers to first stiffening and final set refers to an attainment of significant rigidity.

setting shrinkage—*in grouting*, a reduction in volume of grout prior to the final set of cement caused by bleeding, by the decrease in volume due to the chemical combination of water with cement, and by syneresis.

set time—(*1*) the hardening time of portland cement; or (*2*) the gel time for a chemical grout.

shaft—generally a vertical or near vertical excavation driven downward from the surface as access to tunnels, chambers, or other underground workings. (ISRM)

shaking test—a test used to indicate the presence of significant amounts of rock flour, silt, or very fine sand in a fine-grained soil. It consists of shaking a pat of wet soil, having a consistency of thick paste, in the palm of the hand; observing the surface for a glossy or livery appearance; then squeezing the pat; and observing if a rapid apparent drying and subsequent cracking of the soil occurs.

shear failure (failure by rupture)—failure in which movement caused by shearing stresses in a soil or rock mass is of sufficient magnitude to destroy or seriously endanger a structure.

general shear failure—failure in which the ultimate strength of the soil or rock is mobilized along the entire potential surface of sliding before the structure supported by the soil or rock is impaired by excessive movement.

local shear failure—failure in which the ultimate shearing strength of the soil or rock is mobilized only locally along the potential surface of sliding at the time the structure supported by the soil or rock is impaired by excessive movement.

[3] *Annual Book of ASTM Standards*, Vol 04.03.

shear force—a force directed parallel to the surface element across which it acts. (ISRM)

shear plane—a plane along which failure of material occurs by shearing. (ISRM)

shear resistance—see **internal friction.**

shear strain—the change in shape, expressed by the relative change of the right angles at the corner of what was in the undeformed state an infinitesimally small rectangle or cube. (ISRM)

shear strength, *s*, T_f (FL^{-2})—the maximum resistance of a soil or rock to shearing stresses. See **peak shear strength.**

shear stress—stress directed parallel to the surface element across which it acts. (ISRM)

shear stress (*shearing stress*) (*tangential stress*)—see **stress.**

shear wave (rotational, equivoluminal)—wave in which medium changes shape without change of volume (shear-plane wave in isotropic medium is transverse wave).

shelf life—maximum time interval during which a material may be stored and remain in a usable condition; usually related to storage conditions.

shock pulse—a substantial disturbance characterized by a rise of acceleration from a constant value and decay of acceleration to the constant value in a short period of time.

shock wave—a wave of finite amplitude characterized by a shock front, a surface across which pressure, density, and internal energy rise almost discontinuously, and which travels with a speed greater than the normal speed of sound. (ISRM)

shotcrete—mortar or concrete conveyed through a hose and pneumatically projected at high velocity onto a surface. Can be applied by a "wet" or "dry" mix method. (ISRM)

shrinkage-compensating—*in grouting*, a characteristic of grout made using an expansive cement in which volume increase, if restrained, induces compressive stresses that are intended to offset the tendency of drying shrinkage to induce tensile stresses. See also **self-stressing grout.**

shrinkage index, *SI* (D)—the numerical difference between the plastic and shrinkage limits.

shrinkage limit, *SL*, w_s (D)—the maximum water content at which a reduction in water content will not cause a decrease in volume of the soil mass.

shrinkage ratio, *R* (D)—the ratio of: (*1*) a given volume change, expressed as a percentage of the dry volume, to (*2*) the corresponding change in water content above the shrinkage limit, expressed as a percentage of the weight of the oven-dried soil.

sieve analysis—determination of the proportions of particles lying within certain size ranges in a granular material by separation on sieves of different size openings.

silt (inorganic silt) (rock flour)—material passing the No. 200 (75-μm) U.S. standard sieve that is nonplastic or very slightly plastic and that exhibits little or no strength when air-dried.

silt size—that portion of the soil finer than 0.02 mm and coarser than 0.002 mm (0.05 mm and 0.005 mm in some cases).

simple shear—shear strain in which displacements all lie in one direction and are proportional to the normal distances of the displaced points from a given reference plane. The dilatation is zero. (ISRM)

single-grained structure—see **soil structure.**

size effect—influence of specimen size on its strength or other mechanical parameters. (ISRM)

skin friction, *f* (FL^{-2})—the frictional resistance developed between soil and an element of structure.

slabbing—the loosening and breaking away of relatively large flat pieces of rock from the excavated surface, either immediately after or some time after excavation. Often occurring as tensile breaks which can be recognized by the subconchoidal surfaces left on remaining rock surface. (ISRM)

slabjacking—*in grouting*, injection of grout under a concrete slab in order to raise it to a specified grade.

slaking—deterioration of rock on exposure to air or water.

slaking—the process of breaking up or sloughing when an indurated soil is immersed in water.

sleeved grout pipe—see **tube A manchette.**

sliding—relative displacement of two bodies along a surface, without loss of contact between the bodies. (ISRM)

slope—the excavated rock surface that is inclined to the vertical or horizontal, or both, as in an open-cut. (ISRM)

slow test—see **consolidated-drain test.**

slump—a measure of consistency of freshly mixed concrete or grout. See also **slump test.**

slump test—the procedure for measuring slump (Test Method C 143).[2]

slurry cutoff wall—a vertical barrier constructed by excavating a vertical slot under a bentonite slurry and backfilling it with materials of low permeability for the purpose of the containment of the lateral flow of water and other fluids.

slurry grout—a fluid mixture of solids such as cement, sand, or clays in water.

slurry trench—a trench that is kept filled with a bentonite slurry during the excavation process to stabilize the walls of the trench.

slush grouting—application of cement slurry to surface rock as a means of filling cracks and surface irregularities or to prevent slaking; it is also applied to riprap to form grouted riprap.

smooth (-wall) blasting—a method of accurate perimeter blasting that leaves the remaining rock practically undamaged. Narrowly spaced and lightly charged blastholes, sometimes alternating with empty dummy holes, located along the breakline and fired simultaneously as the last round of the excavation. (ISRM)

soil (earth)—sediments or other unconsolidated accumulations of solid particles produced by the physical and chemical disintegration of rocks, and which may or may not contain organic matter.

soil binder—see **binder.**

soil-forming factors—factors, such as parent material, climate, vegetation, topography, organisms, and time involved in the transformation of an original geologic deposit into a soil profile.

soil horizon—see **horizon.**

soil mechanics—the application of the laws and principles of mechanics and hydraulics to engineering problems dealing with soil as an engineering material.

soil physics—the organized body of knowledge concerned with the physical characteristics of soil and with the methods employed in their determinations.

time, while at any instant the displacement at point is function of position of point.

wave front—moving surface in a medium at which a propagated disturbance first occurs.

wave front—(*1*) a continuous surface over which the phase of a wave that progresses in three dimensions is constant, or (*2*) a continuous line along which the phase of a surface wave is constant. (ISRM)

wave length—normal distance between two wave fronts with periodic characteristics in which amplitudes have phase difference of one complete cycle.

weathering—the process of disintegration and decomposition as a consequence of exposure to the atmosphere, to chemical action, and to the action of frost, water, and heat. (ISRM)

wetland—land which has the water table at, near, or above the land surface, or which is saturated for long enough periods to promote hydrophylic vegetation and various kinds of biological activity which are adapted to the wet environment.

wetting agent—a substance capable of lowering the surface tension of liquids, facilitating the wetting of solid surfaces, and facilitating the penetration of liquids into the capillaries.

wet unit weight—see **unit weight**.

working pressure—the pressure adjudged best for any particular set of conditions encountered during grouting.

Discussion—Factors influencing the determination are size of voids to be filled, depth of zone to be grouted, lithology of area to be grouted, grout viscosity, and resistance of the formation to fracture.

yield—*in grouting*, the volume of freshly mixed grout produced from a known quantity of ingredients.

yielding arch—type of support of arch shape, the joints of which deform plastically beyond a certain critical load, that is, continue to deform without increasing their resistance. (ISRM)

yield stress—the stress beyond which the induced deformation is not fully annulled after complete destressing. (ISRM)

Young's modulus—the ratio of the increase in stress on a test specimen to the resulting increase in strain under constant transverse stress limited to materials having a linear stress-strain relationship over the range of loading. Also called elastic modulus.

zero air voids curve (saturation curve)—the curve showing the zero air voids unit weight as a function of water content.

zero air voids density (*zero air voids unit weight*)—see **unit weight**.

APPENDIX

(Nonmandatory Information)

X1. ISRM SYMBOLS RELATING TO SOIL AND ROCK MECHANICS

NOTE—These symbols may not correlate with the symbols appearing in the text.

X1.1 Space

Ω, ω	solid angle
l	length
b	width
h	height or depth
r	radius
A	area
V	volume
t	time
v	velocity
ω	angular velocity
g	gravitational acceleration

X1.2 Periodic and Related Phenomena

T	periodic time
f	frequency
ω	angular frequency
λ	wave length

X1.3 Statics and Dynamics

m	mass
ρ	density (mass density)
G_m	mass specific gravity
G_s	specific gravity of solids
G_w	specific gravity of water
F	force
T	tangential force
W	weight
γ	unit weight
γ_d	dry unit weight
γ_w	unit weight of water
γ'	buoyant unit weight
γ_s	unit of solids
T	torque
I	moment of inertia
W	work
W	energy

X1.4 Applied Mechanics

e	void ratio
n	porosity
w	water content
S_r	degree of saturation
p	pressure
u	pore water pressure
σ	normal stress
σ_x, σ_y, σ_z	stress components in rectangular coordinates
σ_1, σ_2, σ_3	principal stresses
S_1, S_2, S_3	applied stresses (and reactions)
σ_h	horizontal stress
σ_v	vertical stress
τ	shear stress
τ_{xy}, τ_{yz}, τ_{zx}	shear stress components in rectangular coordinates
ϵ	strain
ϵ_x, ϵ_y, ϵ_z	strain components in rectangular coordinates
γ_{xy}, γ_{yz}, γ_{zx}	shear strain components in rectangular coordinates
θ	volume strain
E	Young's modulus; modulus of elasticity $E = \sigma/\epsilon$
ϵ_1, ϵ_2, ϵ_3	principal strains
G	shear modulus; modulus of rigidity $G = \tau/\gamma$
c	cohesion
ϕ_s	angle of friction between solid bodies
ϕ	angle of shear resistance (angle of internal friction)
h	hydraulic head
i	hydraulic gradient
j	seepage force per unit volume or seepage pressure per unit length
k	coefficient of permeability
η	viscosity
η_{pl}	plasticity (viscosity of Bingham body)
t_{ret}	retardation time
t_{rel}	relaxation time
T_s	surface tension
q	quantity rate of flow; rate of discharge
Q	quantity of flow
FS	safety factor

X1.5 Heat

T	temperature
β	coefficient of volume expansion

X1.6 Electricity

I	electric current
Q	electric charge
C	capacitance
L	self-inductance
R	resistance
ρ	resistivity

REFERENCES

(1) Terzaghi, *Theoretical Soil Mechanics*, John Wiley & Sons, Inc., New York, NY (1943).
(2) Terzaghi and Peck, *Soil Mechanics in Engineering Practice*, John Wiley & Sons, Inc., New York, NY (1948).
(3) Taylor, D. W., *Fundamentals of Soil Mechanics*, John Wiley & Sons, Inc., New York, NY (1948).
(4) Krynine, D. P., *Soil Mechanics*, 2nd Edition, McGraw-Hill Book Co., Inc., New York, NY (1947).
(5) Plummer and Dore, *Soil Mechanics and Foundations*, Pitman Publishing Corp., New York, NY (1940).
(6) Tolman, C. F., *Ground Water*, McGraw-Hill Book Co., Inc., New York, NY (1937).
(7) Stewart Sharpe, C. F., *Land Slides and Related Phenomena*, Columbia University Press, New York, NY (1938).
(8) "Letter Symbols and Glossary for Hydraulics with Special Reference to Irrigation," Special Committee on Irrigation Hydraulics, *Manual of Engineering Practice*, Am. Soc. Civil Engrs., No. 11 (1935).
(9) "Soil Mechanics Nomenclature," Committee of the Soil Mechanics and Foundations Division on Glossary of Terms and Definitions and on Soil Classification, *Manual of Engineering Practice*, Am. Soc. Civil Engrs., No. 22 (1941).
(10) "Pile Foundations and Pile Structures," Joint Committee on Bearing Value of Pile Foundations of the Waterways Division, Construction Division, and Soil Mechanics and Foundations Division, *Manual of Engineering Practice*, Am. Soc. Civil Engrs., No. 27 (1956).
(11) Webster's New International Dictionary of the English Language, unabridged, 2nd Edition, G. and C. Merriam Co., Springfield, MA (1941).
(12) Baver, L. D., *Soil Physics*, John Wiley & Sons, Inc., New York, NY (1940).
(13) Longwell, Knopf and Flint, "Physical Geology," *Textbook of Geology*, Part I, 2nd Edition, John Wiley & Sons, Inc., New York, NY (1939).
(14) Runner, D. G., *Geology for Civil Engineers*, Gillette Publishing Co., Chicago, IL (1939).
(15) Leggett, R. F., *Geology and Engineering*, McGraw-Hill Book Co., Inc., New York, NY (1939).
(16) Holmes, A., *The Nomenclature of Petrology*, Thomas Murby and Co., London, England (1920).
(17) Meinzer, O. E., "Outline of Ground Water Hydrology with Definitions," *U. S. Geological Survey Water Supply Paper 494* (1923).
(18) "Reports of the Committee on Sedimentation of the Division of Geology and Geography of the National Research Council," Washington, DC (1930–1938).
(19) Twenhofel, W. H., *A Treatise on Sedimentation*, 2nd Edition, Williams & Wilkins Co., Baltimore, MD (1932).
(20) Hogentogler, C. A., *Engineering Properties of Soils*, McGraw Hill Book Co., Inc., New York, NY (1937).
(21) *Special Procedures for Testing Soil and Rock for Engineering Purposes, ASTM STP 479*, ASTM, 1970.
(22) "Glossary of Terms and Definitions," Preliminary Report of Subcommittee G-3 on Nomenclature and Definitions of ASTM Committee D-18 on Soils for Engineering Purposes.
(23) Sowers and Sowers, *Introductory Soil Mechanics and Foundations*, The Macmillan Co., New York, NY (1951).
(24) Lambe, T. William, *Soil Testing for Engineers*, John Wiley & Sons, Inc., New York, NY (1951).
(25) Capper and Cassie, *The Mechanics of Engineering Soils*, McGraw-Hill Book Co., Inc., New York, NY (1949).
(26) Dunham, C. W., *Foundations of Structures*, McGraw-Hill Book Co., Inc., New York, NY (1950).
(27) Casagrande, A., "Notes on Soil Mechanics," Graduate School of Engineering, Harvard University (1938).
(28) Tschebotarioff, G. P., *Soil Mechanics, Foundations, and Earth Structures*, McGraw-Hill Book Co., Inc., New York, (1951).
(29) Rice, C. M., "Dictionary of Geological Terms," Edwards Bros., Inc., Ann Arbor, MI (1940).
(30) Creager, Justin and Hinds, *Engineering for Dams*, John Wiley & Sons, Inc., New York, NY (1945).
(31) Krumbein and Sloss, *Stratigraphy and Sedimentation*, W. H. Freeman and Co., San Francisco, CA (1951).
(32) Pettijohn, F. J. *Sedimentary Rocks*, Harper and Bros., New York, NY (1949).
(33) Reiche, Parry, *A Survey of Weathering Processes and Products*, University of New Mexico Press, Albuquerque, NM (1945).
(34) Garrels, R. M., *A Textbook of Geology*, Harper and Bros., New York, NY (1951).
(35) Ries and Watson, *Engineering Geology*, John Wiley & Sons, Inc., New York, NY (1936).
(36) Ross and Hendricks, *Minerals of the Montmorillonite Group*, U. S. Geological Survey Professional Paper 205-B (1945).
(37) Hartman, R. J., *Colloid Chemistry*, Houghton Mifflin Co., New York, NY (1947).
(38) "Frost Investigations," Corps of Engineers, Frost Effects Laboratory, Boston, MA, June 1951.
(39) "Standard Specifications for Highway Materials and Methods of Sampling and Testing," Parts I and II, adopted by the American Association of State Highway Officials (1950).
(40) Coates, D. G., "Rock Mechanics Principles," rev ed. Mines Br., Dept. Mines and Tech. Surv., Ottawa, Mines Br. Mon. 874 (1970).
(41) Gary, M., McAfee, R., Jr., and Wolf, C. L., (eds.), *Glossary of Geology*, American Geological Institute (1972).
(42) International Society for Rock Mechanics, Commission on Terminology, Symbols and Graphic Representation, Final Document on Terminology, English Version, 1972, and List of Symbols, 1970.
(43) Jaeger, J. C., and Cook, N. G. W., *Fundamentals of Rock Mechanics*, Methuen, London (1969).
(44) Nelson, A., and Nelson, K. D., *Dictionary of Applied Geology* (Mining and Civil Engineering), Philosophical Library, Inc., New York, NY (1967).
(45) Obert, L. A., and Duvall, W. I., *Rock Mechanics and the Design of Structures in Rock*, John Wiley & Sons, New York, NY (1967).
(46) SME Mining Engineering Handbook: Society Mining Engineers, Vol 2, New York, NY (1973).
(47) Thrush, R. P. (ed), et al., *A Dictionary of Mining, Mineral and Related Terms*, U. S. Bureau of Mines (1968).
(48) Lohman, W. W., and others, *Definitions of Selected Ground-Water Terms—Revisions and Conceptual Refinements*, U. S. Geological Survey Water-Supply Paper 1988, 21 pp., 1972.
(49) *Glossary of Soil Science Terms*, Madison, WI, Soil Science Society of America, 34 pp. (1975).
(50) *International Glossary of Hydrology*, Geneva, Switzerland, World Meterological Organization, WMO No. 385, 393 pp., (1974).

Designation: D 1452 – 80 (Reapproved 1990)[ε1]

Standard Practice for Soil Investigation and Sampling by Auger Borings[1]

This standard is issued under the fixed designation D 1452; the number immediately following the designation indicates the year of original adoption or, in the case of revision, the year of last revision. A number in parentheses indicates the year of last reapproval. A superscript epsilon (ε) indicates an editorial change since the last revision or reapproval.

This standard has been approved for use by agencies of the Department of Defense. Consult the DoD Index of Specifications and Standards for the specific year of issue which has been adopted by the Department of Defense.

[ε1] NOTE—Section 6 was added editorially in July 1990.

1. Scope

1.1 This practice covers equipment and procedures for the use of earth augers in shallow geotechnical exploration. This practice does not apply to sectional continuous flight augers.

1.2 *This standard does not purport to address the safety problems associated with its use. It is the responsibility of the user of this standard to establish appropriate safety and health practices and determine the applicability of regulatory limitations prior to use.*

2. Significance and Use

2.1 Auger borings often provide the simplest method of soil investigation and sampling. They may be used for any purpose where disturbed samples can be used and are valuable in connection with ground water level determination and indication of changes in strata and advancement of hole for spoon and tube sampling. Equipment required is simple and readily available. Depths of auger investigations are, however, limited by ground water conditions, soil characteristics, and the equipment used.

3. Apparatus

3.1 *Hand-Operated Augers*:

3.1.1 *Helical Augers*—Small lightweight augers generally available in sizes from 1 through 3 in. (25.4 through 76.2 mm).

3.1.1.1 *Spiral-Type Auger*, consisting of a flat thin metal strip, machine twisted to a spiral configuration of uniform pitch; having at one end, a sharpened or hardened point, with a means of attaching a shaft or extension at the opposite end.

3.1.1.2 *Ship-Type Auger*—Similar to a carpenter's wood bit. It is generally forged from steel and machined to the desired size and configuration. It is normally provided with sharpened and hardened nibs at the point end and with an integral shaft extending through its length for attachment of a handle or extension at the opposite end.

3.1.2 *Open Tubular Augers*, ranging in size from 1.5 through 8 in. (38.1 through 203.2 mm) and having the common characteristic of appearing essentially tubular when viewed from the digging end.

3.1.2.1 *Orchard-Barrel Type*, consisting essentially of a tube having cutting lips or nibs hardened and sharpened to penetrate the formation on one end and an adaptor fitting for an extension or handle on the opposite end.

3.1.2.2 *Open-Spiral Type*, consisting of a flat thin metal strip that has been helically wound around a circular mandrel to form a spiral in which the flat faces of the strip are parallel to the axis of the augered hole. The lower helix edges are hard-faced to improve wear characteristics. The opposite end is fitted with an adaptor for extension.

3.1.2.3 *Closed-Spiral Type*—Nearly identical to the open-spiral type except, the pitch of the helically wound spiral is much less than that of the open-spiral type.

3.1.3 *Post-Hole Augers*, generally 2 through 8 in. (50.8 through 203.2 mm), and having in common a means of blocking the escape of soil from the auger.

3.1.3.1 *Clam-Shell Type*, consisting of two halves, hinged to allow opening and closing for alternately digging and retrieving. It is not usable deeper than about 3.5 ft (1.07 m).

3.1.3.2 *Iwan Type*, consisting of two tubular steel segments, connected at the top to a common member to form a nearly complete tube, but with diametrically opposed openings. It is connected at the bottom by two radial blades pitched to serve as cutters which also block the escape of contained soil. Attachment of handle or extension is at the top connector.

3.2 *Machine-Operated Augers*:

3.2.1 *Helical Augers*, generally 8 through 48 in. (203.2 through 1219 mm), consisting essentially of a center shaft fitted with a shank or socket for application of power, and having one to three complete 360° (6.28-rad) spirals for conveyance and storage of cut soil. Cutter bits and pilot bits are available in moderate and hard formation types and normally replaceable in the field. They are normally operated by heavy-duty, high-torque machines, designed for heavy construction work.

3.2.2 *Stinger Augers*, generally 6 through 30 in. (152.4 through 762 mm), are similar to the helical auger in 3.2.1, but lighter and generally smaller. They are commonly operated by light-duty machines for post and power pole holes.

3.2.3 *Disk Augers*, generally 10 through 30 in. (254 through 762 mm), consisting essentially of a flat, steel disk with diametrically opposed segments removed and having a

[1] This practice is under the jurisdiction of ASTM Committee D-18 on Soil and Rock and is the direct responsibility of Subcommittee D18.02 on Sampling and Related Field Testing for Soil Investigations.

Current edition approved June 12, 1980. Published August 1980. Originally published as D 1452 – 57 T. Last previous edition D 1452 – 65 (1972).

shank or socket located centrally for application of power. Replaceable cutter bits, located downward from the leading edges of the remaining disk, dig and load soil that is held on the disk by valves or shutters hinged at the disk in order to close the removed segments. The disk auger is specifically designed to be operated by machines having limited vertical clearance between spindle and ground surface.

3.2.4 *Bucket Auger*, generally 12 through 48 in. (304.8 through 1219 mm), consisting essentially of a disk auger, without shank or socket, but hinge-mounted to the bottom of a steel tube or bucket of approximately the same diameter as the disk auger. A socket or shank for power application is located in the top center of the bucket diametral cross piece provided for the purpose.

3.3 *Casing* (when needed), consisting of pipe of slightly larger diameter than the auger used.

3.4 *Accessory Equipment*—Labels, field log sheets, sample jars, sealing wax, sample bags, and other necessary tools and supplies.

4. Procedure

4.1 Make the auger boring by rotating and advancing the desired distance into the soil. Withdraw the auger from the hole and remove the soil for examination and test. Return the empty auger to the hole and repeat the procedure. Continue the sequence until the required depth is reached.

4.2 Casing is required in unstable soil in which the bore hole fails to stay open and especially when the boring is extended below the ground-water level. The inside diameter of the casing must be slightly larger than the diameter of the auger used. The casing shall be driven to a depth not greater than the top of the next sample and shall be cleaned out by means of the auger. The auger can then be inserted into the bore hole and turned below the bottom of the casing to obtain a sample.

4.3 The soil auger can be used both for boring the hole and for bringing up disturbed samples of the soil encountered. The structure of a cohesive soil is completely destroyed and the moisture may be changed by the auger. Seal all samples in a jar or other airtight container and label appropriately. If more than one type of soil is picked up in the sample, prepare a separate container for each type of soil.

4.4 *Field Observations*—Record complete ground water information in the field logs. Where casing is used, measure ground water levels, both before and after the casing is pulled. In sands, determine the water level at least 30 min after the boring is completed; in silts, at least 24 h. In clays, no accurate water level determination is possible unless pervious seams are present. As a precaution, however, water levels in clays shall be taken after at least 24 h.

5. Report

5.1 The data obtained in boring shall be recorded in the field logs and shall include the following:

5.1.1 Date of start and completion of boring,

5.1.2 Identifying number of boring,

5.1.3 Reference datum including direction and distance of boring relative to reference line of project or other suitable reference points,

5.1.4 Type and size of auger used in boring,

5.1.5 Depth of changes in strata,

5.1.6 Description of soil in each major stratum,

5.1.7 Ground water elevation and location of seepage zones, when found, and

5.1.8 Condition of augered hole upon removal of auger, that is, whether the hole remains open or the sides cave, when such can be observed.

6. Keywords

Auger borings; sampling; soil investigations

Standard Test Method for Penetration Test and Split-Barrel Sampling of Soils[1]

This standard is issued under the fixed designation D 1586; the number immediately following the designation indicates the year of original adoption or, in the case of revision, the year of last revision. A number in parentheses indicates the year of last reapproval. A superscript epsilon (ε) indicates an editorial change since the last revision or reapproval.

This standard has been approved for use by agencies of the Department of Defense. Consult the DOD Index of Specifications and Standards for the specific year of issue which has been adopted by the Department of Defense.

ε1 NOTE—Editorial changes were made throughout October 1992.

1. Scope

1.1 This test method describes the procedure, generally known as the Standard Penetration Test (SPT), for driving a split-barrel sampler to obtain a representative soil sample and a measure of the resistance of the soil to penetration of the sampler.

1.2 *This standard does not purport to address all of the safety problems, if any, associated with its use. It is the responsibility of the user of this standard to establish appropriate safety and health practices and determine the applicability of regulatory limitations prior to use.* For a specific precautionary statement, see 5.4.1.

1.3 The values stated in inch-pound units are to be regarded as the standard.

2. Referenced Documents

2.1 *ASTM Standards:*

D 2487 Test Method for Classification of Soils for Engineering Purposes[2]

D 2488 Practice for Description and Identification of Soils (Visual-Manual Procedure)[2]

D 4220 Practices for Preserving and Transporting Soil Samples[2]

D 4633 Test Method for Stress Wave Energy Measurement for Dynamic Penetrometer Testing Systems[2]

3. Terminology

3.1 *Descriptions of Terms Specific to This Standard*

3.1.1 *anvil*—that portion of the drive-weight assembly which the hammer strikes and through which the hammer energy passes into the drill rods.

3.1.2 *cathead*—the rotating drum or windlass in the rope-cathead lift system around which the operator wraps a rope to lift and drop the hammer by successively tightening and loosening the rope turns around the drum.

3.1.3 *drill rods*—rods used to transmit downward force and torque to the drill bit while drilling a borehole.

3.1.4 *drive-weight assembly*—a device consisting of the hammer, hammer fall guide, the anvil, and any hammer drop system.

3.1.5 *hammer*—that portion of the drive-weight assembly consisting of the 140 ± 2 lb (63.5 ± 1 kg) impact weight which is successively lifted and dropped to provide the energy that accomplishes the sampling and penetration.

3.1.6 *hammer drop system*—that portion of the drive-weight assembly by which the operator accomplishes the lifting and dropping of the hammer to produce the blow.

3.1.7 *hammer fall guide*—that part of the drive-weight assembly used to guide the fall of the hammer.

3.1.8 *N-value*—the blowcount representation of the penetration resistance of the soil. The *N*-value, reported in blows per foot, equals the sum of the number of blows required to drive the sampler over the depth interval of 6 to 18 in. (150 to 450 mm) (see 7.3).

3.1.9 ΔN—the number of blows obtained from each of the 6-in. (150-mm) intervals of sampler penetration (see 7.3).

3.1.10 *number of rope turns*—the total contact angle between the rope and the cathead at the beginning of the operator's rope slackening to drop the hammer, divided by 360° (see Fig. 1).

3.1.11 *sampling rods*—rods that connect the drive-weight assembly to the sampler. Drill rods are often used for this purpose.

3.1.12 *SPT*—abbreviation for Standard Penetration Test, a term by which engineers commonly refer to this method.

4. Significance and Use

4.1 This test method provides a soil sample for identification purposes and for laboratory tests appropriate for soil obtained from a sampler that may produce large shear strain disturbance in the sample.

4.2 This test method is used extensively in a great variety of geotechnical exploration projects. Many local correlations and widely published correlations which relate SPT blowcount, or *N*-value, and the engineering behavior of earthworks and foundations are available.

5. Apparatus

5.1 *Drilling Equipment*—Any drilling equipment that provides at the time of sampling a suitably clean open hole before insertion of the sampler and ensures that the penetration test is performed on undisturbed soil shall be acceptable. The following pieces of equipment have proven to be

---

[1] This method is under the jurisdiction of ASTM Committee D-18 on Soil and Rock and is the direct responsibility of Subcommittee D18.02 on Sampling and Related Field Testing for Soil Investigations.

Current edition approved Sept. 11, 1984. Published November 1984. Originally published as D 1586 – 58 T. Last previous edition D 1586 – 67 (1974).

[2] *Annual Book of ASTM Standards*, Vol 04.08.

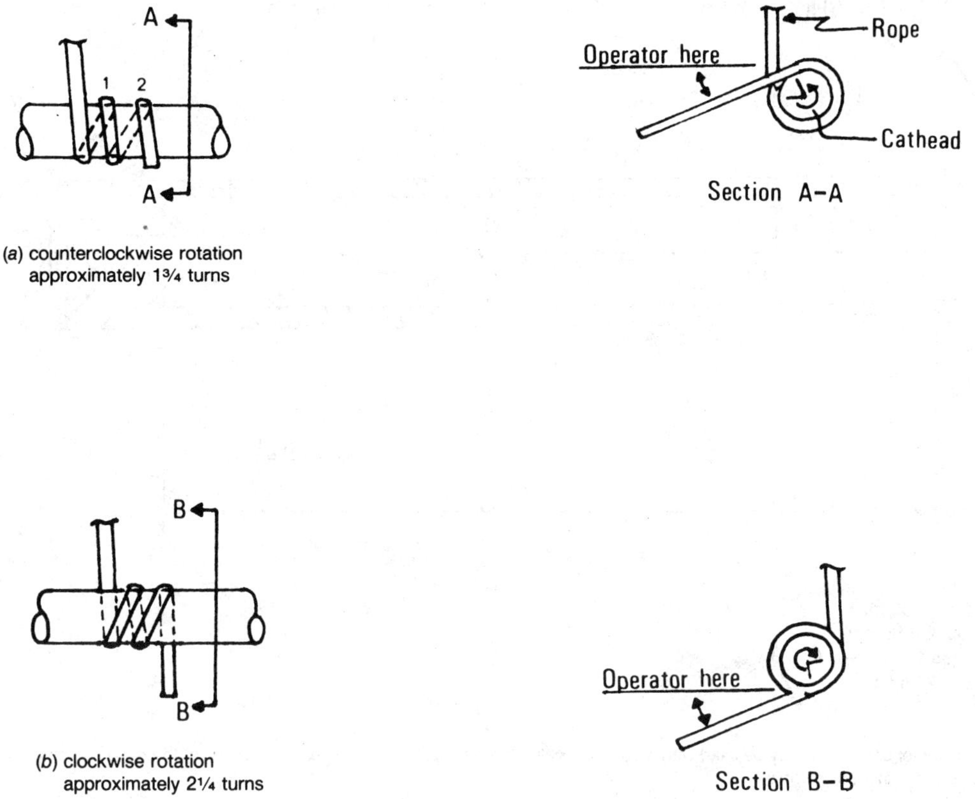

(*a*) counterclockwise rotation approximately 1¾ turns

(*b*) clockwise rotation approximately 2¼ turns

FIG. 1 Definitions of the Number of Rope Turns and the Angle for (*a*) Counterclockwise Rotation and (*b*) Clockwise Rotation of the Cathead

suitable for advancing a borehole in some subsurface conditions.

5.1.1 *Drag, Chopping, and Fishtail Bits*, less than 6.5 in. (162 mm) and greater than 2.2 in. (56 mm) in diameter may be used in conjuction with open-hole rotary drilling or casing-advancement drilling methods. To avoid disturbance of the underlying soil, bottom discharge bits are not permitted; only side discharge bits are permitted.

5.1.2 *Roller-Cone Bits*, less than 6.5 in. (162 mm) and greater than 2.2 in. (56 mm) in diameter may be used in conjunction with open-hole rotary drilling or casing-advancement drilling methods if the drilling fluid discharge is deflected.

5.1.3 *Hollow-Stem Continuous Flight Augers*, with or without a center bit assembly, may be used to drill the boring. The inside diameter of the hollow-stem augers shall be less than 6.5 in. (162 mm) and greater than 2.2 in. (56 mm).

5.1.4 *Solid, Continuous Flight, Bucket and Hand Augers*, less than 6.5 in. (162 mm) and greater than 2.2 in. (56 mm) in diameter may be used if the soil on the side of the boring does not cave onto the sampler or sampling rods during sampling.

5.2 *Sampling Rods*—Flush-joint steel drill rods shall be used to connect the split-barrel sampler to the drive-weight assembly. The sampling rod shall have a stiffness (moment of inertia) equal to or greater than that of parallel wall "A" rod (a steel rod which has an outside diameter of 1⅝ in. (41.2 mm) and an inside diameter of 1⅛ in. (28.5 mm).

NOTE 1—Recent research and comparative testing indicates the type rod used, with stiffness ranging from "A" size rod to "N" size rod, will usually have a negligible effect on the *N*-values to depths of at least 100 ft (30 m).

5.3 *Split-Barrel Sampler*—The sampler shall be constructed with the dimensions indicated in Fig. 2. The driving shoe shall be of hardened steel and shall be replaced or repaired when it becomes dented or distorted. The use of liners to produce a constant inside diameter of 1⅜ in. (35 mm) is permitted, but shall be noted on the penetration record if used. The use of a sample retainer basket is permitted, and should also be noted on the penetration record if used.

NOTE 2—Both theory and available test data suggest that *N*-values may increase between 10 to 30 % when liners are used.

5.4 *Drive-Weight Assembly:*

5.4.1 *Hammer and Anvil*—The hammer shall weigh 140 ± 2 lb (63.5 ± 1 kg) and shall be a solid rigid metallic mass. The hammer shall strike the anvil and make steel on steel contact when it is dropped. A hammer fall guide permitting a free fall shall be used. Hammers used with the cathead and rope method shall have an unimpeded overlift capacity of at least 4 in. (100 mm). For safety reasons, the use of a hammer assembly with an internal anvil is encouraged.

NOTE 3—It is suggested that the hammer fall guide be permanently marked to enable the operator or inspector to judge the hammer drop height.

5.4.2 *Hammer Drop System*—Rope-cathead, trip, semi-automatic, or automatic hammer drop systems may be used, providing the lifting apparatus will not cause penetration of

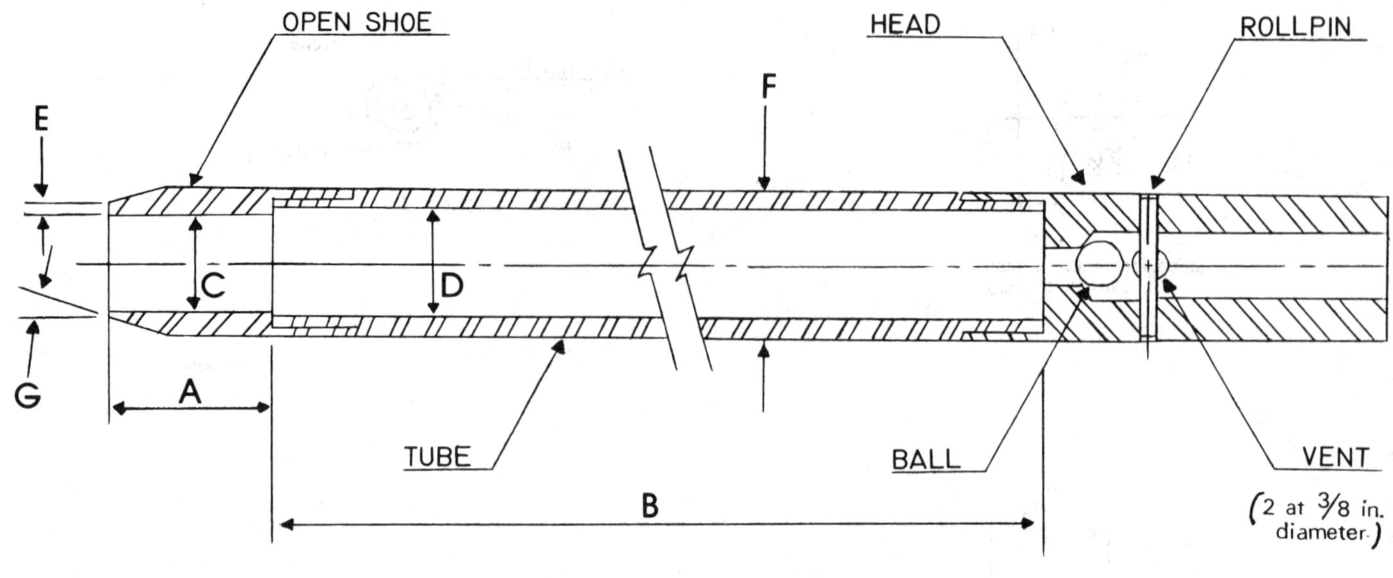

A = 1.0 to 2.0 in. (25 to 50 mm)
B = 18.0 to 30.0 in. (0.457 to 0.762 m)
C = 1.375 ± 0.005 in. (34.93 ± 0.13 mm)
D = 1.50 ± 0.05 − 0.00 in. (38.1 ± 1.3 − 0.0 mm)
E = 0.10 ± 0.02 in. (2.54 ± 0.25 mm)
F = 2.00 ± 0.05 − 0.00 in. (50.8 ± 1.3 − 0.0 mm)
G = 16.0° to 23.0°

The 1½ in. (38 mm) inside diameter split barrel may be used with a 16-gage wall thickness split liner. The penetrating end of the drive shoe may be slightly rounded. Metal or plastic retainers may be used to retain soil samples.

FIG. 2 Split-Barrel Sampler

the sampler while re-engaging and lifting the hammer.

5.5 *Accessory Equipment*—Accessories such as labels, sample containers, data sheets, and groundwater level measuring devices shall be provided in accordance with the requirements of the project and other ASTM standards.

6. Drilling Procedure

6.1 The boring shall be advanced incrementally to permit intermittent or continuous sampling. Test intervals and locations are normally stipulated by the project engineer or geologist. Typically, the intervals selected are 5 ft (1.5 mm) or less in homogeneous strata with test and sampling locations at every change of strata.

6.2 Any drilling procedure that provides a suitably clean and stable hole before insertion of the sampler and assures that the penetration test is performed on essentially undisturbed soil shall be acceptable. Each of the following procedures have proven to be acceptable for some subsurface conditions. The subsurface conditions anticipated should be considered when selecting the drilling method to be used.

6.2.1 Open-hole rotary drilling method.

6.2.2 Continuous flight hollow-stem auger method.

6.2.3 Wash boring method.

6.2.4 Continuous flight solid auger method.

6.3 Several drilling methods produce unacceptable borings. The process of jetting through an open tube sampler and then sampling when the desired depth is reached shall not be permitted. The continuous flight solid auger method shall not be used for advancing the boring below a water table or below the upper confining bed of a confined non-cohesive stratum that is under artesian pressure. Casing may not be advanced below the sampling elevation prior to sampling. Advancing a boring with bottom discharge bits is not permissible. It is not permissible to advance the boring for subsequent insertion of the sampler solely by means of previous sampling with the SPT sampler.

6.4 The drilling fluid level within the boring or hollow-stem augers shall be maintained at or above the in situ groundwater level at all times during drilling, removal of drill rods, and sampling.

7. Sampling and Testing Procedure

7.1 After the boring has been advanced to the desired sampling elevation and excessive cuttings have been removed, prepare for the test with the following sequence of operations.

7.1.1 Attach the split-barrel sampler to the sampling rods and lower into the borehole. Do not allow the sampler to drop onto the soil to be sampled.

7.1.2 Position the hammer above and attach the anvil to the top of the sampling rods. This may be done before the sampling rods and sampler are lowered into the borehole.

7.1.3 Rest the dead weight of the sampler, rods, anvil, and drive weight on the bottom of the boring and apply a seating blow. If excessive cuttings are encountered at the bottom of the boring, remove the sampler and sampling rods from the boring and remove the cuttings.

7.1.4 Mark the drill rods in three successive 6-in. (0.15-m) increments so that the advance of the sampler under the impact of the hammer can be easily observed for each 6-in. (0.15-m) increment.

7.2 Drive the sampler with blows from the 140-lb (63.5-

kg) hammer and count the number of blows applied in each 6-in. (0.15-m) increment until one of the following occurs:

7.2.1 A total of 50 blows have been applied during any one of the three 6-in. (0.15-m) increments described in 7.1.4.

7.2.2 A total of 100 blows have been applied.

7.2.3 There is no observed advance of the sampler during the application of 10 successive blows of the hammer.

7.2.4 The sampler is advanced the complete 18 in. (0.45 m) without the limiting blow counts occurring as described in 7.2.1, 7.2.2, or 7.2.3.

7.3 Record the number of blows required to effect each 6 in. (0.15 m) of penetration or fraction thereof. The first 6 in. is considered to be a seating drive. The sum of the number of blows required for the second and third 6 in. of penetration is termed the "standard penetration resistance," or the "*N*-value." If the sampler is driven less than 18 in. (0.45 m), as permitted in 7.2.1, 7.2.2, or 7.2.3, the number of blows per each complete 6-in. (0.15-m) increment and per each partial increment shall be recorded on the boring log. For partial increments, the depth of penetration shall be reported to the nearest 1 in. (25 mm), in addition to the number of blows. If the sampler advances below the bottom of the boring under the static weight of the drill rods or the weight of the drill rods plus the static weight of the hammer, this information should be noted on the boring log.

7.4 The raising and dropping of the 140-lb (63.5-kg) hammer shall be accomplished using either of the following two methods:

7.4.1 By using a trip, automatic, or semi-automatic hammer drop system which lifts the 140-lb (63.5-kg) hammer and allows it to drop 30 ± 1.0 in. (0.76 m ± 25 mm) unimpeded.

7.4.2 By using a cathead to pull a rope attached to the hammer. When the cathead and rope method is used the system and operation shall conform to the following:

7.4.2.1 The cathead shall be essentially free of rust, oil, or grease and have a diameter in the range of 6 to 10 in. (150 to 250 mm).

7.4.2.2 The cathead should be operated at a minimum speed of rotation of 100 RPM, or the approximate speed of rotation shall be reported on the boring log.

7.4.2.3 No more than 2¼ rope turns on the cathead may be used during the performance of the penetration test, as shown in Fig. 1.

NOTE 4—The operator should generally use either 1¾ or 2¼ rope turns, depending upon whether or not the rope comes off the top (1¾ turns) or the bottom (2¼ turns) of the cathead. It is generally known and accepted that 2¾ or more rope turns considerably impedes the fall of the hammer and should not be used to perform the test. The cathead rope should be maintained in a relatively dry, clean, and unfrayed condition.

7.4.2.4 For each hammer blow, a 30-in. (0.76-m) lift and drop shall be employed by the operator. The operation of pulling and throwing the rope shall be performed rhythmically without holding the rope at the top of the stroke.

7.5 Bring the sampler to the surface and open. Record the percent recovery or the length of sample recovered. Describe the soil samples recovered as to composition, color, stratification, and condition, then place one or more representative portions of the sample into sealable moisture-proof containers (jars) without ramming or distorting any apparent stratification. Seal each container to prevent evaporation of soil moisture. Affix labels to the containers bearing job designation, boring number, sample depth, and the blow count per 6-in. (0.15-m) increment. Protect the samples against extreme temperature changes. If there is a soil change within the sampler, make a jar for each stratum and note its location in the sampler barrel.

8. Report

8.1 Drilling information shall be recorded in the field and shall include the following:

8.1.1 Name and location of job,

8.1.2 Names of crew,

8.1.3 Type and make of drilling machine,

8.1.4 Weather conditions,

8.1.5 Date and time of start and finish of boring,

8.1.6 Boring number and location (station and coordinates, if available and applicable),

8.1.7 Surface elevation, if available,

8.1.8 Method of advancing and cleaning the boring,

8.1.9 Method of keeping boring open,

8.1.10 Depth of water surface and drilling depth at the time of a noted loss of drilling fluid, and time and date when reading or notation was made,

8.1.11 Location of strata changes,

8.1.12 Size of casing, depth of cased portion of boring,

8.1.13 Equipment and method of driving sampler,

8.1.14 Type sampler and length and inside diameter of barrel (note use of liners),

8.1.15 Size, type, and section length of the sampling rods, and

8.1.16 Remarks.

8.2 Data obtained for each sample shall be recorded in the field and shall include the following:

8.2.1 Sample depth and, if utilized, the sample number,

8.2.2 Description of soil,

8.2.3 Strata changes within sample,

8.2.4 Sampler penetration and recovery lengths, and

8.2.5 Number of blows per 6-in. (0.15-m) or partial increment.

9. Precision and Bias

9.1 *Precision*—A valid estimate of test precision has not been determined because it is too costly to conduct the necessary inter-laboratory (field) tests. Subcommittee D18.02 welcomes proposals to allow development of a valid precision statement.

9.2 *Bias*—Because there is no reference material for this test method, there can be no bias statement.

9.3 Variations in *N*-values of 100 % or more have been observed when using different standard penetration test apparatus and drillers for adjacent borings in the same soil formation. Current opinion, based on field experience, indicates that when using the same apparatus and driller, *N*-values in the same soil can be reproduced with a coefficient of variation of about 10 %.

9.4 The use of faulty equipment, such as an extremely massive or damaged anvil, a rusty cathead, a low speed cathead, an old, oily rope, or massive or poorly lubricated rope sheaves can significantly contribute to differences in *N*-values obtained between operator-drill rig systems.

9.5 The variability in N-values produced by different drill rigs and operators may be reduced by measuring that part of the hammer energy delivered into the drill rods from the sampler and adjusting N on the basis of comparative energies. A method for energy measurement and N-value adjustment is given in Test Method D 4633.

10. Keywords

10.1 blow count; in-situ test; penetration resistance; split-barrel sampling; standard penetration test

The American Society for Testing and Materials takes no position respecting the validity of any patent rights asserted in connection with any item mentioned in this standard. Users of this standard are expressly advised that determination of the validity of any such patent rights, and the risk of infringement of such rights, are entirely their own responsibility.

This standard is subject to revision at any time by the responsible technical committee and must be reviewed every five years and if not revised, either reapproved or withdrawn. Your comments are invited either for revision of this standard or for additional standards and should be addressed to ASTM Headquarters. Your comments will receive careful consideration at a meeting of the responsible technical committee, which you may attend. If you feel that your comments have not received a fair hearing you should make your views known to the ASTM Committee on Standards, 1916 Race St., Philadelphia, PA 19103.

ASTM Designation: D 1587 – 83

Standard Practice for Thin-Walled Tube Sampling of Soils[1]

This standard is issued under the fixed designation D 1587; the number immediately following the designation indicates the year of original adoption or, in the case of revision, the year of last revision. A number in parentheses indicates the year of last reapproval. A superscript epsilon (ϵ) indicates an editorial change since the last revision or reapproval.

This practice has been approved for use by agencies of the Department of Defense and for listing in the DOD Index of Specifications and Standards.

1. Scope

1.1 This practice covers a procedure for using a thin-walled metal tube to recover relatively undisturbed soil samples suitable for laboratory tests of structural properties. Thin-walled tubes used in piston, plug, or rotary-type samplers, such as the Denison or Pitcher, must comply with the portions of this practice which describe the thin-walled tubes (5.3).

NOTE 1—This practice does not apply to liners used within the above samplers.

2. Referenced Documents

2.1 *ASTM Standards:*

D 2488 Practice for Description and Identification of Soils (Visual-Manual Procedure)[2]

D 3550 Practice for Ring-Lined Barrel Sampling of Soils[2]

D 4220 Practices for Preserving and Transporting Soil Samples[2]

3. Summary of Practice

3.1 A relatively undisturbed sample is obtained by pressing a thin-walled metal tube into the in-situ soil, removing the soil-filled tube, and sealing the ends to prevent the soil from being disturbed or losing moisture.

4. Significance and Use

4.1 This practice, or Practice D 3550, is used when it is necessary to obtain a relatively undisturbed specimen suitable for laboratory tests of structural properties or other tests that might be influenced by soil disturbance.

5. Apparatus

5.1 *Drilling Equipment*—Any drilling equipment may be used that provides a reasonably clean hole; that does not disturb the soil to be sampled; and that does not hinder the penetration of the thin-walled sampler. Open borehole diameter and the inside diameter of driven casing or hollow stem auger shall not exceed 3.5 times the outside diameter of the thin-walled tube.

5.2 *Sampler Insertion Equipment,* shall be adequate to provide a relatively rapid continuous penetration force. For hard formations it may be necessary, although not recommended, to drive the thin-walled tube sampler.

5.3 *Thin-Walled Tubes,* should be manufactured as shown in Fig. 1. They should have an outside diameter of 2 to 5 in. and be made of metal having adequate strength for use in the soil and formation intended. Tubes shall be clean and free of all surface irregularities including projecting weld seams.

5.3.1 *Length of Tubes*—See Table 1 and 6.4.

5.3.2 *Tolerances,* shall be within the limits shown in Table 2.

5.3.3 *Inside Clearance Ratio,* should be 1 % or as specified by the engineer or geologist for the soil and formation to be sampled. Generally, the inside clearance ratio used should increase with the increase in plasticity of the soil being sampled. See Fig. 1 for definition of inside clearance ratio.

5.3.4 *Corrosion Protection*—Corrosion, whether from galvanic or chemical reaction, can damage or destroy both the thin-walled tube and the sample. Severity of damage is a function of time as well as interaction between the sample and the tube. Thin-walled tubes should have some form of protective coating. Tubes which will contain samples for more than 72 h shall be coated. The type of coating to be used may vary depending upon the material to be sampled. Coatings may include a light coat of lubricating oil, lacquer, epoxy, Teflon, and others. Type of coating must be specified by the engineer or geologist if storage will exceed 72 h. Plating of the tubes or alternate base metals may be specified by the engineer or geologist.

5.4 *Sampler Head,* serves to couple the thin-walled tube to the insertion equipment and, together with the thin-walled tube, comprises the thin-walled tube sampler. The sampler head shall contain a suitable check valve and a venting area to the outside equal to or greater than the area through the check valve. Attachment of the head to the tube shall be concentric and coaxial to assure uniform application of force to the tube by the sampler insertion equipment.

6. Procedure

6.1 Clean out the borehole to sampling elevation using whatever method is preferred that will ensure the material to be sampled is not disturbed. If groundwater is encountered, maintain the liquid level in the borehole at or above ground water level during the sampling operation.

6.2 Bottom discharge bits are not permitted. Side discharge bits may be used, with caution. Jetting through an open-tube sampler to clean out the borehole to sampling elevation is not permitted. Remove loose material from the center of a casing or hollow stem auger as carefully as

[1] This practice is under the jurisdiction of ASTM Committee D-18 on Soil and Rock and is the direct responsibility of Subcommittee D18.02 on Sampling and Related Field Testing for Soil Investigations.

Current edition approved Aug. 17, 1983. Published October 1983. Originally published as D 1587 – 58 T. Last previous edition D 1587 – 74.

[2] *Annual Book of ASTM Standards,* Vol 04.08.

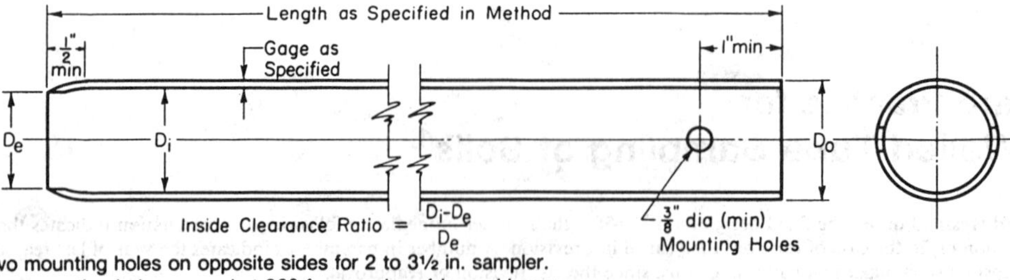

NOTE 1—Minimum of two mounting holes on opposite sides for 2 to 3½ in. sampler.

NOTE 2—Minimum of four mounting holes spaced at 90° for samplers 4 in. and larger.

NOTE 3—Tube held with hardened screws.

NOTE 4—Two-inch outside-diameter tubes are specified with an 18-gage wall thickness to comply with area ratio criteria accepted for "undisturbed samples." Users are advised that such tubing is difficult to locate and can be extremely expensive in small quantities. Sixteen-gage tubes are generally readily available.

Metric Equivalents

in.	mm
⅜	6.77
½	12.7
1	25.4
2	50.8
3½	88.9
4	101.6

FIG. 1 Thin-Walled Tube for Sampling

TABLE 1 Suitable Thin-Walled Steel Sample Tubes[A]

Outside diameter:			
in.	2	3	5
mm	50.8	76.2	127
Wall thickness:			
Bwg	18	16	11
in.	0.049	0.065	0.120
mm	1.24	1.65	3.05
Tube length:			
in.	36	36	54
m	0.91	0.91	1.45
Clearance ratio, %	1	1	1

[A] The three diameters recommended in Table 1 are indicated for purposes of standardization, and are not intended to indicate that sampling tubes of intermediate or larger diameters are not acceptable. Lengths of tubes shown are illustrative. Proper lengths to be determined as suited to field conditions.

TABLE 2 Dimensional Tolerances for Thin-Walled Tubes

Size Outside Diameter	Nominal Tube Diameters from Table 1[A] Tolerances, in. 2	3	5
Outside diameter	+0.007 −0.000	+0.010 −0.000	+0.015 −0.000
Inside diameter	+0.000 −0.007	+0.000 −0.010	+0.000 −0.015
Wall thickness	±0.007	±0.010	±0.015
Ovality	0.015	0.020	0.030
Straightness	0.030/ft	0.030/ft	0.030/ft

[A] Intermediate or larger diameters should be proportional. Tolerances shown are essentially standard commercial manufacturing tolerances for seamless steel mechanical tubing. Specify only two of the first three tolerances; that is, O.D. and I.D., or O.D. and Wall, or I.D. and Wall.

possible to avoid disturbance of the material to be sampled.

NOTE 2—Roller bits are available in downward-jetting and diffused-jet configurations. Downward-jetting configuration rock bits are not acceptable. Diffuse-jet configurations are generally acceptable.

6.3 Place the sample tube so that its bottom rests on the bottom of the hole. Advance the sampler without rotation by a continuous relatively rapid motion.

6.4 Determine the length of advance by the resistance and condition of the formation, but the length shall never exceed 5 to 10 diameters of the tube in sands and 10 to 15 diameters of the tube in clays.

NOTE 3—Weight of sample, laboratory handling capabilities, transportation problems, and commercial availability of tubes will generally limit maximum practical lengths to those shown in Table 1.

6.5 When the formation is too hard for push-type insertion, the tube may be driven or Practice D 3550 may be used. Other methods, as directed by the engineer or geologist, may be used. If driving methods are used, the data regarding weight and fall of the hammer and penetration achieved must be shown in the report. Additionally, that tube must be prominently labeled a "driven sample."

6.6 In no case shall a length of advance be greater than the sample-tube length minus an allowance for the sampler head and a minimum of 3 in. for sludge-end cuttings.

NOTE 4—The tube may be rotated to shear bottom of the sample after pressing is complete.

6.7 Withdraw the sampler from the formation as carefully as possible in order to minimize disturbance of the sample.

7. Preparation for Shipment

7.1 Upon removal of the tube, measure the length of sample in the tube. Remove the disturbed material in the upper end of the tube and measure the length again. Seal the upper end of the tube. Remove at least 1 in. of material from the lower end of the tube. Use this material for soil description in accordance with Practice D 2488. Measure the overall sample length. Seal the lower end of the tube. Alternatively, after measurement, the tube may be sealed without removal of soil from the ends of the tube if so directed by the engineer or geologist.

NOTE 5—Field extrusion and packaging of extruded samples under the specific direction of a geotechnical engineer or geologist is permitted.

NOTE 6—Tubes sealed over the ends as opposed to those sealed with expanding packers should contain end padding in end voids in order to prevent drainage or movement of the sample within the tube.

7.2 Prepare and immediately affix labels or apply markings as necessary to identify the sample. Assure that the markings or labels are adequate to survive transportation and storage.

8. Report

8.1 The appropriate information is required as follows:

8.1.1 Name and location of the project,

8.1.2 Boring number and precise location on project,

8.1.3 Surface elevation or reference to a datum,

8.1.4 Date and time of boring—start and finish,

8.1.5 Depth to top of sample and number of sample,

8.1.6 Description of sampler: size, type of metal, type of coating,

8.1.7 Method of sampler insertion: push or drive,

8.1.8 Method of drilling, size of hole, casing, and drilling fluid used,

8.1.9 Depth to groundwater level: date and time measured,

8.1.10 Any possible current or tidal effect on water level,

8.1.11 Soil description in accordance with Practice D 2488,

8.1.12 Length of sampler advance, and

8.1.13 Recovery: length of sample obtained.

9. Precision and Bias

9.1 This practice does not produce numerical data; therefore, a precision and bias statement is not applicable.

Standard Specification for Poly(Vinyl Chloride) (PVC) Plastic Pipe, Schedules 40, 80, and 120[1]

This standard is issued under the fixed designation D 1785; the number immediately following the designation indicates the year of original adoption or, in the case of revision, the year of last revision. A number in parentheses indicates the year of last reapproval. A superscript epsilon (ε) indicates an editorial change since the last revision or reapproval.

This standard has been approved for use by agencies of the Department of Defense. Consult the DoD Index of Specifications and Standards for the specific year of issue which has been adopted by the Department of Defense.

1. Scope

1.1 This specification covers poly(vinyl chloride) (PVC) pipe made in Schedule 40, 80, and 120 sizes and pressure-rated for water (see Appendix). Included are criteria for classifying PVC plastic pipe materials and PVC plastic pipe, a system of nomenclature for PVC plastic pipe, and requirements and test methods for materials, workmanship, dimensions, sustained pressure, burst pressure, flattening, and extrusion quality. Methods of marking are also given.

1.2 The values stated in inch-pound units are to be regarded as the standard. The values given in parentheses are for information only.

1.3 The following safety hazards caveat pertains only to the test methods portion, Section 8, of this specification: *This standard does not purport to address all of the safety problems, if any, associated with its use. It is the responsibility of the user of this standard to establish appropriate safety and health practices and determine the applicability of regulatory limitations prior to use.* A specific precautionary statement is given in Note 7.

NOTE 1—CPVC plastic pipes, Schedules 40 and 80, which were formerly included in this specification, are now covered by Specification F 441.

NOTE 2—The sustained and burst pressure test requirements, and the pressure ratings in the Appendix, are calculated from stress values obtained from tests made on pipe 4 in. (100 mm) and smaller. However, tests conducted on pipe as large as 24-in. (600-mm) diameter have shown these stress values to be valid for larger diameter PVC pipe.

NOTE 3—PVC pipe made to this specification is often belled for use as line pipe. For details of the solvent cement bell, see Specification D 2672 and for details of belled elastomeric joints, see Specifications D 3139 and D 3212.

2. Referenced Documents

2.1 *ASTM Standards:*

D 618 Practice for Conditioning Plastics and Electrical Insulating Materials for Testing[2]

D 1598 Test Method for Time-to-Failure of Plastic Pipe Under Constant Internal Pressure[3]

D 1599 Test Method for Short-Time Hydraulic Failure Pressure of Plastic Pipe, Tubing, and Fittings[3]

D 1600 Terminology for Abbreviated Terms Relating to Plastics[4]

D 1784 Specification for Rigid Poly(Vinyl Chloride) (PVC) Compounds and Chlorinated Poly(Vinyl Chloride) (CPVC) Compounds[5]

D 2122 Test Method for Determining Dimensions of Thermoplastic Pipe and Fittings[3]

D 2152 Test Method for Degree of Fusion of Extruded Poly(Vinyl Chloride) (PVC) Pipe and Molded Fittings by Acetone Immersion[3]

D 2672 Specification for Joints for IPS PVC Pipe Using Solvent Cement[3]

D 2837 Test Method for Obtaining Hydrostatic Design Basis for Thermoplastic Pipe Materials[3]

D 3139 Specification for Joints for Plastic Pressure Pipes Using Flexible Elastomeric Seals[3]

D 3212 Specification for Joints for Drain and Sewer Plastic Pipes Using Flexible Elastomeric Seals[3]

F 412 Terminology Relating to Plastic Piping Systems[3]

F 441 Specification for Chlorinated Poly(Vinyl Chloride) (CPVC) Plastic Pipe, Schedules 40 and 80[3]

2.2 *Federal Standard:*

Fed. Std. No. 123 Marking for Shipment (Civil Agencies)[6]

2.3 *Military Standard:*

MIL-STD-129 Marking for Shipment and Storage[6]

2.4 *NSF Standards:*

Standard No. 14 for Plastic Piping Components and Related Materials[7]

Standard No. 61 for Drinking Water System Components—Health Effects[7]

3. Terminology

3.1 *Definitions*—Definitions are in accordance with Terminology F 412 and abbreviations are in accordance with Terminology D 1600, unless otherwise specified. The abbreviation for poly(vinyl chloride) plastic is PVC.

3.2 *Descriptions of Terms Specific to This Standard:*

3.2.1 *hydrostatic design stress*—the estimated maximum tensile stress in the wall of the pipe in the circumferential orientation due to internal hydrostatic water pressure that can be applied continuously with a high degree of certainty

---

[1] This specification is under the jurisdiction of ASTM Committee F-17 on Plastic Piping Systems and is the direct responsibility of Subcommittee F17.25 on Vinyl Based Pipe.

Current edition approved April 15, 1993. Published July 1993. Originally published as D 1785 – 60. Last previous edition D 1785 – 91.

[2] *Annual Book of ASTM Standards*, Vol 08.01.

[3] *Annual Book of ASTM Standards*, Vol 08.04.

[4] *Annual Book of ASTM Standards*, Vols 08.01 and 08.04.

[5] *Annual Book of ASTM Standards*, Vols 08.02 and 08.04.

[6] Available from Standardization Documents Order Desk, Bldg. 4 Section D, 700 Robbins Ave., Philadelphia, PA 19111-5094, Attn: NPODS.

[7] Available from the National Sanitation Foundation, P.O. Box 1468, Ann Arbor, MI 48106.

that failure of the pipe will not occur.

3.2.2 *pressure rating* (*PR*)—the estimated maximum pressure that water in the pipe can exert continuously with a high degree of certainty that failure of the pipe will not occur.

3.2.3 *relation between dimensions, design stress, and pressure rating*—the following expression, commonly known as the ISO equation,[8] is used in this specification to relate dimensions, hydrostatic design stress, and pressure rating:

$$2S/P = (D_0/t) - 1$$

where:

S = hydrostatic design stress, psi (or MPa),
P = pressure rating, psi (or MPa),
D_0 = average outside diameter, in. (or mm), and
t = minimum wall thickness, in. (or mm).

3.2.4 *standard thermoplastic pipe materials designation code*—the pipe materials designation code shall consist of the abbreviation PVC for the type of plastic, followed by the ASTM type and grade in Arabic numerals and the design stress in units of 100 psi (0.7 MPa) with any decimal figures dropped. When the design stress code contains less than two figures, a cipher shall be used before the number. Thus a complete material code shall consist of three letters and four figures for PVC plastic pipe materials (see Section 5).

4. Classification

4.1 *General*—This specification covers PVC pipe made to and marked with one of six type/grade/design stress designations (see X1.2) in Schedule 40, 80, and 120 wall sizes.

4.2 *Hydrostatic Design Stresses*—This specification covers pipe made from PVC plastics as defined by four hydrostatic design stresses which have been developed on the basis of long-term tests (Appendix).

5. Materials and Manufacture

5.1 *General*—Poly(vinyl chloride) plastics used to make pipe meeting the requirements of this specification are categorized by means of two criteria, namely, (*1*) short-term strength tests and (*2*) long-term strength tests.

NOTE 4—The PVC pipe intended for use in the transport of potable water should be evaluated and certified as safe for this purpose by a testing agency acceptable to the local health authority. The evaluation should be in accordance with requirements for chemical extraction, taste, and odors that are no less restrictive than those included in NSF Standard No. 14. The seal or mark of the laboratory making the evaluation should be included on the pipe.

5.2 *Basic Materials*—This specification covers pipe made from PVC plastics having certain physical and chemical properties as described in Specification D 1784.

5.3 *Compound*—The PVC compounds used for this pipe shall equal or exceed the following classes described in Specification D 1784; PVC 12454-B, 12454-C, or 14333-D.

5.4 *Rework Material*—Clean, rework material of the same type and grade (cell classification), generated from the manufacturer's own pipe production, may be used by the same manufacturer, as long as the pipe produced meets all the requirements of this specification.

6. Requirements

6.1 *Dimensions and Tolerances:*

6.1.1 Dimensions and tolerances shall be as shown in Tables 1 and 2 when measured in accordance with Test Method D 2122. The tolerances for out-of-roundness shall apply only to pipe prior to shipment.

6.2 *Sustained Pressure*—The pipe shall not fail, balloon, burst, or weep as defined in Test Method D 1598, at the test pressures given in Tables 3, 4, or 5 when tested in accordance with 8.4.

6.2.1 *Accelerated Regression Test*—At the option of the manufacturer, the accelerated regression test may be used as a substitute for both pressure tests—sustained and burst. The test shall be conducted in accordance with 8.4.1. The pipe shall demonstrate a hydrostatic design basis projection at the 100 000-h intercept that meets the hydrostatic design basis category requirement (see Table 1 and Test Method D 2837) for the PVC material used in its manufacture. (*Example:* PVC 1120 pipe must have a minimum 100 000-h projection of 3830 psi (26.40 MPa) and 85 % lower confidence limit (LCL).

6.3 *Burst Pressure*—The minimum burst pressures for PVC plastic pipe shall be as given in Table 6, when determined in accordance with Test Method D 1599.

NOTE 5—Times greater than 60 s may be needed to bring large size specimens to burst pressure. The test is more difficult to pass using greater pressurizing times.

6.4 *Flattening*—There shall be no evidence of splitting, cracking, or breaking when the pipe is tested in accordance with 8.5.

6.5 *Extrusion Quality*—The pipe shall not flake or disintegrate when tested in accordance with Test Method D 2152.

7. Workmanship, Finish, and Appearance

7.1 The pipe shall be homogeneous throughout and free of visible cracks, holes, foreign inclusions, or other defects. The pipe shall be as uniform as commercially practicable in color, opacity, density, and other physical properties.

NOTE 6—Color and transparency or opacity should be specified in the contract or purchase order.

8. Test Methods

8.1 *Conditioning*—Condition the test specimens at 73.4 ± 3.6°F (23 ± 2°C) and 50 ± 5 % relative humidity for not less than 40 h prior to test in accordance with Procedure A of Practice D 618, for those tests where conditioning is required.

8.2 *Test Conditions*—Conduct tests in the standard laboratory atmosphere of 73.4 ± 3.6°F (23 ± 2°C) and 50 ± 5 % relative humidity, unless otherwise specified in the test methods or in this specification.

8.3 *Sampling*—The selection of the sample or samples of pipe shall be as agreed upon by the purchaser and seller. In case of no prior agreement, any sample selected by the testing laboratory shall be deemed adequate.

8.3.1 *Test Specimens*—Not less than 50 % of the test specimens required for any pressure test shall have at least a part of the marking in their central sections. The central section is that portion of pipe which is at least one pipe diameter away from an end closure.

[8] ISO R161-1960, Pipes of Plastics Materials for the Transport of Fluids (Outside Diameters and Nominal Pressures) Part 1, Metric Series.

8.4 *Sustained Pressure Test*—Select the test specimens at random. Test individually with water at the internal pressures given in Tables 3, 4, and 5, six specimens of pipe, each specimen at least ten times the nominal diameter in length, but not less than 10 in. (250 mm) or more than 3 ft (1 m) between end closures and bearing the permanent marking on the pipe. Maintain the specimens at the pressure indicated for a period of 1000 h. Hold the pressure as closely as possible, but within ±10 psi (±70 kPa). Condition the specimens at the test temperature of 73.4°F (23°C) to within 3.6°F (±2°C). Test in accordance with Test Method D 1598, except maintain the pressure at the values given in Tables 3, 4, or 5 for 1000 h. Failure of one of the six specimens tested is cause for retest of six additional specimens. Failure of one of the six specimens tested in retest shall constitute failure in the test. Evidence of failure of the pipe shall be as defined in Test Method D 1598.

8.4.1 *Accelerated Regression Test*—Test in accordance with procedures in Test Method D 1598, except that restrained-end fittings may be employed. A minimum of six samples will be tested at pressures selected to yield data points as follows:

0.010 to 0.099 h (36 s to 6 min)
0.10 to 0.999 h (6 min to 1 h)
1.00 to 9.999 h
10.0 to 99.999 h
100 to 100+ h
0 to 100+ h (random point)

Additional points may be added if necessary to improve projection or LCL, or both. No points shall be excluded unless an obvious defect is detected in the failure area of the test sample. Characterize the results using the least squares extrapolation described in Test Method D 2837.

NOTE 7—**Caution:** Since rupture of the test specimen is expected in quick burst and high-stress regression testing, well-shielded test equipment and protective personal equipment should be used when conducting the tests.

8.5 *Burst Pressure*—Determine the minimum burst pressure with at least five specimens in accordance with Test Method D 1599. The time of testing of each specimen shall be between 60 and 70 s.

8.6 *Flattening*—Flatten three specimens of the pipe, 2 in. (50 mm) long, between parallel plates in a suitable press until the distance between the plates is 40 % of the outside diameter of the pipe or the walls of the pipe touch, whichever occurs first. The rate of loading shall be uniform and such that the compression is completed within 2 to 5 min. On removal of the load examine the specimens for evidence of splitting, cracking, or breaking.

9. Retest and Rejection

9.1 If the results of any test(s) do not meet the requirements of this specification, the test(s) may be conducted again in accordance with an agreement between the purchaser and the seller. There shall be no agreement to lower the minimum requirement of the specification by such means as omitting tests that are a part of the specification, substituting or modifying a test method, or by changing the specification limits. In retesting, the product requirements of this specification shall be met, and the test methods designated in the specification shall be followed. If, upon retest, failure occurs, the quantity of product represented by the test(s) shall be rejected.

10. Product Marking

10.1 *Quality of Marking*—The marking shall be applied to the pipe in such a manner that it remains legible (easily read) after installation and inspection.

10.2 *Content of Marking:*

10.2.1 Marking on the pipe shall include the following, spaced at intervals of not more than 5 ft (1.5 m):

10.2.1.1 Nominal pipe size (for example, 2 in. (50 mm)),

10.2.1.2 Type of plastic pipe material in accordance with the designation code prescribed in 3.2.4, for example, PVC1120,

10.2.1.3 Schedule (40, 80, or 120, whichever is applicable) and the pressure rating in pounds per square inch (megapascals) for water at 73°F (23°C) shown as the number followed by psi (for example, 200 psi (1.4 MPa)). When the indicated pressure rating is lower than that calculated in accordance with 3.2.3 (see appendix), this shall be indicated by placing a star after the pressure rating,

10.2.1.4 ASTM designation D 1785, with which the pipe complies,

10.2.1.5 Manufacturer's name (or trademark) and code (see Note 3), and

10.2.1.6 Pipe intended for the transport of potable water shall also include the seal or mark of the laboratory making the evaluation for this purpose, spaced at intervals specified by the laboratory.

NOTE 8—Manufacturers using the seal or mark of a laboratory must obtain prior authorization from the laboratory concerned.

NOTE 9—It is common practice to dual mark Schedule 40 piping for potable water and DWV usage in which compliance with each applicable standard is met.

11. Quality Assurance

11.1 *Quality Assurance*—When the product is marked with this designation "ASTM D 1785," the manufacturer affirms that this product was inspected, sampled, and tested in accordance with this specification and has been found to meet the requirements of this specification.

12. Keywords

12.1 pressure pipe; PVC pipe; Schedule 40 pipe; Schedule 80 pipe; Schedule 120 pipe

TABLE 1 Outside Diameters and Tolerances for PVC Plastic Pipe Schedules 40, 80, and 120, in. (mm)

Nominal Pipe Size	Outside Diameter	Tolerances		
		Average	For Maximum and Minimum Diameter (Out-of-Roundness)	
			Schedule 40 sizes 3½ in. and over; Schedule 80 sizes 8 in. and over	Schedule 40 sizes 3 in. and less; Schedule 80 sizes 6 in. and less; Schedule 120 sizes all
⅛	0.405 (10.29)	±0.004 (±0.10)	. . .	±0.008 (±0.20)
¼	0.540 (13.72)	±0.004 (±0.10)	. . .	±0.008 (±0.20)
⅜	0.675 (17.14)	±0.004 (±0.10)	. . .	±0.008 (±0.20)
½	0.840 (21.34)	±0.004 (±0.10)	. . .	±0.008 (±0.20)
¾	1.050 (26.67)	±0.004 (±0.10)	. . .	±0.010 (±0.25)
1	1.315 (33.40)	±0.005 (±0.13)	. . .	±0.010 (±0.25)
1¼	1.660 (42.16)	±0.005 (±0.13)	. . .	±0.012 (±0.30)
1½	1.900 (48.26)	±0.006 (±0.15)	. . .	±0.012 (±0.30)
2	2.375 (60.32)	±0.006 (±0.15)	. . .	±0.012 (±0.30)
2½	2.875 (73.02)	±0.007 (±0.18)	. . .	±0.015 (±0.38)
3	3.500 (88.90)	±0.008 (±0.20)	. . .	±0.015 (±0.38)
3½	4.000 (101.60)	±0.008 (±0.20)	±0.050 (±1.27)	±0.015 (±0.38)
4	4.500 (114.30)	±0.009 (±0.23)	±0.050 (±1.27)	±0.015 (±0.38)
5	5.563 (141.30)	±0.010 (±0.25)	±0.050 (±1.27)	±0.030 (±0.76)
6	6.625 (168.28)	±0.011 (±0.28)	±0.050 (±1.27)	±0.035 (±0.89)
8	8.625 (219.08)	±0.015 (±0.38)	±0.075 (±1.90)	±0.045 (±1.14)
10	10.750 (273.05)	±0.015 (±0.38)	±0.075 (±1.90)	±0.050 (±1.27)
12	12.750 (323.85)	±0.015 (±0.38)	±0.075 (±1.90)	±0.060 (±1.52)
14	14.000 (355.60)	±0.015 (±0.38)	±0.100 (±2.54)	. . .
16	16.000 (406.40)	±0.019 (±0.48)	±0.160 (±4.06)	. . .
18	18.000 (457.20)	±0.019 (±0.48)	±0.180 (±4.57)	. . .
20	20.000 (508.00)	±0.023 (±0.58)	±0.200 (±5.08)	. . .
24	24.000 (609.60)	±0.031 (±0.79)	±0.240 (±6.10)	. . .

TABLE 2 Wall Thicknesses and Tolerances for PVC Plastic Pipe, Schedules 40, 80, and 120,[A,B] in. (mm)

Nominal Pipe Size	Wall Thickness[A]					
	Schedule 40		Schedule 80		Schedule 120	
	Minimum	Tolerance	Minimum	Tolerance	Minimum	Tolerance
⅛	0.068 (1.73)	+0.020 (+0.51)	0.095 (2.41)	+0.020 (+0.51)	. . .	. . .
¼	0.088 (2.24)	+0.020 (+0.51)	0.119 (3.02)	+0.020 (+0.51)	. . .	. . .
⅜	0.091 (2.31)	+0.020 (+0.51)	0.126 (3.20)	+0.020 (+0.51)	. . .	. . .
½	0.109 (2.77)	+0.020 (+0.51)	0.147 (3.73)	+0.020 (+0.51)	0.170 (4.32)	+0.020 (+0.51)
¾	0.113 (2.87)	+0.020 (+0.51)	0.154 (3.91)	+0.020 (+0.51)	0.170 (4.32)	+0.020 (+0.51)
1	0.133 (3.38)	+0.020 (+0.51)	0.179 (4.55)	+0.021 (+0.53)	0.200 (5.08)	+0.024 (+0.61)
1¼	0.140 (3.56)	+0.020 (+0.51)	0.191 (4.85)	+0.023 (+0.58)	0.215 (5.46)	+0.026 (+0.66)
1½	0.145 (3.68)	+0.020 (+0.51)	0.200 (5.08)	+0.024 (+0.61)	0.225 (5.72)	+0.027 (+0.68)
2	0.154 (3.91)	+0.020 (+0.51)	0.218 (5.54)	+0.026 (+0.66)	0.250 (6.35)	+0.030 (+0.76)
2½	0.203 (5.16)	+0.024 (+0.61)	0.276 (7.01)	+0.033 (+0.84)	0.300 (7.62)	+0.036 (+0.91)
3	0.216 (5.49)	+0.026 (+0.66)	0.300 (7.62)	+0.036 (+0.91)	0.350 (8.89)	+0.042 (+1.07)
3½	0.226 (5.74)	+0.027 (+0.68)	0.318 (8.08)	+0.038 (+0.96)	0.350 (8.89)	+0.042 (+1.07)
4	0.237 (6.02)	+0.028 (+0.71)	0.337 (8.56)	+0.040 (+1.02)	0.437 (11.10)	+0.052 (+1.32)
5	0.258 (6.55)	+0.031 (+0.79)	0.375 (9.52)	+0.045 (+1.14)	0.500 (12.70)	+0.060 (+1.52)
6	0.280 (7.11)	+0.034 (+0.86)	0.432 (10.97)	+0.052 (+1.32)	0.562 (14.27)	+0.067 (+1.70)
8	0.322 (8.18)	+0.039 (+0.99)	0.500 (12.70)	+0.060 (+1.52)	0.718 (18.24)	+0.086 (+2.18)
10	0.365 (9.27)	+0.044 (+1.12)	0.593 (15.06)	+0.071 (+1.80)	0.843 (21.41)	+0.101 (+2.56)
12	0.406 (10.31)	+0.049 (+1.24)	0.687 (17.45)	+0.082 (+2.08)	1.000 (25.40)	+0.120 (+3.05)
14	0.437 (11.10)	+0.053 (+1.35)	0.750 (19.05)	+0.090 (+2.29)	. . .	. . .
16	0.500 (12.70)	+0.060 (+1.52)	0.843 (21.41)	+0.101 (+2.57)	. . .	. . .
18	0.562 (14.27)	+0.067 (+1.70)	0.937 (23.80)	+0.112 (+2.84)	. . .	. . .
20	0.593 (15.06)	+0.071 (+1.80)	1.031 (26.19)	+0.124 (+3.15)	. . .	. . .
24	0.687 (17.45)	+0.082 (+2.08)	1.218 (30.94)	+0.146 (+3.71)	. . .	. . .

[A] The minimum is the lowest wall thickness of the pipe at any cross section. The maximum permitted wall thickness, at any cross section, is the minimum wall thickness plus the stated tolerance. All tolerances are on the plus side of the minimum requirement.

[B] These dimensions conform to nominal IPS dimensions, with the exception that Schedule 120 wall thickness for pipe sizes ½ to 3½ in. (12.5 to 87.5 mm), inclusive, are special PVC plastic pipe sizes.

TABLE 3 Sustained Pressure Test Conditions for Water at 73°F (23°C) for PVC Plastic Pipe, Schedule 40

Nominal Pipe Size	Pressure Required for Test[A]			
	PVC1120 PVC1220 PVC2120	PVC2116	PVC2112	PVC2110
in.	psi			
1/8	1690	1360	1130	930
1/4	1640	1310	1090	900
3/8	1310	1050	870	720
1/2	1250	1000	840	690
3/4	1010	810	680	550
1	950	760	630	520
1 1/4	770	620	520	420
1 1/2	690	560	460	380
2	580	470	390	320
2 1/2	640	510	430	350
3	590	440	370	300
3 1/2	500	400	340	280
4	470	370	310	260
5	410	330	270	220
6	370	300	250	200
8	330	260	220	180
10	300	240	200	160
12	280	220	180	150
14	270	220	180	150
16	270	220	180	150
18	270	220	180	150
20	260	210	170	140
24	250	200	170	140
in.	MPa			
1/8	11.65	9.38	7.79	6.41
1/4	11.31	9.03	7.52	6.21
3/8	9.03	7.24	6.00	4.96
1/2	8.62	6.89	5.79	4.76
3/4	6.96	5.58	4.69	3.79
1	6.55	5.24	4.34	3.59
1 1/4	5.31	4.27	3.59	2.90
1 1/2	4.76	3.86	3.17	2.62
2	4.00	3.24	2.69	2.21
2 1/2	4.41	3.52	2.96	2.41
3	4.07	3.03	2.55	2.07
3 1/2	3.45	2.76	2.34	1.93
4	3.24	2.55	2.14	1.79
5	2.83	2.28	1.86	1.52
6	2.55	2.07	1.72	1.38
8	2.28	1.79	1.52	1.24
10	2.07	1.65	1.38	1.10
12	1.93	1.52	1.24	1.03
14	1.89	1.54	1.26	1.05
16	1.89	1.54	1.26	1.05
18	1.89	1.54	1.26	1.05
20	1.82	1.47	1.19	0.98
24	1.75	1.40	1.19	0.98

[A] The fiber stresses used to derive these test pressures are as follows:

	psi	MPa
PVC1120	4200	29.0
PVC1220	4200	29.0
PVC2120	4200	29.0
PVC2116	3360	23.2
PVC2112	2800	19.3
PVC2110	2300	15.9

TABLE 4 Sustained Pressure Test Conditions for Water at 73°F (23°C) for PVC Plastic Pipe, Schedule 80

Nominal Pipe Size	Pressure Required for Test[A]			
	PVC1120 PVC1220 PVC2120	PVC2116	PVC2112	PVC2110
in.	psi			
1/8	2570	2060	1720	1410
1/4	2370	1900	1580	1300
3/8	1930	1540	1290	1060
1/2	1780	1430	1190	980
3/4	1440	1160	960	790
1	1320	1060	880	720
1 1/4	1090	870	730	600
1 1/2	990	790	660	540
2	850	680	570	460
2 1/2	890	710	590	490
3	790	630	520	430
3 1/2	730	580	480	400
4	680	540	450	370
5	610	490	400	330
6	590	470	390	320
8	520	410	340	280
10	490	390	330	270
12	480	380	320	260
14	470	380	320	260
16	470	370	310	260
18	460	370	310	250
20	460	370	300	250
24	450	360	300	250
in.	MPa			
1/8	17.72	14.21	11.86	9.72
1/4	16.34	13.10	10.90	8.96
3/8	13.31	10.62	8.89	7.31
1/2	12.27	9.86	8.20	6.76
3/4	9.93	8.00	6.62	5.45
1	9.10	7.31	6.07	4.96
1 1/4	7.52	6.00	5.03	4.14
1 1/2	6.83	4.96	4.55	3.72
2	5.86	4.69	3.93	3.17
2 1/2	6.14	4.90	4.07	3.38
3	5.45	4.34	3.59	2.96
3 1/2	5.03	4.00	3.31	2.76
4	4.69	3.72	3.10	2.55
5	4.21	3.38	2.76	2.28
6	4.07	3.24	2.69	2.21
8	3.59	2.83	2.34	1.93
10	3.38	2.69	2.28	1.86
12	3.31	2.62	2.21	1.79
14	3.29	2.66	2.24	1.82
16	3.29	2.59	2.17	1.82
18	3.22	2.59	2.17	1.75
20	3.22	2.59	2.10	1.75
24	3.15	2.52	2.10	1.75

[A] The fiber stresses used to derive these test pressures are as follows:

	psi	MPa
PVC1120	4200	29.0
PVC1220	4200	29.0
PVC2120	4200	29.0
PVC2116	3360	23.2
PVC2112	2800	19.3
PVC2110	2300	15.9

TABLE 5 **Sustained Pressure Test Conditions for Water at 73°F (23°C) for PVC Plastic Pipe, Schedule 120**

Nominal Pipe Size	Pressure Required for Test[A]			
	PVC1120 PVC1220 PVC2120	PVC2116	PVC2112	PVC2110
in.	psi			
½	2130	1710	1420	1170
¾	1620	1300	1080	890
1	1510	1200	1000	830
1¼	1250	1000	830	680
1½	1130	900	750	620
2	990	790	660	540
2½	980	780	650	540
3	930	750	620	510
3½	810	640	540	440
4	900	720	600	490
5	830	660	550	450
6	780	620	520	430
8	760	610	510	420
10	770	620	510	420
12	710	570	480	390
in.	MPa			
½	14.69	11.79	9.79	8.07
¾	11.17	8.96	7.45	6.14
1	10.41	8.27	6.89	5.72
1¼	8.62	6.89	5.72	4.69
1½	7.79	6.21	5.17	4.27
2	6.83	5.45	4.55	3.72
2½	6.76	5.38	4.48	3.72
3	6.41	5.17	4.27	3.52
3½	5.58	4.41	3.72	3.03
4	6.21	4.96	4.14	3.38
5	5.72	4.55	3.79	3.10
6	5.38	4.27	3.59	2.96
8	5.24	4.21	3.52	2.90
10	5.31	4.27	3.52	2.90
12	4.90	3.93	3.31	2.69

[A] The fiber stresses used to derive these test pressures are as follows:

	psi	MPa
PVC1120	4200	29.0
PVC1220	4200	29.0
PVC2120	4200	29.0
PVC2116	3360	23.2
PVC2112	2800	19.3
PVC2110	2300	15.9

TABLE 6 Burst Pressure Requirements for Water at 73°F (23°C) for PVC Plastic Pipe, Schedules 40, 80, and 120

Nominal Pipe Size	Min Burst Pressures[A]					
	Schedule 40		Schedule 80		Schedule 120	
	PVC1120 PVC1220 PVC2120	PVC2112 PVC2116 PVC2110	PVC1120 PVC1220 PVC2120	PVC2112 PVC2116 PVC2110	PVC1120 PVC1220 PVC2120	PVC2112 PVC2116 PVC2110
in.	psi					
1/8	2580	2020	3920	3060	...	...
1/4	2490	1950	3620	2830	...	...
3/8	1990	1560	2940	2300	...	...
1/2	1910	1490	2720	2120	3250	2540
3/4	1540	1210	2200	1720	2470	1930
1	1440	1130	2020	1580	2300	1790
1 1/4	1180	920	1660	1300	1900	1490
1 1/2	1060	830	1510	1180	1720	1340
2	890	690	1290	1010	1510	1180
2 1/2	970	760	1360	1060	1490	1170
3	840	660	1200	940	1420	1110
3 1/2	770	600	1110	860	1230	960
4	710	560	1040	810	1380	1080
5	620	390	930	720	1260	990
6	560	440	890	700	1190	930
8	500	390	790	620	1160	910
10	450	350	750	580	1170	920
12	420	330	730	570	1090	850
14	410	320	720	570	...	...
16	410	320	710	560	...	...
18	410	320	700	550	...	...
20	390	310	700	540	...	...
24	380	300	680	530	...	...
in.	MPa					
1/8	17.79	13.93	27.03	21.10	...	...
1/4	17.17	13.45	24.96	19.52	...	...
3/8	13.72	10.76	20.27	15.86	...	...
1/2	13.17	10.27	18.76	14.62	22.41	17.52
3/4	10.62	8.34	15.17	11.86	17.03	13.31
1	9.93	7.79	13.93	10.89	15.86	12.34
1 1/4	8.14	6.34	11.45	8.96	13.10	10.27
1 1/2	7.31	5.72	10.41	8.14	11.86	9.24
2	6.14	4.76	8.89	6.96	10.41	8.14
2 1/2	6.69	5.24	9.38	7.31	10.27	8.07
3	5.79	4.55	8.27	6.48	9.79	7.65
3 1/2	5.31	4.14	7.65	5.93	8.48	6.62
4	4.90	3.86	7.17	5.58	9.51	7.45
5	4.27	2.69	6.41	4.96	8.69	6.83
6	3.86	3.03	6.14	4.83	8.20	6.41
8	3.45	2.69	5.45	4.27	8.00	6.27
10	3.10	2.41	5.17	4.00	8.07	6.34
12	2.90	2.28	5.03	3.93	7.52	5.86
14	2.87	2.24	5.04	3.99	...	...
16	2.87	2.24	4.97	3.92	...	...
18	2.87	2.24	4.90	3.85	...	...
20	2.73	2.17	4.90	3.78	...	...
24	2.66	2.10	4.76	3.71	...	...

[A] The fiber stresses used to derive these test pressures are as follows:

	psi	MPa
PVC1120	6400	44.1
PVC1220	6400	44.1
PVC2120	6400	44.1
PVC2116	5000	34.5
PVC2112	5000	34.5
PVC2110	5000	34.5

SUPPLEMENTARY REQUIREMENTS

This requirement applies whenever a regulatory authority or user calls for product to be used to convey or to be in contact with potable water.

S1. *Potable Water Requirement*—Products intended for contact with potable water shall be evaluated, tested, and certified for conformance with ANSI/NSF Standard No. 61 or the health effects portion of NSF Standard No. 14 by an acceptable certifying organization when required by the regulatory authority having jurisdiction.

GOVERNMENT/MILITARY PROCUREMENT

These requirements apply *only* to federal/military procurement, not domestic sales or transfers.

S2. *Responsibility for Inspection*—Unless otherwise specified in the contract or purchase order, the producer is responsible for the performance of all inspection and test requirements specified herein. The producer may use his own or any other suitable facilities for the performance of the inspection and test requirements specified herein, unless the purchaser disapproves. The purchaser shall have the right to perform any of the inspections and tests set forth in this specification where such inspections are deemed necessary to ensure that material conforms to prescribed requirements.

NOTE S2.1—In U.S. Federal contracts, the contractor is responsible for inspection.

S3. *Packaging and Marking for U.S. Government Procurement:*

S3.1 *Packaging*—Unless otherwise specified in the contract, the materials shall be packaged in accordance with the supplier's standard practice in a manner ensuring arrival at destination in satisfactory condition and which will be acceptable to the carrier at lowest rates. Containers and packing shall comply with Uniform Freight Classification rules or National Motor Freight Classification rules.

S3.2 *Marking*—Marking for shipment shall be in accordance with Fed. Std. No. 123 for civil agencies and MIL-STD-129 for military agencies.

NOTE S3.1—The inclusion of U.S. Government procurement requirements should not be construed as an indication that the U.S. Government uses or endorses the products described in this specification.

APPENDIX

(Nonmandatory Information)

X1. SOURCE OF HYDROSTATIC DESIGN STRESSES

X1.1 The hydrostatic design stresses recommended by the Plastics Pipe Institute are used to pressure rate PVC plastic pipe. These hydrostatic design stresses are 2000 psi (14 MPa), 1600 psi (11 MPa), 1250 psi (9 MPa), and 1000 psi (7 MPa) for water at 73°F (23°C). These hydrostatic design stresses apply only to pipe meeting all the requirements of this specification.

X1.2 Six PVC pipe materials are included based on the requirements of Specification D 1784 and the PPI-recommended hydrostatic design stresses as follows:

X1.2.1 Type I, Grade 1 (12454-B), with a hydrostatic design stress of 2000 psi (14 MPa), designated as PVC1120.

X1.2.2 Type I, Grade 2 (12454-C), with a hydrostatic design stress of 2000 psi (14 MPa), designated as PVC1220.

X1.2.3 Type II, Grade 1 (14333-D), with a hydrostatic design stress of 2000 psi (14 MPa), designated as PVC2120.

X1.2.4 Type II, Grade 1 (14333-D), with a hydrostatic design stress of 1600 psi (11.2 MPa), designated as PVC2116.

X1.2.5 Type II, Grade 1 (14333-D), with a hydrostatic design stress of 1250 psi (8.7 MPa), designated as PVC2112.

X1.2.6 Type II, Grade 1 (14333-D), with a hydrostatic design stress of 1000 psi (7.0 MPa), designated as PVC2110.

X1.3 The standard method for obtaining hydrostatic basis for thermoplastic pipe materials is Test Method D 2837. Additional information regarding the test method and other criteria used in developing these hydrostatic design stresses may be obtained from the Plastics Pipe Institute, a division of The Society of the Plastics Industry, 355 Lexington Ave., New York, NY 10017. These hydrostatic design stresses may not be suitable for materials that show a wide departure from a straight-line plot of log stress versus log time to failure. All the data available to date on PVC pipe materials made in the United States exhibit a straight-line plot under these plotting conditions.

X1.4 The pipe is rated for use with water at 73°F (23°C) at the maximum internal pressures shown in Tables X1.1, X1.2, and X1.3. Lower pressure ratings than those calculated in accordance with 3.2.3 may be recommended, at the option of the pipe manufacturer, in which case the SDR shall be included in the marking. Experience of the industry indicates that PVC plastic pipe meeting the requirements of this specification gives satisfactory service under normal conditions for a long period at these pressure ratings. The sustained pressure requirements are related to these ratings through the slopes of the strength-time plots of these materials in pipe form.

X1.5 The hydrostatic design stresses recommended by the Plastics Pipe Institute are based on tests made on pipe ranging in size from ½ to 2½ in. (12.5 to 63.5 mm).

TABLE X1.1 Water Pressure Ratings at 73°F (23°C) for Schedule 40 PVC Plastic Pipe

Nominal Pipe Size	Pressure Ratings[A]			
	PVC1120[B] PVC1220 PVC2120	PVC2116	PVC2112[B]	PVC2110[B]
in.	psi			
1/8	810	650	500	400
1/4	780	620	490	390
3/8	620	500	390	310
1/2	600	480	370	300
3/4	480	390	300	240
1	450	360	280	220
1 1/4	370	290	230	180
1 1/2	330	260	210	170
2	280	220	170	140
2 1/2	300	240	190	150
3	260	210	160	130
3 1/2	240	190	150	120
4	220	180	140	110
5	190	160	120	100
6	180	140	110	90
8	160	120	100	80
10	140	110	90	70
12	130	110	80	70
14	130	100	80	60
16	130	100	80	60
18	130	100	80	60
20	120	100	80	60
24	120	90	70	60
in.	MPa			
1/8	5.58	4.48	3.45	2.76
1/4	5.38	4.27	3.38	2.69
3/8	4.27	3.45	2.69	2.14
1/2	4.14	3.31	2.55	2.07
3/4	3.31	2.69	2.07	1.65
1	3.10	2.48	1.93	1.52
1 1/4	2.55	2.04	1.59	1.24
1 1/2	2.28	1.79	1.45	1.17
2	1.93	1.52	1.17	0.97
2 1/2	2.07	1.65	1.31	1.03
3	1.79	1.45	1.10	0.90
3 1/2	1.65	1.31	1.03	0.83
4	1.52	1.24	0.97	0.76
5	1.31	1.10	0.83	0.69
6	1.24	0.97	0.76	0.62
8	1.10	0.83	0.69	0.55
10	0.97	0.76	0.62	0.48
12	0.90	0.76	0.55	0.48
14	0.91	0.70	0.56	0.42
16	0.91	0.70	0.56	0.42
18	0.91	0.70	0.56	0.42
20	0.84	0.70	0.56	0.42
24	0.84	0.63	0.49	0.42

[A] These pressure ratings apply only to unthreaded pipe. The industry does not recommend threading PVC plastic pipe in Schedule 40 dimensions in nominal pipe sizes 6 in. (150 mm) and smaller.

[B] See Appendix for code designation.

TABLE X1.2 Water Pressure Ratings at 73°F (23°C) for Schedule 80 PVC Plastic Pipe

Nominal Pipe Size, in.	PVC1120, PVC1220, PVC2120		PVC2116		PVC2112		PVC2110	
	Unthreaded	Threaded	Unthreaded	Threaded	Unthreaded	Threaded	Unthreaded	Threaded
	psi							
1/8	1230	610	980	490	770	380	610	310
1/4	1130	570	900	450	710	350	570	280
3/8	920	460	730	370	570	290	460	230
1/2	850	420	680	340	530	260	420	210
3/4	690	340	550	280	430	210	340	170
1	630	320	500	250	390	200	320	160
1 1/4	520	260	420	210	320	160	260	130
1 1/2	470	240	380	190	290	150	240	120
2	400	200	320	160	250	130	200	100
2 1/2	420	210	340	170	260	130	210	110
3	370	190	300	150	230	120	190	90
3 1/2	350	170	280	140	220	110	170	90
4	320	160	260	130	200	100	160	80
5	290	140	230	120	180	90	140	70
6	280	140	220	110	170	90	140	70
8	250	120	200	100	150	80	120	60
10	230	120	190	90	150	70	120	60
12	230	110	180	90	140	70	110	60
14	220	...	180	...	140	...	110	...
16	220	...	180	...	140	...	110	...
18	220	...	180	...	140	...	110	...
20	220	...	170	...	140	...	110	...
24	210	...	170	...	130	...	110	...
	MPa							
1/8	8.48	4.21	6.76	3.38	5.31	2.62	4.21	2.14
1/4	7.79	3.93	6.21	3.10	4.90	2.41	3.93	1.93
3/8	6.34	3.17	5.03	2.55	3.93	2.00	3.17	1.59
1/2	5.86	2.90	4.69	2.34	3.65	1.79	2.90	1.45
3/4	4.76	2.34	3.79	1.93	2.96	1.45	2.34	1.17
1	4.34	2.21	3.45	1.72	2.69	1.38	2.21	1.10
1 1/4	3.59	1.79	2.90	1.45	2.21	1.10	1.79	0.90
1 1/2	3.24	1.65	2.62	1.31	2.0	1.03	1.65	0.83
2	2.76	1.38	2.21	1.10	1.72	0.90	1.38	0.69
2 1/2	2.90	1.45	2.34	1.17	1.79	0.90	1.45	0.76
3	2.55	1.31	2.07	1.03	1.59	0.83	1.31	0.62
3 1/2	2.41	1.17	1.93	0.97	1.52	0.76	1.17	0.62
4	2.21	1.10	1.79	0.90	1.38	0.69	1.10	0.55
5	2.00	0.97	1.59	0.83	1.24	0.62	0.97	0.48
6	1.93	0.97	1.52	0.76	1.17	0.62	0.97	0.48
8	1.72	0.83	1.38	0.69	1.03	0.55	0.83	0.41
10	1.59	0.83	1.31	0.62	1.03	0.48	0.83	0.41
12	1.59	0.76	1.24	0.62	0.97	0.48	0.76	0.41
14	1.54	...	1.26	...	0.98	...	0.77	...
16	1.54	...	1.26	...	0.98	...	0.77	...
18	1.54	...	1.26	...	0.98	...	0.77	...
20	1.54	...	1.19	...	0.98	...	0.77	...
24	1.47	...	1.19	...	0.91	...	0.77	...

TABLE X1.3 Water Pressure Ratings at 73°F (23°C) for Schedule 120 PVC Plastic Pipe

Nominal Pipe Size, in.	PVC1120, PVC1220, PVC2120		PVC2116		PVC2112		PVC2110	
	Unthreaded	Threaded	Unthreaded	Threaded	Unthreaded	Threaded	Unthreaded	Threaded
	psi							
½	1010	510	810	410	630	320	510	250
¾	770	390	620	310	480	240	390	190
1	720	360	570	290	450	220	360	180
1¼	600	300	480	240	370	190	300	150
1½	540	270	430	210	340	170	270	130
2	470	240	380	190	290	150	240	120
2½	470	230	370	190	290	150	230	120
3	440	220	360	180	280	140	220	110
3½	380	190	310	150	240	120	190	100
4	430	220	340	170	270	130	220	110
5	400	200	320	160	250	120	200	100
6	370	190	300	150	230	120	190	90
8	380	180	290	140	230	110	180	90
10	370	180	290	140	230	110	180	90
12	340	170	270	140	210	110	170	80
	MPa							
½	6.96	3.52	5.58	2.83	4.34	2.21	3.52	1.72
¾	5.31	2.69	4.27	2.14	3.31	1.65	2.69	1.31
1	4.96	2.48	3.93	2.00	3.10	1.52	2.48	1.24
1¼	4.14	2.07	3.31	1.65	2.55	1.31	2.07	1.03
1½	3.72	1.86	2.96	1.45	2.34	1.17	1.86	0.90
2	3.24	1.65	2.62	1.31	2.00	1.03	1.65	0.83
2½	3.24	1.59	2.55	1.31	2.00	1.03	1.59	0.83
3	3.03	1.52	2.48	1.24	1.93	0.97	1.52	0.76
3½	2.62	1.31	2.14	1.03	1.65	0.83	1.31	0.69
4	2.96	1.52	2.34	1.17	1.86	0.90	1.52	0.76
5	2.76	1.38	2.21	1.10	1.72	0.83	1.38	0.69
6	2.55	1.31	2.07	1.03	1.59	0.83	1.31	0.62
8	2.62	1.24	2.00	0.97	1.59	0.76	1.24	0.62
10	2.55	1.24	2.00	0.97	1.59	0.76	1.24	0.62
12	2.34	1.17	1.86	0.97	1.45	0.76	1.17	0.55

Standard Practice for Diamond Core Drilling for Site Investigation[1]

This standard is issued under the fixed designation D 2113; the number immediately following the designation indicates the year of original adoption or, in the case of revision, the year of last revision. A number in parentheses indicates the year of last reapproval. A superscript epsilon (ε) indicates an editorial change since the last revision or reapproval.

This standard has been approved for use by agencies of the Department of Defense. Consult the DoD Index of Specifications and Standards for the specific year of issue which has been adopted by the Department of Defense.

1. Scope

1.1 This practice describes equipment and procedures for diamond core drilling to secure core samples of rock and some soils that are too hard to sample by soil-sampling methods. This method is described in the context of obtaining data for foundation design and geotechnical engineering purposes rather than for mineral and mining exploration.

2. Referenced Documents

2.1 *ASTM Standards:*

D 1586 Method for Penetration Test and Split-Barrel Sampling of Soils[2]

D 1587 Practice for Thin-Walled Tube Sampling of Soils[2]

D 3550 Practice for Ring-Lined Barrel Sampling of Soils[2]

3. Significance and Use

3.1 This practice is used to obtain core specimens of superior quality that reflect the in-situ conditions of the material and structure and which are suitable for standard physical-properties tests and structural-integrity determination.

4. Apparatus

4.1 *Drilling Machine,* capable of providing rotation, feed, and retraction by hydraulic or mechanical means to the drill rods.

4.2 *Fluid Pump or Air Compressor,* capable of delivering sufficient volume and pressure for the diameter and depth of hole to be drilled.

4.3 *Core barrels,* as required:

4.3.1 *Single Tube Type, WG Design,* consisting of a hollow steel tube, with a head at one end threaded for drill rod, and a threaded connection for a reaming shell and core bit at the other end. A core lifter, or retainer located within the core bit is normal, but may be omitted at the discretion of the geologist or engineer.

4.3.2 *Double Tube, Swivel-Type, WG Design*—An assembly of two concentric steel tubes joined and supported at the upper end by means of a ball or roller-bearing swivel arranged to permit rotation of the outer tube without causing rotation of the inner tube. The upper end of the outer tube, or removable head, is threaded for drill rod. A threaded connection is provided on the lower end of the outer tube for a reaming shell and core bit. A core lifter located within the core bit is normal but may be omitted at the discretion of the geologist or engineer.

4.3.3 *Double-Tube, Swivel-Type, WT Design,* is essentially the same as the double tube, swivel-type, WG design, except that the WT design has thinner tube walls, a reduced annular area between the tubes, and takes a larger core from the same diameter bore hole. The core lifter is located within the core bit.

4.3.4 *Double Tube, Swivel Type, WM Design,* is similar to the double tube, swivel-type, WG design, except that the inner tube is threaded at its lower end to receive a core lifter case that effectively extends the inner tube well into the core bit, thus minimizing exposure of the core to the drilling fluid. A core lifter is contained within the core lifter case on the inner tube.

4.3.5 *Double Tube Swivel-Type, Large-Diameter Design,* is similar to the double tube, swivel-type, WM design, with the addition of a ball valve, to control fluid flow, in all three available sizes and the addition of a sludge barrel, to catch heavy cuttings, on the two larger sizes. The large-diameter design double tube, swivel-type, core barrels are available in three core per hole sizes as follows: 2¾ in. (69.85 mm) by 3⅞ in. (98.43 mm), 4 in. (101.6 mm) by 5½ in. (139.7 mm), and 6 in. (152.4 mm) by 7¾ in. (196.85 mm). Their use is generally reserved for very detailed investigative work or where other methods do not yield adequate recovery.

4.3.6 *Double Tube, Swivel-Type, Retrievable Inner-Tube Method,* in which the core-laden inner-tube assembly is retrieved to the surface and an empty inner-tube assembly returned to the face of the borehole through the matching, large-bore drill rods without need for withdrawal and replacement of the drill rods in the borehole. The inner-tube assembly consists of an inner tube with removable core lifter case and core lifter at one end and a removable inner-tube head, swivel bearing, suspension adjustment, and latching device with release mechanism on the opposite end. The inner-tube latching device locks into a complementary recess in the wall of the outer tube such that the outer tube may be rotated without causing rotation of the inner tube and such that the latch may be actuated and the inner-tube assembly transported by appropriate surface control. The outer tube is threaded for the matching, large-bore drill rod and internally configured to receive the inner-tube latching device at one end and threaded for a reaming shell and bit, or bit only, at the other end.

4.4 *Longitudinally Split Inner Tubes*—As opposed to

[1] This practice is under the jurisdiction of ASTM Committee D-18 on Soil and Rock and is the direct responsibility of Subcommittee D18.02 on Sampling and Related Field Testing for Soil Investigations.

Current edition approved June 24, 1983. Published August 1983. Originally published as D 2113 – 62 T. Last previous edition D 2113 – 70 (1976).

[2] *Annual Book of ASTM Standards,* Vol 04.08.

conventional cylindrical inner tubes, allow inspection of, and access to, the core by simply removing one of the two halves. They are not standardized but are available for most core barrels including many of the retrievable inner-tube types.

4.5 *Core Bits*—Core bits shall be surface set with diamonds, impregnated with small diamond particles, inserted with tungsten carbide slugs, or strips, hard-faced with various hard surfacing materials or furnished in saw-tooth form, all as appropriate to the formation being cored and with concurrence of the geologist or engineer. Bit matrix material, crown shape, water-way type, location and number of water ways, diamond size and carat weight, and bit facing materials shall be for general purpose use unless otherwise approved by the geologist or engineer. Nominal size of some bits is shown in Table 1.

NOTE 1—Size designation (letter symbols) used throughout the text and in Tables 1, 2, and 3 are those standardized by the Diamond Core Drill Manufacturers' Assoc. (DCDMA). Inch dimensions in the tables have been rounded to the nearest hundredth of an inch.

4.6 *Reaming Shells,* shall be surface set with diamonds, impregnated with small diamond particles, inserted with tungsten carbide strips or slugs, hard faced with various types of hard surfacing materials, or furnished blank, all as appropriate to the formation being cored.

4.7 *Core Lifters*—Core lifters of the split-ring type, either plain or hard-faced, shall be furnished and maintained, along with core-lifter cases or inner-tube extensions or inner-tube shoes, in good condition. Basket or finger-type lifters, together with any necessary adapters, shall be on the job and available for use with each core barrel if so directed by the geologist or engineer.

TABLE 1 Core Bit Sizes

Size Designation	Outside Diameter		Inside Diameter	
	in.	mm	in.	mm
RWT	1.16	29.5	0.375	18.7
EWT	1.47	37.3	0.905	22.9
EWG, EWM	1.47	37.3	0.845	21.4
AWT	1.88	47.6	1.281	32.5
AWG, AWM	1.88	47.6	1.185	30.0
BWT	2.35	59.5	1.750	44.5
BWG, BWM	2.35	59.5	1.655	42.0
NWT	2.97	75.3	2.313	58.7
NWG, NWM	2.97	75.3	2.155	54.7
2¾ × 3⅞	3.84	97.5	2.69	68.3
HWT	3.89	98.8	3.187	80.9
HWG, . . .	3.89	98.8	3.000	76.2
4 × 5½	5.44	138.0	3.97	100.8
6 × 7¾	7.66	194.4	5.97	151.6

4.8 *Casings:*

4.8.1 *Drive Pipe or Drive Casing,* shall be standard weight (schedule 40), extra-heavy (schedule 80), double extra-heavy (schedule 160) pipe or W-design flush-joint casing as required by the nature of the overburden or the placement method. Drive pipe or W-design casing shall be of sufficient diameter to pass the largest core barrel to be used, and it shall be driven to bed rock or to firm seating at an elevation below water-sensitive formation. A hardened drive shoe is to be used as a cutting edge and thread protection device on the bottom of the drive pipe or casing. The drive shoe inside diameter shall be large enough to pass the tools intended for use, and the shoe and pipe or casing shall be free from burrs or obstructions.

4.8.2 *Casing*—When necessary to case through formations already penetrated by the borehole or when no drive casing has been set, auxiliary casing shall be provided to fit inside the borehole to allow use of the next smaller core barrel. Standard sizes of telescoping casing are shown in Table 2. Casing bits have an obstruction in their interior and will not pass the next smaller casing size. Use a casing shoe if additional telescoping is anticipated.

4.8.3 *Casing Liner*—Plastic pipe or sheet-metal pipe may be used to line an existing large-diameter casing. Liners, so used, should not be driven, and care should be taken to maintain true alignment throughout the length of the liner.

4.8.4 *Hollow Stem Auger*—Hollow stem auger may be used as casing for coring.

4.9 *Drill Rods:*

4.9.1 *Drill Rods of Tubular Steel Construction* are normally used to transmit feed, rotation, and retraction forces from the drilling machine to the core barrel. Drill-rod sizes that are presently standardized are shown in Table 3.

4.9.2 Large bore drill rods used with retrievable inner-tube core barrels are not standardized. Drill rods used with retrievable inner-tube core barrels should be those manufactured by the core-barrel manufacturer specifically for the core barrel.

4.9.3 *Composite Drill Rods* are specifically constructed from two or more materials intended to provide specific properties such as light weight or electrical nonconductivity.

4.9.4 *Nonmagnetic Drill Rods* are manufactured of nonferrous materials such as aluminum or brass and are used primarily for hole survey work. Some nonmagnetic rods have left-hand threads in order to further their value in survey work. No standard exists for nonmagnetic rods.

4.10 *Auxiliary Equipment,* shall be furnished as required by the work and shall include: roller rock bits, drag bits,

TABLE 2 Casing Sizes

Size Designation	Outside Diameter		Inside Diameter		Threads per in.	Will Fit Hole Drilled with Core Bit Size
	in.	mm	in.	mm		
RW	1.144	36.5	1.19	30.1	5	EWT, EWG, EWM
EW	1.81	46.0	1.50	38.1	4	AWT, AWG, AWM
AW	2.25	57.1	1.91	48.4	4	BWT, BWG, BWM
BW	2.88	73.0	2.38	60.3	4	NWT, NWG, NWM
NW	3.50	88.9	3.00	76.2	4	HWT, HWG
HW	4.50	114.3	4.00	101.6	4	4 × 5½
PW	5.50	139.7	5.00	127.0	3	6 × 7¾
SW	6.63	168.2	6.00	152.4	3	6 × 7¾
UW	7.63	193.6	7.00	177.8	2	. . .
ZW	8.63	219.0	8.00	203.2	2	. . .

TABLE 3 Drill Rods

Size Designation	Rod and Coupling Outside Diameter		Rod Inside Diameter		Coupling Bore, Threads		
	in.	mm	in.	mm	in.	mm	per in.
RW	1.09	27.7	0.72	18.2	0.41	10.3	4
EW	1.38	34.9	1.00	25.4	0.44	11.1	3
AW	1.72	43.6	1.34	34.1	0.63	15.8	3
BW	2.13	53.9	1.75	44.4	0.75	19.0	3
NW	2.63	66.6	2.25	57.1	1.38	34.9	3
HW	3.50	88.9	3.06	77.7	2.38	60.3	3

chopping bits, boulder busters, fishtail bits, pipe wrenches, core barrel wrenches, lubrication equipment, core boxes, and marking devices. Other recommended equipment includes: core splitter, rod wicking, pump-out tools or extruders, and hand sieve or strainer.

5. Transportation and Storage of Core Containers

5.1 *Core Boxes,* shall be constructed of wood or other durable material for the protection and storage of cores while enroute from the drill site to the laboratory or other processing point. All core boxes shall be provided with longitudinal separators and recovered cores shall be laid out as a book would read, from left to right and top to bottom, within the longitudinal separators. Spacer blocks or plugs shall be marked and inserted into the core column within the separators to indicate the beginning of each coring run. The beginning point of storage in each core box is the upper left-hand corner. The upper left-hand corner of a hinged core box is the left corner when the hinge is on the far side of the box and the box is right-side up. All hinged core boxes must be permanently marked on the outside to indicate the top and the bottom. All other core boxes must be permanently marked on the outside to indicate the top and the bottom and additionally, must be permanently marked internally to indicate the upper-left corner of the bottom with the letters UL or a splotch of red paint not less than 1 in.^{2} Lid or cover fitting(s) for core boxes must be of such quality as to ensure against mix up of the core in the event of impact or upsetting of the core box during transportation.

5.2 Transportation of cores from the drill site to the laboratory or other processing point shall be in durable core boxes so padded or suspended as to be isolated from shock or impact transmitted to the transporter by rough terrain or careless operation.

5.3 Storage of cores, after initial testing or inspection at the laboratory or other processing point, may be in cardboard or similar less costly boxes provided all layout and marking requirements as specified in 5.1 are followed. Additional spacer blocks or plugs shall be added if necessary at time of storage to explain missing core. Cores shall be stored for a period of time specified by the engineer but should not normally be discarded prior to completion of the project for which they were taken.

6. Procedure

6.1 Use core-drilling procedures when formations are encountered that are too hard to be sampled by soil-sampling methods. A 1-in. (25.4-mm) or less penetration for 50 blows in accordance with Method D 1586 or other criteria established by the geologist or engineer, shall indicate that soil-sampling methods are not applicable.

6.1.1 Seat the casing on bedrock or in a firm formation to prevent raveling of the borehole and to prevent loss of drilling fluid. Level the surface of the rock or hard formation at the bottom of the casing when necessary, using the appropriate bits. Casing may be omitted if the borehole will stand open without the casing.

6.1.2 Begin the core drilling using an N-size double-tube swivel-type core barrel or other size or type approved by the engineer. Continue core drilling until core blockage occurs or until the net length of the core barrel has been drilled in. Remove the core barrel from the hole and disassemble it as necessary to remove the core. Reassemble the core barrel and return it to the hole. Resume coring.

6.1.3 Place the recovered core in the core box with the upper (surface) end of the core at the upper-left corner of the core box as described in 5.1. Continue boxing core with appropriate markings, spacers, and blocks as described in 5.1. Wrap soft or friable cores or those which change materially upon drying in plastic film or seal in wax, or both, when such treatment is required by the engineer. Use spacer blocks or slugs properly marked to indicate any noticeable gap in recovered core which might indicate a change or void in the formation. Fit fracture, bedded, or jointed pieces of core together as they naturally occurred.

6.1.4 Stop the core drilling when soft materials are encountered that produce less than 50 % recovery. If necessary, secure samples of soft materials in accordance with the procedures described in Method D 1586, Practice D 1587, or Practice D 3550, or by any other method acceptable to the geologist or engineer. Resume diamond core drilling when refusal materials as described in 6.1 are again encountered.

6.2 Subsurface structure, including the dip of strata, the occurrence of seams, fissures, cavities, and broken areas are among the most important items to be detected and described. Take special care to obtain and record information about these features. If conditions prevent the continued advance of the core drilling, the hole should be cemented and redrilled, or reamed and cased, or cased and advanced with the next smaller-size core barrel, as required by the geologist or engineer.

6.3 Drilling mud or grouting techniques must be approved by the geologist or engineer prior to their use in the borehole.

6.4 *Compatibility of Equipment:*

6.4.1 Whenever possible, core barrels and drill rods should be selected from the same letter-size designation to ensure maximum efficiency. See Tables 1 and 3.

6.4.2 Never use a combination of pump, drill rod, and core barrel that yields a clear-water up-hole velocity of less than 120 ft/min.

6.4.3 Never use a combination of air compressor, drill

rod, and core barrel that yields a clear-air up-hole velocity of less than 3000 ft/min.

7. Boring Log

7.1 The boring log shall include the following:

7.1.1 Project identification, boring number, location, date boring began, date boring completed, and driller's name.

7.1.2 Elevation of the ground surface.

7.1.3 Elevation of or depth to ground water and raising or lowering of level including the dates and the times measured.

7.1.4 Elevations or depths at which drilling fluid return was lost.

7.1.5 Size, type, and design of core barrel used. Size, type, and set of core bit and reaming shell used. Size, type, and length of all casing used. Description of any movements of the casing.

7.1.6 Length of each core run and the length or percentage, or both, of the core recovered.

7.1.7 Geologist's or engineer's description of the formation recovered in each run.

7.1.8 Driller's description, if no engineer or geologist is present, of the formation recovered in each run.

7.1.9 Subsurface structure description, including dip of strata and jointing, cavities, fissures, and any other observations made by the geologist or engineer that could yield information regarding the formation.

7.1.10 Depth, thickness, and apparent nature of the filling of each cavity or soft seam encountered, including opinions gained from the feel or appearance of the inside of the inner tube when core is lost. Record opinions as such.

7.1.11 Any change in the character of the drilling fluid or drilling fluid return.

7.1.12 Tidal and current information when the borehole is sufficiently close to a body of water to be affected.

7.1.13 Drilling time in minutes per foot and bit pressure in pound-force per square inch gage when applicable.

7.1.14 Notations of character of drilling, that is, soft, slow, easy, smooth, etc.

8. Precision and Bias

8.1 This practice does not produce numerical data; therefore, a precision and bias statement is not applicable.

NOTE 2—Inclusion of the following tables and use of letter symbols in the foregoing text is not intended to limit the practice to use of DCDMA tools. The table and text references are included as a convenience to the user since the vast majority of tools in use do meet DCDMA dimensional standards. Similar equipment of approximately equal size on the metric standard system is acceptable unless otherwise stipulated by the engineer or geologist.

Designation: D 2217 – 85

Standard Practice for
Wet Preparation of Soil Samples for Particle-Size Analysis and Determination of Soil Constants[1]

This standard is issued under the fixed designation D 2217; the number immediately following the designation indicates the year of original adoption or, in the case of revision, the year of last revision. A number in parentheses indicates the year of last reapproval. A superscript epsilon (ε) indicates an editorial change since the last revision or reapproval.

This practice has been approved for use by agencies of the Department of Defense and for listing in the DoD Index of Specifications and Standards.

1. Scope

1.1 This practice covers the wet preparation of soil samples as received from the field for particle-size analysis and determination of soil constants.

1.2 Procedure A provides for drying the field sample at a temperature not exceeding 140°F (60°C), making a wet separation on the No. 10 (2.00-mm) sieve, or No. 40 (425-μm) sieve, or both, as needed, and finally drying at a temperature not exceeding 140°F. Procedure B provides that the sample shall be kept at a moisture content equal to or greater than the natural water content. The procedure to be used should be indicated in the specification for the material being tested. If no procedure is specified, the provisions of Procedure B shall govern.

1.3 *This standard may involve hazardous materials, operations, and equipment. This standard does not purport to address all of the safety problems associated with its use. It is the responsibility of whoever uses this standard to consult and establish appropriate safety and health practices and determine the applicability of regulatory limitations prior to use.*

2. Referenced Documents

2.1 *ASTM Standards:*

D 421 Practice for Dry Preparation of Soil Samples for Particle-Size Analysis and Determination of Soil Constants[2]

D 422 Method for Particle-Size Analysis of Soils[2]

E 11 Specification for Wire-Cloth Sieves for Testing Purposes[3]

3. Significance and Use

3.1 Procedure A is used to prepare soil samples for plasticity tests and particle-size analysis when the coarse-grained particles of a sample are soft and pulverize readily, as in Practice D 421, or when the fine particles are very cohesive and tend to resist removal from the coarse particles.

3.2 Some soils never dry out in nature and may change their characteristics greatly when dried. If the true natural gradation and plasticity characteristics of such soils are desired, these soils should be shipped to the laboratory in sealed containers and processed in accordance with Procedure B of this practice.

3.3 Liquid limit and plasticity index values derived from samples containing their natural moisture are usually, but not always, equal to or higher than values derived from similar samples of the dried soil. In the case of fine-grained organic soil, there is a radical drop in plasticity due to oven drying.

4. Apparatus

4.1 *Balance*, sensitive to 0.1 g.

4.2 *Mortar and Rubber-Covered Pestle*, suitable for breaking up the aggregations of soil particles.

4.3 *Sieves*, No. 10 (2.00-mm) and No. 40 (425-μm), of square mesh woven-wire cloth, conforming to Specification E 11.

4.4 *Sampler*—A riffle sampler or sample splitter for quartering the samples.

4.5 *Drying Apparatus*—Thermostatically controlled drying oven for use at 140°F (60°C) or below and at 230°F (110°C), infrared lamps; air drier; or other suitable device for drying samples.

4.6 *Filter Funnels or Candles*—Büchner funnels 10 in. (254 mm) in diameter and filter paper or filter candles.

4.7 *Miscellaneous Equipment*—Pans 12 in. (304.8 mm) in diameter and 3 in. (76.2 mm) in depth; a suitable container that will prevent loss of moisture during storage of the moist test sample prepared in Procedure B.

PROCEDURE A

5. Sampling

5.1 Dry the soil sample as received from the field, using one of the following methods: (*1*) in air at room temperature, (*2*) in a drying oven at a temperature not exceeding 140°F (60°C), or (*3*) using any warming device that will not raise the temperature of the sample above 140°F. Break up thoroughly any aggregations of particles using the mortar and rubber-covered pestle or other suitable device (Note 1). Select a representative portion by the method of quartering or by use of the sampler. This portion must be sufficient to provide samples for particle-size analyses of material retained on and passing the No. 10 (2.00-mm) sieve, and to provide an adequate amount of material passing the No. 40 (425-μm) sieve for the tests to determine soil constants. The amounts of material required to perform the individual tests are as follows:

---

[1] This practice is under the jurisdiction of ASTM Committee D-18 on Soil and Rock and is the direct responsibility of Subcommittee D18.03 on Texture, Plasticity, and Density Characteristics of Soils.

Current edition approved July 26, 1985. Published September 1985. Originally published as D 2217 – 63 T. Last previous edition D 2217 – 66 (1978)$^{\epsilon 1}$.

[2] *Annual Book of ASTM Standards*, Vol 04.08.

[3] *Annual Book of ASTM Standards*, Vol 14.02.

Particle-Size Analysis of Material Retained on No. 10 (2.00-mm) Sieve:	
Gravelly soils, g	4 000 to 10 000
Sandy soils, g	1 500
Silty or clayey soils, g	400
Particle-Size Analysis of Material Passing No. 10 (2.00-mm) Sieve:	
Sandy soils, g	115
Silty or clayey soils, g	65
Tests for Determination of Soil Constants:	
Liquid limit, g	100
Plastic limit, g	15
Centrifuge moisture equivalent, g	10
Shrinkage factors, g	30
Check tests, g	65

NOTE 1—When the sample contains particles of soft shale or sandstone or similar weak material, proper care must be exercised to avoid excessive reduction in the size of the particles.

6. Preparation of Test Samples

6.1 *For Particle-Size Analysis:*

6.1.1 Weigh the portion of the test sample selected for particle-size analysis and record as the weight of test sample uncorrected for hygroscopic moisture. Separate this material into two portions using the No. 10 (2.00-mm) sieve. Set aside the portion passing for later recombination with additional material washed from the portion retained on the No. 10 (2.00-mm) sieve.

6.1.2 Place the material retained on the No. 10 (2.00-mm) sieve in a pan, cover with water, and allow to soak until the particle aggregations become soft. After soaking, wash the material on a No. 10 (2.00-mm) sieve in the following manner: Place an empty No. 10 (2.00-mm) sieve on the bottom of a clean pan and pour the water from the soaked sample into the sieve. Add sufficient water to bring the level approximately ½ in. (12.7 mm) above the mesh of the sieve. Transfer the soaked material to the sieve in increments not exceeding 1 lb (0.45 kg), stirring each increment with the fingers while agitating the sieve up and down. Crumble or mash any lumps that have not slaked, using the thumb and fingers. Raise the sieve above the water in the pan and complete the washing operation using a small amount of clean water. Transfer the washed material on the sieve to a clean pan before placing another increment of soaked material on the sieve.

6.1.3 Dry the material retained on the No. 10 (2.00-mm) sieve at a temperature of 230 ± 9°F (110 ± 5°C), sieve on the No. 10 (2.00-mm) sieve, and add the material passing the sieve to similar material obtained in 6.1.1. Set aside the material retained on the sieve for use in the particle-size analysis.

6.1.4 Set aside the pan containing the washings for a period of several hours or until the water above the particles is clear. Decant, pipet, or siphon off as much of the clear water as possible (Note 2). Dry the soil remaining in the pan at a temperature not exceeding 140°F (60°C). Grind the dried soil in the mortar with the rubber-covered pestle or other suitable device, and combine with similar material obtained in 6.1.1.

6.1.5 Alternatively, after all the soaked material has been washed, remove most of the water by filtering the wash water on one or more Büchner funnels fitted with filter paper or by using filter candles. Remove the moist soil from the filter paper or filter candles, combine with any sediment remaining in the pan, and dry at a temperature not exceeding 140°F (60°C). Grind the dried soil in the mortar with a rubber-covered pestle or other suitable device and combine with similar material obtained in 6.1.1.

NOTE 2—In some instances, the wash water will not become clear in a reasonable length of time; in this case the entire volume must be evaporated.

6.2 *For Determination of Soil Constants*—Proceed in accordance with 6.1, substituting a No. 40 (425-μm) sieve for the No. 10 (2.00-mm) sieve.

NOTE 3—In some areas it is possible that the cations of salts present in the tap water may exchange with the natural cations in the soil and alter significantly the values of the soil constants should tap water be used in the soaking and washing operations. Unless it is known that such cations are not present in the tap water, distilled or demineralized water should be used. The soaking and washing operation will remove soluble salts contained in the soil. When soluble salts are present in the soil, the wash water should be saved and evaporated, and the salts returned to the soil sample.

7. Test Samples

7.1 Keeping each portion separate from the other portion, mix thoroughly the portions of the soil sample passing the No. 10 (2.00-mm) sieve and the No. 40 (425-μm) sieve. By the method of quartering or by the use of the sampler, select and weigh out test samples of the weights indicated in Section 5, as may be needed to make the required tests.

PROCEDURE B

8. Samples

8.1 Samples prepared in accordance with this procedure must be shipped from the field to the laboratory in sealed containers and must contain all their natural moisture. Samples obviously containing only particles passing the No. 10 (2.00-mm) sieve may be tested in the particle-size analysis without first washing on the No. 10 (2.00-mm) sieve. Samples obviously containing only particles passing the No. 40 (425-μm) sieve may be used in the tests to determine soil constants without first washing on the No. 40 (425-μm) sieve.

9. Preparation of Test Samples

9.1 *For Particle-Size Analysis:*

9.1.1 Select and weigh a representative portion of the moist sample estimated to contain 50 g of particles passing the No. 10 (2.00-mm) sieve for silty and clayey soil, or 100 g for sandy soil. For samples containing particles not passing the No. 10 (2.00-mm) sieve for which a particle-size analysis is required, select and weigh a representative sample estimated to contain the required amounts of particles both passing and not passing the No. 10 (2.00-mm) sieve. Determine the moisture content at 230 ± 9°F (110 ± 5°C) using an auxiliary sample, for use in Method D 422.

9.1.2 Soak the moist sample and wash on a No. 10 (2.00-mm) sieve as described in 6.1.2. After washing, dry the material retained on the No. 10 (2.00-mm) sieve in an oven at a temperature of 230 ± 9°F (110 ± 5°C), weigh, and retain for the particle-size analysis. If the volume of the wash water and soil is too large for use in the sedimentation procedure of the test for particle-size analysis, evaporate excess water by exposure to air at room temperature, by heating in an oven at a temperature not exceeding 230°F (110°C), or by boiling. Regardless of the method of evaporation used, the following precautions must be taken: (*1*) stir the slurry from time to time to prevent a dry soil ring from forming on the walls of the evaporation vessel, and (*2*) return the temperature of the sample to room temperature before testing.

9.2 *For Determination of Soil Constants*—Select a representative portion of the moist sample estimated to contain sufficient particles passing the No. 40 (425-μm) sieve to make the required tests for determination of soil constants. Soak this selected portion of the moist sample and wash on the No. 40 (425-μm) sieve as described in 6.2 (Note 2). Reduce the moisture content of the material passing the No. 40 (425-μm) sieve until the mass reaches a putty-like consistency (such as 30 to 35 drops of the cup in the liquid limit test) but never below the natural moisture content. Reduction of moisture content may be accomplished as follows: by exposure to air at ordinary room temperature, by heating in an oven at a temperature not exceeding 230°F (110°C), by boiling, by filtering on a Büchner funnel, or by use of filter candles. During evaporation and cooling, stir the sample often enough to prevent overdrying of the fringes and soil pinnacles on the surface. Cool the heated samples to normal room temperature before testing. For soil samples containing soluble salts, use a method of water reduction that will not eliminate the soluble salts from the test sample. Protect the prepared sample in a suitable container from further drying until all required tests have been performed.

Designation: D 2487 – 92

Standard Classification of Soils for Engineering Purposes (Unified Soil Classification System)[1]

This standard is issued under the fixed designation D 2487; the number immediately following the designation indicates the year of original adoption or, in the case of revision, the year of last revision. A number in parentheses indicates the year of last reapproval. A superscript epsilon (ε) indicates an editorial change since the last revision or reapproval.

This standard has been approved for use by agencies of the Department of Defense. Consult the DOD Index of Specifications and Standards for the specific year of issue which has been adopted by the Department of Defense.

1. Scope

1.1 This standard describes a system for classifying mineral and organo-mineral soils for engineering purposes based on laboratory determination of particle-size characteristics, liquid limit, and plasticity index and shall be used when precise classification is required.

NOTE 1—Use of this standard will result in a single classification group symbol and group name except when a soil contains 5 to 12 % fines or when the plot of the liquid limit and plasticity index values falls into the crosshatched area of the plasticity chart. In these two cases, a dual symbol is used, for example, GP-GM, CL-ML. When the laboratory test results indicate that the soil is close to another soil classification group, the borderline condition can be indicated with two symbols separated by a slash. The first symbol should be the one based on this standard, for example, CL/CH, GM/SM, SC/CL. Borderline symbols are particularly useful when the liquid limit value of clayey soils is close to 50. These soils can have expansive characteristics and the use of a borderline symbol (CL/CH, CH/CL) will alert the user of the assigned classifications of expansive potential.

1.2 The group symbol portion of this sytem is based on laboratory tests performed on the portion of a soil sample passing the 3-in. (75-mm) sieve (see Specification E 11).

1.3 As a classification system, this standard is limited to naturally occurring soils.

NOTE 2—The group names and symbols used in this test method may be used as a descriptive system applied to such materials as shale, claystone, shells, crushed rock, etc. See Appendix X2.

1.4 This standard is for qualitative application only.

NOTE 3—When quantitative information is required for detailed designs of important structures, this test method must be supplemented by laboratory tests or other quantitative data to determine performance characteristics under expected field conditions.

1.5 This standard is the ASTM version of the Unified Soil Classification System. The basis for the classification scheme is the Airfield Classification System developed by A. Casagrande in the early 1940's.[2] It became known as the Unified Soil Classification System when several U.S. Government Agencies adopted a modified version of the Airfield System in 1952.

1.6 *This standard does not purport to address all of the safety problems, if any, associated with its use. It is the responsibility of the user of this standard to establish appropriate safety and health practices and determine the applicability of regulatory limitations prior to use.*

2. Referenced Documents

2.1 *ASTM Standards:*

C 117 Test Method for Materials Finer Than 75-μm (No. 200) Sieve in Mineral Aggregates by Washing[3]

C 136 Method for Sieve Analysis of Fine and Coarse Aggregates[3]

C 702 Practice for Reducing Field Samples of Aggregate to Testing Size[3]

D 420 Guide for Investigating and Sampling Soil and Rock[4]

D 421 Practice for Dry Preparation of Soil Samples for Particle-Size Analysis and Determination of Soil Constants[4]

D 422 Test Method for Particle-Size Analysis of Soils[4]

D 653 Terminology Relating to Soil, Rock, and Contained Fluids[4]

D 1140 Test Method for Amount of Material in Soils Finer than the No. 200 (75-μm) Sieve[4]

D 2216 Test Method for Laboratory Determination of Water (Moisture) Content of Soil and Rock[4]

D 2217 Practice for Wet Preparation of Soil Samples for Particle-Size Analysis and Determination of Soil Constants[4]

D 2488 Practice for Description and Identification of Soils (Visual-Manual Procedure)[4]

D 4083 Practice for Description of Frozen Soils (Visual-Manual Procedure)[4]

D 4318 Test Method for Liquid Limit, Plastic Limit, and Plasticity Index of Soils[4]

D 4427 Classification of Peat Samples by Laboratory Testing[4]

E 11 Specification for Wire-Cloth Sieves for Testing Purposes[3]

3. Terminology

3.1 *Definitions*—Except as listed below, all definitions are in accordance with Terminology D 653.

[1] This standard is under the jurisdiction of ASTM Committee D-18 on Soil and Rock and is the direct responsibility of Subcommittee D18.07 on Identification and Classification of Soils.

Current edition approved May 15, 1992. Published July 1992. Originally published as D 2487 – 66 T. Last previous edition D 2487 – 90$^{\epsilon 1}$.

[2] Casagrande, A., "Classification and Identification of Soils," *Transactions*, ASCE, 1948, p. 901.

[3] *Annual Book of ASTM Standards*, Vol 04.02.

[4] *Annual Book of ASTM Standards*, Vol 04.08.

NOTE 4—For particles retained on a 3-in. (75-mm) U.S. standard sieve, the following definitions are suggested:

Cobbles—particles of rock that will pass a 12-in. (300-mm) square opening and be retained on a 3-in. (75-mm) U.S. standard sieve, and

Boulders—particles of rock that will not pass a 12in. (300-mm) square opening

3.1.1 *gravel*—particles of rock that will pass a 3-in. (75-mm) sieve and be retained on a No. 4 (4.75-mm) U.S. standard sieve with the following subdivisions:

Coarse—passes 3-in. (75-mm) sieve and retained on ¾-in. (19-mm) sieve, and

Fine—passes ¾-in. (19-mm) sieve and retained on No. 4 (4.75-mm) sieve.

3.1.2 *sand*—particles of rock that will pass a No. 4 (4.75-mm) sieve and be retained on a No. 200 (75-µm) U.S. standard sieve with the following subdivisions:

Coarse—passes No. 4 (4.75-mm) sieve and retained on No. 10 (2.00-mm) sieve,

Medium—passes No. 10 (2.00-mm) sieve and retained on No. 40 (425-µm) sieve, and

Fine—passes No. 40 (425-µm) sieve and retained on No. 200 (75-µm) sieve.

3.1.3 *clay*—soil passing a No. 200 (75-µm) U.S. standard sieve that can be made to exhibit plasticity (putty-like properties) within a range of water contents and that exhibits considerable strength when air dry. For classification, a clay is a fine-grained soil, or the fine-grained portion of a soil, with a plasticity index equal to or greater than 4, and the plot of plasticity index versus liquid limit falls on or above the "A" line.

3.1.4 *silt*—soil passing a No. 200 (75-µm) U.S. standard sieve that is nonplastic or very slightly plastic and that exhibits little or no strength when air dry. For classification, a silt is a fine-grained soil, or the fine-grained portion of a soil, with a plasticity index less than 4 or if the plot of plasticity index versus liquid limit falls below the "A" line.

3.1.5 *organic clay*—a clay with sufficient organic content to influence the soil properties. For classification, an organic clay is a soil that would be classified as a clay except that its liquid limit value after oven drying is less than 75 % of its liquid limit value before oven drying.

3.1.6 *organic silt*—a silt with sufficient organic content to influence the soil properties. For classification, an organic silt is a soil that would be classified as a silt except that its liquid limit value after oven drying is less than 75 % of its liquid limit value before oven drying.

3.1.7 *peat*—a soil composed of vegetable tissue in various stages of decomposition usually with an organic odor, a dark-brown to black color, a spongy consistency, and a texture ranging from fibrous to amorphous.

3.2 *Descriptions of Terms Specific to This Standard:*

3.2.1 *coefficient of curvature*, Cc—the ratio $(D_{30})^2/(D_{10} \times D_{60})$, where D_{60}, D_{30}, and D_{10} are the particle diameters corresponding to 60, 30, and 10 % finer on the cumulative particle-size distribution curve, respectively.

3.2.2 *coefficient of uniformity*, Cu—the ratio D_{60}/D_{10}, where D_{60} and D_{10} are the particle diameters corresponding to 60 and 10 % finer on the cumulative particle-size distribution curve, respectively.

4. Summary

4.1 As illustrated in Table 1, this classification system identifies three major soil divisions: coarse-grained soils, fine-grained soils, and highly organic soils. These three divisions are further subdivided into a total of 15 basic soil groups.

4.2 Based on the results of visual observations and prescribed laboratory tests, a soil is catalogued according to the basic soil groups, assigned a group symbol(s) and name, and thereby classified. The flow charts, Fig. 1 for fine-grained soils, and Fig. 2 for coarse-grained soils, can be used to assign the appropriate group symbol(s) and name.

5. Significance and Use

5.1 This standard classifies soils from any geographic location into categories representing the results of prescribed laboratory tests to determine the particle-size characteristics, the liquid limit, and the plasticity index.

5.2 The assigning of a group name and symbol(s) along with the descriptive information required in Practice D 2488 can be used to describe a soil to aid in the evaluation of its significant properties for engineering use.

5.3 The various groupings of this classification system have been devised to correlate in a general way with the engineering behavior of soils. This standard provides a useful first step in any field or laboratory investigation for geotechnical engineering purposes.

5.4 This standard may also be used as an aid in training personnel in the use of Practice D 2488.

5.5 This standard may be used in combination with Practice D 4083 when working with frozen soils.

6. Apparatus

6.1 In addition to the apparatus that may be required for obtaining and preparing the samples and conducting the prescribed laboratory tests, a plasticity chart, similar to Fig. 3, and a cumulative particle-size distribution curve, similar to Fig. 4, are required.

NOTE 5—The "U" line shown on Fig. 3 has been empirically determined to be the approximate "upper limit" for natural soils. It is a good check against erroneous data, and any test results that plot above or to the left of it should be verified.

7. Sampling

7.1 Samples shall be obtained and identified in accordance with a method or methods, recommended in Recommended Guide D 420 or by other accepted procedures.

7.2 For accurate identification, the minimum amount of test sample required for this test method will depend on which of the laboratory tests need to be performed. Where only the particle-size analysis of the sample is required, specimens having the following minimum dry weights are required:

Maximum Particle Size, Sieve Opening	Minimum Specimen Size, Dry Weight
4.75 mm (No. 4)	100 g (0.25 lb)
9.5 mm (⅜ in.)	200 g (0.5 lb)
19.0 mm (¾ in.)	1.0 kg (2.2 lb)
38.1 mm (1½ in.)	8.0 kg (18 lb)
75.0 mm (3 in.)	60.0 kg (132 lb)

Whenever possible, the field samples should have weights two to four times larger than shown.

TABLE 1 Soil Classification Chart

Criteria for Assigning Group Symbols and Group Names Using Laboratory Tests[A]				Soil Classification	
				Group Symbol	Group Name[B]
COARSE-GRAINED SOILS More than 50 % retained on No. 200 sieve	Gravels More than 50 % of coarse fraction retained on No. 4 sieve	Clean Gravels Less than 5 % fines[C]	$Cu \geq 4$ and $1 \leq Cc \leq 3$[E]	GW	Well-graded gravel[F]
			$Cu < 4$ and/or $1 > Cc > 3$[E]	GP	Poorly graded gravel[F]
		Gravels with Fines More than 12 % fines[C]	Fines classify as ML or MH	GM	Silty gravel[F,G,H]
			Fines classify as CL or CH	GC	Clayey gravel[F,G,H]
	Sands 50 % or more of coarse fraction passes No. 4 sieve	Clean Sands Less than 5 % fines[D]	$Cu \geq 6$ and $1 \leq Cc \leq 3$[E]	SW	Well-graded sand[I]
			$Cu < 6$ and/or $1 > Cc > 3$[E]	SP	Poorly graded sand[I]
		Sands with Fines More than 12 % fines[D]	Fines classify as ML or MH	SM	Silty sand[G,H,I]
			Fines classify as CL or CH	SC	Clayey sand[G,H,I]
FINE-GRAINED SOILS 50 % or more passes the No. 200 sieve	Silts and Clays Liquid limit less than 50	inorganic	$PI > 7$ and plots on or above "A" line[J]	CL	Lean clay[K,L,M]
			$PI < 4$ or plots below "A" line[J]	ML	Silt[K,L,M]
		organic	$\frac{\text{Liquid limit – oven dried}}{\text{Liquid limit – not dried}} < 0.75$	OL	Organic clay[K,L,M,N] / Organic silt[K,L,M,O]
	Silts and Clays Liquid limit 50 or more	inorganic	PI plots on or above "A" line	CH	Fat clay[K,L,M]
			PI plots below "A" line	MH	Elastic silt[K,L,M]
		organic	$\frac{\text{Liquid limit – oven dried}}{\text{Liquid limit – not dried}} < 0.75$	OH	Organic clay[K,L,M,P] / Organic silt[K,L,M,Q]
HIGHLY ORGANIC SOILS		Primarily organic matter, dark in color, and organic odor		PT	Peat

[A] Based on the material passing the 3-in. (75-mm) sieve.

[B] If field sample contained cobbles or boulders, or both, add "with cobbles or boulders, or both" to group name.

[C] Gravels with 5 to 12 % fines require dual symbols:
GW-GM well-graded gravel with silt
GW-GC well-graded gravel with clay
GP-GM poorly graded gravel with silt
GP-GC poorly graded gravel with clay

[D] Sands with 5 to 12 % fines require dual symbols:
SW-SM well-graded sand with silt
SW-SC well-graded sand with clay
SP-SM poorly graded sand with silt
SP-SC poorly graded sand with clay

[E] $Cu = D_{60}/D_{10}$ $\quad Cc = \frac{(D_{30})^2}{D_{10} \times D_{60}}$

[F] If soil contains ≥ 15 % sand, add "with sand" to group name.

[G] If fines classify as CL-ML, use dual symbol GC-GM, or SC-SM.

[H] If fines are organic, add "with organic fines" to group name.

[I] If soil contains ≥ 15 % gravel, add "with gravel" to group name.

[J] If Atterberg limits plot in hatched area, soil is a CL-ML, silty clay.

[K] If soil contains 15 to 29 % plus No. 200, add "with sand" or "with gravel," whichever is predominant.

[L] If soil contains ≥ 30 % plus No. 200, predominantly sand, add "sandy" to group name.

[M] If soil contains ≥ 30 % plus No. 200, predominantly gravel, add "gravelly" to group name.

[N] $PI \geq 4$ and plots on or above "A" line.

[O] $PI < 4$ or plots below "A" line.

[P] PI plots on or above "A" line.

[Q] PI plots below "A" line.

7.3 When the liquid and plastic limit tests must also be performed, additional material will be required sufficient to provide 150 g to 200 g of soil finer than the No. 40 (425-μm) sieve.

7.4 If the field sample or test specimen is smaller than the minimum recommended amount, the report shall include an appropriate remark.

8. Classification of Peat

8.1 A sample composed primarily of vegetable tissue in various stages of decomposition and has a fibrous to amorphous texture, a dark-brown to black color, and an organic odor should be designated as a highly organic soil and shall be classified as peat, PT, and not subjected to the classification procedures described hereafter.

8.2 If desired, classification of type of peat can be performed in accordance with Classification D 4427.

9. Preparation for Classification

9.1 Before a soil can be classified according to this standard, generally the particle-size distribution of the minus 3-in. (75-mm) material and the plasticity characteristics of the minus No. 40 (425-μm) sieve material must be determined. See 9.8 for the specific required tests.

9.2 The preparation of the soil specimen(s) and the testing for particle-size distribution and liquid limit and plasticity index shall be in accordance with accepted standard procedures. Two procedures for preparation of the soil specimens for testing for soil classification purposes are given in Appendixes X3 and X4. Appendix X3 describes the wet preparation method and is the preferred method for cohesive soils that have never dried out and for organic soils.

9.3 When reporting soil classifications determined by this standard, the preparation and test procedures used shall be reported or referenced.

9.4 Although the test procedure used in determining the particle-size distribution or other considerations may require a hydrometer analysis of the material, a hydrometer analysis is not necessary for soil classification.

9.5 The percentage (by dry weight) of any plus 3-in. (75-mm) material must be determined and reported as auxiliary information.

9.6 The maximum particle size shall be determined (measured or estimated) and reported as auxiliary information.

9.7 When the cumulative particle-size distribution is required, a set of sieves shall be used which include the

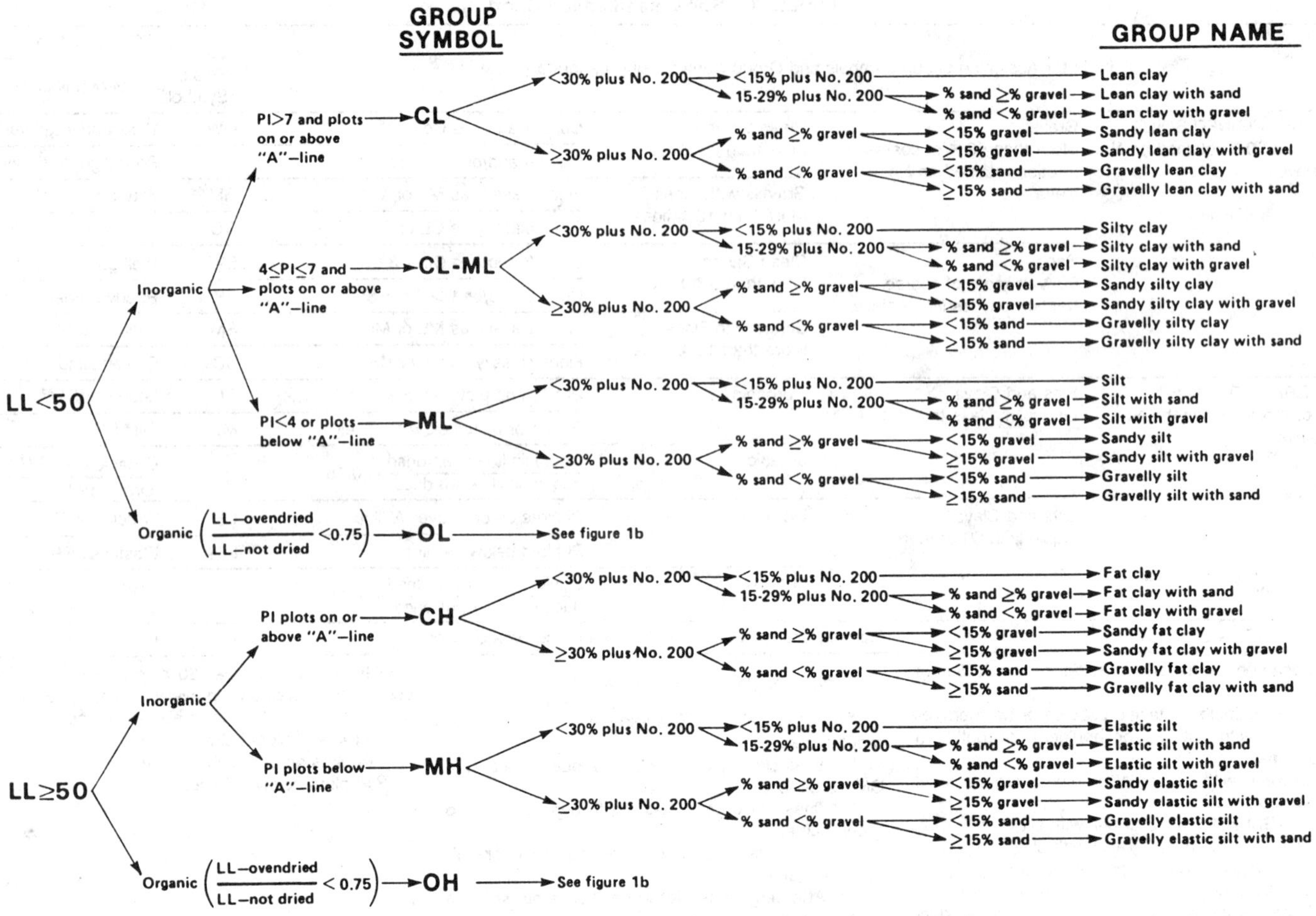

FIG. 1a Flow Chart for Classifying Fine-Grained Soil (50 % or More Passes No. 200 Sieve)

following sizes (with the largest size commensurate with the maximum particle size) with other sieve sizes as needed or required to define the particle-size distribution:

3-in. (75-mm)
¾-in.(19.0-mm)
No. 4 (4.75-mm)
No. 10 (2.00-mm)
No. 40 (425-μm)
No. 200 (75-μm)

9.8 The tests required to be performed in preparation for classification are as follows:

9.8.1 For soils estimated to contain less than 5 % fines, a plot of the cumulative particle-size distribution curve of the fraction coarser than the No. 200 (75-μm) sieve is required. The cumulative particle-size distribution curve may be plotted on a graph similar to that shown in Fig. 4.

9.8.2 For soils estimated to contain 5 to 15 % fines, a cumulative particle-size distribution curve, as described in 9.8.1, is required, and the liquid limit and plasticity index are required.

9.8.2.1 If sufficient material is not available to determine the liquid limit and plasticity index, the fines should be estimated to be either silty or clayey using the procedures described in Practice D 2488 and so noted in the report.

9.8.3 For soils estimated to contain 15 % or more fines, a determination of the percent fines, percent sand, and percent gravel is required, and the liquid limit and plasticity index are required. For soils estimated to contain 90 % fines or more, the percent fines, percent sand, and percent gravel may be estimated using the procedures described in Practice D 2488 and so noted in the report.

10. Preliminary Classification Procedure

10.1 Class the soil as fine-grained if 50 % or more by dry weight of the test specimen passes the No. 200 (75-μm) sieve and follow Section 11.

10.2 Class the soil as coarse-grained if more than 50 % by dry weight of the test specimen is retained on the No. 200 (75-μm) sieve and follow Section 12.

11. Procedure for Classification of Fine-Grained Soils (50 % or more by dry weight passing the No. 200 (75-μm) sieve)

11.1 The soil is an inorganic clay if the position of the plasticity index versus liquid limit plot, Fig. 3, falls on or above the "A" line, the plasticity index is greater than 4, and the presence of organic matter does not influence the liquid limit as determined in 11.3.2.

NOTE 6—The plasticity index and liquid limit are determined on the minus No. 40 (425 μm) sieve material.

11.1.1 Classify the soil as a *lean clay*, CL, if the liquid limit is less than 50. See area identified as CL on Fig. 3.

11.1.2 Classify the soil as a *fat clay*, CH, if the liquid limit

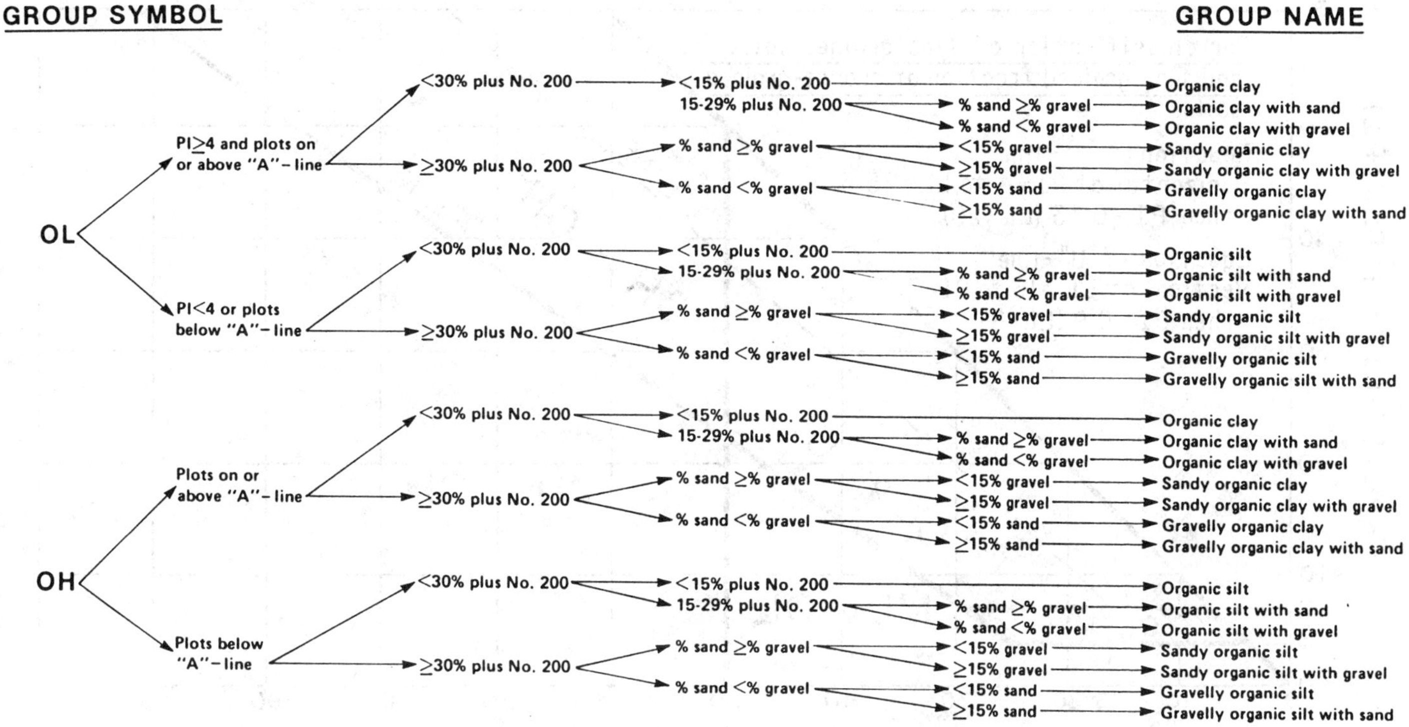

FIG. 1b Flow Chart for Classifying Organic Fine-Grained Soil (50 % or More Passes No. 200 Sieve)

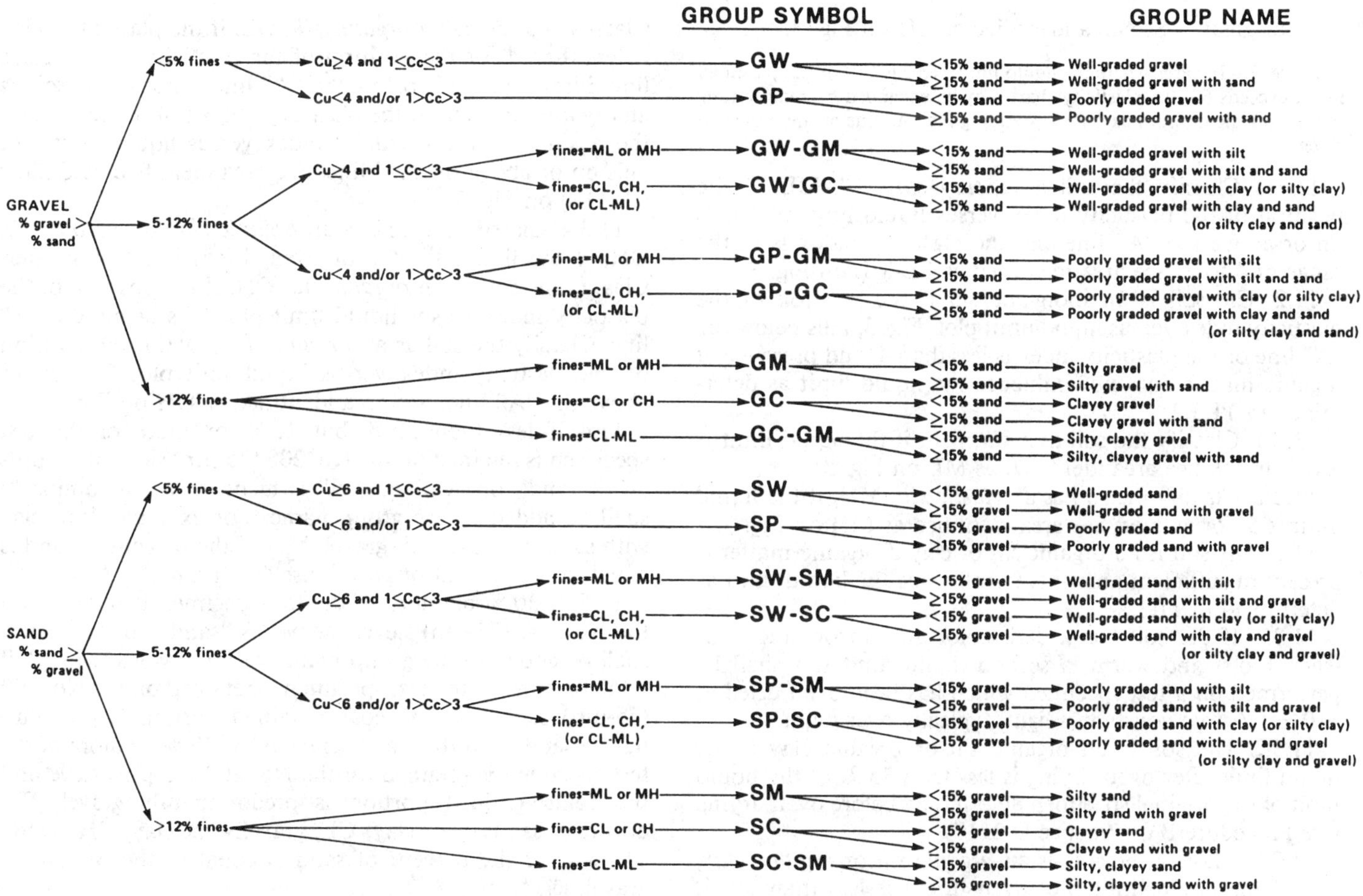

FIG. 2 Flow Chart for Classifying Coarse-Grained Soils (More Than 50 % Retained on No. 200 Sieve)

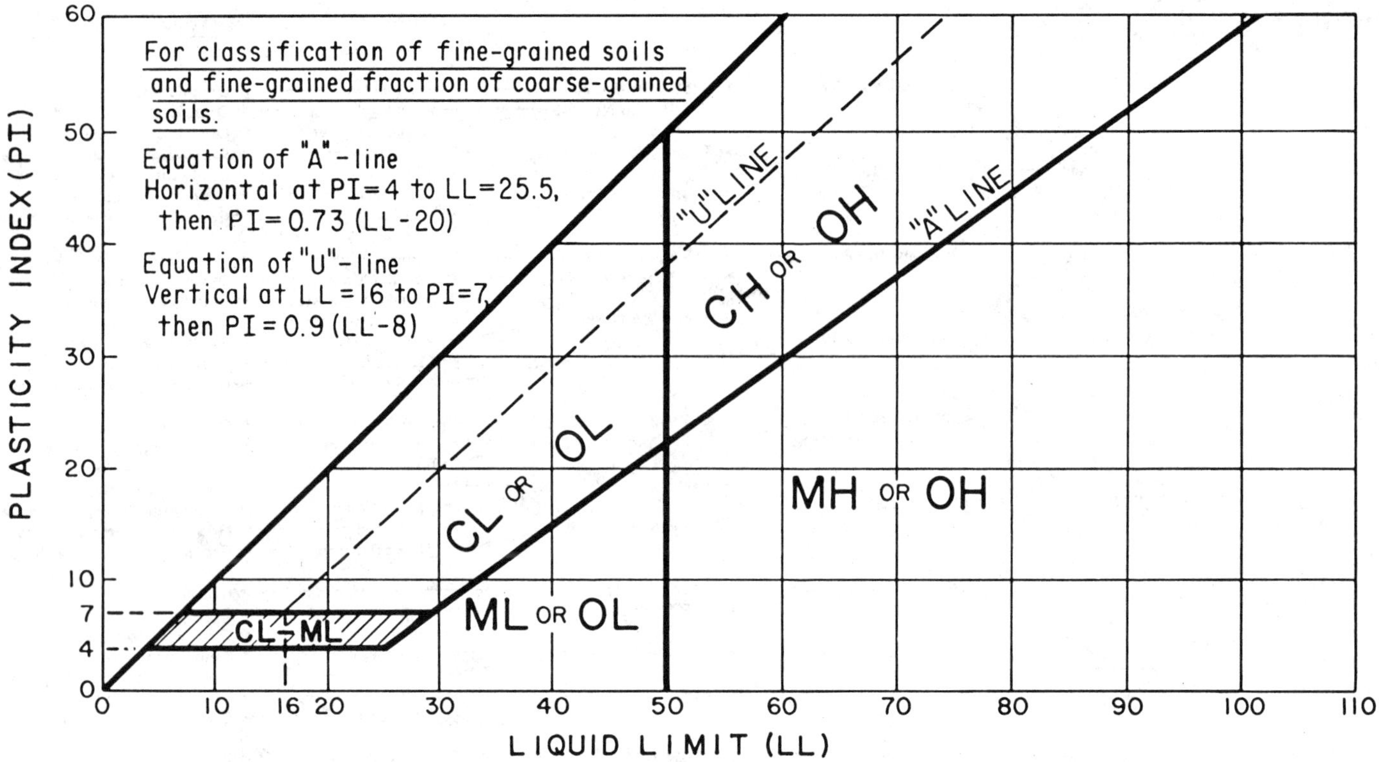

FIG. 3 Plasticity Chart

is 50 or greater. See area identified as CH on Fig. 3.

NOTE 7—In cases where the liquid limit exceeds 110 or the plasticity index exceeds 60, the plasticity chart may be expanded by maintaining the same scale on both axes and extending the "A" line at the indicated slope.

11.1.3 Classify the soil as a *silty clay*, CL-ML, if the position of the plasticity index versus liquid limit plot falls on or above the "A" line and the plasticity index is in the range of 4 to 7. See area identified as CL-ML on Fig. 3.

11.2 The soil is an inorganic silt if the position of the plasticity index versus liquid limit plot, Fig. 3, falls below the "A" line or the plasticity index is less than 4, and presence of organic matter does not influence the liquid limit as determined in 11.3.2.

11.2.1 Classify the soil as a *silt*, ML, if the liquid limit is less than 50. See area identified as ML on Fig. 3.

11.2.2 Classify the soil as an *elastic silt*, MH, if the liquid limit is 50 or greater. See area identified as MH on Fig. 3.

11.3 The soil is an organic silt or clay if organic matter is present in sufficient amounts to influence the liquid limit as determined in 11.3.2.

11.3.1 If the soil has a dark color and an organic odor when moist and warm, a second liquid limit test shall be performed on a test specimen which has been oven dried at 110 ± 5°C to a constant weight, typically over night.

11.3.2 The soil is an organic silt or organic clay if the liquid limit after oven drying is less than 75 % of the liquid limit of the original specimen determined before oven drying (see Procedure B of Practice D 2217).

11.3.3 Classify the soil as an *organic silt* or *organic clay*, OL, if the liquid limit (not oven dried) is less than 50 %. Classify the soil as an *organic silt*, OL, if the plasticity index is less than 4, or the position of the plasticity index versus liquid limit plot falls below the "A" line. Classify the soil as an *organic clay*, OL, if the plasticity index is 4 or greater and the position of the plasticity index versus liquid limit plot falls on or above the "A" line. See area identified as OL (or CL-ML) on Fig. 3.

11.3.4 Classify the soil as an *organic clay* or *organic silt*, OH, if the liquid limit (not oven dried) is 50 or greater. Classify the soil as an *organic silt*, OH, if the position of the plasticity index versus liquid limit plot falls below the "A" line. Classify the soil as an *organic clay*, OH, if the position of the plasticity index versus liquid-limit plot falls on or above the "A" line. See area identified as OH on Fig. 3.

11.4 If less than 30 % but 15 % or more of the test specimen is retained on the No. 200 (75-μm) sieve, the words "with sand" or "with gravel" (whichever is predominant) shall be added to the group name. For example, lean clay with sand, CL; silt with gravel, ML. If the percent of sand is equal to the percent of gravel, use "with sand."

11.5 If 30 % or more of the test specimen is retained on the No. 200 (75-μm) sieve, the words "sandy" or "gravelly" shall be added to the group name. Add the word "sandy" if 30 % or more of the test specimen is retained on the No. 200 (75-μm) sieve and the coarse-grained portion is predominantly sand. Add the word "gravelly" if 30 % or more of the test specimen is retained on the No. 200 (75-μm) sieve and the coarse-grained portion is predominantly gravel. For example, sandy lean clay, CL; gravelly fat clay, CH; sandy silt, ML. If the percent of sand is equal to the percent of gravel, use "sandy."

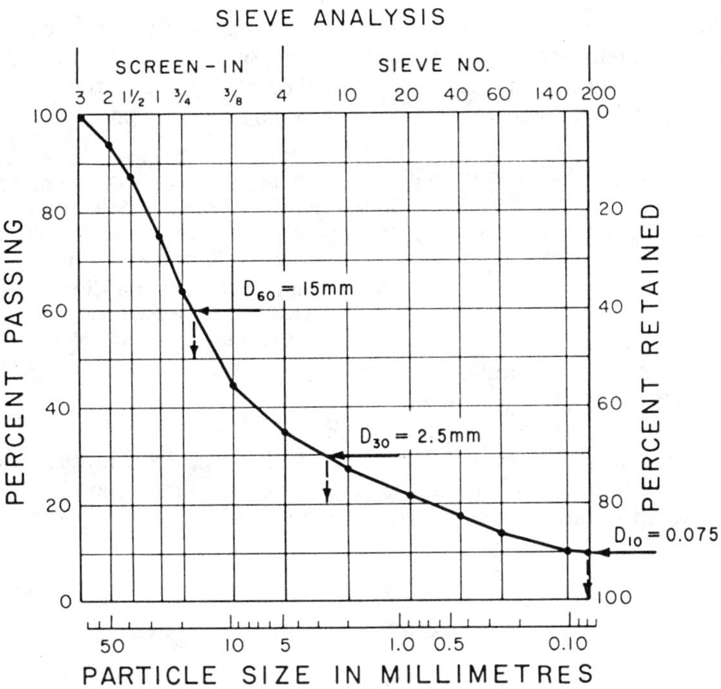

$$Cu = \frac{D_{60}}{D_{10}} = \frac{15}{0.075} = 200 \qquad Cc = \frac{(D_{30})^2}{D_{10} \times D_{60}} = \frac{(2.5)^2}{0.075 \times 15} = 5.6$$

FIG. 4 Cumulative Particle-Size Plot

12. Procedure for Classification of Coarse-Grained Soils (more than 50 % retained on the No. 200 (75-μm) sieve)

12.1 Class the soil as gravel if more than 50 % of the coarse fraction [plus No. 200 (75-μm) sieve] is retained on the No. 4 (4.75-mm) sieve.

12.2 Class the soil as sand if 50 % or more of the coarse fraction [plus No. 200 (75-μm) sieve] passes the No. 4 (4.75-mm) sieve.

12.3 If 12 % or less of the test specimen passes the No. 200 (75-μm) sieve, plot the cumulative particle-size distribution, Fig. 4, and compute the coefficient of uniformity, Cu, and coefficient of curvature, Cc, as given in Eqs 1 and 2.

$$Cu = D_{60}/D_{10} \quad (1)$$

$$Cc = (D_{30})^2/(D_{10} \times D_{60}) \quad (2)$$

where:

D_{10}, D_{30}, and D_{60} = the particle-size diameters corresponding to 10, 30, and 60 %, respectively, passing on the cumulative particle-size distribution curve, Fig. 4.

NOTE 8—It may be necessary to extrapolate the curve to obtain the D_{10} diameter.

12.3.1 If less than 5 % of the test specimen passes the No. 200 (75-μm) sieve, classify the soil as a *well-graded gravel*, GW, or *well-graded sand*, SW, if Cu is greater than 4.0 for gravel or greater than 6.0 for sand, and Cc is at least 1.0 but not more than 3.0.

12.3.2 If less than 5 % of the test specimen passes the No. 200 (75-μm) sieve, classify the soil as *poorly graded gravel*, GP, or *poorly graded sand*, SP, if either the Cu or the Cc criteria for well-graded soils are not satisfied.

12.4 If more than 12 % of the test specimen passes the No. 200 (75-μm) sieve, the soil shall be considered a coarse-grained soil with fines. The fines are determined to be either clayey or silty based on the plasticity index versus liquid limit plot on Fig. 3. (See 9.8.2.1 if insufficient material available for testing). (See NOTE 6)

12.4.1 Classify the soil as a *clayey gravel*, GC, or *clayey sand*, SC, if the fines are clayey, that is, the position of the plasticity index versus liquid limit plot, Fig. 3, falls on or above the "A" line and the plasticity index is greater than 7.

12.4.2 Classify the soil as a *silty gravel*, GM, or *silty sand*, SM, if the fines are silty, that is, the position of the plasticity index versus liquid limit plot, Fig. 3, falls below the "A" line or the plasticity index is less than 4.

12.4.3 If the fines plot as a silty clay, CL-ML, classify the soil as a *silty, clayey gravel*, GC-GM, if it is a gravel or a *silty, clayey sand*, SC-SM, if it is a sand.

12.5 If 5 to 12 % of the test specimen passes the No. 200 (75-μm) sieve, give the soil a dual classification using two group symbols.

12.5.1 The first group symbol shall correspond to that for a gravel or sand having less than 5 % fines (GW, GP, SW, SP), and the second symbol shall correspond to a gravel or sand having more than 12 % fines (GC, GM, SC, SM).

12.5.2 The group name shall correspond to the first group symbol plus "with clay" or "with silt" to indicate the plasticity characteristics of the fines. For example, well-graded gravel with clay, GW-GC; poorly graded sand with silt, SP-SM (See 9.8.2.1 if insufficient material available for testing).

NOTE 9—If the fines plot as a *silty clay*, CL-ML, the second group symbol should be either GC or SC. For example, a poorly graded sand with 10 % fines, a liquid limit of 20, and a plasticity index of 6 would be classified as a poorly graded sand with silty clay, SP-SC.

12.6 If the specimen is predominantly sand or gravel but contains 15 % or more of the other coarse-grained constituent, the words "with gravel" or "with sand" shall be added to the group name. For example, poorly graded gravel with sand, clayey sand with gravel.

12.7 If the field sample contained any cobbles or boulders or both, the words "with cobbles," or "with cobbles and boulders" shall be added to the group name. For example, silty gravel with cobbles, GM.

13. Report

13.1 The report should include the group name, group symbol, and the results of the laboratory tests. The particle-size distribution shall be given in terms of percent of gravel, sand, and fines. The plot of the cumulative particle-size distribution curve shall be reported if used in classifying the soil. Report appropriate descriptive information according to the procedures in Practice D 2488. A local or commercial name or geologic interpretation for the material may be added at the end of the descriptive information if identified as such. The test procedures used shall be referenced.

NOTE 10—*Example: Clayey Gravel with Sand and Cobbles* (GC)—46 % fine to coarse, hard, subrounded gravel; 30 % fine to coarse, hard, subrounded sand; 24 % clayey fines, LL = 38, PI = 19; weak reaction with HCl; original field sample had 4 % hard, subrounded cobbles; maximum dimension 150 mm.

In-Place Conditions—firm, homogeneous, dry, brown,

Geologic Interpretation—alluvial fan.

NOTE 11—Other examples of soil descriptions are given in Appendix X1.

14. Keywords

14.1 Atterberg limits; classification; clay; gradation; gravel; laboratory classification; organic soils; sand; silt; soil classification; soil tests

APPENDIXES

(Nonmandatory Information)

X1. EXAMPLES OF DESCRIPTIONS USING SOIL CLASSIFICATION

X1.1 The following examples show how the information required in 13.1 can be reported. The appropriate descriptive information from Practice D 2488 is included for illustrative purposes. The additional descriptive terms that would accompany the soil classification should be based on the intended use of the classification and the individual circumstances.

X1.1.1 *Well-Graded Gravel with Sand (GW)*—73 % fine to coarse, hard, subangular gravel; 23 % fine to coarse, hard, subangular sand; 4 % fines; Cc = 2.7, Cu = 12.4.

X1.1.2 *Silty Sand with Gravel (SM)*—61 % predominantly fine sand; 23 % silty fines, LL = 33, PI = 6; 16 % fine, hard, subrounded gravel; no reaction with HCl; (field sample smaller than recommended). *In-Place Conditions*—Firm, stratified and contains lenses of silt 1 to 2 in. thick, moist, brown to gray; in-place density = 106 lb/ft^3 and in-place moisture = 9 %.

X1.1.3 *Organic Clay (OL)*—100 % fines, LL (not dried) = 32, LL (oven dried) = 21, PI (not dried) = 10; wet, dark brown, organic odor, weak reaction with HCl.

X1.1.4 *Silty Sand with Organic Fines (SM)*—74 % fine to coarse, hard, subangular reddish sand; 26 % organic and silty dark-brown fines, LL (not dried) = 37, LL (oven dried) = 26, PI (not dried) = 6, wet, weak reaction with HCl.

X1.1.5 *Poorly Graded Gravel with Silt, Sand, Cobbles and Boulders (GP-GM)*—78 % fine to coarse, hard, subrounded to subangular gravel; 16 % fine to coarse, hard, subrounded to subangular sand; 6 % silty (estimated) fines; moist, brown; no reaction with HCl; original field sample had 7 % hard, subrounded cobbles and 2 % hard, subrounded boulders with a maximum dimension of 18 in.

X2. USING SOIL CLASSIFICATION AS A DESCRIPTIVE SYSTEM FOR SHALE, CLAYSTONE, SHELLS, SLAG, CRUSHED ROCK, ETC.

X2.1 The group names and symbols used in this standard may be used as a descriptive system applied to materials that exist in situ as shale, claystone, sandstone, siltstone, mudstone, etc., but convert to soils after field or laboratory processing (crushing, slaking, etc.).

X2.2 Materials such as shells, crushed rock, slag, etc., should be identified as such. However, the procedures used in this standard for describing the particle size and plasticity characteristics may be used in the description of the material. If desired, a classification in accordance with this standard may be assigned to aid in describing the material.

X2.3 If a classification is used, the group symbol(s) and group names should be placed in quotation marks or noted with some type of distinguishing symbol. See examples.

X2.4 Examples of how soil classifications could be incorporated into a description system for materials that are not naturally occurring soils are as follows:

X2.4.1 *Shale Chunks*—Retrieved as 2 to 4-in. pieces of shale from power auger hole, dry, brown, no reaction with HCl. After laboratory processing by slaking in water for 24 h, material classified as "Sandy Lean Clay (CL)"—61 % clayey fines, LL = 37, PI = 16; 33 % fine to medium sand; 6 % gravel-size pieces of shale.

X2.4.2 *Crushed Sandstone*—Product of commercial crushing operation; "Poorly Graded Sand with Silt (SP-SM)"—91 % fine to medium sand; 9 % silty (estimated) fines; dry, reddish-brown, strong reaction with HCl.

X2.4.3 *Broken Shells*—62 % gravel-size broken shells;

31 % sand and sand-size shell pieces; 7 % fines; would be classified as "Poorly Graded Gravel with Sand (GP)".

X2.4.4 *Crushed Rock*—Processed gravel and cobbles from Pit No. 7; "Poorly Graded Gravel (GP)"—89 % fine, hard, angular gravel-size particles; 11 % coarse, hard, angular sand-size particles, dry, tan; no reaction with HCl; Cc = 2.4, Cu = 0.9.

X3. PREPARATION AND TESTING FOR CLASSIFICATION PURPOSES BY THE WET METHOD

X3.1 This appendix describes the steps in preparing a soil sample for testing for purposes of soil classification using a wet-preparation procedure.

X3.2 Samples prepared in accordance with this procedure should contain as much of their natural water content as possible and every effort should be made during obtaining, preparing, and transportating the samples to maintain the natural moisture.

X3.3 The procedures to be followed in this standard assume that the field sample contains fines, sand, gravel, and plus 3-in. (75-mm) particles and the cumulative particle-size distribution plus the liquid limit and plasticity index values are required (see 9.8). Some of the following steps may be omitted when they are not applicable to the soil being tested.

X3.4 If the soil contains plus No. 200 (75-μm) particles that would degrade during dry sieving, use a test procedure for determining the particle-size characteristics that prevents this degradation.

X3.5 Since this classification system is limited to the portion of a sample passing the 3-in. (75-mm) sieve, the plus 3-in. (75-mm) material shall be removed prior to the determination of the particle-size characteristics and the liquid limit and plasticity index.

X3.6 The portion of the field sample finer than the 3-in. (75-mm) sieve shall be obtained as follows:

X3.6.1 Separate the field sample into two fractions on a 3-in. (75-mm) sieve, being careful to maintain the natural water content in the minus 3-in. (75-mm) fraction. Any particles adhering to the plus 3-in. (75-mm) particles shall be brushed or wiped off and placed in the fraction passing the 3-in. (75-mm) sieve.

X3.6.2 Determine the air-dry or oven-dry weight of the fraction retained on the 3-in. (75-mm) sieve. Determine the total (wet) weight of the fraction passing the 3-in. (75-mm) sieve.

X3.6.3 Thoroughly mix the fraction passing the 3-in. (75-mm) sieve. Determine the water content, in accordance with Test Method D 2216, of a representative specimen with a minimum dry weight as required in 7.2. Save the water-content specimen for determination of the particle-size analysis in accordance with X3.8.

X3.6.4 Compute the dry weight of the fraction passing the 3-in. (75-mm) sieve based on the water content and total (wet) weight. Compute the total dry weight of the sample and calculate the percentage of material retained on the 3-in. (75-mm) sieve.

X3.7 Determine the liquid limit and plasticity index as follows:

X3.7.1 If the soil disaggregates readily, mix on a clean, hard surface and select a representative sample by quartering in accordance with Practice C 702.

X3.7.1.1 If the soil contains coarse-grained particles coated with and bound together by tough clayey material, take extreme care in obtaining a representative portion of the No. 40 (425-μm) fraction. Typically, a larger portion than normal has to be selected, such as the minimum weights required in 7.2.

X3.7.1.2 To obtain a representative specimen of a basically cohesive soil, it may be advantageous to pass the soil through a ¾-in. (19-mm) sieve or other convenient size so the material can be more easily mixed and then quartered or split to obtain the representative specimen.

X3.7.2 Process the representative specimen in accordance with Procedure B of Practice D 2217.

X3.7.3 Perform the liquid-limit test in accordance with Test Method D 4318, except the soil shall not be air dried prior to the test.

X3.7.4 Perform the plastic-limit test in accordance with Test Method D 4318, except the soil shall not be air dried prior to the test, and calculate the plasticity index.

X3.8 Determine the particle-size distribution as follows:

X3.8.1 If the water content of the fraction passing the 3-in. (75-mm) sieve was required (X3.6.3), use the water-content specimen for determining the particle-size distribution. Otherwise, select a representative specimen in accordance with Practice C 702 with a minimum dry weight as required in 7.2.

X3.8.2 If the cumulative particle-size distribution including a hydrometer analysis is required, determine the particle-size distribution in accordance with Test Method D 422. See 9.7 for the set of required sieves.

X3.8.3 If the cumulative particle-size distribution without a hydrometer analysis is required, determine the particle-size distribution in accordance with Method C 136. See 9.7 for the set of required sieves. The specimen should be soaked until all clayey aggregations have softened and then washed in accordance with Test Method C 117 prior to performing the particle-size distribution.

X3.8.4 If the cumulative particle-size distribution is not required, determine the percent fines, percent sand, and percent gravel in the specimen in accordance with Test Method C 117, being sure to soak the specimen long enough to soften all clayey aggregations, followed by Method C 136 using a nest of sieves which shall include a No. 4 (4.75-mm) sieve and a No. 200 (75-μm) sieve.

X3.8.5 Calculate the percent fines, percent sand, and percent gravel in the minus 3-in. (75-mm) fraction for classification purposes.

X4. AIR-DRIED METHOD OF PREPARATION OF SOILS FOR TESTING FOR CLASSIFICATION PURPOSES

X4.1 This appendix describes the steps in preparing a soil sample for testing for purposes of soil classification when air-drying the soil before testing is specified or desired or when the natural moisture content is near that of an air-dried state.

X4.2 If the soil contains organic matter or mineral colloids that are irreversibly affected by air drying, the wet-preparation method as described in Appendix X3 should be used.

X4.3 Since this classification system is limited to the portion of a sample passing the 3-in. (75-mm) sieve, the plus 3-in. (75-mm) material shall be removed prior to the determination of the particle-size characteristics and the liquid limit and plasticity index.

X4.4 The portion of the field sample finer than the 3-in. (75-mm) sieve shall be obtained as follows:

X4.4.1 Air dry and weigh the field sample.

X4.4.2 Separate the field sample into two fractions on a 3-in. (75-mm) sieve.

X4.4.3 Weigh the two fractions and compute the percentage of the plus 3-in. (75-mm) material in the field sample.

X4.5 Determine the particle-size distribution and liquid limit and plasticity index as follows (see 9.8 for when these tests are required):

X4.5.1 Thoroughly mix the fraction passing the 3-in. (75-mm) sieve.

X4.5.2 If the cumulative particle-size distribution including a hydrometer analysis is required, determine the particle-size distribution in accordance with Test Method D 422. See 9.7 for the set of sieves that is required.

X4.5.3 If the cumulative particle-size distribution without a hydrometer analysis is required, determine the particle-size distribution in accordance with Test Method D 1140 followed by Method C 136. See 9.7 for the set of sieves that is required.

X4.5.4 If the cumulative particle-size distribution is not required, determine the percent fines, percent sand, and percent gravel in the specimen in accordance with Test Method D 1140 followed by Method C 136 using a nest of sieves which shall include a No. 4 (4.75-mm) sieve and a No. 200 (75-μm) sieve.

X4.5.5 If required, determine the liquid limit and the plasticity index of the test specimen in accordance with Test Method D 4318.

X5. RATIONALE

X5.1 Significant revisions were made to the standard which appeared as D 2487 – 83 from the previous version of D 2487 – 69 (1975). The changes are documented in the literature.[5]

X5.2 The 1992 edition differs from the previous edition in that the title was changed to better indicate the use of the standard and identifying it as the ASTM version of the Unified Soil Classification System.

[5] Howard, A.K., "The Revised ASTM Standard on the Unified Soil Classification System," Geotechnical Testing Journal, GTJODJ Vol 7, No. 4, December 1984.

Standard Practice for Description and Identification of Soils (Visual-Manual Procedure)[1]

This standard is issued under the fixed designation D 2488; the number immediately following the designation indicates the year of original adoption or, in the case of revision, the year of last revision. A number in parentheses indicates the year of last reapproval. A superscript epsilon (ε) indicates an editorial change since the last revision or reapproval.

This standard has been approved for use by agencies of the Department of Defense. Consult the DoD Index of Specifications and Standards for the specific year of issue which has been adopted by the Department of Defense.

1. Scope

1.1 This practice covers procedures for the description of soils for engineering purposes.

1.2 This practice also describes a procedure for identifying soils, at the option of the user, based on the classification system described in Test Method D 2487. The identification is based on visual examination and manual tests. It must be clearly stated in reporting an identification that it is based on visual-manual procedures.

1.2.1 When precise classification of soils for engineering purposes is required, the procedures prescribed in Test Method D 2487 shall be used.

1.2.2 In this practice, the identification portion assigning a group symbol and name is limited to soil particles smaller than 3 in. (75 mm).

1.2.3 The identification portion of this practice is limited to naturally occurring soils (disturbed and undisturbed).

NOTE 1—This practice may be used as a descriptive system applied to such materials as shale, claystone, shells, crushed rock, etc. (See Appendix X2).

1.3 The descriptive information in this practice may be used with other soil classification systems or for materials other than naturally occurring soils.

1.4 *This standard does not purport to address all of the safety problems, if any, associated with its use. It is the responsibility of the user of this standard to establish appropriate safety and health practices and determine the applicability of regulatory limitations prior to use.* For specific precautionary statements see Section 8.

1.5 The values stated in inch-pound units are to be regarded as the standard.

2. Referenced Documents

2.1 *ASTM Standards:*

D 653 Terminology Relating to Soil, Rock, and Contained Fluids[2]

D 1452 Practice for Soil Investigation and Sampling by Auger Borings[2]

D 1586 Method for Penetration Test and Split-Barrel Sampling of Soils[2]

D 1587 Practice for Thin-Walled Tube Sampling of Soils[2]

D 2113 Practice for Diamond Core Drilling for Site Investigation[2]

D 2487 Test Method for Classification of Soils for Engineering Purposes[2]

D 4083 Practice for Description of Frozen Soils (Visual-Manual Procedure)[2]

3. Terminology

3.1 *Definitions:*

3.1.1 Except as listed below, all definitions are in accordance with Terminology D 653.

NOTE 2—For particles retained on a 3-in. (75-mm) US standard sieve, the following definitions are suggested:

Cobbles—particles of rock that will pass a 12-in. (300-mm) square opening and be retained on a 3-in. (75-mm) sieve, and

Boulders—particles of rock that will not pass a 12-in. (300-mm) square opening.

3.1.1.2 *clay*—soil passing a No. 200 (75-μm) sieve that can be made to exhibit plasticity (putty-like properties) within a range of water contents, and that exhibits considerable strength when air-dry. For classification, a clay is a fine-grained soil, or the fine-grained portion of a soil, with a plasticity index equal to or greater than 4, and the plot of plasticity index versus liquid limit falls on or above the "A" line (see Fig. 3 of Test Method D 2487).

3.1.1.3 *gravel*—particles of rock that will pass a 3-in. (75-mm) sieve and be retained on a No. 4 (4.75-mm) sieve with the following subdivisions:

coarse—passes a 3-in. (75-mm) sieve and is retained on a ¾-in. (19-mm) sieve.

fine—passes a ¾-in. (19-mm) sieve and is retained on a No. 4 (4.75-mm) sieve.

3.1.1.4 *organic clay*—a clay with sufficient organic content to influence the soil properties. For classification, an organic clay is a soil that would be classified as a clay, except that its liquid limit value after oven drying is less than 75 % of its liquid limit value before oven drying.

3.1.1.5 *organic silt*—a silt with sufficient organic content to influence the soil properties. For classification, an organic silt is a soil that would be classified as a silt except that its liquid limit value after oven drying is less than 75 % of its liquid limit value before oven drying.

3.1.1.6 *peat*—a soil composed primarily of vegetable tissue in various stages of decomposition usually with an organic odor, a dark brown to black color, a spongy consistency, and a texture ranging from fibrous to amorphous.

3.1.1.7 *sand*—particles of rock that will pass a No. 4

[1] This practice is under the jurisdiction of ASTM Committee D-18 on Soil and Rock and is the direct responsibility of Subcommittee D18.07 on Identification and Classification of Soils.

Current edition approved June 29, 1990. Published August 1990. Originally published as D 2488 – 66 T. Last previous edition D 2488 – 84$^{\epsilon 1}$.

[2] *Annual Book of ASTM Standards*, Vol 04.08.

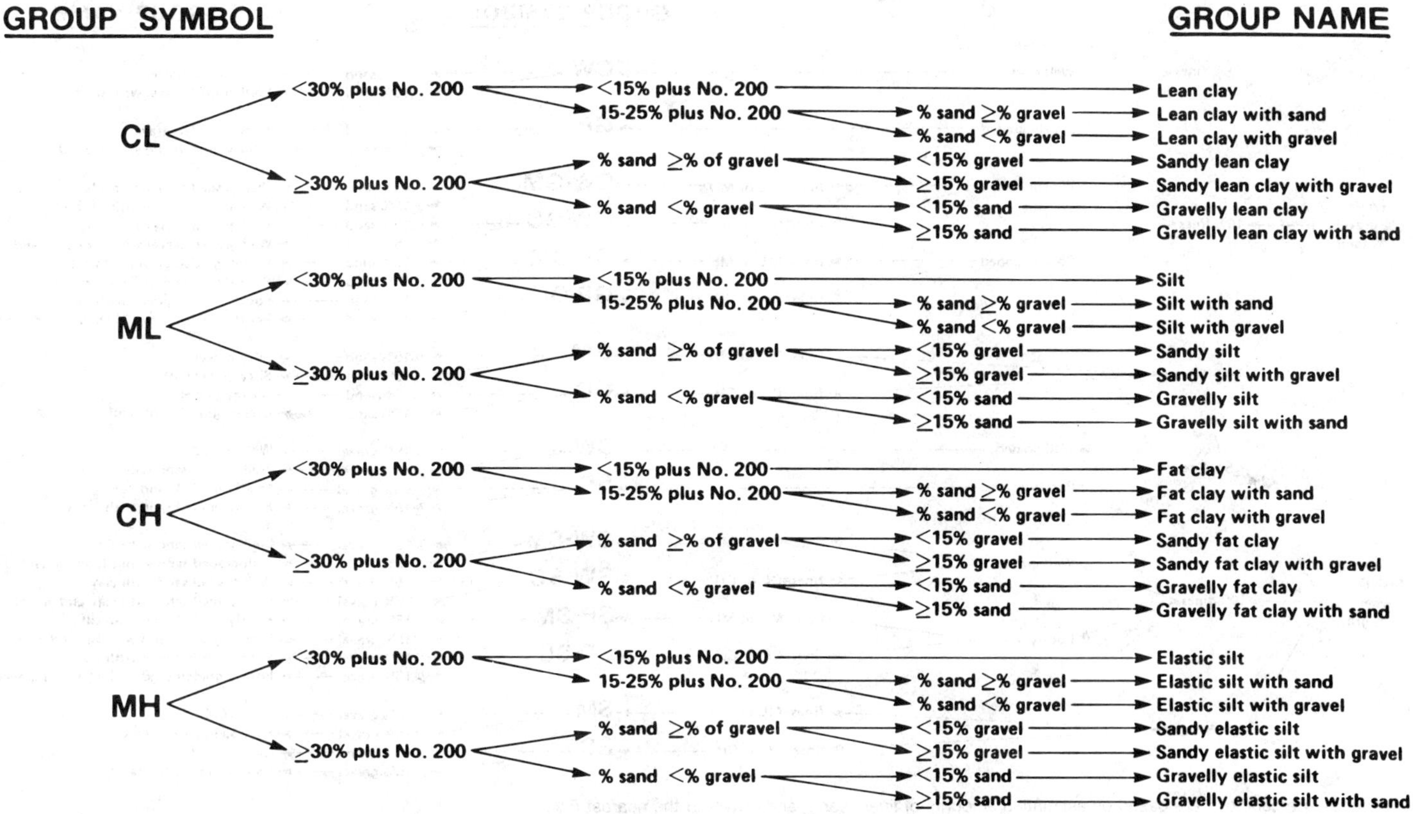

NOTE—Percentages are based on estimating amounts of fines, sand, and gravel to the nearest 5 %.

FIG. 1a Flow Chart for Identifying Inorganic Fine-Grained Soil (50 % or more fines)

(4.75-mm) sieve and be retained on a No. 200 (75-μm) sieve with the following subdivisions:

coarse—passes a No. 4 (4.75-mm) sieve and is retained on a No. 10 (2.00-mm) sieve.

medium—passes a No. 10 (2.00-mm) sieve and is retained on a No. 40 (425-μm) sieve.

fine—passes a No. 40 (425-μm) sieve and is retained on a No. 200 (75-μm) sieve.

3.1.1.8 *silt*—soil passing a No. 200 (75-μm) sieve that is nonplastic or very slightly plastic and that exhibits little or no strength when air dry. For classification, a silt is a fine-grained soil, or the fine-grained portion of a soil, with a plasticity index less than 4, or the plot of plasticity index versus liquid limit falls below the "A" line (see Fig. 3 of Test Method D 2487).

4. Summary of Practice

4.1 Using visual examination and simple manual tests, this practice gives standardized criteria and procedures for describing and identifying soils.

4.2 The soil can be given an identification by assigning a group symbol(s) and name. The flow charts, Figs. 1a and 1b for fine-grained soils, and Fig. 2, for coarse-grained soils, can be used to assign the appropriate group symbol(s) and name. If the soil has properties which do not distinctly place it into a specific group, borderline symbols may be used, see Appendix X3.

NOTE 3—It is suggested that a distinction be made between *dual symbols* and *borderline symbols.*

Dual Symbol—A dual symbol is two symbols separated by a hyphen, for example, GP-GM, SW-SC, CL-ML used to indicate that the soil has been identified as having the properties of a classification in accordance with Test Method D 2487 where two symbols are required. Two symbols are required when the soil has between 5 and 12 % fines or

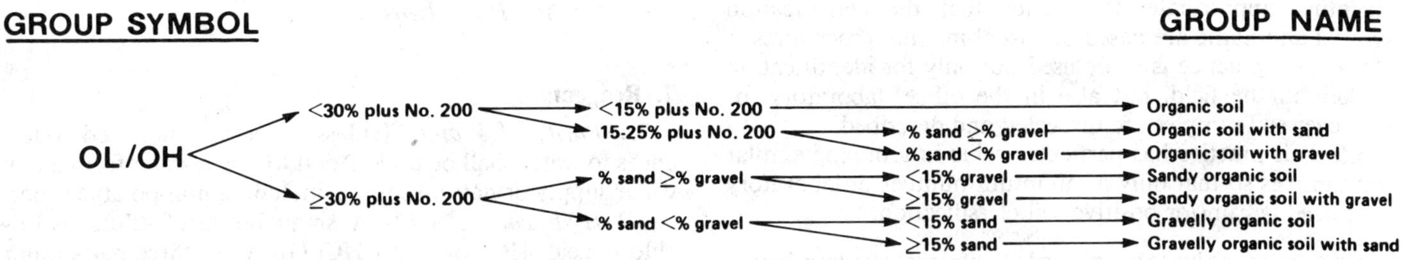

NOTE—Percentages are based on estimating amounts of fines, sand, and gravel to the nearest 5 %.

FIG. 1b Flow Chart for Identifying Organic Fine-Grained Soil (50 % or more fines)

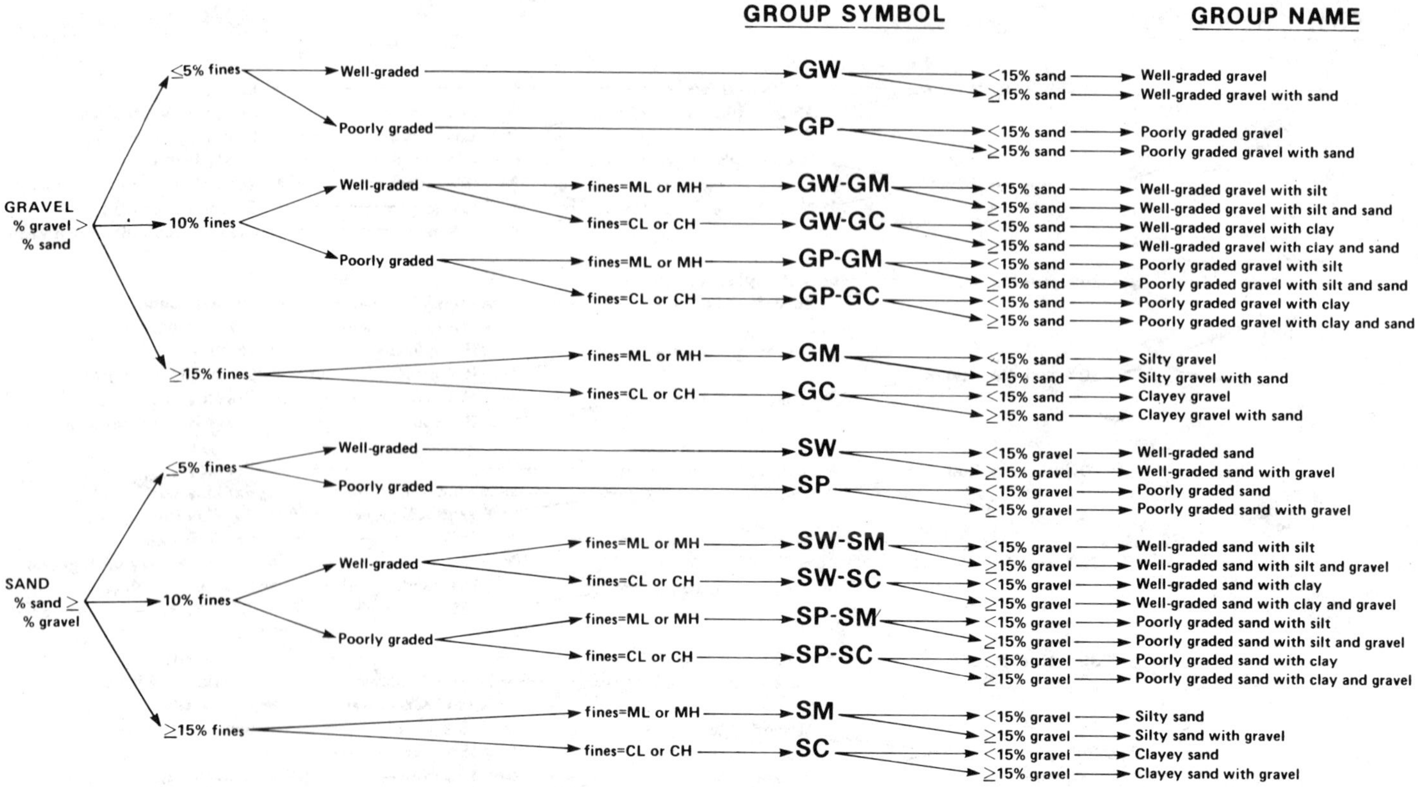

NOTE—Percentages are based on estimating amounts of fines, sand, and gravel to the nearest 5 %.

FIG. 2 Flow Chart for Identifying Coarse-Grained Soils (less than 50 % fines)

when the liquid limit and plasticity index values plot in the CL-ML area of the plasticity chart.

Borderline Symbol—A borderline symbol is two symbols separated by a slash, for example, CL/CH, GM/SM, CL/ML. A borderline symbol should be used to indicate that the soil has been identified as having properties that do not distinctly place the soil into a specific group (see Appendix X3).

5. Significance and Use

5.1 The descriptive information required in this practice can be used to describe a soil to aid in the evaluation of its significant properties for engineering use.

5.2 The descriptive information required in this practice should be used to supplement the classification of a soil as determined by Test Method D 2487.

5.3 This practice may be used in identifying soils using the classification group symbols and names as prescribed in Test Method D 2487. Since the names and symbols used in this practice to identify the soils are the same as those used in Test Method D 2487, it shall be clearly stated in reports and all other appropriate documents, that the classification symbol and name are based on visual-manual procedures.

5.4 This practice is to be used not only for identification of soils in the field, but also in the office, laboratory, or wherever soil samples are inspected and described.

5.5 This practice has particular value in grouping similar soil samples so that only a minimum number of laboratory tests need be run for positive soil classification.

NOTE 4—The ability to describe and identify soils correctly is learned more readily under the guidance of experienced personnel, but it may also be acquired systematically by comparing numerical laboratory test results for typical soils of each type with their visual and manual characteristics.

5.6 When describing and identifying soil samples from a given boring, test pit, or group of borings or pits, it is not necessary to follow all of the procedures in this practice for every sample. Soils which appear to be similar can be grouped together; one sample completely described and identified with the others referred to as similar based on performing only a few of the descriptive and identification procedures described in this practice.

5.7 This practice may be used in combination with Practice D 4083 when working with frozen soils.

6. Apparatus

6.1 *Required Apparatus:*

6.1.1 *Pocket Knife or Small Spatula.*

6.2 *Useful Auxiliary Apparatus:*

6.2.1 *Small Test Tube and Stopper* (or jar with a lid).

6.2.2 *Small Hand Lens.*

7. Reagents

7.1 *Purity of Water*—Unless otherwise indicated, references to water shall be understood to mean water from a city water supply or natural source, including non-potable water.

7.2 *Hydrochloric Acid*—A small bottle of dilute hydrochloric acid, HCl, one part HCl (10 *N*) to three parts water (This reagent is optional for use with this practice). See Section 8.

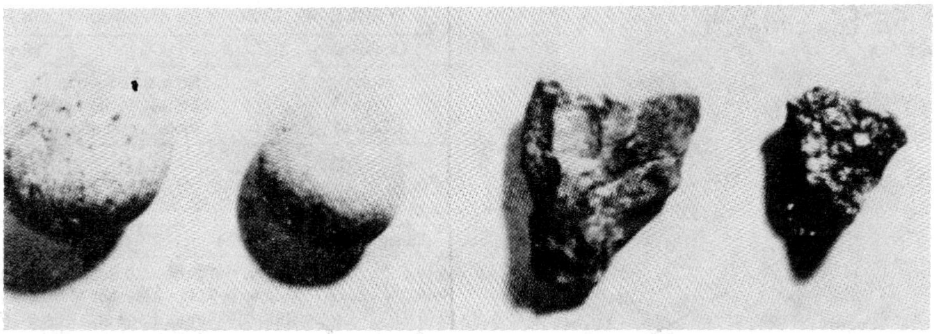

(a) Rounded (b) Angular

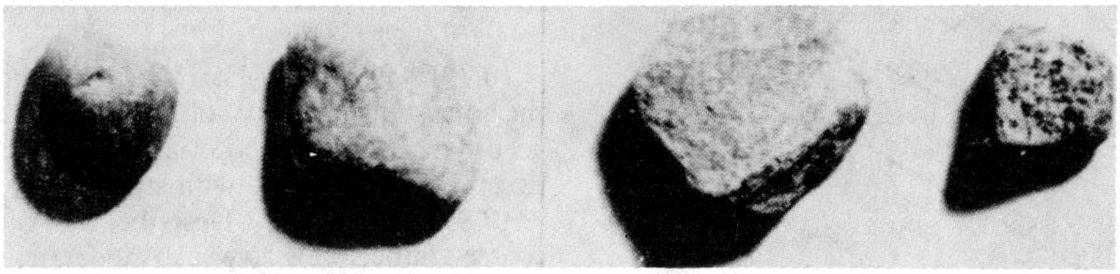

(c) Subrounded (d) Subangular

FIG. 3 Typical Angularity of Bulky Grains

8. Safety Precautions

8.1 When preparing the dilute HCl solution of one part concentrated hydrochloric acid (10 *N*) to three parts of distilled water, slowly add acid into water following necessary safety precautions. Handle with caution and store safely. If solution comes into contact with the skin, rinse thoroughly with water.

8.2 **Caution**—Do not add water to acid.

9. Sampling

9.1 The sample shall be considered to be representative of the stratum from which it was obtained by an appropriate, accepted, or standard procedure.

NOTE 5—Preferably, the sampling procedure should be identified as having been conducted in accordance with Practices D 1452, D 1587, or D 2113, or Method D 1586.

9.2 The sample shall be carefully identified as to origin.

NOTE 6—Remarks as to the origin may take the form of a boring number and sample number in conjunction with a job number, a geologic stratum, a pedologic horizon or a location description with respect to a permanent monument, a grid system or a station number and offset with respect to a stated centerline and a depth or elevation.

9.3 For accurate description and identification, the minimum amount of the specimen to be examined shall be in accordance with the following schedule:

Maximum Particle Size, Sieve Opening	Minimum Specimen Size, Dry Weight
4.75 mm (No. 4)	100 g (0..5 lb)
9.5 mm (3/8 in.)	200 g (0.5 lb)
19.0 mm (3/4 in.)	1.0 kg (2.2 lb)
38.1 mm (1 1/2 in.)	8.0 kg (18 lb)
75.0 mm (3 in.)	60.0 kg (132 lb)

NOTE 7—If random isolated particles are encountered that are significantly larger than the particles in the soil matrix, the soil matrix can be accurately described and identified in accordance with the preceeding schedule.

9.4 If the field sample or specimen being examined is smaller than the minimum recommended amount, the report shall include an appropriate remark.

10. Descriptive Information for Soils

10.1 *Angularity*—Describe the angularity of the sand (coarse sizes only), gravel, cobbles, and boulders, as angular, subangular, subrounded, or rounded in accordance with the criteria in Table 1 and Fig. 3. A range of angularity may be stated, such as: subrounded to rounded.

10.2 *Shape*—Describe the shape of the gravel, cobbles, and boulders as flat, elongated, or flat and elongated if they meet the criteria in Table 2 and Fig. 4. Otherwise, do not mention the shape. Indicate the fraction of the particles that have the shape, such as: one-third of the gravel particles are flat.

10.3 *Color*—Describe the color. Color is an important property in identifying organic soils, and within a given

TABLE 1 Criteria for Describing Angularity of Coarse-Grained Particles (see Fig. 3)

Description	Criteria
Angular	Particles have sharp edges and relatively plane sides with unpolished surfaces
Subangular	Particles are similar to angular description but have rounded edges
Subrounded	Particles have nearly plane sides but have well-rounded corners and edges
Rounded	Particles have smoothly curved sides and no edges

TABLE 2 Criteria for Describing Particle Shape (see Fig. 4)

The particle shape shall be described as follows where length, width, and thickness refer to the greatest, intermediate, and least dimensions of a particle, respectively.

Flat	Particles with width/thickness > 3
Elongated	Particles with length/width > 3
Flat and elongated	Particles meet criteria for both flat and elongated

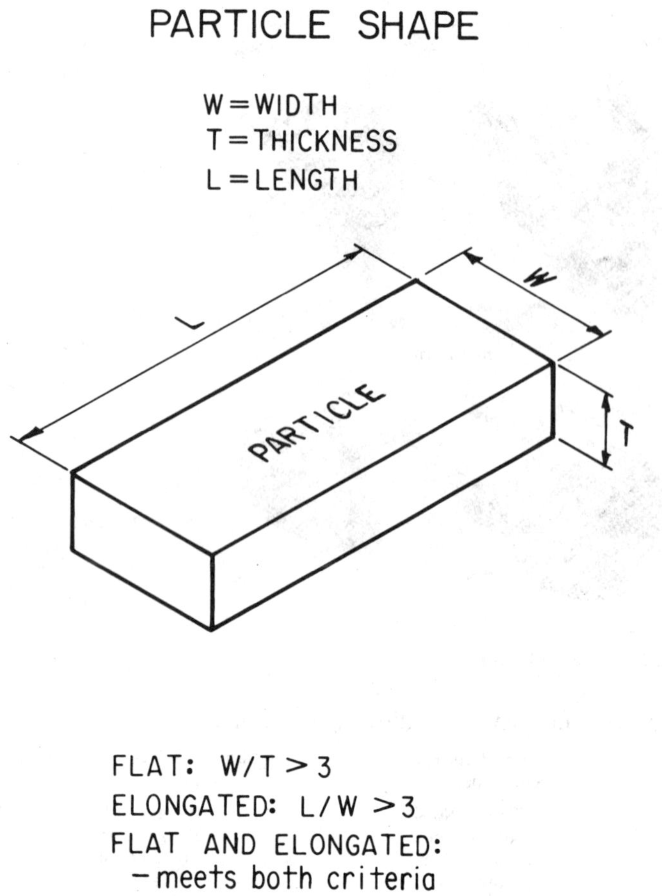

FIG. 4 Criteria for Particle Shape

TABLE 3 Criteria for Describing Moisture Condition

Description	Criteria
Dry	Absence of moisture, dusty, dry to the touch
Moist	Damp but no visible water
Wet	Visible free water, usually soil is below water table

locality it may also be useful in identifying materials of similar geologic origin. If the sample contains layers or patches of varying colors, this shall be noted and all representative colors shall be described. The color shall be described for moist samples. If the color represents a dry condition, this shall be stated in the report.

10.4 *Odor*—Describe the odor if organic or unusual. Soils containing a significant amount of organic material usually have a distinctive odor of decaying vegetation. This is especially apparent in fresh samples, but if the samples are dried, the odor may often be revived by heating a moistened sample. If the odor is unusual (petroleum product, chemical, and the like), it shall be described.

10.5 *Moisture Condition*—Describe the moisture condition as dry, moist, or wet, in accordance with the criteria in Table 3.

10.6 *HCl Reaction*—Describe the reaction with HCl as none, weak, or strong, in accordance with the critera in Table 4. Since calcium carbonate is a common cementing agent, a report of its presence on the basis of the reaction with dilute hydrochloric acid is important.

TABLE 4 Criteria for Describing the Reaction With HCl

Description	Criteria
None	No visible reaction
Weak	Some reaction, with bubbles forming slowly
Strong	Violent reaction, with bubbles forming immediately

TABLE 5 Criteria for Describing Consistency

Description	Criteria
Very soft	Thumb will penetrate soil more than 1 in. (25 mm)
Soft	Thumb will penetrate soil about 1 in. (25 mm)
Firm	Thumb will indent soil about ¼ in. (6 mm)
Hard	Thumb will not indent soil but readily indented with thumbnail
Very hard	Thumbnail will not indent soil

10.7 *Consistency*—For intact fine-grained soil, describe the consistency as very soft, soft, firm, hard, or very hard, in accordance with the criteria in Table 5. This observation is inappropriate for soils with significant amounts of gravel.

10.8 *Cementation*—Describe the cementation of intact coarse-grained soils as weak, moderate, or strong, in accordance with the criteria in Table 6.

10.9 *Structure*—Describe the structure of intact soils in accordance with the criteria in Table 7.

10.10 *Range of Particle Sizes*—For gravel and sand components, describe the range of particle sizes within each component as defined in 3.1.2 and 3.1.6. For example, about 20 % fine to coarse gravel, about 40 % fine to coarse sand.

10.11 *Maximum Particle Size*—Describe the maximum particle size found in the sample in accordance with the following information:

10.11.1 *Sand Size*—If the maximum particle size is a sand size, describe as fine, medium, or coarse as defined in 3.1.6. For example: maximum particle size, medium sand.

10.11.2 *Gravel Size*—If the maximum particle size is a gravel size, describe the maximum particle size as the smallest sieve opening that the particle will pass. For example, maximum particle size, 1½ in. (will pass a 1½-in. square opening but not a ¾-in. square opening).

10.11.3 *Cobble or Boulder Size*—If the maximum particle size is a cobble or boulder size, describe the maximum dimension of the largest particle. For example: maximum dimension, 18 in. (450 mm).

10.12 *Hardness*—Describe the hardness of coarse sand and larger particles as hard, or state what happens when the particles are hit by a hammer, for example, gravel-size particles fracture with considerable hammer blow, some gravel-size particles crumble with hammer blow. "Hard" means particles do not crack, fracture, or crumble under a hammer blow.

10.13 Additional comments shall be noted, such as the presence of roots or root holes, difficulty in drilling or augering hole, caving of trench or hole, or the presence of mica.

10.14 A local or commercial name or a geologic interpre-

TABLE 6 Criteria for Describing Cementation

Description	Criteria
Weak	Crumbles or breaks with handling or little finger pressure
Moderate	Crumbles or breaks with considerable finger pressure
Strong	Will not crumble or break with finger pressure

TABLE 7 Criteria for Describing Structure

Description	Criteria
Stratified	Alternating layers of varying material or color with layers at least 6 mm thick; note thickness
Laminated	Alternating layers of varying material or color with the layers less than 6 mm thick; note thickness
Fissured	Breaks along definite planes of fracture with little resistance to fracturing
Slickensided	Fracture planes appear polished or glossy, sometimes striated
Blocky	Cohesive soil that can be broken down into small angular lumps which resist further breakdown
Lensed	Inclusion of small pockets of different soils, such as small lenses of sand scattered through a mass of clay; note thickness
Homogeneous	Same color and appearance throughout

tation of the soil, or both, may be added if identified as such.

10.15 A classification or identification of the soil in accordance with other classification systems may be added if identified as such.

11. Identification of Peat

11.1 A sample composed primarily of vegetable tissue in various stages of decomposition that has a fibrous to amorphous texture, usually a dark brown to black color, and an organic odor, shall be designated as a highly organic soil and shall be identified as peat, PT, and not subjected to the identification procedures described hereafter.

12. Preparation for Identification

12.1 The soil identification portion of this practice is based on the portion of the soil sample that will pass a 3-in. (75-mm) sieve. The larger than 3-in. (75-mm) particles must be removed, manually, for a loose sample, or mentally, for an intact sample before classifying the soil.

12.2 Estimate and note the percentage of cobbles and the percentage of boulders. Performed visually, these estimates will be on the basis of volume percentage.

NOTE 8—Since the percentages of the particle-size distribution in Test Method D 2487 are by dry weight, and the estimates of percentages for gravel, sand, and fines in this practice are by dry weight, it is recommended that the report state that the percentages of cobbles and boulders are by volume.

12.3 Of the fraction of the soil smaller than 3 in. (75 mm), estimate and note the percentage, by dry weight, of the gravel, sand, and fines (see Appendix X4 for suggested procedures).

NOTE 9—Since the particle-size components appear visually on the basis of volume, considerable experience is required to estimate the percentages on the basis of dry weight. Frequent comparisons with laboratory particle-size analyses should be made.

12.3.1 The percentages shall be estimated to the closest 5 %. The percentages of gravel, sand, and fines must add up to 100 %.

12.3.2 If one of the components is present but not in sufficient quantity to be considered 5 % of the smaller than 3-in. (75-mm) portion, indicate its presence by the term *trace*, for example, trace of fines. A trace is not to be considered in the total of 100 % for the components.

13. Preliminary Identification

13.1 The soil is *fine grained* if it contains 50 % or more fines. Follow the procedures for identifying fine-grained soils of Section 14.

13.2 The soil is *coarse grained* if it contains less than 50 % fines. Follow the procedures for identifying coarse-grained soils of Section 15.

14. Procedure for Identifying Fine-Grained Soils

14.1 Select a representative sample of the material for examination. Remove particles larger than the No. 40 sieve (medium sand and larger) until a specimen equivalent to about a handful of material is available. Use this specimen for performing the dry strength, dilatancy, and toughness tests.

14.2 *Dry Strength:*

14.2.1 From the specimen, select enough material to mold into a ball about 1 in. (25 mm) in diameter. Mold the material until it has the consistency of putty, adding water if necessary.

14.2.2 From the molded material, make at least three test specimens. A test specimen shall be a ball of material about ½ in. (12 mm) in diameter. Allow the test specimens to dry in air, or sun, or by artificial means, as long as the temperature does not exceed 60°C.

14.2.3 If the test specimen contains natural dry lumps, those that are about ½ in. (12 mm) in diameter may be used in place of the molded balls.

NOTE 10—The process of molding and drying usually produces higher strengths than are found in natural dry lumps of soil.

14.2.4 Test the strength of the dry balls or lumps by crushing between the fingers. Note the strength as none, low, medium, high, or very high in accorance with the criteria in Table 8. If natural dry lumps are used, do not use the results of any of the lumps that are found to contain particles of coarse sand.

14.2.5 The presence of high-strength water-soluble cementing materials, such as calcium carbonate, may cause exceptionally high dry strengths. The presence of calcium carbonate can usually be detected from the intensity of the reaction with dilute hydrochloric acid (see 10.6).

14.3 *Dilatancy:*

14.3.1 From the specimen, select enough material to mold into a ball about ½ in. (12 mm) in diameter. Mold the material, adding water if necessary, until it has a soft, but not sticky, consistency.

14.3.2 Smooth the soil ball in the palm of one hand with the blade of a knife or small spatula. Shake horizontally, striking the side of the hand vigorously against the other hand several times. Note the reaction of water appearing on

TABLE 8 Criteria for Describing Dry Strength

Description	Criteria
None	The dry specimen crumbles into powder with mere pressure of handling
Low	The dry specimen crumbles into powder with some finger pressure
Medium	The dry specimen breaks into pieces or crumbles with considerable finger pressure
High	The dry specimen cannot be broken with finger pressure. Specimen will break into pieces between thumb and a hard surface
Very high	The dry specimen cannot be broken between the thumb and a hard surface

TABLE 9 Criteria for Describing Dilatancy

Description	Criteria
None	No visible change in the specimen
Slow	Water appears slowly on the surface of the specimen during shaking and does not disappear or disappears slowly upon squeezing
Rapid	Water appears quickly on the surface of the specimen during shaking and disappears quickly upon squeezing

TABLE 10 Criteria for Describing Toughness

Description	Criteria
Low	Only slight pressure is required to roll the thread near the plastic limit. The thread and the lump are weak and soft
Medium	Medium pressure is required to roll the thread to near the plastic limit. The thread and the lump have medium stiffness
High	Considerable pressure is required to roll the thread to near the plastic limit. The thread and the lump have very high stiffness

the surface of the soil. Squeeze the sample by closing the hand or pinching the soil between the fingers, and note the reaction as none, slow, or rapid in accordance with the criteria in Table 9. The reaction is the speed with which water appears while shaking, and disappears while squeezing.

14.4 *Toughness:*

14.4.1 Following the completion of the dilatancy test, the test specimen is shaped into an elongated pat and rolled by hand on a smooth surface or between the palms into a thread about ⅛ in. (3 mm) in diameter. (If the sample is too wet to roll easily, it should be spread into a thin layer and allowed to lose some water by evaporation.) Fold the sample threads and reroll repeatedly until the thread crumbles at a diameter of about ⅛ in. The thread will crumble at a diameter of ⅛ in. when the soil is near the plastic limit. Note the pressure required to roll the thread near the plastic limit. Also, note the strength of the thread. After the thread crumbles, the pieces should be lumped together and kneaded until the lump crumbles. Note the toughness of the material during kneading.

14.4.2 Describe the toughness of the thread and lump as low, medium, or high in accordance with the criteria in Table 10.

14.5 *Plasticity*—On the basis of observations made during the toughness test, describe the plasticity of the material in accordance with the criteria given in Table 11.

14.6 Decide whether the soil is an *inorganic* or an *organic* fine-grained soil (see 14.8). If inorganic, follow the steps given in 14.7.

14.7 *Identification of Inorganic Fine-Grained Soils:*

TABLE 11 Criteria for Describing Plasticity

Description	Criteria
Nonplastic	A ⅛-in. (3-mm) thread cannot be rolled at any water content
Low	The thread can barely be rolled and the lump cannot be formed when drier than the plastic limit
Medium	The thread is easy to roll and not much time is required to reach the plastic limit. The thread cannot be rerolled after reaching the plastic limit. The lump crumbles when drier than the plastic limit
High	It takes considerable time rolling and kneading to reach the plastic limit. The thread can be rerolled several times after reaching the plastic limit. The lump can be formed without crumbling when drier than the plastic limit

14.7.1 Identify the soil as a *lean clay*, CL, if the soil has medium to high dry strength, no or slow dilatancy, and medium toughness and plasticity (see Table 12).

14.7.2 Identify the soil as a *fat clay*, CH, if the soil has high to very high dry strength, no dilatancy, and high toughness and plasticity (see Table 12).

14.7.3 Identify the soil as a *silt*, ML, if the soil has no to low dry strength, slow to rapid dilatancy, and low toughness and plasticity, or is nonplastic (see Table 12).

14.7.4 Identify the soil as an *elastic silt*, MH, if the soil has low to medium dry strength, no to slow dilatancy, and low to medium toughness and plasticity (see Table 12).

NOTE 11—These properties are similar to those for a lean clay. However, the silt will dry quickly on the hand and have a smooth, silky feel when dry. Some soils that would classify as MH in accordance with the criteria in Test Method D 2487 are visually difficult to distinguish from lean clays, CL. It may be necessary to perform laboratory testing for proper identification.

14.8 *Identification of Organic Fine-Grained Soils:*

14.8.1 Identify the soil as an *organic soil*, OL/OH, if the soil contains enough organic particles to influence the soil properties. Organic soils usually have a dark brown to black color and may have an organic odor. Often, organic soils will change color, for example, black to brown, when exposed to the air. Some organic soils will lighten in color significantly when air dried. Organic soils normally will not have a high toughness or plasticity. The thread for the toughness test will be spongy.

NOTE 12—In some cases, through practice and experience, it may be possible to further identify the organic soils as organic silts or organic clays, OL or OH. Correlations between the dilatancy, dry strength, toughness tests, and laboratory tests can be made to identify organic soils in certain deposits of similar materials of known geologic origin.

14.9 If the soil is estimated to have 15 to 25 % sand or gravel, or both, the words "with sand" or "with gravel" (whichever is more predominant) shall be added to the group name. For example: "lean clay with sand, CL" or "silt with gravel, ML" (see Figs. 1a and 1b). If the percentage of sand is equal to the percentage of gravel, use "with sand."

14.10 If the soil is estimated to have 30 % or more sand or gravel, or both, the words "sandy" or "gravelly" shall be added to the group name. Add the word "sandy" if there appears to be more sand than gravel. Add the word "gravelly" if there appears to be more gravel than sand. For example: "sandy lean clay, CL", "gravelly fat clay, CH", or "sandy silt, ML" (see Figs. 1a and 1b). If the percentage of sand is equal to the percent of gravel, use "sandy."

15. Procedure for Identifying Coarse-Grained Soils (Contains less than 50 % fines)

15.1 The soil is a *gravel* if the percentage of gravel is estimated to be more than the percentage of sand.

TABLE 12 Identification of Inorganic Fine-Grained Soils from Manual Tests

Soil Symbol	Dry Strength	Dilatancy	Toughness
ML	None to low	Slow to rapid	Low or thread cannot be formed
CL	Medium to high	None to slow	Medium
MH	Low to medium	None to slow	Low to medium
CH	High to very high	None	High

15.2 The soil is a *sand* if the percentage of gravel is estimated to be equal to or less than the percentage of sand.

15.3 The soil is a *clean gravel* or *clean sand* if the percentage of fines is estimated to be 5 % or less.

15.3.1 Identify the soil as a *well-graded gravel*, GW, or as a *well-graded sand*, SW, if it has a wide range of particle sizes and substantial amounts of the intermediate particle sizes.

15.3.2 Identify the soil as a *poorly graded gravel*, GP, or as a *poorly graded sand*, SP, if it consists predominantly of one size (uniformly graded), or it has a wide range of sizes with some intermediate sizes obviously missing (gap or skip graded).

15.4 The soil is either a *gravel with fines* or a *sand with fines* if the percentage of fines is estimated to be 15 % or more.

15.4.1 Identify the soil as a *clayey gravel*, GC, or a *clayey sand*, SC, if the fines are clayey as determined by the procedures in Section 14.

15.4.2 Identify the soil as a *silty gravel*, GM, or a *silty sand*, SM, if the fines are silty as determined by the procedures in Section 14.

15.5 If the soil is estimated to contain 10 % fines, give the soil a dual identification using two group symbols.

15.5.1 The first group symbol shall correspond to a clean gravel or sand (GW, GP, SW, SP) and the second symbol shall correspond to a gravel or sand with fines (GC, GM, SC, SM).

15.5.2 The group name shall correspond to the first group symbol plus the words "with clay" or "with silt" to indicate the plasticity characteristics of the fines. For example: "well-graded gravel with clay, GW-GC" or "poorly graded sand with silt, SP-SM" (see Fig. 2).

15.6 If the specimen is predominantly sand or gravel but contains an estimated 15 % or more of the other coarse-grained constituent, the words "with gravel" or "with sand" shall be added to the group name. For example: "poorly graded gravel with sand, GP" or "clayey sand with gravel, SC" (see Fig. 2).

15.7 If the field sample contains any cobbles or boulders, or both, the words "with cobbles" or "with cobbles and boulders" shall be added to the group name. For example: "silty gravel with cobbles, GM."

16. Report

16.1 The report shall include the information as to origin, and the items indicated in Table 13.

NOTE 13—*Example: Clayey Gravel with Sand and Cobbles, GC*—About 50 % fine to coarse, subrounded to subangular gravel; about 30 % fine to coarse, subrounded sand; about 20 % fines with medium plasticity, high dry strength, no dilatancy, medium toughness; weak reaction with HCl; original field sample had about 5 % (by volume) subrounded cobbles, maximum dimension, 150 mm.

In-Place Conditions—Firm, homogeneous, dry, brown

Geologic Interpretation—Alluvial fan

NOTE 14—Other examples of soil descriptions and identification are given in Appendixes X1 and X2.

NOTE 15—If desired, the percentages of gravel, sand, and fines may be stated in terms indicating a range of percentages, as follows:

Trace—Particles are present but estimated to be less than 5 %
Few—5 to 10 %
Little—15 to 25 %
Some—30 to 45 %
Mostly—50 to 100 %

TABLE 13 Checklist for Description of Soils

1. Group name
2. Group symbol
3. Percent of cobbles or boulders, or both (by volume)
4. Percent of gravel, sand, or fines, or all three (by dry weight)
5. Particle-size range:
   Gravel—fine, coarse
   Sand—fine, medium, coarse
6. Particle angularity: angular, subangular, subrounded, rounded
7. Particle shape: (if appropriate) flat, elongated, flat and elongated
8. Maximum particle size or dimension
9. Hardness of coarse sand and larger particles
10. Plasticity of fines: nonplastic, low, medium, high
11. Dry strength: none, low, medium, high, very high
12. Dilatancy: none, slow, rapid
13. Toughness: low, medium, high
14. Color (in moist condition)
15. Odor (mention only if organic or unusual)
16. Moisture: dry, moist, wet
17. Reaction with HCl: none, weak, strong

For intact samples:

18. Consistency (fine-grained soils only): very soft, soft, firm, hard, very hard
19. Structure: stratified, laminated, fissured, slickensided, lensed, homogeneous
20. Cementation: weak, moderate, strong
21. Local name
22. Geologic interpretation
23. Additional comments: presence of roots or root holes, presence of mica, gypsum, etc., surface coatings on coarse-grained particles, caving or sloughing of auger hole or trench sides, difficulty in augering or excavating, etc.

16.2 If, in the soil description, the soil is identified using a classification group symbol and name as described in Test Method D 2487, it must be distinctly and clearly stated in log forms, summary tables, reports, and the like, that the symbol and name are based on visual-manual procedures.

17. Precision and Bias

17.1 This practice provides qualitative information only, therefore, a precision and bias statement is not applicable.

18. Keywords

18.1 classification; clay; gravel; organic soils; sand; silt; soil classification; soil description; visual classification

APPENDIXES

(Nonmandatory Information)

X1. EXAMPLES OF VISUAL SOIL DESCRIPTIONS

X1.1 The following examples show how the information required in 16.1 can be reported. The information that is included in descriptions should be based on individual circumstances and need.

X1.1.1 *Well-Graded Gravel with Sand (GW)*—About 75 % fine to coarse, hard, subangular gravel; about 25 % fine to coarse, hard, subangular sand; trace of fines; maximum size, 75 mm, brown, dry; no reaction with HCl.

X1.1.2 *Silty Sand with Gravel (SM)*—About 60 % predominantly fine sand; about 25 % silty fines with low plasticity, low dry strength, rapid dilatancy, and low toughness; about 15 % fine, hard, subrounded gravel, a few gravel-size particles fractured with hammer blow; maximum size, 25 mm; no reaction with HCl (Note—Field sample size smaller than recommended).

In-Place Conditions—Firm, stratified and contains lenses of silt 1 to 2 in. (25 to 50 mm) thick, moist, brown to gray; in-place density 106 lb/ft^3; in-place moisture 9 %.

X1.1.3 *Organic Soil (OL/OH)*—About 100 % fines with low plasticity, slow dilatancy, low dry strength, and low toughness; wet, dark brown, organic odor; weak reaction with HCl.

X1.1.4 *Silty Sand with Organic Fines (SM)*—About 75 % fine to coarse, hard, subangular reddish sand; about 25 % organic and silty dark brown nonplastic fines with no dry strength and slow dilatancy; wet; maximum size, coarse sand; weak reaction with HCl.

X1.1.5 *Poorly Graded Gravel with Silt, Sand, Cobbles and Boulders (GP-GM)*—About 75 % fine to coarse, hard, subrounded to subangular gravel; about 15 % fine, hard, subrounded to subangular sand; about 10 % silty nonplastic fines; moist, brown; no reaction with HCl; original field sample had about 5 % (by volume) hard, subrounded cobbles and a trace of hard, subrounded boulders, with a maximum dimension of 18 in. (450 mm).

X2. USING THE IDENTIFICATION PROCEDURE AS A DESCRIPTIVE SYSTEM FOR SHALE, CLAYSTONE, SHELLS, SLAG, CRUSHED ROCK, AND THE LIKE

X2.1 The identification procedure may be used as a descriptive system applied to materials that exist in-situ as shale, claystone, sandstone, siltstone, mudstone, etc., but convert to soils after field or laboratory processing (crushing, slaking, and the like).

X2.2 Materials such as shells, crushed rock, slag, and the like, should be identified as such. However, the procedures used in this practice for describing the particle size and plasticity characteristics may be used in the description of the material. If desired, an identification using a group name and symbol according to this practice may be assigned to aid in describing the material.

X2.3 The group symbol(s) and group names should be placed in quotation marks or noted with some type of distinguishing symbol. See examples.

X2.4 Examples of how group names and symbols can be incororated into a descriptive system for materials that are not naturally occurring soils are as follows:

X2.4.1 *Shale Chunks*—Retrieved as 2 to 4-in. (50 to 100-mm) pieces of shale from power auger hole, dry, brown, no reaction with HCl. After slaking in water for 24 h, material identified as "Sandy Lean Clay (CL)"; about 60 % fines with medium plasticity, high dry strength, no dilatancy, and medium toughness; about 35 % fine to medium, hard sand; about 5 % gravel-size pieces of shale.

X2.4.2 *Crushed Sandstone*—Product of commercial crushing operation; "Poorly Graded Sand with Silt (SP-SM)"; about 90 % fine to medium sand; about 10 % nonplastic fines; dry, reddish-brown, strong reaction with HCl.

X2.4.3 *Broken Shells*—About 60 % gravel-size broken shells; about 30 % sand and sand-size shell pieces; about 10 % fines; "Poorly Graded Gravel with Sand (GP)."

X2.4.4 *Crushed Rock*—Processed from gravel and cobbles in Pit No. 7; "Poorly Graded Gravel (GP)"; about 90 % fine, hard, angular gravel-size particles; about 10 % coarse, hard, angular sand-size particles; dry, tan; no reaction with HCl.

X3. SUGGESTED PROCEDURE FOR USING A BORDERLINE SYMBOL FOR SOILS WITH TWO POSSIBLE IDENTIFICATIONS.

X3.1 Since this practice is based on estimates of particle size distribution and plasticity characteristics, it may be difficult to clearly identify the soil as belonging to one category. To indicate that the soil may fall into one of two possible basic groups, a borderline symbol may be used with the two symbols separated by a slash. For example: SC/CL or CL/CH.

X3.1.1 A borderline symbol may be used when the

percentage of fines is estimated to be between 45 and 55 %. One symbol should be for a coarse-grained soil with fines and the other for a fine-grained soil. For example: GM/ML or CL/SC.

X3.1.2 A borderline symbol may be used when the percentage of sand and the percentage of gravel are estimated to be about the same. For example: GP/SP, SC/GC, GM/SM. It is practically impossible to have a soil that would have a borderline symbol of GW/SW.

X3.1.3 A borderline symbol may be used when the soil could be either well graded or poorly graded. For example: GW/GP, SW/SP.

X3.1.4 A borderline symbol may be used when the soil could either be a silt or a clay. For example: CL/ML, CH/MH, SC/SM.

X3.1.5 A borderline symbol may be used when a fine-grained soil has properties that indicate that it is at the boundary between a soil of low compressibility and a soil of high compressibility. For example: CL/CH, MH/ML.

X3.2 The order of the borderline symbols should reflect similarity to surrounding or adjacent soils. For example: soils in a borrow area have been identified as CH. One sample is considered to have a borderline symbol of CL and CH. To show similarity, the borderline symbol should be CH/CL.

X3.3 The group name for a soil with a borderline symbol should be the group name for the first symbol, except for:

CL/CH lean to fat clay
ML/CL clayey silt
CL/ML silty clay

X3.4 The use of a borderline symbol should not be used indiscriminately. Every effort shall be made to first place the soil into a single group.

X4. SUGGESTED PROCEDURES FOR ESTIMATING THE PERCENTAGES OF GRAVEL, SAND, AND FINES IN A SOIL SAMPLE

X4.1 *Jar Method*—The relative percentage of coarse- and fine-grained material may be estimated by thoroughly shaking a mixture of soil and water in a test tube or jar, and then allowing the mixture to settle. The coarse particles will fall to the bottom and successively finer particles will be deposited with increasing time; the sand sizes will fall out of suspension in 20 to 30 s. The relative proportions can be estimated from the relative volume of each size separate. This method should be correlated to particle-size laboratory determinations.

X4.2 *Visual Method*—Mentally visualize the gravel size particles placed in a sack (or other container) or sacks. Then, do the same with the sand size particles and the fines. Then, mentally compare the number of sacks to estimate the percentage of plus No. 4 sieve size and minus No. 4 sieve size present. The percentages of sand and fines in the minus sieve size No. 4 material can then be estimated from the wash test (X4.3).

X4.3 *Wash Test (for relative percentages of sand and fines)*—Select and moisten enough minus No. 4 sieve size material to form a 1-in (25-mm) cube of soil. Cut the cube in half, set one-half to the side, and place the other half in a small dish. Wash and decant the fines out of the material in the dish until the wash water is clear and then compare the two samples and estimate the percentage of sand and fines. Remember that the percentage is based on weight, not volume. However, the volume comparison will provide a reasonable indication of grain size percentages.

X4.3.1 While washing, it may be necessary to break down lumps of fines with the finger to get the correct percentages.

X5. RATIONALE

X5.1 This practice was significantly revised in the D 2488 – 84 version from the previous version D 2488 – 69 (1975). The revisions are documented in the literature.[3]

X5.2 Changes in this version from the previous version include rewording of 1.2.3 to say (disturbed and undisturbed), the addition of 5.7 to refer to the practice for describing frozen soils, and the addition of Appendix X5 on Rationale.

[3] Howard, A. K. "The Revised ASTM Standard on the Description and Identification of Soils (Visual-Manual Procedure)," *Geotechical Testing Journal*, GTJODJ Vol. 10, No. 4, December 1987.

Designation: D 2777 – 86

Standard Practice for Determination of Precision and Bias of Applicable Methods of Committee D-19 on Water[1]

This standard is issued under the fixed designation D 2777; the number immediately following the designation indicates the year of original adoption or, in the case of revision, the year of last revision. A number in parentheses indicates the year of last reapproval. A superscript epsilon (ε) indicates an editorial change since the last revision or reapproval.

1. Scope

1.1 This practice establishes uniform standards for expressing the precision and bias of applicable test methods for Committee D-19 on Water.

1.2 Except as specified in 1.3, 1.4, 1.5, and 1.6, this practice requires that the task group proposing a new test method must carry out a collaborative study from which statements for precision (multiple laboratory operational and single-operator standard deviation estimates) and observed bias can be developed. This practice provides general guidance to task groups in planning and conducting such determinations of precision and bias.

1.3 Where, due to the nature of the test or instability of the sample, a full-scale collaborative test is not technically feasible, a statement of single-operator precision (Section 10) will meet the requirements of this practice. Whenever possible, this statement should be developed from data generated by multiple operators, each doing replicate analyses on independent samples which generally fall within specified concentration ranges. This practice is not applicable to methodology involving continuous sampling and measurement of specific constituents and properties.

1.4 Many of the test methods adopted prior to 1980 do not have adequate precision and bias statements, and there is insufficient information to satisfactorily improve the statements. These test methods are being used, and it is generally agreed that they should be continued as standards until 1985, while a proper collaborative study is conducted. In the interim, a warning must be placed in the scope and precision sections to alert the user that there has not been adequate collaborative testing to validate the test method.

1.5 Existing test methods that do have adequate precision and bias statements based on a minimum of three laboratories for a total of six operators do not require further collaborative testing, and such deficiencies must be listed in the precision and bias statements.

1.6 With the recommendation of the Results Advisor and the concurrence of the Technical Operations Section of the Executive Subcommittee, a statement giving a compelling reason why all or specific points of this practice cannot be fully complied with will meet both ASTM requirements **(1)**[2] and the related requirements of this practice.

1.7 In principle, all test methods are covered by this practice.

1.8 In Section 11, this practice shows exemplary precision and bias statement formats for: (*1*) test methods yielding a numerical measure, (*2*) test methods yielding a non-numerical report of success or failure based on criteria specified in the procedure, and (*3*) test methods specifying that procedures in another ASTM method are to be used with only insignificant modifications.

2. Referenced Documents

2.1 *ASTM Standards:*

D 596 Practice for Reporting Results of Analysis of Water[3]
D 858 Test Methods for Manganese in Water[3]
D 1129 Terminology Relating to Water[3]
D 1193 Specification for Reagent Water[3]
E 177 Practice for Use of the Terms Precision and Bias in ASTM Test Methods[4]
E 178 Practice for Dealing with Outlying Observations[4]

3. Terminology

3.1 *Definitions*—For definitions of terms used in this practice, refer to Terminology D 1129.

3.2 *Descriptions of Terms Specific to This Standard:*

3.2.1 *accuracy*—a measure of the degree of conformity of a value generated by a specific procedure to the assumed or accepted true value and includes both precision and bias.

3.2.2 *bias*—the persistent positive or negative deviation of the method average value from the assumed or accepted value.

3.2.3 *precision*—the degree of agreement of repeated measurements of the same property, expressed in terms of dispersion of test results about the arithmetical mean result obtained by repetitive testing of a homogeneous sample under specified conditions. The precision of a test method is expressed quantitatively as the standard deviation computed from the results of a series of controlled determinations.

4. Summary of Practice

4.1 After the task group has assured itself that the test method has had all preliminary evaluation work completed, it should prepare the method writeup in final form. The plan for the collaborative test is developed in accordance with this practice and submitted along with method writeup to the Results Advisor for concurrence except as specified in 1.3, 1.4, 1.5, and 1.6. Upon receipt of concurrence, the collabo-

---

[1] This practice is under the jurisdiction of ASTM Committee D-19 on Water and is the direct responsibility of Subcommittee D19.02 on General Specifications and Technical Resources.

Current edition approved Aug. 29, 1986. Published October 1986. Originally published as D 2777 – 69 T. Last previous edition D 2777 – 85.

[2] The boldface numbers in parentheses refer to the list of references at the end of this practice.

[3] *Annual Book of ASTM Standards*, Vol 11.01.

[4] *Annual Book of ASTM Standards*, Vol 14.02.

rative test is conducted, data analyzed, and precision and bias statements formulated by the task group. The statements, with backup data including the reported results summary, the calculations leading up to the statements, and the method writeup with precision and bias statements included, are submitted to the subcommittee vice-chairman who in turn sends it to the Results Advisor for concurrence before balloting. This assures having a copy of the results to send along to ASTM for items on the main committee ballot. In most instances, the collaborative test must be complete before a subcommittee ballot. If the collaborative test is not complete, then the test method goes on the ballot as a "proposed" method rather than a "standard" method. Copies of the test data and approved calculations must be filed at ASTM Headquarters when the test method is submitted by the subcommittee chairman as an item for the main committee ballot.

4.1.1 Appendix X1 shows a sample of Form A, "Approval of Plans for Interlaboratory Testing."

4.1.2 Appendix X2 shows a sample of Form B, "Data Report from Individual Laboratories."

4.1.3 Appendix X3 shows a sample calculation of precision and bias from real collaborative test data.

4.1.4 Appendix X4 shows a sample of information required for research report from real collaborative test data.

5. Significance and Use

5.1 The significance of this practice to the user of the test method is the assurance that a cross section of interested laboratories have performed the test method, using the method as written, and achieved satisfactory results. The user is provided guidance as to what levels of precision and bias can be expected in application of the test method.

5.2 The test method specifies the media for which the test method is appropriate. The collaborative test corroborates the writeup within the limitations of the test design. An extensive test can only use representative media so that universal applicability cannot be implied from the results.

5.3 The fundamental assumption of the collaborative test is that the media tested, the concentrations tested, and the precipitating laboratories are a representative and fair evaluation of the scope and applicability of the test method as written.

6. Preliminary Studies

6.1 Considerable pilot work on a test method must precede the determination of its precision and bias **(2, 3)**. This pilot work should explore such variables as reaction time and temperature, concentration of reagents, interferences, calibration, and sample size. Potentially significant factors must be investigated and controlled closely in advance because cooperative identification of them can be difficult. Disregard of such factors may introduce so much variation among cooperators that results are misleading or inconclusive **(4)**.

6.2 Only after a proposed test method has been tried, proved, and reduced to unequivocal written form should a determination of its precision and bias be attempted.

7. Planning the Collaborative Test

7.1 Based upon the task group's knowledge of a test method and having the unequivocal writeup, several factors must be considered in planning the collaborative test. The testing variables that must be considered in planning are discussed below. The task group will also have to decide whether to use replicates to estimate precision and assure sufficient data or use the "unit block" or "paired sample" design of Youden **(5)** to estimate precision. Finally, the task group must determine within the resources available, how best to estimate the observed bias of the test method.

7.2 *Testing Variables:*

7.2.1 It is desirable to develop a statement of precision of a test method that indicates the contribution to overall variation of selected causes such as concentration, laboratory, operators, apparatus, and other factors that may or have been shown to have strong effects on the results. Since any test method can be tried in only a limited number of applications, the standard deviation calculated from the results of any planned program can be only an estimate of the universe standard deviation, Σ. For this reason the symbol s (sample standard deviation) is used here. The multiple laboratory operational standard deviation, s_{mlo}, of the planned sample is the result of many sources of variation, both recognized and unknown. The mathematical relationship is:

$$s_{mlo}^2 = s_t^2 + s_m^2$$

$$s_t^2 = s_a^2 + s_l^2 + s_{oo}^2$$

$$s_{oo}^2 = s_p^2 + s_o^2$$

$$s_o^2 = s_c^2 + s_e^2$$

where:

mlo = multiple laboratory operational,
t = overall,
oo = single operator operational,
o = single operator,
a = apparatus,
l = laboratory,
c = concentration of material,
m = sample container material and its preparation,
p = preservation, storage, and transportation, and
e = inherent random error.

Ordinarily, only multiple laboratory operational and single-operator standard deviations are identified in the collaborative test.

7.2.2 Four variables must be included in virtually every planned determination of precision: apparatus, laboratory, operator, and level, because these items cannot be identical for all users of a test method. Definition of *apparatus* and *laboratory* is superfluous. If the apparatus required is unique or of such a kind that no laboratory could be expected to have more than one, laboratory and apparatus variables cannot be separated but must be pooled in the laboratory variable. *Operator* designates the individual analysts who participate in the program in the several cooperating laboratories. *Concentration of material* refers to the amount or concentration measured by the test method, since it is important to know whether precision is constant throughout the range of applicability or if it is related to level. Any other variables that pilot work has shown cannot be controlled closely enough to be insignificant must be added to the above basic ones. Storage time, day of reaction, quality of atmosphere, and sample container material are variables that may be significant.

7.2.2.1 These are the only sources of variability represented in the precision and bias statements resulting from the usual collaborative study. They may not represent the additional influence that can arise from differences in sample splitting, field preservation, transportation, etc., all of which may influence routine analytical results as shown in the general precision definitions in Definitions D 1129. These latter variables should also be included in collaborative tests whenever precision is to be determined at concentrations near the estimated limit of detection in real samples and whenever there is reason to believe that they may contribute significantly to the multiple laboratory operational precision.

7.2.3 *Apparatus*—When the variables to be explored have been defined, decisions must be made as to the number of each variable to be employed. Each must be employed at least in duplicate. Larger numbers, while statistically desirable, increase the labor and expense of the project. All alternative apparatus in each laboratory must be used in the evaluation of a test method. If only usual laboratory apparatus (glassware, oven, balance, etc.) is required, at least two such arrays should be segregated in each cooperating laboratory and each employed by each operator for a series of determinations. (This may seem extreme, but it is a sensible precaution against unsuspected error from a familiar source.)

7.2.4 *Laboratories*—A minimum of six independent laboratories producing six usable sets of data must be involved.

7.2.5 *Operators*—If only single-operator precision is to be determined, there must be a minimum of six operators providing usable data.

7.2.6 *Concentrations of Material*—If pilot work has shown that precision is linear with increasing concentrations of material, at least three concentrations covering the range of the test method should be included. If pilot work suggests that precision is other than constant or linear, more levels should be analyzed. They should also be rather uniformly distributed over the range of the test method.

7.2.6.1 Generally, samples in a collaborative study should not be near the limits of detection (a subjective judgment of the lowest concentration that can be determined with useful accuracy and precision for the intended application of the analytical test method) unless one of the intended uses of the method is at its limit of detection. A proposed method for calculating the limit of detection based on single-operator precision is given by Glaser **(6)**.

7.2.7 *Days*—Generally, day will be a random variable and should be treated as such in the planning of tests and in the statistical analysis of data. However, readings within each laboratory at two quite different times or days are preferable. If time or day has been shown to influence the measurements, close control over the date and time of analysis will be essential. Similar decisions on numbers of other agreed-upon variables must be made up by the task group.

7.2.8 *Media*—The collaborative test must be conducted with at least one representative sample matrix which is the same for all laboratories. This reference matrix should be reproducible by user-laboratories so that they can compare their results with the results of the collaborative study.

7.2.8.1 Typically, a reagent water or synthetic medium is used as the reference matrix. Analytes may be supplied separately as concentrates for addition to this matrix by each laboratory. The information on the reference matrix, its preparation, addition of analytes and analyses should be described briefly in the statements on precision and bias.

7.2.8.2 Additional collaborative testing should also be conducted using other matrices specified in the scope of the test method. It is strongly recommended that at least one environmental sample matrix (other than the reference matrix in 7.2.8.1) should be prepared or obtained from a single source and distributed to all laboratories, to permit calculation of overall and single analyst precision and bias on the same sample matrix. The chairperson of the collaborative study should confirm that each study participant can obtain similar samples of each of the remaining matrix types to be tested. He or she can supply a concentrate of the analyte(s) to be added to these sample matrices sample by each laboratory prior to analyses. By analysis of these sample matrices before and after addition of the analyte-increment, test method bias and recoveries can be measured by difference calculation and compared with recovery from the reference sample matrix, so as to obtain separate measures of test method bias and matrix interferences.

7.2.8.3 The same study design as detailed in Section 9, should be used for all sample matrices. A separate precision and bias statement should be generated for each sample matrix with a brief description of the matrix tested.

7.2.8.4 When studies are available indicating the applicability of the method for matrices untested in 7.2.8.1 and 7.2.8.2 and not meeting the other requirements of this practice, at the discretion of the task group responsible for the test method and the results advisor, and providing the data is analyzed in accordance with Section 10 of this practice, this supporting data may be included in a separate section of the precision and bias statement. A clear but brief description must be included of the matrices, and the test protocol employed. It is the intent of this practice that ultimately, data concerning the precision and bias of the test method in the full range of matrices covered in the scope and analyzed in accordance with Section 10, be made available to the users of the test method.

7.2.9 If pilot work has shown that the sample container must be of a specified material prepared in a specified manner prior to use, the variations in the container as obtained from different sources and prepared by the participants will be a random variable and should be treated as such in planning of the test and in the statistical analysis of the data.

7.2.10 The manner of preservation of the sample, if known to affect the precision and bias of the results, is necessarily incorporated in the method of test. While preservation up to a stipulated maximum period of time is treated as a random variable in planning the tests, it is preferable, as noted in 7.2.7, the readings be made within each laboratory after quite different lengths of storage time.

7.3 *Measurement of Precision:*

7.3.1 *Replicates Studies*—After the variables and number of each have been determined, the number of determinations to be made under each set of conditions can be calculated. This number, called the number of replicates, must satisfy the following inequality:

$$n > 1 + (30/P)$$

where:
n = the number of replicates, and
P = the product of the number of the several variables.
For example, to test at three levels, without duplicate apparatus, with a single operator per laboratory and six laboratories:

$$P = 3 \times 1 \times 1 \times 6 = 18$$
$$1 + 30/18$$
$$n > 2.7$$

Thus for this example, the minimum number of replicates is three.

7.3.2 *Unit Block Studies*—Youden and Steiner **(3)** have summarized the concepts, assumptions, sample preparation, and data analysis techniques associated with the "unit-block" or "paired-sample" approach to collaborative testing (each test determination consists of two similar but not identical concentrations). The minimum number of laboratories using the "paired-sample" concept is higher since only one analysis per sample is made. This single analysis has led to a "nonreplicate" description being given to the design. It is both feasible and practical to determine both multiple laboratory operational and single-operator precision using this approach.

7.3.2.1 The only real difference in treatment of data from a unit block study is the calculations used to estimate the means and standard deviations; these calculations may be found in Youden and Steiner **(7)**. Once developed, these mean and standard deviation estimates may be treated the same as statistics from a study with the usual replicate design. A detailed example with and without raw experimental data is given in Refs. **(8)** and **(9)**, respectively.

7.3.2.2 The value of the nonreplicate design is that the single-operator standard deviation estimates are clearly free of any influence that may cause results from known replicate analyses to be artificially closer together.

7.4 *Measurement of Bias:*

7.4.1 The concept of accuracy comprises both precision and bias (see Definitions D 1129 and Practice E 177). As discussed in Practice E 177, there is not a single form for statement of accuracy that can be universally recommended. Since the accuracy of a measurement process is affected by both random and systematic sources of error, measures of both kinds of errors are needed. The standard deviation is a universal measure of random sources of error (or precision). Bias is a measure of the systematic errors of a test method.

7.4.2 A collaborative test evaluation of bias results in a set of laboratory averages that differ among themselves by more than the within-laboratory precision would predict. The average of the laboratory averages or grand average is characteristic of the procedure. The difference between the true value (however defined) and the grand average is an estimate of the systematic error of the procedure. Each laboratory average differs from the grand average due to calibration differences and other departures from the norm. Therefore, the difference between a laboratory average and the true value is a composite of two errors; the systematic error of the test method as modified by the systematic errors of the individual laboratory as it interprets the writeup and applies to the test method. Even if the true value of a property (concentration) is not known, the systematic error of a laboratory, relative to the consensus of all laboratories, can be determined.

7.4.3 There are three major devices commonly used to test a measurement procedure: (*1*) measurement of known materials, (*2*) comparison with other measurement procedures, and (*3*) comparison with modifications of the given procedure **(10)**. The third device or technique may involve the standard addition technique or the simultaneous analysis of several aliquots of different sizes (for example, 0.5, 1, 1.5, 2, 2.5 units). The task group will select the approach that best suits its needs within the resources available to it.

7.4.4 The most likely task-group approach will be the use of "known" materials. Since reference standards are unlikely to be available, the task group will prepare its samples with added (therefore known) quantities of the constituent(s) being tested. The best available chemical and analytical techniques for preparing, stabilizing, if necessary, storing and shipping the prepared samples should be known within the task group and will not be addressed in this practice. However, if the sample preparation and handling techniques employed for the test are different from those expected to be employed for operational samples subjected to the procedure, those differences and their possible effect, if known, on the precision and bias of the test method as determined by the test must be pointed out in the presentation of the test results.

8. Collaborative Test Design Approval

8.1 The task group chairman (or designee) will summarize the design of the collaborative test except as specified in 1.3, 1.4, 1.5, and 1.6. This summary will include: (*1*) the testing variables included; (*2*) the number of replicates in accordance with 7.3.1 or the number of samples in accordance with the paired sample plan of 7.3.2; (*3*) approach for determining the bias of the test method as exemplified in the collaborative test; (*4*) range of concentration covered and approximate concentration of material in each sample or set; (*5*) the approximate number of laboratories and analysts; and (*6*) the media being tested. The summary should be presented in the form of a letter to the Results Advisor.

8.1.1 As an aid, the task group chairman may use, "Approval of Plans for Interlaboratory Testing," Form A, in Appendix X1 (a completed example is shown in X1.1 to X1.1.1.9).

8.2 Upon review of the plan, the Results Advisor will advise the task group chairman whether the plan meets the requirements of this practice.

8.3 Upon receipt of approval of the collaborative test plan, the task group chairman will conduct the collaborative test.

9. Conducting the Collaborative Test

9.1 The task group will prepare the samples for the collaborative test and ship them to the participants with instructions for the test, a copy of the exact test (if not already supplied), and the participant reporting form (or reporting instructions).

9.1.1 The instructions for collaborative testing must require preliminary work by potential collaborators to familiarize them with the procedure prior to test measurements. This is necessary to ensure that the collaborative tests are made by peer groups and that a "learning experience" is not included in the statistics of the collaborative test. The task group may also develop procedures to qualify prospective collaborators, and this approach is strongly recommended.

9.1.2 As an aid, the task group chairman may use Form B, "Data Report from Individual Laboratories," in Appendix X2 (a completed example is shown in X2.1 to X2.2).

9.2 Sample containers should be clearly marked with a code, informative to the distributors but not informative to the operator. Samples should be sized to supply more than the predetermined number of replicates (with reasonable allowance for pipetting, rinsing, etc.) to allow trial runs. There should be a separate set of samples for each operator so that operators within a laboratory do not draw from a common supply nor consult effectively with one another concerning results. Sample concentrations should not be easily surmised values (1, 5, etc.) or near limits of detection. The assignment of samples to the participating laboratories should be randomized within each concentration level. The above are requirements for statistical independence of results.

9.3 A copy of the test method under investigation, the written instructions for carrying out his part of the program, and the necessary samples for test should be supplied to each operator. No supplementary instructions or explanations such as by telephone or from a task group member within a cooperating laboratory, should be supplied to one participant, if not to all. Distribution should be made from one location, and operator's reports should be submitted to one place, preferably to the same location.

9.4 The written instructions should cover such items as: (*1*) directives for storing and subdividing the sample; (*2*) preparation of sample prior to using the test method; (*3*) number of replicate determinations to be made; (*4*) order of analysis of samples if desired; (*5*) details required on the reporting form; and (*6*) the time limit for return of results.

9.4.1 Laboratories shall be required to report all figures obtained in making measurements, instead of rounding results before recording them. This may result in recording in one or more significant figures than is usual in the report section of test method. Negative numbers should be included. A decision about rounding all data can be made when the final analysis is performed.

9.4.2 The participating laboratories should analyze a minimum of two typical samples in replicate to familiarize themselves with the methodology and demonstrate their attainment of a state of statistical control prior to analyzing the test samples. These data should be sent to the task group chairman along with the data of the collaborative test.

9.5 The task group chairman (or designee) should monitor the collaborative test to assure that the participants who take part, should report the results back within the agreed upon time limit, and should be sure that the results are free of obvious transcription, clerical, or calculational errors. Careful design of the reporting form (or reporting instructions) will facilitate this task.

10. Collaborative Test Data Analysis

10.1 The steps involved by the task group chairman in the data analysis consist of: (*1*) tabulating the data; (*2*) eliminating any laboratories that are so consistently high or low that their results are unreasonable; (*3*) eliminating any individual "outlier" data points; (*4*) calculating the multiple laboratory operational precision, single-operator precision, and bias for each concentration of material in each media studied; (*5*) tabulating the statistics; (*6*) assembling information required for the research report; and (*7*) summarizing these results in a graph for the test method statement.

10.1.1 As an aid, the task group chairman may use sample calculation of precision and bias from real collaborative test data shown in Appendix X3 (a completed example is shown in X3.1 to X3.8).

10.2 *Tabulation of Data*—The data reported by the laboratories may have to be made consistent in reporting units or in the number of reported values per operator or laboratory **(11)**.

10.2.1 A completed example is shown in X3.1 to X3.2.

10.3 *Rejection of Outlier Laboratories*—If one or more laboratory's data are so consistently high or low that there must be a large systematic error specific to that laboratory, all the data from the laboratory should be rejected. Youden **(12)** describes a laboratory ranking test appropriate for such decisions. This test requires calculation of the arithmetic mean from each of the various sets of replicate analyses reported in the study.

10.3.1 If there is another suspected outlier laboratory, it is permissible to repeat this. However, the precision and bias estimate with two or more laboratories deleted may not be representative and reasons for failure should be pursued.

10.3.2 A completed example is shown in X3.3.

10.4 *Initial Statistical Calculations*—For each concentration level in the collaborative study, calculate initial values for the mean recovery ($\bar{x}$) and multiple laboratory operational (s_{mlo}), or overall standard deviation (s_t) as follows:

$$\bar{x} = \left(\sum_{i=1}^{m} \sum_{j=1}^{n_i} x_{ij} \right) \Big/ n$$

$$s_t = \sqrt{\frac{\sum_{i=1}^{m} \sum_{j=1}^{n_i} (x_{ij} - \bar{x})^2}{n-1}}$$

$$= \sqrt{\frac{\sum_{i=1}^{m} \sum_{j=1}^{n_i} x_{ij}^2 - \left(\sum_{i=1}^{m} \sum_{j=1}^{n_i} x_{ij} \right)^2 \Big/ n}{n-1}}$$

where:

i = the index identifying operator and varies from 1 to m at this concentration of material,

j = the replication index and varies from 1 to n_i for operator i at this level,

x_{ij} = the jth retained value reported by operator i at this level, and

n = $\sum_{j=1}^{m} n_i$ or the total number of values retained so far for this level.

10.4.1 A completed example is shown in X3.4 with detailed use of the formulas shown in X5.1 (specifically X5.2.1 to X5.2.3 and X5.2.3.1 to X5.2.3.2).

10.5 *Rejection of Individual Outliers*—Absolutely no data should be discarded unless valid statistical criteria show them clearly to be erroneous or aberrant. Control charts and variance analyses may be misleading in some cases. Statis-

TABLE 1 Critical Values for T (Two-Sided Test) When Standard Deviation is Calculated from the Same Samples (for Outliers)[A]

Number of Observations, n	10 % Significance Level	5 % Significance Level	2 % Significance Level	1 % Significance Level
3	1.15	1.15	1.15	1.15
4	1.46	1.48	1.49	1.50
5	1.67	1.71	1.75	1.76
6	1.82	1.89	1.94	1.97
7	1.94	2.02	2.10	2.14
8	2.03	2.13	2.22	2.27
9	2.11	2.21	2.32	2.39
10	2.18	2.29	2.41	2.48
11	2.23	2.36	2.48	2.56
12	2.29	2.41	2.55	2.64
13	2.33	2.46	2.61	2.70
14	2.37	2.51	2.66	2.75
15	2.41	2.55	2.70	2.81
16	2.44	2.58	2.75	2.85
17	2.47	2.62	2.78	2.89
18	2.50	2.65	2.82	2.93
19	2.53	2.68	2.85	2.97
20	2.56	2.71	2.88	3.00
21	2.58	2.73	2.91	3.03
22	2.60	2.76	2.94	3.06
23	2.62	2.78	2.96	3.09
24	2.64	2.80	2.99	3.11
25	2.66	2.82	3.01	3.13
30	2.75	2.91	3.10	3.24
35	2.82	2.98	3.18	3.32
40	2.87	3.04	3.24	3.38
45	2.92	3.08	3.29	3.43
50	2.96	3.13	3.34	3.48
60	3.03	3.20	3.41	3.56
70	3.09	3.26	3.47	3.62
80	3.14	3.30	3.52	3.67
90	3.18	3.35	3.56	3.72
100	3.21	3.38	3.60	3.75

[A] Values of T for $n \leq 25$ are based on those given in Ref. **(13)**. For $n > 25$, the values of T are approximated. All values have been adjusted for division by $n - 1$ instead of n in calculating s. Tabulated values come from Practice E 178 and may also be found in Grubbs **(14)**. Levels of significance shown in Practice E 178 were doubled, since this is a two-sided test for significance instead of a one-sided test.

tical references provide valid criteria for excluding data from precision evaluations. When the experimenter is aware that a gross deviation from prescribed experimental procedure has occurred, the resultant observation should be discarded. Otherwise the most extreme value among the data at each concentration of material may be tested by calculating its T value.

10.5.1 If the T value is greater than the critical value recorded in Table 1 at the selected significance level (see Practice E 178), the outlier may be rejected (the selection of the significance level is left up to the collaborative study chairman). The test criterion for the suspected outlier, x_{ij}, is as follows:

$$T = (x_{ij} - \bar{x})/s_t$$

where $\bar{x}$ and s_t are the current estimates of the mean and overall standard deviation for all retained data at the concentration of material associated with x_{ij}.

10.5.1.1 Because the direction of the value being tested is not known beforehand and because the experimenter is interested in detecting values that could be either high or low, a two-sided test is being performed (see page 5 in treatise by Grubbs **(15)** for detailed explanation).

10.5.2 If the suspected outlier fails the test in 10.5.1, recalculate the $\bar{x}$ and s_t from the remaining data at this concentration of material.

10.5.3 If there is another suspected value among the remaining data for a specific concentration level a second iteration of steps 10.5.1 may be justified but is generally not recommended.

10.5.4 A completed example is shown in X3.5.

10.6 *Calculation of Single-Operator Precision*—For each concentration level in the collaborative study, calculate the single-operator standard deviation (s_o) from the data after removal of all outliers, as follows:

$$s_o = \sqrt{\frac{\sum_{i=1}^{m} (n_i - 1)s_{oi}^2}{n - m}}$$

where:

$$s_{oi}^2 = \frac{\sum_{j=1}^{n_i} x_{ij}^2 - \left(\sum_{j=1}^{n_i} x_{ij}\right)^2 \Big/ n_i}{n_i - 1}$$

and all the other variables are defined as in 10.4.

10.6.1 A completed example is shown in X3.6 with detailed use of the formulas shown in X5.1 (specifically X5.2.1 to X5.2.3 and X5.2.3.3).

10.7 Since precision will usually vary significantly at the different concentrations of materials, make plots of overall precision, s_t, and single-operator precision, s_o, versus the mean recovery, $\bar{x}$.

10.7.1 A plot of s_t versus $\bar{x}$ will indicate quickly any significant relationship. Unless several plotted points fit a smooth curve of considerable slope, there is no significant relationship requiring mention.

10.7.2 A completed example is shown in X3.7.

10.7.3 The results of a single operator should agree more closely than those between operators or laboratories. If the precision of a single-operator (s_o) is relatively small compared to that contributed by other sources, this fact will be helpful to users of the method and must be conveyed to them. Having only s_t as a guide, such a user might be satisfied with a degree of agreement among his own results that would be mediocre or suspicious. The difference is worth mentioning in the method, however, only if the value for s_o is less than 0.8 that for s_t.

10.8 *Calculation of Bias:*

10.8.1 The calculation of the bias of a test method will logically follow the collaborative test design (7.3). The usual collaborative test technique will involve reporting the recovery of added (therefore known) amounts of the constituent being analyzed.

10.8.2 The calculation of bias from recovery study is made as follows:

$$\text{Bias} = \text{average amount found} - \text{amount added}$$

10.8.3 Where other types of studies are used to develop a value for the test method bias, logical calculational techniques should be used. Consultation with the Results Advisor may be appropriate in such cases.

10.8.4 The reporting of a test method bias should be preceded by a two-tailed t-test for significance at the 5 % level ($\alpha = 0.05$), such as given by Natrella **(16)**.

10.8.5 A completed example is given in X3.8.

10.9 *Information Required for Research Report*—Refer to Appendix X4 for instructions.

11. Statement Format

11.1 *Precision Statement:*

11.1.1 Having the precision values versus the concentrations of material, draw a smooth curve through the points. The precision will be found to be (*1*) constant and not vary with level, (*2*) vary directly with level linearly, or (*3*) vary with level in a curvilinear manner.

11.1.2 If the precision is constant over the tested concentration range, state:

The precision of this test method within its designated range may be expressed as follows:

$$s_t = \_\_\_\_$$

where s_t = overall precision expressed in mg/L, etc., of ____.

11.1.3 Consult Method D 596 for other appropriate units.

11.1.4 If the precision varies linearly or curvilinearly with the concentrations of material, state:

The overall precision of this test method within its designated range varies with the quantity being tested according to the following figure: (plot precision, s_t, on the *y* axis and test level on the *x* axis).

11.1.5 The format for statements of single-operator precision is the same as for overall precision and will be included in the same statement where both are to be reported.

11.1.6 The format for statements of precision for special types of tests such as radioactivity measurements and monitoring methods shall be consistent with techniques for measuring appropriate to those methods.

11.1.7 A brief note should give the reader of the test method an understanding of the size of the collaborative test. It should point out the number of laboratories, concentrations, replicates, and other highly significant aspects of the test. The note should conclude with reference as to which version of Practice D 2777 was used to develop the precision and bias estimates.

11.1.8 A brief note must be included to give the reader a description of the kinds of water that were used to obtain the precision. An example follows:

11.1.8.1 These collaborative test data were obtained on reagent grade water and ____ waters. For other matrices, these data may not apply.

11.1.9 A brief note should give the reader a description of the kind of total precision obtained. Two examples follow:

11.1.9.1 The precision obtained by the interlaboratory study conducted for this test method was multiple laboratory operational, s_{mlo}, since natural samples, after collection, (with or without spiking before splitting) were split in the field under conditions normal for the type of sample. These estimates include the increase in variability and the potential change in the bias that may be attributed to the field splitting and transportation activity customarily common to the application of the method to samples originating outside of, and destined for more than one laboratory.

11.1.9.2 The precision obtained by the interlaboratory study conducted for this test method was overall, s_t, since standardized samples of the matrix of choice, prepared and split in the laboratory were used rather than natural samples collected and split at the source. These estimates *do not* include the increase in variability and the potential change in bias that may be attributed to the field splitting and transportation activity customarily common to the application of the method to samples originating outside of, and destined for more than one laboratory.

11.2 *Bias Statement:*

11.2.1 The bias statement will be developed from the data generated in the collaborative test. Some illustrative examples follow:

11.2.2 *Prepared Standards*—Recoveries of known amounts of ____ in a series of prepared standards were as follows:

Amount Added, mg/L	Amount Found, mg/L	± Bias	± % Bias	Statistically Significant
10.0	...	...	...	...
50.0	...	...	...	...
100.0	...	...	...	...

The bias will be either positive or negative depending upon whether more material or less material is recovered than added. The blank or background value will have already been subtracted prior to making this calculation. The percentage (± % bias) value may be useful in normalizing the data due to the range of concentrations.

11.2.2.1 A graph of amount found on the *y*-axis against that added on the *x*-axis shall be given with the overall precision clearly shown. This graph will be useful in estimating the bias of the test method throughout its applicable concentration range.

11.2.2.2 A brief note should give the reader a description of the kinds of water that were used to obtain the bias. An example follows:

11.2.2.3 These collaborative test data were obtained on reagent grade water and ____ waters. For other matrices, these data may not apply.

11.2.3 *Reference Method*—The determined values for ____ content averages ____ percent lower than the expected values based on the ____ content of the spiked samples of ____ representative waters (see scope of test method), as determined by Test Method D ____. This was true for all three levels (____, ____, and ____ mg/L) used in the interlaboratory test.

11.2.4 *Standard Sample*—The average value obtained in the analysis of a reference sample was ____ versus a value of ____. Or, the average value obtained in the analysis of a sample of ____ was ____ versus a theoretical value of ____.

11.3 *Test Methods Approved Before or During the Year This Practice is Adopted:*

11.3.1 For these test methods, add the following statement:

11.3.1.1 This test method meets requirements specified in Practice D 2777 – 77.

11.4 *Test Methods with Numerical Measure:*

11.4.1 When a test method results in a numerical measure, prepare precision and bias statements as directed in 11.1 and 11.2 using data from six or more laboratories.

11.5 *Test Methods with Nonnumerical Report:*

11.5.1 When a test method specifies that a test result is a nonnumerical report of success or failure based on criteria in the procedure, the statement on precision and bias should read as follows.

11.5.1.1 *Precision and Bias*—No statement is made about either the precision or the bias of Test Method D ____ for measuring (insert here the name of property) since the result merely states whether there is conformance to the criteria for success specified in the procedure.

11.6 *Test Methods Specifying Other Procedures:*

11.6.1 When a test method specifies that the procedures in another ASTM test method are to be used, a statement such as the following should be used to assure the user that precision and bias statements apply.

11.6.1.1 *Precision and Bias*—The precision and bias of this test method of measuring (insert here the name of the property) are as specified in Test Method (insert here the designation of the other test method).

12. Approval of Data Analysis and Statements

12.1 Approval of the precision and bias statement must be obtained from the Results Advisor, at the same time providing him with a copy of:

12.1.1 All test data resulting from the collaborative tests,

12.1.2 All statistical calculations,

12.1.3 Any background information that may have influenced the results, and

12.1.4 Information required for research report.

12.2 *Experimental Data*—When a proposed test method is submitted, the collaborative data used in determining its precision and bias must be placed in the Research Reports file at ASTM. The test method must then carry a footnote indicating where the supporting data can be found. The footnote must read as in the following example:

> Supporting data for the precision and bias statements have been filed at ASTM Headquarters. Request RR:D ____-____.

APPENDIX

(Nonmandatory Information)

X1. APPROVAL OF STUDY DESIGN

X1.1 *Sample of Form A—Approval of Plans for Interlaboratory Testing*[5]

X1.1.1 The task group chairman shall submit the following information to the Results Advisor for approval before proceeding with interlaboratory testing of test methods for chemical analysis of water.

X1.1.1.1 Task group-approved draft of the test method in ASTM format *Determination of Chromium in Water by Atomic Absorption Spectrophotometry, D 2576.*

NOTE X1.1—This test method is not included in this example.

X1.1.1.2 Copy of plans for developing study samples and instructions for the participating laboratories. *See attached instructions.*

NOTE X1.2—The attached directions are not included in this example.

X1.1.1.3 Number of laboratories involved in the study __7__.

NOTE X1.3—Practice D 2777 calls for a minimum of six usable sets of data from six laboratories, each having one operator perform the analysis at three levels with triplicate analyses at each level.

X1.1.1.4 Number of operators involved in the study (six minimum) __8__.

X1.1.1.5 Number of times (three is minimum for six operators) each operator is to replicate the analyses at each level __3__.

NOTE X1.4—For 10 or more laboratories at three concentration levels, each operator shall be required to do a minimum of two replicate analyses.

X1.1.1.6 Number of concentration levels (three minimum) __4__.

X1.1.1.7 Number of samples issued __32__.

X1.1.1.8 Concentration ranges covered __0.2 to 1.0__, __1.0 to 4.0__, __4.0 to 7.0__, __7.0 to 10.0__. Units of concentration (spelled out) *milligrams per litre.*

X1.1.1.9 Kinds of water (two minimum) *Type II* reagent water and *wastewater.*

NOTE X1.5—Only the Type II reagent water data are shown in the following example.

[5] Full size, 8½ by 11 in. copies of blank forms A and B, and the form for information required for Research Reports (Appendix X4) are available from ASTM Headquarters, 1916 Race St., Philadelphia, PA 19103.

X2. REPORTING OF STUDY DATA

X2.1 *Sample of Form B—Data Report from Individual Laboratories*

X2.2 After the collaborative test plans have been approved by the Results Advisor, the Task Group Chairman shall conduct the study and have each participating laboratory report their data on a form similar to Form B.

TABLE X2.1 Sample of Completed Form B For Data Report from Individual Laboratories

Complete Title of Test Method, Draft Number, and Date of Draft: Determination of Chromium in Water by Atomic Absorption Spectrophotometry, D 2576, Draft No. 5, August 5, 1975

	Reagent Water[A]							Water Matrix of Your Choice[A]					
	Chromium, milligrams per litre			Found[B]								Found[B]	
	Operator							Operator					
Sample	1			2			Sample	1			2		
Number	Replicate Analysis						Number	Replicate Analysis					
	1	2	3	1	2	3		1	2	3	1	2	3
1	0.70	0.70	0.70	0.75	0.75	0.75	1						
2	2.30	2.30	2.4	2.4	2.4	2.4	2						
3	5.40	5.40	5.50	5.60	5.60	5.60	3						
4	8.5	8.6	8.7	8.7	8.7	8.7	4						

[A] Specify type of reagent water: Type II
Specify water matrix used:
Specify background concentration of constituent being tested in water matrix used: 0.02 milligrams per litre of chromium

[B] Insert concentration units and material sought: milligrams per litre of chromium found
Insert laboratory code no. A1; A2
Do not list laboratory identification on this sheet.

X3. SAMPLE CALCULATION OF PRECISION AND BIAS

X3.1 The following is a sample calculation of precision and bias from collaborative (round-robin) test data *for one water* (see 10.1). These procedures must be followed for *each* water requiring precision and bias statements.

X3.1.1 The example shows calculations for triplicate results on a single day from six of seven scheduled laboratories analyzing for known concentrations of chromium in Type II Reagent Water. One laboratory used two operators. For simplicity, these two operators are considered as separate laboratories.

X3.2 Arrange the test data in tabular form as shown in Table X3.1. Test data should be reported to the same number of significant figures (see 10.2).

X3.3 Test the data for outlying laboratories (see 10.3). The laboratory ranking test is commonly used if visual inspection of the data shows an apparent large discrepancy.

X3.3.1 Calculate the arithmetic mean from each of the various sets of triplicate analyses. Within each concentration, rank these means from highest to lowest. When ties occur, the same average rank is assigned to each laboratory in the tie. Sum the ranks for each laboratory and compare these ranks (Table X3.2) with the appropriate critical values obtained from Table X3.3. If the rank for any laboratory falls

TABLE X3.1 Tabulation of Collaborative Test Data (Data for Determination of Chromium in Water by Atomic Absorption Spectrophotometry, D 2576) Concentrations, mg/L

Laboratory	Amount Added			
	0.70	2.6	5.8	8.75
	Found			
A1	0.70	2.30	5.40	8.5
	0.70	2.30	5.40	8.6
	0.70	2.4	5.50	8.7
A2	0.75	2.4	5.60	8.7
	0.75	2.4	5.60	8.7
	0.75	2.4	5.60	8.7
B	0.65	2.40	5.40	8.70
	0.62	2.50	5.50	8.70
	0.63	2.50	5.50	8.80
C	no data reported			
D	0.56	2.06	4.93	8.38
	0.56	2.04	4.98	8.20
	0.54	2.04	4.95	8.29
E	0.60	2.60	6.0	9.7
	0.60	2.50	6.0	9.8
	0.50	2.50	6.1	9.8
F	0.65	2.6	7.0	10.9
	0.65	2.8	6.8	10.3
	0.65	2.7	6.8	10.3
G	0.50	2.00	5.00	8.20
	0.53	2.00	5.00	8.00
	0.52	2.10	5.00	8.00

TABLE X3.2 Rejection of Outlier Laboratories

Laboratory	Average Results, mg/Litre by Concentration				Ranked Results by Concentration				Cumulative Score for each Laboratory
	1	2	3	4	1	2	3	4	
A1	0.70	2.33	5.43	8.60	2	5	5	5	17
A2	0.75	2.40	5.60	8.70	1	4	3	4	12
B	0.63	2.47	5.47	8.73	4	3	4	3	14
D	0.55	2.05	4.95	8.29	6	6	7	6	25
E	0.57	2.53	6.03	9.77	5	2	2	2	11
F	0.65	2.70	6.87	10.50	3	1	1	1	6[A]
G	0.52	2.03	5.00	8.07	7	7	6	7	27[A]

[A] Since none of the cumulative scores fall outside the limits from Table X3.3, test data from all laboratories are accepted for further analysis.

TABLE X3.3 Approximate Limits (Two-Sided Test) for Ranking Scores at a 5 % Level of Significance

No. of Laboratories	Number of Materials 3	4	5	6	7	8	9	10	11	12	13	14	15
3	...	4	5	7	8	10	12	13	15	17	19	20	22
		12	15	17	20	22	24	27	29	31	33	36	38
4	...	4	6	8	10	12	14	16	18	20	22	24	26
		16	19	22	25	28	31	34	37	40	43	46	49
5	...	5	7	9	11	13	16	18	21	23	26	28	31
		19	23	27	31	35	38	42	45	49	52	56	59
6	3	5	7	10	12	15	18	21	23	26	29	32	35
	18	23	28	32	37	41	45	49	54	58	62	66	70
7	3	5	8	11	14	17	20	23	26	29	32	36	39
	21	27	32	37	42	47	52	57	62	67	72	76	81
8	3	6	9	12	15	18	22	25	29	32	36	39	43
	24	30	36	42	48	54	59	65	70	76	81	87	92
9	3	6	9	13	16	20	24	27	31	35	39	43	47
	27	34	41	47	54	60	66	73	79	85	91	97	103
10	4	7	10	14	17	21	26	30	34	38	43	47	51
	29	37	45	52	60	67	73	80	87	94	100	107	114
11	4	7	11	15	19	23	27	32	36	41	46	51	55
	32	41	49	57	65	73	81	88	96	103	110	117	125
12	4	7	11	15	20	24	29	34	39	44	49	54	59
	35	45	54	63	71	80	88	96	104	112	120	128	136
13	4	8	12	16	21	26	31	36	42	47	52	58	63
	38	48	58	68	77	86	95	104	112	121	130	138	147
14	4	8	12	17	22	27	33	38	44	50	56	61	67
	41	52	63	73	83	93	102	112	121	130	139	149	158
15	4	8	13	18	23	29	35	41	47	53	59	65	71
	44	56	67	78	89	99	109	119	129	139	149	159	169

outside these critical values, data from that laboratory are not included in any subsequent calculations.

X3.4 Calculate initial estimates of the mean recoveries ($\bar{x}$'s) and overall standard deviations (s_t's) from the retained data in Table X3.1 as specified in 10.4. Enter the results into Table X3.4.

X3.5 Test for individual outliers within the data at each level by following 10.5. For sets of 21 observations, the critical value from Table 1 of the practice is 2.73 at the 5 % significance level. The T value calculated from the most extreme value at each level must be larger than this critical value before the extreme value may be rejected as an individual outlier. The remainder of Table X3.4 summarizes the process.

X3.5.1 If any of the extreme values had been discarded, the mean ($\bar{x}$) and overall standard deviation (s_t) for that level would have been recalculated from the remaining 20 observations. It would also be permissible to test the next most extreme value as a possible individual outlier by repeating X3.5.

TABLE X3.4 Initial Statistics and Rejection of Individual Outliers

Amount Added	0.70	2.6	5.8	8.75
Observations	21	21	21	21
Standard deviation, (s_t)	0.082	0.238	0.628	0.833
Mean ($\bar{x}$)	0.624	2.36	5.62	8.95
Most different value from Table X3.1	0.75	2.8	7.0	10.9
Largest difference	0.13	0.44	1.38	2.05
T = (difference)/s_t	1.59	1.85	2.20	2.46
Critical T value from Table 1	2.73	2.73	2.73	2.73
Reject	no	no	no	no

TABLE X3.5 Final Statistics

Concentration ($\bar{x}$)	0.624	2.36	5.62	8.95
s_t	0.082	0.238	0.628	0.833
s_o	0.025	0.059	0.061	0.152

TABLE X3.6 Determination of Bias

Amount Added, mg/L	Mean Recovery ($\bar{x}$), mg/L	± Bias	± % Bias	Statistically Significant at the 5 % Level ($\alpha = 0.05$)
0.70	0.624	−0.076	−11%	yes
2.6	2.36	−0.24	−9%	yes
5.8	5.62	−0.18	−3%	no
8.75	8.95	+0.20	+3%	no

TABLE X3.7 Calculations to Determine Significance of the Bias in Table X3.6

Line (l)	Factor	Data			
1	Amount added, A	0.70	2.6	5.8	8.75
2	Amount found, $\bar{x}$	0.624	2.36	5.62	8.95
3	Number of observations, n	21	21	21	21
4	Standard deviation, s_t	0.082	0.238	0.628	0.833
5	Standard deviation of mean, $S_{\bar{x}} = (s_t/\sqrt{n})$ or $l4 \div \sqrt{l3}$	0.018	0.052	0.137	0.182
6	Critical Student's t value from Table X3.8 for 21 observations or 20 degrees of freedom	2.09	2.09	2.09	2.09
7	Observed difference, $(A - \bar{x})$ or $(l1 - l2)$	0.076	0.24	0.18	0.20
8	Observed Student's t value, $(A - \bar{x})/s_{\bar{x}}$	4.22	4.61	1.31	1.10
9	Significant: Yes if $l_8 > l_6$ No if $l_8 < l_6$	yes	yes	no	no

X3.5.2 Since no individual value can be rejected, the entire set of data for each concentration level is used for subsequent calculations and the mean ($\bar{x}$) and overall standard deviation (s_t) estimates in Table X3.4 can be entered among the final statistics in Table X3.5.

X3.6 Calculate the single-operator precision (s_o) from the retained data in Table X3.1, in accordance with 10.6, and enter the results in Table X3.5.

X3.7 Since the overall and single-operator standard deviations in Table X3.5 both obviously increase with concentration, plot s_t and s_o versus $\bar{x}$ as shown in Fig. X3.1 (in accordance with 10.7).

X3.7.1 To introduce the precision plot, prepare an appropriate precision statement in accordance with 11.1.

X3.8 In accordance with 10.8, calculate the bias of the method as shown in Table X3.6 and test the significance of these bias estimates as demonstrated in Table X3.7.

TABLE X3.8 Critical Student's "t" Test (Two-Sided Test) Values at a 5 % Level of Significance

Number of Observations	Degrees of Freedom	Critical t Value at $\alpha = 0.05$
2	1	12.706
3	2	4.303
4	3	3.182
5	4	2.776
6	5	2.571
7	6	2.447
8	7	2.365
9	8	2.306
10	9	2.262
11	10	2.228
12	11	2.201
13	12	2.179
14	13	2.160
15	14	2.145
16	15	2.131
17	16	2.120
18	17	2.110
19	18	2.101
20	19	2.093
21	20	2.086
22	21	2.080
23	22	2.074
24	23	2.069
25	24	2.064
26	25	2.060
27	26	2.056
28	27	2.052
29	28	2.048
30	29	2.045
Infinite	Infinite	1.960

X3.8.1 Prepare an appropriate bias statement as directed in 11.2. For these data, it would be sufficient to briefly introduce a table just like Table X3.6. In addition, it might be of interest to plot $\bar{x}$ versus the amount added, such as in Fig. X3.2.

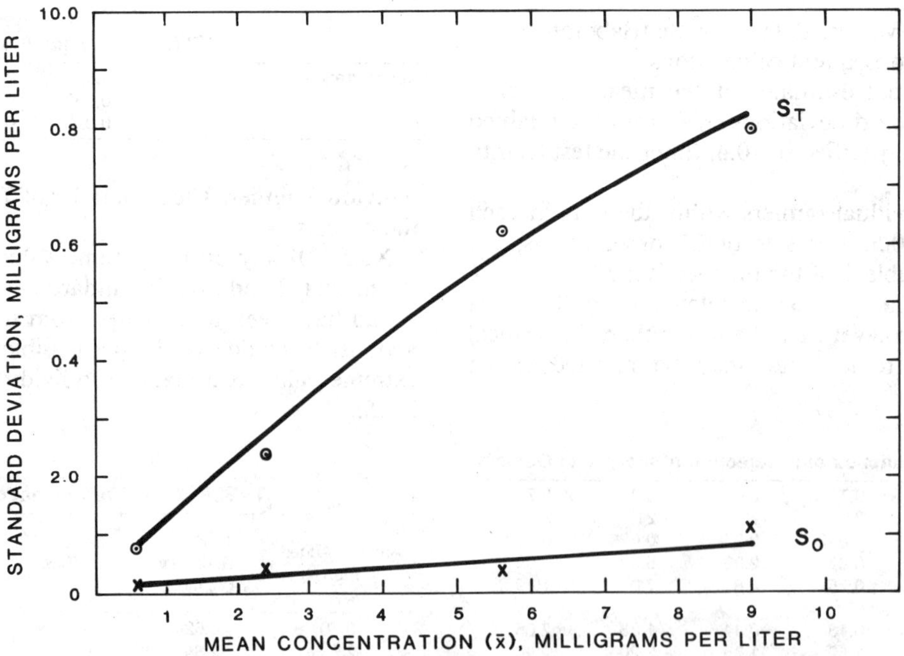

FIG. X3.1 Overall and Single-Operator Standard Deviations Against Mean Concentration in Milligrams per Litre for Interlaboratory Experimental Recovery by Atomic Absorption, Test Method D 2576, of Chromium from Reagent Water

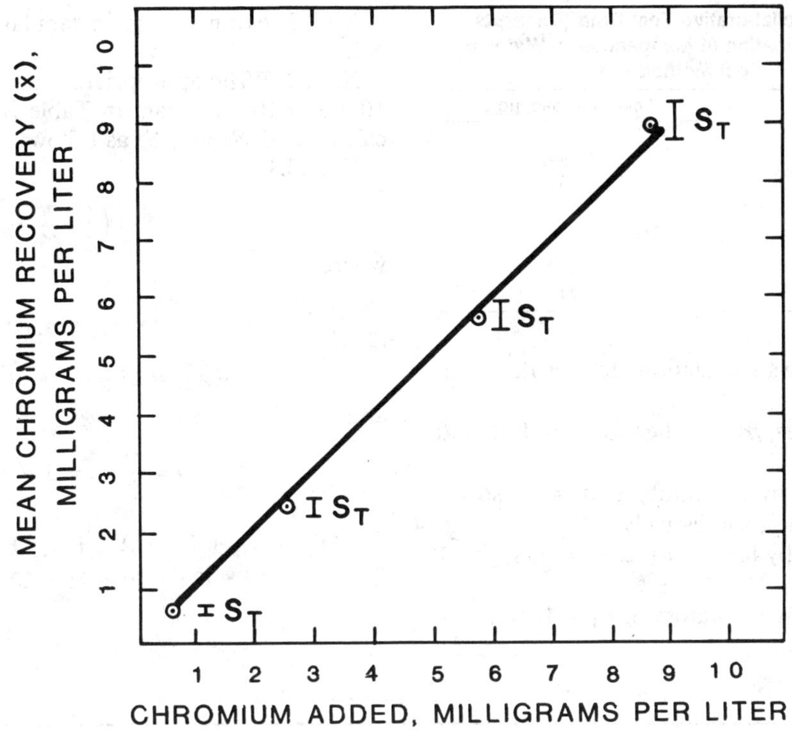

FIG. X3.2 Mean Chromium Recovery Against Concentration added with Overall Standard Deviation Shown in Milligrams per Litre for Interlaboratory Experimental Recovery by Atomic Absorption, Test Method D 2576, of Chromium from Reagent Water

X4. SAMPLE INFORMATION REQUIRED FOR RESEARCH REPORT

X4.1 After calculating the precision and bias results, the task group chairman shall assemble and complete information required for research report in a format similar to that shown as follows.

X4.2 Draft of standard for which research report is being submitted is attached.

X4.3 Signed approval by Results Advisor that advice on statistical analyses outlined in Form A of Practice D 2777 were followed and that round robin tests meet Practice D 2777 requirements.

X4.4 Details on preparation of round robin samples: Preparation of standard solutions:

Source[A] ___ Composition ___ Concentra tion ___
Final concentration ___ Dilutions ___
Preservative ___ Concentration ___

[A] Attached page lists particulars concerning composition of commercial source, date, lot no., and address of supplier.

X4.5 Letter sent to participants, which includes instructions, number of samples, replicates, dilutions and any special instructions.

X4.6 Name of laboratories which participated:

____

____

____

____

____

____

X4.7 Report Form B, showing data from individual laboratories.

X4.8 Report Form C, showing completion of results from the different laboratories (Tables X3.1, X3.5, and X3.6).

X4.9 Precision and bias statements in the standard, including figures if required.

X5. SAMPLE CALCULATION OF OVERALL STANDARD DEVIATION (S_t) AND SINGLE-OPERATOR STANDARD DEVIATION (S_o)

X5.1 The following is a sample calculation of the overall standard deviation (S_t) and single-operator standard deviation (S_o) from excerpts of collaborative (round-robin) test data for one water to illustrate the use of the compact formulas for S_t in 10.4 and S_o in 10.6 in their respective expanded forms.

TABLE X5.1 Tabulation of Collaborative Test Data (Excerpts of Collaborative Data for Determination of Manganese in Water by Chelation-Extraction, Test Method D 858)

Laboratory		Amount Added, μg/L
i	n_i	20
		Found
1	3	25
		20
		21
2	2	20
		20
3	1	24

X5.2.1 The example shows calculations for results (with terms defined in 10.4):

X5.2.1.1 For 3 laboratories, $m = 3$, therefore $i = 1, 2$, and 3, successively

X5.2.1.2 For 3 replicates by laboratory 1, $n_1 = 3$, so for this laboratory, $j = 1, 2$, and 3, successively

X5.2.1.3 For 2 replicates by laboratory 2, $n_2 = 2$, so $j = 1$ and 2, successively

X5.2.1.4 For 1 replicate by laboratory 3, $n_3 = 1$, so $j = 1$

X5.2.2 Arrange data in tabular form as shown in Table X5.1.

X5.2.3 Write appropriate expanded forms of equations in 10.4 and 10.6 for data in Table X5.1, insert the values and calculate $\bar{X}$, S_t, and S_o as follows:

X5.2.3.1

$$\bar{X} = \left(\sum_{i=1}^{m}\sum_{j=1}^{n_i} X_{ij}\right) \Big/ n \quad \text{(Eq X5.1)}$$

where:

$$\sum_{i=1}^{m}\sum_{j=1}^{n_i} X_{ij} = X_{11} + X_{12} + X_{13} + X_{21} + X_{22} + X_{31} = 25 + 20 + 21 + 20 + 20 + 24 \quad \text{(Eq X5.2)}$$

and:

$$n = \sum_{j=1}^{m} n_i = 3 + 2 + 1 = 6 \quad \text{(Eq X5.3)}$$

$$\bar{X} = [X_{11} + X_{12} + X_{13} + X_{21} + X_{22} + X_{31}]/6 = [25 + 20 + 21 + 20 + 20 + 24]/6 = [130]/6 = 21.6667$$

X5.2.3.2 (Eq X5.4)

$$S_t = \sqrt{\left[\sum_{i=1}^{m}\sum_{j=1}^{n_i}(X_{ij} - \bar{X})^2\right]} \Big/ \sqrt{(n-1)} = \frac{\sqrt{\begin{bmatrix}(X_{11} - \bar{X})^2 + (X_{12} - \bar{X})^2 \\ + \cdots + (X_{31} - \bar{X})^2\end{bmatrix}}}{\sqrt{[6-1]}} = \frac{\sqrt{\left[\sum_{1=1}^{m}\sum_{j=1}^{n_i} X_{ij}^2 - \left(\sum_{i=1}^{m}\sum_{j=1}^{n_i} X_{ij}\right)^2 \Big/ n\right]}}{\sqrt{(n-1)}}$$

$$= \frac{\sqrt{\begin{bmatrix}(X_{11})^2 + (X_{12})^2 + \cdots + (X_{31})^2 \\ -(X_{11} + X_{12} + \cdots + X_{31})^2/(6)\end{bmatrix}}}{\sqrt{[6-1]}} = \frac{\sqrt{\begin{bmatrix}(25)^2 + (20)^2 + (21)^2 + (20)^2 + (20)^2 \\ + (24)^2 - (25 + 20 + 21 + 20 + 24)^2/6\end{bmatrix}}}{\sqrt{5}} = \frac{\sqrt{\begin{bmatrix}(625) + (400) + (441) + (400) \\ + (400) + (576) + (130)^2/6\end{bmatrix}}}{\sqrt{5}}$$

$$= \frac{\sqrt{[(2842) - (2816.667)]}}{2.236068} = \frac{5.03319}{2.236068} = 2.251 \simeq 2.3$$

X5.2.3.3 (Eq X5.5)

$$S_{oi}^2 = \frac{\sum_{j=1}^{n_i} X_{ij}^2 - \left(\sum_{j=1}^{n_i} X_{ij}\right)^2 \Big/ n_i}{n_i - i}$$

$$S_{o1}^2 = \frac{[(X_{11})^2 + (X_{12})^2 + (X_{13})^2 - (X_{11} + X_{12} + X_{13})^2/(3)]}{[3-1]} = \frac{[(25)^2 + (20)^2 + (21)^2 - (25 + 20 + 21)^2\,(3)]}{[2]} = \frac{[1466 - 1452]}{2} = 7.00$$

$$S_{o2}^2 = \frac{[(X_{21})^2 + (X_{22})^2 - [(X_{21} + X_{22})^2/(2)]}{[2-1]} = \frac{[(20)^2 + (20)^2 - (20 + 20)^2/(2)]}{[1]} = \frac{0}{1} = 0$$

$$S_{o3}^2 = \frac{[(X_{31})^2 - (X_{31})^2]/[(1)]}{[1-1]} = \frac{[(24)^2 - (24)^2]/[(1)]}{[1-1]} = \frac{0}{0} = \text{not defined}$$

X5.2.3.4 (Eq X5.6)

$$S_o = \frac{\sqrt{\left[\sum_{i=1}^{m}(n_i - 1)\,S_{oi}^2\right]}}{\sqrt{(n-m)}} = \frac{\sqrt{[(3-1)\,S_{o1}^2 + (2-1)\,S_{o2}^2]}}{\sqrt{(5-2)}} = \frac{\sqrt{[(2)\,(7) + (1)\,(0)]}}{\sqrt{3}} = \frac{3.74166}{1.73205} = 2.160 \simeq 2.2$$

REFERENCES

(1) ASTM Circular Letter No. 587 dated May 20, 1975.

(2) *ASTM Manual for Conducting an Interlaboratory Study of a Test Method, ASTM STP 355*, ASTM, 1964, p. 1.

(3) Youden, W. J., and Steiner, E. H., *Statistical Manual of the Association of Official Analytical Chemists*, Association of Official Analytical Chemists, Washington, DC, 1975, pp. 33–36.

(4) *ASTM Manual on Quality Control of Materials*, ASTM *STP 15-C*, ASTM, 1951, pp. 55–118.

(5) Youden, W. J., and Steiner, E. H., *ibid.*, p. 27.

(6) Glaser, J. A., Foerst, D. L., McKee, G. D., Quave, S. A., and Budde, W. L. "Trace Analyses for Wastewaters," *Environmental Science and Technology*, Vol 15 (No. 12), December 1981, pp. 1426–1435.

(7) Youden, W. J., and Seiner, E. H. *op. cit.*, pp. 21–26.

(8) Winter, J., Britton, P., Clements, H., and Kroner, R., "EPA Method Study 8, Total Mercury in Water," *EPA-600/4-77-012*, Environmental Monitoring and Support Laboratory, Office of Research and Development, U.S. Environmental Protection Agency, Cincinnati, OH, February 1977, p. 76.

(9) Winter, J. A., and Clements, H. A., "Interlaboratory Study of the Cold Vapor Technique for Total Mercury in Water," *Symposium on Water Quality Parameters, ASTM STP 573*, ASTM, 1975, pp. 556–580.

(10) Youden, W. J. "How to Evaluate Accuracy," *Precision Measurement and Calibration—Statistical Concepts and Procedures*, NBS Special Publication 300, Vol 1, Edited by H. H. Ku, Superintendent of Documents, U.S. G.P.O. Washington, DC, 1969, p. 363.

(11) Steiner, E. H., "Planning and Analysis of Results of Collaborative Results," Youden, W. J., and Steiner, E. H., *op. cit.*, pp. 73–74.

(12) Youden, W. J., *op. cit.*, pp. 31–33.

(13) Grubbs, F. E., "Sample Criteria for Testing Outlying Observations," *Annals of Mathematical Statistics*, Vol 21, March 1950, pp. 27–58.

(14) Grubbs, F. E., and Beck, G., "Extension of Sample Sizes and Percentage Points for Significance Tests of Outlying Observations," *Technometrics*, Vol 14, No. 4, November 1972, pp 847–854.

(15) Grubbs, F. E., "Procedures for Detecting Outlying Observations in Samples," *Technometrics*, Vol. 11 (No. 1), February 1969, p. 4.

(16) Natrella, M. G., *Experimental Statistics, National Bureau of Standards Handbook 91*, Superintendent of Documents, U.S. G.P.O., Washington, DC 20025, 1963, pp. 3-1–3-8.

(17) Steiner, E. H., *op. cit.* p. 33.

(18) Miller, I. and Freund, J. E., *Probability and Statistics for Engineers*, Prentice-Hall, Inc., Englewood Cliffs, NJ, 1965, p. 168.

Standard Guide for Measuring Matric Potential in the Vadose Zone Using Tensiometers[1]

This standard is issued under the fixed designation D 3404; the number immediately following the designation indicates the year of original adoption or, in the case of revision, the year of last revision. A number in parentheses indicates the year of last reapproval. A superscript epsilon (ϵ) indicates an editorial change since the last revision or reapproval.

1. Scope

1.1 This guide covers the measurement of matric potential in the vadose zone using tensiometers. The theoretical and practical considerations pertaining to successful onsite use of commercial and fabricated tensiometers are described. Measurement theory and onsite objectives are used to develop guidelines for tensiometer selection, installation, and operation.

1.2 The values stated in SI units are to be regarded as the standard. The inch-pound units given in parentheses are for information only.

1.2 *This standard does not purport to address all of the safety problems, if any, associated with its use. It is the responsibility of the user of this standard to establish appropriate safety and health practices and determine the applicability of regulatory limitations prior to use.*

2. Terminology

2.1 *Descriptions of Terms Specific to This Standard:*

2.1.1 *accuracy of measurement*—the difference between the value of the measurement and the true value.

2.1.2 *hysteresis*—that part of inaccuracy attributable to the tendency of a measurement device to lag in its response to environmental changes. Parameters affecting pressure-sensor hysteresis are temperature and measured pressure.

2.1.3 *precision (repeatability)*—the variability among numerous measurements of the same quantity.

2.1.4 *resolution*—the smallest division of the scale used for a measurement and it is a factor in determining precision and accuracy.

3. Summary of Guide

3.1 The measurement of matric potential in the vadose zone can be accomplished using tensiometers that create a saturated hydraulic link between the soil water and a pressure sensor. A variety of commercial and fabricated tensiometers are commonly used. A saturated porous ceramic material that forms an interface between the soil water and bulk water inside the instrument is available in many shapes, sizes, and pore diameters. A gage, manometer, or electronic pressure transducer is connected to the porous material with small- or large-diameter tubing. Selection of these components allows the user to optimize one or more characteristics, such as accuracy, versatility, response time, durability, maintenance, extent of data collection, and cost.

4. Significance and Use

4.1 Movement of water in the unsaturated zone is of considerable interest in studies of hazardous-waste sites **(1, 2, 3, 4)**[2]; recharge studies **(5, 6)**; irrigation management **(7, 8, 9)**; and civil-engineering projects **(10, 11)**. Matric-potential data alone can be used to determine direction of flow **(11)** and, in some cases, quantity of water flux can be determined using multiple tensiometer installations. In theory, this technique can be applied to almost any unsaturated-flow situation whether it is recharge, discharge, lateral flow, or combinations of these situations.

4.2 If the moisture-characteristic curve is known for a soil, matric-potential data can be used to determine the approximate water content of the soil **(10)**. The standard tensiometer is used to measure matric potential between the values of 0 and −867 cm of water; this range includes most values of saturation for many soils **(12)**.

4.3 Tensiometers directly and effectively measure soil-water tension, but they require care and attention to detail. In particular, installation needs to establish a continuous hydraulic connection between the porous material and soil, and minimal disturbance of the natural infiltration pattern are necessary for successful installation. Avoidance of errors caused by air invasion, nonequilibrium of the instrument, or pressure-sensor inaccuracy will produce reliable values of matric potential.

4.4 Special tensiometer designs have extended the normal capabilities of tensiometers, allowing measurement in cold or remote areas, measurement of matric potential as low as −153 m of water (−15 bars), measurement at depths as deep as 6 m (recorded at land surface), and automatic measurement using as many as 22 tensiometers connected to a single pressure transducer, but these require a substantial investment of effort and money.

4.5 Pressure sensors commonly used in tensiometers include vacuum-gages, mercury manometers, and pressure transducers. Only tensiometers equipped with pressure transducers allow for the automated collection of large quantities of data. However, the user needs to be aware of the pressure-transducer specifications, particularly temperature sensitivity and longterm drift. Onsite measurement of known zero and "full-scale" readings probably is the best calibration procedure; however, onsite temperature measurement or periodic recalibration in the laboratory may be sufficient.

[1] This guide is under the jurisdiction of ASTM Committee D-18 on Soil and Rock and is the direct responsibility of Subcommittee D18.21 on Ground Water and Vadose Zone Investigations.

Current edition approved May 15, 1991. Published October 1991.

[2] The boldface numbers in parentheses refer to a list of references at the end of the text.

5. Measurement Theory

5.1 In the absence of osmotic effects, unsaturated flow obeys the same laws that govern saturated flow: Darcy's Law and the Equation of Continuity, that were combined as the Richards' Equation **(13)**. Baver *et al.* **(14)** presents Darcy's Law for unsaturated flow as follows:

$$q = -K\nabla(\psi + Z) \tag{1}$$

where:

q = the specific flow, $\left[\frac{L}{T}\right]$,
K = the unsaturated hydraulic conductivity, $\left[\frac{L}{T}\right]$,
ψ = the matric potential of the soil water at a point, $[L]$,
Z = the elevation at the same point, relative to some datum, $[L]$, and
∇ = the gradient operator, $[L^{-1}]$.

The sum of $\psi + Z$ commonly is referred to as the hydraulic head.

5.2 Unsaturated hydraulic conductivity, K, can be expressed as a function of either matric potential, ψ, or water content, $\theta[L^3$ of water$/L^3$ of soil$]$, although both functions are affected by hysteresis **(5)**. If the wetting and drying limbs of the $K(\psi)$ function are known for a soil, time series of onsite matric-potential profiles can be used to determine: which limb is more appropriate to describe the onsite $K(\psi)$; the corresponding values of the hydraulic-head gradient; and an estimate of flux using Darcy's Law. If, instead, K is known as a function of θ, onsite moisture-content profiles (obtained, for example, from neutron-scattering methods) can be used to estimated K, and combined with matric-potential data to estimate flux. In either case, the accuracy of the flux estimate needs to be assessed carefully. For many porous media, $\frac{dK}{d\psi}$ and $\frac{dK}{d\theta}$ are large, within certain ranges of ψ or θ, making estimates of K particularly sensitive to onsite-measurement errors of ψ or θ. (Onsite-measurement errors of ψ also have direct effect on $\nabla(\psi + Z)$ in Darcy's Law). Other sources of error in flux estimates can result from: inaccurate data used to establish the $K(\psi)$ or $K(\theta)$ functions (accurate measurement of very small permeability values is particularly difficult) **(16)**; use of an analytical expression for $K(\psi)$ or $K(\theta)$ that facilitates computer simulation, but only approximates the measured data; an insufficient density of onsite measurements to define adequately the θ or ψ profile, which can be markedly nonlinear; onsite soil parameters that are different from those used to establish $K(\psi)$ or $K(\theta)$; and invalid assumptions about the state of onsite hysteresis. Despite the possibility of large errors, certain flow situations occur where these errors are minimized and fairly accurate estimates of flux can be obtained **(6, 17)**. The method has a sound theoretical basis and refinement of the theory to match measured data markedly would improve reliability of the estimates.

5.3 The concept of fluid tension refers to the difference between standard atmospheric pressure and the absolute fluid pressure. Values of tension and pressure are related as follows:

$$T_F = P_{AT} - P_F \tag{2a}$$

where:

T_F = the tension of an elemental volume of fluid, $\left[\frac{M}{LT^2}\right]$,
P_{AT} = the absolute pressure of the standard atmosphere, $\left[\frac{M}{LT^2}\right]$, and
P_F = the absolute pressure of the same elemental volume or fluid $\left[\frac{M}{LT^2}\right]$.

Soil-water tension (or soil-moisture tension) similarly is equal to the difference between soil-gas pressure and soil-water pressure. Thus:

$$T_W + P_G = P_W \tag{2b}$$

where:

T_W = the tension of an elemental volume of soil water, $\left[\frac{M}{LT^2}\right]$,
P_G = the absolute pressure of the surrounding soil gas, $\left[\frac{M}{LT^2}\right]$, and
P_W = the absolute pressure of the same elemental volume of soil water, $\left[\frac{M}{LT^2}\right]$.

In this guide, for simplicity, soil-gas pressure is assumed to be equal to 1 atmosphere, except as noted. Various units are used to express tension or pressure of soil water, and are related to each other by the equation:

$$\begin{aligned}&1.000 \text{ bar} = 100.0 \text{ kPa} = 0.9869 \text{ atm} =\\&1020 \text{ cm of water at } 4°\text{C} =\\&1020 \text{ g per cm}^2 \text{ in a standard}\\&\text{gravitational field.}\end{aligned} \tag{3}$$

A standard gravitational field is assumed in this guide; thus, centimetres of water at 4°C are used interchangeably with grams per square centimetre.

5.4 The negative of soil-water tension is known formally as matric potential. The matric potential of water in an unsaturated soil arises from the attraction of the soil-particle surfaces for water molecules (adhesion), the attraction of water molecules for each other (cohesion), and the unbalanced forces across the air-water interface. The unbalanced forces result in the concave water films typically found in the interstices between soil particles. Baver *et al.* **(14)** present a thorough discussion of matric potential and the forces involved.

5.5 The tensiometer, formally named by Richards and Gardner **(18)**, has undergone many modifications for use in specific problems **(1, 11, 19–31)**. However, the basic components have remained unchanged. A tensiometer comprises a porous surface (usually a ceramic cup) connected to a pressure sensor by a water-filled conduit. The porous cup, buried in a soil, transmits the soil-water pressure to a manometer, a vacuum gage, or an electronic-pressure transducer (referred to in this guide as a pressure transducer). During normal operation, the saturated pores of the cup prevent bulk movement of soil gas into the cup.

5.6 An expanded cross-sectional view of the interface between a porous cup and soil is shown in Fig. 1. Water held by the soil particles is under tension; absolute pressure of the soil water, P_W, is less than atmospheric. This pressure is transmitted through the saturated pores of the cup to the water inside the cup. Conventional fluid statics relates the

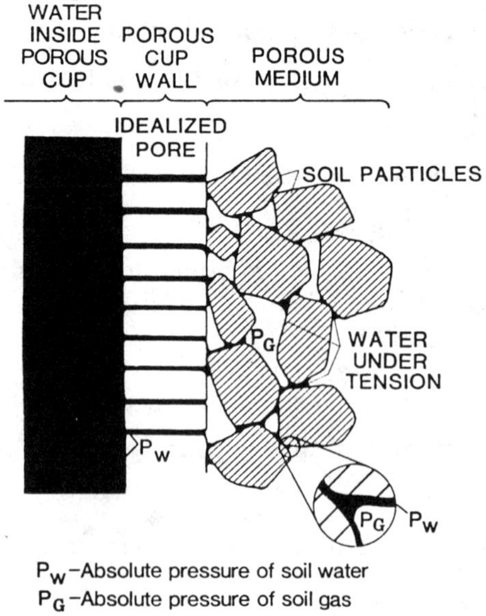

FIG. 1 Enlarged Cross Section of Porous Cup-Porous Medium Interface

pressure in the cup to the reading obtained at the manometer, vacuum gage, or pressure transducer.

5.6.1 In the case of a mercury manometer (see Fig. 1(*a*)):

$$T_W = P_A - P_W = (\rho_{Hg} - \rho_{H_2O})r - \rho_{H_2O}(h + d) \quad (4)$$

where:

T_W = the soil-water tension relative to atmospheric pressure, in centimetres of water at 4°C,
P_A = the atmospheric pressure, in centimetres of water at 4°C,
P_W = the average pressure in the porous cup and soil, in centimetres of water at 4°C,
ρ_{Hg} = the average density of the mercury column, in grams per cubic centimetre,
ρ_{H_2O} = the average density of the water column, in grams per cubic centimetre,
r = the reading, or height of mercury column above the mercury-reservoir surface, in centimetres,
h = the height of the mercury-reservoir surface above land surface, in centimetres, and
d = the depth of the center of the cup below land surface, in centimetres.

5.7 Although the density of mercury and water both vary about 1 % between 0 and 45°C, Eq. 4 commonly is used with ρ_{Hg} and ρ_{H_2O} constant.

5.7.1 Using ρ_{Hg} = 13.54 and ρ_{H_2O} = 0.995 (the median values for this temperature range) yields about a 0.25 % error (1.5 cm H_2O) at 45°C, for Tw ≈ 520 cm H_2O. This small, but needless, error can be removed by using the following density functions:

$$\rho_{Hg} = 13.595 - 2.458 \times 10^{-3}\,(T) \quad (5)$$

and

$$\rho_{H_2O} = 0.9997 + 4.879 \times 10^{-5}\,(T) - 5.909 \times 10^{-6}\,(T)^2 \quad (6)$$

where: ρ_{Hg} and ρ_{H_2O} are as defined above, and
T = average temperature of the column, in °C.

5.7.2 Average temperature of the buried segment of water column can be estimated with a thermocouple or thermistor in contact with the tubing, buried at about 45 % of the depth of the porous cup. Air temperature is an adequate estimate for exposed segments.

5.8 Most vacuum gages used with tensiometers are graduated in bars (and centibars) and have an adjustable zero-reading. The zero adjustment is used to offset the effects of altitude, the height of the gage above the porous cup (see Fig. 3(*b*)), and changes in the internal characteristics of the gage with time. The adjustment is set by filling the tensiometer with water and then setting the gage to zero while immersing the porous cup to its midpoint in a container of water. This setting is done at the altitude at which the tensiometer will be used and it needs to be repeated periodically after installation either by removing the tensiometer from the soil or by unscrewing the gage and measuring a tension equal to that used in the original calibration. The gage then reads directly the tension in the porous cup. Use of a vacuum gage without an adjustable zero reading could result in inaccurate measurements because the zero-reading could become negative and, therefore, would be indeterminate.

5.9 Pressure transducers convert pressure, or pressure difference, into a voltage (or current) signal. The pressure transducer can be connected remotely to the porous cup with tubing **(22, 24)** attached directly to the cup **(19, 32)**, or transported between sites **(24)**. An absolute pressure transducer measures the absolute pressure (P_P) in its port. A gage pressure transducer measures the difference between ambient-atmospheric pressure (P_A) and the pressure in its port (P_P), known as gage pressure. When $P_P < P_A$, gage pressure is identical to tension. A differential pressure transducer measures the difference between two pressures; one in each of its two ports. When used with tensiometers, the second port usually is connected to the atmosphere; the unit is used as a gage pressure transducer and it measures tension.

5.10 A calibration equation supplied by the manufacturer, or determined by the user, is used to convert the measured signal into pressure or tension at the pressure-transducer port. The tension in the porous cup and soil is then (see Fig. 3(*c*)):

$$T_W = T_P - t\,(\rho_{H_2O}) \quad (7)$$

where:

T_W = the average tension in the porous cup and soil, in centimetres of water at 4°C,
T_P = the tension in the pressure-transducer port, in centimetres of water at 4°C,
t = the difference in elevation between the pressure-transducer port and the center of the porous cup, in centimetres, and
ρ_{H_2O} = the average density of the water column connecting the porous cup and transducer, in grams per cubic centimetre.

5.11 At 15°C, pure liquid water begins to cavitate (vaporize) if its tension exceeds 969 cm H_2O. If cavitation happens in a tensiometer, liquid continuity is interrupted and tension readings are invalid. Water used in tensiometers is deaerated as completely as practicable, but some impurities and dissolved gases remain that decrease the tension sustainable by liquid water to about 867 cm H_2O **(33)**. Thus, the operating range of tensiometers is described by the following equations:

$$T_C + \Delta h < 867 \text{ cm} \tag{8}$$

and

$$T_C < 867 \text{ cm} \tag{9}$$

where:

T_C = the tension in the porous cup, in centimetres of water at 4°C, and

Δh = the elevation of the highest point in the hydraulic connection between the porous cup and the pressure sensor, minus the elevation of the porous cup, in centimetres.

Equation 8 indicates that a "trade-off" occurs between depth of installation of the porous cup and the maximum tension measurable; Eq. 9 sets the upper limit of that tension. Eqs. 8 and 9 are approximate; if the water is insufficiently deaerated, the value 867 would be replaced with a smaller value.

5.12 The only tensiometer described thus far that measures absolute soil-water pressure (P_W) directly is the absolute-pressure-transducer type. The others, differential tensiometers, measure the quantity $P_A - P_W$, where P_A is ambient atmospheric pressure. The driving forces for liquid water in the unsaturated zone (ignoring osmotic potential) are the absolute pressure gradient in the liquid-water phase and gravity (see Eq. 1). If the pressure wave propagates easily through the unsaturated zone, then differential tensiometers can be used directly to determine pressure gradients. However, if a barometric-pressure change is transmitted readily to one differential tensiometer porous cup and not to another (because of an intermediate confining layer), the calculated gradient between the two porous cups would be in error. If a porous cup is isolated from the atmosphere by a confining layer, then a time series of soil-water pressure at the porous cup, calculated with P_A constant, will indicate fluctuations that correlate well with barometric fluctuations. In this case, a recording barometer will provide a record of ambient atmospheric pressure from which absolute soil-water pressure and pressure gradients can be determined. The resulting series of absolute soil-water pressure at the isolated porous cup will be a smoother curve, that will indicate real pressure changes in the water phase.

5.13 Richards **(12)** defined the time constant of a tensiometer as follows:

$$\tau = \frac{1}{K_c S} \tag{10}$$

where:

τ = the time constant, or time required for 63.2 % of a step change in pressure to be recorded by a tensiometer, when the cup is surrounded by water, in seconds,

K_c = the conductance of the saturated porous cup, or the volume of water passing through the cup wall per unit of time per unit of hydraulic-head difference, in centimetre^2 second^{-1}, and

S = the tensiometer sensitivity, or change in pressure reading per unit volume of water passing through the porous-cup wall, in centimetre^{-2}.

Also, the porous cup conductance may be expressed as:

$$K_c = \frac{kA}{W}; \tag{11}$$

where:

K_c = the cup conductance, in centimetre^2 second^{-1},

k = the permeability of the cup material to water at the prevailing temperature, in centimetre second^{-1},

A = the average surface area of porous-cup material, estimated as the mean of the inside area and the outside area, in centimetre^2, and

W = the average wall thickness of the porous cup, in centimetres.

5.14 Richards' **(12)** definition does not apply to a tensiometer buried in a soil because soil conductance (K_s) is in series with K_c and usually $K_s \ll K_c$. In fact, an onsite time constant cannot be defined **(19)** because the response is not logarithmic due to a varying K_s during equilibration. However, the phrase "response time" is used to describe the rate of onsite response to pressure changes **(33)**. The term is not to be confused with the time constant because two tensiometers with equal time constants emplaced in the same soil can have different response times. For example, if $K_{c_1} = 10\ K_{c_2}$ and $S_2 = 10\ S_1$, then $\tau_1 = \tau_2$; but if $K_s \approx K_{c_2}$, then *response time*$_1$ > *response time*$_2$. Nonetheless, τ as defined here can be used comparatively to help evaluate tensiometer design. Greater sensitivity, large porous-cup surface area and permeability, and thin porous-cup walls are characteristics of a tensiometer with a short response time. Use of a sensitive pressure transducer is the most effective way to decrease response time in a soil of low hydraulic conductivity.

5.15 A bubble that interrupts hydraulic continuity between the porous cup and the pressure sensor will cause a change in the calculated value of P_W as follows:

$$\Delta = (E_P - E_C)\ \rho_{H_2O} \tag{12}$$

where:

Δ = the change in the calculated value of P_W, in centimetres of water at 4°C,

E_P = the elevation of the end of the bubble nearest the pressure sensor, in centimetres,

E_C = is the elevation of the end of the bubble nearest the cup, in centimetres, and

ρ_{H_2O} = the density of water adjacent to the air bubble, in grams centimetre^{-3}.

If bubbles are detected and measured, the above correction(s) can be made to P_W as calculated in Eqs. 4 or 7. Small bubbles that cling to the wall of the tubing and do not block the entire cross-section do not affect the calculated value of P_W.

6. Procedure

6.1 *Construction and Applications:*

6.1.1 The definitions used to describe the quality of a measurement and used in Table 1 to compare types of tensiometers are given in Section 2.

6.1.2 The operating characteristics of commonly available tensiometers vary (see Table 1) and they need to be matched to the specific installation, cost constraints, and the desired quality of data collection. Complete tensiometers may be purchased from soils and agricultural research companies, made entirely from parts, or made from parts of commercial units modified to suit the user's needs. The advantages and

TABLE 1 Tensiometer Characteristics

Characteristic	Commercial		Constructed			
			Manometer		Pressure Transducer	
	Vacuum Gage	Manometer (Hybrid)	Small Diameter	Hybrid	Small Diameter	Hybrid
Accuracy	Poor	Excellent	Excellent	Excellent	Good to excellent	Good to excellent
Precision[A]	Poor	Good	Good	Good	Excellent	Excellent
Hysteresis	Poor	Excellent	Excellent	Excellent	Fair to excellent	Fair to excellent
Response time	Poor to excellent	Fair	Fair	Fair	Excellent	Excellent
Versatility of application	Fair	Fair	Excellent	Fair	Excellent	Fair
Durability	Good	Good	Good to excellent	Good	Good	Good
Purging	Seldom	Occasionally	Often	Occasionally	Often	Occasionally
Recalibration	Occasionally	Never	Never	Never	Often	Often
Data-collection method	Manual	Manual	Manual	Manual	Manual or automatic	Manual or automatic
Cost of Five[B]	$260.00	$200.00	$120.00[C]	$150.00	$410.00[D]	$440.00[D]

[A] Precision (repeatability) is rated for either a wetting or drying cycle to distinguish from hysteresis effects.
[B] Estimated for five 0.914 m (3-ft) deep tensiometers.
[C] Does not include cost of deaerating water.
[D] Does not include cost of deaerating water or recording equipment.

disadvantages of some of the different types are discussed in the following sub-sections and in Table 1.

6.1.3 Commercially available vacuum-gage-type units (see Fig. 2(*b*)) usually have a large diameter porous cup cemented to a rigid plastic tube of equal diameter (19 or 22 mm). A vacuum gage that indicates from 0 to 100 centibars of tension is screwed into the side of the tube, several centimetres below the top. The space between the vacuum gage and the top of the tube is a reservoir for air (the water may or may not be deaerated beforehand) to collect. When the water level inside the tube approaches the vacuum-gage inlet, the tube cap is unscrewed and the air space is refilled with water. Some vacuum-gage tensiometers have a large water reservoir connected to the top of the tube with a spring-loaded valve to simplify refilling.

6.1.4 The advantages of vacuum-gage tensiometers include simplicity of use, relatively low cost, and the maintenance of a hydraulic connection between the porous cup and gage, even with large quantities of air present. However, this last advantage is typically offset by the use of a vacuum gage with a resolution of 0.5 centibar (5 cm H_2O) and an overall accuracy of 3 centibars (31 cm H_2O). Response time is excellent immediately after removing all air, but it slows rapidly as the air reservoir fills up. Efforts can be made to minimize thermal effects on the air column by shielding it from the sun. The construction is fairly durable, but its rigidity can transfer shock and actually damage the porous cup, cup-tube bond, or hydraulic connection with the soil if the top is impacted after installation. Although the tube usually is installed vertically, it can be inclined to a nearly horizontal orientation as long as the zero adjustment of the vacuum gage is made at the same inclination. Installations greater than 45° from the vertical are more likely to have air accumulation problems.

6.1.5 A vacuum-gage tensiometer is used predominantly for irrigation scheduling where extreme accuracy is not necessary. It is not recommended for measurement of unsaturated hydraulic gradients **(33)**. However, replacement of a standard vacuum gage with a more accurate, higher-resolution gage, or with an accurate pressure transducer, would improve the usability of the tensiometer.

6.1.6 In this guide, a tensiometer with a large diameter cup-tube assembly connected to the pressure sensor with small-diameter (3.2 mm, for example) tubing is referred to as a hybrid tensiometer (see Fig. 2(*a*)). Hybrid tensiometers, like vacuum-gage tensiometers, have a space at the top of the large tube to collect air. Hydraulic continuity is not broken, unless air bubbles block an entire cross-section of the small-diameter tubing.

6.1.7 Commercial manometer-type tensiometers commonly are hybrid types. Almost all of the air that enters the tensiometer through the porous cup collects harmlessly at the top. However, air also tends to be liberated from solution near the top of the manometer, where the maximum tension occurs; use of deaerated water minimizes air production.

6.1.8 A mercury manometer is probably the most accurate pressure-sensor commonly used in tensiometers and it never needs calibrating (a water manometer, usable only in special cases, is more accurate because of a better resolution). Hysterisis in a manometer tensiometer (from surface tension at the interface) or pressure-transducer tensiometer is usually much less than that in a vacuum-gage tensiometer. Thus, the hybrid-manometer tensiometer and pressure-transducer tensiometer combine fairly maintenance-free operation with excellent accuracy.

6.1.9 The major advantages of constructing a manometer tensiometer with small-diameter tubing are the versatility of onsite application, accuracy, and low cost. Flexible nylon tubing can be routed around obstructions, connecting a porous cup to a gage hundreds of feet away. Installation orientation is limited only by backfilling capabilities. A typical design **(30)** employs two 3.2 mm-diameter (nominal ⅛-in.-diameter) nylon tubes cemented with epoxy directly to a 9.5 mm-diameter (nominal ⅜-in.-diameter) porous cup (see Fig. 3(*a*)). The water-supply tube is connected via a shutoff valve to a deaerated water supply and the measurement tube is routed directly to the mercury reservoir. Manometer sensitivity (S, see Eq. 10) is the reciprocal of the cross-sectional area of the manometer tubing; response time is minimized by using small-diameter tubing. The design is simple, but the epoxy-nylon-tube bond is somewhat susceptible to rupture, from differential movement of the nylon tubes. A more robust design (see Fig. 3(*b*)) uses a larger porous cup and a metal plate to separate the nylon tubes,

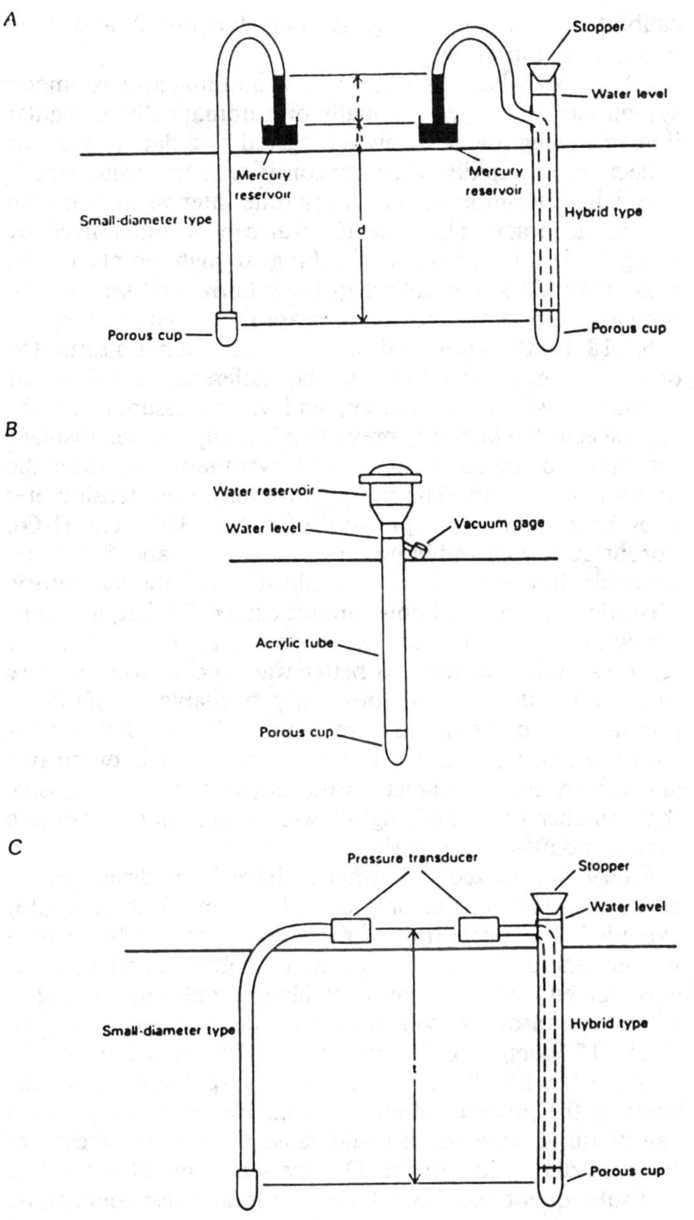

FIG. 2 Three Common Types of Tensiometers: (a) Manometer; (b) Vacuum Gage; and (c) Pressure Transducer

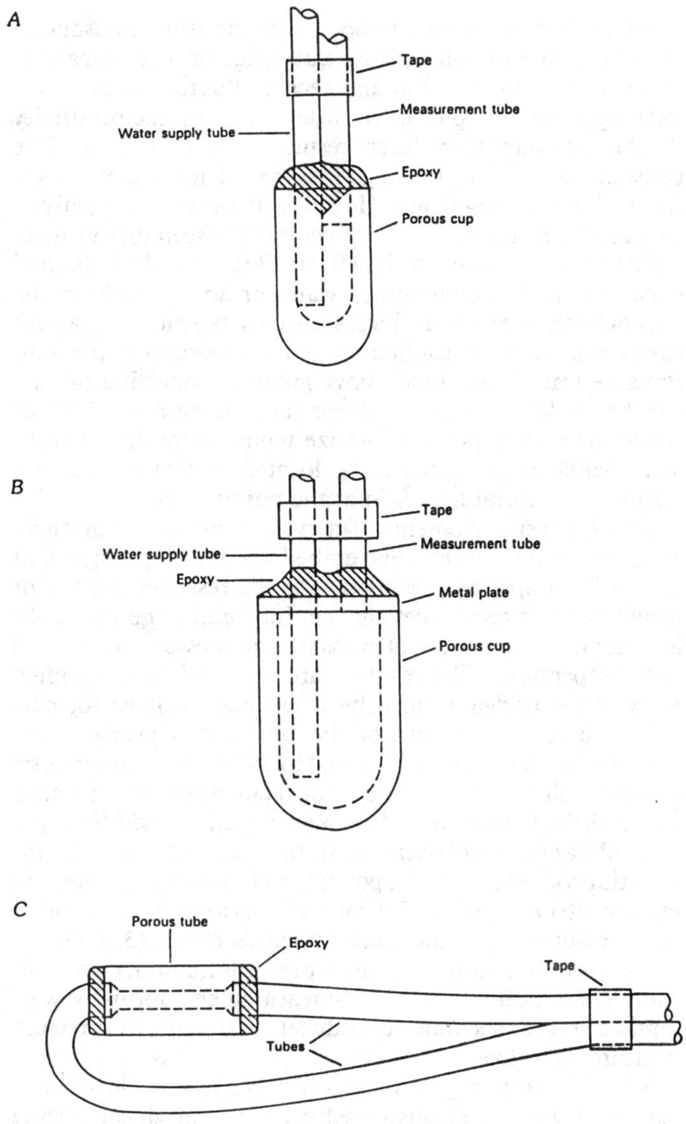

FIG. 3 Porous-Cup and Tube Designs

allowing the epoxy to form a stronger bead around the tubes. A third design (see Fig. 3(*c*)) uses a porous tube, made by cutting the rounded end off a 6.4 mm-diameter (nominal ¼-in.-diameter) porous cup. One of the nylon tubes is molded into a "U" shape with heat, to produce the design in Fig. 3(*c*). The third design is extremely durable and versatile. The porous tube can be purged of air in any orientation, which is not entirely true of designs (*a*) and (*b*).

6.1.10 Tensiometers tend to collect air bubbles onsite that originate from the following: insufficiently deaerated water; air that diffuses through the water-filled pores of the porous cup; dissolved gases in the soil moisture that flow into the porous cup during a wetting cycle; and air that diffuses through the tubing material.

6.1.11 The major disadvantage of the small-diameter designs is that air bubbles can easily block the entire cross-section of the tubing, interrupting hydraulic continuity. Thus, small-diameter designs require frequent purging of air—especially at large tensions. Use of a thick-walled tubing decreases diffusion through the tube wall. Some plastics (such as polyethylene) are relatively permeable to air and are unsatisfactory tubing material. Use of metal tubing almost eliminates air diffusion through the tube wall, however, other adverse affects can occur from the use of materials that have a high thermal conductivity.

6.1.12 A hybrid-manometer tensiometer can be constructed for two-thirds to three-fourths the price of a commercial unit. The cost in Table 1 was determined for five "cup-tube kits" purchased at a supply company and outfitted with a manometer made from parts. A cup-tube kit consists of a porous cup cemented to a stoppered acrylic tube with a fitting to accept 1.6 mm-diameter (nominal 1/16-in.-diameter) manometer tubing. The manometer tubing extends to the bottom of the acrylic tube. The cost could be decreased further by assembling the entire unit from basic parts.

6.1.13 Tensiometers equipped with pressure transducers are well-suited to collect large quantities of data. Measurements can be made often and recorded automatically by a data logger or a strip-chart recorder. The extreme sensitivity of the pressure transducer results in the shortest time constant obtainable, making this type of tensiometer ideal for tracking wetting fronts. However, the extreme sensitivity also results in the transducer becoming susceptible to measuring transient-temperature effects **(34)**, caused by thermal expansion and contraction of water or air, or both, in the tensiometer. Water that freezes inside a pressure-transducer cavity will affect the calibration, and it can rupture the unit. Pressure transducers (and above-ground connecting tubing) can be enclosed in an insulated (and, if need be, heated) shelter or surface pit to minimize temperature fluctuations. The shelter or pit needs to be located so that it does not disturb the natural flow field at the porous cup(s).

6.1.14 Sensing elements of a typical pressure transducer are semiconductive resistors, embedded in a diaphragm that moves from applied pressure(s). As the resistors shorten or lengthen, their resistances change. Inherently, the resistance is a nonlinear function of pressure (or pressure difference) and temperature. The resistors are included in a modified Wheatstone bridge, excited by a regulated voltage (or current) source. The output of the bridge is a nearly linear function of pressure that is independent of temperature; however, all pressure transducers retain some nonlinearity and a slight temperature dependence **(35)**. In addition, the zero offset and, to a lesser extent, the sensitivity may change with time (known as drift), possibly requiring recalibration at regular intervals **(35)**. Other sources of error are repeatability and pressure and temperature hysteresis effects **(35)**. Onsite application determines required pressure-transducer specifications; for example, gradient-measurement normally warrants a more accurate transducer than simple pressure measurement does.

6.1.15 A large degree of accuracy may be achieved in a variety of ways. A sophisticated pressure transducer costing seven to eight hundred dollars typically has an overall accuracy of 1 cm H_2O, at 23°C, and a temperature coefficient of 0.07 cm H_2O °C^{-1}. Without temperature correction, a worst-case error (at 0°C) of 2.6 cm H_2O could result. At the other extreme, a simple pressure transducer may be purchased for about fifty dollars and its output may be corrected for nonlinearity and temperature dependence by measuring temperature and applying a second-order polynomial fit to the measured data **(36)**. The decreased linearity and temperature errors, combined with a typical repeatability and hysteresis error of 1.5 cm H_2O, produce a root-sum-square error of 1.8 cm H_2O. Such a transducer is listed in Table 1. Of course, the same curve-fitting procedure can be applied to a pressure transducer with better repeatability and hysteresis to achieve greater accuracy.

6.1.16 Accuracies determined in the two examples discussed in 6.1.15 further are degraded by drift or lack of long-term-stability. If a long-term-stability specification cannot be supplied by the pressure-transducer manufacturer, much of the data collected may be inaccurate. A long-term-stability specification is used to determine how often a transducer needs to be recalibrated to maintain desired accuracy. A hanging column of water or mercury (or a calibrated vacuum source) is used for initial and then periodic recalibration.

6.1.17 Air that collects in an automated-tensiometer system can be purged manually or automatically at regular intervals using solenoid valves triggered by a data logger. Air collects more rapidly when tension is greater; onsite experience will determine the necessary time interval to maintain desired accuracy. The time interval can be maximized by using horizontal sections of tubing at high points in the system. Bubbles that collect in these horizontal sections do not cause errors but they do increase the response time.

6.1.18 Porous cups used in tensiometers remain saturated during normal operation. If the difference between air pressure outside the porous cup and water pressure inside the cup exceeds the bubbling pressure of the cup, air will displace the water in the largest pores and eventually will enter the interior of the cup. The more commonly used tensiometer cups have a bubbling pressure of 1 bar (1020 cm H_2O). Porous cups with bubbling pressures of 2, 3, and 5 bars are available, but they only have applications in the laboratory. If onsite tensions are known to not exceed 0.5 bar, a porous cup with a 0.5-bar bubbling pressure can be used to decrease response time. However, a better way to decrease response time is by the use of commonly available, "high-flow" porous cups, made with a pore-size distribution that emphasizes the larger pores. (Cost of a porous cup is between two and fifteen dollars. Larger porous cups are more expensive than smaller ones, and high-flow-cups are more expensive than standard-flow-cups.)

6.1.19 Specialized tensiometers have been developed to address specific onsite problems. Peck and Rabbidge **(35)** extended the upper limit of measurable tension by using a reference solution with a low osmotic potential and a porous cup coated with a semipermeable membrane. Using a pressure transducer, soil-water tensions as large as 153 m H_2O (15 bars) can be measured. The solution in the tensiometer usually is under positive pressure, thus decreasing the problem of air invasion. However, the pressure can be quite large and it could cause permanent "creep" of the transducer diaphragm. Depolymerization of the solute and subsequent loss through the membrane also could cause creep. Ambient temperature affects the osmotic potential directly (as qualitatively predicted by van't Hoff's law) and indirectly by causing flow of water through the membrane. Thus, an osmotic tensiometer is valuable for measuring extremely low matric potentials, but some sources of measurement error are unique to it.

6.1.20 The U.S. Forest Service developed an inexpensive recording tensiometer for use in remote areas where electric power is unavailable **(31)**. This instrument can record as much as 1 month of continuous data on a battery-driven raingage chart. Oaksford **(27)** designed a unit based on a coaxial water manometer that provides maximum sensitivity at depths as deep as about 6 m. This unit uses a calibrated wire and ohmmeter to sense a below-ground, free-water surface inside the unit.

6.1.21 Fluid-scanning switches have been used successfully **(20, 34, 37, 38)** to connect as many as 22 tensiometers sequentially to a single pressure transducer. The approach minimizes cost and removes the bias between tensiometers caused by using different pressure transducers. Equally relevant, this network has the capability of measuring a zero and a full-scale calibration tension before each scan. This capability removes measurement errors from hysteresis, temperature dependence, and long-term drift. However, if the pressure transducer fails, data from all its tensiometers are lost. Also, the scanning switch is made with precise tolerances and it may develop leaks over time.

6.1.22 Another approach to efficient data collection with large numbers of tensiometers **(25, 39)** is to connect a portable pressure transducer to each tensiometer using a hypodermic needle and septum. The needle tip is inserted in an air space above the tensiometer fluid. A small change in tensiometer pressure is caused by the connection, probably affecting measurements by a few cm H_2O. Also, changes in the fluid-surface elevation, although small, affect measurements directly unless the changes are accounted for.

6.1.23 A small proportion of water in a frozen soil may remain liquid and, therefore, mobile. Measurement of tension in frozen soils can be accomplished using a pressure transducer and ethylene glycol **(26)** as a tensiometer fluid. The pressure transducer minimizes exchange of fluid between the soil and tensiometer, but slight bulk flow and diffusive flux do occur that decrease the freezing point of the soil fluid. The osmotic tensiometer probably will work well in frozen soils.

6.1.24 Porous cups have been connected directly to pressure transducers; the internal cavity has been filled with deaerated fluid and the entire assembly has been buried in a soil, with no provisions for purging of the fluid **(19, 32)**. Although this approach provides a stable temperature environment, the pressure transducer cannot be recalibrated readily for drift. Also, soil gas diffusing through the porous cup creates an air pocket in the cavity. Because this pocket is at a pressure less than that of the soil gas, air will continue to diffuse through the porous-cup wall, eventually emptying the cup of water. A purging system is needed for extended undisturbed operation.

6.1.25 The specialized tensiometers developed thus far have resulted from insight and persistence of researchers faced with particular problems. Most of these solutions require extra effort, care, and expense; review of the cited report(s) is needed before implementation.

6.2 *Installation:*

6.2.1 Continuous hydraulic connection between the porous cup and soil **(18)**, and minimal disturbance of the natural flow field are essential to collection of accurate tensiometric data. When a hole is made to accept the tensiometer, the cuttings need to be preserved in the order they were removed if the hole is to be backfilled.

6.2.2 Hydraulic connection can be established in several ways. Commercial tensiometers fit snugly into holes made with a coring tool (available as a tensiometer accessory) or made with standard iron pipe. The porous cup is forced against the hole bottom and no backfill is used. When the soil is rocky, or when a small-diameter tensiometer is installed, a hole larger in diameter than the porous cup is excavated. If the soil at the bottom of the hole is soft, the porous cup may be forced into the soil. A hard soil sometimes can be softened with water. If the porous cup will not penetrate the moistened soil, the last cuttings from the hole need to be used to backfill around the cup, either by making them into a slurry or by careful tamping. A tremie pipe ensures clean delivery of cuttings or slurry to the bottom of the hole. Gaps between the porous cup and soil increase the tensiometer response time by reducing the effective area of the porous cup. In the worst case, no hydraulic connection occurs and the tensiometer will not indicate the soil-water tension.

6.2.3 If water is used to establish hydraulic connection, the tension adjacent to the porous cup will be reduced, and it will recover asymptotically as the added water is dispersed in the soil. The rate of dispersal will depend on the K (see Eq. 1) of the soil and a time series of tension will indicate when natural conditions are restored sufficiently.

6.2.4 If vertical profiles or gradients of pressure are to be measured, multiple small-diameter tensiometers can be installed in a single hole. Ideally, the original lithology is duplicated (except that gravel larger than 6 mm in diameter and cobbles and boulders can be removed) by using the cuttings in reverse order for backfilling. A backfill that is more compacted than the undisturbed soil will tend to shed infiltrating water, but, after a short time, the tension in the undisturbed soil and in the backfill will be in equilibrium. A backfill that is less compacted than the surrounding soil, or one with excessive gaps, will be a conduit for infiltrating water, resulting in abnormally low tensions in the backfill and in the undisturbed soil. Therefore, compaction of the backfill to a slightly greater bulk density than that of the undisturbed soil is desirable. Less permeable layers (such as clay lenses) need to be reproduced or even exaggerated by importing a fine-grained material.

6.2.5 Deep installation of tensiometers can be accomplished by drilling horizontal holes radially from a central caisson hole. This method also preserves undisturbed conditions above and below each porous cup. Backfilling horizontal holes with a tamping rod is painstaking. An alternative is to backfill in the vicinity of the porous cup and to fill the remainder of the hole with an expanding insulation foam. Pressure transducers work particularly well in caisson holes. If the entire length of tubing connecting the porous cup to the pressure transducer is horizontal, air bubbles do not cause errors. Also, the problem of transducer sensitivity to temperature change is minimized, because temperature in the caisson hole remains relatively constant. A caisson hole allows use of water manometers for improved precision because the entire manometer can be placed below the level of the porous cup.

6.3 *Operation:*

6.3.1 Testing the porous cup, the porous cup-tubing interface, all fittings, and the measurement and recording device(s), before installation, is desirable. After saturating the porous cup, apply air pressure to the interior of the tensiometer while the parts to be tested are submersed. If bubbles appear at a gage pressure substantially less than the bubbling pressure of the porous cup, the unit is faulty and the appropriate parts need to be repaired or replaced.

6.3.2 Water used in tensiometers is deaerated in a carboy by applying a vacuum or heat, or both. Excellent results have been obtained using a pump that generates 970 cm H_2O vacuum with a heated magnetic stirrer for 48 h. Insufficiently deaerated water requires frequent onsite purging, or, in the worst case, allows bubbles to form before a tensiometer has reached equilibrium, preventing accurate data collection. The deaerated water is siphoned (to minimize reaeration) from the carboy to a collapsible plastic container (available at a sporting goods store) for onsite use. Although the air above the water is forced out of the spout immediately, the water reaerates slowly by diffusion of air through the container wall. Proper fittings should be used to connect the container spout to the tensiometer supply tube. Use of water native to the soil being measured will minimize the effect of osmotic potential across the cup on the measured matric potential.

6.3.3 Purge a small-diameter tensiometer by connecting the water supply container to the water supply tube and raising it above the top of the tensiometer, while opening the supply valve. Several tensiometers can be connected in a "tee" network to simplify multiple purging. Purging instructions for vacuum-gage and hybrid tensiometers are supplied by the manufacturer. These instructions can be modified if use of deaerated water is desired. Purging time needs to be short to minimize wetting of the soil immediately surrounding the porous cup. When purging is complete, the system is closed and the soil draws water through the porous cup until equilibrium is established. The pressure inside the porous cup approaches the soil-water pressure asymptotically at a rate determined by the time constant and the unsaturated hydraulic conductivity of the soil. When equilibrium is reached, make the measurement. Record a single value of pressure when using a vacuum-gage or pressure transducer; a manometer requires measurement of the mercury column and reservoir elevations.

6.3.4 The most reliable data are obtained by purging a tensiometer and allowing it to equilibrate before recording the measurement. However, wet soils, lack of wetting fronts, low permeability tubing, or thoroughly deaerated water tend to prevent air accumulation for long periods; these conditions, either singly or in combination permit reliable data collection without purging.

6.3.5 Dry soils or inadequate manometer design, or both, occasionally result in mercury being pulled over the top of the manometer and into the porous cup. The porous cups shown in Fig. 3 may be purged of mercury by applying pressure to the measurement tube, forcing mercury out the supply tube. Pressure applied to the top of a hybrid tensiometer will force the mercury out the measurement tube.

6.3.6 Porous cups that are removed from a soil to be reused need to be washed with warm water to prevent plugging of the pores. A porous cup with plugged pores possibly can be restored by sanding or rinsing in a weak HCl solution **(33)**.

REFERENCES

(1) Healy, R. W., Peters, C. A., DeVries, M. P., Mills, P. C., and Moffett, D. L., "Study of the Unsaturated Zone at a Low-Level Radioactive-Waste Disposal Site near Sheffield, Ill.," *Proceedings*, National Water Well Association Conference on Characterization and Monitoring of the Vadose Zone, Los Vegas, NV, Dec. 1983, pp. 820–831.

(2) McMahon, P. B., and Dennehy, K. F., "Water Movement in the Vadose Zone at Two Experimental Waste-Burial Trenches in South Carolina," *Proceedings*, National Water Well Association Conference on Characterization and Monitoring of the Unsaturated (Vadose) Zone, Denver, CO, Nov. 1985, pp. 34–54.

(3) Ripp, J. A., and Villaume, J. F., "A Vadose Zone Monitoring System for a Flyash Landfill," *Proceedings*, National Water Well Association Conference on Characterization and Monitoring of the Unsaturated (Vadose) Zone, Denver, CO, Nov. 1985, pp. 73–96.

(4) Ryan, B. J., DeVries, M. P., Garklavs, George, Gray, J. R., Healy, R. W., Mills, P. C., Peters, C. A., and Striegl, R. G., "Results of Hydrologic Research at a Low-Level Radioactive-Waste Disposal Site Near Sheffield, Illinois," *U.S. Geological Survey Water-Supply Paper 2367*, 1991.

(5) Lichtler, W. F., Stannard, D. I., and Kouma, E., "Investigation of Artificial Recharge of Aquifers in Nebraska," *U.S. Geological Survey Water-Resources Investigations Report 80-93*, 1980.

(6) Sophocleous, Marios, and Perry, C. A., "Experimental Studies in Natural Groundwater Recharge Dynamics: Analysis of Observed Recharge Events," *Journal of Hydrology*, Vol 81, 1985, pp. 297–332.

(7) Anderson, E. M., "Tipburn of Lettuce. Effect of Maturity, Air and Soil Temperature, and Soil Moisture Tension," Cornell Agricultural Experimental Station, *Bulletin*, No. 829, Cornell University, New York, 1946.

(8) Richards, L. A., and Neal, O. R., "Some Field Observations with Tensiometers," *Proceedings*, Soil Science Society of America (1936), Vol 1, 1937, pp. 71–91.

(9) Richards, S. J., Willardson, L. S., Davis, Sterling, and Spencer, J. R., "Tensiometer Use in Shallow Ground-Water Studies," *Proceedings*, American Society of Civil Engineers, Vol 99, No. IR4, 1973, pp. 457–464.

(10) McKim, H. L., Walsh, J. E., and Arion, D. N., "Review of Techniques for Measuring Soil Moisture In Situ," *U.S. Army Corps of Engineers, Cold Regions Research and Engineering Lab Special Report 80-31*, 1980.

(11) Richards, L. A., Russell, M. B., and Neal, O. R., "Further Developments on Apparatus for Field Moisture Studies," *Proceedings*, Soil Science Society of America, Vol 2, 1938, pp. 55–64.

(12) Richards, L. A., "Methods of Measuring Soil Moisture Tension," *Soil Science*, Vol 68, 1949, pp. 95–112.

(13) Richards, L. A., "Capillary Conduction of Liquids through Porous Mediums," *Physics 1*, 1931, pp. 318–333.

(14) Baver, L. D., Gardner, W. H., and Gardner, W. R., *Soil Physics*, Wiley, New York, 1972, pp. 291–296.

(15) Brooks, R. H., and Corey, A. T., "Hydraulic Properties of Porous Media," *Hydrology Papers*, No. 3, Colorado State University, Fort Collins, CO, 1964.

(16) Marshall, T. J., "Relations Between Water and Soil," Commonwealth Bureau of Soil Science, England, *Technical Communication* No. 50, 1960.

(17) Haverkamp, R., Vauclin, M., Touma, J., Wierenga, P. J., and Vachaud, G., "A Comparison of Numerical Simulation Models for One-Dimensional Infiltration," *Soil Science Society of America Journal*, Vol 41, 1977, pp. 285–294.

(18) Richards, L. A., and Gardner, W., "Tensiometers for Measuring the Capillary Tension of Soil Water," *Journal of American Society of Agronomy*, Vol 28, 1936, pp. 352–358.

(19) Bianchi, W. C., "Measuring Soil Moisture Tension Changes," *Agricultural Engineering*, Vol 43, 1962, pp. 398–404.

(20) Bianchi, W. C., and Tovey, Rhys, "Continuous Monitoring of Soil Moisture Tension Profiles," *Transactions*, American Society of Agricultural Engineers, Vol 11, No. 3, 1968, pp. 441–447.

(21) Colman, E. A., Hanawalt, W. B., and Burck, C. R., "Some Improvements in Tensiometer Design," *Journal of the American Society of Agronomy*, Vol 38, 1946, pp. 455–458.

(22) Healy, R. W., deVries, M. P., and Striegl, R. G., "Concepts and Data-Collection Techniques Used in a Study of the Unsaturated Zone at a Low-Level Radioactive-Waste Disposal Site Near Sheffield, Illinois," *U.S. Geological Survey Water-Resources Investigations Report 85-4228*, 1986.

(23) Hunter, A. S., and Kelley, O. J., "Changes in the Construction of Soil Moisture Tensiometers for Field Use," *Soil Science*, Vol 61, 1946, pp. 215–218.

(24) Klute, A., and Peters, D. B., "A Recording Tensiometer with a Short Response Time," *Proceedings*, Soil Science Society of America, Vol 26, 1962, pp. 87–88.

(25) Marthaler, H. P., Vogelsanger, W., Richard, F., and Wierenga, P. J., "A Pressure Transducer for Field Tensiometers," *Soil Science Society of America Journal*, Vol 47, 1983, pp. 624–627.

(26) McKim, H. L., Berg, R. L., McGraw, R. W., Atkins, R. T., and Ingersoll, J., "Development of a Remote-Reading Tensiometer Transducer System for Use in Subfreezing Temperatures," *Proceedings*, American Geophysical Union Second Conference on Soil-Water Problems in Cold Regions, Edmonton, Alberta, 1976, pp. 31–45.

(27) Oaksford, E. T., "Water-Manometer Tensiometers Installed and Read from the Land Surface," *Geotechnical Testing Journal*, Vol 1, No. 4, 1978, pp. 199–202.

(28) Peck, A. J., and Rabbidge, R. M., "Design and Performance of an Osmotic Tensiometer for Measuring Capillary Potential," *Proceedings*, Soil Science Society of America, Vol 33, 1969, pp. 196–202.

(29) Richards, L. A., "Soil Moisture Tensiometer Materials and Construction," *Soil Science*, Vol 53, 1942, pp. 241–248.

(30) Stannard, D. I., "Theory, Construction, and Operation of Simple Tensiometers," *Ground Water Monitoring Review*, Vol 6, No. 3, 1986, pp. 70–78.

(31) Walkotten, W. J., "A Recording Soil Moisture Tensiometer," *U.S.D.A. Forest Service Research Note PNW-180*, 1972.

(32) Watson, K. K., "A Recording Field Tensiometer with Rapid Response Characteristics," *Journal of Hydrology*, Vol 5, 1967, pp. 33–39.

(33) Cassel, D. K., and Klute, A., "Water Potential: Tensiometry," American Society of Agronomy—Soil Science Society of America, *Agronomy Monograph* No. 9 (Second Edition). *Methods of Soil Analysis, Part 1. Physical and Mineralogical Methods*, 1986, pp. 563–596.

(34) Rice, Robert, "A Fast-Response, Field Tensiometer System," *Transactions*, American Society of Agricultural Engineers, Vol 12, 1969, pp. 48–50.

(35) Omega Engineering, Inc., *Omega 1987 Complete Pressure, Strain and Force Measurement Handbook and Encyclopedia*, Dec 1986.

(36) Micro Switch, Freeport, Illinois, *Pressure Sensors, Catalog 15*, 1985.

(37) Fitzsimmons, D. W., and Young, N. C., "Tensiometer-Pressure Transducer System for Studying Unsteady Flow Through Soils," *Transactions*, American Society of Agricultural Engineers, Vol 15, No. 2, 1972, pp. 272–275.

(38) Williams, T. H. Lee, "An Automatic Scanning and Recording Tensiometer System," *Journal of Hydrology*, Vol 39, 1978, pp. 175–183.

(39) Haldorsen, Sylvi, Petsonk, A. M., and Torstensson, B. A., "An Instrument for In Situ Monitoring of Water Quality and Movement in the Vadose Zone," *Proceedings*, National Water Well Association Conference on Characterization and Monitoring of the Unsaturated (Vadose) Zone, Denver, CO, Nov. 1985, pp. 158–172.

(40) Mualem, Yechezkel, "A New Model for Predicting the Hydraulic Conductivity of Unsaturated Porous Media," *Water Resources Research*, Vol 12, No. 3, 1976, pp. 513–522.

Standard Practices for Preparation of Sample Containers and for Preservation of Organic Constituents[1]

1. Scope

1.1 These practices cover the various means of (*1*) preparing sample containers used for collection of waters to be analyzed for organic constituents and (*2*) preservation of such samples from the time of sample collection until the time of analysis.

1.2 The sample preservation practice is dependent upon the specific analysis to be conducted. See Section 9 for preservation practices listed with the corresponding applicable general and specific constituent test method. The preservation method for waterborne oils is given in Practice D 3325. Use of the information given herein will make it possible to choose the minimum number of sample preservation practices necessary to ensure the integrity of a sample designated for multiple analysis. For further considerations of sample preservation, see the *Manual on Water.*[2]

1.3 *This standard does not purport to address all of the safety problems, if any, associated with its use. It is the responsibility of the user of this standard to establish appropriate safety and health practices and determine the applicability of regulatory limitations prior to use.* For specific hazard statements, see 6.7, 6.24, and 8.1.3.

2. Referenced Documents

2.1 *ASTM Standards:*

D 1129 Terminology Relating to Water[3]
D 1252 Test Methods for Chemical Oxygen Demand (Dichromate Oxygen Demand) of Water[4]
D 1783 Test Methods for Phenolic Compounds in Water[4]
D 2036 Test Methods for Cyanides in Water[4]
D 2330 Test Method for Methylene Blue Active Substances[4]
D 2579 Test Methods for Total and Organic Carbon in Water[4]
D 2580 Test Method for Phenols in Water by Gas-Liquid Chromatography[4]
D 2908 Practice for Measuring Volatile Organic Matter in Water by Aqueous-Injection Gas Chromatography[4]
D 3086 Test Method for Organochlorine Pesticides in Water[4]
D 3113 Test Methods for Sodium Salts of EDTA in Water[4]
D 3325 Practice for Preservation of Waterborne Oil Samples[4]
D 3371 Test Method for Nitriles in Aqueous Solution by Gas-Liquid Chromatography[4]
D 3478 Test Method for Chlorinated Phenoxy Acid Herbicides in Water[4]
D 3534 Test Method for Polychlorinated Biphenyls (PCBs) in Water[4]
D 3590 Test Methods for Total Kjeldahl Nitrogen in Water[3]
D 3695 Test Method for Volatile Alcohols in Water by Direct Aqueous-Injection Gas Chromatography[4]
D 3871 Test Method for Purgeable Organic Compounds in Water Using Headspace Sampling[4]
D 3921 Test Method for Oil and Grease and Petroleum Hydrocarbons in Water[4]
D 3973 Test Method for Low-Molecular Weight Halogenated Hydrocarbons in Water[4]
D 4129 Test Method for Total and Organic Carbon in Water by High-Temperaure Oxidation and Coulometric Detection[4]
D 4165 Test Method for Cyanogen Chloride in Water[4]
D 4193 Test Method for Thiocyanate in Water[4]
D 4281 Test Method for Oil and Grease (Fluorocarbon Extractable Substances) by Gravimetric Determination[4]
D 4282 Test Method for Determination of Free Cyanide in Water and Wastewater by Microdiffusion[4]
D 4374 Test Methods for Cyanide in Water—Automated Methods for Total Cyanide, Acid Dissociable Cyanide, and Thiocyanate[4]
D 4515 Practice for Estimation of Holding Time for Water Samples Containing Organic Constituents[4]
D 4657 Test Method for Polynuclear Aromatic Hydrocarbons in Water[4]
D 4744 Test Method for Organic Halides in Water by Carbon Adsorption Microcoulometric Detection[4]
D 4763 Practice for Identification of Chemicals in Water by Fluorescence Spectroscopy[4]
D 4779 Test Method for Total, Organic, and Inorganic Carbon in High Purity Water by Ultraviolet (UV) or Persulfate Oxidation, or Both, and Infrared Detection[4]
D 4839 Test Method for Total Carbon and Inorganic Carbon in Water by Ultraviolet, or Persulfate Oxidation, or Both, and Infrared Detection[4]
D 4841 Practice for Estimation of Holding Time for Water Samples Containing Organic and Inorganic Constituents[3]
D 4983 Test Method for Cyclohexylamine, Morpholine, and Diethylaminoethanol in Water and Condensed Steam by Direct Aqueous Injection Gas Chromatography[4]

---

[1] These practices are under the jurisdiction of ASTM Committee D-19 on Water and are the direct responsibilities of Subcommittee D19.06 on Methods for Analysis for Organic Substances in Water.

Current edition approved Feb. 12, 1992. Published April 1992. Originally published as D 3694 – 78. Last previous edition D 3694 – 89$^{\epsilon 1}$.

[2] *Manual on Water, ASTM STP 442*, ASTM, 1969.

[3] *Annual Book of ASTM Standards*, Vol 11.01.

[4] *Annual Book of ASTM Standards*, Vol 11.02.

3. Terminology

3.1 *Definitions*—For definitions of terms used in this practice, refer to Terminology D 1129.

4. Significance and Use

4.1 There are four basic steps necessary to obtain meaningful analytical data: preparation of the sample container, sampling, sample preservation, and analysis. In fact these four basic steps comprise the analytical method and for this reason no step should be overlooked. Although the significance of preservation is dependent upon the time between sampling and the analysis, unless the analysis is accomplished within 2 h after sampling, preservation is preferred and usually required.

5. Apparatus

5.1 *Forced Draft Oven*, capable of operating at 275 to 325°C.

5.2 *Sample Bottle*, borosilicate or flint glass.

NOTE 1—High-density polyethylene (HDPE) bottles and caps have been demonstrated to be of sufficient quality to be compatible for all tests except pesticides, herbicides, polychlorinated biphenyls, and volatile organics. However, this type of bottle cannot be recycled.

5.3 *Sample Bottle Cap*, TFE-fluorocarbon or aluminum foil-lined.

NOTE 2—Even these liners have some disadvantages. TFE is known to collect some organic constituents, for example, PCBs. Aluminum foil will react with samples that are strongly acid or alkaline. Clean TFE liners as described in 7.1. Replace aluminum foil with new foil after each use.

5.4 *Sample Vial*, glass.

5.5 *Septa*, PTFE-faced with screw cap lid and matching aluminum foil disks.

6. Reagents and Materials

6.1 *Purity of Reagents*—Reagent grade chemicals shall be used in all tests. Unless otherwise indicated, it is intended that all reagents shall conform to the specifications of the Committee on Analytical Reagents of the American Chemical Society, where such specifications are available.[5] Other grades may be used, provided it is first ascertained that the reagent is of sufficiently high purity to permit its use without lessening the accuracy of the determination.

6.2 *Purity of Water*—Unless otherwise indicated, reference to water shall be understood to mean reagent water conforming to Specification D 1193, Type II and demonstrated to be free of specific interference for the test being performed.

6.3 *Acetic Acid Buffer Solution* (pH 4)—Dissolve 6.0 g of sodium acetate ($NaC_2H_3O_2$) in 75 mL of water. Add 30 mL of glacial acetic acid, with stirring.

6.4 *Acetone.*

6.5 *Acid Buffer Solution* (pH 3.75)—Dissolve 125 g of potassium chloride (KCl) and 70 g of sodium acetate trihydrate ($NaC_2H_3O_2 \cdot 3H_2O$) in 500 mL of water. Add 300 mL of glacial acetic acid and dilute to 1 L.

6.6 *Ascorbic Acid.*

6.7 *Chromic Acid Cleaning Solution*—To a 2-L beaker, add 35 mL of saturated sodium dichromate ($Na_2Cr_2O_7$) solution followed by 1-L of sulfuric acid (H_2SO_4) (sp gr 1.84) with stirring. **Warning**—Use rubber gloves, safety goggles, and protective clothing when preparing and handling this corrosive cleaning agent that is a powerful oxidant. Store the reagent in a glass bottle with a glass stopper.

6.8 *Detergent*, formulated for cleaning laboratory glassware.

6.9 *Hydrochloric Acid*—Concentrated HCl (sp gr 1.19).

6.10 *Hydrochloric Acid* (1+2)—To 200 mL of water, carefully add 100 mL of hydrochloric acid (6.9). Store in a glass-stoppered reagent bottle.

6.11 *Ice*, crushed wet.

6.12 *Lead Acetate Test Paper.*

6.13 *Lead Acetate Solution*—Dissolve 50 g of lead acetate $[Pb(C_2H_3O_2)_2 \cdot 3H_2O]$ in water and dilute to 1 L.

6.14 *Lead Carbonate*, powdered.

6.15 *Lime, Hydrated*, powdered.

6.16 *Nitric Acid*—Concentrated HNO_3 (sp gr 1.42).

6.17 *Phosphate Buffer*—Dissolve 138 g of sodium dihydrogen phosphate ($NaH_2PO_4 \cdot H_2O$) in water and dilute to 1 L. Refrigerate this solution.

6.18 *Phosphate Solution*—Dissolve 33.8 g of potassium dihydrogen phosphate (KH_2PO_4) in 250 mL of water.

6.19 *Phosphoric Acid*—Concentrated H_3PO_4 (sp gr 1.83).

6.20 *Phosphoric Acid Solution* (1+1)—Dilute 1 volume of phosphoric acid (sp gr 1.83).

6.21 *pH Paper*, narrow range for pH < 2, pH > 12, and pH 5 to 7.

6.22 *Potassium Iodide-Starch Test Paper.*

6.23 *Sodium Bisulfate* ($NaHSO_4$).

6.24 *Sodium Bisulfite Solution*—Dissolve 2 g of sodium bisulfite ($NaHSO_4$) in 1-L of water and adjust to pH 2 by the slow addition of H_2SO_4 (1+1). **Warning**—Prepare and use this reagent in a well-ventilated hood to avoid exposure to SO_2 fumes.

6.25 *Sodium Sulfite Solution* (0.1 *M*)—Transfer approximately 10.3 g of sodium sulfite to a 1-L volumetric flask. Dilute to volume with water.

6.26 *Sodium Thiosulfate* ($Na_2S_2O_3$).

6.27 *Sodium Hydroxide Pellets* (NaOH).

6.28 *Sulfuric Acid*—Concentrated H_2SO_4 (sp gr 1.84).

6.29 *Sulfuric Acid* (1+1)—Slowly and carefully add 1 volume of sulfuric acid (see 6.27) to 1 volume of water, stirring and cooling the solution during addition.

7. Preparation of HDPE Sample Bottles

7.1 Wash the bottles with two 100-mL portions of HCl (1+2) and rinse with three 100-mL portions of water. These volumes of wash and rinse portions are recommended for 1-L sample bottles; therefore, use proportionate volumes for washing and rinsing sample bottles of a different volume.

[5] "Reagent Chemicals, American Chemical Society Specifications," Am. Chemical Soc., Washington, DC. For suggestions on the testing of reagents not listed by the American Chemical Society, see "Analar Standards for Laboratory Chemicals," BDH Ltd., Poole, Dorset, U.K., and the "United States Pharmacopeia."

TABLE 1 Recommended Preservation Practice for General Organic Constituent Test Methods

NOTE—The container preparation procedures described in Sections 7 and 8 should yield bottles of sufficient quality to be compatible with the test methods listed in this table. However, a sample bottle blank should be obtained to establish the fact.

ASTM Test Method	Recommended Practice
D 2579, Organic carbon, total, by combustion-infrared or reduction-FID	9.1.3 refrigeration and 9.1.1.2[A] hydrochloric acid
D 4129, Total and organic carbon, oxidation coulometric	9.1.3 refrigeration and 9.1.4 hermetically sealing, zero headspace
D 4281, Oil and grease, gravimetric	9.1.1.1 sulfuric acid or 9.1.1.2 hydrochloric acid
D 3921, Oil and grease, petroleum hydrocarbons	9.1.1.1 sulfuric acid (1+1) or 9.1.1.3 sodium bisulfate
D 1252, Oxygen demand, chemical	9.1.1.1 sulfuric acid or 9.1.1.3 sodium bisulfate or 9.1.3 refrigeration
D 4744, Organic halides, by carbon absorption-microcoulometry	9.1.1.8 nitric acid and 9.1.2 sodium sulfite and 9.1.3 refrigeration
D 4763, Identification by fluorescence	9.1.3 refrigeration
D 4779, Carbon, total, organic and inorganic in high-purity water	9.1.1.8 nitric acid and 9.1.3 refrigeration
D 4839, Carbon, total and organic in water	9.1.3 refrigeration and either 9.1.1.1[A] sulfuric acid or 9.1.1.5 phosphoric acid or 9.1.1.8 nitric acid

[A] Acidification can be used only when organic carbon alone is being determined. If total carbon is of interest, the sample must not be acidified; refrigeration is the only appropriate preservation technique.

8. Preparation of Glass Sample Bottles and Vials

8.1 *Solvent-Detergent/Chromic Acid Preparation of Glass Sample Bottles:*

8.1.1 Rinse the container with 100 mL of dilute detergent or acetone. For some residues, a few alternative detergent and acetone rinses may be more satisfactory. Then rinse at least three times with tap water followed by a reagent water rinse to remove the residual detergent or acetone, or both.

8.1.2 Rinse the container with 100 mL of chromic acid solution, returning the chromic acid to its original container after use. Then rinse with at least three 100-mL portions of tap water followed by a reagent water rinse.

8.1.3 Rinse the container with 100 mL of $NaHSO_3$ solution to remove residual hexavalent chromium. **Warning—** Carry out this step in a hood to prevent exposure to SO_2 fumes.

8.1.4 Rinse the container with water until sulfurous acid and its vapors have been removed. Test rinsings for acid with a pH meter or an appropriate narrow range pH paper. Rinsings should have a pH approximately the same as the water used for rinsing.

8.1.5 When the last trace of $NaHSO_3$ has been removed, wash with three additional 100-mL portions of water. Allow to drain. This procedure is for 1-L sample containers, therefore, use proportionate volumes for washing and rinsing sample containers of a different volume.

8.1.6 Heat for a minimum of 4 h (mouth up) in a forced draft oven at 275 to 325°C. Upon cooling, fit the bottles with caps and the vials with septa.

NOTE 3—For some tests, heating may not be required. Refer to the individual method to determine the necessity for this treatment.

8.2 *Machine Washing Glass Sample Bottles and Vials:*

NOTE 4—Machine washing of narrow mouth sample bottles may not yield acceptable results.

8.2.1 Rinse the container with 100 mL of chromic acid solution, returning the chromic acid to its original container after use. Then rinse with at least three 100-mL portions of tap water.

8.2.2 Machine wash in accordance with the machine manufacturer's instructions using a detergent and 90°C water.

8.2.3 Remove the bottles from the machine and rinse them with two 100-mL portions of HCl (1+2), followed with three 100-mL portions of water.

8.2.4 Heat for a minimum of 4 h (mouth up) in a forced draft oven at 275 to 325°C. Upon cooling, fit the bottles with caps and the vials with septa (see Note 3).

9. Sample Preservation

9.1 Depending upon the type of analysis required, use any one or a combination of the following methods of sample preservation (see Tables 1, 2, and 3 and Annexes A1 and A2).

9.1.1 Adjust the pH. An adjustment to neutral pH is usually prescribed when chemical reactions such as hydrolysis are to be avoided. Adjustment to an extreme pH, for example, <2, is usually prescribed to inhibit biological activity for biodegradable organic chemicals.

NOTE 5—To confirm the adjustment of the pH of samples to the proper value, place a drop of sample on an appropriate pH test paper or measure the pH with a pH meter.

9.1.1.1 *Sulfuric Acid*—To the sample bottle partially filled with sample, slowly add 2 mL of H_2SO_4 (sp gr 1.84) and mix thoroughly. Confirm that the pH is less than 2. If the pH is greater than 2, add additional acid until the pH is less than 2. This procedure is based on a 1-L sample bottle; therefore, use proportionate volumes for sample bottles with a different volume.

TABLE 2 Recommended Preservation Practice for Specific Organic Constituent Test Methods

ASTM Test Method	Recommended Practice
D 2330 Alkyl benzene sulfonate	9.1.1.1 sulfuric acid or 9.1.1.3 sodium bisulfate
D 4983 Cyclohexamine, morpholine, diethanolamine by DAI	9.1.1.9 phosphate/phosphoric acid and 9.1.3 refrigeration.
D 2036[A] Cyanide	9.1.1.4 sodium hydroxide and in presence of chlorine 9.1.2 chlorine removal
D 4282 Free cyanide	9.1.3 refrigeration, 9.1.1.4 sodium hydroxide, and 9.1.5 in dark
D 4374 Total dissociable cyanide, automated	9.1.3 refrigeration and 9.1.1.4 sodium hydroxide, in dark
D 4165 Cyanogen chloride	9.1.1.6 phosphate buffer to ph 8.0 to 8.5
D 4193 Thiocyanate	9.1.1.2 acid or 9.1.1.4 base
D 3478 Chlorophenoxy acid herbicides	9.1.3 refrigeration and 9.1.1.1 sulfuric acid
D 3086 Pesticides, organochlorine	9.1.3 refrigeration
D 3534 Polychlorinated biphenyls	9.1.3 refrigeration
D 3113 Ethylenediaminetetraacetate	9.1.3 refrigeration
D 3590 Total nitrogen, Kjeldahl	9.1.3 refrigeration and 9.1.1.1 sulfuric acid
D 1783 Phenolic compounds by 4-AAP	9.1.3 refrigeration and 9.1.1.1 sulfuric acid or 9.1.1.3 sodium bisulfate or 9.1.1.2 hydrochloric acid or 9.1.1.5 phosphoric acid
D 2580 Phenols by gas liquid chromatography	9.1.3 refrigeration and 9.1.1.1 sulfuric acid or 9.1.1.3 sodium bisulfate
D 4657 Polynuclear aromatic hydrocarbons	9.1.3 refrigeration, 9.1.1.2 acid, or 9.1.1.4 sodium hydroxide (to pH 6.0 to 8.0), 9.1.2 chlorine removal
D 2908 Volatile organic matter in water by aqueous injection gas chromatography (DAIGC)	9.1.3 refrigeration and 9.1.1.1 sulfuric acid or 9.1.1.3 sodium bisulfate
D 3695 Volatile alcohols by DAIGC	9.1.3 refrigeration and 9.1.1.1 sulfuric acid or 9.1.1.3 sodium bisulfate
D 3371 Nitriles by DAIGC	9.1.3 refrigeration and 9.1.1.1 sulfuric acid or 9.1.1.3 sodium bisulfate
D 3871 Purgeable organic compounds	9.1.2 chlorine removal, 9.1.3 refrigeration, 9.1.4 hermetically sealing
D 3973 Low molecular weight hydrocarbons	9.1.3 refrigeration, and in presence of chlorine 9.1.2 chlorine removal and 9.1.4 hermetically sealing, zero headspace

[A] See Annex A1 for alternative treatment if the sample is suspected to contain sulfide or a high concentration of carbonate.

9.1.1.2 *Hydrochloric Acid*—To a sample bottle partially filled with sample, add 6 mL of HCl (sp gr 1.19) while swirling the bottle. After the acid addition, confirm that the pH is less than 2. If the pH is greater than 2, add additional acid to lower the pH to less than 2. This procedure is for a 1-L sample bottle, therefore, use proportionate volumes for sample bottles with a different volume.

9.1.1.3 *Sodium Bisulfate*—To a sample bottle partially filled with sample, add approximately 9 g of $NaHSO_4$. Mix to dissolve and confirm that the pH is less than 2. If the pH is greater than 2, add additional $NaHSO_4$ until the pH is less than 2. This procedure is based on a 1-L sample bottle, therefore, use proportionate amounts for sample bottles with a different volume.

9.1.1.4 *Sodium Hydroxide*—Adjust the sample pH to above 12 using NaOH (pellets). Store the sample away from light.

9.1.1.5 *Phosphoric Acid*—To a sample bottle partially filled with sample, slowly add 2 mL of phosphoric acid (sp gr 1.83) and mix thoroughly. Confirm that the pH is less than 2. If the pH is greater than 2, add additional acid until the pH is less than 2. This procedure is based on a 1-L sample bottle; therefore, use proportionate volumes for sample bottles with a different volume.

9.1.1.6 *Phosphate Buffer*—Reduce pH to 8.0 to 8.5 range with careful additions of phosphate buffer.

9.1.1.7 *Acid Buffer*—Add 4 mL per 100 mL of sample.

9.1.1.8 *Nitric Acid*—To a sample bottle partially filled with sample, slowly add 2 mL of nitric acid (sp gr 1.42) and mix thoroughly. Confirm that the pH is less than 2. If the pH is greater than 2, add additional acid until the pH is less than 2. This procedure is based on a 1-L sample bottle; therefore, use proportionate volumes for sample bottles with a different volume.

9.1.1.9 *Phosphate/Phosphoric Acid*—Add approximately 1 mL phosphate solution followed by a few drops of 1+1 phosphoric acid solution to 115 mL of water sample to bring the pH approximately to 3.

9.1.2 *Chlorine Removal*—Chlorine is added to water supplies and discharges as a disinfectant and oxidant for organic compounds. If the chlorine is not eliminated at the time of sampling, chlorination of organics present in the sample may occur; that is, trihalomethanes and chlorophenols will form, causing a positive interference for these analytes. Test a drop of the sample with potassium iodide-starch paper; a blue color indicates the need for the following treatment: Add ascorbic acid, dissolving a few crystals at a time, until a drop of sample produces no color on the indicator paper. Then add an additional 0.05 g of ascorbic acid[6] ($C_6H_8O_6$) per litre of sample. Alternatively, sodium sulfite solution (0.1 *M*) is used; typically, 0.2 mL/100 mL of sample is sufficient.

9.1.3 *Refrigeration at 4°C*— Samples are cooled to reduce biological activity on the organic chemicals. Cool the sample to 4°C immediately after sampling using a wet ice water bath. During storage or shipment, or both, maintain the sample at 4°C. Prior to the analysis, raise the sample temperature to room temperature using a water bath with a temperature no more than 5°C above room temperature. If the sample temperature is not adjusted, then an appropriate temperature-volume correction must be made.

[6] Do not use ascorbic acid when organic carbon is to be determined.

TABLE 3 Maximum Holding Times Allowed by ASTM Test Methods and U.S. EPA Regulations

ASTM Test Method	Holding Times	
	In Standard	In 40 CFR 136[A]
D 2330 Alkyl benzene sulfonate	7 days	NA[B]
D 2036 Cyanide	None stated	NA
D 4282 Free cyanide	Immediate analysis recommended	14 days
D 4374 Total dissociable cyanide, automated	Immediate analysis recommended	14 days
D 4165 Cyanogen chloride	Immediate analysis recommended	NA
D 4193 Thiocyanate	None stated	NA
D 3478 Chlorophenoxy acid herbicides	24 h	7 days for extraction, 40 days after extraction
D 3086 Pesticides, organochlorine	None stated	7 days for extraction, 40 days after extraction
D 3534 Polychlorinated biphenyls	None stated	7 days for extraction, 40 days after extraction
D 3113 Ethylenediaminetetracetate	15 min recommended	NA
D 3590 Total nitrogen, Kjeldahl	28 days	28 days
D 4281 Oil and grease, gravimetric	Up to 2 months	28 days
D 3921 Oil and grease, petroleum hydrocarbons	None stated	28 days
D 2579 Organic carbon, total by combustion-infrared or reduction-FID	None stated	28 days
D 4129 Total and organic carbon, oxidation coulometry	None stated	28 days
D 1252 Oxygen demand, chemical	24 h if not acidified	28 days
D 1783 Phenolic compounds by 4-AAP	48 h	28 days
D 2580 Phenols by gas liquid chromatography	Keep to minimum	28 days
D 4657 Polynuclear aromatic hydrocarbons	7 days for extraction, 30 days for analysis	7 days for extraction, 40 days after extraction
D 2908 Volatile organic matter in water by aqueous injection gas chromatography (DAIGC)	Keep to minimum	NA
D 3695 Volatile alcohols by DAIGC	None stated	NA
D 3371 Nitriles by gas liquid chromatography by DAIGC	Keep to minimum	NA
D 3871 Purgeable organic compounds	None stated	14 days
D 3973 Low molecular weight hydrocarbons	15 days	14 days
D 4744 Organic halides, by carbon absorption-microcoulometry	7 days	NA
D 4763 Identification by fluorescence	None stated	NA
D 4779 Carbon, total, organic and inorganic, in high-purity water	None stated	28 days
D 4839 Carbon, total and organic, in water	None stated	28 days
D 4983 Cyclohexamine, morpholine diethanolamine by DAI	None stated	NA

[A] Title 40, Code of Federal Regulations, Part 136 (40 CFR 136), by the U.S. Environmental Protection Agency.
[B] NA = Not applicable.

9.1.4 *Hermetically Sealing (Purgeable Organics)*—Add 1 to 2 mg of sodium thiosulfate ($Na_2S_2O_3$) or ascorbic acid[6] ($C_6H_8O_6$) to a sample vial. Fill the vial to overflowing so that a convex meniscus forms at the top. Place a septum, PTFE side down, carefully on the opening of the vial, displacing the excess water. Place aluminum foil disk over the septum. Seal the vial with the screw cap and invert to verify the seal by demonstrating the absence of air bubbles (zero headspace).

NOTE 6—The purpose of the aluminum foil disk is to prevent contamination of the sample through permeation of the septum by ambient organic vapors. Alternative methods of preventing sample contamination that have been demonstrated to be equally effective may be used.

9.1.5 *Minimize Photodecomposition*—Collect samples in dark bottles and store in the dark.

10. Sample Holding Times

10.1 Table 3 lists maximum holding times prescribed in ASTM standards for measuring organic compounds in water. The applicable holding times cited in the U.S. EPA "Guidelines Establishing Test Procedures for the Analysis of Pollutants Under the Clean Water Act"[7] are also included in Table 3.

10.2 Samples that exceed holding times should be discarded rather than analyzed.

10.3 Holding times for organic constituents are highly dependent upon the chemical and biological composition of the sample. Sample holding times for a specific matrix may be determined by using the procedure described in Practices D 4515 or D 4841. These practices are particularly useful when sampling, transporting, or scheduling complications make it desirable to have holding times beyond those prescribed in the test method.

[7] "Guidelines Establishing Test Procedures for the Analysis of Pollutants Under the Clean Water Act," Title 40, Code of Federal Regulations, Part 136.3(e), written by the U.S. Environmental Protection Agency, available from the Superintendent of Documents, U.S. Government Printing Office, Washington, DC 20401.

ANNEXES

(Mandatory Information)

A1. ALTERNATIVE TREATMENT STEPS

A1.1 The following is a list of additional or alternative treatment steps that are required in the presence of specific interfering constituents:

ASTM Test Method	Interfering Constituent	Recommended Treatment for Preservation
D 2036 Cyanide	sulfide	Sulfide in the sample can convert CN^- to SCN^-, especially at high pH. Before stabilizing the sample by raising the pH for the preservation of the cyanide content, test for sulfide by placing a drop of sample on lead acetate test paper previously moistened with acetic acid buffer solution (pH 4). Darkening of the paper indicates the presence of sulfide. (The simultaneous presence of both sulfide and oxidizing agents is not anticipated. Oxidizing agents should be also removed before sample preservation.) Sulfide is removed by adding lead acetate solution (50 g/L) a drop at a time; retest on the test paper and continue until a negative paper test has been read, add 1 drop in excess and immediately filter out the black lead sulfide precipitate. If sulfide content is high, add instead powdered lead carbonate to avoid significant reduction of the pH. After sulfide removal, continue with 9.1.1.5.
	high carbonate content	When sampling effluents such as coal gasification wastes, atmospheric emission scrub waters, and other high carbonate content wastes, use hydrated lime in the powder form to stabilize the sample instead of NaOH. Add slowly with stirring to raise the pH to 12 to 12.5. Decant the sample into the sample bottle after the precipitate has been settled. (High carbonate content affects the distillation procedure by causing excessive gasing when the acid is added. The CO_2 released also may significantly reduce the NaOH content in the adsorbent.)

A2. PRESERVATION OF COMPOSITE SAMPLES

A2.1 When composite samples are collected, the appropriate preservation reagents must be added to the compositing vessel prior to collection. If the preservation requirements call for refrigeration, the sample must be refrigerated during collection. The collection time for a single composite sample shall not exceed 24 h. If longer sampling periods are necessary, a series of composite samples shall be collected.

Designation: D 3856 – 88

Standard Guide for Good Laboratory Practices in Laboratories Engaged in Sampling and Analysis of Water[1]

This standard is issued under the fixed designation D 3856; the number immediately following the designation indicates the year of original adoption or, in the case of revision, the year of last revision. A number in parentheses indicates the year of last reapproval. A superscript epsilon (ε) indicates an editorial change since the last revision or reapproval.

1. Scope

1.1 This guide addresses the laboratory practices considered essential to assure high quality performance by laboratories that provide services in the sampling and analysis of water. The utilization of these practices will provide the data whose quality can be substantiated through written field and laboratory records.

1.2 This guide is designed to be used by those responsible for the selection, operation, or control of laboratory organizations engaged in sampling and analysis of water.

1.3 This guide presents features of organization, facilities, resources, and operations which by their selection and control affect the usefulness of the data generated.

1.4 Moreover, criteria are presented for the selection and control of the features described in 1.3 so that acceptable performance may be attained and sustained.

1.5 This guide also makes recommendations for the correction of unacceptable performance.

2. Referenced Documents

2.1 *ASTM Standards:*

D 1193 Specification for Reagent Water[2]

D 1129 Terminology Relating to Water[2]

D 2777 Practice for Determination of Precision and Bias of Applicable Methods of Committee D-19 on Water[2]

D 3370 Practices for Sampling Water[2]

D 3694 Practices for Preparation of Sample Containers and for Preservation of Organic Constituents[3]

D 4210 Practice for Intralaboratory Quality Control Procedures and a Discussion on Reporting Low-Level Data[2]

D 4375 Terminology for Basic Statistics in Committee D-19 on Water[2]

D 4840 Practice for Sampling Chain of Custody Procedures[2]

D 4841 Practice for Estimation of Holding Time for Water Samples Containing Organic and Inorganic Constituents[2]

E 456 Terminology Relating to Quality and Statistics[4]

E 548 Guide for General Criteria Used for Evaluating Laboratory Competence[4]

3. Terminology

3.1 For definitions of terms used in this guide refer to Terminology D 1129 and E 456, Practices D 4375 and E 548, and ASTM STP 15D **(1)**.[5]

4. Summary of Guide

4.1 This guide describes the criteria, guidelines, and recommendations for physical and human resources and data validation for the operation of a laboratory.

5. Significance and Use

5.1 Data on the composition and characteristics of water are frequently used to evaluate the health and safety to humans and the environment.

5.2 Moreover, such data are frequently used to ascertain compliance with regulatory statutes that place limits on acceptable compositions and characteristics of waters.

5.3 Laboratories that conduct water sampling and generate analytical data, and those persons who have the responsibility for selecting a laboratory to perform water quality studies need to use criteria, guidelines, and recommendations that have been developed by consensus and are well accepted in making this selection.

5.4 Demonstration and documentation by a laboratory that there was judicious selection and control of organization, facilities, resources, and operations will enhance the credibility of the data produced and promote its acceptance.

6. Aspects of Quality Assurance

6.1 *General*—The function of a laboratory is to provide analytical results and related information which are adequate for the intended use. This function is achieved through effective use of a quality assurance program. The elements of such a program are:

6.2 *Organizational Structure*—A table of the organization should be available which shows the lines of authority, areas of responsibility, and job functions. The laboratory should also provide a description of its capabilities. Laboratory management should demonstrate and foster a positive quality assurance attitude and provide the analytical staff with a written policy to carry out a defined quality assurance program.

[1] This guide is under the jurisdiction of ASTM Committee D-19 on Water and is the direct responsibility of Subcommittee D19.02 on General Specifications and Technical Resources.

Current edition approved Sept. 13, 1988. Published November 1988. Last previous edition D 3856 – 80$^{\epsilon 1}$.

[2] *Annual Book of ASTM Standards*, Vol 11.01.

[3] *Annual Book of ASTM Standards*, Vol 11.02.

[4] *Annual Book of ASTM Standards*, Vol 14.02.

[5] The boldface numbers in parentheses refer to the references at the end of this guide.

6.2.1 *Human Resources*—The key personnel of the organization should be described by means of personal resumes presenting the education and work experience appropriate to the table of organization and the qualifications of the position.

6.2.2 *Physical Resources*—The laboratory facilities should provide a working environment that is clean, comfortable, and safe. The instrumentation and equipment must be suitable for the operational needs of the laboratory.

6.3 *Methodology*—Written procedures should be readily available to personnel.

6.3.1 Written sample collection, handling and storage requirements should be followed.

6.3.2 Analytical procedures should be written.

6.3.3 There should be a document control system to track the currency and completeness of procedures.

6.4 *Metrology*—Systems for making measurements should have:

6.4.1 Written calibration procedures, including traceability and schedule.

6.4.2 Written preventative maintenance procedures with scheduled intervals.

6.4.3 Records available to document any repair or service of equipment, replacement or change of reagents, or modification of procedures.

6.5 *Data Recording*—The laboratory should keep records of sample source disposition, and analyses to provide information on sample collection and preservation, analytical procedures and results, and the persons responsible for the sampling and analysis. All laboratory data sheets should be dated and signed by the analyst and his supervisor or a suitable colleague.

6.6 *Data Validation*—To assess and demonstrate a suitable level of analytical performance the laboratory should keep records of audit procedures, reference sample programs, and interlaboratory tests. One criterion for measurement of analytical performance should be the precision and bias data contained in the analytical method being used. Where applicable, quality control charts as described in Practice D 4210 should be used.

6.7 *Trouble Shooting*—The organization should provide the authority and the responsibility to a designated person or persons for investigation of out-of-control results and for informing the laboratory management of any problems that occur.

6.7.1 A current log should be kept of analytical deficiencies and the action taken to correct them.

7. Organization

7.1 *General*—The production of reliable data is effected through the effort of everyone involved with the service. It is paramount, therefore, that personnel have a clear understanding of their duties and responsibilities and their relationship to the product produced. Management has the responsibility for defining function and goals as applied to the individual. A formal document describing objectives, staff functions and responsibilities, should be distributed and explained to all staff members.

7.1.1 The personnel in a laboratory will vary with the specific functions that are to be served, but minimal qualifications and duties generally will be as described in 7.2 through 7.3.5.11.

7.2 *Human Resources/Personnel Duties and Responsibilities:*

7.2.1 *The Director*—The director should have a working knowledge of the laboratory and its scope of activities. The director should have earned a baccalaureate degree in science or engineering from an accredited college or university or the equivalent (Note 1) and have at least five years experience in laboratory work.

NOTE 1—The purpose of the "equivalent" requirement is to allow the assignment of persons who have comparable skills obtained through qualified training which did not result in the award of a baccalaureate degree. Interpretation of the term "equivalent" will necessarily require careful judgment by the user of these guidelines. Certification by professional boards is to be encouraged.

The laboratory director or manager should be a full-time employee that operates the laboratory with at least the responsibilities outlined in 7.2.1.1 through 7.2.1.6.

7.2.1.1 Establishment of long-term program plans and shorter term work plans and assignments to meet the program objectives.

7.2.1.2 Operation and maintenance of the physical plant (building, equipment, instrumentation, services, etc.).

7.2.1.3 Selection, training, and development of personnel.

7.2.1.4 Overview and approval of methods of sampling and analyses.

7.2.1.5 Development and implementation of a quality assurance (QA) program to monitor and maintain the quality of laboratory performance. This includes insuring staff participation in appropriate interlaboratory quality control activities, intercalibration checks, performance audit programs, etc. Such interlaboratory checks are the most effective measure of comparative performance and should demonstrate the worth of a good quality assurance program worth to upper management or regulator agencies. A QA program also provides each laboratory staff member with a copy of the quality assurance plan for the laboratory, which document responsibilities, and kinds and frequency of quality control checks. The plan should also specify the monitoring and overview responsibilities of management.

7.2.1.6 Establishment of a development and operational performance appraisal system for the staff and an individual career development plan for each staff member. Performance standards should be developed and agreed to jointly by each staff member and their supervisor. The director should be responsible for assuring a periodic review of performance of all staff members by supervisors, for rewarding good quality performance, and for implementing and encouraging on-the-job or off-site training. This joint development of performance standards is key to obtaining an understanding between the worker and the supervisor, as to what is expected for satisfactory performance. It also paves the way for rewarding outstanding performance or identifying unsatisfactory performance. These standards should be used to evaluate performance frequently but informally, and formally on a less frequent (annual or semiannual) basis.

7.2.2 *Senior Staff*—The senior professional staff of the laboratory conduct the difficult and non-routine sampling and analyses, resolve analytical problems, and modify and develop analytical procedures.

7.2.2.1 Senior staff supervise and assist the technical staff in analyses, other laboratory operations and training.

7.2.2.2 Senior staff members should have earned a baccalaureate degree in science or engineering from an accredited college or the equivalent (see Note 1) and have at least two years experience at the bench level in a water laboratory.

7.2.3 *Technical Staff*—The technical staff are personnel who perform routine and specialized sampling and analyses.

7.2.3.1 Where appropriate, technical staff members should have formal training in the sampling and analytical methodology, and quality control, as applied to the specific sample types and concentration levels of analytes which are of interest to the laboratory.

7.2.3.2 Technical staff may be required to satisfactorily complete analytical tests to qualify initially and to periodically requalify throughout their work career. Qualification should be based on the generation of analytical results with precision and bias recovery within limits known to be possible for the particular method and which meet the data user's requirements.

7.2.4 *Laboratory Support Staff*—The support staff are nontechnical workers who perform routine field and laboratory services in support of the professional and technical staff. In the field, they may collect samples, and they may transport, handle and maintain records of collection and storage, including chain of custody. The support staff also maintains sampling and other field equipment.

7.2.4.1 In the laboratory, they wash glassware, operate laboratory reagent water systems, autoclaves, drying ovens, and incubators. The support staff also receives, stores, and ships samples, materials, and laboratory equipment.

7.2.5 *Office Support Staff*—The office staff are nontechnical clerical or secretarial personnel who are trained either on the job or by formal schooling in typing, filing, recordkeeping, communications by telephone or personal visits, payroll, travel, or both.

7.2.5.1 The laboratory or office support staff may be an integral part of the laboratory, or may be provided as part of the administrative function in a larger organization.

7.3 *Physical Resources and Related Operating Procedures:*

7.3.1 *General*—The laboratory environment can significantly affect the results of water analyses; therefore, the laboratory facility should be carefully designed and periodically inspected and reevaluated. In general, the physical conditions in the laboratory should comply with the applicable U.S. OSHA requirements or their equivalent, and other safety and legal requirements. For further information see "Laboratory Planning for Chemistry and Chemical Engineering" **(2)**.

7.3.2 The specific instrumentation, equipment, materials, and supplies needed for the performance of a standard test method are usually described in the written standard. If the laboratory proposes to perform a new analytical procedure, it must be prepared to acquire the necessary instrumentation, supplies and space, and to conduct an appropriate training period prior to its routine use.

7.3.3 *Laboratory Environment*—The laboratory should be kept as free from environmental contamination as possible in order to protect the samples and instrumentation. By doing so, the laboratory ensures the opportunity to produce quality data. The production of valid data not only depends on the collection of representative samples, but also on maintaining such samples as closely as possible to their original condition through careful handling and storage. If the sample cannot be analyzed at once, it should be preserved and stored as required for the analytes of interest. Recommended procedures for collecting, transporting and handling water and wastewater samples are described in this guide or in Practices D 3370 and D 3694. Recommended chain of custody procedures are described in Practice D 4840. Whenever sample holding times must be determined, recommended procedures are described in Practice D 4841.

7.3.4 *Ventilation System*—Laboratories should be well ventilated and free of dust, drafts, and extreme temperature changes. Central air conditioning is recommended because: 1) incoming air is filtered, reducing the likelihood of airborne laboratory contamination; 2) uniform temperature is conducive to stable operation of instrumentation and equipment; and 3) low humidity reduces moisture problems with hygroscopic chemicals, samples, and corrosion problems with analytical balances and other instrumentation.

7.3.4.1 In order for the hoods to be effective in removing fumes and aerosols from the laboratory environment, they must be operating at their designed capacity. Proper hood performance cannot be assumed. Hoods should be tested periodically for proper air flow by qualified support staff or a professional maintenance contractor. Hoods should not be located in areas of countervailing drafts, such as between two open doors. Under usual operating conditions, hoods require from 50 to 125 CFM/sq ft of face area. For a more detailed treatment of ventilation consult "Industrial Ventilation—A Manual of Recommended Practice" **(3)**.

7.3.5 *Facilities*—Ideally the areas provided for cleaning of glassware and portable equipment should be separated from the laboratory working area but located close enough for convenience.

7.3.5.1 Laboratories conducting trace organic analyses which use organic solvents in extraction and clean-up procedures must separate these activities from analytical instrumentation rooms to avoid contamination and reduce hazards.

7.3.5.2 Laboratories conducting analyses with a wide range of concentrations must take care to avoid cross contamination among samples in storage or analysis. Relatively clean samples, highly polluted samples and reagents should be stored separately from each other in vented cabinets and hoods to avoid cross-contamination.

7.3.5.3 Calibration standards should be stored separately from all samples.

7.3.5.4 *Laboratory Design*—Limited facilities and restricted work space may affect the quality and validity of results. Visitors and incidental traffic should be discouraged in work areas. Through traffic can be prevented by good laboratory design.

7.3.5.5 Bench tops should be set at heights of 36 to 38 in. with a depth of 28 to 30 in. This height is comfortable for work in a standing or sitting position. Desk tops or sit-down benches should be set at 30 to 31 in. height to accommodate calculations and writing activities. Bench tops should be constructed of stainless steel, epoxy plastic, or another smooth impervious material that is inert, corrosion-resistant, and has minimal seams.

7.3.5.6 Walls should be covered with waterproof paint, enamel, tile, or other surface materials that provide a smooth finish that is easily cleaned or disinfected, or both. Floors should be covered with good quality tiles or other heavy duty material, which can be maintained by wet-mopping or scrubbing machine and application of a skid-proof wax. Bacteriostatic agents contained in some wall or floor finishes increase the effectiveness of disinfection. Floors should never be swept or dry-mopped.

7.3.5.7 High standards of cleanliness should be maintained and monitored for contamination in work areas and the laboratory. If there is any doubt about the effects of the surrounding laboratory facility upon the analytical results, blanks that have been protected against the laboratory environment should be compared periodically against sample blanks that have been exposed to the laboratory environment.

7.3.5.8 To minimize the effects of harmful or interfering fumes in the laboratory environment, samples, reagents, and solvents must be used in a properly operating hood or glove box. Smoking, eating, and drinking should not be allowed in the work area. Soiled hands should be washed before handling analytical materials. Sinks must not be used for sample or reagent disposal. Under the current Resource Conservation and Recovery Act (RCRA) regulations, laboratories handling hazardous or toxic reagent chemicals, solvents, or samples must dispose of such waste through a licensed hauler to a permitted disposal facility. The disposal area must meet applicable environmental and safety standards. The specific disposal methods must be treated separately. Procedures for handling, storage, and disposal of hazardous reagents and samples must be defined. When relevant information is not supplied in an ASTM standard, useful information may be obtained from guides for the control of hazardous chemical spills and manuals of laboratory safety available from various laboratory supply firms and from "The Firefighters Handbook of Hazardous Materials" **(4)**. Short courses on handling hazardous and toxic chemicals are available also from chemical companies and others.

7.3.5.9 *Working Area*—Adequate bench space or working areas should be provided for each analyst. As a general guideline, there should be 150 to 300 ft^2 per analyst or 12 to 25 lineal ft of bench space per analyst. The space requirement per analyst will vary with the instrumentation or equipment in use, the sample load, and the number of operations that are to be performed by a single analyst. The laboratory may require specialized facilities such as a perchloric acid hood. The lighting level may vary between 50 footcandles (fc) (54 lux) and 100 fc (108 lux) depending upon the closeness of the tasks being performed in the area.

7.3.5.10 *Electric Power Supply*—The reliability of the instruments is affected by the electrical power supply. Some instruments require separate circuits or a regulated power supply for stable operation. The line voltage and stability should be monitored periodically and not assumed as based on records. Surge suppressors should be installed for any sensitive instrumentation or computers. An isolated ground for individual instruments and antistatic pads are helpful in eliminating stray currents.

7.3.5.11 *Safety Considerations*—The laboratory should be supplied with fire extinguishers suitable for Class *A, B*, or *C* fires; spill control materials for acids, bases, and flammables; eye wash and safety shower facilities; and other safety devices that may be consistent with the particular laboratory operation. The facilities should provide for the safe disposal of reagents and samples with written instructions for the utilization of these procedures by all personnel. Wearing of safety glasses or masks, or both, should be required for everyone entering the laboratory. A senior staff member should be assigned the responsibility for monitoring laboratory safety including periodic inspection of facilities and fire extinguishers.

8. Methodology

8.1 *General*—The use of written procedures is an essential element in providing useful data. The use of published procedures is usually not sufficient to guarantee reliable and reproducible analytical performance. Each laboratory performing an analysis should reduce the methodology to a protocol, which is an exactly specified series of steps that guarantee the same behavior on the part of the independent analysts.

8.1.1 The quality assurance program should clearly identify the duties of: the laboratory staff for routine quality control (QC) checks, the senior staff for continuing review of the quality control activities, and the laboratory director for periodic quality control reviews..

8.2 For ease of cross referencing methods and ensuring completeness of methods, all protocols should follow the same format.

8.3 Strict adherence to the method protocol must be maintained and checked using a system of method performance audits. When deviations are necessary, the protocol should be rewritten to reflect the changes. If time does not permit a rewriting of the protocol, the necessary deviations from the protocol must be approved in writing by supervision before proceeding with the analysis.

8.4 Whenever a deviation from a written protocol is necessary, it should be noted in the final data report along with identification of the person who authorized the change.

8.5 A file of method protocols should be maintained by the laboratory director or manager. Each new protocol or revised protocol should carry its effective date so that recorded data may be related to the appropriate protocol.

9. Metrology

9.1 A set of Class *S* weights or better must be available to make periodic checks on balances. An NBS certified thermometer should be used periodically to check temperature measurement devices. A set of color standards may be used to check the wavelength calibration and the stray light characteristics of a spectrophotometer or colorimeter. Systems such as balances and spectrometers can be maintained and certified under an annual service contract.

9.2 The laboratory providing analytical services on the characterization of water may wish to make additional measurements, such as water flow. The calibration of the flow measurement system, used in the field, should be traceable to standard weights and measures.

9.3 All metrology systems should have a record of calibration and maintenance schedules, and should note configuration changes that may have occurred in such a system. Records of significant changes in calibration should be noted and reviewed periodically.

10. Data Recording

10.1 Any system is acceptable as long as it meets the requirements outlined in 6.5. A commonly used method of recording laboratory data is a notebook specifically designed for this purpose. The pages are serially numbered in pairs with a carbon between the pages to provide duplicate serial numbered copies of the data. These books are permanently bound, but the duplicate page is perforated for easy removal. The duplicate page then can be filed separately. The pages of the notebook are generally lined in a grid pattern with provision for information such as, project identification, date, and signatures of the analyst and supervisor. The analyst should record information on the analyte, method of analysis, analytical conditions and results, and remarks. There should be an example of the calculations. Entries must be in ink with no erasures or whiteout. Revisions should be indicated by a single line through the original entry with the correction along side or referenced. Changes or corrections must be dated and initialed.

10.2 When data are generated using an automated or semiautomated instrumentation, it is generally displayed on a strip chart recorder or printed tape. Chart paper should contain the signature of the analyst, the date, the sample identification, and the operational parameters of the instrument. Chart papers and tapes should be retained as a part of the permanent record. Some laboratories prefer to use microfilm for record retention.

10.3 Laboratory sampling, analyses, and data recording are increasingly automated and computerized. Projections into the future indicate an even more rapid transition to computerized operations of almost all sampling and analyses in water chemistry as has begun in clinical chemistry. Universities, colleges, and technical schools now require chemists to take courses in computer logic, programming, and operation so that graduated chemists and technicians are computer literate and able to utilize the advances already available. Instrument responses to samples, quality control checks, and blanks are now directly entered into the computer. Results and limits are calculated, and data acceptability is indicated immediately. Results are reviewed by the analyst on a video screen, and a hard copy is printed out only as desired. Results, evaluations, and summaries are archived off-line. Use of hard disks, floppy disks, and streamer tape back-ups provide the necessary redundancy to avoid loss from system crashes. A wealth of versatile software programs for personal computers permits statistical evaluations, spread sheets, scheduling, complete record-keeping for cost monitoring, supply management, quality control monitoring, report writing, and laboratory management.

10.4 The recording of the data and the analytical results should be in a format that is agreed to by the laboratory and the data user. The laboratory should have a written protocol regarding the number of significant figures, detection limits, reporting convention for nondetection, analytical range, etc.

11. Data Validation

11.1 *General*—The validation of data will require a variety of techniques due to the variety of ways in which data are produced. If the data is collected manually, the validation procedures should take into account the sampling, the sample handling, the calibration and performance of the analytical system, and the calculations. The sampling, the calibration, and the instrument performance should be taken into account if the data are generated by instrumental means.

11.2 *Sampling*—Because the sampling of water (Practice D 3370), whether performed manually or by instrumental means, involves operations upon a heterogeneous mass under uncontrolled conditions, reliable conclusions can seldom be drawn from one or a few samples. The sampling plan must provide an adequate number and volume of samples to permit statistical evaluation of the data produced. The reasons for obtaining the information, the methods of obtaining it, and the desired levels of confidence in the information, cannot be addressed within this practice for all situations. For further information, see the USEPA "Handbook for Sampling and Sample Preservation" **(5)**.

11.3 *Sample Handling and Identification*—To ensure that proper procedures are observed, to track sample collection, transportation, storage, and analysis, and to protect against loss, misidentification, tampering, or other errors that may be introduced, the sampler is responsible for providing the following information for every sample collected:

11.3.1 Collection date, time, and location;

11.3.2 Weather conditions and other remarks considered appropriate;

11.3.3 Sample identification number and the name of sampler;

11.3.4 The analytes to be tested and the sample preservation techniques utilized, if applicable; and

11.3.5 Appropriate warnings whenever the samples are time, light, or temperature sensitive, coupled with an indication of the allowable holding time. If it is necessary to estimate appropriate holding times, refer to Practice D 4841.

11.4 *Chain of Custody*—The laboratory should record the history of every sample received, including its collection, preservation, transportation, transfers, analysis, and final disposal. This record will assist the laboratory in the investigation of any problems regarding the sample. If the sample is to meet regulatory or legal requirements, a formal chain of custody is a must. For details regarding chain of custody procedures see Practice D 4840.

11.5 *Analytical Quality Control* (*AQC*)—Items stated in 11.6 through 11.12 are recommended as the basis for a routine within laboratory Analytical Quality Control Program. For further information, see the USEPA "Handbook for Analytical Quality Control in Water and Wastewater Laboratories" **(6)**.

11.6 *Calibration*— For each analyte, prepare a calibration curve which covers the entire working range of the method. Construct the curve using at least three points, including one near the upper limit of the concentration range and one near the lower limit with a reasonably equitable distribution of the remaining points. The actual minimum number of calibration points depends upon the width of analytical range and the shape of the calibration curve. For example, a broad range or a curve not known to be linear might require calibration at five to seven points. The difference (D) between the value observed for each calibration standard (X_{OBS}) and the value predicted for that standard using the fitted calibration curve (X_{PRED}), can be evaluated as follows:

11.6.1 After 20 to 25 initial values of $D = X_{OBS} - X_{PRED}$ are available, calculate their mean ($\bar{D}$) and standard deviation (S_D).

11.6.2 If a subsequent D value is not within the interval $\bar{D} \pm 3$ (S_D), the system calibration is out of control and the system must be recalibrated before continuing routine analyses.

11.6.3 If the calibration curve was fitted by minimizing the sum-of-squares of the absolute deviations of each calibration point from the curve, then the D values should also be in absolute units. However, if *percentage* deviations were minimized, then the D's also should be percentages of the predicted value.

11.6.4 If there is no obvious reason to require recalibration, instead of following steps in accordance with 11.6, an existing calibration curve may be verified at the beginning of the analytical run by measuring two calibration standards, one in the lowest quarter of the total range of the existing calibration curve and one in the highest quarter of the range.

11.6.5 Evaluate the difference $D = X_{OBS} - X_{PRED}$ for each calibration standard using the criteria developed in 11.6 and recalibrate if results are not within the control limits. As in the step described in 11.6, whether the D values are percentages or absolute units depends on how the calibration curve was fitted.

11.6.6 Record the D values for all acceptable calibration analyses and after 20 to 25 additional results, revise related control limits by recalculating $\bar{D}$ and S_D from the new data.

11.6.7 It is important to preserve the sign of all D values.

11.7 *Method Blank:*

11.7.1 Determine reagent water, reagents, and solvent blanks for each set of samples analyzed, when there is a change in the reagent water system, or whenever a new source (newly prepared reagent or solvent) is introduced into the analytical system. Reagents or solvents, or both, should also be checked for purity prior to use.

11.7.2 Carry each method blank through the entire procedure.

11.7.3 Response to a nonzero method blank depends a great deal upon the method, but the associated data must certainly be corrected or discarded and every effort should be made to resolve or minimize system interferences.

11.8 *Field Blank:*

11.8.1 Analyze a field blank with each set of samples from a given source.

11.8.2 Carry each field blank through the entire procedure.

11.8.3 When interferences occur, it is best to discard the associated analytical results, investigate the cause so such losses may be avoided in the future, and resample. In situations where it is impossible to resample, however, it may be necessary to report the available results along with a note explaining the extent of the interference and its affect upon the data.

11.9 *Precision*—Precision is the closeness of agreement between the results of repeated analysis on the same sample.

11.9.1 *General*—Develop the necessary initial data by randomly selecting routine samples to be analyzed twice in order to provide duplicate analyses. Consider the steps in 11.9.1.1 through 11.9.8.

11.9.1.1 Develop these data over a reasonable period of time to reflect day-to-day operations.

11.9.1.2 Choose the samples that are most representative of the interference potential of the sample type. If the laboratory handles multiple sample types with different precision characteristics, it will be necessary to establish and maintain separate background data and evaluation criteria for each sample type.

11.9.1.3 Ultimately, samples representing the entire concentration range and interference potential should be included, within each sample type if necessary.

11.9.2 From each pair of duplicate analytes (X_1 and X_2) calculate their relative range value, (R):

$$R = \frac{X_1 - X_2}{(X_1 + X_2)/2}$$

where:

$X_1 - X_2$ means the unsigned difference between X_1 and X_2.

11.9.3 After 50 to 100 R values are available for an analyte, order the R values by their related sample concentration estimates, organize the values into concentration ranges that seem to have a similar underlying R value and calculate the average R value ($\bar{R}$) for each of these concentration ranges. Minimize the number of concentration ranges as much as practical.

11.9.4 Calculate the upper limit (UCL) for each concentration range as:

$$\text{UCL} = 3.27\ (\bar{R})$$

(Note 2).

NOTE 2—This factor may be found in Table 2, p. 83 of Ref. **(1)**.

11.9.5 Review the initial data for R values greater than the UCL value for the appropriate concentration range. If such values are found, they should be discarded and the related UCL value should be recalculated from the remaining R values within that concentration range.

11.9.6 Within each set of 20 or fewer samples to be analyzed together, evaluate system precision by conducting duplicate analyses on one of the samples selected at random. If the relative range value (R) calculated from these duplicates is greater than the appropriate UCL value, system precision is judged to be out-of-control and analyses must stop until the problem has been resolved. Problems with these data may indicate the need for stricter adherence to accepted laboratory practices.

11.9.7 After obtaining 20 to 25 additional acceptable pairs of data within each concentration level for a sample type, periodically update the table of critical relative range values by repeating the step described in 11.9.4 using the new data. Review the criteria being maintained and combine any which are very similar for related concentrations or sample types. If the criteria for adjacent concentration ranges are quite different, further subdivision by concentration may be necessary.

11.9.8 Table A1.1 gives an example of precision estimates from duplicate analyses within specific concentration ranges for three analytes.

11.10 *Accuracy Check Using Standard Solutions:*

11.10.1 Analyze at least one standard through the complete method for every subset of 20 or fewer routine samples to be analyzed together.

11.10.2 To provide a complete record of the calibration and recovery for each analytical run, one of these standard samples should be the last sample analyzed.

11.10.3 Use concentrations that approximate those found in the related routine samples. Calculate percent recovery (P) as:

$$P = \frac{100 \text{ (observed value)}}{\text{(true value)}}$$

11.10.4 After 20 to 25 standards are analyzed over time, calculate average percent recovery ($\bar{P}$) and standard deviation (S_P) of the resulting P values.

11.10.5 If subsequent standards for percent recovery are not within the interval $\bar{P} \pm 3\ S_P$, the analytical system should be checked for problems. If problems exist, resolve them, before continuing the analyses. Problems with these data often require greater care in sample processing prior to actual measurement.

11.10.6 Runs of seven or more successive points, all either above or below $\bar{P}$, also indicate the system is out-of-control. Use of a Shewhart $\bar{X}$-chart is recommended to facilitate evaluation of percent recovery results. An example of the calculation of percent recovery and development of a Shewhart $\bar{X}$-chart is given in Table A1.2 and Fig. A1.1.

11.10.7 Record recovery of all acceptable check standards and after 20 to 25 additional results, revise the related control limits by recalculating P and S_P from the new data. As in 11.9.7, the criteria subdivisions by sample type and concentration range should be periodically reviewed to judge their appropriateness.

11.11 *Accuracy Check Using Recovery of Spikes:*

11.11.1 Do essentially the same thing for recovery as was done in 11.10, except that a concentrate spike is added to a randomly selected *routine environmental sample* from the current analytical run rather than to *reagent water*, and P values for the recovery data are calculated as:

$$P = \frac{100\ [X - BV/(V + v)]}{T}$$

where:

X = analytical result from the spiked sample,
B = background concentration naturally present in the sample,
T = known concentration of the spike,
V = volume of sample used, and
v = volume of spike added.

11.11.2 In spiking samples, make sure that:

11.11.2.1 Sufficient spike is added to at least double the background concentration or to reach a concentration for which the calibration curve has been established. If the background concentration is higher than the midpoint of the standard curve, the background water should be diluted into the lower half of the calibration range and reanalyzed before spiking.

11.11.2.2 The volume of a spike should be kept to a minimum and not exceed 5 % of the sample volume. In organic analyses, the volume of spike should be no greater than 150 mL so that the solubility of the standard in the water will not be affected.

11.11.3 Resulting P values must fall within $\bar{P} \pm 3\ S_P$ calculated from previous related spike recovery data. If not, the system is out-of-control, and the cause must be found and corrected before continuing the analyses. Problems with these data often indicate sample matrix interferences.

11.11.4 As in 11.10.6, runs of seven or more results on the same side of X indicate the system is out-of-control and the use of a Shewhart X-chart is recommended to facilitate evaluation of results.

11.11.5 By simply calculating P from P values calculated as specified in either 11.11.1 or 11.10.3, percent recoveries of a spike can be treated as shown in the example given in Table A1.2 and Fig. A1.1.

11.11.6 Periodically review and update the recovery criteria similarly to 11.10.7.

11.12 *Summary of the Analytical Quality Control*—The following recommended analytical quality control program should be the standard practice in any laboratory.

11.12.1 Three or more standards are needed to develop a calibration curve in concentrations covering the working range, as necessary or measurement of two calibration standards to verify the existing calibration curve.

11.12.2 One method blank per run.

11.12.3 One field blank per set of samples.

11.12.4 One duplicate for precision check (at least one every 20 routine samples).

11.12.5 One standard sample for recovery and calibration check (at least one every 20 routine samples). A standard should be the last sample analyzed, in each run.

11.12.6 One spiked sample for recovery check in the presence of a sample matrix (at least one every 20 routine samples).

11.12.7 *Total*—Seven to ten Analytical Quality Control analyses may be required for runs of up to 20 routine samples, 10 to 13 Analytical Quality Control analyses may be required for runs of 21 to 40 routine samples, etc.

11.12.8 *Minimal Analytical Quality Control*—For very small operations or small sample loads, the described Analytical Quality Control Program may not be practical or necessary for all analytes. Whenever Analytical Quality Control must be reduced below the level recommended, the following minimal Analytical Quality Control Program should be maintained.

11.12.9 Continue calibration or calibration checks as described in 11.6.

11.12.10 Analyze one field blank per set of samples to check for contamination. If an out-of-control situation is indicated, a method blank should be run to find out whether the contamination problem is in the laboratory or the field.

11.12.11 Analyze one spiked sample at the end of each analytical run to check for recovery or precision problems. If an out-of-control situation is indicated, analyze a standard to find out whether the problem is basic recovery or calibration, or both. Successful recovery of the standard would suggest either a matrix problem or a precision problem. A precision problem would produce random out-of-control indications, probably caused by poor or inconsistent analytical techniques or instrumentation.

11.13 *Performance Review*—Analysts should maintain a permanent record of the quality control checks which are performed. The laboratory supervisor should hold frequent meetings to review the quality control program with analysts to discuss the quality control checks performed and the resolution of any problems which are detected. Deficiencies which are detected should be documented in the record book indicating the analytes involved, the problem, the action taken, and the date of the action.

12. Trouble Shooting

12.1 Extreme, unexpected or questionable results are normally detected and reported by the analyst, or are noted by the supervisor in the daily reviews of results. When a deviation is noted, the train of sampling and analytical methods and quality control, must be investigated. The documented intralaboratory quality control checks provide the primary means for the investigation.

12.1.1 Review the records of the sample collection. Check the preservation technique used, the chain of custody record, the time in transit, and comments on the conditions of the samples upon arrival at the laboratory, for example, temperature upon arrival, etc.

12.1.2 *Analytical Procedure*—Check calculations for transposition of numbers and mathematical error. Any significant positive blank result indicates field or laboratory contamination of sample, sampling device, sample container, reagents, reagent water, etc. Check monitoring data on reagent water. Check reagents for changes in bottle and lot, and expiration dates. Analyze or reanalyze samples to confirm source and resolution of problem. Confirm recoveries with analyses of known reference samples.

12.2 The investigation may indicate good field and laboratory practices were not followed, such as:

12.2.1 The field sampling team should keep a bound field logbook for recording field measurements, time, temperature, sampling location, weather conditions, and other pertinent information.

12.2.2 The analyst should keep records on incoming chemicals and reagents, and the preparation of reagents, with estimated shelf lives. Reagent containers should be properly labeled and dated. A mechanism established for reorder of chemicals within the estimated shelf lives.

12.2.3 Reagent blanks should be carried through all sampling and analytical procedures. In colorimetry, the reagent blank should be compared with reagent water to detect an unusual reagent blank response.

12.2.4 When the data are obtained through the use of a standard curve, the points on the curve should be treated statistically and a regression line should be developed for the analytical method.

12.2.5 Use of reference materials of known quality from sources such as the U.S. National Institute of Standards and Technology (NIST), U.S. Environmental Protection Agency (USEPA), or others, should be used to confirm the adequacy of the technique and the analyst.

12.3 Senior analysts must maintain a permanent log of quality control checks performed. The laboratory supervisor must hold frequent quality assurance review meetings with senior analysts to discuss the quality control checks performed, and the resolution of problems detected. Deficiencies and corrective actions must be recorded in a log indicating the analytes involved, the problem, the action taken, and the date of the action. Only when the deficiencies have been discovered, corrected, and confirmed as corrected is the real benefit of a quality assurance plan realized, that is, improved data quality.

ANNEX

(Mandatory Information)

A1. QUALITY CONTROL EXAMPLES

A1.1 Table A1.1 presents the results of carrying out the steps described in 11.5.4.3 and 11.5.4.4 for three different analytes to illustrate use of the UCL values. If duplicate chromium results of 31.2 and 33.7 were obtained, the system precision would be checked as follows:

$$R = \frac{|31.2 - 33.7|}{(31.2 + 33.7)/2} = \frac{|-2.5|}{64.9/2} = \frac{2.5}{32.45} = 0.0770 \quad \text{(A1.1)}$$

Since the appropriate upper control limit (UCL) from Table A1.1 is 0.109 and the current R values are not greater, precision of the analytical system is judged to be within control.

A1.2 The following calculations result from carrying out the step described in 11.5.5.5 using the data in Table A1.2:

$$\bar{P} = \left(\sum_{i=1}^{n} P_i\right) \Big/ n = 2104/21 = 100.2 \quad \text{(A1.2)}$$

$$S_p = \sqrt{1/n - 1\left(\sum_{i=1}^{n} P_i^2 - \left(\sum_{i=1}^{n} P_i\right)^2 \Big/ n\right)} \quad \text{(A1.3)}$$

$$= \sqrt{1/20\,(211\,504 - (2104)^2/21)} = \sqrt{1/20\,(703.24)}$$

$$= \sqrt{35.162} = 5.93$$

$$\bar{P} \pm 3S_p = 100.2 \pm 17.8 = 82.4 \text{ to } 118.0 \quad \text{(A1.4)}$$

A1.3 Therefore, as specified in 11.5.5.6, percent recovery values for total PO_4-P standards roughly within the concentration range from 0.34 to 4.9, that occurred below 82.4 % or above 118 % would indicate that the accuracy of the analytical system is under control. The related Shewhart $\bar{X}$-Chart is illustrated in Fig. A1.1.

TABLE A1.1 Precision Estimates from Duplicate Analyses Within Specific Concentration Ranges for Three Analytes

Analytes	Concentration Range	No. of Sets of Duplicates	Average Concentration of Data	Average Relative Range (R)	R for Combined Concentration Ranges	Final UCL Results
BOD, 5-Day (mg/L) L	1 to <10	21	5.85	0.1776		
	10 to <25	30	17.6	0.1104	0.1381	0.452
	25 to <50	27	36.1	0.0924		
	50 to <150	29	102.	0.0638		
	150 to <300	17	197.	0.0564	0.0652	0.213
	300 to <1,000	12	520.	0.0232		
	1,000 up	3	3341.	0.0528		
Chromium (μg/L) L	5 to <10	32	6.15	0.0612	0.0612	0.200
	10 to <25	15	16.7	0.0340		
	25 to <50	16	36.2	0.0310		
	50 to <150	15	85.1	0.0446	0.0334	0.109
	150 to <500	8	240.	0.0218		
	500 up	5	3170.	0.0240		
Copper (μg/L) L	5 to <15	16	11.1	0.1234		
	15 to <25	23	19.1	0.0736	0.0940	0.307
	25 to <50	21	35.4	0.0338		
	50 to <100	26	65.9	0.0354		
	100 to <200	10	134.	0.0210	0.0313	0.102
	200 up	3	351.	0.0130		

TABLE A1.2 Analysis[A] of Total Phosphate-Phosphorus Standards, in mg/L Total PO_4-P

Point	Known	Obtained	% Recovery P_1	P_1^2
1	0.34	0.33	97	9 409
2	0.34	0.34	100	10 000
3	0.40	0.40	100	10 000
4	0.49	0.49	100	10 000
5	0.49	0.49	100	10 000
6	0.50	0.47	94	8 836
7	0.50	0.53	106	11 236
8	0.50	0.56	112	12 544
9	0.52	0.59	113	12 769
10	0.66	0.70	106	11 236
11	0.66	0.60	91	8 281
12	0.67	0.65	97	9 409
13	0.68	0.65	96	9 216
14	0.83	0.80	96	9 216
15	1.3	1.2	92	8 464
16	1.3	1.3	100	10 000
17	1.6	1.7	106	11 236
18	2.3	2.3	100	10 000
19	2.3	2.4	104	10 816
20	3.3	3.3	100	10 000
21	4.9	4.6	94	8 836
		TOTALS:	2 104	211 504

[A] Using a colorimetric method with persulfate digestion.

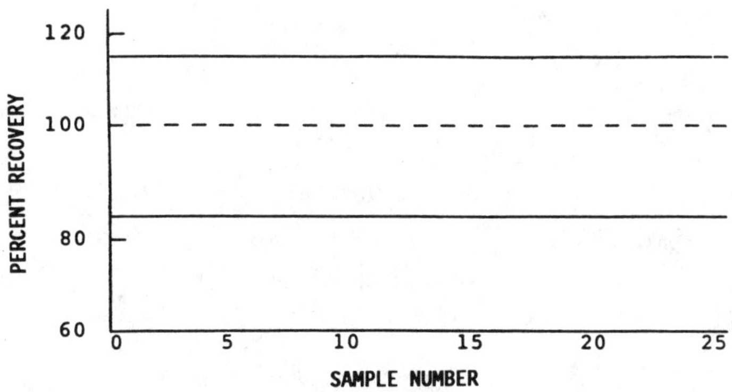

FIG. A1.1 Shewart Control Chart for Percent Recovery Data

REFERENCES

(1) "ASTM Manual on Presentation of Data and Control Chart Analysis," ASTM STP 15D, ASTM, 1976.

(2) Lewis, H. F. (ed.), "Laboratory Planning for Chemistry and Chemical Engineering," Reinhold Publishing Corp., NY, 1962.

(3) "Industrial Ventilation—A Manual of Recommended Practice," Available from the American Conference of Governmental Industrial Hygienists, 6500 Glenway Ave., Building D-5, Cincinnati, OH 45211.

(4) Baker, C. J. (ed.), "The Firefighters Handbook of Hazardous Materials," P. O. Box 31009, Indianapolis, IN, 46231.

(5) "Handbook for Sampling and Sample Preservation of Water and Wastewater," Document No. EPA-600/4-82-029. Available from The National Technical Information Service (NTIS), 5285 Port Royal Rd., Springfield, VA 22165.

(6) "Handbook for Analytical Quality Control in Water and Wastewater Laboratories," Document No. EPA-600/4-79-019. Available from USEPA, Environmental Monitoring and Support Laboratory, 26 W. King Dr., Cincinnati, OH 45268.

Designation: D 3864 – 79 (Reapproved 1990)

Standard Guide for Continual On-Line Monitoring Systems for Water Analysis[1]

This standard is issued under the fixed designation D 3864; the number immediately following the designation indicates the year of original adoption or, in the case of revision, the year of last revision. A number in parentheses indicates the year of last reapproval. A superscript epsilon (ε) indicates an editorial change since the last revision or reapproval.

1. Scope

1.1 This guide covers the selection, establishment, and application of monitoring systems for determining water characteristics by continual sampling, automatic analysis, and recording or otherwise signaling of output data. The system chosen will depend on the purpose for which it is intended: whether it is for regulatory compliance, process monitoring, or to alert the user of adverse trends. If it is to be used for regulatory compliance, the method published or referenced in the regulations should be used in conjunction with this guide and other ASTM methods.

1.2 *This standard does not purport to address all of the safety problems associated with its use. It is the responsibility of the user of this standard to establish appropriate safety and health practices and determine the applicability of regulatory limitations prior to use.* Specific precautionary statements are given in Section 7.

2. Referenced Documents

2.1 *ASTM Standards:*

D 1129 Terminology Relating to Water[2]

D 1193 Specification for Reagent Water[2]

3. Terminology

3.1 *Definitions:*

3.1.1 For definitions of terms used in this guide, refer to Terminology D 1129.

3.2 *Descriptions of Terms Specific to This Standard:*

3.2.1 *monitoring system*—the integrated equipment package comprising sampling system, analyzer, and data output equipment, required to perform water quality analysis automatically.

3.2.1.1 *sampling system*—equipment necessary to deliver a continual representative sample to the analyzer.

3.2.1.2 *analyzer*—a device that measures continually the chemical composition or a specific physical, chemical, or biological property of a sample.

3.2.1.3 *data output equipment*—analog or digital devices for recording, acquiring, or processing the output signals from the analyzer.

3.2.2 *cycle time*—the interval between repetitive sample introductions in a monitoring system with discrete sampling.

3.2.3 *detection limit*—the smallest incremental input signal that can be determined reliably, usually considered as twice the noise level.

3.2.4 *drift*—the change in system output, with constant input over a stated time period of unadjusted, continuous operation; usually expressed as percentage of full scale over a 24-h period.

3.2.4.1 *span drift*—drift when the input is at a constant, stated upscale value.

3.2.4.2 *zero drift*—drift when the input is at zero.

3.2.5 *full scale*—the maximum measuring limit of the system for a given range.

3.2.6 *input*—the value of the parameter being measured at the inlet to the analyzer.

3.2.7 *interference*—an undesired output caused by a substance or substances other than the one being measured. (The effect of interfering substance(s) on the measured parameter of interest shall be expressed as a percentage change (±) in the measured component as the interference varies from 0 to 100 % of the measuring scale. If the interference is nonlinear, an algebraic expression should be developed (or curve plotted) to show the varying effect.)

3.2.8 *linearity*—the extent to which an actual analyzer reading agrees with the reading predicted by a straight line drawn between upper and lower calibration points—generally zero and full-scale. (The maximum deviation from linearity is frequently expressed as a percentage of full-scale.)

3.2.9 *output*—a signal, usually electrical, that is related to the parametric measurement and is the intended input to data output equipment.

3.2.10 *range*—the region defined by the minimum and maximum measurable limits.

3.2.11 *repeatability*—a measure of the precision of one analyzer to repeat its results on independent introduction of the same sample at different time intervals.

3.2.12 *reproducibility*—a measure of the precision of different analyzers to repeat results on the same sample.

3.2.13 *response time*—the time interval from a step change in the input or output reading to 90 % of the ultimate reading.

3.2.13.1 *lag time*—the time interval from a step change in input to the first corresponding change in readout.

3.2.13.2 *rise time*—response minus lag time.

3.2.13.3 *total time*—the time interval from a step change in the input to a constant analyzer signal output.

4. Summary of Guide

4.1 This guide provides a unified approach to the use of on-line monitoring systems for water quality analysis. It presents definitions of terms, safety precautions, system design and installation considerations, calibration techniques, general operating procedures, and comments relating

---

[1] This guide is under the jurisdiction of ASTM Committee D-19 on Water and is the direct responsibility of Subcommittee D 19.03 on Sampling of Water and Water-Formed Deposits, and Surveillance of Water.

Current edition approved Nov. 30, 1979. Published March 1980. Last previous edition D 3864 – 79$^{\epsilon 1}$.

[2] *Annual Book of ASTM Standards*, Vol 11.01.

to the validity and reliability of results.

5. Significance and Use

5.1 Many of the manual and automated laboratory methods for measurement of physical, chemical, and biological parameters in water and waste water are adaptable to on-line sampling and analysis. The resulting real-time data output can have a variety of uses, including confirming regulatory compliance, controlling process operations, or detecting leaks or spills.

5.2 This guide is intended to be a common reference that can be applied to all water quality monitoring systems. However, calibration and validation sections may be inappropriate for certain tests since the act of removing a sample from a flowing stream may change the sample.

5.3 Technical details of the specific methodology are contained in the pertinent ASTM standard test methods, which will reference this guide for guidance in selection of systems and their proper implementation.

5.4 This guide complements descriptive information on this subject found in the *ASTM Manual on Water, STP 442.*

6. Reagents

6.1 *Purity of Reagents*—Reagent grade chemicals shall be used in all tests. Unless otherwise indicated, it is intended that all reagents shall conform to the specifications of the Committee on Analytical Reagents of the American Chemical Society.[3] Other grades may be used, provided it is first ascertained that the reagent is of sufficiently high purity to permit its use without lessening the accuracy of the determination.

6.2 *Purity of Water*—Unless otherwise indicated, references to water shall be understood to mean reagent water conforming to Specification D 1193, Type II.

7. Safety Precautions

7.1 Each analyzer installation shall be given a thorough safety engineering study **(1)**.[4]

7.2 Electrically the monitoring system as well as the individual components shall meet all code requirements for the particular area classification.

7.2.1 All analyzers using 120-V, a-c, 60-Hz, 3-wire systems shall observe polarity and shall not use mechanical adapters for 2-wire outlets.

7.2.2 Check the neutral side of the power supply at the analyzer to see that it is at ground potential.

7.2.3 Connect the analyzer's ground connection to earth ground and checked for proper continuity.

7.2.4 The metallic framework of the analyzer shall be at ground potential.

7.2.5 Consider additional protection in the form of ground fault interrupters for each individual application.

7.2.6 Analyzers containing electrically heated sections shall have a temperature-limit device.

7.2.7 The analyzer, and any related electrical equipment (the system), shall have a power cut-off switch, and a fuse or breaker, on the "hot" side of the line(s) of each device.

7.3 Give full consideration to safe disposal of the analyzer's spent samples and reagents.

7.4 Provide pressure relief valves, if applicable, to protect both the analyzer and monitoring system.

7.5 Take precautions when using cylinders containing gases or liquids under pressure. Helpful guidance may be obtained from Refs **(2)**, **(3)**, **(4)**, and **(5)**.

7.5.1 Gas cylinders must be handled by trained personnel only.

7.5.2 Fasten gas cylinders to a rigid structure.

7.5.3 Take special safety precautions when using or storing combustible or toxic gases to ensure that the system is safe and free of leaks.

7.6 Gas piping, where possible, shall be metallic, especially inside the analyzer housing.

8. Measurement Objectives

8.1 Carefully define the measurement objective for the monitoring system before selecting components of the system and set specifications realistically, to meet the objective. Terms used as specifications shall be consistent with the terminology in Section 3.

8.2 If the monitoring system is intended primarily to determine compliance with regulatory standards, the accuracy, precision, frequency of sampling, and response time may be dictated by the requirements of the regulations. A high degree of stability and on-line reliability is generally required. The analyzer response for a specific parameter must be referenced to a recognized or specified laboratory method approved by the regulatory agency.

8.3 Monitoring systems intended to detect leaks and uncontrolled discharges, that is, spills, to protect treatment plants or receiving waters, require short sampling cycles and rapid response. Typically, these systems will activate alarms to alert operating personnel. They then may cause flow to be diverted from normal channels until the upset has passed or has been corrected. Frequently, the monitoring system is used in some way to locate and identify the source of the spill.

8.4 Systems that monitor performance of process operations, such as waste treatment, may have varying degrees of sophistication and complexity, depending on the specific nature of the application.

8.4.1 Simple, inexpensive, and low-precision analyzers with indicating or recording devices and alarms are acceptable for monitoring trends in operating parameters and for alerting operating personnel to off-standard performance.

8.4.2 Monitoring systems that provide data to be used to manually control process operations or to manually set automatic controllers are generally more complex and frequently require that outputs be transmitted long distances.

8.4.3 Monitoring systems intended to process data for operating guidance or management presentation and to provide varying degrees of automatic process control must be compatible with digital computers or telemetering systems. The reliability and stability of such systems, particularly the data output equipment, shall be high.

[3] "Reagent Chemicals, American Chemical Society Specifications," Am. Chemical Soc., Washington, DC. For suggestions on the testing of reagents not listed by the American Chemical Society, see "Reagent Chemicals and Standards," by Joseph Rosin, D. Van Nostrand Co., Inc., New York, NY, and the "United States Pharmacopeia."

[4] The boldface numbers in parentheses refer to the references at the end of this practice.

9. Sample System Design Considerations

9.1 Carefully examine the measurement objectives of the monitoring system and select a sampling system that matches these requirements.

9.2 Review all sample requirements with the equipment supplier. Be sure to define accurately all conditions of intended operation, the components in the sample and expected variations in the measured parameters.

9.3 Choose materials of construction that will be in contact with the sample so as to avoid reaction of these materials with the sample and subsequent contamination, corrosion, or other damage to critical parts or sorption of measurable components.

9.4 Select the sampling point(s) so as to provide a representative and measurable sample as close as possible to the sample system and analyzer, and as outlined in Practices D 3370.

9.5 Design the sample probe to be consistent with the measurement objective and to require a minimum of maintenance.

9.6 Select the sample transfer system, including pumps and transfer lines, so that the integrity of the sample is maintained from sampling point to analyzer, especially with respect to suspension of solids and biological growth.

9.7 Provide necessary sample conditioning equipment (for example, filters, diluters, homogenizers, stream splitters), that is consistent with the defined measurement objective.

9.8 Provide a connection, when necessary, for introducing standard samples or withdrawing check samples immediately upstream of the analyzer.

9.9 Keep single- or multiple-sample streams that interface a single analyzer flowing all the time. Keep the manifold close to the analyzer to minimize cross-contamination.

9.10 Always keep sample lines as short as possible.

9.11 Provide appropriate protection of sample lines from extremely hot or freezing temperatures.

10. Considerations for Analyzer Selection

10.1 The analyzer selected must meet the measurement objective of the system over the complete range of application.

10.1.1 Precision and accuracy of measurement and response time for the parameter of interest shall coincide with system specifications at all levels of measurement.

10.1.2 Interference shall be insignificant relative to the measured component or shall be controllable. When used for regulatory compliance, known interferences shall affect the reading of more than 5 % from the true value.

10.1.3 If required for compliance, the analyzer shall be capable of validation by calibration with approved and certified standard reference materials using standard ASTM (or equivalent) tests.

10.2 In choosing a specific analyzer for a specific application, on-line reliability of the instrument is of prime concern.

10.2.1 Downtime for maintenance because of component failures or other malfunction shall be minimal. Ease, promptness, and minimal cost of repair or replacement are essential.

10.2.2 The analyzer shall be stable. Drift and changes in response with changes in conditions such as flow and temperature shall be insignificant or means for compensation shall be provided.

10.2.3 The analyzer shall be relatively simple and easy to operate and maintain at a satisfactory level of performance.

11. Data Output Equipment Considerations

11.1 Equipment for the acquisition of output data from the analyzer shall meet the requirements of the measuring objectives for the monitoring system.

11.2 Visual or audible alarms and simple output meters are acceptable and desirable in many applications.

11.3 The analyzer output can be recorded locally at the field location. The digital or analog signal is frequently transmitted to a centralized location, such as a control room, often by a data line shared with other instruments.

11.4 Recorded or real-time data can be transferred to computers for storage, process control, or report generation.

11.5 Process equipment such as valves and pumps can be actuated by output generated by analyzers in a number of ways:

11.5.1 Recorders and output meters can have setpoints as integral parts of their design which actuate the equipment directly for either on-off or proportional control.

11.5.2 Controllers can be manually adjusted in response to analyzer signals read from a recorder or from output presented in a data report, typed or displayed on a cathode ray tube.

11.5.3 Direct digital process control is possible in more complicated and sophisticated systems, where real-time analyzer output is integrated with other process data and used to maintain desirable process conditions.

12. Installation of Monitoring System

12.1 Obtain information required for installation and operation of the monitoring system from the supplier.

12.2 Study operational data and design parameters furnished by the supplier before installation.

12.3 Choose materials of construction and components of the monitoring system to withstand the environment in which it is installed.

12.4 Select a location for the analyzer that is as close as possible to the sample intake and which provide adequate protection from extremes of temperature and humidity, where this is essential to proper performance.

12.5 Provide a convenient access to the entire monitoring system.

12.6 Provide proper outlets for the analyzer's exit streams so that no liquid or gas pressure buildup occurs (see 7.4).

12.7 After the installation has been completed, allow the analyzer to stabilize and calibrate before testing performance specifications.

13. Calibration

13.1 Establish a written calibration procedure and frequency consistent with the parameter being measured and the accuracy and reliability demanded by the measurement or control objectives.

13.1.1 Consult the analyzer supplier to determine the best calibration procedure to use with the specific analyzer in a particular application.

13.1.2 When required for regulatory compliance, use

calibration procedures specified by the appropriate agency.

13.1.3 Refer to ASTM standards, where applicable, to determine appropriate calibration standards.

13.1.4 Provide calibration standards at concentrations and compositions as close as possible to those of the sample stream being analyzed.

13.1.5 Before calibration, assure that the sampling system and output instrumentation are functioning properly and that all preliminary adjustments to the analyzer required by the procedure have been made.

13.2 *Spot Sample Calibration:*

13.2.1 With the sample flowing through the analyzer sampling line uniformly and the analyzer readout as close as possible to an equilibrium value, withdraw a sample from the inlet stream, as described in Practices D 3370.

13.2.2 Record time, sample number, date, and the corresponding analyzer readout and immediately analyze the sample, in duplicate, using the appropriate test method.

13.2.3 Average the results of the duplicate analyses. Discard tests *that* fail to meet repeatability requirements. Do *not* use that average.

13.2.4 Determine the analyzer calibration adjustment required so that results of analysis and the analyzer readout coincide. Adjust the analyzer with the analyzer controls accordingly.

13.2.5 Repeat this procedure until no further change is needed, consistent with the quality of data required.

13.3 *Multiple Standard Calibration:*

13.3.1 Prepare a series of calibration standards covering the range of measurements for the sample being analyzed, following instructions in the test method or in the analyzer supplier's instructions.

13.3.2 Check all operating conditions of the system in accordance with analyzer specifications and allow sufficient time for instrument equilibrium.

13.3.3 Introduce a calibration standard of a concentration level recommended by the instrument supplier into the analyzer using the recommended instrument operating procedure. Activate the readout equipment.

13.3.4 After sufficient sample has been allowed to flow through the analyzer, adjust the readout to conform to the desired value.

13.3.5 Repeat 13.3.3 for the remaining standards from the calibration series, recording the equilibrium readout value each time.

13.3.6 Plot a calibration curve of standard value versus readout response from the above data.

13.3.7 Discard any standard when any change of composition is detected.

13.4 *Laboratory Check Sample for Flow-Through System:*

13.4.1 Withdraw from the spot sampling line or otherwise obtain directly from the sample stream sufficient sample for calibration, representative of one concentration within the range of measurement of the analyzer (see Practices D 3370).

13.4.2 Analyze the sample for the parameter of interest, using the appropriate test method.

13.4.3 If necessary, prepare additional standards to cover the range of interest by dilution with reagent water or by "spiking" with known amounts of an appropriate standard.

13.4.4 Serially, introduce the standards into the analyzer, using the recommended instrument operating procedures, and allow the analyzer readout to reach equilibrium.

13.4.5 Plot a calibration curve of concentration of parameter being determined vesus readout response from the readout data.

13.5 *Probe Calibration:*

13.5.1 Provide special calibration procedure for analyzers for which the instrumental measuring technique utilizes a sensor that is inserted directly into the sample, for example, pH, dissolved oxygen, conductivity.

13.5.2 Prepare two calibration solutions in accordance with the appropriate test method, selecting them to bracket the anticipated value of measurement.

13.5.3 Remove the probe from the sample stream, clean if appropriate and perform any necessary maintenance.

13.5.4 Fill a test container with the first calibration solution. The container shall have the means for monitoring temperature and, where appropriate, provide and maintain an adequate flow of sample past the sensor.

13.5.5 Insert the probe in the container containing the calibration solution and, using the procedure provided by the suppliers, adjust controls so that the analyzer output coincides with the accepted value of the standard. Make necessary adjustments for temperature compensation.

13.5.6 Rinse the probe thoroughly, place it in a second container containing the other calibration solution and readjust the controls, if necessary, so that the output agrees with the value of this standard.

13.5.7 Recheck with both solutions at least once. If either point differs from the true value by a significant amount, as determined by the quality of measurement required, perform necessary maintenance and recalibrate.

13.5.8 Alternatively, insert a second probe, with independent readout equipment and previously calibrated, into the sample alongside the probe and calibrate in situ, by adjusting its controls until the outputs of the two probes coincide.

13.6 After initial calibration with standard solutions or actual samples, as in 13.2 through 13.5, analyzer calibrations can be rechecked with secondary standards.

13.6.1 An electrical signal may be imposed to produce an analyzer output corresponding to a specific value produced by the parameters being analyzed.

13.6.2 A solution containing material other than the component of interest, but producing the same analyzer output as that component, may be used in place of the standard solution.

13.6.3 An optical filter may be placed in the beam of a photometric analyzer to produce an output equivalent to that produced by the component of interest.

14. Procedure

14.1 Begin introducing the sample to the analyzer or insert the analyzer sensor into sample.

14.2 Check all operating conditions for the monitoring system in accordance with the application engineering data and method for specific analysis.

14.3 Observe the sample analysis as indicated by the readout equipment after the analyzer has been thoroughly flushed or otherwise equilibrated with the sample.

14.4 At a frequency dictated by the type of analyzer and the measurement objective, check calibration of the system by applying the appropriate technique from Section 13.

Generally, a single-point calibration at some "benchmark" value shall be made.

14.5 If the analyzer output and the "benchmark" value agree, within the limits of precision and reliability established for the system, reintroduce sample to the analyzer and proceed.

14.6 If the deviation between analyzer output and the "benchmark" value is outside the established limits, make appropriate adjustments to restore the output to an acceptable value and reintroduce sample to the analyzer.

14.7 If the offset between the instrument and "benchmark" readings is significant, and cannot be easily corrected by instrumental adjustments, perform the necessary maintenance of the analyzer or other portion of the system and recalibrate.

14.8 Record all calibration check values, adjustments made, and maintenance performed. If the record indicates unsatisfactory performance, investigate the system thoroughly to determine the cause of the problem.

14.9 Perform all calibration checks and maintenance as quickly as possible to minimize the time the analyzer is off-line.

14.10 Set up a sound and thorough preventive maintenance program. Successful operation of the monitoring system depends to a large extent on the amount of maintenance provided. Type of analyzer, complexity of the system, and condition of the sample stream usually determine the maintenance requirements.

15. Calculation

15.1 Each individual monitoring system and ASTM test method chosen determines the calculations necessary to perform on the output signal. Most analyses are recorded as direct readouts based on instrument calibration. However, in some cases, the measurement sensitivity range is involved and scale factors are necessary to determine the final results. This is usually a simple multiplication step.

16. Precision and Bias

16.1 Preferably, each laboratory standard test method that is applied to on-line monitoring shall include its own precision section based on cooperative test program results.

16.2 If it is desirable to validate the monitoring system results relative to the laboratory method, statistical equivalency of the data generated by the two techniques shall be demonstrated by applying the "t" test for mean differences of the paired observations at the 95 % ($P \leq 0.05$) confidence level ($t_{0.025}$–two tailed). This shows whether any significant difference exists between the mean values of the differences and zero for paired observations having different values. The procedure for this test is given in the annex.

REFERENCES

(1) The user, equipment, supplier, and installer should be familiar with requirements of the National Electrical Code, any local applicable, electrical code, U.L. Safety Codes, and the Occupational Safety and Health Standards (*Federal Register*, Vol 36, No. 105, Part II, May 29, 1971). Helpful guidance may also be obtained from API RP500, "Classification of Areas for Electrical Installations in Petroleum Refineries;" ISA RP12.1, "Electrical Instruments in Hazardous Atmospheres;" ISA RP12.2, "Intrinsically Safe and Nonincendive Electrical Instruments," ISA RP12.4, "Instrument Purging for Reduction of Hazardous Area Classification;" and API RP550, "Installation of Refinery Instruments and Control Systems, Part II."

(2) "Safe Handling of Compressed Gases," Pamphlet P-1, Compressed Gas Association, Inc., New York, NY.

(3) "Compressed Gases, Safe Practices," Pamphlet No. 95, National Safety Council, Chicago, Ill.

(4) "Chemical Safety Data Sheets", Manufacturing Chemists Assn., 1825 Connecticut Ave., N.W., Washington, DC 20009.

(5) Sax, N. I., *Dangerous Properties of Industrial Materials,* 3rd edition, 1968, Reinhold Book Corp., New York, NY.

ANNEX

(Mandatory Information)

A1. VALIDATION OF RESULTS

A1.1 Obtain a set of at least seven pairs of data comparing analyzer response (A_i) with determinations, by laboratory analysis (L_i), of a simultaneously obtained "benchmark" sample.

A1.2 Determine the difference (d) between each matched pair of observations, $d_i = A_i - L_i$.

A.1.3 Establish the validity of the data by comparing the mean difference for pairs of analyzer and laboratory results using Student's "t" test, as follows:

$$t_c = \bar{d}\sqrt{n}/S_d$$

where:

$\bar{d}$ = mean difference, $(1/n) \sum_{i=1}^{n} d_i$,

n = number of paired measurements comparing analyzer and laboratory results, and

S_d = the standard deviation for the differences,

$$\sqrt{\frac{\sum_{i=1}^{n} d_i^2 - (1/n)\left(\sum_{i=1}^{n} d_i\right)^2}{n-1}}$$

A1.4 Compare t_c with the value of "t" given in Table A1.1 for the number of degrees of freedom (df), $n - 1$, used in the calculation for the 95 % ($P \leq 0.05$) confidence level ($t_{0.025}$–two tailed).

A1.5 If t_c is equal to or less than the tabulated value for "t" the analyzer results can be considered valid because

TABLE A1.1 Student's "t" Test for 95 % Confidence Level

Degrees of Freedom	"t"	Degrees of Freedom	"t"
1	12.706	16	2.120
2	4.303	17	2.110
3	3.182	18	2.101
4	2.776	19	2.093
5	2.571	20	2.086
6	2.447	21	2.080
7	2.365	22	2.074
8	2.306	23	2.069
9	2.262	24	2.064
10	2.228	25	2.060
11	2.201	30	2.042
12	2.179	40	2.021
13	2.160	60	2.000
14	2.145	120	1.980
15	2.131	∞	1.960

there is no significant difference between the analyzer and laboratory values. If t_c is greater than the tabulated "t", the analyzer and laboratory values differ significantly and the analyzer results cannot be considered valid.

APPENDIX

(Nonmandatory Information)

X1. SAMPLE DETERMINATION OF VALIDITY

X1.1 The following is a sample calculation for determining the validity of monitoring system results relative to a laboratory method, in accordance with the procedure described in the annex.

X1.2 Tabulate analyzer response and laboratory analytical results for at least seven matched pairs of data, as in Table X1.1.

X1.3 Calculate the differences (d) for each pair of observations (see A1.2), square the differences (d^2) and include both values in the tabulation.

X1.4 Calculate the mean difference ($\bar{d}$) and the standard deviation of the differences (S_d) in accordance with A1.3.

$$\bar{d} = (1/n)\Sigma\, d_i = (1/7) \times 7 = 1$$

$$S_d = \sqrt{\frac{\Sigma\, d_i^2 - (1/7)(\Sigma\, d_i)^2}{n-1}} = \sqrt{\frac{227 - ((7)^2/7)}{6}} = 6.06$$

X1.5 Compute t_c for this set of data (see A1.3) as follows:

$$t_c = \bar{d}\sqrt{n}/S_d = 1\sqrt{7}/6.06 = 0.435$$

X1.6 Compare t_c with the tabulated value for "t" for six degrees of freedom (see Table A1.1).

$$\text{"}t\text{"}_{6\text{df}} = 2.447;\ t_c = 0.435$$

$$t_c < \text{"}t\text{"}_{6\text{df}}$$

X1.7 Since t_c is less than "t", the analyzer and laboratory results are not significantly different and the analyzer results can be considered valid.

TABLE X1.1 Comparison of Results from On-Line Analyzer and Laboratory Analysis for Total Organic Carbon (Test Methods D 2579)

Sample No.	Total Organic Carbon, mg/L		d_i	d^2
	Analyzer (A_i)	Laboratory (L_i)		
1	84	76	8	64
2	68	72	−4	16
3	88	81	7	49
4	75	69	6	36
5	80	86	−6	36
6	78	77	1	1
7	73	78	−5	25
Σ			7	227

Standard Guide for Selection of Aquifer-Test Method in Determining of Hydraulic Properties by Well Techniques[1]

This standard is issued under the fixed designation D 4043; the number immediately following the designation indicates the year of original adoption or, in the case of revision, the year of last revision. A number in parentheses indicates the year of last reapproval. A superscript epsilon (ε) indicates an editorial change since the last revision or reapproval.

1. Scope

1.1 This guide is an integral part of a series of standards that are being prepared on the in situ determination of hydraulic properties of aquifer systems by single- or multiple-well tests. This guide provides guidance for development of a conceptual model of a field site and selection of an analytical test method for determination of hydraulic properties. This guide does not establish a fixed procedure for determination of hydrologic properties.

1.2 The values stated in SI units are to be regarded as standard.

1.3 *Limitations*—Well techniques have limitations in the determination of hydraulic properties of ground-water flow systems. These limitations are related primarily to the simplifying assumptions that are implicit in each test method. The response of an aquifer system to stress is not unique; therefore, the system must be known sufficiently to select the proper analytical method.

1.4 *This standard does not purport to address all of the safety problems, if any, associated with its use. It is the responsibility of the user of this standard to establish appropriate safety and health practices and determine the applicability of regulatory limitations prior to use.*

2. Referenced Documents

2.1 *ASTM Standards:*

D 653 Terminology Relating to Soil, Rock, and Contained Fluids[2]

D 4044 Test Method (Field Procedure) for Instantaneous Change in Head (Slug Tests) for Determining Hydraulic Properties of Aquifers[2]

D 4104 Test Method (Analytical Procedure) for Determining Transmissivity of Nonleaky Confined Aquifers by Overdamped Well Response to Instantaneous Change in Head (Slug Test)[2]

D 4105 Test Method (Analytical Procedure) for Determining Transmissivity and Storage Coefficient of Nonleaky Confined Aquifers by the Modified Theis Nonequilibrium Method[1]

D 4106 Test Method (Analytical Procedure) for Determining Transmissivity and Storage Coefficient of Nonleaky Confined Aquifers by the Theis Nonequilibrium Method[2]

3. Terminology

3.1 *Definitions:*

3.1.1 *aquifer, confined*—an aquifer bounded above and below by confining beds and in which the static head is above the top of the aquifer.

3.1.2 *unconfined aquifer*—an aquifer that has a water table.

3.1.3 *barometric efficiency*—the ratio of the change in depth to water in a well to the inverse of water-level change in barometric pressure, expressed in length of water.

3.1.4 *conceptual model*—a simplified representation of the hydrogeologic setting and the response of the flow system to stress.

3.1.5 *confining bed*—a hydrogeologic unit of less permeable material bounding one or more aquifers.

3.1.6 *control well*—well by which the aquifer is stressed, for example, by pumping, injection, or change of head.

3.1.7 *hydraulic conductivity (field aquifer tests)*—the volume of water at the existing kinematic viscosity that will move in a unit time under unit hydraulic gradient through a unit area measured at right angles to the direction of flow.

3.1.8 *observation well*—a well open to all or part of an aquifer.

3.1.9 *piezometer*—a device so constructed and sealed as to measure hydraulic head at a point in the subsurface.

3.1.10 *specific capacity*—the rate of discharge from a well divided by the drawdown of the water level within the well at a specific time since pumping started.

3.1.11 *specific storage*—the volume of water released from or taken into storage per unit volume of the porous medium per unit change in head.

3.1.12 *specific yield*—the ratio of the volume of water that the saturated rock or soil will yield by gravity to the volume of the rock or soil. In the field, specific yield is generally determined by tests of unconfined aquifers and represents the change that occurs in the volume of water in storage per unit area of unconfined aquifer as the result of a unit change in head. Such a change in storage is produced by the draining or filling of pore space and is, therefore, mainly dependent on particle size, rate of change of the water table, and time of drainage.

3.1.13 *storage coefficient*—the volume of water an aquifer releases from or takes into storage per unit surface area of the aquifer per unit change in head. For a confined aquifer the storage coefficient is equal to the product of specific storage and aquifer thickness. For an unconfined aquifer, the storage coefficient is approximately equal to the specific yield.

3.1.14 *transmissivity*—the volume of water at the existing kinematic viscosity that will move in a unit time under a unit

---

[1] This guide is under the jurisdiction of ASTM Committee D-18 on Soil and Rock and is the direct responsibility of Subcommittee D18.21 on Ground Water and Vadose Zone Investigations.

Current edition approved Aug. 15, 1991. Published December 1991.

[2] *Annual Book of ASTM Standards*, Vol 04.08.

hydraulic gradient through a unit width of the aquifer.

3.1.15 For definitions of other terms used in this guide, see Terminology D 653.

4. Significance and Use

4.1 An aquifer test is a controlled field experiment made to determine the approximate hydraulic properties of water-bearing material. The hydraulic properties determined are specific to the test method. The hydraulic properties that can be determined are also dependent upon the instrumentation of the field test, the knowledge of the aquifer system at the field site, and conformance of the hydrogeologic conditions at the field site to the assumptions of the test method. Hydraulic conductivity and storage coefficient of the aquifer are the basic properties determined by most test methods. Some tests are designed to determine vertical and horizontal anisotropy, aquifer discontinuities, and vertical hydraulic conductivity of confining beds.

5. Procedure

5.1 The procedure for selection of an aquifer test method or methods is primarily based on selection of a test method that is compatible with the hydrogeology of the proposed test site. Secondarily, the test method is selected on the basis of the testing conditions specified by the test method, such as the method of stressing or causing water-level changes in the aquifer and the requirements of a test method for observations of water level response in the aquifer. The decision tree in Fig. 1 is designed to assist, first, in selecting test methods applicable to specific hydrogeologic site characteristics. Secondly, the decision tree will assist in selecting a test method on the basis of the nature of the stress on the aquifer imposed by the control well. The decision tree references the sections in this guide where the test methods are cited.

5.2 *Pre-test-selection procedures*—Aquifer test methods are highly specific to the assumptions of the analytical solution of the test method. Reliability of determination of hydraulic properties depends upon conformance of the hydrologic site characteristics to the assumptions of the test method. A prerequisite for selecting an aquifer test method is knowledge of the hydrogeology of the test site. A conceptual understanding of the hydrogeology of the aquifer system at the prospective test site should be gained in as much detail as possible from existing literature and data, and a site reconnaissance. In developing a site characterization, incorporate geologic mapping, driller's logs, geophysical logs, records of existing wells, water-level and water-quality data, and results of geophysical surveys. Include information on the thickness, lithology, stratification, depth, attitude, continuity, and extent of the aquifer and confining beds.

5.3 *Select Applicable Aquifer Test Methods*—Select a test method based on conformation of the site hydrogeology to assumptions of the test model and the parameters to be determined. A summary of principal aquifer-test methods and their applicability to hydrogeologic site conditions is given in the following paragraphs. The decision tree for aquifer test selection, Table 1, provides a graphic display of the hydrogeologic site conditions for each test method and references to the section where each test method is cited.

5.3.1 *Extensive, Isotropic, Homogeneous, Confined, Nonleaky Aquifer:*

5.3.1.1 *Constant Discharge*—Methods for tests in which the discharge or injection rate in the control well is constant are given by the nonequilibrium method of Theis **(1)**[3] for the drawdown and recovery phases. The Theis method is the most widely referenced and applied aquifer-test method and is the basis for the solution to other more complicated boundary condition problems. The Theis test method for the pumping or injection phase is given in Test Method D 4106. Cooper and Jacob **(2)** and Jacob **(3)** recognized that for large values of time and small values of distance from the control well the Theis solution yields a straight line on semilogarithmic plots of various combinations of drawdown and distance from the control well. The solution of the Theis equation can therefore be simplified by the use of semilogarithmic plots. The modified Theis nonequilibrium method is given in Test Method D 4105.

5.3.1.2 *Variable Discharge*—Test methods for a variably discharging control well have been presented by Stallman **(4)** and Moench **(5)**. These test methods simulate pumpage as a sequence of constant-rate stepped changes in discharge. The test methods utilize the principle of superposition in constructing type curves by summing the effects of successive changes in discharge. The type curves may be derived for control wells discharging from extensive, leaky, and nonleaky confined aquifers or any situation where the response to a unit stress is known. Hantush **(6)** developed drawdown functions for three types of decreases in control-well discharge. Abu-Zied and Scott **(7)** presented a general solution for drawdown in an extensive confined aquifer in which the discharge of the control well decreases at an exponential rate. Aron and Scott **(8)** proposed an approximate test method of determining transmissivity and storage from an aquifer test in which discharge decreases with time during the early part of the test. Lai et al **(9)** presented test methods for determining the drawdown in an aquifer taking into account storage in the control well and having an exponentially and linearly decreasing discharge.

5.3.1.3 *Constant Drawdown*—Test methods have been presented to determine hydraulic-head distribution around a discharging well in a confined aquifer with near constant drawdown. Such conditions are most commonly achieved by shutting in a flowing well long enough for the head to fully recover, then opening the well. The solutions of Jacob and Lohman **(10)** and Hantush **(6)** apply to aerially extensive, nonleaky aquifers. Rushton and Rathod **(11)** used a numerical model to analyze aquifer-test data.

5.3.1.4 *Slug Test Methods*—Test methods for estimating transmissivity by injecting a given quantity or *slug* of water into a well were introduced by Hvorslev **(12)** and Ferris and Knowles **(13)**. Solutions to overdamped well response to slug tests have also been presented by Cooper et al **(14)**. The solution presented by Cooper et al **(14)** is given in Test Method D 4104. Solutions for slug tests in wells that exhibit oscillatory water-level fluctuations caused by a sudden injection or removal of a volume of water have been presented by Krauss **(15)**, van der Kamp **(16)** and Shinohara and Ramey **(17)**. Kipp **(18)** analyzed the complete range of response of

[3] The boldface numbers in parentheses refer to the list of references at the end of this guide.

TABLE 1 Decision Tree for Selection of Aquifer Test Method

Decision	Branch	Test Methods
FRACTURED AQUIFER	—yes→	**Fractured Aquifer, Sec. 5.2.6** Gringarten and Ramey (33) Barenblatt et al (34) Boulton and Streltsova (35) Moench (36)
no ↓		
SINGLE AQUIFER	—no→	**Multiple Aquifers, Sec. 5.2.5** Bennett and Patten (28) Hantush (29) Neuman and Witherspoon (30, 31) Javendel and Witherspoon (32)
yes ↓		
ISOTROPIC AQUIFER	—no→	**Radial-Vertical Anisotropy, Sec. 5.2.3.1** Hantush (21) Weeks (22, 23) **Horizontal Anisotropy, Sec. 5.2.3.2** Papadopulos (24)
yes ↓		
LATERAL BOUNDARY	—yes→	**Bounded Aquifer, Sec. 5.2.4** Ferris et al (25) Stallman (26) Lohman (27)
yes ↓		
CONFINED AQUIFER	—no→	**Unconfined Aquifer, Sec. 5.3** Constant Discharge, Sec.5.3.1: Boulton (37, 38, 39); Neuman (40, 41, 42) Slug Test, Sec. 5.2.3: Bouwer and Rice (43); Bouwer (44)
yes ↓		
LEAKY AQUIFER	—no→	**Confined Aquifer, Sec. 5.2.1** Constant Discharge, Sec. 5.2.1.1: Theis (1); Cooper and Jacob (2); Jacob (3) Slug Test, Sec.5.2.1.4: Hvorslev (12); Cooper et al (14) Variable Discharge, Sec. 5.2.1.2: Stallman (4); Hantush (6); Abu-Zied and Scott (8); Aron and Scott (8); Lai et al (9); Moench (5) Constant Drawdown, Sec. 5.2.1.3: Jacob and Lohman (10); Hantush (6); Rushton and Rathod (11)
yes ↓		
STORAGE IN CONFINING BED	—no→	**Without Storage, Sec. 5.2.2.1** Hantush and Jacob (19)
	yes →	**With Storage, Sec. 5.2.2.2** Hantush (20)

wells ranging from those having negligible inertial effects through full oscillatory behavior and developed type curves for the analysis of slug-test data. The field procedure for slug test methods is given in Test Method D 4044.

5.3.2 *Extensive, Isotropic, Homogeneous, Confined, Leaky Aquifers*—Confining beds above or below the aquifer commonly allow transmission of water to the aquifer by leakage. Test methods that account for this source of water have been presented for several aquifer-confining bed situations.

5.3.2.1 *Leaky Confining Bed, Without Storage*—Hantush and Jacob **(19)** presented a solution for the situation in which a confined aquifer is overlain, or underlain, by a leaky confining layer having uniform properties. Radial flow is assumed in a uniform aquifer. The hydraulic properties of the aquifer and confining bed are determined by matching logarithmic plots of aquifer-test data to a family of type curves.

5.3.2.2 *Leaky Confining Bed, With Storage*—Solutions for determining the response of a leaky confined aquifer where the release of water in the confining bed is taken into account were presented by Hantush **(20)**. Flow in the uniform confined aquifer is assumed to be radial and flow in the leaky confining beds is assumed to be vertical.

5.3.3 *Extensive, Confined, Anisotropic Aquifer:*

5.3.3.1 *Radial-Vertical Anisotropy*—Solutions to the head distribution in a homogeneous confined aquifer with radial-vertical anisotropy in response to constant discharge of a partially penetrating well are presented by Hantush **(21)**. Weeks **(22, 23)** presented test methods to determine the ratio of horizontal to vertical hydraulic conductivity.

5.3.3.2 *Horizontal Anisotropy*—Papadopulos **(24)** presented a test method for determination of horizontal plane anisotropy in an aerially extensive homogeneous confined aquifer.

5.3.4 *Aerially Bounded Aquifers*—Aquifer test methods discussed previously are based on the assumption that the aquifer is aerially extensive. Effects of limitations in aerial extent of aquifers by impermeable boundaries or by source boundaries, such as hydraulically connected streams, may preclude the direct application of an aquifer-test method. The method of images, described by Ferris et al **(25)**, Stallman **(26)** and Lohman **(27)**, provide solutions to head distribution in aerially finite aquifers.

5.3.5 *Multiple Aquifers*—Test methods for multiple aquifers, that is, two or more aquifers separated by a leaky confining bed and penetrated by a control well, require special methods for analysis. Bennett and Patten **(28)** presented a method for testing a multiaquifer system using downhole metering and constant drawdown. Hantush **(29)** presented solutions for two aquifers separated by a leaky confining bed. Neuman and Witherspoon **(30)** provided solutions for drawdown in leaky confining beds above and below an aquifer being pumped. Neuman and Witherspoon **(31)** developed an analytical solution for the flow in a leaky confined system of two aquifers separated by a leaky confining bed with storage. Javendel and Witherspoon **(32)** presented a finite-element method of analyzing anisotropic multiaquifer systems.

5.3.6 *Fractured Media*—Solutions for the flow in a single finite fracture are presented by Gringarten and Ramey **(33)**. Barenblatt et al **(34)** presented a test method for solving a double-porosity model. Boulton and Streltsova **(35)** presented a solution for a system of porous layers separated by fractures. Moench **(36)** developed type curves for a double-porosity model with a fracture skin that may be present at the fracture-block interface as a result of mineral deposition or alteration.

5.4 *Extensive, Isotropic, Homogeneous, Unconfined Aquifer*—Conditions governing drawdown due to discharge from an unconfined aquifer differ markedly from those due to discharge from a nonleaky confined aquifer. Difficulties in deriving analytical solutions to the hydraulic-head distribution in an unconfined aquifer result from the following characteristics: (*1*) transmissivity varies in space and time as the water table is drawn down and the aquifer is dewatered, (*2*) water is derived from storage in an unconfined aquifer mainly at the free water surface and, to a lesser degree, from each discrete point within the aquifer, and (*3*) vertical components of flow exist in the aquifer in response to withdrawal of water from a well in an unconfined aquifer.

5.4.1 Boulton **(37, 38, 39)** introduced a mathematical solution to the head distribution in response to discharge at a constant rate from an unconfined aquifer. Boulton's solution invokes the use of a semi-empirical delay index that was not defined on a physical basis. Neuman **(40, 41, 42)** presented solutions for unconfined aquifer tests utilizing fully penetrating and partially penetrating control and observation wells hypothesized on well-defined physical properties of the aquifer.

5.4.2 Bouwer and Rice **(43)** and Bouwer **(44)** present a slug test method for unconfined aquifer conditions.

6. Keywords

6.1 aquifers; aquifer tests; confining beds; control wells; discharging wells; hydraulic conductivity; observation wells; piezometers; storage coefficient; storativity; transmissivity

REFERENCES

(1) Theis, C. V., "The Relation Between The Lowering of the Piezometric Surface and the Rate and Duration of Discharge of a Well Using Ground-Water Storage," *American Geophysical Union Transactions*, Vol 16, Pt. 2, 1935, pp. 519–524.

(2) Cooper, H. H., Jr., and Jacob, C. E., "Generalized Graphical Method for Evaluating Formation Constants and Summarizing Well-Field History," *American Geophysical Union Transactions*, Vol 27, No. 4, 1946, pp. 526–534.

(3) Jacob, C. E., "Flow of Ground Water," *Engineering Hydraulics*, Proceedings of the Fourth Hydraulics Conference, June 12–15, 1949, John Wiley & Sons, Inc., New York, NY, 1950, pp. 321–386.

(4) Stallman, R. W., "Variable Discharge without Vertical Leakage (Continuously Varying Discharge)," *Theory of Aquifer Tests*, U. S. Geological Survey Water-Supply Paper 1536-E, 1962, pp. 118–122.

(5) Moench, A. E., "Ground-Water Fluctuations in Response to Arbitrary Pumpage," Ground Water Vol 9, No. 2, 1971, pp. 4–8.

(6) Hantush, M. S., "Hydraulics of Wells," *Advances in Hydroscience*, Vol 1, Academic Press, Inc., New York, 1964, pp. 281–442.

(7) Abu-Zied, M., and Scott, V. H., "Nonsteady Flow for Wells with Decreasing Discharge," *Proceedings*, American Society of Civil Engineers, Vol 89, No. HY3, 1963, pp. 119–132.

(8) Aron, G., and Scott, V. H., "Simplified Solutions for Decreasing Flow in Wells," *Proceedings*, American Society of Civil Engineers Vol 91, No. HY5, 1965, pp. 1–12.

(9) Lai, R. Y., Karadi, G. M., and Williams, R. A., "Drawdown at Time-Dependent Flowrate," *Water Resources Bulletin*, Vol 9, No. 5, 1973, pp. 854–859.

(10) Jacob, C. E., and Lohman, S. W., "Nonsteady Flow to a Well of Constant Drawdown in an Extensive Aquifer," *American Geophysical Union Transactions*, Vol 33, No. 4, 1952, pp. 552–569.

(11) Rushton, K. R., and Rathod, K. S., "Overflow Tests Analyzed by Theoretical and Numerical Methods," *Ground Water*, Vol 18, No. 1, 1980, pp. 61–69.

(12) Hvorslev, M. J., "Time Lag and Soil Permeability in Ground-Water Observations," U. S. Army Corps of Engineers, 1951, p. 49.

(13) Ferris, J. G., and Knowles, D. B., "The Slug Test for Estimating Transmissibility," *Ground Water Note 26*, U. S. Geological Survey, 1954, p. 26.

(14) Cooper, H. H., Jr., Bredehoeft, J. D., and Papdopulos, I. S., "Response of a Finite-Diameter Well to an Instantaneous Charge of Water," *Water Resources Research*, Vol 3, 1967, pp. 263–269.

(15) Krauss, I., "Die Bestimmung der Transmissivitat von Grundwasserleitern aus dem Einschwingverhalten des Brunnen-Grundwasserleitersystems," *Journal of Geophysics*, Vol 40, 1974, pp. 381–400.

(16) van der Kamp, Garth, "Determining Aquifer Transmissivity by Means of Well Response Tests—The Underdamped Case," *Water Resources Research*, Vol 12, No. 1, 1976, pp. 71–77.

(17) Shinohara, K., and Ramey, H. J., "Slug Test Data Analysis, including the Inertial Effect of the Fluid in the Well Bore," *54th Annual Fall Conference and Exhibition of the Society of Petroleum Engineers*, 1979.

(18) Kipp, K. L., "Type Curve Analysis of Inertial Effects in the Response of a Well to a Slug Test," *Water Resources Research*, Vol 21, No. 9, 1985, pp. 1399–1408.

(19) Hantush, M. S., and Jacob, C. E., "Non-Steady Radial Flow in an Infinite Leaky Aquifer," *American Geophysical Union Transactions*, Vol 36, No. 1, 1955, pp. 95–100.

(20) Hantush, M. S., "Modification of the Theory of Leaky Aquifers," *Journal of Geophysical Research*, Vol 65, No. 11, 1960, pp. 3713-3725.

(21) Hantush, M. S., "Drawdown Around a Partially Penetrating Well," *Proceedings*, American Society of Civil Engineers, Vol 87, No. HY4, 1961, pp. 83-98.

(22) Weeks, E. P., "Field Methods for Determining Vertical Permeability and Aquifer Anisotropy," *U. S. Geological Survey Professional Paper 501-D*, 1964, pp. D193–D198.

(23) Weeks, E. P., "Determining the Ratio of Horizontal to Vertical Permeability by Aquifer-Test Analysis" *Water Resources Research*, Vol 5, No. 1, 1969, pp. 196–214.

(24) Papadopulos, I. S., "Nonsteady Flow to a Well in an Infinite Anisotropic Aquifer," *Symposium of Dubrovnik*, International Association of Scientific Hydrology, 1965, pp. 21–31.

(25) Ferris, J. G., Knowles, D. B., Brown, R. H., and Stallman, R. W., "Theory of Aquifer Tests," *U. S. Geological Survey Water-Supply Paper 1536-E*, 1962, pp. 69–174.

(26) Stallman, R. W., "Type Curves for the Solution of Single-Bound Problems," in Bentall, Ray, Compiler, *Short Cuts and Special Problems in Aquifer Tests, U. S. Geological Survey Water-Supply Paper 1545-C*, 1963, pp. 45–47.

(27) Lohman, S. W., "Ground-Water Hydraulics," *Professional Paper 708*, U. S. Geological Survey, 1972, p. 70.

(28) Bennett, G. D., and Patten, E. P., Jr., "Constant-Head Pumping Test of a Multiaquifer Well to Determine Characteristics of Individual Aquifers," *U. S. Geological Survey Water-Supply Paper 1536-G*, 1962, p. 203.

(29) Hantush, M. S., "Flow to Wells Separated by Semipervious Layer," *Journal of Geophysical Research*, Vol 72, No. 6, 1967, pp. 1709–1720.

(30) Neuman, S. P., and Witherspoon, P. A., "Field Determination of the Hydraulic Properties of Leaky Multiple Aquifer Systems," *Water Resources Research*, Vol 8, No. 5, 1972, pp. 1284–1298.

(31) Neuman, S. P., and Witherspoon, P. A., "Theory of Flow in a Confined Two Aquifer System," *Water Resources Research*, Vol 5, No. 4, 1969, pp. 803–816.

(32) Javandel, I., and Witherspoon, P. A., "Method of Analyzing Transient Fluid Flow in Multilayered Aquifers," *Water Resources Research*, Vol 5, No. 4, 1969, pp. 856–869.

(33) Gringarten, A. C., and Ramey, H. J., Jr., "Unsteady-State Pressure Distributions Created by a Well with a Single Horizontal Fracture, Partial Penetration, or Restricted Entry," *Society of Petroleum Engineers Journal*, Vol 14, No. 4, 1974, pp. 413–426.

(34) Barenblatt, G. I., Zheltov, Iu P., and Kochina, I. N., "Basic Concepts in the Theory of Seepage of Homogeneous Liquids in Fissured Rocks [Strata]," *Journal of Applied Mathematics and Mechanics*, Vol 24, 1960, pp. 1286–1301.

(35) Boulton, N. S., and Streltsova, T. D., "Unsteady Flow to a Pumped Well in a Fissured Water-Bearing Formation," *Journal of Hydrology*, Vol 35, 1977, pp. 257–269.

(36) Moench, A. F., "Double Porosity Model for a Fissured Groundwater Reservoir with Fracture Skin," *Water Resources Research*, Vol 20, No. 7, 1984, pp. 831–846.

(37) Boulton, N. S., "Drawdown of the Water Table Under Non-Steady Conditions Near a Pumped Well in an Unconfined Formation," *Proceedings*, Institution of Civil Engineers, Vol 3, Pt. 3, 1954, pp. 564–579.

(38) Boulton, N. S., "Unsteady Flow to a Pumped Well Allowing for Delayed Yield from Storage," *International Association of Scientific Hydrology*, Publication 37, 1954, pp. 472–477.

(39) Boulton, N. S., "Analysis of Data from Non-Equilibrium Pumping Tests Allowing for Delayed Yield from Storage," *Proceedings*, Institution of Civil Engineers, Vol 26, 1963, pp. 469–482.

(40) Neuman, S. P., "Theory of Flow in Unconfined Aquifers Considering Delayed Response of the Water Table," *Water Resources Research*, Vol 8, No. 4, 1972, pp. 1031–1045.

(41) Neuman, S. P., "Supplementary Comments on 'Theory of Flow in Unconfined Aquifers Considering Delayed Response of the Water Table'," *Water Resources Research*, Vol 9, No. 4, 1973, pp. 1102.

(42) Neuman, S. P., "Analysis of Pumping Test Data from Anisotropic Unconfined Aquifers Considering Delayed Gravity Response," *Water Resources Research*, Vol 11, No. 2, 1975, pp. 329–342.

(43) Bouwer, H., and Rice, R. C., "A Slug Test for Determining Hydraulic Conductivity of Unconfined Aquifers with Completely or Partially Penetrating Wells," *Water Resources Research*, Vol 12, No. 3, pp. 423–428.

(44) Bouwer, H., "The Bouwer-Rice Slug Test—An Update," *Ground Water*, Vol 27, No. 3, 1989, pp. 304–309.

Standard Test Method for (Field Procedure) for Instantaneous Change in Head (Slug Tests) for Determining Hydraulic Properties of Aquifers[1]

This standard is issued under the fixed designation D 4044; the number immediately following the designation indicates the year of original adoption or, in the case of revision, the year of last revision. A number in parentheses indicates the year of last reapproval. A superscript epsilon (ε) indicates an editorial change since the last revision or reapproval.

1. Scope

1.1 This test method covers the field procedure for performing an in situ instantaneous change in head (slug) test.

1.2 This test method is used in conjunction with an analytical procedure such as Test Method D 4104 to determine aquifer properties.

1.3 The values stated in the SI units are to be regarded as standard.

1.4 *This standard does not purport to address all of the safety problems, if any, associated with its use. It is the responsibility of the user of this standard to establish appropriate safety and health practices and determine the applicability of regulatory limitations prior to use.*

2. Referenced Documents

2.1 *ASTM Standards:*

D 653 Terminology Relating to Soil, Rock, and Contained Fluids[2]

D 4043 Guide for Selection of Aquifer-Test Method in Determination of Hydraulic Properties by Well Techniques[2]

D 4104 Test Method (Analytical Procedure) for Determining Transmissivity of Confined Nonleaky Aquifers by Overdamped Well Response to Instantaneous Change in Head (Slug Test)[2]

D 4750 Test Method for Determining Subsurface Liquid Levels in a Borehole or Monitoring Well (Observation Well)[2]

3. Terminology

3.1 *Definitions:*

3.1.1 *control well*—well by which the aquifer is stressed, for example, by pumping, injection, or imposing a constant change of head.

3.1.2 *hydraulic conductivity—(field aquifer tests)*, the volume of water at the existing kinematic viscosity that will move in a unit time under a unit hydraulic gradient through a unit area measured at right angles to the direction of flow.

3.1.3 *observation well*—a well open to all or part of an aquifer.

3.1.4 *overdamped-well response*—characterized by the water level returning to the static level in an approximately exponential manner following a sudden change in water level. (See for comparison *underdamped well.*)

3.1.5 *slug*—a volume of water or solid object used to induce a sudden change of head in a well.

3.1.6 *storage coefficient*—the volume of water an aquifer releases from or takes into storage per unit surface area of the aquifer per unit change in head. For a confined aquifer, it is equal to the product of specific storage and aquifer thickness. For an unconfined aquifer, the storage coefficient is approximately equal to the specific yield.

3.1.7 *transmissivity*—the volume of water at the existing kinematic viscosity that will move in a unit time under a unit hydraulic gradient through a unit width of the aquifer.

3.1.8 *underdamped-well response*—characterized by the water level oscillating about the static water level following a sudden change in water level. (See for comparison *overdamped well.*)

3.1.9 For definitions of other terms used in this test method, refer to Terminology D 653.

4. Summary of Test Method

4.1 This test method describes the field procedures involved in conducting an instantaneous head (slug) test. The slug test method involves causing a sudden change in head in a control well and measuring the water level response within that control well. Head change may be induced by suddenly injecting or removing a known quantity or "slug" of water into the well, rapid removal of a mechanical "slug" from below the water level, increasing or decreasing the air pressure in the well casing, or emplacement of a mechanical slug into the water column.

4.2 The water-level response in the well is a function of the mass of water in the well and the transmissivity and coefficient of storage of the aquifer. One method of analysis of the data from this field practice is described in Test Method D 4104.

5. Significance and Use

5.1 This slug test field procedure is used in conjunction with a slug test analytical procedure, such as Test Method D 4104 to provide quick and relatively inexpensive estimates of transmissivity. The value of coefficient of storage determined from this test method is considered less reliable than the value of transmissivity.

5.2 The slug test provides an advantage over pumping tests in that it does not require the disposal of the large quantities of water that may be produced. This is of special importance when testing a potentially contaminated aquifer. However, slug tests reflect conditions near the well, therefore

---

[1] This test method is under the jurisdiction of ASTM Committee D-18 on Soil and Rock and is the direct responsibility of Subcommittee D18.21 on Ground Water and Vadose Zone Investigations.

Current edition approved June 15, 1991. Published August 1991.

[2] *Annual Book of ASTM Standards*, Vol 04.08.

are influenced by near-well conditions, such as gravel pack, poor well development, and skin effects.

5.3 Slug tests may be made in aquifer materials of lower hydraulic conductivity than generally considered suitable for hydraulic testing with pumping tests.

5.4 The method of data analysis (analytical procedure) should be known prior to the field testing to ensure that all appropriate dimensions and measurements are properly recorded. Selection of the analytical procedure can be aided by using Guide D 4043.

6. Apparatus

6.1 *Slug-Inducing Equipment*—This test method describes the types of equipment that can be used. Because of the infinite variety of testing conditions and because similar results can be achieved with different apparatus, engineering specifications for apparatus are not appropriate. This test method specifies the results to be achieved by the equipment to satisfy the requirements of this practice.

6.2 *Water-Level Measurement Equipment*—The method of water level measurement may be dependent on the method selected for injection or withdrawal of water, and the nature of the response of the well. For an open-well test, that is, where access to the water level is open to the surface, measure water levels manually as described in Test Method D 4750, by an automatic recording device linked to a float, or with a pressure transducer linked to a data logger or display device. A pressure transducer linked to a data logger will be necessary for a test in a closed well in which water-level changes are induced by vacuum or pressure on the control well and where manual measurements do not provide measurements of adequate frequency (see 9.3).

7. Conditioning

7.1 *Pre-Test Procedure:*

7.1.1 *Measuring Pre-Test Water Levels*—Measure the water level in the control well before beginning the test for a period of about the duration of the test to determine the pre-test water level fluctuations and to determine a reference static water level.

8. Procedure

8.1 Cause a change in water level, either a rise or decline, by one of the following methods:

8.1.1 *Water Slug*—Inject or withdraw water of a known quantity into or from the control well.

8.1.2 *Mechanical Slug*—Inject or withdraw a mechanical slug below or above the water level. The water within the control well will then rise or decline an amount equal to the volume of the mechanical slug.

8.1.3 *Release Vacuum or Pressure*—A method of simulating the injection or withdrawal of a slug of water is by the release of a vacuum or pressure on a tightly capped (shut-in) control well. After the release, the vacuum or pressure is held constant.

NOTE—There is no fixed requirement for the magnitude of the change in water level. Similar results can be achieved with a wide range in induced head change. Some considerations include a magnitude of change that can be readily measured with the apparatus selected, for example the head change should be such that the method of measurement should be accurate to 1 % of the maximum head change. Generally, an induced head change of from one-third to one meter is adequate.

The mechanical model for the test assumes the head change is induced instantaneously. Practically, a finite time is required to effect a head change. Selection of time zero can be selected experimentally. Refer to the method of analysis (such as Test Method D 4104) to determine time zero and to evaluate the suitability of the change effected in the well.

8.2 Measure water-level response to the change in water level. The frequency of water-level measurement during the test is dependent upon the hydraulic conductivity of the material being tested. During the early portions of the test, measure water levels at closely-spaced intervals. Measurements of water level made manually with a tape should be made as frequently as possible until the water level has recovered about 60 to 80 %. Increase the length of time between measurements with increasing duration of the test. Since most methods of data analysis are curve-fitting techniques, it is essential that water levels are measured frequently enough to define the water-level response curve (see Test Method D 4104).

8.2.1 In aquifer-well systems where water-level changes are rapid, it may be necessary to use a pressure transducer linked to an electronic data logger to measure and record the water levels frequently enough to adequately define the water-level response. The use of transducers and data loggers generally provides a greater than adequate frequency of measurements, ranging from several measurements per second in the early part of the test to a specified frequency in the later portions of a test. With such equipment, the test analysis may use a reduced data set of measurements to calculate the hydraulic properties (see Test Method D 4104 for analysis of water level data).

8.3 *Post-Test Procedure*—Make preliminary analysis of data before leaving the field and evaluate the test regarding the criteria given in this test method and the method of analysis, such as Test Method D 4104 to determine if the test should be rerun.

9. Report

9.1 Include the information listed below in the report of the field procedure:

9.2 All test reports should include the following:

9.2.1 Date, time, and well identification,

9.2.2 Method of slug withdrawal or injection, as well as whether the test is a falling head (injection) or a rising head (withdrawal) test,

9.2.3 Inside diameter of well screen and well casing above screen,

9.2.4 Depth of well,

9.2.5 Length and depth setting of screen,

9.2.6 Volume of mechanical slug or pressure change imposed on water level, and

9.2.7 Pre-testing water-level trend.

9.3 Establish and record the measurement point from which all measurements of water level are made. Record date, time, and depth to water level below measurement point of all water levels.

9.4 Water levels measured during the test should be recorded with information on date, clock time, and time since test started. If the water levels are measured with a pressure transducer and recorded with an electronic data logger, record the name of the data file on the data logger.

10. Precision and Bias

10.1 It is not practical to specify the precision of this test method because the response of aquifer systems during aquifer tests is dependent upon ambient system stresses. No statement can be made about bias because no true reference values exist.

11. Keywords

11.1 aquifers; aquifer tests; ground water; hydraulic conductivity; hydraulic properties; instantaneous head test; slug tests; storage coefficient; transmissivity

Standard Test Method (Field Procedure) for Withdrawal and Injection Well Tests for Determining Hydraulic Properties of Aquifer Systems[1]

This standard is issued under the fixed designation D 4050; the number immediately following the designation indicates the year of original adoption or, in the case of revision, the year of last revision. A number in parentheses indicates the year of last reapproval. A superscript epsilon (ε) indicates an editorial change since the last revision or reapproval.

1. Scope

1.1 This test method describes the field procedure for selecting well locations, controlling discharge or injection rates, and measuring water levels used to analyze the hydraulic properties of an aquifer or aquifers and adjacent confining beds.

1.2 This test method is used in conjunction with an analytical procedure such as Test Method D 4105 or D 4106 to determine aquifer properties.

1.3 The appropriate field and analytical procedures are selected as described in Guide D 4043.

1.4 The values stated in SI units are to be regarded as standard.

1.5 *This standard does not purport to address all of the safety problems, if any, associated with its use. It is the responsibility of the user of this standard to establish appropriate safety and health practices and determine the applicability of regulatory limitations prior to use.*

2. Referenced Documents

2.1 *ASTM Standards:*

D 653 Terminology Relating to Soil, Rock, and Contained Fluids[2]

D 2488 Practice for Description and Identification of Soils (Visual-Manual Procedure)[2]

D 4043 Guide for Selection of Aquifer-Test Field and Analytical Procedures in Determination of Hydraulic Properties by Well Techniques[2]

D 4105 Test Method (Analytical Procedure) for Determining Transmissivity and Storativity of Nonleaky Confined Aquifers by the Modified Theis Nonequilibrium Method[2]

D 4106 Test Method (Analytical Procedure) for Determining Transmissivity and Storativity of Confined Nonleaky Aquifers by the Theis Nonequilibrium Method[2]

D 4750 Test Method for Determining Subsurface Liquid Levels in a Borehole or Monitoring Well (Observation Well)[2]

3. Terminology

3.1 *Definitions:*

3.1.1 *aquifer, confined*—an aquifer bounded above and below by confining beds and in which the static head is above the top of the aquifer.

3.1.2 *confining bed*—a hydrogeologic unit of less permeable material bounding one or more aquifers.

3.1.3 *control well*—well by which the head and flow in the aquifer is changed, for example, by pumping, injection, or imposing a constant change of head.

3.1.4 *hydraulic conductivity (field aquifer tests)*—the volume of water at the existing kinematic viscosity that will move in a unit time under a unit hydraulic gradient through a unit area measured at right angles to the direction of flow.

3.1.5 *observation well*—a well open to all or part of an aquifer.

3.1.6 *piezometer*—a device so constructed and sealed as to measure hydraulic head at a point in the subsurface.

3.1.7 *specific storage*—the volume of water released from or taken into storage per unit volume of the porous medium per unit change in head.

3.1.8 *storage coefficient* —the volume of water an aquifer releases from or takes into storage per unit surface area of the aquifer per unit change in head. For a confined aquifer, the storage coefficient is equal to the product of the specific storage and aquifer thickness. For an unconfined aquifer, the storage coefficient is approximately equal to the specific yield.

3.1.9 *transmissivity*—the volume of water at the existing kinematic viscosity that will move in a unit time under a unit hydraulic gradient through a unit width of the aquifer.

3.1.10 For definitions of other terms used in this test method, see Terminology D 653.

4. Summary of Test Method

4.1 This test method describes the field practices in conducting withdrawal and injection well tests. These methods involve withdrawal of water from or injection of water to an aquifer through a control well and measurement of the water-level response in the aquifer. The analysis of the data from this field practice is described in standards such as Test Methods D 4105 and D 4106.

5. Significance and Use

5.1 Withdrawal and injection well test field procedures are used with appropriate analytical procedures in appropriate hydrogeological sites to determine transmissivity and storage coefficient of aquifers and hydraulic conductivity of confining beds.

6. Apparatus

6.1 Various types of equipment can be used to withdraw or inject water into the control well, measure withdrawal and

---

[1] This test method is under the jurisdiction of ASTM Committee D-18 on Soil and Rock and is the direct responsibility of Subcommittee D18.21 on Ground Water and Vadose Zone Investigations.

Current edition approved June 15, 1991. Published August 1991.

[2] *Annual Book of ASTM Standards*, Vol 04.08.

injection rates, and measure water levels. The test procedure may be conducted with different types of equipment to achieve similar results. The objectives to be achieved by the use of the equipment are given in this section and in Sections 7 and 8.

6.2 *Control Well*—Discharge or injection well test methods require that water be withdrawn from or injected into a single well. This well, known as the control well, must be drilled and completed such that it transmits water to or from the aquifer (usually the entire thickness of the aquifer) at rates such that a measurable water level change will occur at observation wells. The control well should be as efficient as possible, to reduce the head loss between the aquifer and the well. Well development should be as complete as possible to eliminate additional production of sand or silt and consequent changes in well efficiency and pumping water levels during the test. The cuttings from the control well should be described and recorded according to Practice D 2488. The analytical method selected for analysis of the data may specify certain dimensions of the control well such as screen length and depth of screen placement. Specific requirements for control wells may be given in standards for specific analytical methods (see, for example, Test Methods D 4105 and D 4106).

6.3 *Observation Wells or Piezometers*—Numbers of observation wells and their distance from the control well and their screened interval may be dependent upon the test method to be employed. Refer to the analytical test method to be used for specifications of observation wells (see, for example, Test Methods D 4105 and D 4106).

6.4 *Control Well Pump*—A pump capable of withdrawal of a constant or predetermined variable rate of water from the control well. The pump and motor should be adequately sized for the designed pumping rate and lift. The pump or motor must be equipped with a control mechanism to adjust discharge rate. In the case of diesel-, gasoline-, or natural-gas-fueled engines, throttle settings should allow for small adjustments in pumping rates. Pumps equipped with electric motors are usually controlled by adjusting backpressure on the pump through a gate valve in the discharge line. Take care to select a discharge rate small enough such that the rate can be maintained throughout the test without fully opening the gate valve. If neither method of control is practical, split the discharge and route part of the discharge back to the well through a separate discharge line.

6.5 Many aquifer tests are made at "sites of opportunity," that is, using existing production wells as the control well and using other existing wells for observation of water level. In such cases the locations and screened intervals of the wells should be compatible with the requirements of the method of test analysis.

6.6 *Water-Level Measurement Equipment*—Manual measurements can be made with a steel tape or electric tape as described in Test Method D 4750, with a mechanical recorder linked to a float, or combination of pressure transducer and electronic data logger.

6.6.1 *Mechanical Recorders*—Mechanical recorders employ a float in the well to produce a graphic record of water level changes. Early in the test, it may be difficult to distinguish small increments of time on the recorder chart, therefore the recorder should be supplemented with additional early time measurements or by marking the trace of an automatic water-level recorder chart and recording the time by the mark. Check the mechanical recorder periodically throughout the test using the steel tape.

6.6.2 *Pressure Transducers and Electronic Data Loggers*—A combination of a pressure transducer and electronic data logger can provide rapid measurements of water-level change, and can be programmed to sample at reduced frequency late in the test. Select the pressure transducer to measure pressure changes equivalent to the range of expected water level changes. Check the transducer in the field by raising and lowering the transducer a measured distance in the well. Also check the transducer readings periodically with a steel tape.

7. Conditioning

7.1 *Pre-Test Procedures:*

7.1.1 *Selecting Aquifer-Test Method*—Develop a conceptual model of the site hydrogeology and select the appropriate aquifer test method according to Guide D 4043. Observe the requirements of the selected test method with regard to specifications for the control well and observations wells.

7.1.2 *Field Reconnaissance*—Make a field reconnaissance of the site before conducting the test to include as much detail as possible on depth, continuity, extent, and preliminary estimates of the hydrologic properties of the aquifers and confining beds. Note the location of existing wells and water-holding or conveying structures that might interfere with the test. The control should be equipped with a pipeline or conveyance structure adequate to transmit the water away from the test site, so that recharge is not induced near the site. Make arrangements to ensure that nearby wells are turned off well before the test, and automatic pump controls are disabled throughout the anticipated test period. Alternately, it may be necessary to pump some wells throughout the test. If so, they should be pumped at a constant rate, and not started and stopped for a duration equal to that of the test before nor should they be started and stopped during the test.

7.1.3 *Testing of Control Well*—Conduct a short term preliminary test of the control well to estimate hydraulic properties of the aquifer, estimate the duration of the test and establish a pumping rate for the field procedure.

7.1.4 *Testing Observation Wells*—Test the observation wells or piezometers prior to the aquifer test to ensure that they are hydraulically connected to the aquifer. Accomplish this by adding or withdrawing a known volume of water (slug) and measure the water-level response in the well. The resultant response should be rapid enough to ensure that the water level in the piezometer will reflect the water level in the aquifer during the test. Redevelop piezometers with unusually sluggish response.

7.1.5 *Measuring Pre-Testing Water-Level Trends*—Measure water levels in all observation wells prior to start of pumping for a period long enough to establish the pre-pumping trend. This period is at least equal to the length of the test. The trend in all observation wells should be similar. A well with an unusual trend may reflect effects of local disturbances in the hydrologic system, or may be inadequately developed.

7.1.6 *Selecting of Pumping Rate*—Select the pumping rate, on the basis of the preliminary test (see 7.1.3), at which the well is to be pumped, such that, the rate can be sustained by the pump for the duration of the test. The rate should not be so large that the water level is drawn down below the perforations in the control well, causing cascading water and entrained air in the well. Under no circumstances should the rate be so large that the water level is drawn down to the water-entry section of the pump or tailpipe.

8. Procedure

8.1 *Withdrawing or Injecting Water from the Aquifer*—Regulate the rate at which water is withdrawn from, or injected into, the control well throughout the test. The short-term discharge should not vary more than 10 % about the mean discharge. For constant-discharge tests, long-term variation of discharge from the beginning to end of test generally should be less than 5 %.

8.2 Measure discharge frequently, for example every 5 min, and if necessary adjust discharge during the beginning of the test. When the discharge becomes more stable, reduce the frequency of adjustments and check discharge at least once every 2 h throughout the test. Variations in electric line load throughout the day will cause variations in discharge of pumps equipped with electric motors. Changes in air temperature and barometric pressure will likewise affect diesel motors. Late in a lengthy test, measure and adjust discharge much more frequently than the water levels are measured.

8.3 *Measuring Water Level; Frequency of Measurement*—Measure water levels in each observation well at approximately logarithmic intervals of time. Measure at least ten data points throughout each logarithmic interval. A typical measurement schedule is listed in Table 1.

8.4 *Duration of Pumping Phase of Test*—Make preliminary analysis of the aquifer-test data during the test using the appropriate test method (such as Test Methods D 4105 and D 4106). Continue the test until the analysis shows adequate test duration.

8.5 *Measuring Recovery of Water Levels:*

8.5.1 The recovery of water levels following pumping phase should be measured and recorded for a period of time equal to the pumping time. Analyze the recovery data to determine the hydraulic parameters of the system. The frequency of measuring water levels should be similar to the frequency during the pumping phase (see Table 1).

8.5.2 If water level data during the early part of the recovery phase are to be used from the control well, the pump should be equipped with a foot valve to prevent the column pipe fluid from flowing back into the well when the pump is turned off.

8.6 *Post-Testing Procedures:*

8.6.1 Tabulate water levels, including, pre-pumping water levels, for each well or piezometer, date, clock time, time since pumping started or stopped, and measurement point (Test Method D 4750).

TABLE 1 Typical Measurement Frequency

Frequency, One Measurement Every:	Elapsed Time, For the First:
30 s	3 min
1 min	3 to 15 min
5 min	15 to 60 min
10 min	60 to 120 min
20 min	2 to 3 h
1 h	3 to 15 h
5 h	15 to 60 h

8.6.2 Tabulate measurements of the rate of discharge or injection at the control well, date, clock time, time since pumping started, and method of measurement.

8.6.3 Prepare a written description of each well, describing the measuring point, giving its altitude and the method of obtaining the altitude, and the distance of the measuring point above the mean land surface.

8.6.4 Make plots of water-level changes and discharge measurements as follows:

8.6.4.1 Plot water levels in the control well and each observation well against the logarithm of time since pumping began. Plot the rate of discharge, Q, of the control well on arithmetic paper.

8.6.4.2 Prepare a plot of the log of drawdown, s, versus the log of the ratio of time since pumping began, t, to the square of the distance from the control well to the observation well, r, that is $\log_{10}s$ versus $\log_{10}t/r^2$, on a single graph and maintain the graph as the test progresses. Unexpected, rapid deviations of the data from the type curves may be caused by variations in discharge of the control well, or by other wells in the vicinity starting, stopping or changing discharge rates, or by other changes in field conditions. Such interfering effects may need to be measured, and adjustments made in the final data, or it may be necessary to abort the test.

8.6.4.3 *Plot Recovery of Water Levels*—Plot recovery data, consisting of plots of water level versus log of the ratio of time since pumping started (t) to the time since pumping stopped (t'). Prepare mass plots of log of recovery versus log of the quantity: ratio of time since pumping stopped (t') to the square of the distance from the control well to the observation well (r^2), that is $\log_{10}t$ versus $\log_{10}t'/r^2$.

9. Report

9.1 Prepare a report containing field data including a description of the field site, plots of water level and discharge with time, and preliminary analysis of data.

9.1.1 An introduction stating purpose of the test, dates and times water-level measurements were begun, dates and times discharge or injection was begun and ended, and the average rate of discharge or injection.

9.1.2 The "as built" description and diagrams of all control wells, observation wells, and piezometers.

9.1.3 A map of the site showing all well locations, the distances between wells, and location of all geologic boundaries or surface-water bodies which might effect the test.

9.1.3.1 The locations of wells and boundaries that would affect the aquifer tests need to be known with sufficient accuracy to provide a valid analysis. For most analyses, this means the locations must provide data points within plotting accuracy on the semilog or log-log graph paper used in the analysis. Radial distances from the control well to the observation wells usually need to be known within ±0.5 %. For prolonged, large-scale testing it may be sufficient to locate wells from maps or aerial photographs. However, for small-scale tests, the well locations should be surveyed. All faults, streams, and canals or other potential boundaries should be located. When test wells are deep relative to their spacing it may be necessary to conduct well-deviation surveys to determine the true horizontal distance between well screens in the aquifer.

9.1.4 Include tabulated field data collected during the test.

10. Precision and Bias

10.1 It is not practicable to specify the precision of this test method because the response of aquifer systems during aquifer tests is dependent upon ambient system stresses. No statement can be made about bias because no true reference values exist.

11. Keywords

11.1 aquifers; aquifer tests; discharging wells; drawdown; ground water; hydraulic conductivity; injection wells; recovery; storage coefficient; transmissivity

Standard Test Method (Analytical Procedure) for Determining Transmissivity of Nonleaky Confined Aquifers by Overdamped Well Response to Instantaneous Change in Head (Slug Test)[1]

This standard is issued under the fixed designation D 4104; the number immediately following the designation indicates the year of original adoption or, in the case of revision, the year of last revision. A number in parentheses indicates the year of last reapproval. A superscript epsilon (ε) indicates an editorial change since the last revision or reapproval.

1. Scope

1.1 This test method covers the determination of transmissivity from the measurement of force-free (overdamped) response of a well-aquifer system to a sudden change of water level in a well. Force-free response of water level in a well to a sudden change in water level is characterized by recovery to initial water level in an approximate exponential manner with negligible inertial effects.

1.2 The analytical procedure in this test method is used in conjunction with the field procedure in Test Method D 4044 for collection of test data.

1.3 *Limitations*—Slug tests are considered to provide an estimate of transmissivity. The determination of aquifer storage coefficient is less reliable. Although the assumptions of this test method prescribe a fully penetrating well (a well open through the full thickness of the aquifer), the slug test method is commonly conducted using a partially penetrating well. Such a practice may be acceptable for application under conditions in which the aquifer is stratified and horizontal hydraulic conductivity is much greater than vertical hydraulic conductivity. In such a case the test would be considered to be representative of the average hydraulic conductivity of the portion of the aquifer adjacent to the open interval of the well.

1.4 The values stated in SI units are to be regarded as standard.

1.5 *This standard does not purport to address all of the safety problems, if any, associated with its use. It is the responsibility of the user of this standard to establish appropriate safety and health practices and determine the applicability of regulatory limitations prior to use.*

2. Referenced Documents

2.1 *ASTM Standards:*

D 653 Terminology Relating to Soil, Rock, and Contained Fluids[2]

D 4043 Guide for Selection of Aquifer-Test Method in Determining of Hydraulic Properties by Well Techniques[3]

D 4044 Test Method (Field Procedure) for Instantaneous Change in Head (Slug Test) for Determining Hydraulic Properties of Aquifers[3]

D 4750 Test Method for Determining Subsurface Liquid Levels in a Borehole or Monitoring Well (Observation Well)[2]

3. Terminology

3.1 *Definitions:*

3.1.1 *aquifer, confined*—an aquifer bounded above and below by confining beds and in which the static head is above the top of the aquifer.

3.1.2 *confining bed*—a hydrogeologic unit of less permeable material bounding one or more aquifers.

3.1.3 *control well*—well by which the aquifer is stressed, for example, by pumping, injection, or change of head.

3.1.4 *head, static*—the height above a standard datum of the surface of a column of water (or other liquid) that can be supported by the static pressure at a given point.

3.1.5 *hydraulic conductivity—(field aquifer tests)*, the volume of water at the existing kinematic viscosity that will move in a unit time under a unit hydraulic gradient through a unit area measured at right angles to the direction of flow.

3.1.6 *observation well*—a well open to all or part of an aquifer.

3.1.7 *overdamped-well response*—characterized by the water level returning to the static level in an approximately exponential manner following a sudden change in water level. (See for comparison *underdamped-well response.*)

3.1.8 *slug*—a volume of water or solid object used to induce a sudden change of head in a well.

3.1.9 *specific storage*—the volume of water released from or taken into storage per unit volume of the porous medium per unit change in head.

3.1.10 *storage coefficient*—the volume of water an aquifer releases from or takes into storage per unit surface area of the aquifer per unit change in head. For a confined aquifer, the storage coefficient is equal to the product of specific storage and aquifer thickness. For an unconfined aquifer, the storage coefficient is approximately equal to the specific yield.

3.1.11 *transmissivity*—the volume of water at the existing kinematic viscosity that will move in a unit time under a unit hydraulic gradient through a unit width of the aquifer.

3.1.12 *underdamped-well response*—response characterized by the water level oscillating about the static water level following a sudden change in water level. (See for comparison *overdamped-well response.*)

3.1.13 For definitions of other terms used in this test method, see Terminology D 653.

3.2 *Symbols and Dimensions:*

3.2.1 J_0 [nd]—zero-order Bessel function of the first kind.

[1] This test method is under the jurisdiction of ASTM Committee D-18 on Soil and Rock and is the direct responsibility of Subcommittee D18.21 on Ground Water and Vadose Zone Investigations.

Current edition approved June 15, 1991. Published August 1991.

[2] *Annual Book of ASTM Standards*, Vol 04.08.

[3] *Annual Book of ASTM Standards*, Vol

3.2.2 J_1 [nd]—first-order Bessel function of the first kind.
3.2.3 K [LT^{-1}]—hydraulic conductivity.
3.2.4 T [L^2T^{-1}]—transmissivity.
3.2.5 S [nd]—storage coefficient.
3.2.6 Y_0 [nd]—zero order Bessel function of the second kind.
3.2.7 Y_1 [nd]—first order Bessel function of the second kind.
3.2.8 r_c [L]—radius of control-well casing or open hole in interval where water level changes.
3.2.9 r_w [L]—radius of control well screen or open hole adjacent to water bearing unit.
3.2.10 u—variable of integration.
3.2.11 H [L]—change in head in control well.
3.2.12 H_o [L]—initial head rise (or decline) in control well.
3.2.13 t—time.
3.2.14 β—Tt/r_c^2.
3.2.15 α—r_w^2S/r_c^2.

4. Summary of Test Method

4.1 This test method describes the analytical procedure for analyzing data collected during an instantaneous head (slug) test using an overdamped well. The field procedures in conducting a slug test are given in Test Method D 4044. The analytical procedure consists of analyzing the recovery of water level in the well following the change in water level induced in the well.

4.2 *Solution*—The solution given by Cooper et al (**1**)[4] is as follows:

$$H = \frac{2H_o}{\pi}\int_0^\infty [[\exp(-\beta u^2/\alpha)[J_0(ur/r_w)[uY_0(u) - 2\alpha Y_1(u)] - Y_0(ur/r_w)[uJ_0(u) - 2\alpha J_1(u)]]/\Delta(u)]]du \quad (1)$$

where:

$$\alpha = r_w^2S/r_c^2,$$
$$\beta = Tt/r_c^2,$$

and:

$$\Delta(u) = [uJ_0(u) - 2\alpha J_1(u)]^2 + [uY_0(u) - 2\alpha Y_1(u)]^2$$

NOTE 1—Other analytical solutions are described by Hvorslev (**2**) Bouwer and Rice (**3**), and Bouwer (**4**).

5. Significance and Use

5.1 *Assumptions of Solution of Cooper et al* (**1**):
5.1.1 The head change in the control well is instantaneous at time $t = 0$.
5.1.2 Well is of finite diameter and fully penetrates the aquifer.
5.1.3 Flow in the nonleaky aquifer is radial.
5.2 *Implications of Assumptions:*
5.2.1 The mathematical equations applied ignore inertial effects and assume the water level returns the static level in an approximate exponential manner. The geometric configuration of the well and aquifer are shown in Fig. 1.
5.2.2 Assumptions are applicable to artesian or confined conditions and fully penetrating wells. However, this test method is commonly applied to partially penetrating wells and in unconfined aquifers where it may provide estimates of hydraulic conductivity for the aquifer interval adjacent to the open interval of the well if the horizontal hydraulic conductivity is significantly greater than the vertical hydraulic conductivity.

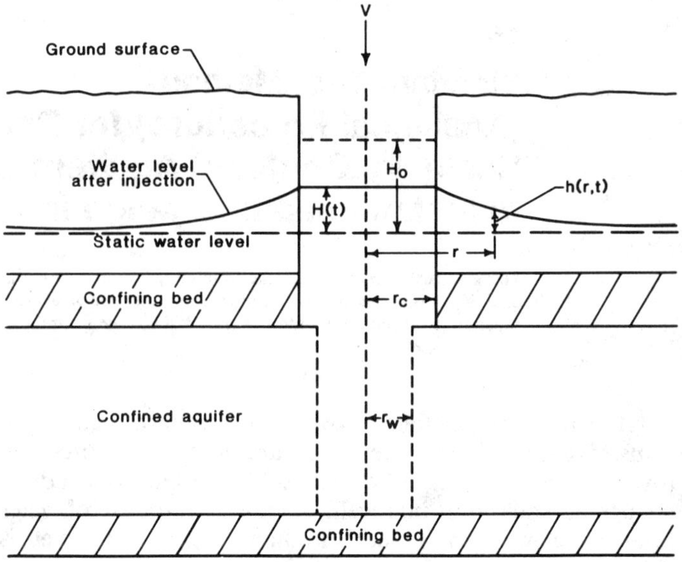

FIG. 1 Cross Section Through a Well in Which a Slug of Water is Suddenly Injected

5.2.3 As pointed out by Cooper et al (**1**) the determination of storage coefficient by this test method has questionable reliability because of the similar shape of the curves, whereas, the determination of transmissivity is not as sensitive to choosing the correct curve. However, the curve selected should not imply a storage coefficient unrealistically large or small.

6. Procedure

6.1 The overall procedure consists of conducting the slug test field procedure (see Test Method D 4044) and analysis of the field data, that is addressed in this test method.

6.2 The integral expression in the solution given in Eq (1) cannot be evaluated analytically. A graphical solution for determination of transmissivity and coefficient of storage can be made using a set of type curves that can be drawn from the values in Table 1.

7. Calculation

7.1 Plot, on semilogarithmic paper, a set of type curves of values of $F(\beta, \alpha) = H/H_o$, on the arithmetic scale, as a function of β, on the logarithmic scale from the values of the functions in Table 1.

7.2 Plot, on semilogarithmic paper of the same scale as that of the type-curve plot, the water level data in the control well, expressed as a fraction, H/H_o, on the arithmetic scale, versus time, t, on the logarithmic scale.

NOTE 2—If the water level rise is very rapid with a small disparity between the calculated and measured change in water level, then time = 0 can be used as the instant the change was initiated and H_o can be the calculated rise. If there is a significant time lag between initiation of the head change and the peak rise or decline is significantly less than the

[4] The boldface numbers in parentheses refer to a list of references at the end of the text.

TABLE 1 Values of H/H_o

$\beta = Tt/r_c^2$	α	10^{-1}	10^{-2}	10^{-3}	10^{-4}	10^{-5}
	From Cooper, Bredehoeft, and Papadopulos (1)					
	1.00	0.9771	0.9920	0.9969	0.9985	0.9992
10^{-3}	2.15	0.9658	0.9876	0.9949	0.9974	0.9985
	4.64	0.9490	0.9807	0.9914	0.9954	0.9970
	1.00	0.9238	0.9693	0.9853	0.9915	0.9942
10^{-2}	2.15	0.8860	0.9505	0.9744	0.9841	0.9883
	4.64	0.8293	0.9187	0.9545	0.9701	0.9781
	1.00	0.7460	0.8655	0.9183	0.9434	0.9572
10^{-1}	2.15	0.6289	0.7782	0.8538	0.8935	0.9167
	4.64	0.4782	0.6436	0.7436	0.8031	0.8410
	1.00	0.3117	0.4598	0.5729	0.6520	0.7080
10^{0}	2.15	0.1665	0.2597	0.3543	0.4364	0.5038
	4.64	0.07415	0.1086	0.1554	0.2082	0.2620
	7.00	0.04625	0.06204	0.08519	0.1161	0.1521
	1.00	0.03065	0.03780	0.04821	0.06355	0.08378
	1.40	0.02092	0.02414	0.02844	0.03492	0.04426
10^{1}	2.15	0.01297	0.01414	0.01545	0.01723	0.01999
	3.00	0.009070	0.009615	0.01016	0.01083	0.01169
	4.64	0.005711	0.004919	0.006111	0.006319	0.006554
	7.00	0.003722	0.003809	0.003884	0.003962	0.004046
	1.00	0.002577	0.002618	0.002653	0.002688	0.002725
10^{2}	2.15	0.001179	0.001187	0.001194	0.001201	0.001208

$\beta = Tt/r_c^2$	α	10^{-6}	10^{-7}	10^{-8}	10^{-9}	10^{-10}
	From Papadopulos, Bredehoeft, and Cooper (5)					
	1	0.9994	0.9996	0.9996	0.9997	0.9997
	2	0.9989	0.9992	0.9993	0.9994	0.9995
10^{-3}	4	0.9980	0.9985	0.9987	0.9989	0.9991
	6	0.9972	0.9978	0.9982	0.9984	0.9986
	8	0.9964	0.9971	0.9976	0.9980	0.9982
	1	0.9956	0.9965	0.9971	0.9975	0.9978
	2	0.9919	0.9934	0.9944	0.9952	0.9958
10^{-2}	4	0.9848	0.9875	0.9894	0.9908	0.9919
	6	0.9782	0.9819	0.9846	0.9866	0.9881
	8	0.9718	0.9765	0.9799	0.9824	0.9844
	1	0.9655	0.9712	0.9753	0.9784	0.9807
	2	0.9361	0.9459	0.9532	0.9587	0.9631
10^{-1}	4	0.8828	0.8995	0.9122	0.9220	0.9298
	6	0.8345	0.8569	0.8741	0.8875	0.8984
	8	0.7901	0.8173	0.8383	0.8550	0.8686
	1	0.7489	0.7801	0.8045	0.8240	0.8401
	2	0.5800	0.6235	0.6591	0.6889	0.7139
	3	0.4554	0.5033	0.5442	0.5792	0.6096
	4	0.3613	0.4093	0.4517	0.4891	0.5222
10^{0}	5	0.2893	0.3351	0.3768	0.4146	0.4487
	6	0.2337	0.2759	0.3157	0.3525	0.3865
	7	0.1903	0.2285	0.2655	0.3007	0.3337
	8	0.1562	0.1903	0.2243	0.2573	0.2888
	9	0.1292	0.1594	0.1902	0.2208	0.2505
	1	0.1078	0.1343	0.1620	0.1900	0.2178
	2	0.02720	0.03343	0.04129	0.05071	0.06149
	3	0.01286	0.01448	0.01667	0.01956	0.02320
10^{1}	4	0.008337	0.008898	0.009637	0.01062	0.01190
	5	0.006209	0.006470	0.006789	0.007192	0.007709
	6	0.004961	0.005111	0.005283	0.005487	0.005735
	8	0.003547	0.003617	0.003691	0.003773	0.003863
	1	0.002763	0.002803	0.002845	0.002890	0.002938
10^{2}	2	0.001313	0.001322	0.001330	0.001339	0.001348

calculated change use $t = 0$ as the time of maximum observed change and take H_o as the maximum observed change.

7.3 Overlay the data plot on the set of type curve plots and, with the arithmetic axes coincident, shift the data plot to match one curve or an interpolated curve of the type curve set.

7.4 Using the coordinates of the point, determine the transmissivity and storage coefficient from the following equations:

$$T = \beta r_c^2/t$$

and:

$$S = \alpha r_c^2/r_w^2$$

8. Report

8.1 Prepare a report including the information described in this section. The final report of the analytical procedure will include information from the report on test method selection (see Guide D 4043) and the field testing procedure (see Test Method D 4044).

8.1.1 *Introduction*—The introductory section is intended to present the scope and purpose of the slug test method for determining transmissivity and storage coefficient. Summarize the field hydrogeologic conditions and the field equipment and instrumentation including the construction of the control well, and the method of measurement and of effecting a change in head. Discuss the rationale for selecting the method used (see Guide D 4043).

8.1.2 *Hydrogeologic Setting*—Review information available on the hydrogeology of the site; interpret and describe the hydrogeology of the site as it pertains to the method selected for conducting and analyzing an aquifer test. Compare hydrogeologic characteristics of the site as it conforms and differs from assumptions made in the solution to the aquifer test method.

8.1.3 *Equipment*—Report the field installation and equipment for the aquifer test. Include in the report, well construction information, diameter, depth, and open interval to the aquifer, and location of control well.

8.1.3.1 Report the techniques used for observing water levels, pumping rate, barometric changes, and other environmental conditions pertinent to the test. Include a list of measuring devices used during the test, the manufacturers name, model number, and basic specifications for each major item, and the name and date of the last calibration, if applicable.

8.1.4 *Testing Procedures*—Report the steps taken in conducting the pretest and test phases. Include the frequency of head measurements made in the control well, and other environmental data recorded before and during the testing procedure.

8.1.5 *Presentation and Interpretation of Test Results:*

8.1.5.1 *Data*—Present tables of data collected during the test.

8.1.5.2 *Data Plots*—Present data plots used in analysis of the data. Show overlays of data plots and type curve with match points and corresponding values of parameters at match points.

8.1.5.3 Show calculation of transmissivity and storage coefficient.

8.1.5.4 Evaluate the overall quality of the test on the basis of the adequacy of instrumentation and observations of stress and response and the conformance of the hydrogeologic conditions and the performance of the test to the assumptions (see 5.1).

9. Precision and Bias

9.1 It is not practical to specify the precision of this test method because the response of aquifer systems during aquifer tests is dependent upon ambient system stresses. No statement can be made about bias because no true reference values exist.

10. Keywords

10.1 Aquifers; aquifer tests; control wells; ground water; hydraulic conductivity; observation wells; storage coefficient storativity; transmissivity

REFERENCES

(1) Cooper, H. H., Jr., Bredehoeft, J. D., and Papadopulos, I. S., "Response of a Finite-Diameter Well to an Instantaneous Charge of Water," *Water Resources Research*, Vol 3, No. 1, 1967, pp. 263–269.

(2) Hvorslev, M. J., "Time Lag and Soil Permeability in Ground-Water Observations," Waterways Experiment Station, Corps of Engineers, U.S. Army, Bulletin No. 36, 1951, p. 50.

(3) Bouwer, H., and Rice, R. C., "A Slug Test for Determining Hydraulic Conductivity of Unconfined Aquifers with Completely or Partially Penetrating Wells," *Water Resources Research*, Vol 12, No. 3, 1976, pp. 423–423.

(4) Bouwer, H., "The Bouwer-Rice Slug Test—An Update," *Ground Water*, Vol 27, No. 3, 1989, pp. 304–309.

(5) Papadopulos, I. S., Bredehoeft, J. D., and Cooper, H. H., Jr., "On the Analysis of Slug Test Data," *Water Resources Research*, Vol 9, No. 4, 1973, pp. 1087–1089.

Designation: D 4105 – 91

Standard Test Method (Analytical Procedure) for Determining Transmissivity and Storage Coefficient of Nonleaky Confined Aquifers by the Modified Theis Nonequilibrium Method[1]

This standard is issued under the fixed designation D 4105; the number immediately following the designation indicates the year of original adoption or, in the case of revision, the year of last revision. A number in parentheses indicates the year of last reapproval. A superscript epsilon (ε) indicates an editorial change since the last revision or reapproval.

1. Scope

1.1 This test method covers an analytical procedure for determining transmissivity and storage coefficient of a nonleaky confined aquifer under conditions of radial flow to a fully penetrating well of constant flux. This test method is a shortcut procedure used to apply the Theis nonequilibrium method. The Theis method is described in Test Method D 4106.

1.2 This test method is used in conjunction with the field procedure given in Test Method D 4050.

1.3 *Limitations*—The limitations of this test method are primarily related to the correspondence between the field situation and the simplifying assumptions of this test method (see 5.1). Furthermore, application is valid only for values of u less than 0.01 (u is defined in Eq 2, in 8.6).

1.4 The values stated in SI units are to be regarded as standard.

1.5 *This standard does not purport to address all of the safety problems, if any, associated with its use. It is the responsibility of the user of this standard to establish appropriate safety and health practices and determine the applicability of regulatory limitations prior to use.*

2. Referenced Documents

2.1 *ASTM Standards:*

D 653 Terminology Relating to Soil, Rock, and Contained Fluids[2]

D 4043 Guide for Selection of Aquifer-Test Method in Determining Hydraulic Properties by Well Techniques[2]

D 4050 Test Method (Field Procedure) for Withdrawal and Injection Well Tests for Determining Hydraulic Properties of Aquifer Systems[2]

D 4106 Test Method (Analytical Procedure) for Determining Transmissivity and Storage Coefficient of Nonleaky Confined Aquifers by the Theis Nonequilibrium Method[2]

3. Terminology

3.1 *Definitions:*

3.1.1 *aquifer, confined*—an aquifer bounded above and below by confining beds and in which the static head is above the top of the aquifer.

3.1.2 *confining bed*—a hydrogeologic unit of less permeable material bounding one or more aquifers.

3.1.3 *control well*—well by which the aquifer is stressed, for example, by pumping, injection, or change of head.

3.1.4 *drawdown*—vertical distance the static head is lowered due to the removal of water.

3.1.5 *hydraulic conductivity—(field aquifer tests),* the volume of water at the existing kinematic viscosity that will move in a unit time under unit hydraulic gradient through a unit area measured at right angles to the direction of flow.

3.1.6 *observation well*—a well open to all or part of an aquifer.

3.1.7 *piezometer*—a device so constructed and sealed as to measure static head at a point in the subsurface.

3.1.8 *specific storage*—the volume of water released from or taken into storage per unit volume of the porous medium per unit change in head.

3.1.9 *storage coefficient*—the volume of water an aquifer releases from or takes into storage per unit surface area of the aquifer per unit change in head. For a confined aquifer, it is equal to the product of specific storage and aquifer thickness. For an unconfined aquifer, the storage coefficient is approximately equal to the specific yield.

3.1.10 *transmissivity*—the volume of water at the existing kinematic viscosity that will move in a unit time under a unit hydraulic gradient through a unit width of the aquifer.

3.1.11 *unconfined aquifer*—an aquifer that has a water table.

3.1.12 For definitions of other terms used in this test method, see Terminology D 653.

3.2 *Symbols and Dimensions:*

3.2.1 K [LT^{-1}]—hydraulic conductivity.

3.2.2 K_{xy}—hydraulic conductivity in the horizontal direction.

3.2.3 K_z—hydraulic conductivity in the vertical direction.

3.2.4 T [L^2T^{-1}]—transmissivity.

3.2.5 S [nd]—storage coefficient.

3.2.6 Ss [L^{-1}]—specific storage.

3.2.7 s [L]—drawdown.

3.2.8 Q [L^3T^{-1}]—discharge.

3.2.9 r [L]—radial distance from control well.

3.2.10 t [T]—time.

3.2.11 b [L]—thickness of the aquifer.

4. Summary of Test Method

4.1 This test method describes an analytical procedure for

[1] This test method is under the jurisdiction of ASTM Committee D-18 on Soil and Rock and is the direct responsibility of Subcommittee D18.21 on Ground Water and Vadose Zone Investigations.

Current edition approved June 15, 1991. Published August 1991.

[2] *Annual Book of ASTM Standards*, Vol 04.08.

analyzing data collected during a withdrawal or injection well test. The field procedure (see Test Method D 4050) involves pumping a control well at a constant rate and measuring the water level response in one or more observation wells or piezometers. The water-level response in the aquifer is a function of the transmissivity and coefficient of storage of the aquifer. Alternatively, the test can be performed by injecting water at a constant rate into the aquifer through the control well. Analysis of buildup of water level in response to injection is similar to analysis of drawdown of water level in response to withdrawal in a confined aquifer. Drawdown of water level is analyzed by plotting drawdown against factors incorporating either time or distance from the control well, or both, and matching the drawdown response with a type curve.

4.2 *Solution*—The solution given by Theis **(1)**[3] can be expressed as follows:

$$s = \frac{Q}{4\pi T}\int_u^\infty \frac{e^{-y}}{y}\,dy \tag{1}$$

where:

$$u = \frac{r^2 S}{4Tt} \tag{2}$$

and:

$$\int_u^\infty \frac{e^{-y}}{y}\,dy = W(u) = -0.577216 - \log_e u + u - \frac{u^2}{2!2} + \frac{u^3}{3!3} - \frac{u^4}{4!4} + \ldots \tag{3}$$

4.3 The sum of the terms to the right of $\log_e u$ in the series of Eq 3 is not significant when u becomes small, that is, equal or less than 0.01. The value of u decreases with time, t, and decreases as the radial distance, r, decreases. Therefore, for large values of t and reasonably small values of r, the terms to the right of $\log_e u$ in Eq 3 may be neglected as recognized by Theis **(1)** and Jacob **(2)**. The Theis equation can then be written as follows:

$$s = \frac{Q}{4\pi T}\left[-0.577216 - ln\left(r^2 \frac{S}{4Tt}\right)\right] \tag{4}$$

from which it has been shown by Lohman **(3)** that

$$T = \frac{2.3Q}{4\pi\Delta s/\Delta\log_{10} t} \tag{5}$$

and:

$$T = -\frac{2.3Q}{2\pi\Delta s/\Delta\log_{10} r} \tag{6}$$

where:

$\Delta s/\Delta\log_{10} t$ = the drawdown (measured or projected) over one log cycle of time, and

$\Delta s/\Delta\log_{10} r$ = the drawdown (measured or projected) over one log cycle of radial distance from the control well.

[3] The boldface numbers in parentheses refer to a list of references at the end of the text.

5. Significance and Use

5.1 *Assumptions:*

5.1.1 Well discharges at a constant rate, Q.

5.1.2 Well is of infinitesimal diameter and fully penetrates the aquifer, that is, the well is open to the full thickness of the aquifer.

5.1.3 The nonleaky aquifer is homogeneous, isotropic, and aerially extensive. A nonleaky aquifer receives insignificant contribution of water from confining beds.

5.1.4 Discharge from the well is derived exclusively from storage in the aquifer.

5.1.5 The geometry of the assumed aquifer and well conditions are shown in Fig. 1.

5.2 *Implications of Assumptions:*

5.2.1 Implicit in the assumptions are the conditions of radial flow. Vertical flow components are induced by a control well that partially penetrates the aquifer, that is, not open to the aquifer through its full thickness. If the control well does not fully penetrate the aquifer, the nearest piezometer or partially penetrating observation well should be located at a distance, r, beyond which vertical flow components are negligible, where according to Reed **(4)**

$$r = \frac{1.5b}{\sqrt{\dfrac{K_z}{K_{xy}}}} \tag{7}$$

Fully penetrating observation wells may be placed at less than distance r from the control well. Observation wells may be on the same or on various radial lines from the control well.

5.2.2 The Theis method assumes the control well is of infinitesimal diameter. Also, it assumes that the water level in the control well is the same as in the aquifer contiguous to the well. In practice these assumptions may cause a difference between the theoretical drawdown and field measurements of drawdown in the early part of the test and in and near the control well. Control well storage is negligible after a time, t, given by the following equation after Papadopulos and Cooper **(5)**.

$$t = \frac{2.5 \times 10^2 r_c^{\,2}}{T} \tag{8}$$

where:

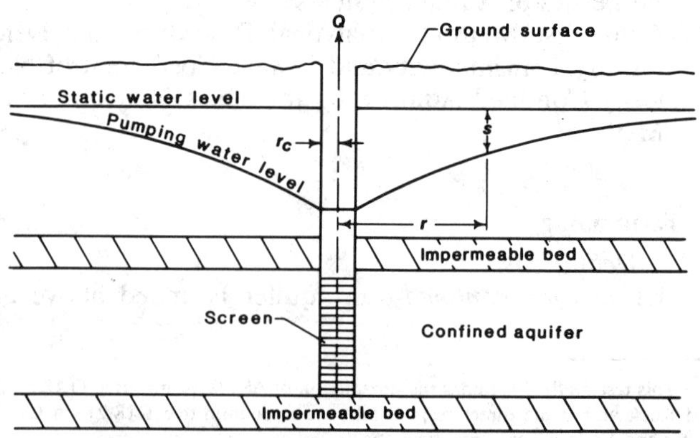

FIG. 1 Cross Section Through a Discharging Well in a Nonleaky Confined Aquifer

r_c = the radius of the control well in the interval that includes the water level changes.

5.2.3 *Application of Theis Nonequilibrium Method to Unconfined Aquifers*:

5.2.3.1 Although the assumptions are applicable to artesian or confined conditions, the Theis solution may be applied to unconfined aquifers if drawdown is small compared with the saturated thickness of the aquifer or if the drawdown is corrected for reduction in thickness of the aquifer and the effects of delayed gravity yield are small.

5.2.3.2 *Reduction in Aquifer Thickness*—In an unconfined aquifer, dewatering occurs when the water levels decline in the vicinity of a pumping well. Corrections in drawdown need to be made when the drawdown is a significant fraction of the aquifer thickness as shown by Jacob **(6)**. The drawdown, s, needs to be replaced by s', the drawdown that would occur in an equivalent confined aquifer, where:

$$s' = s - \frac{s^2}{2b} \tag{9}$$

5.2.3.3 *Gravity Yield Effects*—In unconfined aquifers, delayed gravity yield effects may invalidate measurements of drawdown during the early part of the test for application to the Theis method. Effects of delayed gravity yield are negligible in partially penetrating observation wells at a distance, r, from the control well, where:

$$r = \frac{b}{\sqrt{\dfrac{K_z}{K_{xy}}}} \tag{7}$$

after the time, t, as given in the following equation from Neuman **(7)**:

$$t = 10 S_y \frac{r^2}{T} \tag{11}$$

where:

S_y = the specific yield.

For fully penetrating observation wells, the effects of delayed yield are negligible at the distance, r, in Eq 10 after one tenth of the time given in the Eq 11.

6. Apparatus

6.1 Analysis of data from the field procedure (see Test Method D 4050) by this test method requires that the control well and observation wells meet the requirements specified in 6.2 through 6.4.

6.2 *Control Well*—Screen the control well in the aquifer and equip with a pump capable of discharging water from the well at a constant rate for the duration of the test. Preferably, screen the control well throughout the full thickness of the aquifer. If the control well partially penetrates the aquifer, take special precaution in the placement or design of observation wells (see 5.2.1).

6.3 *Construction of Observation Wells*—Construct one or more observation wells or piezometers at a distance from the control well. Observation wells may be partially open or fully open throughout the thickness of the aquifer.

6.4 *Location of Observation Wells*—Locate observation wells at various distances from the control well within the area of influence of pumping. However, if radial flow components are significant and if partially penetrating observation wells are used, locate them at a distance beyond the effect of vertical flow components (see 5.2.1). If the aquifer is unconfined, constraints are imposed on the distance to partially penetrating observation wells and the validity of early time measurements (see 5.2.3).

7. Procedure

7.1 The overall procedure consists of conducting the field procedure for withdrawal or injection well tests described in Test Method D 4050 and analysis of the field data, that is addressed in this test method.

7.2 Use a graphical procedure to solve for transmissivity and coefficient of storage as described in 8.2.

8. Calculation

8.1 Plot drawdown, s, at a specified distance on the arithmetic scale and time, t, on the logarithmic scale.

8.2 Plot drawdown, s, for several observation wells at a specified time on the arithmetic scale and distance on the logarithmic scale.

8.3 For convenience in calculations, by choosing drawdown, Δs_t, as that which occurs over one log cycle of time:

$$\Delta \log_{10} t = \log_{10}\left(\frac{t_2}{t_1}\right) = 1 \tag{12}$$

and, similarly for convenience in calculations, by choosing the drawdown, Δs_r, as that which occurs over one log cycle of distance,

$$\Delta \log_{10} r = \log_{10}\left(\frac{r_2}{r_1}\right) = 1 \tag{13}$$

8.4 Calculate transmissivity using the semilog plot of drawdown versus time by the following equation derived from Eq 5:

$$t = 2.3Q/2\pi\Delta s_r \tag{14}$$

or calculate transmissivity using the semilog plot of drawdown versus radial distance from control well by the following equation derived from Eq 6:

$$T = -\frac{2.3Q}{2\pi\Delta s_r} \tag{15}$$

8.5 Determine the coefficient of storage from these semilog plots of drawdown versus time or distance by a method proposed by Jacob **(2)** where:

$$s = \frac{2.3Q}{4\pi T}\log_{10}\left(\frac{2.25Tt}{r^2 S}\right) \tag{16}$$

Taking $s = 0$ at the zero-drawdown intercept of the straight-line semilog plot of time or distance versus drawdown,

$$S = \frac{2.25Tt}{r^2} \tag{17}$$

where:

either r or t = the value at the zero-drawdown intercept.

8.6 To apply the modified Theis nonequilibrium method to thin unconfined aquifers, where the drawdown is a significant fraction of the initial saturated thickness, apply a correction to the drawdown in solving for T and S (see 5.2.3.2).

8.7 This method is applicable only for values of $u < 0.01$, that is:

$$u = \frac{r^2S}{4Tt} < 0.01 \quad (18)$$

It is seen from Eq 12 that u decreases as time increases, other things being equal. Because S is in the numerator, the value of u is much smaller for a confined aquifer, whose storage coefficient may range from only about 10^{-5} to 10^{-3}, than for an unconfined aquifer, whose specific yield may be from 0.1 to 0.3. To compensate for this, t must be greater by several orders of magnitude in testing an unconfined aquifer than testing a confined aquifer.

8.7.1 In a drawdown-time test (s versus $\log_{10}t$ or $\log_{10}t/r^2$), data points for any particular distance will begin to fall on a straight line only after the time is sufficiently long to satisfy the above criteria. In a drawdown-distance test (s versus $\log_{10}r$), the well must be pumped long enough that the data for the most distant observation well satisfy the requirements; then only the drawdowns at or after this value of t may be analyzed on a semilogarithmic plot for one particular value of t.

NOTE—The analyst may also find it useful to analyze the data using the Theis nonequilibrium procedure (see Test Method D 4106).

9. Report

9.1 Report the information described below. The report of the analytical procedure will include information from the report on test method selection (see Guide D 4043) and the field testing procedure (see Test Method D 4050).

9.1.1 *Introduction*—The introductory section is intended to present the scope and purpose of the recovery method for determining transmissivity and storativity in a nonleaky confined aquifer. Summarize the field hydrogeologic conditions and the field equipment and instrumentation including the construction of the control well and observation wells and piezometers, the method of measurement of discharge and water levels, and the duration of the test and pumping rate. Discuss rationale for selecting the modified Theis method.

9.1.2 *Hydrogeologic Setting*—Review the information available on the hydrogeology of the site; interpret and describe the hydrogeology of the site as it pertains to the selection of this method for conducting and analyzing an aquifer test. Compare the hydrogeologic characteristics of the site as it conforms and differs from the assumptions in the solution to the aquifer test method.

9.1.3 *Equipment*—Report the field installation and equipment for the aquifer test, including the construction, diameter, depth of screened interval, and location of control well and pumping equipment, and the construction, diameter, depth, and screened interval of observation wells.

9.1.4 Describe the methods of observing water levels, pumping rate, barometric changes, and other environmental conditions pertinent to the test. Include a list of measuring devices used during the test, the manufacturers name, model number, and basic specifications for each major item, and the name and date and method of the last calibration, if applicable.

9.1.5 *Testing Procedures*—State the steps taken in conducting pre-test, drawdown, and recovery phases of the test. Include the date, clock time, and time since pumping started or stopped for measurements of discharge rate, water levels, and other environmental data recorded during the testing procedure.

9.1.6 *Presentation and Interpretation of Test Results:*

9.1.6.1 *Data*—Present tables of data collected during the test. Show methods of adjusting water levels for barometric changes and calculation of drawdown and residual drawdown.

9.1.6.2 *Data Plots*—Present data plots used in analysis of the data.

9.1.6.3 Evaluate qualitatively the determinations of transmissivity and coefficient of storage on the basis of the adequacy of instrumentation, observations of stress and response, and the conformance of the hydrogeologic conditions, and the performance of the test to the assumptions of the method.

10. Precision and Bias

10.1 It is not practicable to specify the precision of this test method because the response of aquifer systems during aquifer tests is dependent upon ambient system stresses. No statement can be made about bias because no true reference values exist.

11. Keywords

11.1 aquifers; aquifer tests; confined aquifers; control wells; ground water; hydraulic properties; observation wells; storage coefficient; transmissivity; unconfined aquifers

REFERENCES

(1) Theis C. V., "The Relation Between the Lowering of the Piezometric Surface and the Rate and Duration of Discharge of a Well Using Ground-Water Storage," *American Geophysical Union Transactions*, Vol 16, Part 2, 1935, pp. 519–524.

(2) Jacob, C. E., "Flow of Ground Water," in *Engineering Hydraulics*, Proceedings of the Fourth Hydraulics Conference, June 12–15, 1949, New York, John Wiley and Sons, Inc., 1950, pp. 321–386.

(3) Lohman, S. W., "Ground-Water Hydraulics," U.S. Geological Survey Professional Paper 708, 1972.

(4) Reed, J. E., "Type Curves for Selected Problems of Flow to Wells in Confined Aquifers," *U.S. Geological Survey Techniques of Water-Resources Investigations*, Book 3, Chapter B3, 1980.

(5) Papadopulos, S. S., and Cooper, H. H., Jr., "Drawdown in a Well of Large Diameter," *Water Resources Research*, Vol 1, 1967, pp. 241–244.

(6) Jacob, C. E., "Determining Permeability of Water-Table Aquifers," in Bentall, Ray, compiler, *Methods of Determining Permeability, Transmissibility, and Drawdown*, U.S. Geological Survey Water-Supply Paper 1536-I, 1963, pp. 272–292.

(7) Neuman, S. P., "Effect of Partial Penetration on Flow in Unconfined Aquifers Considering Delayed Gravity Response," *Water Resources Research*, Vol 10, No. 2, 1974, pp. 303–312.

Standard Test Method (Analytical Procedure) for Determining Transmissivity and Storage Coefficient of Nonleaky Confined Aquifers by the Theis Nonequilibrium Method[1]

This standard is issued under the fixed designation D 4106; the number immediately following the designation indicates the year of original adoption or, in the case of revision, the year of last revision. A number in parentheses indicates the year of last reapproval. A superscript epsilon (ε) indicates an editorial change since the last revision or reapproval.

1. Scope

1.1 This test method covers an analytical procedure for determining the transmissivity and storage coefficient of a nonleaky confined aquifer. It is used to analyze data on water-level response collected during radial flow to or from a well of constant discharge or injection.

1.2 This analytical procedure is used in conjunction with the field procedure given in Test Method D 4050.

1.3 *Limitations*—The limitations of this test method for determination of hydraulic properties of aquifers are primarily related to the correspondence between the field situation and the simplifying assumptions of this test method (see 5.1).

1.4 The values stated in SI units are to be regarded as standard.

1.5 *This standard does not purport to address all of the safety problems, if any, associated with its use. It is the responsibility of the user of this standard to establish appropriate safety and health practices and determine the applicability of regulatory limitations prior to use.*

2. Referenced Documents

2.1 *ASTM Standards:*

D 653 Terminology Relating to Soil, Rock, and Contained Fluids[2]

D 4043 Guide for Selection of Aquifer Test Method in Determining of Hydraulic Properties by Well Techniques[2]

D 4050 Test Method (Field Procedure) for Withdrawal and Injection Well Tests for Determining Hydraulic Properties of Aquifer Systems[2]

3. Terminology

3.1 *Definitions:*

3.1.1 *aquifer, confined*—an aquifer bounded above and below by confining beds and in which the static head is above the top of the aquifer.

3.1.2 *confining bed*—a hydrogeologic unit of less permeable material bounding one or more aquifers.

3.1.3 *control well*—well by which the head and flow in the aquifer is changed, for example, by pumping, injection, or imposing a constant change of head.

3.1.4 *drawdown*—vertical distance the static head is lowered due to the removal of water.

3.1.5 *head*—see *head, static.*

3.1.6 *head, static*—the height above a standard datum of the surface of a column of water (or other liquid) that can be supported by the static pressure at a given point.

3.1.7 *hydraulic conductivity (field aquifer tests)*—the volume of water at the existing kinematic viscosity that will move in a unit time under a unit hydraulic gradient through a unit area measured at right angles to the direction of flow.

3.1.8 *observation well*—a well open to all or part of an aquifer.

3.1.9 *piezometer*—a device so constructed and sealed as to measure hydraulic head at a point in the subsurface.

3.1.10 *specific storage*—the volume of water released from or taken into storage per unit volume of the porous medium per unit change in head.

3.1.11 *storage coefficient*—the volume of water an aquifer releases from or takes into storage per unit surface area of the aquifer per unit change in head. For a confined aquifer, the storage coefficient is equal to the product of the specific storage and aquifer thickness. For an unconfined aquifer, the storage coefficient is approximately equal to the specific yield.

3.1.12 *transmissivity*—the volume of water at the existing kinematic viscosity that will move in a unit time under a unit hydraulic gradient through a unit width of the aquifer.

3.1.13 *unconfined aquifer*—an aquifer that has a water table.

3.1.14 For definitions of other terms used in this test method, see Terminology D 653.

3.2 *Symbols and Dimensions:*

3.2.1 K [LT^{-1}]—hydraulic conductivity.

3.2.2 K_{xy}—hydraulic conductivity in the horizontal plane, radially from the control well.

3.2.3 K_z—hydraulic conductivity in the vertical direction.

3.2.4 Q [L^3T^{-1}]—discharge.

3.2.5 S [nd]—storage coefficient.

3.2.6 S_s [L^{-1}]—specific storage.

3.2.7 T [L^2T^{-1}]—transmissivity.

3.2.8 $W(u)$ [nd]—well function of u.

3.2.9 b [L]—thickness of aquifer.

3.2.10 r [L]—radial distance from control well.

3.2.11 s [L]—drawdown.

4. Summary of Test Method

4.1 This test method describes an analytical procedure for analyzing data collected during a withdrawal or injection well test. The field procedure (see Test Method D 4050)

[1] This test method is under the jurisdiction of ASTM Committee D-18 on Soil and Rock and is the direct responsibility of Subcommittee D18.21 on Ground Water and Vadose Zone Investigations.

Current edition approved June 15, 1991. Published August 1991.

[2] *Annual Book of ASTM Standards*, Vol 04.08.

involves pumping a control well at a constant rate and measuring the water level response in one or more observation wells or piezometers. The water-level response in the aquifer is a function of the transmissivity and storage coefficient of the aquifer. Alternatively, this test method can be performed by injecting water at a constant rate into the aquifer through the control well. Analysis of buildup of water level in response to injection is similar to analysis of drawdown of water level in response to withdrawal in a confined aquifer. Drawdown of water level is analyzed by plotting drawdown against factors incorporating either time or distance from the control well, or both, and matching the drawdown response with a type curve.

4.2 *Solution*—The solution given by Theis **(1)**[3] may be expressed as follows:

$$s = \frac{Q}{4\pi T} \int_u^\infty \frac{e^{-y}}{y} dy \tag{1}$$

where:

$$u = \frac{r^2 S}{4Tt} \tag{2}$$

$$\int_u^\infty \frac{e^{-y}}{y} dy = W(u)$$

$$= -0.577216 - \log_e u + u - \frac{u^2}{2!2} + \frac{u^3}{3!3} - \frac{u^4}{4!4} + \ldots \tag{3}$$

5. Significance and Use

5.1 *Assumptions:*

5.1.1 Well discharges at a constant rate, Q.

5.1.2 Well is of infinitesimal diameter and fully penetrates the aquifer.

5.1.3 The nonleaky aquifer is homogeneous, isotropic, and aerially extensive. A nonleaky aquifer receives insignificant contribution of water from confining beds.

5.1.4 Discharge from the well is derived exclusively from storage in the aquifer.

5.1.5 The geometry of the assumed aquifer and well conditions are shown in Fig. 1.

5.2 *Implications of Assumptions:*

5.2.1 Implicit in the assumptions are the conditions of radial flow. Vertical flow components are induced by a control well that partially penetrates the aquifer, that is, the well is not open to the aquifer through its full thickness. If the control well does not fully penetrate the aquifer, the nearest piezometer or partially penetrating observation well should be located at a distance, r, beyond which vertical flow components are negligible, where according to Reed **(2)**:

$$r = 1.5 \frac{b}{\sqrt{\frac{K_z}{K_{xy}}}} \tag{4}$$

Fully penetrating observation wells may be placed at less than distance r from the control well. Observation wells may be on the same or on various radial lines from the control well.

5.2.2 The Theis method assumes the control well is of infinitesimal diameter. Also, it assumes that the water level in the control well is the same as in the aquifer contiguous to the well. In practice these assumptions may cause a difference between the theoretical drawdown and field measurements of drawdown in the early part of the test and in and near the control well. Control well storage is negligible after a time, t, given by the Eq 5 after Weeks **(3)**.

$$t = 25 \times \frac{r_c^2}{T} \tag{5}$$

where:

r_c = the radius of the control well in the interval in which the water level changes.

5.2.3 *Application of Theis Method to Unconfined Aquifers*:

5.2.3.1 Although the assumptions are applicable to artesian or confined conditions, the Theis solution may be applied to unconfined aquifers if drawdown is small compared with the saturated thickness of the aquifer or if the drawdown is corrected for reduction in thickness of the aquifer, and the effects of delayed gravity yield are small.

5.2.3.2 *Reduction in Aquifer Thickness*—In an unconfined aquifer dewatering occurs when the water levels decline in the vicinity of a pumping well. Corrections in drawdown need to be made when the drawdown is a significant fraction of the aquifer thickness as shown by Jacob **(4)**. The drawdown, s, needs to be replaced by s', the drawdown that would occur in an equivalent confined aquifer, where:

$$s' = s - \left(\frac{s^2}{2b}\right) \tag{6}$$

5.2.3.3 *Gravity Yield Effects*—In unconfined aquifers, delayed gravity yield effects may invalidate measurements of drawdown during the early part of the test for application to the Theis method. Effects of delayed gravity yield are negligible in partially penetrating observation wells at and beyond a distance, r, from the control well, where:

$$r = \frac{b}{\sqrt{\frac{K_z}{K_{xy}}}} \tag{7}$$

After the time, t, as given in Eq 8 from Neuman **(5)**.

$$t = 10 \times S_y(r^2/T) \tag{8}$$

where:

S_y = the specific yield. For fully penetrating observation wells, the effects of delayed yield are negligible at the distance, r, in Eq 7 after one tenth of the time given in the Eq 8.

6. Apparatus

6.1 Analysis of data from the field procedure (see Test Method D 4050) by the method specified in this test method requires that the control well and observation wells meet the specifications in the following paragraphs.

6.2 *Construction of Control Well*—Screen the control well

[3] The boldface numbers in parentheses refer to a list of references at the end of the text.

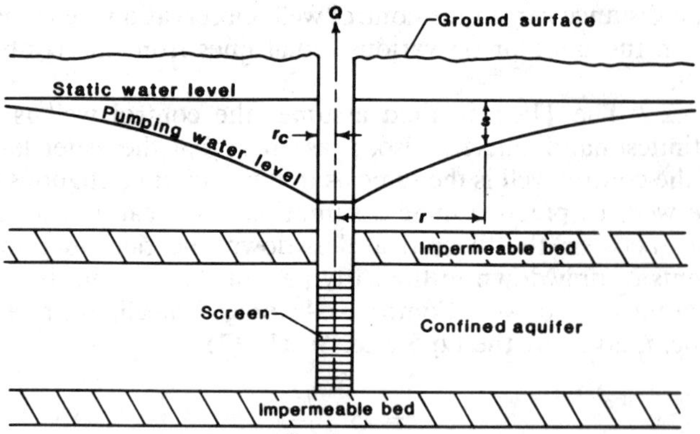

FIG. 1 Cross Section Through a Discharging Well in a Nonleaky Confined Aquifer

in the aquifer to be tested and equip with a pump capable of discharging water from the well at a constant rate for the duration of the test. Preferably, screen the control well throughout the full thickness of the aquifer. If the control well partially penetrates the aquifer, take special precaution in the placement and design of observation wells (see 5.2.1).

6.3 *Construction of Observation Wells*—Construct one or more observation wells at a distance from the control well. Observation wells may be partially open or open throughout the thickness of the aquifer.

6.4 *Location of Observation Wells*—Locate observation wells at various distances from the control well within the area of influence of pumping. However, if vertical flow components are significant and if partially penetrating observation wells are used, locate them at a distance beyond the effect of vertical flow components (see 5.2.1). If the aquifer is unconfined, constraints are imposed on the distance to partially penetrating observation wells and the validity of early time measurements (see 5.2.3).

7. Procedure

7.1 The overall procedure consists of conducting the field procedure for withdrawal or injection well tests (described in Test Method D 4050) and analysis of the field data that is addressed in this test method.

7.2 The integral expression in Eqs 1 and 2 can not be evaluated analytically. A graphical procedure is used to solve for the two unknown parameters transmissivity and storage coefficient where:

$$s = \frac{Q}{4\pi T} \mathrm{W(u)} \tag{9}$$

and:

$$u = \frac{r^2 S}{4Tt} \tag{10}$$

8. Calculation

8.1 The graphical procedure used to calculate test results is based on the functional relations between *W(u)* and *s* and between *u* and *t* or t/r^2.

8.1.1 Plot values of *W(u)* versus $1/u$ on logarithmic-scale paper (see Table 1). This plot is referred to as the type curve plot.

8.1.2 On logarithmic tracing paper of the same scale and size as the *W(u)* versus $1/u$ type curve, plot values of drawdown, *s*, on the vertical coordinate versus either time on the horizontal coordinate if one observation well is used or versus t/r^2 on the horizontal coordinate if more than one observation well is used.

8.1.3 Overlay the data plot on the type curve plot and, while the coordinate axes of the two plots are held parallel,

TABLE 1 Values of Theis Equation *W(u)* for values of $1/u$, From Reed (2)

$1/u$	$1/u \times 10^{-1}$	1	10	10^2	10^3	10^4	10^5	10^6
1.0	0.00000[A]	0.21938	1.82292	4.03793	6.33154	8.63322	10.93572	13.23830
1.2	0.00003	0.29255	1.98932	4.21859	6.51369	8.81553	11.11804	13.42062
1.5	0.00017	0.39841	2.19641	4.44007	6.73667	9.03866	11.34118	13.64376
2.0	0.00115	0.55977	2.46790	4.72610	7.02419	9.32632	11.62886	13.93144
2.5	0.00378	0.70238	2.68126	4.94824	7.24723	9.54945	11.85201	14.15459
3.0	0.00857	0.82889	2.85704	5.12990	7.42949	9.73177	12.03433	14.33691
3.5	0.01566	0.94208	3.00650	5.28357	7.58359	9.88592	12.18847	14.49106
4.0	0.02491	1.04428	3.13651	5.41675	7.71708	10.01944	12.32201	14.62459
5.0	0.04890	1.22265	3.35471	5.63939	7.94018	10.24258	12.54515	14.84773
6.0	0.07833	1.37451	3.53372	5.82138	8.12247	10.42490	12.72747	15.03006
7.0	0.11131	1.50661	3.68551	5.97529	8.27659	10.57905	12.88162	15.18421
8.0	0.14641	1.62342	3.81727	6.10865	8.41011	10.71258	13.01515	15.31774
9.0	0.18266	1.72811	3.93367	6.22629	8.52787	10.83036	13.13294	15.43551
$1/u$	$1/u \times 10^7$	10^8	10^9	10^{10}	10^{11}	10^{12}	10^{13}	10^{14}
1.0	15.54087	17.84344	20.14604	22.44862	24.75121	27.05379	29.36638	31.65897
1.2	15.72320	18.02577	20.32835	22.63094	24.93353	27.23611	29.53870	31.84128
1.5	15.94634	18.24892	20.55150	22.85408	25.15668	27.45926	29.76184	32.06442
2.0	16.23401	18.53659	20.83919	23.14177	25.44435	27.74693	30.04953	32.35211
2.5	16.45715	18.76974	21.06233	23.36491	25.66750	27.97008	30.27267	32.57526
3.0	16.63948	18.94206	21.24464	23.54723	25.84982	28.15240	30.45499	32.75757
3.5	16.79362	19.09621	21.39880	23.70139	26.00397	28.30655	30.60915	32.91173
4.0	16.92715	19.22975	21.53233	23.83492	26.13750	28.44008	30.74268	33.04526
5.0	17.15030	19.45288	21.75548	24.05806	26.36064	28.66322	30.96582	33.26840
6.0	17.33263	19.63521	21.93779	24.24039	26.54297	28.84555	31.14813	33.45071
7.0	17.48677	19.78937	22.09195	24.39453	26.69711	28.99969	31.30229	33.60487
8.0	17.62030	19.92290	22.22548	24.52806	26.83064	29.13324	31.43582	33.73840
9.0	17.73808	20.04068	22.34326	24.64584	26.94843	29.25102	31.55360	33.85619

[A] Value shown as 0.00000 is nonzero but less than 0.000005.

shift the plot to align with the type curve (see Fig. 2).

8.1.4 Select and record the values of $W(u)$, $1/u$, s, and t at an arbitrary point, referred to as the match point (see Fig. 2), anywhere on the overlapping part of the plots. For convenience the point may be selected where $W(u)$ and $1/u$ are integer values.

NOTE 1—Alternatively, the type curve can be constructed by plotting $W(u)$ against u, then plotting the data as s versus r^2/t.

8.1.5 Using the coordinates of the point, determine the transmissivity and storage coefficient from Eqs 11 and 12:

$$T = \frac{QW(u)}{4\pi s} \tag{11}$$

$$S = 4Tu\frac{t}{r^2} \tag{12}$$

8.1.6 To apply the Theis nonequilibrium method to thin unconfined aquifers where the drawdown is a significant fraction of the initial saturated thickness, apply a correction to the drawdown in solving for transmissivity and coefficient of storage (see 5.2.3.2).

9. Report

9.1 Prepare a report including the information described in this section. The report of the analytical procedure will include information from the report on test method selection (see Guide D 4043) and the field testing procedure (see Test Method D 4050).

9.1.1 *Introduction*—The introductory section is intended to present the scope and purpose of the constant discharge method for determining transmissivity and storativity in a confined nonleaky aquifer under constant flux. Summarize the field hydrogeologic conditions and the field equipment and instrumentation including the construction of the control well and observation wells or piezometers, or both, the method of measurement of discharge and water levels, and the duration of the test and pumping rate. Discuss rationale for selecting the Theis nonequilibrium method.

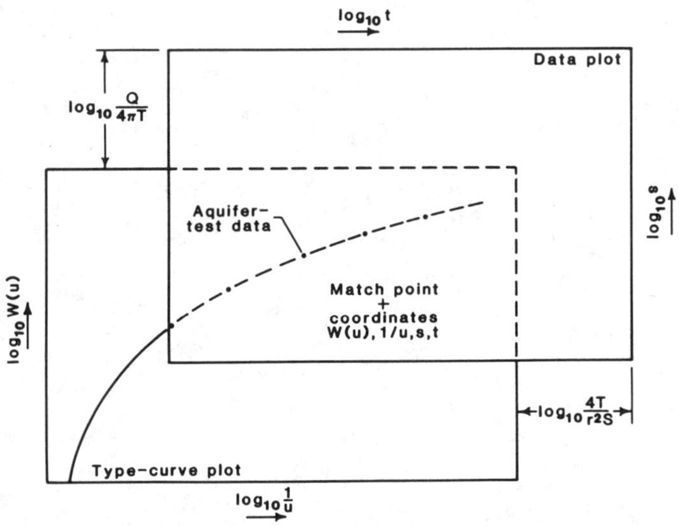

FIG. 2 Relation of $1/u$, $W(u)$ Type Curve and t, s Data Plot

9.1.2 *Hydrogeologic Setting*—Review the information available on the hydrogeology of the site; interpret and describe the hydrogeology of the site as it pertains to the selection of this test method for conducting and analyzing an aquifer test. Compare the hydrogeologic characteristics of the site as it conforms and differs from the assumptions of this test method.

9.1.3 *Equipment*—Report the field installation and equipment for the aquifer test, including the construction, diameter, depth of screened and gravel packed intervals, and location of control well and pumping equipment, and the construction, diameter, depth, and screened interval of observation wells or piezometers.

9.1.4 Describe the methods of observing water levels, pumping rate, barometric changes, and other environmental conditions pertinent to the test. Include a list of measuring devices used during the test, the manufacturers name, model number, and basic specifications for each major item, and the name and date and method of the last calibration, if applicable.

9.1.5 *Testing Procedures*—State the steps taken in conducting pre-test, drawdown, and recovery phases of the test. Include the date, clock time, and time since pumping started or stopped for measurements of discharge rate, water levels, and other environmental data recorded during the testing procedure.

9.2 *Presentation and Interpretation of Test Results*:

9.2.1 *Data*—Present tables of data collected during the test. Show methods of adjusting water levels for background water-level and barometric changes and calculation of drawdown and residual drawdown.

9.2.2 *Data Plots*—Present data plots used in analysis of the data. Show overlays of data plots and type curve with match points and corresponding values of parameters at match points.

9.2.3 Show calculation of transmissivity and storage coefficient.

9.2.4 Evaluate qualitatively the test on the basis of the adequacy of instrumentation, observations of stress and response, the conformance of the hydrogeologic conditions, and the performance of the test to the assumptions of this test method.

10. Precision and Bias

10.1 It is not practicable to specify the precision of this test method because the response of aquifer systems during aquifer tests is dependent upon ambient system stresses. No statement can be made about bias because no true reference values exist.

11. Keywords

11.1 aquifers; aquifer tests; control wells; ground water; hydraulic conductivity; observation wells; storage coefficient; transmissivity

REFERENCES

(1) Theis, C. V., "The Relation Between the Lowering of the Piezometric Surface and the Rate and Duration of Discharge of a Well Using Ground-Water Storage," *American Geophysical Union Transactions*, Vol 16, Part 2, 1935, pp. 519–524.

(2) Reed, J. E., "Type Curves for Selected Problems of Flow to Wells in Confined Aquifers," *U.S. Geological Survey Techniques of Water-Resources Investigations*, Book 3, Chapter B3, 1980.

(3) Weeks, E. P., "*Aquifer Tests—The State of the Art in Hydrology*" in Proceedings of the International Well-Testing Symposium, October 19–21, 1977, Berkeley, California, LBL, 7027, Lawrence Berkeley Laboratory, pp 14–26.

(4) Jacob, C. C., "Determining the Permeability of Water-Table Aquifers," *in* Bentall, Ray, compiler, "Methods of Determining Permeability, Transmissibility, and Drawdown," U.S. Geological Survey Water-Supply Paper 1536-I, 1963, pp. 245–271.

(5) Neuman, S. P., "Effect of Partial Penetration on Flow in Unconfined Aquifers Considering Delayed Gravity Response," *Water Resources Research*, Vol 10, No. 2, 1974, pp. 303–312.

(6) Wenzel, L. K., "Methods for Determining Permeability of Water-Bearing Materials, with Special Reference to Discharging Well Methods," U.S. Geological Survey Water Supply Paper 887, 1942.

Designation: D 4220 – 89

Standard Practices for Preserving and Transporting Soil Samples[1]

This standard is issued under the fixed designation D 4220; the number immediately following the designation indicates the year of original adoption or, in the case of revision, the year of last revision. A number in parentheses indicates the year of last reapproval. A superscript epsilon (ε) indicates an editorial change since the last revision or reapproval.

1. Scope

1.1 These practices cover procedures for preserving soil samples immediately after they are obtained in the field and accompanying procedures for transporting and handling the samples.

1.2 *Limitations*—These practices are not intended to address requirements applicable to transporting of soil samples known or suspected to contain hazardous materials.

1.3 *This standard may involve hazardous materials, operations, and equipment. This standard does not purport to address all of the safety problems associated with its use. It is the responsibility of the user of this standard to establish appropriate safety and health practices and determine the applicability of regulatory limitations prior to use.* See Section 7.

2. Referenced Documents

2.1 *ASTM Standards:*

D 420 Practice for Investigating and Sampling Soil and Rock for Engineering Purposes[2]

D 653 Terminology Relating to Soil, Rock, and Contained Fluids[2]

D 1452 Practice for Soil Investigation and Sampling by Auger Borings[2]

D 1586 Method for Penetration Test and Split-Barrel Sampling of Soils[2]

D 1587 Practice for Thin-Walled Tube Sampling of Soils[2]

D 2488 Practice for Description and Identification of Soils (Visual-Manual Procedure)[2]

D 3550 Practice for Ring-Lined Barrel Sampling of Soils[2]

3. Terminology

3.1 Terminology in these practices is in accordance with Terminology D 653.

4. Summary of Practices

4.1 The various procedures are given under four groupings as follows:

4.1.1 *Group A*—Samples for which only general visual identification is necessary.

4.1.2 *Group B*—Samples for which only water content and classification tests, proctor and relative density, or profile logging is required, and bulk samples that will be remolded or compacted into specimens for swell pressure, percent swell, consolidation, permeability, shear testing, CBR, stabilimeter, etc.

4.1.3 *Group C*—Intact, naturally formed or field fabricated, samples for density determinations; or for swell pressure, percent swell, consolidation, permeability testing and shear testing with or without stress-strain and volume change measurements, to include dynamic and cyclic testing.

4.1.4 *Group D*—Samples that are fragile or highly sensitive for which tests in Group C are required.

4.2 The procedure(s) to be used should be included in the project specifications or defined by the designated responsible person.

5. Significance and Use

5.1 Use of the various procedures recommended in these practices is dependent on the type of samples obtained (Practice D 420), the type of testing and engineering properties required, the fragility and sensitivity of the soil, and the climatic conditions. In all cases, the primary purpose is to preserve the desired inherent conditions.

5.2 The procedures presented in these practices were primarily developed for soil samples that are to be tested for engineering properties, however, they may be applicable for samples of soil and other materials obtained for other purposes.

6. Apparatus

6.1 The type of materials and containers needed depend upon the conditions and requirements listed under the four groupings A to D in Section 4, and also on the climate and transporting mode and distance.

6.1.1 *Sealing Wax*, includes microcrystalline wax, paraffin, beeswax, ceresine, carnaubawax, or combinations thereof.

6.1.2 *Metal Disks*, about 1/16 in. (about 2 mm) thick and having a diameter slightly less than the inside diameter of the tube, liner, or ring and to be used in union with wax or caps and tape, or both.

6.1.3 *Wood Disks*, prewaxed, 1 in. (25 mm) thick and having a diameter slightly less than the inside diameter of the liner or tube.

6.1.4 *Tape*, either waterproof plastic, adhesive friction, or duct tape.

6.1.5 *Cheesecloth*, to be used in union with wax in alternative layers.

6.1.6 *Caps*, either plastic, rubber or metal, to be placed over the end of thin-walled tubes (Practice D 1587), liners and rings (Practice D 3550), in union with tape or wax.

6.1.7 *O'ring (Sealing End Caps)*, used to seal the ends of samples within thin-walled tubes, by mechanically expanding an O'ring against the tube wall.

---

[1] These practices are under the jurisdiction of ASTM Committee D-18 on Soil and Rock and are the direct responsibility of Subcommittee D18.02 on Sampling and Related Field Testing for Soil Investigations.

Current edition approved July 13, 1989. Published September 1989. Originally published as D 4220 – 83. Last previous edition D 4220 – 83.

[2] *Annual Book of ASTM Standards*, Vol 04.08.

NOTE 1—Plastic expandable end caps are preferred. Metal expandable end caps seal equally well; however, long-term storage may cause corrosion problems.

6.1.8 *Jars*, wide mouthed, with rubber-ringed lids or lids lined with a coated paper seal and of a size to comfortably receive the sample, commonly ½ pt (250 mL), 1 pt (500 mL) and quart-sized (1000 mL).

6.1.9 *Bag*, either plastic, burlap with liner, burlap or cloth type (Practice D 1452).

6.1.10 *Packing Material*, to protect against vibration and shock.

6.1.11 *Insulation*, either granule (bead), sheet or foam type, to resist temperature change of soil or to prevent freezing.

6.1.12 *Sample Cube Boxes*, for transporting cube (block) samples. Constructed with ½ to ¾ in. (13 to 19 mm) thick plywood (marine type).

6.1.13 *Cylindrical Sample Containers*, somewhat larger in dimension than the thin-walled tube or liner samples, such as cylindrical frozen food cartons.

6.1.14 *Shipping Containers*, either box or cylindrical type and of proper construction to protect against vibration, shock, and the elements, to the degree required.

NOTE 2—The length, girth and weight restrictions for commercial transportation must be considered.

6.1.15 *Identification Material*—This includes the necessary writing pens, tags, and labels to properly identify the sample(s).

7. Precautions

7.1 Special instructions, descriptions, and marking of containers must accompany any sample that may include radioactive, chemical, toxic, or other contaminant material.

7.2 Interstate transportation containment, storage, and disposal of soil samples obtained from certain areas within the United States and the transportation of foreign soils into or through the United States are subject to regulations established by the U.S. Department of Agriculture, Animal, and Plant Health Service, Plant Protection and Quarantine Programs, and possibly to regulations of other federal, state, or local agencies.

7.2.1 Samples shipped by way of common carrier or U.S. Postal Service must comply with the Department of Transportation Hazardous Materials Regulation, 49CRF Part 172.

7.3 Sample traceability records (see Fig. 1) are encouraged and should be required for suspected contaminated samples.

7.3.1 The possession of all samples must be traceable, from collection to shipment to laboratory to disposition, and should be handled by as few persons as possible.

7.3.2 The sample collector(s) should be responsible for initiating the sample traceability record; recording the project, sample identification and location, sample type, date, and the number and types of containers.

7.3.3 A separate traceability record shall accompany each shipment.

7.3.4 When transferring the possession of samples the person(s) relinquishing and receiving the samples shall sign, date, record the time, and check for completeness of the traceability record.

8. Procedure

8.1 *All Samples*—Properly identify samples with tags, labels, and markings prior to transporting them as follows:

8.1.1 Job name or number, or both,

8.1.2 Sampling date,

8.1.3 Sample/boring number and location,

8.1.4 Depth or elevation, or both,

8.1.5 Sample orientation,

8.1.6 Special shipping or laboratory handling instructions, or both, including sampling orientation, and

8.1.7 Penetration test data, if applicable (Test Method D 1586).

8.1.8 Subdivided samples must be identified while maintaining association to the original sample.

8.1.9 If required, sample traceability record.

8.2 *Group A*—Transport samples in any type of container by way of available transportation. If transported commercially, the container need only meet the minimum requirements of the transporting agency and any other requirements necessary to assure against sample loss.

8.3 *Group B:*

8.3.1 Preserve and transport these samples in sealed, moistureproof containers. All containers shall be of sufficient thickness and strength to ensure against breakage and moisture loss. The container types include: plastic bags or pails, glass or plastic (provided they are waterproof) jars, thin walled tubes, liners, and rings. Wrap cylindrical and cube samples in suitable plastic film or aluminum foil, or both, (Note 3) and coat with several layers of wax, or seal in several layers of cheesecloth and wax.

8.3.2 Transport these samples by any available transportation. Ship these samples as prepared or placed in larger shipping containers, including bags, cardboard, or wooden boxes or barrels.

NOTE 3—Some soils may cause holes to develop in aluminum foil, due to corrosion. Avoid direct contact where adverse affects to sample composition are a concern.

8.3.3 *Plastic Bags*—Place the plastic bags as tightly as possible around the sample, squeezing out as much air as possible. They shall be 3 mil or thicker to prevent leakage.

8.3.4 *Glass-Plastic Jars*—If the jar lids are not rubber ringed or lined with new waxed paper seals, seal the lids with wax.

8.3.5 *Plastic Pails*—If the plastic pail lids are not air tight, seal them with wax or tape.

8.3.6 *Thin-Walled Tubes:*

8.3.6.1 *Expandable Packers*—The preferred method of sealing sample ends within tubes is with plastic, expandable packers.

8.3.6.2 *Wax With Disks*—For short-term sealing, paraffin wax is acceptable. For long term sealing (in excess of 3 days) use microcrystalline waxes or combine with up to 15 % beeswax or resin, for better adherence to the wall of the tube and to reduce shrinkage. Several thin layers of wax are preferred over one thick layer. The minimum final thickness shall be 0.4 in. (10 mm).

8.3.6.3 *End Caps*—Seal metal, rubber, or plastic end caps with tape. For long term storage (longer than 3 days), also dip them in wax, applying two or more layers of wax.

8.3.6.4 *Cheesecloth and Wax*—Use alternating layers (a minimum of two each) of cheesecloth and wax to seal each end of the tube and stabilize the sample.

NOTE 4—Where necessary, spacers or appropriate packing materials, or both, must be placed prior to sealing the tube ends to provide proper confinement. Packing material must be nonabsorbent and must maintain its properties to provide the same degree of continued sample support.

8.3.7 *Liners and Rings*—Refer to 8.3.6.3 or 8.3.6.4.

8.3.8 *Exposed Samples:*

8.3.8.1 *Cylindrical, Cubical or Other Samples Wrapped in Plastic*, such as polyethylene and polypropylene, or foil should be further protected with a minimum of three coats of wax.

8.3.8.2 *Cylindrical and Cube Samples Wrapped in Cheesecloth and Wax*, shall be sealed with a minimum of three layers of each, placed alternatively.

8.3.8.3 *Carton Samples (Frozen Food Cartons)*—Samples placed in these containers must be situated so that wax can be poured completely around the sample. The wax should fill the void between the sample and container wall. The wax should be sufficiently warm to flow, but not so hot that it penetrates the pores of the soil. Generally, the samples should be wrapped in plastic or foil before being surrounded with wax.

8.4 *Group C:*

8.4.1 Preserve and seal these samples in containers as covered in 8.3. In addition, they must be protected against vibration and shock, and protected from extreme heat or cold.

8.4.2 Samples transported by the sampling or testing agency personnel on seats of automobiles and trucks need only be placed in cardboard boxes, or similar containers into which the sealed samples fit snugly, preventing bumping, rolling, dropping, etc.

8.4.3 For all other methods of transporting samples, including automobile trunk, bus, parcel services, truck, boat, air, etc., place the sealed samples in wood, metal, or other type of suitable shipping containers that provide cushioning or insulation, or both, for each sample and container. Avoid transporting by any agency whose handling of containers is suspect.

8.4.4 The cushioning material (sawdust, rubber, polystyrene, urethane foam, or material with similar resiliency) should completely encase each sample. The cushioning between the samples and walls of the shipping containers should have a minimum thickness of 1 in. (25 mm). A minimum thickness of 2 in. (50 mm) shall be provided on the container floor.

8.4.5 When required, the samples should be shipped in the same orientation in which they were sampled. Otherwise, special conditions shall be provided such as freezing, controlled drainage, or sufficient confinement, or a combination thereof, to maintain sample integrity.

8.5 *Group D:*

8.5.1 The requirements of 8.4 must be met, in addition to the following:

8.5.1.1 Samples should be handled in the same orientation in which they were sampled, including during transportation or shipping, with appropriate markings on the shipping container.

8.5.1.2 For all modes of private or commercial transportation, the loading, transporting and unloading of the shipment containers should be supervised as much as possible by a qualified person.

NOTE 5—A qualified person may be an engineer, geologist, soil scientist, soils technician or responsible person designated by the project manager.

8.6 *Shipping Containers* (see Figs. 2 to 7 for typical containers):

8.6.1 The following features should be included in the design of the shipping container for Groups C and D.

8.6.1.1 It should be reuseable,

8.6.1.2 It should be constructed so that the samples can be maintained, at all times, in the same position as when sampled or packed, or both,

8.6.1.3 It should include sufficient packing material to cushion or isolate, or both, the tubes from the adverse effect of vibration and shock, and

8.6.1.4 It should include sufficient insulating material to prevent freezing, sublimation and thawing, or undesirable temperature changes.

8.6.2 *Wood Shipping Containers:*

8.6.2.1 Wood is preferred over metal. Outdoor (marine) plywood having a thickness of ½ and ¾ in. (13 to 19 mm) may be used. The top (cover) should be hinged and latched, or fastened with screws.

8.6.2.2 The cushioning requirements are given in 8.4.4.

8.6.2.3 For protection against freezing or extreme temperature variation, the entire shipping container should be lined with a minimum insulation thickness of 2 in. (50 mm).

8.6.3 *Metal Shipping Containers*—The metal shipping containers must incorporate cushioning and insulation material to minimum thicknesses in accordance with 8.6.2, although slightly greater thicknesses would be appropriate. Alternatively, the cushion effect could be achieved with a spring suspension system, or any other means that would provide similar protection.

8.6.4 *Styrene Shipping Containers*—Bulk styrene with slots cut to the dimensions of the sample tube or liner. A protective outer box of plywood or reinforced cardboard is recommended.

8.6.5 *Other Containers*—Containers constructed with laminated fiberboard, plastic or reinforced cardboard outer walls, and properly lined, may also be used.

9. Reporting

9.1 The data obtained in the field shall be recorded and should include the following:

9.1.1 Job name or number, or both,

9.1.2 Sampling date(s),

9.1.3 Sample/boring number(s) and location(s),

9.1.4 Depth(s) or elevation(s), or both,

9.1.5 Sample orientation,

Sample Identification/Traceability Record
(Controlled Document)

Project: ______________________ W.O. * ______________

Shipped by: ______________________

Shipped to: ______________________ Attention of:

Comments: ______________________ Hazardous materials suspected? (yes/no)

Sampling Point	Location	Field ID *	Date	Sample Type	No. of Containers	Analysis/Test Required	(optional) Lab ID

Sampler(s) (signature) ______________________

Field ID	Relinquished by: (signature)	Date/Time		Received by: (signature)	Date/Time		Comments

Shipment prepared by: (signature) ______________ Date/Time ________ Shipment method: ________

Received for Lab by: (signature) ______________ Date/Time ________ Comments ________

Receiving Laboratory: Please return original form after signing for receipt of samples.

FIG. 1 Example Layout of Record Form

9.1.6 Groundwater observation, if any,

9.1.7 Method of sampling, and penetration test data, if applicable,

9.1.8 Sample dimensions,

9.1.9 Soil description (Practice D 2488),

9.1.10 Names of technician/crewman, engineer, project chief, etc.,

9.1.11 Comments regarding contaminated or possible contaminated samples,

9.1.12 If used, a copy of traceability records,

9.1.13 Weather conditions, and

9.1.14 General remarks.

10. Precision and Bias

10.1 This practice provides qualitative and general information only. Therefore, a precision and bias statement is not applicable.

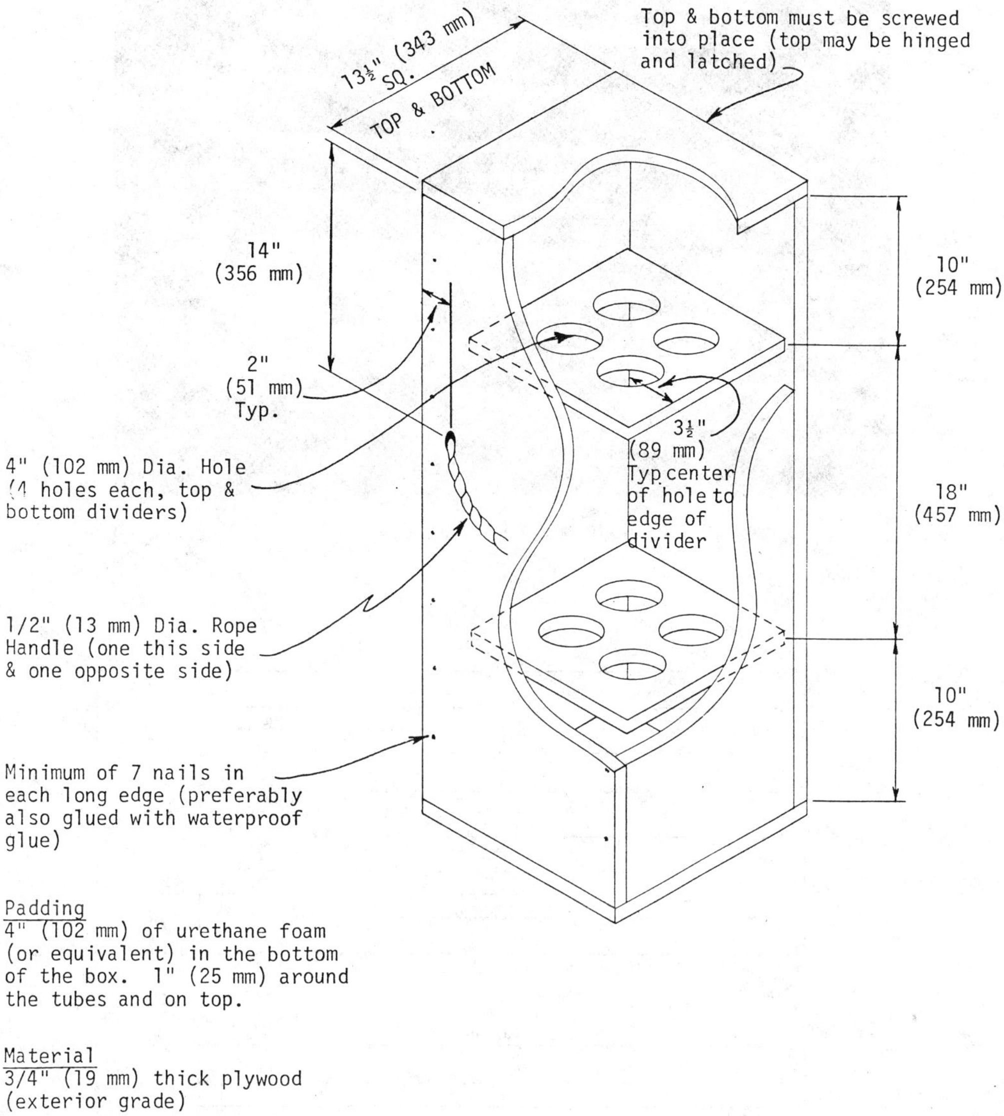

FIG. 2 Shipping Box for 3-in. (76-mm) Thin-Walled Tubes

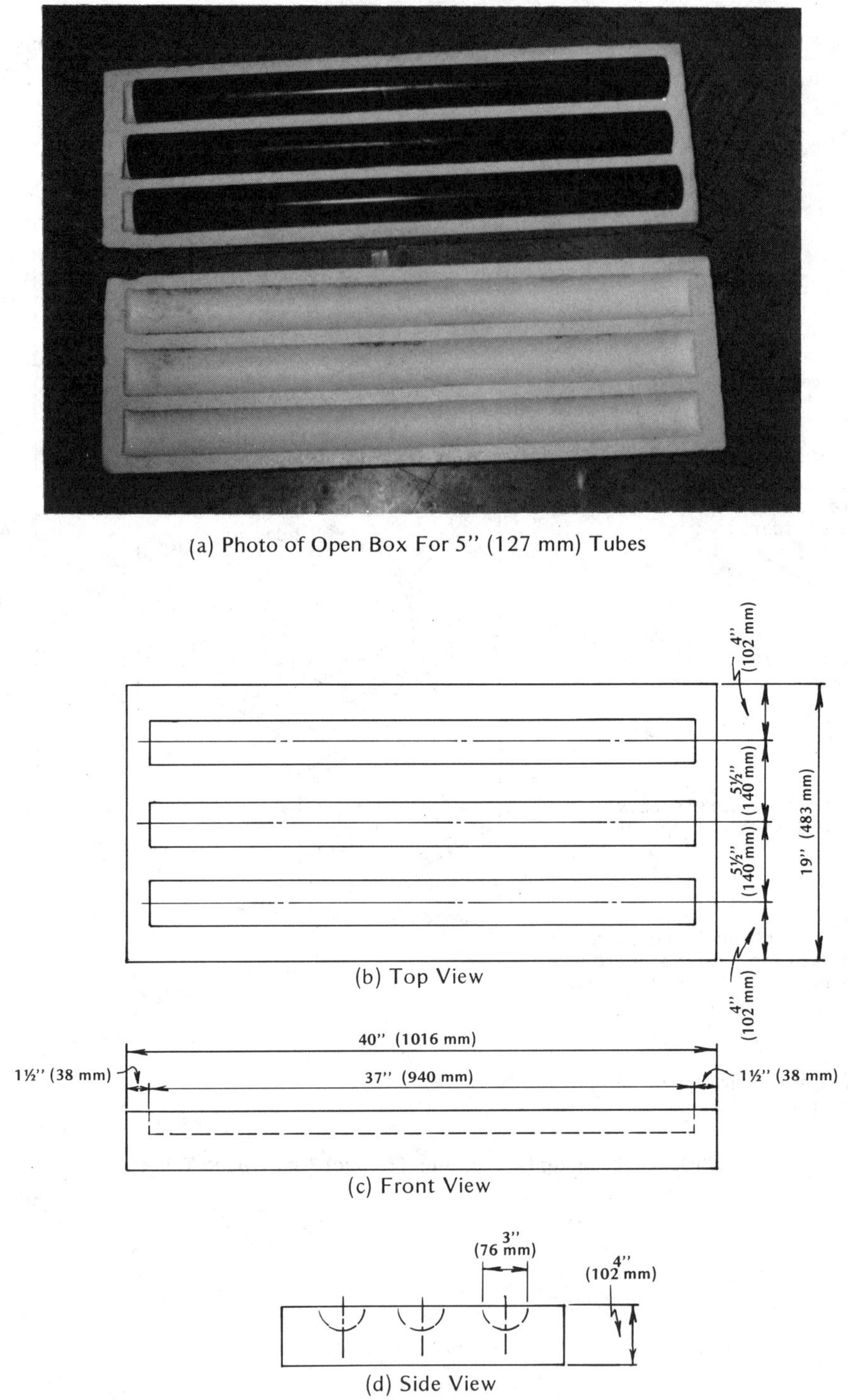

NOTE—Top and bottom halves are identical.

FIG. 3 Styrene Shipping Container for 3-in. (76-mm) Thin-Walled Tubes

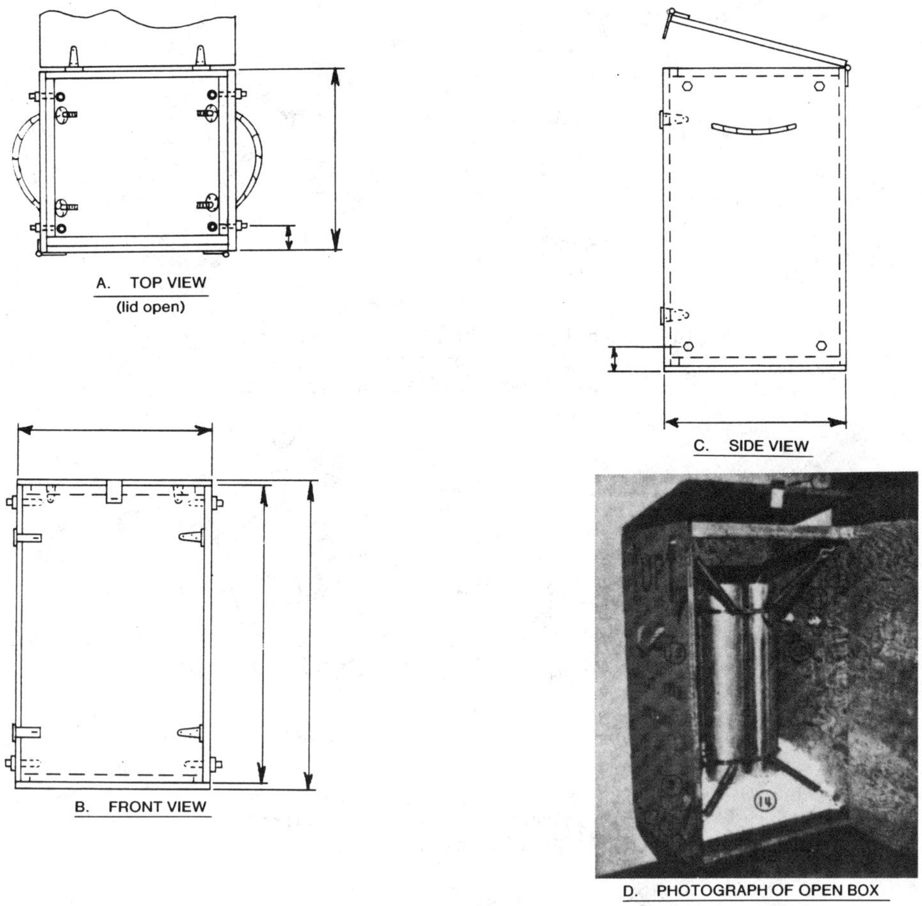

BILL OF MATERIALS

Item No.	Description of Item	Quantity
1	*Plywood*, 4 ft by 8 ft by 3/4 in. (1220 mm by 2440 mm by 19.1 mm) exterior, Grade AC	1 Sheet
2	*Hinge*, strap, 4 in. (102 mm), heavy duty with screws	4 Each
3	*Hasp*, hinged, 4½ in. (114 mm), with screws	3 Each
4	*Screw, Wood, Steel, Flathead*, No. 10 by 1¾ in. (44.5 mm)	72 Each
5	*Bolt, Machine*, 3/8 in. (9.5 mm), with nut to secure hasps	3 Each
6	*Washer*, flat, 3/8 in. (9.5 mm)	3 Each
7	*Eye Bolt*, ½ by 2 in. (6.4 mm by 51 mm), zinc-plated, with nut	8 Each
8	*Washer*, flat, ¼ in. (6.4 mm), for hasp bolt	8 Each
9	*S Hooks*, 2 in. (51 mm), open, zinc-plated	8 Each
10	*Clamp*, adjustable, hose, steel, worm screw adjustment	2 Each
11	*Spring*, expansion	8 Each
12	*Adhesive*, woodworking	1 lb (454 g)
13	*Rope*, nylon, ½-in. (12.7-mm) diameter, solid braided	5 ft (1524 mm)
14	*Cushioning Material*, expanded polystyrene foam	10 ft^3 (0.28 m^3)

NOTES–(a) All wooden components can be sawed from one sheet of plywood.

(b) This shipping box will accommodate approximately three 3-in. (76-mm) diameter tubes or two 5-in (127-mm) diameter tubes up to 30 in. (762 mm) in length. For longer tubes the inside height of the box must be a minimum of 6-in. (152 mm) greater than the length of the tube.

(c) All joints to be glued and fastened with screws.

(d) Stencil all sides as follows (See Views B and C).
TO PROTECT FROM FREEZING

(e) After suspending samples as indicated above, all void space must be filled with a suitable resilient packing material.

FIG. 4 Suspension System Container for Thin-Walled Tubes

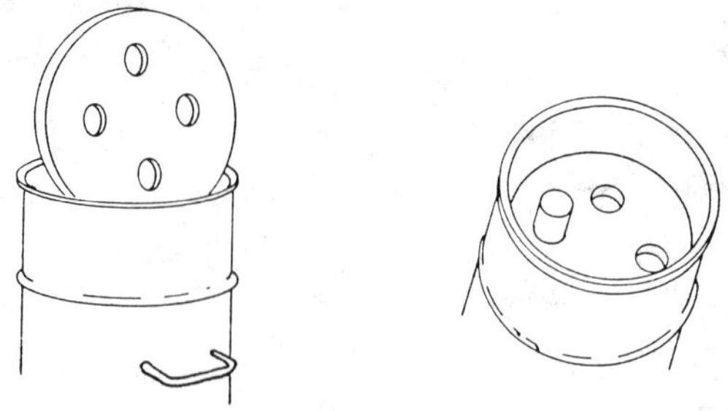

(a) 55-gallon (0.21 m^3) oil barrels with sections of styrofoam insulation; welded handles on each side.

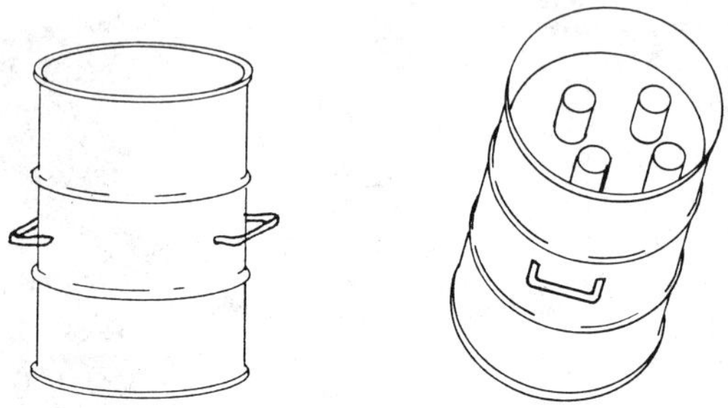

(b) Same as (a) showing barrel ready for shipment. Steel lids bolted on to provide tight seal.

NOTE—Two in. (51 mm) of foam rubber covers 2 in. of styrofoam at the base. One in. (25 mm) of foam rubber overlays the top of the tubes, and the remaining space to the lid is filled with styrofoam.

FIG. 5 Shipping Barrel for Thin-Walled Tubes

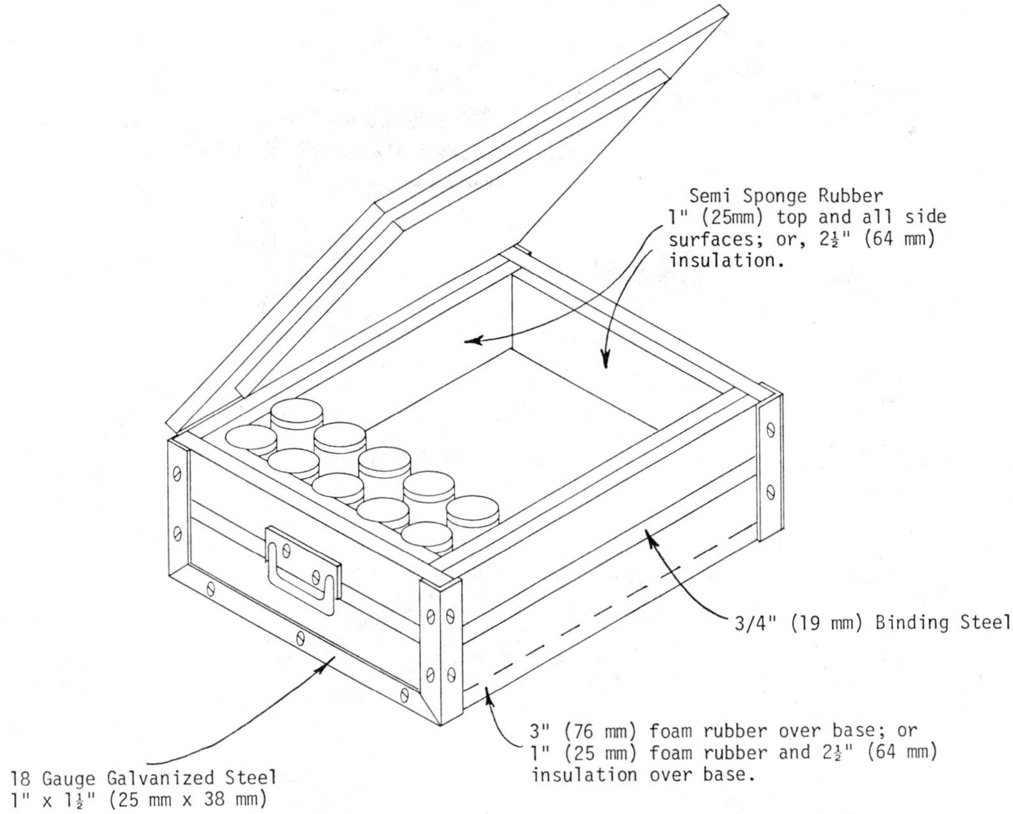

FIG. 6 Shipping Box for Liner (Short Tube) or Ring Samples

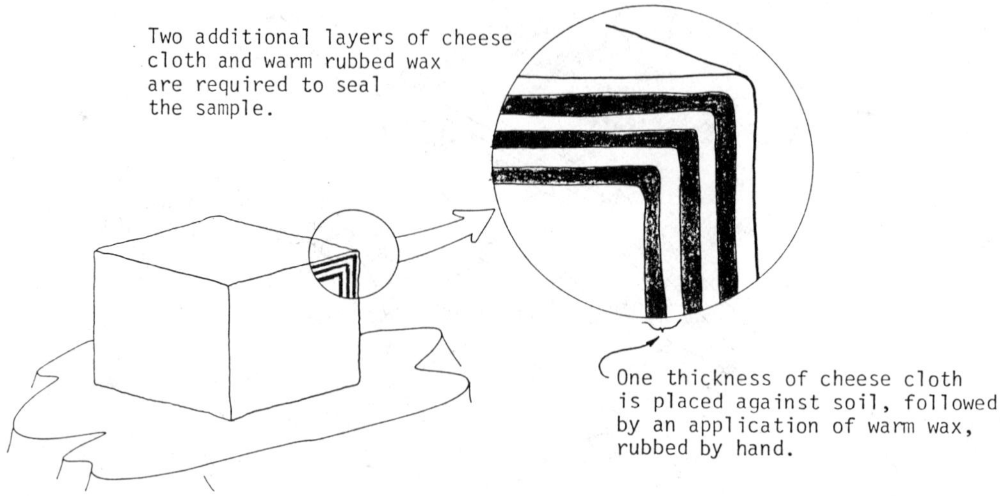

A. METHOD FOR SEALING HAND-CUT UNDISTURBED SAMPLES

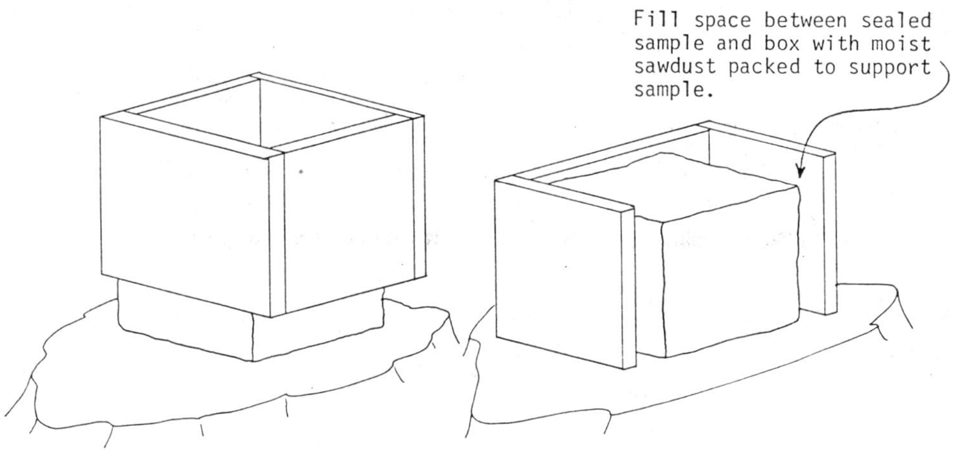

B. ENCASE EASILY DISTURBED SAMPLES IN BOX PRIOR TO CUTTING

Box constructed with 1/2"-3/4" (13 - 19 mm) exterior plywood.

FIG. 7 Preparing and Packaging a Block Sample

The American Society for Testing and Materials takes no position respecting the validity of any patent rights asserted in connection with any item mentioned in this standard. Users of this standard are expressly advised that determination of the validity of any such patent rights, and the risk of infringement of such rights, are entirely their own responsibility.

This standard is subject to revision at any time by the responsible technical committee and must be reviewed every five years and if not revised, either reapproved or withdrawn. Your comments are invited either for revision of this standard or for additional standards and should be addressed to ASTM Headquarters. Your comments will receive careful consideration at a meeting of the responsible technical committee, which you may attend. If you feel that your comments have not received a fair hearing you should make your views known to the ASTM Committee on Standards, 1916 Race St., Philadelphia, PA 19103.

Designation: D 4380 – 84

Standard Test Method for Density of Bentonitic Slurries[1]

This standard is issued under the fixed designation D 4380; the number immediately following the designation indicates the year of original adoption or, in the case of revision, the year of last revision. A number in parentheses indicates the year of last reapproval. A superscript epsilon (ε) indicates an editorial change since the last revision or reapproval.

1. Scope

1.1 This test method covers the determination of the density of slurries used in slurry construction techniques, such as are used for barriers to control the horizontal movement of liquids. This test method is modified from API Recommended Practice 13B.

1.2 *This standard may involve hazardous materials, operations, and equipment. This standard does not purport to address all of the safety problems associated with its use. It is the responsibility of whoever uses this standard to consult and establish appropriate safety and health practices and determine the applicability of regulatory limitations prior to use.*

2. Referenced Documents

2.1 *ASTM Standards:*

D 653 Terminology Relating to Soil, Rock, and Contained Fluids[2]

2.2 *American Petroleum Institute (API) Standard:*

API RP 13B Recommended Practice Standard Procedure for Testing Drilling Fluids (Section 1)[3]

3. Terminology

3.1 For definitions of terms relating to this test method, refer to Terminology D 653.

4. Summary of Method

4.1 The mud balance is the instrument generally used for this test method. The weight of a fixed volume of the slurry is measured by moving a rider counterweight along a graduated scale. The density of the slurry is then read directly off the graduated scale after the instrument is balanced.

5. Significance and Use

5.1 This test method provides for the determination of the density of bentonitic slurries in the laboratory and field. For freshly mixed slurry, this test method may be used as an indicator of mix proportions. For in-trench slurry, a certain value may be specified for maintaining trench stability.

6. Apparatus

6.1 *Mud Balance*—Any instrument of sufficient accuracy to permit measurement within ±0.01 g/cm^3 may be used, however, the mud balance is the instrument generally used (see Fig. 1). The mud balance consists of a mud cup attached to one end of a beam which is balanced on the other end by a fixed counterweight and a rider free to move along a graduated scale. A level bubble is mounted on the beam. Attachments for extending the range of the balance may be used.

7. Calibration

7.1 The instrument should be calibrated frequently with fresh water. Fresh water should give a reading of 1.00 g/cm^3 at 70°F (21.1°C). If it does not, adjust the balancing screw or the amount of lead shot in the well at the end of the graduated arm as required.

8. Procedure

8.1 Set up the instrument base approximately level.

8.2 Fill the clean, dry cup with slurry to be tested, place the cap on the cup, and rotate the cap until firmly seated. Make sure some of the slurry is expelled through the hole in the cap to free trapped air or gas.

8.3 Wash or wipe the excess slurry from the outside of the cup.

8.4 Place the beam on the support and balance it by moving the rider along the graduated scale. The beam is horizontal when the leveling bubble is on the center line.

8.5 Read the density at the side of the rider toward the knife edge. Make appropriate corrections when a range extender is used.

8.6 Clean and dry the instrument thoroughly after each use.

9. Calculations

9.1 To convert the density to other units, use the following relationships:

$$\rho \text{ in g/cm}^3 = \text{specific gravity (numerically), or}$$

$$\rho \text{ in lb/ft}^3 = (\rho \text{ in g/cm}^3)\ 62.43, \text{ or}$$

$$\rho \text{ in lb/gal} = (\rho \text{ in g/cm}^3)\ 8.35$$

10. Report

10.1 Record the density to the nearest 0.01 g/cm^3.

11. Precision and Bias

11.1 The precision and bias of this test method have not been established. Data are being sought that will be suitable for developing a precision and bias statement.

[1] This test method is under the jurisdiction of ASTM Committee D-18 on Soil and Rock and is the direct responsibility of Subcommittee D18.20 on Impermeable Barriers.

Current edition approved June 29, 1984. Published September 1984.

[2] *Annual Book of ASTM Standards*, Vol 04.08.

[3] Available from the American Petroleum Institute, 2101 L St., NW, Washington, DC 20037.

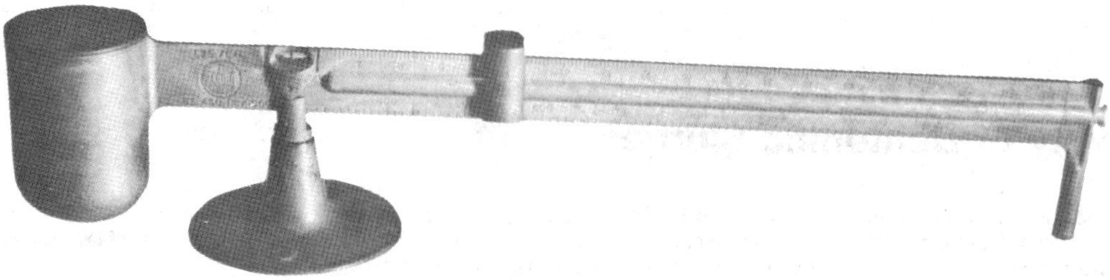

NOTE—Photo courtesy of N. L. Baroid–N. L. Industries, Inc., Houston TX.

FIG. 1 Mud Balance

Designation: D 4448 – 85a (Reapproved 1992)

Standard Guide for Sampling Groundwater Monitoring Wells[1]

This standard is issued under the fixed designation D 4448; the number immediately following the designation indicates the year of original adoption or, in the case of revision, the year of last revision. A number in parentheses indicates the year of last reapproval. A superscript epsilon (ϵ) indicates an editorial change since the last revision or reapproval.

1. Scope

1.1 This guide covers procedures for obtaining valid, representative samples from groundwater monitoring wells. The scope is limited to sampling and "in the field" preservation and does not include well location, depth, well development, design and construction, screening, or analytical procedures.

1.2 This guide is only intended to provide a review of many of the most commonly used methods for sampling groundwater quality monitoring wells and is not intended to serve as a groundwater monitoring plan for any specific application. Because of the large and ever increasing number of options available, no single guide can be viewed as comprehensive. The practitioner must make every effort to ensure that the methods used, whether or not they are addressed in this guide, are adequate to satisfy the monitoring objectives at each site.

1.3 *This standard does not purport to address all of the safety problems, if any, associated with its use. It is the responsibility of the user of this standard to establish appropriate safety and health practices and determine the applicability of regulatory limitations prior to use.*

2. Summary of Guide

2.1 The equipment and procedures used for sampling a monitoring well depend on many factors. These include, but are not limited to, the design and construction of the well, rate of groundwater flow, and the chemical species of interest. Sampling procedures will be different if analyzing for trace organics, volatiles, oxidizable species, or trace metals is needed. This guide considers all of these factors by discussing equipment and procedure options at each stage of the sampling sequence. For ease of organization, the sampling process can be divided into three steps: well flushing, sample withdrawal, and field preparation of samples.

2.2 Monitoring wells must be flushed prior to sampling so that the groundwater is sampled, not the stagnant water in the well casing. If the well casing can be emptied, this may be done although it may be necessary to avoid oxygen contact with the groundwater. If the well cannot be emptied, procedures must be established to demonstrate that the sample represents groundwater. Monitoring an indicative parameter such as pH during flushing is desirable if such a parameter can be identified.

2.3 The types of species that are to be monitored as well as the concentration levels are prime factors for selecting sampling devices (**1, 2**).[2] The sampling device and all materials and devices the water contacts must be constructed of materials that will not introduce contaminants or alter the analyte chemically in any way.

2.4 The method of sample withdrawal can vary with the parameters of interest. The ideal sampling scheme would employ a completely inert material, would not subject the sample to negative pressure and only moderate positive pressure, would not expose the sample to the atmosphere, or preferably, any other gaseous atmosphere before conveying it to the sample container or flow cell for on-site analysis.

2.5 The degree and type of effort and care that goes into a sampling program is always dependent on the chemical species of interest and the concentration levels of interest. As the concentration level of the chemical species of analytical interest decreases, the work and precautions necessary for sampling are increased. Therefore, the sampling objective must clearly be defined ahead of time. For example, to prepare equipment for sampling for mg/L (ppm) levels of Total Organic Carbon (TOC) in water is about an order of magnitude easier than preparing to sample for μg/L (ppb) levels of a trace organic like benzene. The specific precautions to be taken in preparing to sample for trace organics are different from those to be taken in sampling for trace metals. No final Environmental Protection Agency (EPA) protocol is available for sampling of trace organics. A short guidance manual, (**3**) and an EPA document (**4**) concerning monitoring well sampling, including considerations for trace organics are available.

2.6 Care must be taken not to cross contaminate samples or monitoring wells with sampling or pumping devices or materials. All samples, sampling devices, and containers must be protected from the environment when not in use. Water level measurements should be made before the well is flushed. Oxidation-reduction potential, pH, dissolved oxygen, and temperature measurements and filtration should all be performed on the sample in the field, if possible. All but temperature measurement must be done prior to any significant atmospheric exposure, if possible.

2.7 The sampling procedures must be well planned and all sample containers must be prepared and labeled prior to going to the field.

3. Significance and Use

3.1 The quality of groundwater has become an issue of national concern. Groundwater monitoring wells are one of

---

[1] This guide is under the jurisdiction of ASTM Committee D-34 on Waste Management and is the direct responsibility of Subcommittee D34.01 on Sampling and Monitoring.

Current edition approved Aug. 23 and Oct. 25, 1985. Published May 1986.

[2] The boldface numbers in parentheses refer to a list of references at the end of this guide.

TABLE 1 Typical Container and Preservation Requirements for a Ground-Water Monitoring Program

Sample and Measurement	Volume Required (mL)	Container P—Polyethylene G—Glass	Preservative	Maximum Holding Time
Metals As/Ba/Cd/Cr/Fe Pb/Se/Ag/Mn/Na	1000–2000	P/G (special acid cleaning)	high purity nitric acid to pH <2	6 months
Mercury	200–300	P/G (special acid cleaning)	high purity nitric acid to pH <2 +0.05 % $K_2Cr_2O_7$	28 days
Radioactivity alpha/beta/radium	4000	P/G (special acid cleaning)	high purity nitric acid to pH <2	6 months
Phenolics	500–1000	G	cool, 4°C H_2SO_4 to pH <2	28 days
Miscellaneous	1000–2000	P	cool, 4°C	28 days
Fluoride	300–500	P		28 days
Chloride	50–200	P/G		28 days
Sulfate	100–500	P/G		48 hours
Nitrate	100–250	P/G		6 h
Coliform	100	P/G		on site/24 h
Conductivity	100	P/G		on site/6 h
pH	100	P/G		48 h
Turbidity	100	P/G		
Total organic carbon (TOC)	25–100	P/G	cool, 4°C or cool, 4°C HCl or H_2SO_4 to pH <2	24 h 28 days
Pesticides, herbicides and total organic halogen (TOX)	1000–4000	G/TFE-fluorocarbon lined cap solvent rinsed	cool, 4°C	7 days/extraction +30 days/analysis
Extractable organics	1000–2000	G/TFE-fluorocarbon-lined cap solvent rinsed	cool, 4°C	7 days/extraction +30 days/analysis
Organic purgeables acrolein/acrylonitrile	25–120	G/vial TFE-fluorocarbon-lined septum	cool, 4°C	14 days 3 days

the more important tools for evaluating the quality of groundwater, delineating contamination plumes, and establishing the integrity of hazardous material management facilities.

3.2 The goal in sampling groundwater monitoring wells is to obtain samples that are truly representative of the aquifer or groundwater in question. This guide discusses the advantages and disadvantages of various well flushing, sample withdrawal, and sample preservation techniques. It reviews the parameters that need to be considered in developing a valid sampling plan.

4. Well Flushing (Purging)

4.1 Water that stands within a monitoring well for a long period of time may become unrepresentative of formation water because chemical or biochemical change may cause water quality alterations and even if it is unchanged from the time it entered the well, the stored water may not be representative of formation water at the time of sampling, or both. Because the representativeness of stored water is questionable, it should be excluded from samples collected from a monitoring well.

4.2 The surest way of accomplishing this objective is to remove all stored water from the casing prior to sampling. Research with a tracer in a full scale model 2 in. PVC well **(5)** indicates that pumping 5 to 10 times the volume of the well via an inlet near the free water surface is sufficient to remove all the stored water in the casing. The volume of the well may be calculated to include the well screen and any gravel pack if natural flow through these is deemed insufficient to keep them flushed out.

4.3 In deep or large diameter wells having a volume of water so large as to make removal of all the water impractical, it may be feasible to lower a pump or pump inlet to some point well below the water surface, purge only the volume below that point then withdraw the sample from a deeper level. Research indicates this approach should avoid most contamination associated with stored water **(5, 6, 7)**. Sealing the casing above the purge point with a packer may make this approach more dependable by preventing migration of stored water from above. But the packer must be above the top of the screened zone, or stagnant water from above the packer will flow into the purged zone through the well's gravel/sand pack.

4.4 In low yielding wells, the only practical way to remove all standing water may be to empty the casing. Since it is not always possible to remove all water, it may be advisable to let the well recover (refill) and empty it again at least once. If introduction of oxygen into the aquifer may be of concern, it would be best not to uncover the screen when performing the above procedures. The main disadvantage of methods designed to remove all the stored water is that large volumes may need to be pumped in certain instances. The main advantage is that the potential for contamination of samples with stored water is minimized.

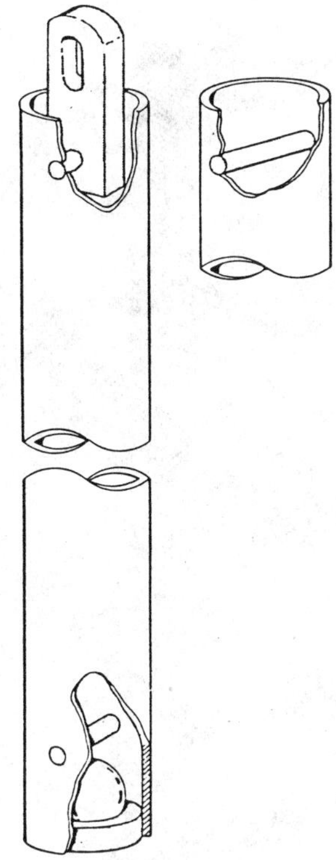

NOTE—Taken from Ref **(15)**.

FIG. 1 Single Check Valve Bailer

4.5 Another approach to well flushing is to monitor one or more indicator parameters such as pH, temperature, or conductivity and consider the well to be flushed when the indicator(s) no longer change. The advantage of this method is that pumping can be done from any location within the casing and the volume of stored water present has no direct bearing on the volume of water that must be pumped. Obviously, in a low yielding well, the well may be emptied before the parameters stabilize. A disadvantage of this approach is that there is no assurance in all situations that the stabilized parameters represent formation water. If significant drawdown has occurred, water from some distance away may be pulled into the screen causing a steady parameter reading but not a representative reading. Also, a suitable indicator parameter and means of continuously measuring it in the field must be available.

4.6 Gibb **(4, 8)** has described a time-drawdown approach using a knowledge of the well hydraulics to predict the percentage of stored water entering a pump inlet near the top of the screen at any time after flushing begins. Samples are taken when the percentage is acceptably low. As before, the advantage is that well volume has no direct effect in the duration of pumping. A current knowledge of the well's hydraulic characteristics is necessary to employ this approach. Downward migration of stored water due to effects other than drawdown (for example density differences) is not accounted for in this approach.

4.7 In any flushing approach, a withdrawal rate that minimizes drawdown while satisfying time constraints should be used. Excessive drawdown distorts the natural flow patterns around a well and can cause contaminants that were not present originally to be drawn into the well.

5. Materials and Manufacture

5.1 The choice of materials used in the construction of sampling devices should be based upon a knowledge of what compounds may be present in the sampling environment and how the sample materials may interact via leaching, adsorption, or catalysis. In some situations, PVC or some other plastic may be sufficient. In others, an all glass apparatus may be necessary.

5.2 Most analytical protocols suggest that the devices used in sampling and storing samples for trace organics analysis (μg/L levels) must be constructed of glass or TFE–fluorocarbon resin, or both. One suggestion advanced by the EPA is that the monitoring well be constructed so that only TFE–fluorocarbon tubing be used in that portion of the sampling well that extends from a few feet above the water table to the bottom of the borehole. **(3, 5)** Although this type of well casing is now commercially available, PVC well casings are currently the most popular. If adhesives are avoided, PVC well casings are acceptable in many cases although their use may still lead to some problems if trace organics are of concern. At present, the type of background presented by PVC and interactions occurring between PVC and groundwater are not well understood. Tin, in the form of an organotin stabilizer added to PVC, may enter samples taken from PVC casing. **(9)**

5.3 Since the most significant problem encountered in trace organics sampling, results from the use of PVC adhesives in monitoring well construction, threaded joints might avoid the problem **(3, 5)**. Milligram per litre (parts per million) levels of compounds such as tetrahydrofuran, methyl-ethyl-ketone, and toluene are found to leach into

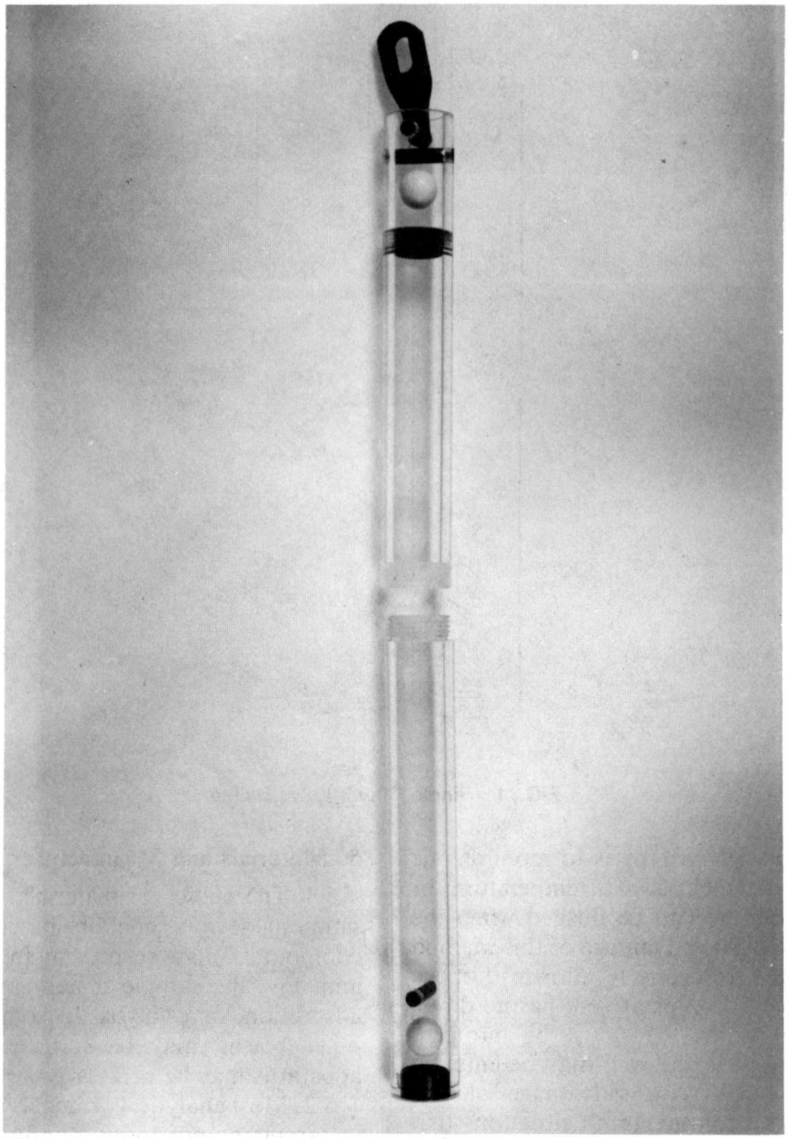

NOTE—Taken from Ref **(17)**.

FIG. 2 Acrylic Point Source Bailer

groundwater samples from monitoring well casings sealed with PVC solvent cement. Pollutant phthalate esters **(8, 10)** are often found in water samples at ppb levels; the EPA has found them on occasion at ppm levels in their samples. The ubiquitous presence of these phthalate esters is unexplained, except to say that they may be leached from plastic pipes, sampling devices, and containers.

5.4 TFE–fluorocarbon resins are highly inert and have sufficient mechanical strength to permit fabrication of sampling devices and well casings. Molded parts are exposed to high temperature during fabrication which destroys any organic contaminants. The evolution of fluorinated compounds can occur during fabrication, will cease rapidly, and does not occur afterwards unless the resin is heated to its melting point.

5.5 Extruded tubing of TFE-fluorocarbon for sampling may contain surface traces of an organic solvent extrusion aid. This can be removed easily by the fabricator and, once removed by flushing, should not affect the sample. TFE-fluorocarbon FEP and TFE-fluorocarbon PFA resins do not require this extrusion aid and may be suitable for sample tubing as well. Unsintered thread-sealant tape of TFE-fluorocarbon is available in an "oxygen service" grade and contains no extrusion aid and lubricant.

5.6 Louneman, et al. **(11)** alludes to problems caused by a lubricating oil used during TFE-fluorocarbon tubing extrusion. This reference also presents evidence that a fluorinated ethylene-propylene copolymer adsorbed acetone to a degree that later caused contamination of a gas sample.

5.7 Glass and stainless steel are two other materials generally considered inert in aqueous environments. Glass is probably among the best choices though it is not inconceivable it could adsorb some constituents as well as release other contaminants (for example, Na, silicate, and Fe). Of course, glass sampling equipment must be handled carefully in the field. Stainless steel is strongly and easily machined to

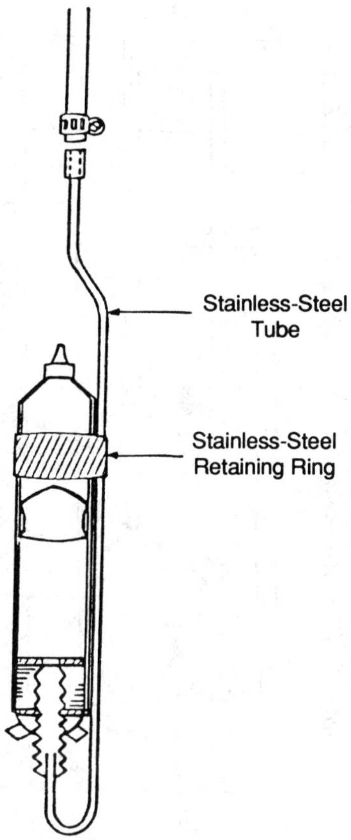

NOTE—Taken from Ref **(21)**.

FIG. 3 Schematic of the Inverted Syringe Sampler

fabricate equipment. Unfortunately, it is not totally immune to corrosion that could release metallic contaminants. Stainless steel contains various alloying metals, some of these (for example Ni) are commonly used as catalysts for various reactions. The alloyed constituents of some stainless steels can be solubilized by the pitting action of nonoxidizing anions such as chloride, fluoride, and in some instances sulfate, over a range of pH conditions. Aluminum, titanium, polyethylene, and other corrosion resistant materials have been proposed by some as acceptable materials, depending on groundwater quality and the constituents of interest.

5.8 Where temporarily installed sampling equipment is used, the sampling device that is chosen should be nonplastic (unless TFE-fluorocarbon), cleanable of trace organics, and must be cleaned between each monitoring well use in order to avoid cross-contamination of wells and samples. The only way to ensure that the device is indeed "clean" and acceptable is to analyze laboratory water blanks and field water blanks that have been soaked in and passed through the sampling device to check for the background levels that may result from the sampling materials or from field conditions. Thus, all samplings for trace materials should be accompanied by samples which represent the field background (if possible), the sampling equipment background, and the laboratory background.

5.9 Additional samples are often taken in the field and spiked (spiked-field samples) in order to verify that the sample handling procedures are valid. The American Chemical Society's committee on environmental improvement has published guidelines for data acquisition and data evaluation which should be useful in such environmental evaluations **(10, 12)**.

6. Sampling Equipment

6.1 There is a fairly large choice of equipment presently available for groundwater sampling from single screened wells and well clusters. The sampling devices can be categorized into the following eight basic types.

6.1.1 *Down-Hole Collection Devices:*

6.1.1.1 Bailers, messenger bailers, or thief samplers **(13, 14)** are examples of down-hole devices that probably provide valid samples once the well has been flushed. They are not practical for removal of large volumes of water. These devices can be constructed in various shapes and sizes from a variety of materials. They do not subject the sample to pressure extremes.

6.1.1.2 Bailers do expose part of the sample to the atmosphere during withdrawal. Bailers used for sampling of volatile organic compounds should have a sample cock or draft valve in or near the bottom of the sampler allowing withdrawal of a sample from the well below the exposed surface of the water or the first few inches of the sample should be discarded. Suspension lines for bailers and other samplers should be kept off the ground and free of other contaminating materials that could be carried into the well. Down-hole devices are not very practical for use in deep

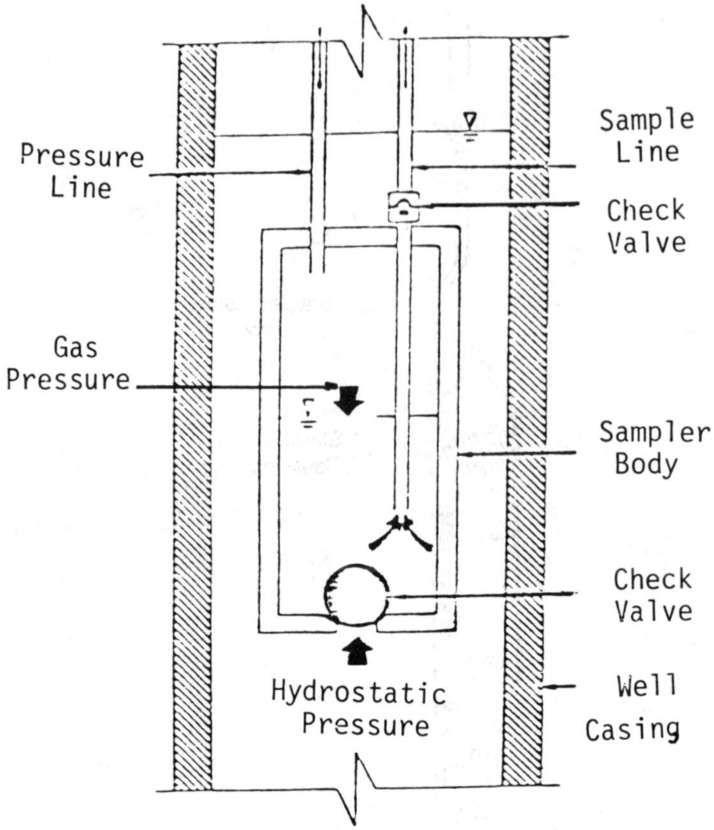

NOTE—Taken from Ref **(5)**.

FIG. 4 The Principal of Gas Displacement Pumping

wells. However, potential sample oxidation during transfer of the sample into a collection vessel and time constraints for lowering and retrieval for deep sampling are the primary disadvantages.

6.1.1.3 Three down-hole devices are the single and double check valve bailers and thief samplers. A schematic of a single check valve unit is illustrated in Fig. 1. The bailer may be threaded in the middle so that additional lengths of blank casing may be added to increase the sampling volume. TFE-fluorocarbon or PVC are the most common materials used for construction **(15)**.

6.1.1.4 In operation, the single check valve bailer is lowered into the well, water enters the chamber through the bottom, and the weight of the water column closes the check valve upon bailer retrieval. The specific gravity of the ball should be about 1.4 to 2.0 so that the ball almost sits on the check valve seat during chamber filling. Upon bailer withdrawal, the ball will immediately seat without any samples loss through the check valve. A similar technique involves lowering a sealed sample container within a weighted bottle into the well. The stopper is then pulled from the bottle via a line and the entire assembly is retrieved upon filling of the container **(14, 16)**.

6.1.1.5 A double check valve bailer allows point source sampling at a specific depth **(15, 17)**. An example is shown in Fig. 2. In this double check valve design, water flows through the sample chamber as the unit is lowered. A venturi tapered inlet and outlet ensures that water passes freely through the

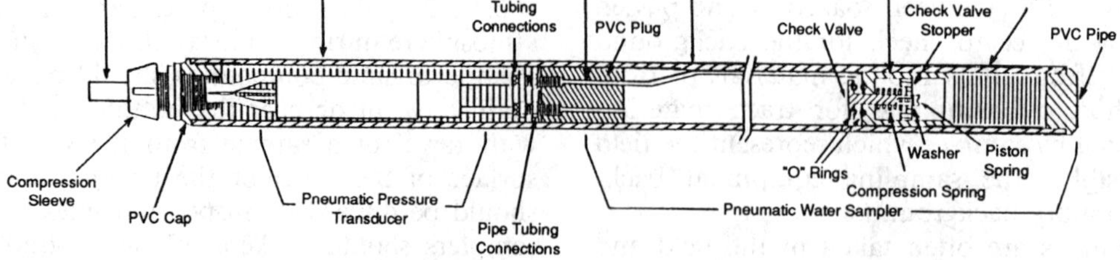

NOTE—Taken from Ref **(41)**.

FIG. 5 Pneumatic Water Sampler With Internal Transducer

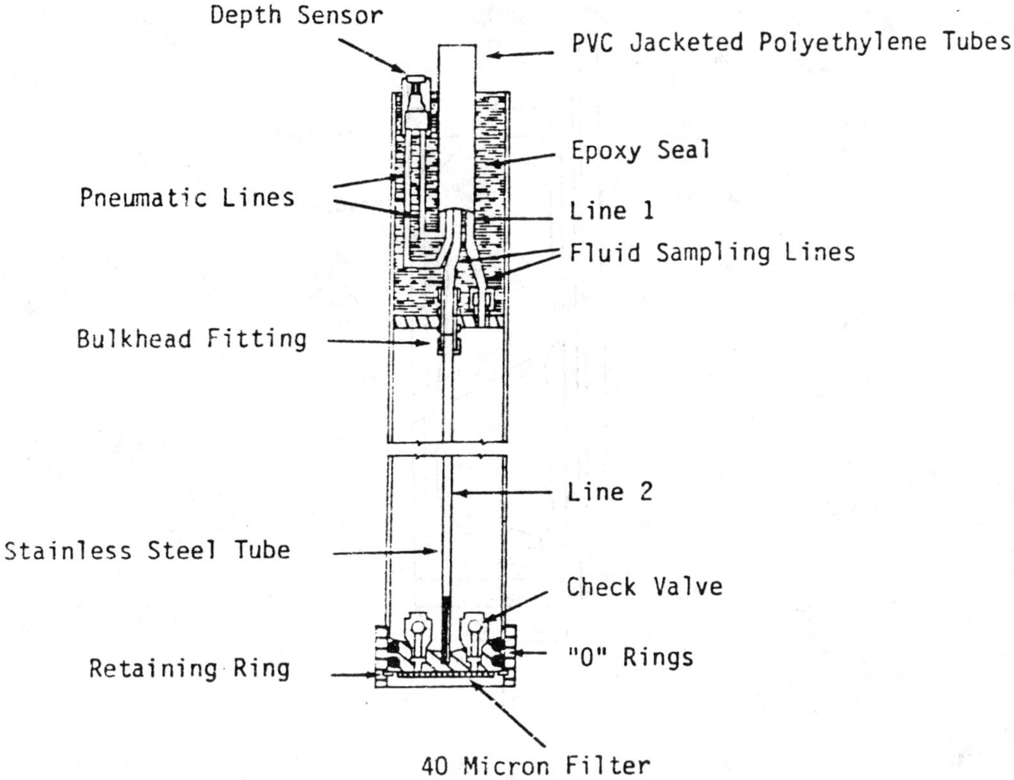

NOTE—Taken from Ref **(42)**.

FIG. 6 Pneumatic Sampler With Externally Mounted Transducer

unit. When a depth where the sample is to be collected is reached, the unit is retrieved. Because the difference between each ball and check valve seat is maintained by a pin that blocks vertical movement of the check ball, both check valves close simultaneously upon retrieval. A drainage pin is placed into the bottom of the bailer to drain the sample directly into a collection vessel to reduce the possibility of air oxidation. The acrylic model in Fig. 2 is threaded at the midsection allowing the addition of threaded casing to increase the sampling volume.

6.1.1.6 Another approach for obtaining point source samples employs a weighted messenger or pneumatic change to "trip" plugs at either end of an open tube (for example, tube water sampler or thief sampler) to close the chamber **(18)**. Foerst, Kemmerer, and Bacon samplers are of this variety **(14, 17, 19)**. A simple and inexpensive pneumatic sampler was recently described by Gillham **(20)**. The device (Fig. 3) consists of a disposable 50 mL plastic syringe modified by sawing off the plunger and the finger grips. The syringe is then attached to a gas-line by means of a rubber stopper assembly. The gas-line extends to the surface, and is used to drive the stem-less plunger, and to raise and lower the syringe into the hole. When the gas-line is pressurized, the rubber plunger is held at the tip of the syringe. The sampler is then lowered into the installation, and when the desired depth is reached, the pressure in the gas-line is reduced to atmospheric (or slightly less) and water enters the syringe. The sampler is then retrieved from the installation and the syringe detached from the gas-line. After the tip is sealed, the syringe is used as a short-term storage container. A number of thief or messenger devices are available in various materials and shapes.

6.1.2 *Suction Lift Pumps:*

6.1.2.1 Three types of suction lift pumps are the direct line, centrifugal, and peristaltic. A major disadvantage of any suction pump is that it is limited in its ability to raise water by the head available from atmospheric pressure. Thus, if the surface of the water is more than about 25 ft below the pump, water may not be withdrawn. The theoretical suction limit is about 34 ft, but most suction pumps are capable of maintaining a water lift of only 25 ft or less.

6.1.2.2 Many suction pumps draw the water through some sort of volute in which impellers, pistons, or other devices operate to induce a vacuum. Such pumps are probably unacceptable for most sampling purposes because they are usually constructed of common materials such as brass or mild steel and may expose samples to lubricants. They often induce very low pressures around rotating vanes or other such parts such that degassing or even cavitation may occur. They can mix air with the sample via small leaks in the casing, and they are difficult to adequately clean between uses. Such pumps are acceptable for purging of wells, but should not generally be used for sampling.

6.1.2.3 One exception to the above statements is a peristaltic pump. A peristaltic pump is a self-priming, low volume suction pump which consists of a rotor with ball bearing rollers **(21)**. Flexible tubing is inserted around the pump rotor and squeezed by heads as they revolve in a circular pattern around the rotor. One end of the tubing is placed into the well while the other end can be connected

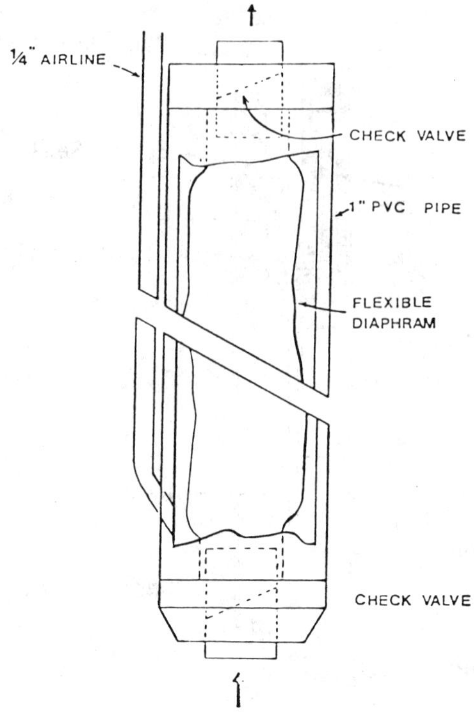

NOTE—Taken from Ref **(4)**.

FIG. 7 Bladder Pump

directly to a receiving vessel. As the rotor moves, a reduced pressure is created in the well tubing and an increased pressure (<40 psi) on the tube leaving the rotor head. A drive shaft connected to the rotor head can be extended so that multiple rotor heads can be attached to a single drive shaft.

6.1.2.4 The peristaltic pump moves the liquid totally within the sample tube. No part of the pump contacts the liquid. The sample may still be degassed (cavitation is unlikely) but the problems due to contact with the pump mechanism are eliminated. Peristaltic pumps do require a fairly flexible section of tubing within the pumphead itself. A section of silicone tubing is commonly used within the peristaltic pumphead, but other types of tubing can be used particularly for the sections extending into the well or from the pump to the receiving container. The National Council of the Paper Industry for Air and Stream Improvement **(22)** recommends using medical grade silicone tubing for organic sampling purposes as the standard grade uses an organic vulcanizing agent which has been shown to leach into samples. Medical grade silicone tube is, however, limited to use over a restricted range of ambient temperatures. Various manufacturers offer tubing lined with TFE-fluorocarbon or Viton[3] for use with their pumps. Gibb **(1, 8)** found little difference between samples withdrawn by a peristaltic pump and those taken by a bailer.

6.1.2.5 A direct method of collecting a sample by suction consists of lowering one end of a length of plastic tubing into the well or piezometer. The opposite end of the tubing is connected to a two way stopper bottle and a hand held or mechanical vacuum pump is attached to a second tubing leaving the bottle. A check valve is attached between the two lines to maintain a constant vacuum control. A sample can then be drawn directly into the collection vessel without contacting the pump mechanism **(5, 23, 24)**.

6.1.2.6 A centrifugal pump can be attached to a length of plastic tubing that is lowered into the well. A foot valve is usually attached to the end of the well tubing to assist in priming the tube. The maximum lift is about 4.6 m (15 ft) for such an arrangement **(23, 25, 26)**.

6.1.2.7 Suction pump approaches offer a simple sample retrieval method for shallow monitoring. The direct line method is extremely portable though considerable oxidation and mixing may occur during collection. A centrifugal pump will agitate the sample to an even greater degree although pumping rates of 19 to 151 Lpm (5 to 40 gpm) can be attained. A peristaltic pump provides a lower sampling rate with less agitation than the other two pumps. The withdrawal rate of peristaltic pumps can be carefully regulated by adjustment of the rotor head revolution.

6.1.2.8 All three systems can be specially designed so that the water sample contacts only the TFE flourocarbon or silicone tubing prior to sample bottle entry. Separate tubing is recommended for each well or piezometer sampled.

6.1.3 *Electric Submersible Pumps:*

6.1.3.1 A submersible pump consists of a sealed electric motor that powers a piston or helical single thread worm at a high rpm. Water is brought to the surface through an access tube. Such pumps have been used in the water well industry for years and many designs exist **(5, 26)**.

6.1.3.2 Submersible pumps provide relatively high discharge rates for water withdrawal at depths beyond suction

[3] Viton is a trademark of E. I. du Pont de Nemours & Co., Wilmington, DE 19898 and has been found suitable for this purpose.

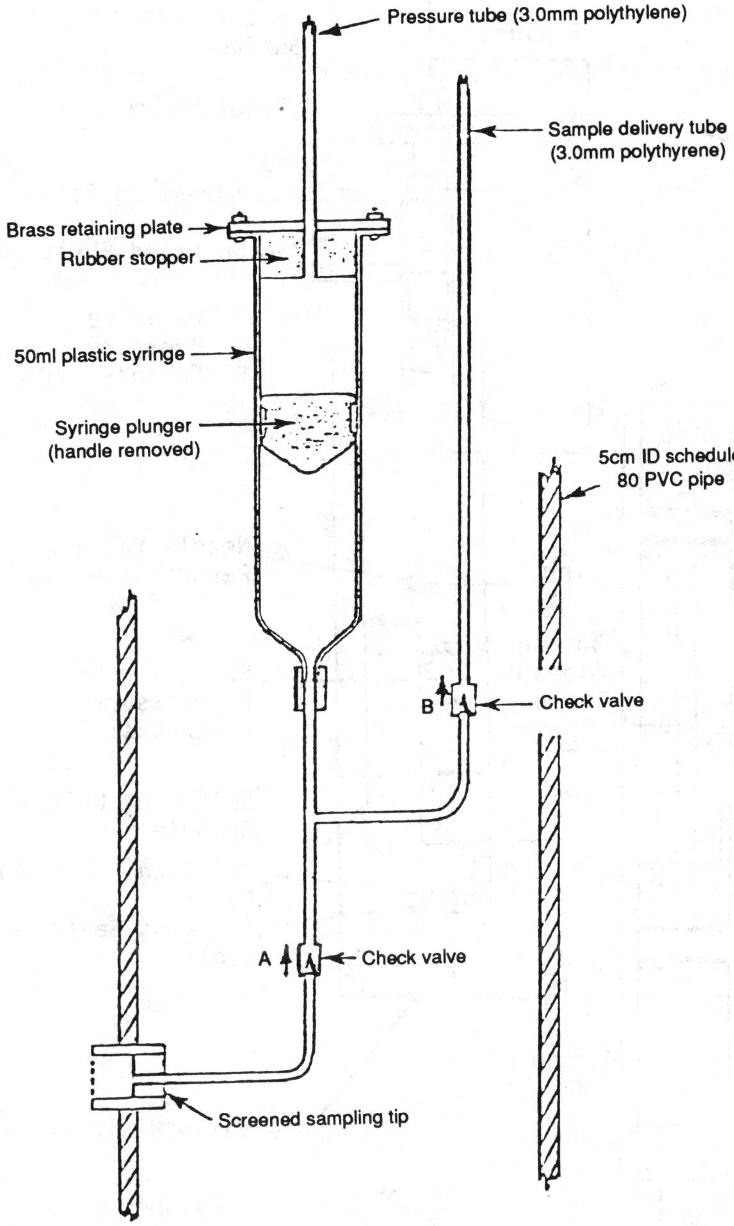

NOTE—Taken from Ref **(48)**.

FIG. 8 Positive Displacement Syringe Pump

lift capabilities. A battery operated unit 3.6 cm (1.4 in.) in diameter and with a 4.5 Lpm (1.2 gpm) flow rate at 33.5 m (110 ft) has been developed **(27)**. Another submersible pump has an outer diameter of 11.4 cm (4.5 in.) and can pump water from 91 m (300 ft). Pumping rates vary up to 53.0 Lpm (14 gpm) depending upon the depth of the pump **(28)**.

6.1.3.3 A submersible pump provides higher extraction rates than many other methods. Considerable sample agitation results, however, in the well and in the collection tube during transport. The possibility of introducing trace metals into the sample from pump materials also exists. Steam cleaning of the unit followed by rinsing with unchlorinated, deionized water is suggested between sampling when analysis for organics in the parts per million (ppm) or parts per billion (ppb) range is required **(29)**.

6.1.4 *Gas-Lift Pumps:*

6.1.4.1 Gas-lift pumps use compressed air to bring a water sample to the surface. Water is forced up an eductor pipe that may be the outer casing or a smaller diameter pipe inserted into the well annulus below the water level **(30, 31)**.

6.1.4.2 A similar principle is used for a unit that consists of a small diameter plastic tube perforated in the lower end. This tube is placed within another tube of slightly larger diameter. Compressed air is injected into the inner tube; the air bubbles through the perforations, thereby lifting the water sample via the annulus between the outer and inner tubing **(32)**. In practice, the eductor line should be submerged to a depth equal to 60 % of the total submerged eductor length during pumping **(26)**. A 60 % ratio is considered optimal although a 30 % submergence ratio is adequate.

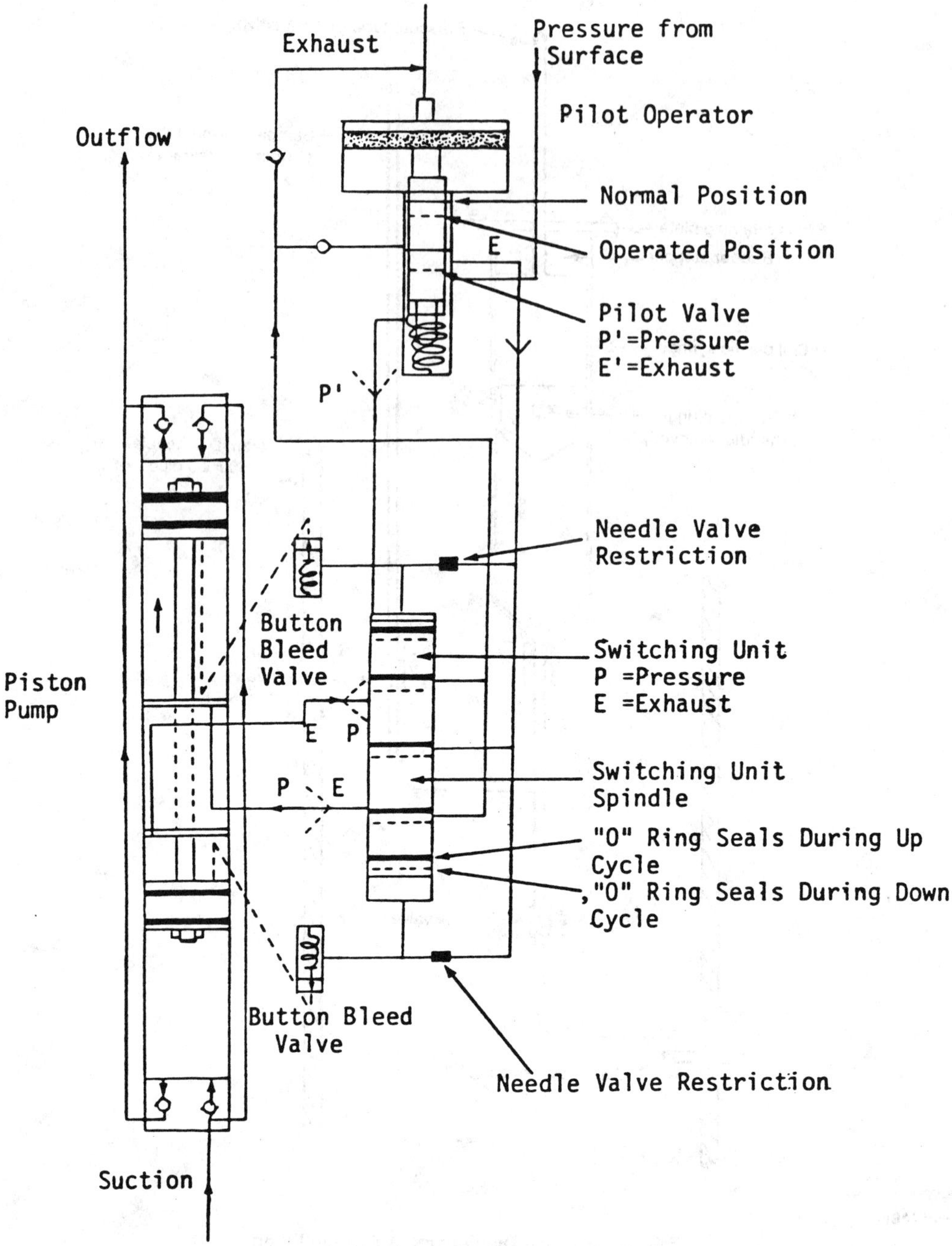

NOTE—Taken from Ref **(49)**.

FIG. 9 Gas Driven Piston Pump

6.1.4.3 The source of compressed gas may be a hand pump for depths generally less than 7.6 m (25 ft). For greater depths, air compressors, pressurized air bottles, and air compressed from an automobile engine have been used.

6.1.4.4 As already mentioned, gas-lift methods result in considerable sample agitation and mixing within the well, and cannot be used for samples which will be tested for volatile organics. The eductor pipe or weighted plastic tubing is a potential source of sample contamination. In addition, Gibb **(8)** uncovered difficulties in sampling for inorganics. These difficulties were attributed to changes in redox, pH, and species transformation due to solubility constant changes resulting from stripping, oxidation, and pressure changes.

6.1.5 *Gas Displacement Pumps:*

6.1.5.1 Gas displacement or gas drive pumps are distinguished from gas-lift pumps by the method of sample transport. Gas displacement pumps force a discrete column of water to the surface via mechanical lift without extensive mixing of the pressurized gas and water as occurs with air-lift equipment. The principle is shown schematically in Fig. 4. Water fills the chamber. A positive pressure is applied to the

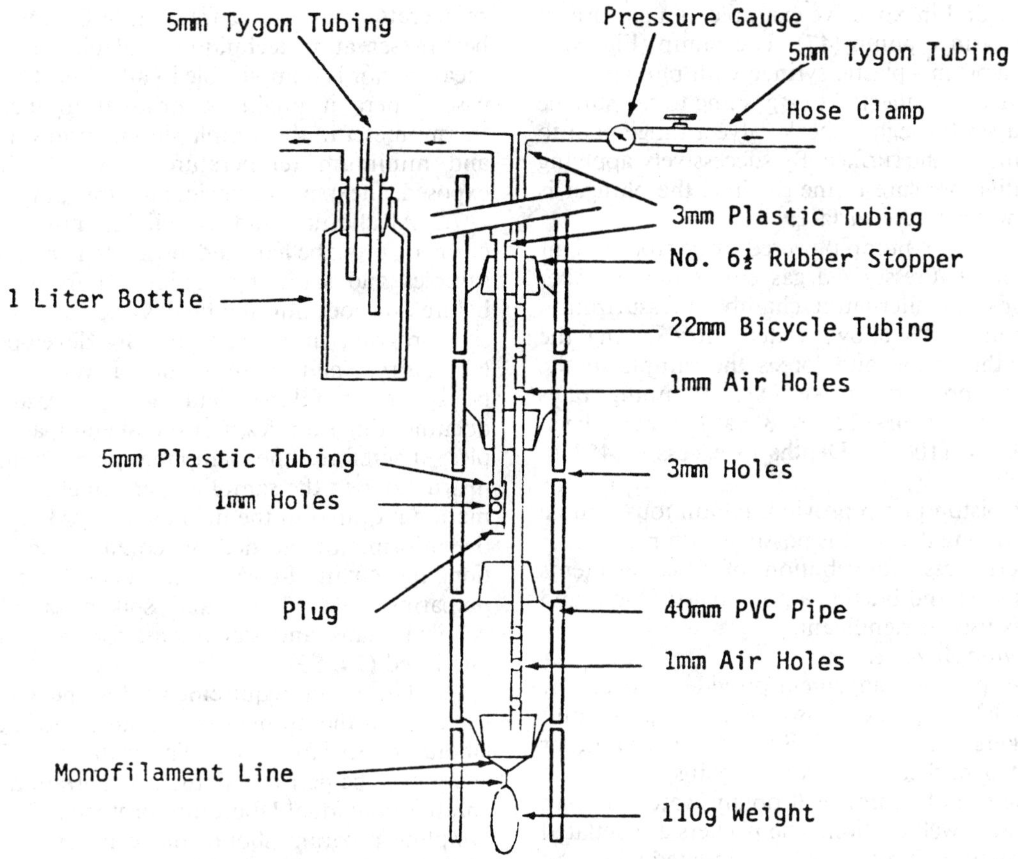

NOTE—Taken from Ref **(53)**.

FIG. 10 Packer Pump Arrangement

gas line closing the sampler check valve and forcing water up the sample line. By removing the pressure the cycle can be repeated. Vacuum can also be used in conjunction with the gas **(30)**. The device can be permanently installed in the well **(33, 34, 35)** or lowered into the well **(36, 37)**.

6.1.5.2 A more complicated two stage design constructed of glass with check valves made of TFE-fluorocarbon has been constructed **(38, 39)**. The unit was designed specifically for sample testing for trace level organics. Continuous flow rates up to 2.3 Lpm (0.6 gpm) are possible with a 5.1 cm (2 in.) diameter unit.

6.1.5.3 Gas displacement pumps have also been developed with multiple functions. The water sample in Fig. 5 provides piezometric data measurements with an internally mounted transducer **(40)**. A sample with its transducer exposed externally for piezometric measurements is illustrated in Fig. 6 **(41)**. The sensor can activate the gas source at the surface to cause sample chamber pressurization at the predetermined depth. Another design can be used as a water sampler or as a tool for injecting brine or other tracers into a well **(42)**.

6.1.5.4 Gas displacement pumps offer reasonable potential for preserving sample integrity because little of the driving gas comes in contact with the sample as the sample is conveyed to the surface by a positive pressure. There is, however, a potential loss of dissolved gasses or contamination from the driving gas and the housing materials.

6.1.6 *Bladder Pumps:*

6.1.6.1 Bladder pumps, also referred to as gas-operated squeeze pumps, consist of a flexible membrane enclosed by a rigid housing. Water enters the membrane through a check valve in the vessel bottom; compressed gas injected into the cavity between the housing and bladder forces the sample through a check valve at the top of the membrane and into a discharge line (Fig. 7). Water is prevented from re-entering the bladder by the top check valve. The process is repeated to cycle the water to the surface. Samples taken from depths of 30.5 m (100 ft) have been reported.

6.1.6.2 A variety of design modifications and materials are available **(43, 44)**. Bladder materials include neoprene, rubber, ethylene propylene terpolymer (E.P.T.), nitrile, and the fluorocarbon Viton.[3] A bladder made of TFE-fluorocarbon is also under development **(45)**. Automated sampling systems have been developed to control the time between pressurization cycles **(46)**.

6.1.6.3 Bladder pumps provide an adaptable sampling tool due primarily to the number of bladder shapes that are feasible. These devices have a distinct advantage over gas displacement pumps in that there is no contact with the driving gas. Disadvantages include the large gas volumes required, low pumping rates, and potential contamination from many of the bladder materials, the rigid housing, or both.

6.1.7 *Gas Driven Piston Pumps:*

6.1.7.1 A simple and inexpensive example of a gas driven piston pump is a syringe pump **(47)**. The pump (Fig. 8) is constructed from a 50 mL plastic syringe with plunger stem removed. The device is connected to a gas line to the surface and the sample passes through a check valve arrangement to a sampling container at the surface. By successively applying positive and negative pressure to the gas-line, the plunger is activated driving water to the surface.

6.1.7.2 A double piston pump powered by compressed air is illustrated in Fig. 9. Pressurized gas enters the chamber between the pistons; the alternating chamber pressurization activates the piston which allows water entry during the suction stroke of the piston and forces the sample to the surface during the pressure stroke **(48)**. Pumping rates between 9.5 and 30.3 L/hr (2.5 to 8 gal/hr) have been reported from 30.5 m (100 ft). Depths in excess of 457 m (1500 ft) are possible.

6.1.7.3 The gas piston pump provides continuous sample withdrawal at depths greater than is possible with most other approaches. Nevertheless, contribution of trace elements from the stainless steel and brass is a potential problem and the quantity of gas used is significant.

6.1.8 *Packer Pump Arrangement:*

6.1.8.1 A packer pump arrangement provides a means by which two expandable "packers" isolate a sampling unit between two packers within a well. Since the hydraulic or pneumatic activated packers are wedged against the casing wall or screen, the sampling unit will obtain water samples only from the isolated well portion. The packers are deflated for vertical movement within the well and inflated when the desired depth is attained. Submersible, gas lift, and suction pumps can be used for sampling. The packers are usually constructed from some type of rubber or rubber compound **(48, 49, 50, 51)**. A packer pump unit consisting of a vacuum sampler positioned between two packers is illustrated in Fig. 10 **(52)**.

6.1.8.2 A packer assembly allows the isolation of discrete sampling points within a well. A number of different samplers can be situated between the packers depending upon the analytical specifications for sample testing. Vertical movement of water outside the well casing during sampling is possible with packer pumps but depends upon the pumping rate and subsequent disturbance. Deterioration of the expandable materials will occur with time with the increased possibility of undesirable organic contaminants contributing to the water sample.

7. Sample Containers and Preservation

7.1 Complete and unequivocal preservation of samples, whether domestic wastewater, industrial wastes, or natural waters, is practically impossible. At best, preservation techniques only retard the chemical and biological changes that inevitably continue after the sample is removed from the source. Therefore, insuring the timely analysis of a sample should be one of the foremost considerations in the sampling plan schedule. Methods of preservation are somewhat limited and are intended to retard biological action, retard hydrolysis of chemical compounds and complexes, and reduce the volatility of constituents. Preservation methods are generally limited to pH control, chemical addition, refrigeration and freezing. For water samples, immediate refrigeration just above freezing (4°C in wet ice) is often the best preservation technique available, but it is not the only measure nor is it applicable in all cases. There may be special cases where it might be prudent to include a recording thermometer in the sample shipment to verify the maximum and minimum temperature to which the samples were exposed. Inexpensive devices for this purpose are available.

7.2 All bottles and containers must be specially pre-cleaned, pre-labelled, and organized in ice-chests (isolating samples and sampling equipment from the environment) before one goes into the field. Otherwise, in any comprehensive program utter chaos usually develops in the field or laboratory. The time in the field is very valuable and should be spent on taking field notes, measurements, and in documenting samples, not on labelling and organizing samples. Therefore, the sampling plan should include clear instructions to the sampling personnel concerning the information required in the field data record logbook (notebook), the information needed on container labels for identification, the chain-of-custody protocols, and the methods for preparing field blanks and spiked samples. Example of detailed plans and documentation procedures have been published **(14, 53)**.

7.3 The exact requirements for the volumes of sample needed and the number of containers to use may vary from laboratory to laboratory. This will depend on the specific analyses to be performed, the concentration levels of interest, and the individual laboratory protocols. The manager of the sampling program should make no assumptions about the laboratory analyses. He should discuss the analytical requirements of the sampling program in detail with the laboratory coordinator beforehand. This is especially the case since some analyses and preservation measures must be performed at the laboratory as soon as possible after the samples arrive. Thus, appropriate arrangements must be made.

7.4 There are a number of excellent references available which list the containers and preservation techniques appropriate for water and soils **(13, 14, 50, 54, 55, 56)**. The "Handbook for Sampling and Sample Preservation of Water and Wastewater" is an excellent reference and perhaps the most comprehensive one **(14)**. Some of this information is summarized in Table 1.

7.5 Sample containers for trace organic samples require special cleaning and handling considerations **(57)**. The sample container for purgeable organics consist of a screw-cap vial (25 to 125 mL) fitted with a TFE-flourocarbon faced silicone septum. The vial is sealed in the laboratory immediately after cleaning and is only opened in the field just prior to pouring sample into it. The water sample then must be sealed into the vial headspace free (no air bubbles) and immediately cooled (4°C) for shipment. Multiple samples (usually about four taken from one large sample container) are taken because leakage of containers may cause losses, may allow air to enter the containers, and may cause erroneous analysis of some constituents. Also, some analyses are best conducted on independent protected samples.

7.6 The purgeable samples must be analyzed by the laboratory within 14 days after collection, unless they are to be analyzed for acrolein or acrylonitrile (in which case they are to be analyzed within 3 days). For samples for solvent extractions (extractable organics-base neutrals, acids and pesticides), the sample bottles are narrow mouth, screw cap quart bottles or half-gallon bottles that have been precleaned, rinsed with the extracting organic solvent and oven dried at 105°C for at least 1 h. These bottles must be sealed with TFE-fluorocarbon lined caps (Note). Samples for organic extraction must be extracted within 7 days and analyzed within 30 days after extraction. Special pre-cleaned, solvent rinsed and oven-dried stainless steel beakers (one for each monitoring well) may be used for transferring samples from the sampling device to the sample containers.

NOTE—When collecting samples, the bottles should not be overfilled or prerinsed with sample before filling because oil and other materials may remain in the bottle. This can cause erroneously high results.

7.7 For a number of groundwater parameters, the most meaningful measurements are those made in the field at the time of sample collection or at least at an on-site laboratory. These include the water level in the well and parameters that sometimes can change rapidly with storage. A discussion of the various techniques for measuring the water level in the well is contained in a NCASI publication **(5)** and detailed procedures are outlined in a U.S. Geological Survey publication **(58)**. Although a discussion of these techniques is beyond the scope of this guide, it is important to point out that accurate measurements must be made before a well is flushed or only after it has had sufficient time to recover. Parameters that can change rapidly with storage include specific conductance, pH, turbidity, redox potential, dissolved oxygen, and temperature. For some of the other parameters, the emphasis in groundwater monitoring is on the concentration of each specific dissolved component, not the total concentration of each. Samples for these types of measurements should be filtered through 0.45 μm membrane filters ideally in the field or possibly at an on-site laboratory as soon as possible. Analyses often requiring filtered samples include all metals, radioactivity parameters, total organic carbon, dissolved orthophosphate (if needed), and total dissolved phosphorous (if needed) **(13, 14)**. If metals are to be analyzed, filter the sample prior to acid preservation. For TOC organics, the filter material should be tested to assure that it does not contribute to the TOC. The type or size of the filter to be used is not well understood. However, if results of metal, TOC or other parameters that could be effected by solids are to be compared, the same filtering procedure must be used in each case. Repeated analytical results should state whether the samples were filtered and how they were filtered.

7.8 Shipment and receipt of samples must be coordinated with the laboratory to minimize time in transit. All samples for organic analysis (and many other parameters), should arrive at the laboratory within one day after it is shipped and be maintained at about 4°C with wet ice. The best way to get them to the laboratory in good condition is to send them in sturdy insulated ice chests (coolers) equipped with bottle dividers. 24-h courier service is recommended, if personal delivery service is not practical.

REFERENCES

(1) Gibb, J. P., Schuller, R. M., Griffin, R. A., *Monitoring Well Sampling and Preservation Techniques*, EPA-600/9-80-101, 1980.

(2) Pettyjohn, W. A., Dunlap, W. J., Cosby, R. L., Keeley, J. W., "Sampling Ground Water for Organic Contaminants," *Ground Water*, Vol 19, (2), March/April 1981, pp. 180–189.

(3) Dunlap, W. J., McNabb, J. F., Scalf, M. R., Cosby, R. L., *Sampling for Organic Chemicals and Microorganisms in the Subsurface*, EPA-600/2-77-176, NTIS PB 276 679, August 1977, 35 pp.

(4) Scalf, M. R., McNabb, J. F., Dunlap, W. J., and Cosby, R. L., *Manual of Ground Water Quality Sampling Procedures*, National Water Well Association, NTIS PB-82 103 045, 1981.

(5) "A Guide to Groundwater Sampling," *NCASI Technical Bulletin*, No. 362, January 1982.

(6) Humenick, M. J., Turk, L. J., Coldrin, M., "Methodology for Monitoring Ground Water at Uranium Solution Mines," *Ground Water*, Vol 18 (3), May–June 1980, p. 262.

(7) Marsh, J. M., and Lloyd, J. W., "Details of Hydrochemical Variations in Flowing Wells," *Ground Water*, Volume 18 (4), July–August 1980, p. 366.

(8) Gibb, J. P., Schuller, R. M., Griffin, R. A., "Collection of Representative Water Quality Data from Monitoring Wells," *Proceeding of the Municipal Solid Waste Resource Recovery Symposium*, EPA-600/9-81-002A, March 1981.

(9) Boettner, E. A., Gwendolyn, L. B., Zand, H., Aquino, R., *Organic and Organotin Compounds Leached from PVC and CPVC Pipe*, NTIS P8 82-108 333, 1982.

(10) Junk, G. A., Svec, H. J., Vick, R. D., Avery, M. J., "Contamination of Water by Synthetic Polymer Tubes," *Environmental Science and Technology*, Vol 8 (13) 1100, December 1974.

(11) Louneman, W. A., Bufalini, J. J., Kuntz, R. L., and Meeks, S. A., "Contamination from Fluorocarbon Films," *Environmental Science and Technology*, Vol 15 (1), January 1981.

(12) ASC Committee on Environmental Improvement, "Guidelines for Data Acquisition and Data Quality Evaluation in Environmental Chemistry." *Analytical Chemistry*, Vol 52, 1980, pp. 2242–2249.

(13) *Procedures Manual for Ground Water Monitoring at Solid Waste Disposal Facilities*, EPA/530/SW-611, August 1977.

(14) *Handbook for Sampling and Sample Preservation of Water and Wastewater*, U.S. Dept. of Commerce NTIS PB-259 946, September 1976.

(15) Timco Manufacturing Co., Inc., "Variable Capacity Bailer," *Timco Geotechnical Catalogue*, Prairie du Sac, WI, 1982.

(16) deVera, E., Simmons, B., Stephens, R., Storm, D., *Samplers and Sampling Procedures for Hazardous Waste Streams*, Environmental Protection Agency, EPA-600/2-80-018, 1980, p. 51.

(17) Morrison, R., *Ground Water Monitoring Technology*, Timco Manufacturing Co., 1982, p. 276.

(18) Eijelkamp, "Equipment for Soil Research," *General Catalogue*, Geisbeek, The Netherlands, 1979, pp. 82–83.

(19) Wood, W., "Guidelines for Collection and Field Analysis of Ground-Water Samples for Selected Unstable Constituents," *Techniques of Water-Resources Investigations of the United States Geological Survey*, Chapter D2, 1976, p. 24.

(20) Gilham, R. W., "Syringe Devices for Groundwater Monitoring," *Ground Water Monitoring Review*, Vol 2 (2), Spring 1982, p. 36.

(21) Masterflex, *Masterflex Pump Catalogue*, Barnant Corp., Barrington, IL, 1981.

(22) "Guidelines for Contracting Sampling and Analyses for Priority Pollutants in Pulp and Paper Industry Effluents," *NCASI Stream Improvement Technical Bulletin*, No. 335, August 1980.

(23) Allison, L., "A Simple Device for Sampling Ground Water in Auger Holes," *Soil Science Society of America Proceedings* 35: 844–45, 1971.

(24) Willardson, L., Meek, B., Huber, M., "A Flow Path Ground Water

Sampler," *Soil Science Society of America Proceedings* 36: 965–66, 1972.

(25) Wilson, L., *Monitoring in the Vadose Zone: A Review of Technical Elements and Methods*, U.S. Environmental Protection Agency, EPA-60017-80-134, 1980, p. 180.

(26) *Ground Water and Wells*, Johnson, E. E., Inc., St. Paul, MN, 1980, p. 440.

(27) Keck, W. G. and Associates, *New "Keck" Submersible Water Sampling Pump for Groundwater Monitoring*, Keck, W. G. and Associates, East Lansing, MI, 1981.

(28) McMillion, L., and Keeley, J. W., "Sampling Equipment for Ground-Water Investigation," *Ground Water*, Vol 6, 1968, pp. 9–11.

(29) Industrial and Environmental Analysts, Inc., *Procedures and Equipment for Groundwater Monitoring*, Industrial and Environmental Analysts, Inc., Essex Junction, VT, 1981.

(30) Trescott, P., and Pinder, G., "Air Pump for Small-Diameter Piezometers," *Ground Water*, Vol 8, 1970, pp. 10–15.

(31) Sommerfeldt, T., and Campbell, D., "A Pneumatic System to Pump Water From Piezometers," *Ground Water*, Vol 13, p. 293.

(32) Smith, A., "Water Sampling Made Easier with New Device," *The Johnson Drillers Journal*, July–August 1976, pp. 1–2.

(33) Morrison, R., and Ross, D., "Monitoring for Groundwater Contamination at Hazardous Waste Disposal Sites," *Proceedings of 1978 National Conference on Control of Hazardous Material Spills*, April 13, Miami Beach, FL, 1968, pp. 281–286.

(34) Morrison, R., and Brewer, P. "Air-Lift Samplers for Zone-of-Saturation Monitoring," *Ground Water Monitoring Review*, Spring 1981, pp. 52–54.

(35) Morrison, R., and Timmons, R., "Groundwater Monitoring II," *Groundwater Digest*, Vol 4, 1981, pp. 21–24.

(36) Bianchi, W. C., Johnson, C., Haskell, E., "A Positive Action Pump for Sampling Small Bore Holes," *Soil Science Society of America Proceedings*, Vol 26, 1961, pp. 86–87.

(37) Timmons, R., Discussion of "An All-Teflon Bailer and An Air-Driven Pump for Evacuating Small-Diameter Ground-Water Wells" by D. Buss and K. Bandt, *Ground Water*, Vol 19, 1981, pp. 666–667.

(38) Timco Manufacturing Co., Inc., "Gas Lift Teflon Pump," *Timco Geotechnical Catalogue*, Prairie du Sac, WI, 1982.

(39) Tomson, M., King, K., Ward, C., "A Nitrogen Powered Continuous Delivery, All Glass Teflon Pumping System for Groundwater Sampling from Below 10 Meters," *Ground Water*, Vol 18, 1980, pp. 444–446.

(40) *Pneumatic Water Sampler*, Slope Indicator Co., Seattle, WA, 1982.

(41) Petur Instrument Co., Inc., *Petur Liquid Sampler*, Petur Instrument Co., Inc., Seattle, WA, 1982.

(42) Idler, G., "Modification of an Electronic Downhole Water Sampler," *Ground Water*, Vol 18, 1980, pp. 532–535.

(43) *Remote Sampler Model 200*, Markland Specialty Engineering, Ltd., Etobicoke, Ontario, Bulletin 200/78, 1978.

(44) Middleburg, R., "Methods for Sampling Small Diameter Wells for Chemical Quality Analysis," *Presented at the National Conference on Quality Assurance of Environmental Measurements*, Nov. 27–29, Denver, CO, 1978.

(45) *Air Squeeze Pump*, Leonard Mold and Die Works, Denver, CO, 1982.

(46) *Automatic Sampler Controller: Markland Model 105 and 2105*, Markland Specialty Engineering, Ltd., Etobicoke, Ontario, Bulletin 105/78, 1981.

(47) Gillham, R. W., and Johnson, P. E., "A Positive Displacement Ground-Water Sampling Device," *Ground Water Monitoring Review*, Vol 1 (2), Summer 1981, p. 33.

(48) Signor, D., "Gas-Driven Pump for Ground-Water Samples," *U.S. Geological Survey, Water Resources Investigation 78–72*, Open File Report, 1978.

(49) *Tigre Tierra HX Pneumatic Packer*, Tigre Tierra, Inc., Puyallup, WA, 1981.

(50) Cherry, R., "A Portable Sampler for Collecting Water Samples from Specific Zones in Uncased or Screened Wells," *U.S. Geological Survey*, Prof. Paper 25-C, 1965, pp. 214–216.

(51) Grisak, G., Merritt, W., Williams, D., "Fluoride Borehole Dilution Apparatus for Groundwater Velocity Measurements," *Canadian Geotechnical Journal*, Vol 14, 1977, pp. 554–561.

(52) Galgowski, C., Wright, W., "A Variable-Depth Ground-Water Sampler," *Soil Science Society of America Proceedings*, Vol 44, 1980, pp. 1120–1121.

(53) *Samplers and Sampling Procedures for Hazardous Waste Streams*, USEPA MERL Laboratory, Cincinnati, OH, EPA-600/2-80-018, January 1980.

(54) *Methods for Chemical Analysis of Water and Wastes*, EPA-600/4-79-020, USEPA EMSL Laboratory, Cincinnati, OH, March 1979.

(55) *Federal Register*, Vol 44, No. 244, Dec. 18, 1979, pp. 75050–75052.

(56) *Standard Methods for the Examination of Water and Wastewater*, APAA, 14th ed., Washington, DC, 1976, pp. 38–45.

(57) *Handbook for Analytical Quality Control in Water and Wastewater Laboratories*, EPA-600/4-79-019, USEPA EMSL Laboratory, Cincinnati, OH, March 1979.

(58) U.S. Department of Interior, "Groundwater," Chapter II, *National Handbook of Recommended Methods for Water Data Acquisition*, 1980.

Designation: D 4630 – 86 (Reapproved 1991)[ε1]

Standard Test Method for Determining Transmissivity and Storativity of Low-Permeability Rocks by In Situ Measurements Using the Constant Head Injection Test[1]

This standard is issued under the fixed designation D 4630; the number immediately following the designation indicates the year of original adoption or, in the case of revision, the year of last revision. A number in parentheses indicates the year of last reapproval. A superscript epsilon (ε) indicates an editorial change since the last revision or reapproval.

ε1 NOTE—Section 10 was added editorially in May 1991.

1. Scope

1.1 This test method covers a field procedure for determining the transmissivity and storativity of geological formations having permeabilities lower than 10^{-3} μm^2 (1 millidarcy) using constant head injection.

1.2 The transmissivity and storativity values determined by this test method provide a good approximation of the capacity of the zone of interest to transmit water, if the test intervals are representative of the entire zone and the surrounding rock is fully water-saturated.

1.3 The values stated in SI units are to be regarded as the standard.

1.4 *This standard does not purport to address all of the safety problems, if any, associated with its use. It is the responsibility of the user of this standard to establish appropriate safety and health practices and determine the applicability of regulatory limitations prior to use.*

2. Terminology

2.1 *Descriptions of Terms Specific to This Standard:*

2.1.1 *transmissivity, T*—the transmissivity of a formation of thickness, *b*, is defined as follows:

$$T = K \cdot b$$

where:
K = equivalent formation hydraulic conductivity (efhc).

The efhc is the hydraulic conductivity of a material if it were homogeneous and porous over the entire interval. The hydraulic conductivity, *K*, is related to the equivalent formation permeability, *k*, as follows:

$$K = k\rho g/\mu$$

where:
ρ = fluid density,
μ = fluid viscosity, and
g = acceleration due to gravity.

2.1.2 *storativity, S*—the storativity (or storage coefficient) of a formation of thickness, *b*, is defined as follows:

$$S = S_s \cdot b$$

where:
S_s = equivalent bulk rock specific storage (ebrss).

The ebrss is the specific storage of a material if it were homogeneous and porous over the entire interval. The specific storage is given as follows:

$$S_s = \rho g\,(C_b + nC_w)$$

where:
C_b = bulk rock compressibility,
C_w = fluid compressibility, and
n = formation porosity.

2.2 *Symbols:*

2.2.1 C_b—bulk rock compressibility ($M^{-1}LT^2$).

2.2.2 C_w—compressibility of water ($M^{-1}LT^2$).

2.2.3 G—dimensionless function.

2.2.4 K—hydraulic conductivity (LT^{-1}).

2.2.5 P—excess test hole pressure ($ML^{-1}T^{-2}$).

2.2.6 Q—excess water flow rate (L^3T^{-1}).

2.2.7 Q_o—maximum excess water flow rate (L^3T^{-1}).

2.2.8 S—storativity (or storage coefficient) (dimensionless).

2.2.9 S_s—specific storage (L^{-1}).

2.2.10 T—transmissivity (L^2T^{-1}).

2.2.11 b—formation thickness (L).

2.2.12 e—fracture aperture (L).

2.2.13 g—acceleration due to gravity (LT^{-2}).

2.2.14 k—permeability (L^2).

2.2.15 n—porosity (dimensionless).

2.2.16 r_w—radius of test hole (L).

2.2.17 t—time elapsed from start of test (T).

2.2.18 α—dimensionless parameter.

2.2.19 μ—viscosity of water ($ML^{-1}T^{-1}$).

2.2.20 ρ—density of water (ML^{-3}).

3. Summary of Test Method

3.1 A borehole is first drilled into the rock mass, intersecting the geological formations for which the transmissivity and storativity are desired. The borehole is cored through potential zones of interest, and is later subjected to geophysical borehole logging over these intervals. During the test, each interval of interest is packed off at top and bottom with inflatable rubber packers attached to high-pressure steel tubing.

3.2 The test itself involves rapidly applying a constant pressure to the water in the packed-off interval and tubing string, and recording the resulting changes in water flow rate.

[1] This test method is under the jurisdiction of ASTM Committee D-18 on Soil and Rock and is the direct responsibility of Subcommittee D18.12 on Rock Mechanics.

Current edition approved Oct. 31, 1986. Published December 1986.

The water flow rate is measured by one of a series of flow meters of different sensitivities located at the surface. The initial transient water flow rate is dependent on the transmissivity and storativity of the rock surrounding the test interval and on the volume of water contained in the packed-off interval and tubing string.

4. Significance and Use

4.1 *Test Method*—The constant pressure injection test method is used to determine the transmissivity and storativity of low-permeability formations surrounding packed-off intervals. Advantages of the method are: (*a*) it avoids the effect of well-bore storage, (*b*) it may be employed over a wide range of rock mass permeabilities, and (*c*) it is considerably shorter in duration than the conventional pump and slug tests used in more permeable rocks.

4.2 *Analysis*—The transient water flow rate data obtained using the suggested test method are evaluated by the curve-matching technique described by Jacob and Lohman **(1)**[2] and extended to analysis of single fractures by Doe *et al.* **(2)**. If the water flow rate attains steady state, it may be used to calculate the transmissivity of the test interval **(3)**.

4.3 *Units:*

4.3.1 *Conversions*—The permeability of a formation is often expressed in terms of the unit darcy. A porous medium has a permeability of 1 darcy when a fluid of viscosity 1 cp (1 mPa·s) flows through it at a rate of 1 cm^3/s (10−6 m^3/s)/1 cm^2 (10−4 m^2) cross-sectional area at a pressure differential of 1 atm (101.4 kPa)/1 cm (10 mm) of length. One darcy corresponds to 0.987 μm^2. For water as the flowing fluid at 20°C, a hydraulic conductivity of 9.66 µm/s corresponds to a permeability of 1 darcy.

4.3.2 *Viscosity of Water*—Table 1 shows the viscosity of water as a function of temperature.

5. Apparatus

NOTE—A schematic of the test equipment is shown in Fig. 1.

5.1 *Source of Constant Pressure*—A pump or pressure intensifier shall be capable of providing an additional amount of water to the water-filled tubing string and packed-off test interval to produce a constant pressure of up to 1 MPA (145 psi) in magnitude, preferably with a rise time of less than 1 % of one half of the flow rate decay (Q/Q_o = 0.5).

5.2 *Packers*—Hydraulically actuated packers are recommended because they produce a positive seal on the borehole wall and because of the low compressibility of water they are also comparatively rigid. Each packer shall seal a portion of the borehole wall at least 0.5 m in length, with an applied pressure at least equal to the excess constant pressure to be applied to the packed-off interval and less than the formation fracture pressure at that depth.

5.3 *Pressure Transducers*—The pressure shall be measured as a function of time, with the transducer located in the packed-off test interval. The pressure transducer shall have an accuracy of at least ±3 kPa (±0.4 psi), including errors introduced by the recording system, and a resolution of at least 1 kPa (0.15 psi).

5.4 *Flow Meters*—Suitable flow meters shall be provided for measuring water flow rates in the range from 10^3 cm^3/s to 10^{-3} cm^3/s. Commercially available flow meters are capable

TABLE 1 Viscosity of Water as a Function of Temperature

Temperature, °C	Absolute Viscosity, mPa · s
0	1.79
2	1.67
4	1.57
6	1.47
8	1.39
10	1.31
12	1.24
14	1.17
16	1.11
18	1.06
20	1.00
22	0.96
24	0.91
26	0.87
28	0.84
30	0.80
32	0.77
34	0.74
36	0.71
38	0.68
40	0.66

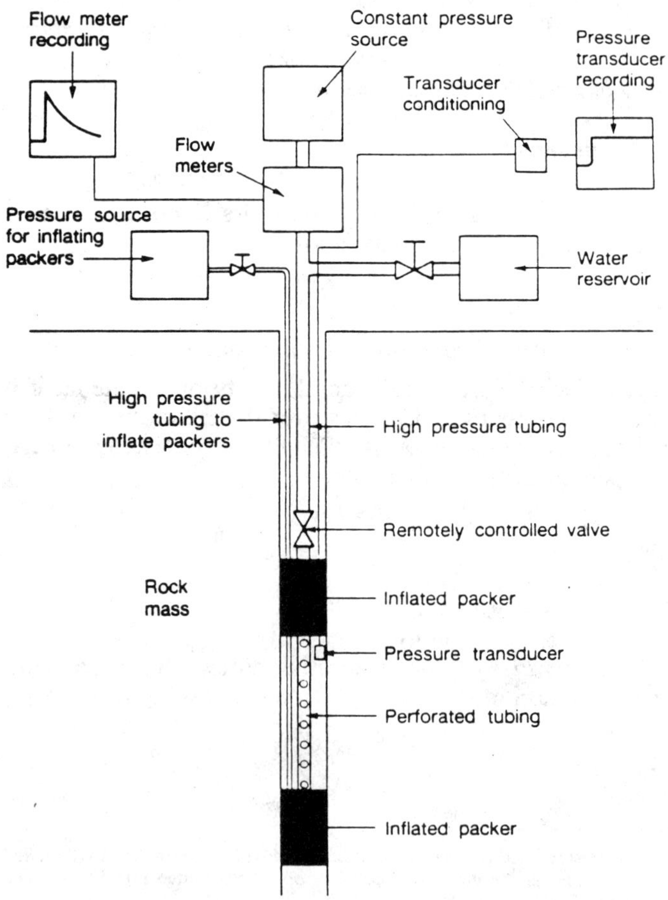

FIG. 1 Equipment Schematic

[2] The **boldface** numbers in parentheses refer to the list of references at the end of this standard.

of measuring flow rates as low as 10^2 cm^3/s with an accuracy of ±1 % and with a resolution of 10^{-5} cm^3/s; these can test permeabilities to 10^{-3} md based on a 10-m packer spacing. Positive displacement flow meters of either the tank type (Haimson and Doe **(4)** or bubble-type (Wilson *et al.* **(3)** are capable of measuring flow rates as low as 10^{-3} cm^3/s; these can test permeabilities to 10^{-4} md based on a 10-m packer spacing.

5.5 *Hydraulic Systems*—The inflatable rubber packers shall be attached to high-pressure steel tubing reaching to the surface. The packers themselves shall be inflated with water using a separate hydraulic system. The pump or pressure intensifier providing the constant pressure shall be attached to the steel tubing at the surface. A remotely controlled down-hole valve, located in the steel tubing immediately above the upper packer, shall be used for shutting in the test interval and for instantaneous starting of tests.

6. Procedure

6.1 *Drilling Test Holes:*

6.1.1 *Number and Orientation*—The number of test holes shall be sufficient to supply the detail required by the scope of the project. The test holes shall be directed to intersect major fracture sets, preferably at right angles.

6.1.2 *Test Hole Quality*—The drilling procedure shall provide a borehole sufficiently smooth for packer seating, shall contain no rapid changes in direction, and shall minimize formation damage.

6.1.3 *Test Holes Cored*—Core the test holes through zones of potential interest to provide information for locating test intervals.

6.1.4 *Core Description*—Describe the rock core from the test holes with particular emphasis on the lithology and natural discontinuities.

6.1.5 *Geophysical Borehole Logging*—Log geophysically the zones of potential interest. In particular, run electrical-induction and gamma-gamma density logs. Whenever possible, also use sonic logs and the acoustic televiewer. Run other logs as required.

6.1.6 *Washing Test Holes*—The test holes must not contain any material that could be washed into the permeable zones during testing, thereby changing the transmissivity and storativity. Flush the test holes with clean water until the return is free from cuttings and other dispersed solids.

6.2 *Test Intervals:*

6.2.1 *Selection of Test Intervals*—Determine test intervals from the core descriptions, geophysical borehole logs, and, if necessary, from visual inspection of the borehole with a borescope or TV camera.

6.2.2 *Changes in Lithology*—Test each major change in lithology that can be isolated between packers.

6.2.3 *Sampling Discontinuities*—Discontinuities are often the major permeable features in hard rock. Test jointed zones, fault zones, bedding planes, and the like, both by isolating individual features and by evaluating the combined effects of several features.

6.2.4 *Redundancy of Tests*—To evaluate variability in transmissivity and storativity, conduct three or more tests in each rock type, if homogeneous. If the rock is not homogeneous, the sets of tests should encompass similar types of discontinuities.

6.3 *Test Water:*

6.3.1 *Quality*—Water used for pressure pulse tests shall be clean, and compatible with the formation. Even small amounts of dispersed solids in the injection water could plug the rock face of the test interval and result in a measured transmissivity value that is erroneously low.

6.3.2 *Temperature*—The lower limit of the test water temperature shall be 5°C below that of the rock mass to be tested. Cold water injected into a warm rock mass causes air to come out of solution, and the resulting bubbles will radically modify the pressure transient characteristics.

6.4 *Testing:*

6.4.1 *Filling and Purging System*—Once the packers have been set, slowly fill the tubing string and packed-off interval with water to ensure that no air bubbles will be trapped in the test interval and tubing. Close the downhole valve to shut in the test interval, and allow the test section pressures (as determined from downhole pressure transducer reading) to dissipate.

6.4.2 *Constant Pressure Test*—Pressurize the tubing, typically to between 300 and 600 kPa (50 to 100 psi) above the shut-in pressure. This range of pressures is in most cases sufficiently low to minimize distortion of fractures adjacent at the test hole, but in no case should the pressure exceed the minimum principal ground stress. It is necessary to provide sufficient volume of pressurized water to maintain constant pressure during testing. Open the down-hole valve, maintain the constant pressure, and record the water flow rate as a function of time. Then close the down-hole valve and repeat the test for a higher value of constant test pressure. A typical record is shown in Fig. 2.

7. Calculation and Interpretation of Test Data

7.1 The solution of the differential equation for unsteady state flow from a borehole under constant pressure located in an extensive aquifer is given by Jacob and Lohman **(1)** as:[3]

$$Q = 2\pi TP\, G(\alpha)/\rho g, \tag{1}$$

where:

Q = water flow rate,
T = transmissivity of the test interval,
P = excess test hole pressure,
ρ = water density,
g = acceleration due to gravity, and
$G(\alpha)$ = function of the dimensionless parameter α:

$$\alpha = Tt/Sr_w^2 \tag{2}$$

where:

t = time elapsed from start of test,
S = storativity, and
r_w = radius of the borehole over the test interval.

7.1.1 In Fig. 2, the flow rate in the shut-in, packed-off interval is considered constant. In those cases where the response of the shut-in interval is time dependent, interpretation of the constant pressure test is unaffected, provided the time dependency is linear.

7.2 To determine the transmissivity, *T*, and storativity, *S*,

[3] For bounded aquifers the reader is referred to Hantush **(5)**.

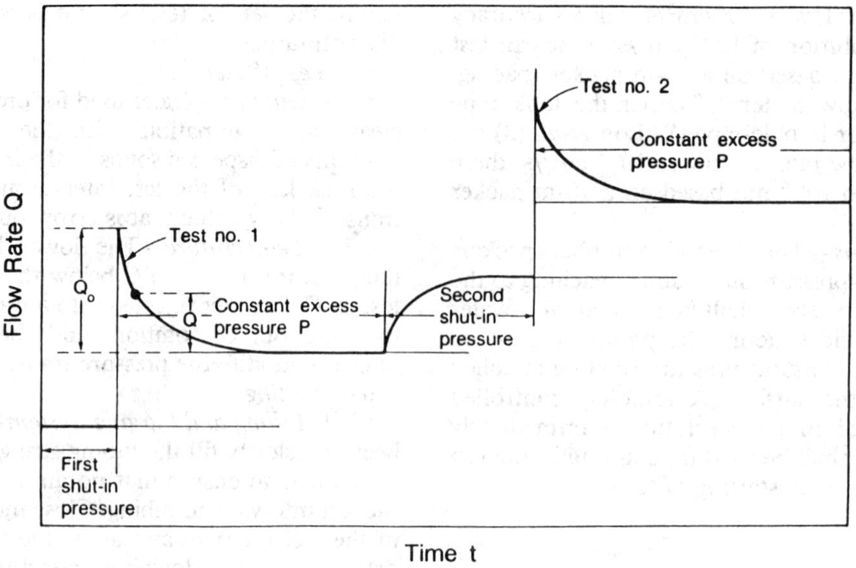

FIG. 2 Typical Flow Rate Record

data on the water flow rate at constant pressure as a function of time are plotted in the following manner **(1)**.

7.2.1 First, plot a type curve on logarithmic paper of the function $G(\alpha)$ versus α where values of $G(\alpha)$ are given in Table 2.

7.2.2 Second, on transparent logarithmic paper to the same scale, plot values of the flow rate, Q, versus values of time, t.

7.2.3 Then, by placing the experimental data over the theoretical curve, the best fit of the data to the curve can be made.

7.2.4 Determine the values of transmissivity, T, and storativity, S, using Eqs. 1 and 2 from the coordinates of any point in both coordinate systems.

8. Report

8.1 The report shall include the following:

8.1.1 *Introduction*—The introductory section is intended to present the scope and purpose of the constant pressure test program, and the characteristics of rock mass tested.

8.1.1.1 *Scope of Testing Program:*

8.1.1.1.1 Report the location and orientation of the boreholes and test intervals. For tests in many boreholes or in a variety of rock types, present the matrix in tabular form.

8.1.1.1.2 Rationale for test location selection, including the reasons for the number, location, and size of test intervals.

8.1.1.1.3 Discuss in general terms limitations of the testing program, stating the areas of interest which are not covered by the testing program and the limitations of the data within the areas of application.

8.1.1.2 *Brief Description of the Test Intervals*—Describe rock type, structure, fabric, grain or crystal size, discontinuities, voids, and weathering of the rock mass in the test

TABLE 2 Values of $G(\alpha)$ for Values of α Between 10^{-4} and 10^{12} [A]

	10^{-4}	10^{-3}	10^{-2}	10^{-1}	1	10	10^{2}	10^{3}
1	56.9	18.34	6.13	2.249	0.985	0.534	0.346	0.251
2	40.4	13.11	4.47	1.716	0.803	0.461	0.311	0.232
3	33.1	10.79	3.74	1.477	0.719	0.427	0.294	0.222
4	28.7	9.41	3.30	1.333	0.667	0.405	0.283	0.215
5	25.7	8.47	3.00	1.234	0.630	0.389	0.274	0.210
6	23.5	7.77	2.78	1.160	0.602	0.377	0.268	0.206
7	21.8	7.23	2.60	1.103	0.580	0.367	0.263	0.203
8	20.4	6.79	2.46	1.057	0.562	0.359	0.258	0.200
9	19.3	6.43	2.35	1.018	0.547	0.352	0.254	0.198
10	18.3	6.13	2.25	0.985	0.534	0.346	0.251	0.196
	10^{4}	10^{5}	10^{6}	10^{7}	10^{8}	10^{9}	10^{10}	10^{11}
1	0.1964	0.1608	0.1360	0.1177	0.1037	0.0927	0.0838	0.0764
2	0.1841	0.1524	0.1299	0.1131	0.1002	0.0899	0.0814	0.0744
3	0.1777	0.1479	0.1266	0.1106	0.0982	0.0883	0.0801	0.0733
4	0.1733	0.1449	0.1244	0.1089	0.0968	0.0872	0.0792	0.0726
5	0.1701	0.1426	0.1227	0.1076	0.0958	0.0864	0.0785	0.0720
6	0.1675	0.1408	0.1213	0.1066	0.0950	0.0857	0.0779	0.0716
7	0.1654	0.1393	0.1202	0.1057	0.0943	0.0851	0.0774	0.0712
8	0.1636	0.1380	0.1192	0.1049	0.0937	0.0846	0.0770	0.0709
9	0.1621	0.1369	0.1184	0.1043	0.0932	0.0842	0.0767	0.0706
10	0.1608	0.1360	0.1177	0.1037	0.0927	0.0838	0.0764	0.0704

[A] From Jacob and Lohman **(1)**.

Data Sheet

Project ______________________

Test Location ______________________

Rock Type ______________________

Date ______________________

Testing by ______________________

Test No. ______________________

Borehole No. ______________________

Borehole Dip and Dip Direction ______________________

Measured Depth of Test to Top Packer, m ______________________

Borehole Diameter, mm ______________________

Rock Temperature, °C ______________________

Equipment Description	Serial No.	Date of last Calibration

Length of Packed-off Interval, m ______________________

Length of Tubing Above Top Packer, m ______________________

Water Temperature, °C ______________________

Packer Pressure, kPa ______________________

Tubing ID, mm ______________________

Shut-in Borehole Pressure, kPa ______________________

Constant Water Pressure, kPa ______________________

FIG. 3 Data Sheet for In Situ Measurement of Transmissivity and Storativity Using the Constant Head Injection Test

intervals. A more detailed description may be needed for certain applications. In a heterogeneous rock mass or for several rock types, many intervals may be described; a tabular presentation is then recommended for clarity.

8.1.2 *Test Method:*

8.1.2.1 *Equipment and Apparatus*—Include a list of the equipment used for the test, the manufacturer's name, model number, and basic specifications for each major item, and the date of last calibration, if applicable.

8.1.2.2 *Procedure*—State the steps actually followed in the procedure for the test.

8.1.2.3 *Variations*—If the actual equipment or procedure deviates from this test method, note each variation and the reasons. Discuss the effects of any deviations upon the test results.

8.1.3 *Theoretical Background:*

8.1.3.1 *Data Reduction Equations*—Clearly present and fully define all equations and type curves used to reduce the data. Note any assumptions inherent in the equations and type curves and any limitations in their applications and discuss their effects on the results.

8.1.3.2 *Site Specific Influences*—Discuss the degree to which the assumptions contained in the data reduction equations pertain to the actual test location and fully explain any factors or methods applied to the data to correct for departures from the assumptions of the data reduction equations.

8.1.4 *Results:*

8.1.4.1 *Summary Table*—Present a table of results, including the types of rock and discontinuities, the average values of the transmissivity and storativity, and their ranges and uncertainties.

8.1.4.2 *Individual Results*—Present a table of results for individual tests, including test number, interval length, rock types, value of constant pressure transmissivity and storativity, and flow rate as a function of time.

8.1.4.3 *Graphic Data*—Present water flow rate versus time curves for each test, together with the appropriate type curves used for their interpretation.

8.1.4.4 *Other*—Other analyses or presentations may be included as appropriate, for example: (*a*) discussion of the characteristic of the permeable zones, (*b*) histograms of results, and (*c*) comparison of results to other studies or previous work.

8.1.5 *Appended Data*—Include in an appendix a completed data form (Fig. 3) for each test.

9. Precision and Bias

9.1 *Error Estimate:*

9.1.1 Analyze the results using standard statistical methods. Calculate all uncertainties using a 95 % confidence interval.

9.1.2 *Measurement Error*—Evaluate the errors in transmissivity and storativity associated with a single test. This includes the combined effects of flow rate determination, measurement of time, and type curve matching.

9.1.3 *Sample Variability*—For each rock or discontinuity type, calculate, as a minimum, the mean transmissivity and storativity and their ranges, standard deviations, and 95 % confidence limits for the means. Compare the uncertainty associated with the transmissivity and storativity for each rock type with the measurement uncertainty to determine whether measurement error or sample variability is the dominant factor in the results.

10. Keywords

10.1 borehole; constant head testing; flow; in situ; faultzones; field testing; flow and flow rate; permeability; pressure testing; rock; saturation; storativity; transmissivity; viscosity; water; water saturation

REFERENCES

(1) Jacob, C. E., and Lohman, S. W., "Non-Steady Flow to a Well of Constant Drawdown in an Extensive Aquifer," Trans. American Geophys. Union, Vol 33, 1952, pp. 559–569.

(2) Doe, T. W., Long, J. C. S., Endo, H. K., and Wilson, C. R., "Approaches to Evaluating the Permeability and Porosity of Fractured Rock Masses," *Proceedings of the 23rd U.S. Symposium on Rock Mechanics,* Berkeley, 1982, pp. 30–38.

(3) Wilson, C. R., Doe, T. W., Long, J. C. S., and Witherspoon, P. A., "Permeability Characterization of Nearly Impermeable Rock Masses for Nuclear Waste Repository Siting," *Proceedings,* Workshop on Low Flow, Low Permeability Measurements in Largely Impermeable Rocks, OECD, Paris, 1979, pp. 13–30.

(4) Haimson, B. C., and Doe, T. W., "State of Stress, Permeability, and Fractures in the Precambrian Granite of Northern Illinois," *Journal of Geophysics Research*, Vol 88, 1983, pp. 7355–7371.

(5) Hantush, M. S., "Non-Steady Flow to Flowing Wells in Leaky Aquifers," *Journal of Geophysics Research*, Vol 64, 1959, pp. 1043–1052.

(6) Zeigler, T., "Determination of Rock Mass Permeability," Tech. Rep. S-76-2, U.S. Army Eng. Waterways Exp. Stn., Vicksburg, MI, 1976, 85 pp.

(7) Earlougher, R. C., "Advances in Well Test Analysis," *Society of Petroleum Engineers of A.I.M.E.,* New York, NY 1977.

(8) Freeze, R. A., and Cherry, J. A., *Groundwater,* Prentice-Hall, Englewood Cliffs, NJ, 1979.

(9) Shuri, F. S., Feves, M. L., Peterson, G. L., Foster, K. M., and Kienle, C. F., Public Draft: "Field and *In Situ* Rock Mechanics Testing Manual," Office of Nuclear Waster Isolation, Document ONWI-310, Section F: "*In Situ* Fluid Properties," GT-F.1 *In Situ* Permeability Measurement of Rock Using Borehole Packers, 1981.

Standard Test Method for Determining Transmissivity and Storativity of Low Permeability Rocks by In Situ Measurements Using the Pressure Pulse Technique[1]

This standard is issued under the fixed designation D 4631; the number immediately following the designation indicates the year of original adoption or, in the case of revision, the year of last revision. A number in parentheses indicates the year of last reapproval. A superscript epsilon (ε) indicates an editorial change since the last revision or reapproval.

[ε1] NOTE—Section 10 was added editorially in May 1991.

1. Scope

1.1 This test method covers a field procedure for determining the transmissivity and storativity of geological formations having permeabilities lower than 10^{-3} μm^2 (1 millidarcy) using the pressure pulse technique.

1.2 The transmissivity and storativity values determined by this test method provide a good approximation of the capacity of the zone of interest to transmit water, if the test intervals are representative of the entire zone and the surrounding rock is fully water saturated.

1.3 The values stated in SI units are to be regarded as the standard.

1.4 *This standard does not purport to address all of the safety problems, if any, associated with its use. It is the responsibility of the user of this standard to establish appropriate safety and health practices and determine the applicability of regulatory limitations prior to use.*

2. Terminology

2.1 *Descriptions of Terms Specific to This Standard:*

2.1.1 *transmissivity, T*—the transmissivity of a formation of thickness, *b*, is defined as follows:

$$T = K \cdot b$$

where:

K = equivalent formation hydraulic conductivity (efhc).

The efhc is the hydraulic conductivity of a material if it were homogeneous and porous over the entire interval. The hydraulic conductivity, *K*, is related to the equivalent formation, *k*, as follows:

$$K = k\rho g/\mu$$

where:

ρ = fluid density,
μ = fluid viscosity, and
g = acceleration due to gravity.

2.1.2 *storativity, S*—the storativity (or storage coefficient) of a formation of thickness, *b*, is defined as follows:

$$S = S_s \cdot b$$

where:

S_s = equivalent bulk rock specific storage (ebrss).

The ebrss is defined as the specific storage of a material if it were homogeneous and porous over the entire interval. The specific storage is given as follows:

$$S_s = \rho g\,(C_b + nC_w)$$

where:

C_b = bulk rock compressibility,
C_w = fluid compressibility, and
n = formation porosity.

2.2 *Symbols:*

2.2.1 C_b—bulk rock compressibility ($M^{-1}LT^2$).
2.2.2 C_w—compressibility of water ($M^{-1}LT^2$).
2.2.3 K—hydraulic conductivity (LT^{-1}).
2.2.4 L—length of packed-off zone (L).
2.2.5 P—excess test hole pressure ($ML^{-1}T^{-2}$).
2.2.6 P_o—initial pressure pulse ($ML^{-1}T^{-2}$).
2.2.7 S—storativity (or storage coefficient) (dimensionless).
2.2.8 S_s—specific storage (L^{-1}).
2.2.9 T—transmissivity (L^2T^{-1}).
2.2.10 V_w—volume of water pulsed (L^3).
2.2.11 b—formation thickness (L).
2.2.12 e—fracture aperture (L).
2.2.13 g—acceleration due to gravity (LT^{-2}).
2.2.14 k—permeability (L^2).
2.2.15 n—porosity (dimensionless).
2.2.16 r_w—radius of test hole (L).
2.2.17 t—time elapsed from pulse initiation (T).
2.2.18 α—dimensionless parameter.
2.2.19 β—dimensionless parameter.
2.2.20 μ—viscosity of water ($ML^{-1}T^{-1}$).
2.2.21 ρ—density of water (ML^{-3}).

3. Summary of Test Method

3.1 A borehole is first drilled into the rock mass, intersecting the geological formations for which the transmissivity and storativity are desired. The borehole is cored through potential zones of interest, and is later subjected to geophysical borehole logging over these intervals. During the test, each interval of interest is packed off at top and bottom with inflatable rubber packers attached to high-pressure steel tubing. After inflating the packers, the tubing string is completely filled with water.

3.2 The test itself involves applying a pressure pulse to the water in the packed-off interval and tubing string, and

[1] This test method is under the jurisdiction of ASTM Committee D-18 on Soil and Rock and is the direct responsibility of Subcommittee D18.12 on Rock Mechanics.

Current edition approved Oct. 31, 1986. Published December 1986.

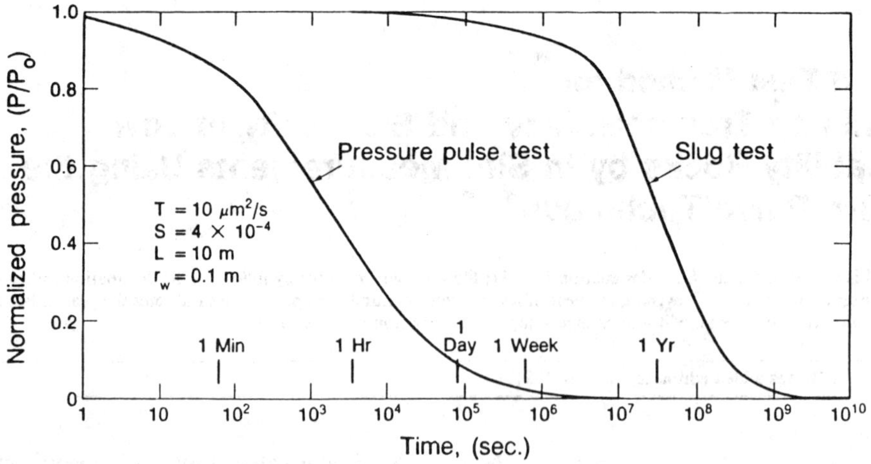

FIG. 1 Comparative Times for Pressure Pulse and Slug Tests

recording the resulting pressure transient. A pressure transducer, located either in the packed-off zone or in the tubing at the surface, measures the transient as a function of time. The decay characteristics of the pressure pulse are dependent on the transmissivity and storativity of the rock surrounding the interval being pulsed and on the volume of water being pulsed. Alternatively, under non-artesian conditions, the pulse test may be performed by releasing the pressure on a shut-in well, thereby subjecting the well to a negative pressure pulse. Interpretation of this test method is similar to that described for the positive pressure pulse.

4. Significance and Use

4.1 *Test Method*—The pulse test method is used to determine the transmissivity and storativity of low-permeability formations surrounding the packed-off intervals. The test method is considerably shorter in duration than the pump and slug tests used in more permeable rocks. To obtain results to the desired accuracy, pump and slug tests in low-permeability formations are too time consuming, as indicated in Fig. 1 (from Bredehoeft and Papadopulos (**1**)).[2]

4.2 *Analysis*—The transient pressure data obtained using the suggested method are evaluated by the curve-matching technique described by Bredehoeft and Papadopulos (**1**), or by an analytical technique proposed by Wang *et al* (**2**). The latter is particularly useful for interpreting pulse tests when only the early-time transient pressure decay data are available.

4.3 *Units:*

4.3.1 *Conversions*—The permeability of a formation is often expressed in terms of the unit darcy. A porous medium has a permeability of 1 darcy when a fluid of viscosity 1 cP (1 mPa·s) flows through it at a rate of 1 cm^3/s (10^{-6} m^3/s)/1 cm^2 (10^{-4} m^2) cross-sectional area at a pressure differential of 1 atm (101.4 kPa)/1 cm (10 mm) of length. One darcy corresponds to 0.987 μm^2. For water as the flowing fluid at 20°C, a hydraulic conductivity of 9.66 μm/s corresponds to a permeability of 1 darcy.

4.3.2 *Viscosity of Water*—Table 1 shows the viscosity of water as a function of temperature.

5. Apparatus

NOTE—A schematic of the test equipment is shown in Fig. 2.

5.1 *Source of Pressure Pulse*—A pump or pressure intensifier shall be capable of injecting an additional amount of water to the water-filled tubing string and packed-off test interval to produce a sharp pressure pulse of up to 1 MPa (145 psi) in magnitude, preferably with a rise time of less than 1 % of one half of the pressure decay ($P/P_o = 0.5$).

5.2 *Packers*—Hydraulically actuated packers are recommended because they produce a positive seal on the borehole wall and because of the low compressibility of water they are also comparatively rigid. Each packer shall seal a portion of the borehole wall at least 0.5 m in length, with an applied pressure at least equal to the excess maximum pulse pressure to be applied to the packed-off interval and less than the formation fracture pressure at that depth.

5.3 *Pressure Transducers*—The test pressure may be measured directly in the packed-off test interval or at the surface

TABLE 1 Viscosity of Water as a Function of Temperature

Temperature, °C	Absolute Viscosity, mPa · s
0	1.79
2	1.67
4	1.57
6	1.47
8	1.39
10	1.31
12	1.24
14	1.17
16	1.11
18	1.06
20	1.00
22	0.96
24	0.91
26	0.87
28	0.84
30	0.80
32	0.77
34	0.74
36	0.71
38	0.68
40	0.66

[2] The **boldface** numbers in parentheses refer to the list of references at the end of this standard.

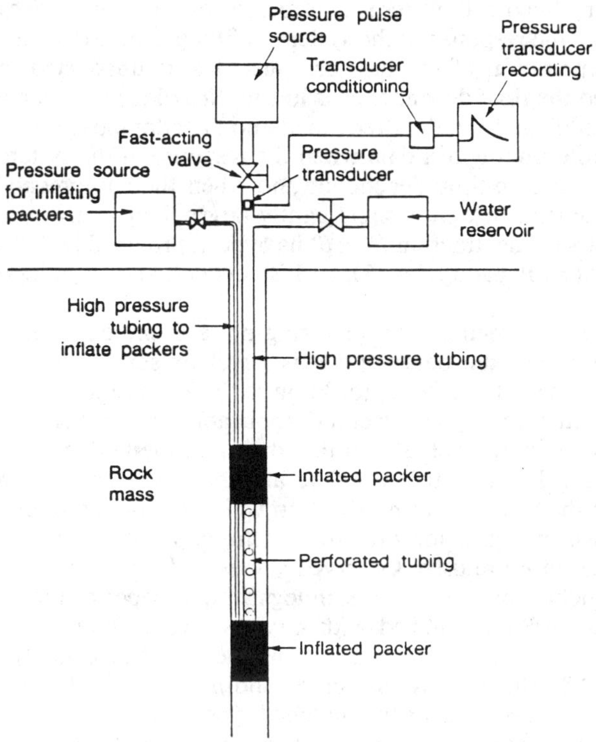

FIG. 2 Schematic of Test Equipment

with an electronic pressure transducer. In either case the pressure shall be recorded at the surface as a function of time. The pressure transducer shall have an accuracy of at least ±3 kPa (±0.4 psi), including errors introduced by the recording system, and a resolution of at least 1 kPa (0.15 psi).

5.4 *Hydraulic Systems*—The inflatable rubber packers shall be attached to high-pressure steel tubing reaching to the surface. The packers themselves shall be inflated with water using a separate hydraulic system. The pump or pressure intensifier providing the pressure pulse shall be attached to the steel tubing at the surface. If the pump is used, a fast-operating valve shall be mounted in the tubing at the surface near the pressure pump.

6. Procedure

6.1 *Drilling Test Holes:*

6.1.1 *Number and Orientation*—The number of test holes shall be sufficient to supply the detail required by the scope of the project. The test holes shall be directed to intersect major fracture sets, preferable at right angles.

6.1.2 *Test Hole Quality*—The drilling procedure shall provide a borehole sufficiently smooth for packer seating, shall contain no rapid changes in direction, and shall minimize formation damage.

6.1.3 *Test Holes Cored*—Core the test holes through zones of potential interest to provide information for locating test intervals.

6.1.4 *Core Description*—Describe the rock core from the test holes with particular emphasis on the lithology and natural discontinuities.

6.1.5 *Geophysical Borehole Logging*—Log geophysically the zones of potential interest. In particular, run electrical-induction and gamma-gamma density logs. Run other logs as required.

6.1.6 *Washing Test Holes*—The test holes must not contain any material that could be washed into the permeable zones during testing, thereby changing the transmissivity and storativity. Flush the test holes with clean water until the return is free from cuttings and other dispersed solids.

6.2 *Test Intervals:*

6.2.1 *Selection of Test Intervals*—Test intervals are determined from the core descriptions, geophysical borehole logs, and, if necessary, from visual inspection of the borehole with a borescope or TV camera.

6.2.2 *Changes in Lithology*—Test each major change in lithology that can be isolated between packers.

6.2.3 *Sampling Discontinuities*—Discontinuities are often the major permeable features in hard rock. Test jointed zones, fault zones, bedding planes, and the like, both by isolating individual features and by evaluating the combined effects of several features.

6.2.4 *Redundancy of Tests*—To evaluate variability in transmissivity and storativity, conduct several tests in each rock type, if homogeneous. If the rock is not homogeneous, each set of tests should encompass similar types of discontinuities.

6.3 *Test Water:*

6.3.1 *Quality*—Water used for pressure pulse tests shall be clean, and compatible with the formation. Even small amounts of dispersed solids in the injection water could plug the rock face of the test interval and result in a measured transmissivity value that is erroneously low.

6.3.2 *Temperature*—The lower limit of the test water temperature shall be 5°C below that of the rock mass to be tested. Cold water injected into a warm rock mass causes air to come out of solution, and the resulting bubbles will radically modify the pressure transient characteristics.

6.4 *Testing:*

6.4.1 *Filling and Purging System*—Allow sufficient time after washing the test hole for any induced formation pressures to dissipate. Once the packers have been set, slowly fill the tubing string and packed-off interval with water to ensure that no air bubbles will be trapped in the test interval and tubing.

6.4.2 *Pressure Pulse Test*—Rapidly pressurize the tubing, typically to between 300 and 600 kPa (50 to 100 psi), and then shut in. This range of pressures is in most cases sufficiently low to minimize distortion of fractures adjacent to the test hole, but in no case should the pressure exceed the minimum principal ground stress. Record the resulting pressure pulse and decay transient detected by the pressure transducer as a function of time. A typical record is shown in Fig. 3.

7. Calculation and Interpretation of Test Data

7.1 The type of matching technique developed by Bredehoeft and Papadopulos **(1)** involves plotting normalized pressure (the ratio of the excess borehole pressure, P, at a given time to the initial pressure pulse, P_o) against the logarithm of time, as indicated in Figs. 2 and 3. The pulse decay is given as follows:

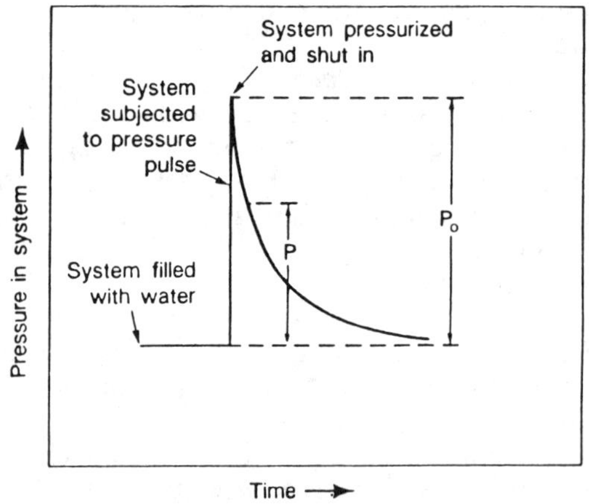

FIG. 3 Typical Pressure Record

$$\frac{P}{P_o} = F(\alpha,\beta) \tag{1}$$

where:
α and β = dimensionless parameters given by:

$$\alpha = \pi r_w^2 \, S/V_w C_w \rho g \tag{2}$$

and

$$\beta = \beta Tt/V_w C_w \rho g \tag{3}$$

where:
V_w = volume of water being pulsed,
r_w = well radius,
t = time elapsed from pulse initiation,
C_w = compressibility of water,
T = transmissivity, and
S = storativity.

7.1.1 In Fig. 3, the pressure of the packed-off interval and tubing filled with water is considered constant. In those cases where the response of the system to filling is time dependent, interpretation of the pressure pulse test is unaffected, provided the time dependency is linear. Neuzil **(3)** has proposed a modification to the Bredehoeft and Papadopulos procedure for interpreting test data when the response of the system to filling is non-time dependent and when the compressibility of the shut-in well is significantly larger than that of water. Tables of the function $F(\alpha, \beta)$ have been provided by Cooper *et al* **(4)**, Papadopulos **(5)**, and Bredehoeft and Papadopulos **(1)**.

7.2 The method for analyzing pulse decay data depends on whether the parameter α is larger or smaller than 0.1. Since the value of α is not known *a priori*, the test data are first analyzed by the method applicable to $\alpha < 0.1$. If this analysis indicates that $\alpha > 0.1$, then that method is used.

7.2.1 For $\alpha < 0.1$, the data are analyzed by the method described by Cooper *et al* **(4)**, in which the family of curves shown in Fig. 4 for $F(\alpha, \beta)$ as a function of β for various values of α are used. Observed values of P/P_o are plotted as a function of time, t, on semilogarithmic paper of the same scale, and are matched with a type curve by keeping the β and t axes coincident and moving the plots horizontally.

7.2.2 The expressions corresponding to α and β in Eqs 1 and 2, the α value of the matched type curve, and the β and t values from a match point are used to determine the transmissivity, T, and the storage coefficient, S, of the tested interval. Bredehoeft and Papadopulos **(1)** indicate that this procedure yields good estimates of the transmissivity when $\alpha \leq 0.1$, but that the storage coefficient could be of questionable reliability for values of $\alpha < 10^{-5}$.

7.2.3 For $\alpha > 0.1$, Bredehoeft and Papadopulos **(1)** recommend the use of the family of curves shown in Fig. 5 for $F(\alpha, \beta)$ as a function of the product $\alpha\beta = \left(\frac{\pi^2 r_w^2 TSt}{(V_w C_w \rho g)^2}\right)$ to

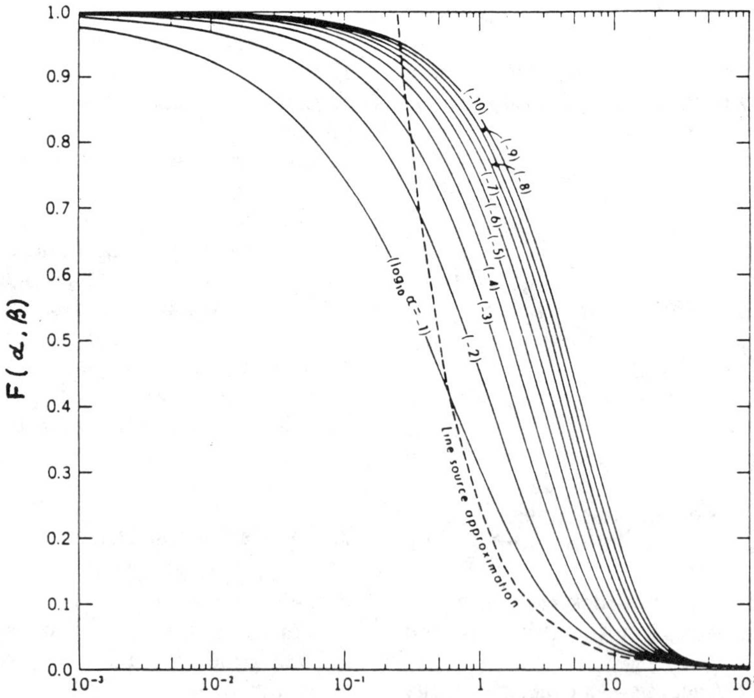

FIG. 4 Type Curves of the Function $F(\alpha, \beta)$ Against the Parameter β for Different Values of α

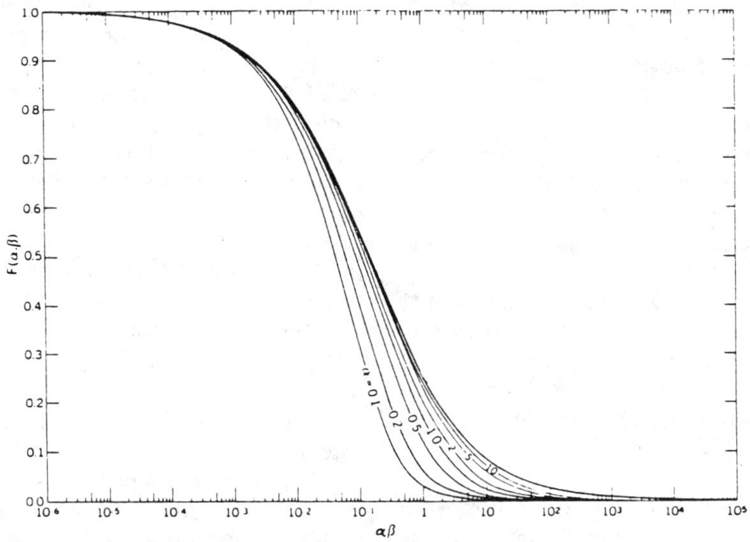

FIG. 5 Type Curves of the Function $F(\alpha, \beta)$ Against the Product Parameter $\alpha\beta$

interpret the data. Matching of the observed values of P/P_o plotted as a function of t with a type curve is performed in the same manner as indicated previously for $\alpha \leq 0.1$. In this way, the product TS and S are determined. Analysis with the type curves shown in Fig. 5 provides an indication as to whether the data are adequate for identifying both α and β and hence, determining both S and T, or whether the data fall in the range where only the product TS can be determined.

7.3 Wang *et al* **(2)** present an alternative method of analyzing pressure pulse data involving analytical solutions for pulse decay in single fractures of both infinite and finite extent. Recognizing that finite fracture geometry introduces errors in the interpretation of the pulse decay data, Wang suggests a method that uses data from elapsed times before the fracture boundaries begin to influence the pressure data. Wang found by linear regression of calculated decay pressure versus time an empirical expression for the fracture aperture of the following form:

$$\begin{aligned} \log(e/10^6) &= -0.32 \log(t) + C \\ &+ 0.32\,[2 \log(r_w/0.04) \\ &+ \log(2.394\mu C_w \times 10^{12})] \\ &+ 0.333 \log(L/2). \end{aligned} \quad (4)$$

where:

e = parallel-plate equivalent aperture, m,
t = time, s,
r_w = borehole radius, m,
μ = water viscosity, mPa·s,
C_w = water compressibility, 1/Pa,
L = length of the packed-off interval, m, and
C = a constant that depends on the fraction of pulse decay, as follows:

Fraction of pulse decay, $(P_o - P)/P_o$	0.05	0.10	0.15
Wang constant, C:	1.09	1.20	1.27

7.3.1 Wang shows that in test zones containing two fractures of different apertures, the wider fracture dominates the early time behavior. The early pressure pulse decay therefore reflects the major fracture only. Doe, *et al* **(6)** view individual fractures as confined aquifers whose transmissivities are given by the cubic relationship:

$$T = \rho g e^3/12\mu \quad (5)$$

Thus, Eq 5 provides transmissivity in terms of a parallel-plate equivalent fracture aperture calculated from Eq 4.

7.3.2 Eqs 4 and 5 can be solved for the early-time pressure pulse decay data to provide a transmissivity value for the test interval from the calculated parallel-plate equivalent aperture.

8. Report

8.1 The report shall include the following:

8.1.1 *Introduction*—The introductory section is intended to present the scope and purpose of the pressure pulse test program, and the characteristics of the rock mass tested.

8.1.1.1 *Scope of Testing Program:*

8.1.1.1.1 Report the location and orientation of the boreholes and test intervals. For tests in many boreholes or in a variety of rock types, present the test matrix in tabular form.

8.1.1.1.2 Rationale for test location selection, including the reasons for the number, location, and size of test intervals.

8.1.1.1.3 Discuss in general terms the limitations of the testing program, stating the areas of interest which are not covered by the testing program and the limitations of the data within the areas of application.

8.1.1.2 *Brief Description of the Test Intervals*—Describe rock type, structure, fabric, grain or crystal size, discontinuities, voids, and weathering of the rock mass in the test intervals. A more detailed description may be needed for certain applications. In a heterogeneous rock mass or for several rock types, many intervals may be described; a tabular presentation is then recommended for clarity.

8.1.2 *Test Method:*

8.1.2.1 *Equipment and Apparatus*—Include a list of the equipment used for the test, the manufacturer's name, model number, and basic specifications for each major item, and the date of last calibration, if applicable.

8.1.2.2 *Procedure*—State the steps actually followed in the procedure for the test.

8.1.2.3 *Variations*—If the actual equipment or procedure deviates from this test method, note each variation and the reasons. Discuss the effects of the deviations upon the test results.

8.1.3 *Theoretical Background:*

8.1.3.1 *Data Reduction Equations*—Clearly present and fully define all equations and type curves used to reduce the data. Note any assumptions inherent in the equations and type curves and any limitations in their applications and discuss their effects on the results.

8.1.3.2 *Site Specific Influences*—Discuss the degree to which the assumptions contained in the data reduction equations pertain to the actual test location and fully explain any factors or methods applied to the data to correct for departures from the assumptions of the data reduction equations.

8.1.4 *Results:*

8.1.4.1 *Summary Table*—Present a table of results, including the types of rock and discontinuities, the average values of the transmissivity and storativity, and their ranges and uncertainties.

8.1.4.2 *Individual Results*—Present a table of results for individual tests, including test number, interval length, rock types, transmissivity and storativity, and pressure pulse amplitude and decay time (or recording time, if the decay is incomplete).

8.1.4.3 *Graphic Data*—Present pressure pulse decay versus time curves for each test, together with the appropriate type curves used for their interpretation.

8.1.4.4 *Other*—Other analysis or presentations may be included as appropriate, for example: (*1*) discussion of the characteristics of the permeable zones, (*2*) histograms of results, and (*3*) comparison of results to other studies or previous work.

8.1.5 *Appended Data*—Include in an appendix a completed data form (Fig. 6) for each test.

9. Precision and Bias

9.1 *Error Estimate:*

9.1.1 Analyze the results using standard statistical methods. Calculate all uncertainties using a 95 % confidence interval.

9.1.2 *Measurement Error*—Evaluate the errors in transmissivity and storativity associated with a single test. This includes the combined effects of pressure determination, measurement of time, and type curve matching or early decay time analysis.

9.1.3 *Sample Variability*—For each rock or discontinuity type, calculate, as a minimum, the mean transmissivity and storativity and their ranges, standard deviations, and 95 % confidence limits for the means. Compare the uncertainty associated with the transmissivity and storativity for each

Data Sheet

Project __________

Test Location __________

Rock Type __________

Date __________

Testing by __________

Test No. __________

Borehole No. __________

Borehole Dip and Dip Direction __________

Measured Depth of Test to Top Packer, m __________

Borehole Diameter, mm __________

Rock Temperature, °C __________

Equipment Description	Serial No.	Date of last Calibration

Length of Packed-off Interval, m __________

Length of Tubing Above Top Packer, m __________

Water Temperature, °C __________

Packer Pressure, kPa __________

Tubing ID, mm __________

Maximum Pulse Pressure, kPa __________

Pulse Decay Time, s __________

FIG. 6 Data Sheet for In Situ Measurement of Transmissivity and Storativity Using the Pressure Pulse Technique

rock type with the measurement uncertainty to determine whether measurement error or sample variability is the dominant factor in the results.

10. Keywords

10.1 borehole drilling; discontinuities; fault zones; field testing flow and flow rate; permeability; pressure testing; pulse testing; rock; saturation; storativity; transmissivity; viscosity; water; water saturation

REFERENCES

(1) Bredehoeft, J. D., and Papadopulos, S. S. "A Method for Determining the Hydraulic Properties of Tight Formations," *Water Resources Research*, Vol 16, 1980, pp. 233–238.

(2) Wang, J. S. Y., Narasimhan, T. N., Tsang, C. F., and Witherspoon, P. A., "Transient Flow in Tight Fractures," *Proceedings of the 1st Invitational Well Testing Symposium*, Berkeley, 1977, pp. 103–116.

(3) Neuzil, C. E., "On Conducting the Modified 'Slug Test' in Tight Formations," *Water Resources Research*, Vol 18, 1982, pp. 439–441.

(4) Cooper, H. H., Bredehoeft, J. D., and Papadopulos, S. S., "Response of a Finite Diameter Well to an Instantaneous Charge of Water," *Water Resources Research*, Vol 3, 1967, pp. 263–269.

(5) Papadopulos, S. S., Bredehoeft, J. D., and Cooper, H. H., "On the Analysis of 'Slug Test' Data," *Water Resources Research*, Vol 9, 1973, pp. 1087–1089.

(6) Doe, T. W., Long, J. C. S., Endo, H. K., and Wilson, C. R., "Approaches to Evaluating the Permeability and Porosity of Fractured Rock Masses," *Proceedings of the 23rd U.S. Symposium on Rock Mechanics*, Berkeley, 1982, pp. 30–38.

(7) Earlougher, R. C., "Advances in Well Test Analysis," *Society of Petroleum Engineers of AIME*, New York, NY, 1977.

(8) Freeze, R. A., and Cherry, J. A., *Groundwater*, Prentice-Hall, Englewood Cliffs, NJ, 1979.

(9) Shuri, F. S., Feves, M. L., Peterson, G. L., Foster, K. M., and Kienle, C. F., Public Draft: "Field and In Situ Rock Mechanics Testing Manual," Office of Nuclear Waste Isolation, Document ONWI-310, Section F: "In Situ Fluid Properties," GT-F.1 In Situ Permeability Measurement of Rock Using Borehole Packers, 1981.

Designation: D 4696 – 92

Standard Guide for Pore-Liquid Sampling from the Vadose Zone[1]

This standard is issued under the fixed designation D 4696; the number immediately following the designation indicates the year of original adoption or, in the case of revision, the year of last revision. A number in parentheses indicates the year of last reapproval. A superscript epsilon (ε) indicates an editorial change since the last revision or reapproval.

1. Scope

1.1 This guide discusses equipment and procedures used for sampling pore-liquid from the vadose zone (unsaturated zone). The guide is limited to in-situ techniques and does not include soil core collection and extraction methods for obtaining samples.

1.2 The term "pore-liquid" is applicable to any liquid from aqueous pore-liquid to oil. However, all of the samplers described in this guide were designed, and are used to sample aqueous pore-liquids only. The abilities of these samplers to collect other pore-liquids may be quite different than those described.

1.3 Some of the samplers described in this guide are not currently commercially available. These samplers are presented because they may have been available in the past, and may be encountered at sites with established vadose zone monitoring programs. In addition, some of these designs are particularly suited to specific situations. If needed, these samplers could be fabricated.

1.4 The values stated in SI units are to be regarded as the standard.

1.5 *This standard does not purport to address all of the safety problems, if any, associated with its use. It is the responsibility of the user of this standard to establish appropriate safety and health practices and determine the applicability of regulatory limitations prior to use.*

2. Referenced Documents

2.1 *ASTM Standards:*

D 653 Terminology Relating to Soil, Rock, and Contained Fluids[2]

3. Terminology

3.1 *Definitions:*

3.1.1 Where reasonable, precise terms and names have been used within this guide. However, certain terms and names with varying definitions are ubiquitous within the literature and industry of vadose zone monitoring. For purposes of recognition, these terms and names have been included in the guide with their most common usage. In these instances, the common definitions have been included in Appendix X1. Examples of such terms are soil, lysimeter, vacuum and pore-liquid tension.

3.2 *Descriptions of Terms Specific to This Standard:*

3.2.1 Appendix X1 is a compilation of those terms used in this guide. More comprehensive compilations, that were used as sources for Appendix X1, are (in decreasing order of their usage):

3.2.1.1 Terminology D 653,

3.2.1.2 *Compilation of ASTM Terminology,*[3]

3.2.1.3 *Glossary of Soil Science Terms,* Soil Science Society of America,[4] and,

3.2.1.4 *Webster's New Collegiate Dictionary.*[5]

4. Summary of Guide

4.1 Pores in the vadose zone can be saturated or unsaturated. Some samplers are designed to extract liquids from unsaturated pores; others are designed to obtain samples from saturated pores (for example, perched ground water) or saturated macropores (for example, fissures, cracks, and burrows). This guide addresses these categories. The sampler types discussed are:

4.1.1 Suction samplers (unsaturated sampling), (see Section 7),

4.1.2 Free drainage samplers (saturated sampling), (see Section 8),

4.1.3 Perched ground water samplers (saturated sampling), (see Section 9), and

4.1.4 Experimental absorption samplers (unsaturated sampling), (see Section 10).

4.2 Most samplers designed for sampling liquid from unsaturated pores may also be used to sample from saturated pores. This is useful in areas where the water table fluctuates, so that both saturated and unsaturated conditions occur at different times. However, samplers designed for sampling from saturated pores cannot be used in unsaturated conditions. This is because the liquid in unsaturated pores is held at less than atmospheric pressures (see *Richard's outflow principle*, in Appendix X1).

4.3 The discussion of each sampler is divided into specific topics that include:

4.3.1 Operating principles,

4.3.2 Description,

4.3.3 Installation,

4.3.4 Operation, and

4.3.5 Limitations.

5. Significance and Use

5.1 Sampling from the vadose zone may be an important component of some ground water monitoring strategies. It can provide information regarding contaminant transport

---

[1] This guide is under the jurisdiction of ASTM Committee D-18 on Soil and Rock and is the direct responsibility of Subcommittee D18.21 on Ground Water Vadose Zone Investigations.

Current edition approved April 15, 1992. Published June 1992.

[2] *Annual Book of ASTM Standards*, Vol 04.08.

[3] *Compilation of ASTM Terminology*, Sixth edition, ASTM, 1916 Race Street, Philadelphia, PA 19103, 1986.

[4] *Glossary of Soil Science Terms*, Soil Science Society of America, 1987.

[5] *Webster's New Collegiate Dictionary*, Fifth edition, 1977.

TABLE 1 Suction Sampler Summary

Sampler Type	Porous Section Material	Maximum[A] Pore Size (µm)	Air Entry Value (cbar)	Operational Suction Range (cbar)	Maximum Operation Depth (m)
Vacuum lysimeters	Ceramic	1.2 to 3.0 (1)[A]	>100	<60 to 80	<7.5
	PTFE	15 to 30 (2)[A]	10 to 21	<10 to 21	<7.5
	Stainless steel	NA[B]	49 to 5	49 to 5	<7.5
Pressure-vacuum lysimeters	Ceramic	1.2 to 3.0 (1) [A]	>100	<60 to 80	<15
	PTFE	15 to 30 (2)[A]	10 to 21	<10 to 21	<15
High pressure-vacuum lysimeters	Ceramic	1.2 to 3.0 (1)[A]	>100	<60 to 80	<91
	PTFE	15 to 30 (2)[A]	10 to 21	<10 to 21	<91
Filter tip samplers	Polyethylene	NA[B]	NA[B]	NA[B]	None
	Ceramic	2 to 3 (1)	>100	<60 to 80	<7.5
	Stainless steel	NA[B]	NA[B]	NA[B]	none
Cellulose-acetate hollow-fiber samplers	Cellulose Acetate	<2.8	>100	<60 to 80	<7.5
	Non cellulosic Polymer	<2.8	>100	<60 to 80	<7.5
Membrane filter samplers	Cellulose Acetate	<2.8	>100	<60 to 80	<7.5
	PTFE	2 to 5	NA[B]	NA[B]	<7.5
Vacuum plate samplers	Alundum	NA[B]	NA[B]	NA[B]	<7.5
	Ceramic	1.2 to 3.0	>100	60 to 80	<7.5
	Fritted glass	4 to 5.5	NA[B]	NA[B]	<7.5
	Stainless steel	NA[B]	49 to 5	49 to 5	<7.5

[A] Pore size determined by bubbling pressure (1) or mercury intrusion (2).
[B] NA = Not available.

and attenuation in the vadose zone. This information can be used for mitigating potential problems prior to degradation of a ground water resource **(1)**.[6]

5.2 The choice of appropriate sampling devices for a particular location is dependent on various criteria. Specific guidelines for designing vadose zone monitoring programs have been discussed by Morrison **(1)**, Wilson **(2)**, Wilson **(3)**, Everett **(4)**, Wilson **(5)**, Everett et al **(6)**, Wilson **(7)**, Everett et al **(8)**, Everett et al **(9)**, Robbins et al **(10)**, Merry and Palmer **(11)**, U.S. EPA **(12)**, Ball **(13)**, and Wilson **(14)**. In general, it is prudent to combine various unsaturated and free drainage samplers into a program, so that the different flow regimes may be monitored.

5.3 This guide does not attempt to present details of installation and use of the equipment discussed. However, an effort has been made to present those references in which the specific techniques may be found.

6. Criteria for Selecting Pore-Liquid Samplers

6.1 Decisions on the types of samplers to use in a monitoring program should be based on consideration of a variety of criteria that include the following:

6.1.1 Required sampling depths,
6.1.2 Required sample volumes,
6.1.3 Soil characteristics,
6.1.4 Chemistry and biology of the liquids to be sampled,
6.1.5 Moisture flow regimes,
6.1.6 Required durability of the samplers,
6.1.7 Required reliability of the samplers,
6.1.8 Climate,
6.1.9 Installation requirements of the samplers,
6.1.10 Operational requirements of the samplers,
6.1.11 Commercial availability, and
6.1.12 Costs.

6.2 Some of these criteria are discussed in this guide. However, the ability to balance many of these factors against one another can only be obtained through field experience.

7. Suction Samplers

7.1 Table 1 presents the various types of suction samplers. The range of operating depths is the major criterion by which suction samplers are differentiated. Accordingly, the categories of suction samplers are as follows:

7.1.1 *Vacuum Lysimeters*—These samplers are theoretically operational at depths less than about 7.5 m. The practical operational depth is 6 m under ideal conditions.

7.1.2 *Pressure-Vacuum Lysimeters*—These samplers are operational at depths less than about 15 m.

7.1.3 *High Pressure-Vacuum Lysimeters*—(Also known as pressure-vacuum lysimeters with transfer vessels.) These samplers are normally operational down to about 46 m, although installations as deep as 91 m have been reported **(15)**.

7.1.4 *Suction Lysimeters With Low Bubbling Pressures (Samplers With PTFE Porous Sections)*—These samplers are available in numerous designs that can be used to maximum depths varying from about 7.5 to 46 m.

NOTE 1—The samplers of 7.1.1, 7.1.2, 7.1.3, and 7.1.4 are referred to collectively as suction lysimeters. Within this standard, lysimeter is defined as a device used to collect percolating water for analyses **(16)**.

7.1.5 *Filter Tip Samplers*—These samplers theoretically have no maximum sampling depth.

7.1.6 *Experimental Suction Samplers*—The samplers have limited field applications at the present time. They include cellulose-acetate hollow-fiber samplers, membrane filter samplers, and vacuum plate samplers. They are generally limited to depths less than about 7.5 m.

7.2 *Operating Principles:*

7.2.1 *General:*

7.2.1.1 Suction lysimeters consist of a hollow, porous section attached to a sample vessel or a body tube. Samples are obtained by applying suction to the sampler and col-

[6] The boldface numbers in parentheses refer to the list of references at the end of this standard.

lecting pore-liquid in the body tube. Samples are retrieved by a variety of methods.

7.2.1.2 Unsaturated portions of the vadose zone consist of interconnecting soil particles, interconnecting air spaces, and interconnecting liquid films. Liquid films in the soil provide hydraulic contact between the saturated porous section of the sampler and the soil (see Fig. 1). When suction greater than the soil pore-liquid tension is applied to the sampler, a pressure potential gradient towards the sampler is created. If the meniscuses of the liquid in the porous segment are able to withstand the applied suction (depending on the maximum pore sizes and hydrophobicity/hydrophilicity), liquid moves into the sampler. The ability of the meniscuses to withstand a suction decreases with increasing pore size and also with increasing hydrophobicity of the porous segment (see 7.6). If the maximum pore sizes are too large and hydrophobicity too great, the meniscuses are not able to withstand the applied suction. As a result, they break down, hydraulic contact is lost, and only air enters the sampler. As described in 7.6, ceramic porous segments are hydrophilic and the maximum pore sizes are small enough to allow meniscuses to withstand the entire range of sampling suctions. Presently available polytetrafluoroethylene (PTFE) porous segments are hydrophobic, the maximum pore sizes are larger, and only a very limited range of sampling suction can be applied before meniscuses break down and sampling ends (see 7.6.1.3). Therefore, samplers made with PTFE porous segments may be used only for sampling soils with low pore-liquid tensions **(12, 17)**.

7.2.1.3 The ability of a sampler to withstand applied suctions can be directly measured by its bubbling pressure. The bubbling pressure is measured by saturating the porous segment, immersing it in water, and pressurizing the inside of the porous segment with air. The pressure at which air starts bubbling through the porous segment into the surrounding water is the bubbling pressure. The magnitude of the bubbling pressure is equal to the magnitude of the maximum suction that can be applied to the sampler before air entry occurs (air entry value). Because the bubbling pressure is a direct measure of how a sampler will perform, it is more useful than measurement of pore size distributions.

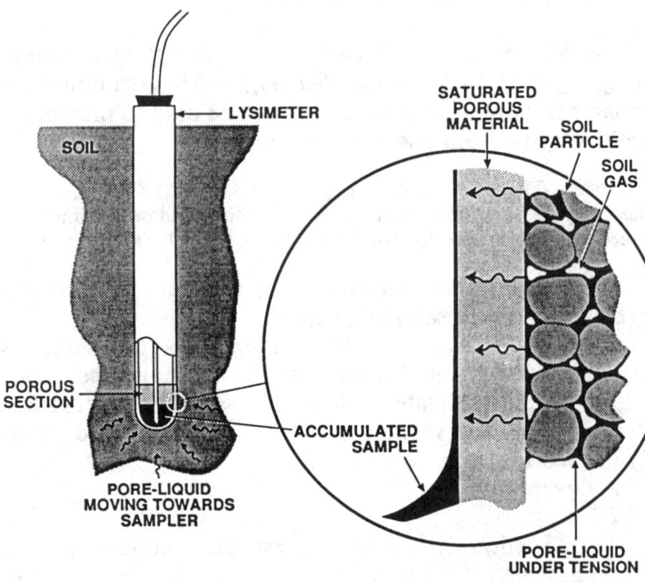

FIG. 1 Porous Section/Soil Interactions

7.2.1.4 As soil pore-liquid tensions increase (low pore-liquid contents), pressure gradients towards the sampler decrease. Also, the soil hydraulic conductivity decreases exponentially. These result in lower flow rates into the sampler. At pore-liquid tensions above about 60 (for coarse grained soils) to 80 cbar (for fine grained soils), the flow rates are effectively zero and samples cannot be collected.

7.2.2 *Suction Lysimeters:*

7.2.2.1 Vacuum lysimeters directly transfer samples to the surface via a suction line. Because the maximum suction lift of water is about 7.5 m, these samplers cannot be operated below this depth. In reality, suction lifts of 6 m should be considered a practical maximum depth.

7.2.2.2 Samples may be retrieved using the same technique as for vacuum lysimeters or, for deeper applications, the sample is retrieved by pressurizing the sampler with one line; this pushes the sample up to the surface in a second line.

7.2.2.3 High pressure-vacuum lysimeters operate in the same manner as pressure-vacuum lysimeters. However, they include an inbuilt check transfer vessel or a chamber between the sampler and the surface. This prevents sample loss through the porous section during pressurization, and prevents possible cup damage due to overpressurization.

7.2.2.4 Suction lysimeters with low bubbling pressures are available in each of the three previous designs. The only difference between these samplers and the three previous designs is that these porous sections are made with PTFE. The low bubbling pressure (and hence large pore size or hydrophobicity, or both) of PTFE constrains these samplers to soils that are nearly saturated (see 7.2.1.2 and 7.6.1.3).

7.2.3 *Filter Tip Samplers*—Samples are collected from a filter tip sampler by lowering an evacuated sample vial down an access tube to a permanently emplaced porous tip. The vial is connected to the porous tip and sample flows through the porous section and into the vial. Once full, the vial is retrieved.

7.2.4 *Experimental Suction Samplers*—Experimental suction samplers generally operate on the same principle as vacuum lysimeters with different combinations of porous materials to enhance hydraulic contact. The samplers are generally fragile and difficult to install. As with vacuum lysimeters, they are generally limited to depths of less than about 7.5 m.

7.3 *Description:*

7.3.1 *Vacuum Lysimeters:*

7.3.1.1 Vacuum lysimeters generally consist of a porous cup mounted on the end of a tube, similar to a tensiometer. The cup is attached to the tube with adhesives **(18[7])** or with "V" shaped flush threading sealed with an "O" ring. A stopper is inserted into the upper end of the body tube and fastened in the same manner as the porous cup or, in the case of rubber stoppers, inserted tightly **(12)**. To recover samples, a suction line is inserted through the stopper to the base of the sampler. The suction line extends to the surface and connects to a sample bottle and suction source in series.

[7] This reference is manufacturer's literature, and it has not been subjected to technical review.

Body tubes up to 1.8 m long have been reported **(15)** (see Fig. 2).

7.3.1.2 Harris and Hansen **(19)** described a vacuum lysimeter with a 6 mm by 65 mm ceramic porous cup designed for intensive sampling in small areas.

7.3.1.3 A variety of materials have been used for the porous segment including nylon mesh **(20)**, fritted glass **(21)**, sintered glass **(22)**, Alundum (manufacturer name), stainless steel **(23**[7]**)**, and ceramics (1.2 to 3.0 μm max pore size) **(18**[7]**)**. The sampler body tube has been made with PVC, ABS, acrylic, stainless steel **(24)** and PTFE **(18**[7]**, 25**[7]**)**. Ceramic porous segments are attached with epoxy adhesives or with flush threading. The stopper is typically made of rubber **(12)**, neoprene, or PTFE. The outlet lines are commonly PTFE, rubber, polyethylene, polypropylene, vinyl, nylon, and historically, copper. Fittings and valves are available in brass or stainless steel.

7.3.2 *Pressure-Vacuum Lysimeters:*

7.3.2.1 These samplers were developed by Parizek and Lane **(26)** for sampling deep moving pollutants in the vadose zone. The porous segment is usually a porous cup at the bottom of a body tube. The porous cup is attached with epoxy adhesives **(18**[7]**)** or with "V" shaped flush threading sealed with an "O" ring **(25**[7]**)**. Two lines are forced through a two-hole stopper sealed into the upper end of the body tube. The discharge line extends to the base of the sampler and the pressure-vacuum line terminates a short distance below the stopper. At the surface, the discharge line connects to a sample bottle and the pressure-vacuum line connects to a pressure-vacuum pump. Designs are available that do not use a stopper but rather an "O" ring sealed, flush threaded top plug **(25**[7]**)**. Tubing lines to the surface are attached to the top plug with threaded tubing fittings of appropriate materials. Body tubes are commonly available with 2.2 and 4.8 cm diameters and in a variety of lengths (see Fig. 3). The sampler and its components have been made out of the same materials used for vacuum lysimeters.

7.3.2.2 These samplers can retrieve samples from depths below 7.5 m because pressure is used for retrieval. However, during pressurization some of the sample is forced back out of the cup. At depths over about 15 m, the volume of sample lost in this fashion may be significant. In addition, at depths over about 15 m, pressures required to bring the sample to the surface may be high enough to damage the cup or to reduce its hydraulic contact with the soil **(27, 28)**. Rapid pressurization causes similar problems. Morrison and Tsai **(29)** developed a tube lysimeter with the porous section located midway up the body tube instead of at the bottom (see Fig. 4). This design mitigates the problem of sample being forced back through the cup. However, it does not prevent problems with porous segment damage due to overpressurization or rapid pressurization. The sleeve lysimeter (that is no longer available) was a modification to this design for use with a monitoring well **(1)** (see Fig. 5). Another modification is the casing lysimeter that consists of several tube lysimeters threaded into one unit (see Fig. 6). This arrangement allows precise spacing between units **(30)**.

7.3.2.3 Nightingale et al **(31)** described a design that

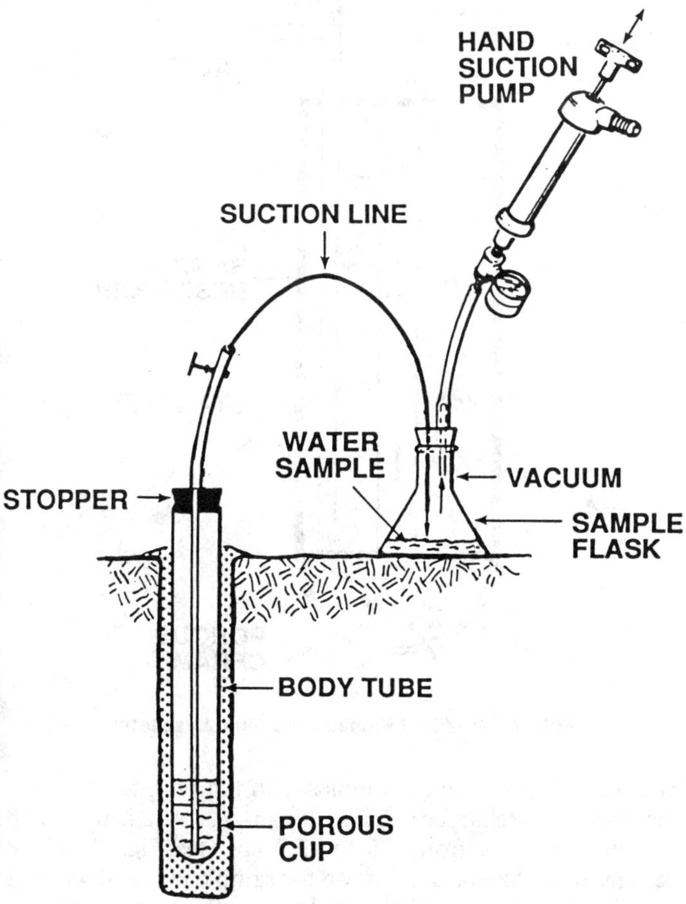

FIG. 2 Vacuum Lysimeter

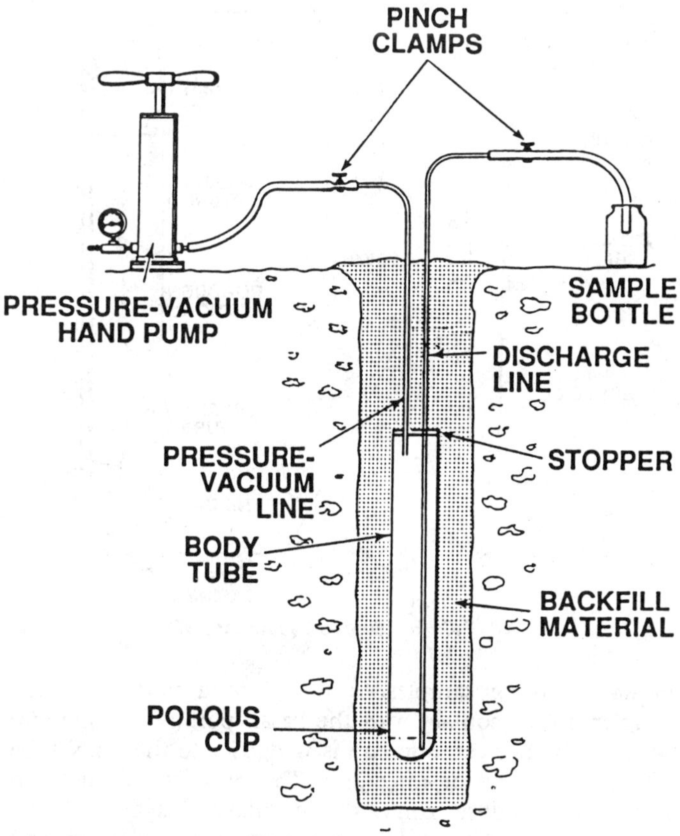

FIG. 3 Pressure-Vacuum Lysimeter

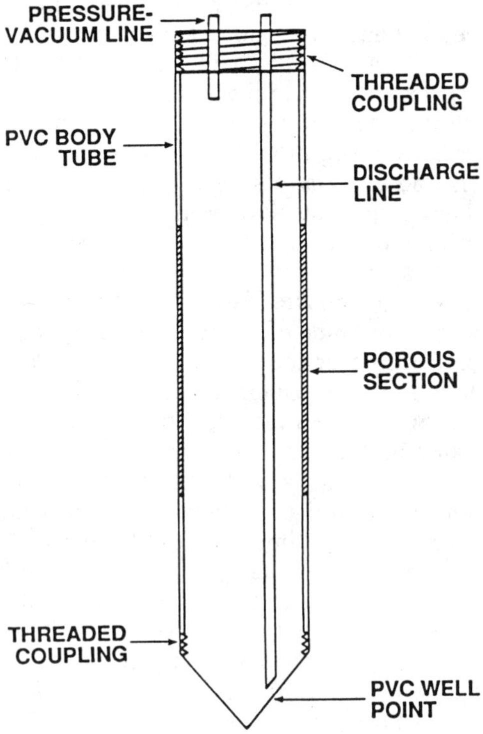

FIG. 4 Tube Pressure-Vacuum Lysimeter

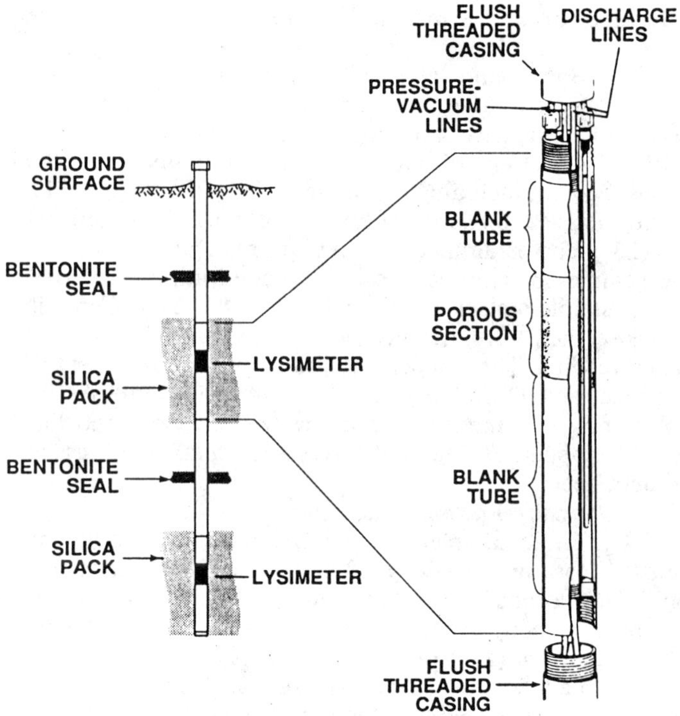

FIG. 6 Casing Lysimeter

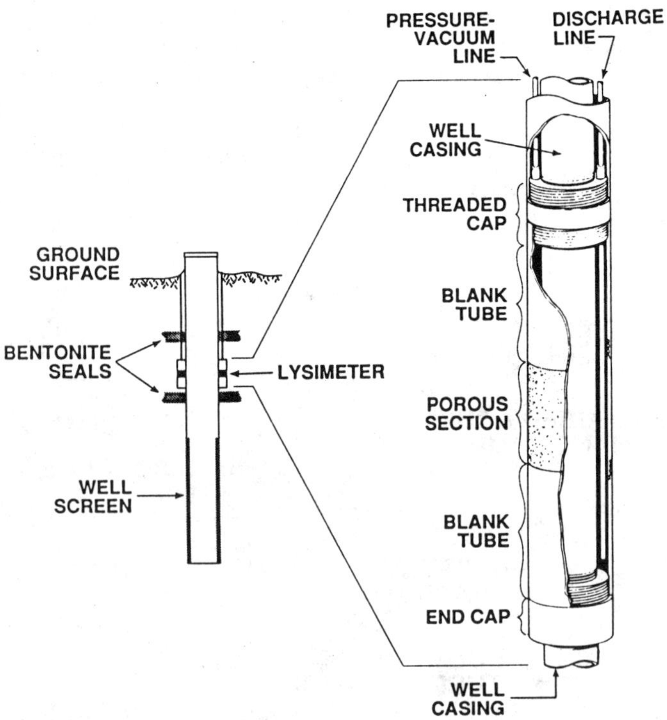

FIG. 5 Sleeve Lysimeter

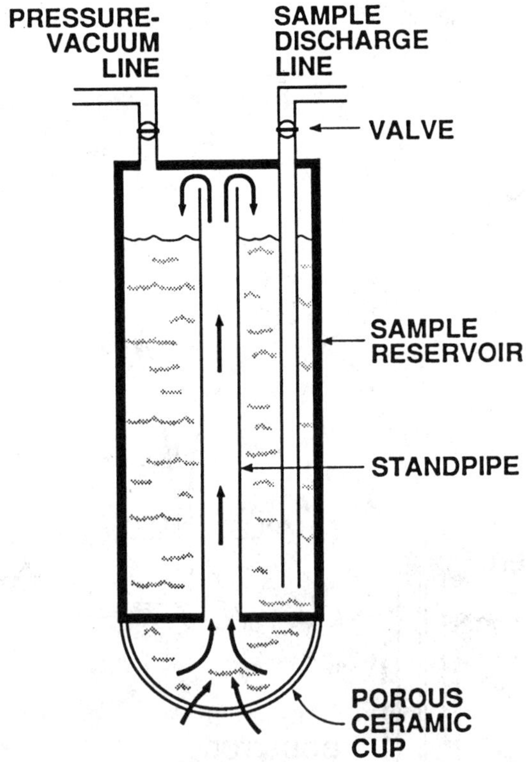

FIG. 7 Modified Pressure-Vacuum Lysimeter

allows incoming samples to flow into a portion of the sampler not in contact with the basal, porous ceramic cup (see Fig. 7). The ceramic cup is wedged into the body tube without adhesives or threading. The sampler was used to sample the vadose zone, the capillary fringe, and the fluctuating water table in a recharge area. Knighton and Streblow (**32**) reported a sampler with the porous cup upon the top of a chamber. This design was used with cup diameters ranging from 7.6 to 12.7 cm (see Fig. 8). These designs also allow pressurization for sample retrieval without significant liquid loss. However, because the porous cups are

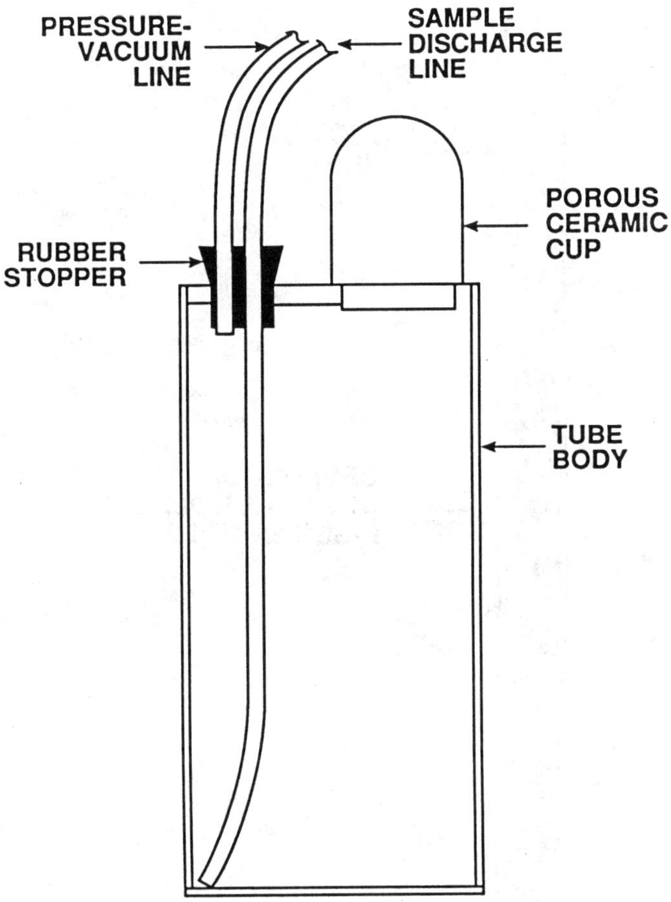

FIG. 8 Knighton and Streblow-Type Vacuum Lysimeter

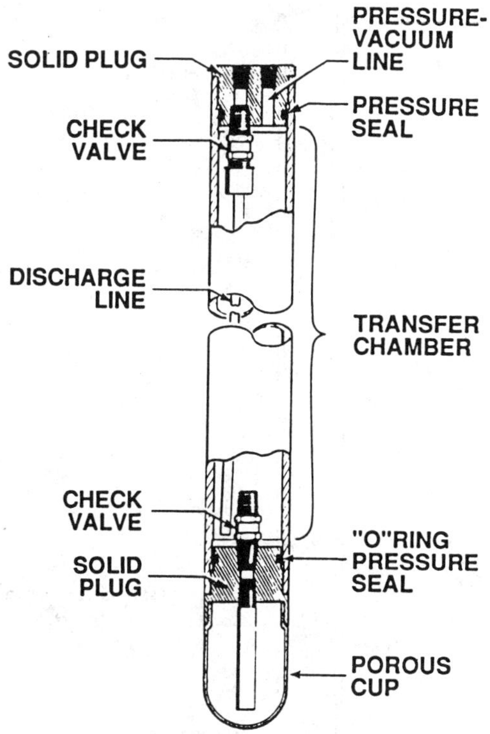

FIG. 9 High Pressure-Vacuum Lysimeter

open to the rest of the samplers, possible damage due to overpressurization or rapid pressurization is still a problem.

7.3.3 *High Pressure-Vacuum Lysimeters (Lysimeters With a Transfer Vessel)*—High pressure-vacuum lysimeters overcome the problems of fluid loss and overpressurization through the use of an attached chamber or a connected transfer vessel (see Fig. 9). The porous segment is usually a porous cup at the bottom of the body tube. The cup is attached with epoxy adhesives (**18**[7]) or with "V" shaped flush threading sealed with an "O" ring (**25**[7]). In the attached chamber design, the body tube is separated into two chambers connected by a one-way check valve. A pressure-vacuum line and a discharge line enter through a two-hole plug at the top of the body tube. The pressure-vacuum line terminates below the plug. The discharge line extends to the bottom of the upper chamber. The transfer vessel design is similar. However, the vessel and body tube are integral components joined by a common double threaded, "O" ring sealed plug containing a check valve. Body tube diameters range from 2.7 to 8.9 cm outside diameter. Total sampler lengths commonly range from 15.2 to 182.9 cm. A threaded top plug allows attachment of casing to the lysimeter. This facilitates accurate placement and provides long-term protection for the tubing lines. The samplers and their components have been made out of the same materials as vacuum lysimeters.

7.3.4 *Suction Lysimeters with Low Bubbling Pressures (Samplers With PTFE Porous Sections)*—Designs are available in each of the three categories described in 7.3.1, 7.3.2, and 7.3.3. The only difference between this group of samplers and the previous three samplers is that PTFE is used for the porous sections of this group of samplers (**25**[7]). The porous PTFE is attached with "V" shaped flush threading sealed with an "O" ring.

7.3.5 *Filter Tip Samplers:*

7.3.5.1 Filter tip samplers consist of two components: a permanently installed filter tip, and a retrievable glass sample vial. The filter tip includes a pointed end to help with installation, a porous section, a nozzle, and a septum. The tip is threaded onto extension pipes that extend to the surface. The sample vial includes a second septum. When in use, the vial is seated in an adaptor that includes a disposable hypodermic needle to penetrate both the septa, allowing sample to flow from the porous segment into the vial (see Fig. 10). Extension pipes vary from 2.5 to 5.1 cm inside diameter. Vial volumes range from 35 to 500 mL (**32**[7]).

7.3.5.2 The body of the filter tip is made of thermoplastic, stainless steel, or brass. The attached porous section is available in high density polyethylene, sintered ceramic, or sintered stainless steel. The septum is made of natural rubber, nitrile rubber, or fluororubber (**32**[7]).

7.3.6 *Experimental Suction Samplers:*

7.3.6.1 Cellulose-acetate, hollow-fiber samplers were described by Jackson et al (**33**) and Wilson (**3**). A sampler consists of a bundle of these flexible, hollow fibers (<2.8 μm max pore size) pinched shut at one end and attached to a suction line at the other end. The suction line leads to the surface and attaches to a sample bottle and source of suction in the same manner as a vacuum lysimeter (see Fig. 11). The fibers, that are analogous to the porous sections of vacuum lysimeters, have outside diameters of up to 250 μm (**33**). Levin and Jackson (**34**) described similar fibers made from a

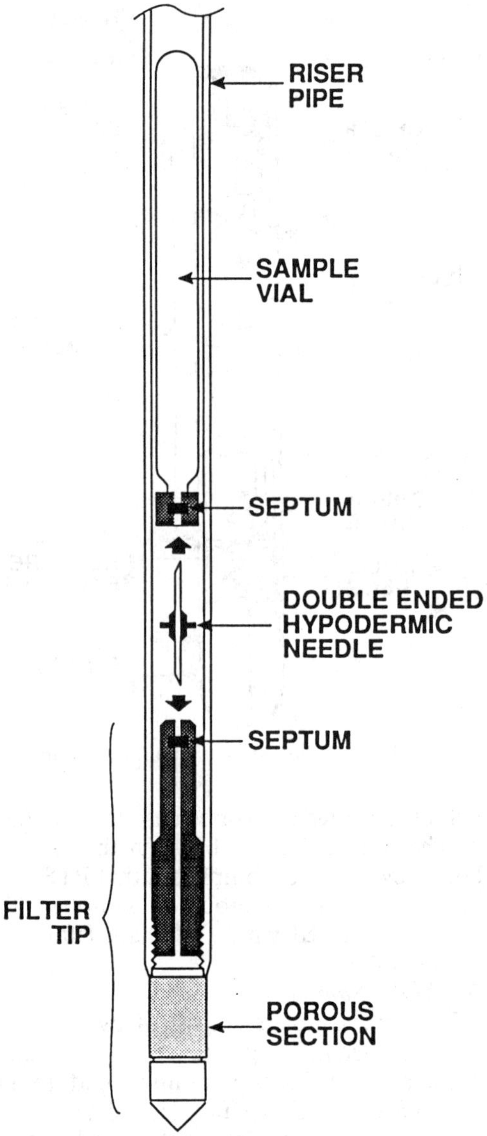

FIG. 10 Filter Tip Sampler

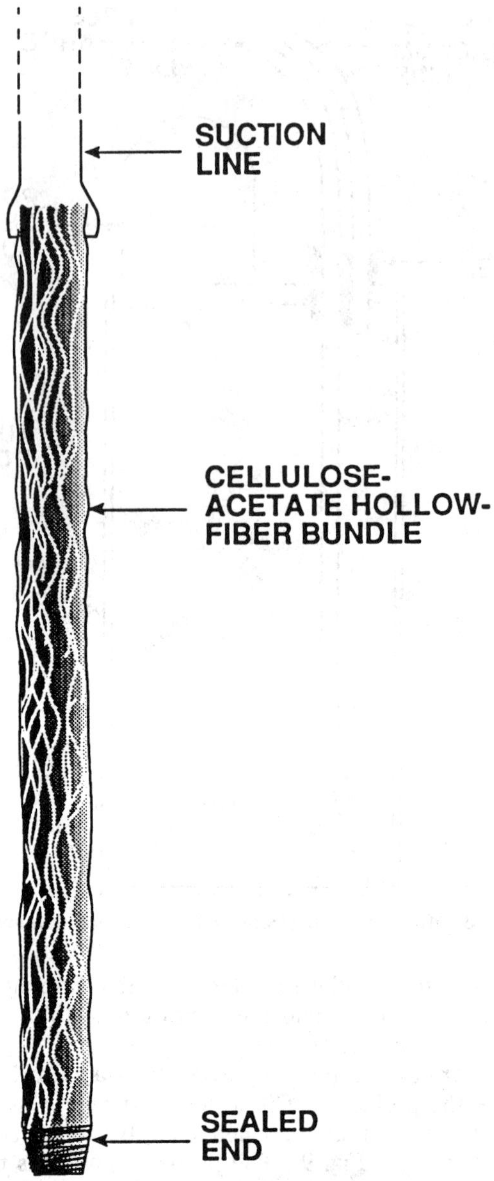

FIG. 11 Cellulose-Acetate Hollow-Fiber Sampler

noncellulosic polymer solution (max pore size <2.8 μm). Those fibers have dense inner layers surrounded by open celled, spongy layers with diameters ranging from 50 to 250 μm.

7.3.6.2 Membrane filter samplers were described by Morrison (**1**), Everett and Wilson (**6**), U.S. EPA (**12**) and Stevenson (**35**). A sampler consists of a membrane filter of polycarbonate, cellulose acetate (<2.8 μm max pore size), cellulose nitrate or PTFE (2 to 5 μm max pore size); mounted in a "swinnex" type filter holder (**35, 36, 37**[7]). The filter rests on a glass fiber prefilter. The prefilter rests on a glass fiber "wick" that in turn sits on a glass fiber collector. The collector is in contact with the soil and extends the sampling area of the small diameter filter (see Fig. 12 and 7.5.1.6). A suction line leads from the filter holder to the surface. At the surface, the suction line is attached to a sample bottle and suction source in a manner similar to vacuum lysimeters.

7.3.6.3 A vacuum plate sampler consists of a flat porous disk fitted with a nonporous backing attached to a suction line that leads to the surface (see Fig. 13). Plates are available in diameters ranging from 4.3 to 25.4 cm and custom designs are easily arranged (**1, 18**[7]). Plates are available in alundum, porous stainless steel (**23**[7]), ceramic (1.2 to 3.0 μm max pore size) or fritted glass (4 to 5.5 μm max pore size) (**38**[7], **6, 39, 40, 41, 42, 43, 44**). The nonpermeable backing can be a fiberglass resin, glass, plastic, or butyl rubber.

7.3.7 *Comments:*

7.3.7.1 When some ceramic cups are glued to the inner wall of the body tube in a suction lysimeter, an inner lip is formed (**45**). As the discharge line is pushed through the stopper at the top of the sampler, it may catch on this lip and the operator may conclude that the line has reached the bottom of the ceramic cup (see Fig. 14). As a result, an 80 mL error can occur in sampling rate determinations. This 80 mL of fluid accumulates in the cup, is not removed during sampling, and will cause cross contamination between sampling events. Soilmoisture (**18**[7]) suggested that the line can be

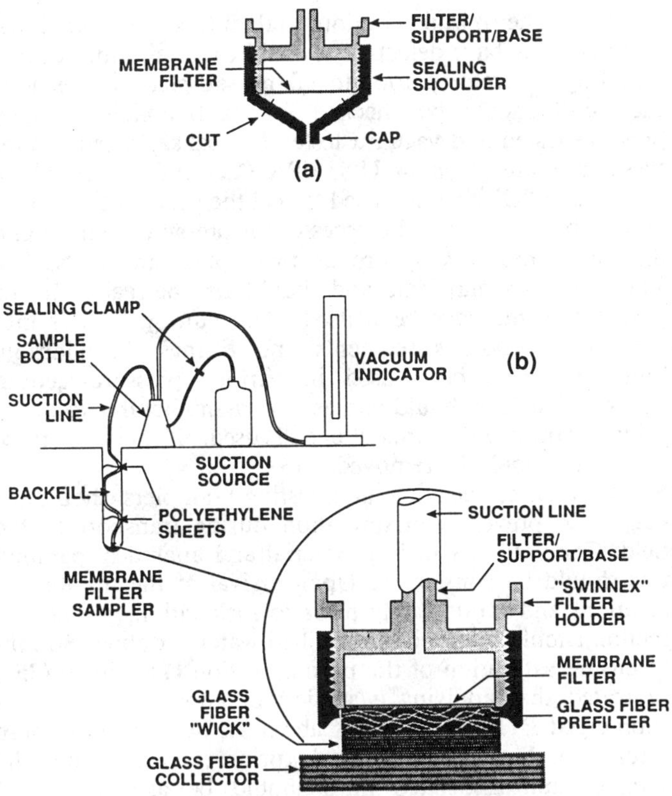

FIG. 12 Membrane Filter Sampler; (*a*) Preparation of Filter Sampler; and (*b*) Installation of Filter Sampler

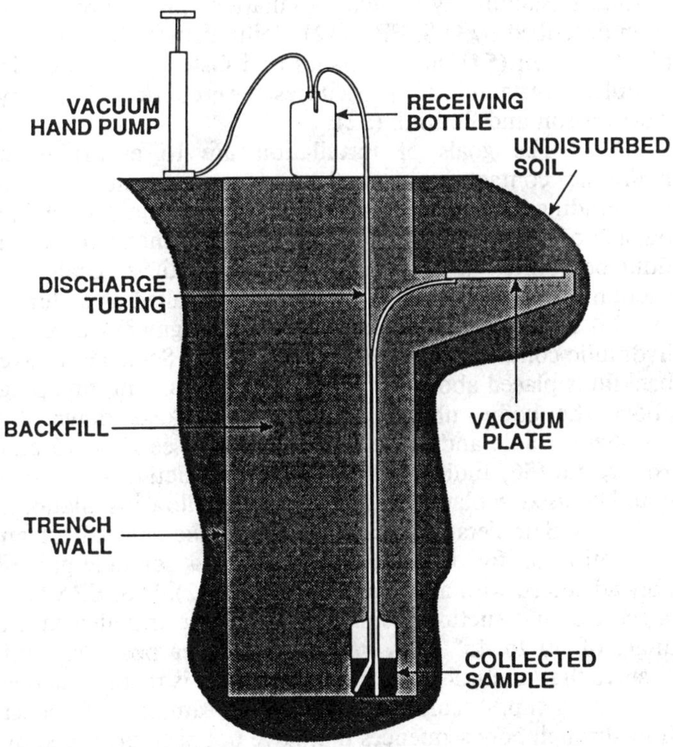

FIG. 13 Vacuum Plate Sampler Installation

kept from catching by cutting its tip at an angle. In all-PTFE suction lysimeters, the discharge line is a rigid PTFE tube

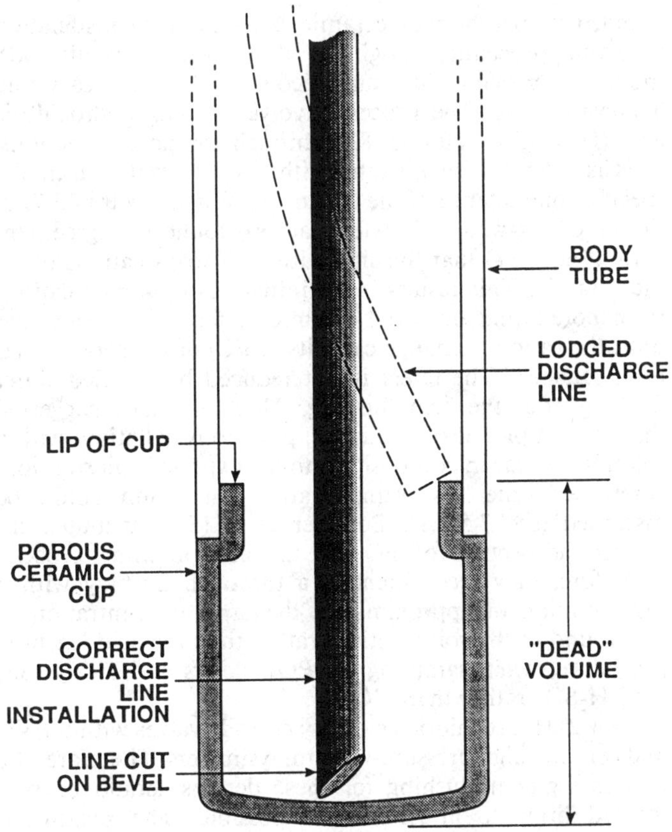

FIG. 14 Location of Potential Dead Volume in Suction Lysimeter

extending to the bottom of the cup. This results in a zero accumulation of fluid. Older samplers with PTFE porous segments and PVC body tubes have a discharge line that does not extend all the way to the bottom. This problem has been corrected in newer PTFE and PVC samplers (**25**[7]). This results in a 34 mL accumulation of fluid (**12**). Filter tip samplers develop an 8 mL accumulation of fluid. Haldorsen et al. (**46**) suggested collecting and discarding an initial sample to purge this accumulated fluid.

7.3.7.2 Because samplers are often handled roughly during installation, durability and ruggedness are important. It has been shown that PTFE has a higher impact strength than ceramics which need to be installed with care (**25**[7]). It has also been found that PTFE threads and ceramic threads (when used) are susceptible to leakage, and must be securely sealed with pipe threading tape (**45**). TFE-fluorocarbon (PTFE) tape is not recommended in square threaded joints since the tape is designed for tapered "V" threaded compression joints.

7.3.7.3 As described above, porous sections can be made from various materials. These materials have physical and chemical limitations that must be considered when designing a monitoring program. Physical limitations are described in 7.6.1. Chemical limitations are described in 7.6.2.

7.4 *Installation Methods:*

7.4.1 *Pre-Installation:*

7.4.1.1 As demonstrated by Neary and Tomassini (**47**), new samplers may be contaminated with dust during manufacturing. In order to reduce chemical interferences from substances on the porous sections, U.S. EPA (**12**) recommended preparation of ceramic units prior to installation following procedures originally developed by Wolff (**48**), modified by Wood (**49**) and recommended by Neary and Tomassini (**47**). The process involves passing hydrochloric acid (HCl) (for example, 8*N*) through the porous sections. This is followed by flushing with distilled water until the specific conductance of the outflowing water is within 2 % of the inflowing water. Debyle et al (**50**) found (in agreement with **49** and **51**) that flushing with HCl strips cations off of the ceramic. This results in an initial adsorption of cations from pore-liquid onto the ceramic surface. This continues until the cation exchange capacity (CEC) of the ceramic has been satisfied. The effect is not reduced by distilled water flushing after the acid flushing. Therefore, they suggested that the sampler also be flushed, prior to installation, with a solution similar in composition to the expected soil solution. Alternately, the first sample after installation could be discarded (see 7.5.2.1). Bottcher et al (**52**) attributed increased adsorption of PO_4 to the acid leaching process. Therefore, they recommended a thorough flushing with a PO_4 solution of approximately the same concentration as that found in the soil solution, rather than the acid leaching procedure, when sampling for PO_4. Peters and Healy (**53**) used H_2SO_4 rather than HCl.

7.4.1.2 Hydrochloric acid may corrode valves within PVC and ceramic high pressure-vacuum lysimeters. Therefore, the porous segment flushing for these designs should be performed prior to attachment if possible. The maximum suction which can be applied is one atmosphere, therefore the flushing process will be slow if suction is used to draw HCl through the porous segment. The flushing can be performed more rapidly if the porous segment is filled with HCl and pressurized to force the acid out of the porous segment since more than one atmosphere of pressure can be applied. This procedure can only be used if the cups are not attached. Care must be taken to prevent overpressurization that might damage the porous section.

7.4.1.3 Corning Laboratories (**38**[7]) recommended washing fritted glass with hot HCl followed by a distilled water rinse. Cleaning procedures for Alundum have not been reported, although an acid and water rinse procedure similar to that for ceramic would appear to be appropriate (**1**). Timco (**25**[7]) described cleaning procedures for PTFE. The method includes passing 0.5 L of distilled water through the material. An I.P.A. bath followed by another in hydrogen peroxide or rinsing with HCl followed by a distilled water rinse.

7.4.1.4 The use of HCl to wash/flush porous segments of lysimeters, that are to be used in sanitary landfills, may cause water quality interpretation problems. Sanitary landfills are notorious generators of methane gas. Reaction of methane with free chloride ion may result in the generation of di- and trichloromethane (also known as methylene chloride and chloroform). Because of the small liquid volumes in lysimeters and the sensitivity of EPA methods (including 601), false positives for one or both of these constituents may occur.

7.4.1.5 Stevenson (**35**) recommended treating cellulose-acetate hollow-fibers with silver nitrate and sodium chloride to prevent biofilm growths. Morrison (**1**) suggested rinsing membrane filters with distilled water.

7.4.1.6 The porous section and fittings of individual samplers may have defects that could cause air entry during sampling. Therefore, prior to taking samplers to the field, each unit should be checked for its bubbling pressure, pressure tested and vacuum tested for leaks. Procedures for these tests are given in U.S. EPA (**12**) and Timco (**25**[7]). Washers or "O" rings are used to seal the plugs at the tops of body tubes. However, the accesses for pressure-vacuum and discharge lines passing through these plugs are not sealed. These accesses may leak, and should also be sealed. In the past, lubricants have been used when cutting threads into body tubes, porous segments and fittings. In addition, lubricants have been used in various pressure-vacuum pumps. The user should contact the manufacturer to determine if these lubricants are still used. If present, these lubricants should be removed.

7.4.1.7 After cleaning and testing, samplers should be bagged to prevent contamination during transport to the field. Compatibility of bag material and analytical parameters should be considered. Upon arrival at the installation location, and immediately prior to installation, the porous section should be placed in distilled water for about 30 min to ensure saturation of the porous section (**1**). Timco (**25**[7]) indicated that applying a suction of about 50 cbar to a submerged PTFE sampler for about an hour would ensure saturation. Finally, immediately prior to installation, the sampler and associated lines should be assembled and inspected for defects (for example, crimped lines).

7.4.2 *Suction Lysimeter and Filter Tip Sampler Installation:*

7.4.2.1 Suction lysimeter installation procedures have been described by U.S. EPA (**12**), Soilmoisture (**18**[7]), Timco (**25**[7]), Linden (**54**), and Rhoades and Oster (**55**). Filter tip sampler installation procedures were described by Torstensson and Petsonk (**32**).

7.4.2.2 The goals of installation are to ensure good hydraulic contact between the porous segment and the surrounding soil, and to minimize leakage of liquid along the outside of the sampler. U.S. EPA (**12**) recommended a silica flour/bentonite clay method to achieve these goals for suction lysimeters. A silica flour layer (installed as a slurry, see 7.4.2.6) placed around the porous segment increases hydraulic contact with the surrounding soil. Screened native backfill is placed above the silica flour, and a bentonite plug above the body tube prevents liquid leakage down the installation hole and along the body tube (see Figs. 15 and 16). Klute (**56**) indicated that a screened native soil slurry could be used in place of silica flour for shallow installations.

7.4.2.3 Samplers may be installed in the sidewall of an excavation or, for deeper applications, in a borehole preferably advanced with a hollow stem auger (**12**). U.S. EPA (**12**) suggested that suction lysimeters should be installed at an angle of 30 to 45° from vertical whenever possible. This ensures that an undisturbed column of soil is retained above the porous cup. Accordingly, pore liquid samples will reflect flow through pore sequences that have not been disturbed by sampler installation. This angular placement also improves the sampler's ability to collect macropore flow. When installed in the sidewall of a trench, the angled emplacement is simple (see Fig. 15). However, when installed in a borehole, angular emplacement entails angled drilling.

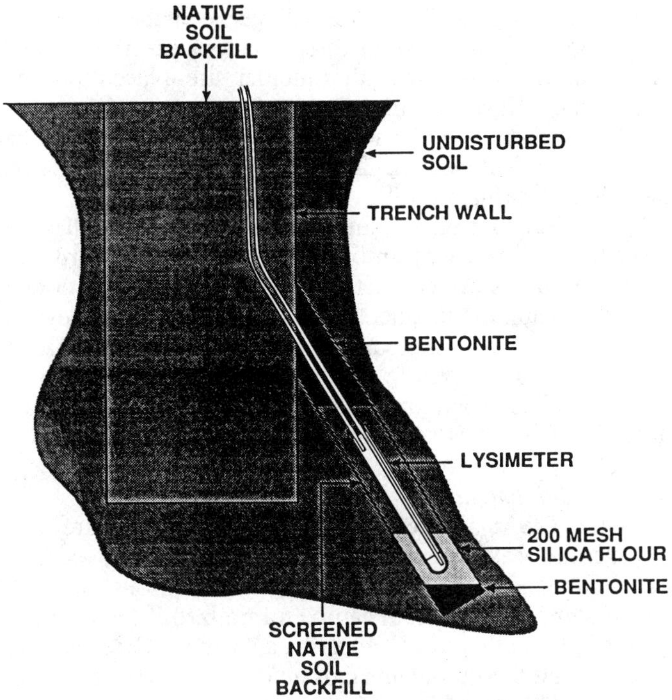

FIG. 15 Pressure-Vacuum Lysimeter Installation in the Sidewall of a Trench

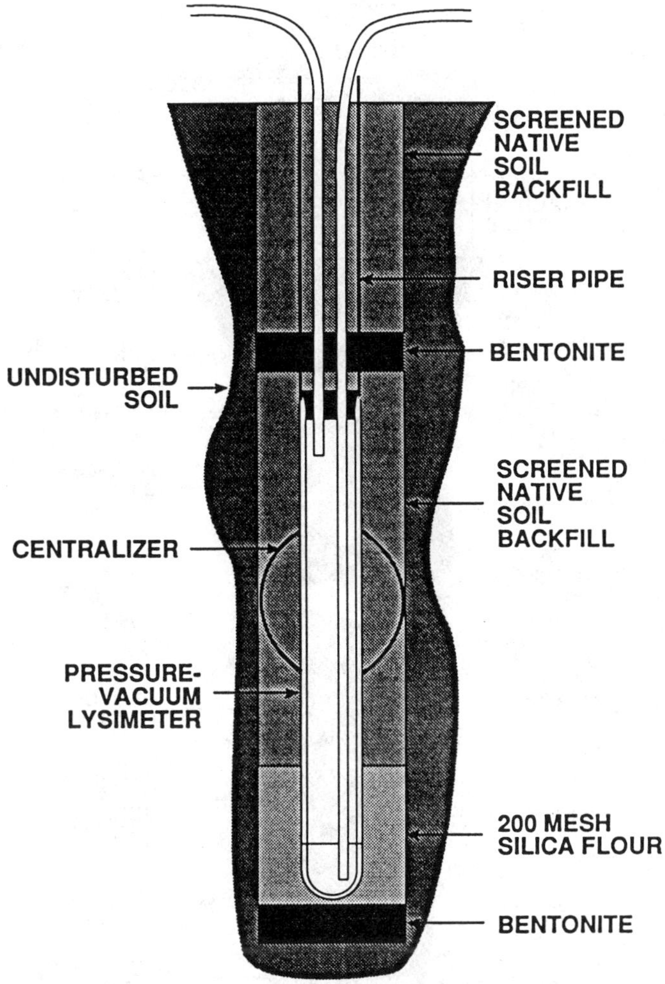

FIG. 16 Pressure-Vacuum Lysimeter Installation in a Borehole

Where soils permit, filter tip samplers can be installed by pushing the filter tip into the ground by applying a static load to the extention pipe **(32)**.

7.4.2.4 When suction lysimeters are installed in a borehole advanced by a drill rig, the hole is usually advanced 15 to 20 cm below the desired location of the porous section. Morrison and Szecsody **(30)** found that the radius of sampling influence is maximized if the borehole diameter is only slightly larger than that of the sampler and if silica flour pack is used. U.S. EPA **(12)** recommended that the hole have a diameter at least 5 cm larger than the sampler. Timco **(25**[7]**)** recommended that the hole have a diameter at least 8 cm greater than that of the sampler to facilitate installation of the silica flour.

7.4.2.5 Suction lysimeters are preferably lowered into place attached to risers. These protect the lines and ensure exact placement at the desired depth. Centralizers are often used to center the sampler in the hole. Suction lysimeters float in the silica flour that is installed as a slurry. Therefore, the samplers should be installed full of distilled water or held in place by rigid risers.

7.4.2.6 The silica flour slurry (for example 200 to 75 μm mesh opening, silica to distilled water ratio of 0.45 kg to 150 mL) is usually installed using the tremie method (side discharge). Alternately, Brose et al **(57)** described a method for freezing the silica slurry around the sampler prior to placement. The sampler and frozen pack are then lowered to the sampling location in the borehole. They cited advantages of this technique as including ensurance of proper sampler placement in the flour pack and elimination of pack contamination by soils which slough down the borehole. U.S. EPA **(12)** recommended filling the borehole to about 30 cm above the suction lysimeter body with the silica. In addition, it was recommended that the powdered bentonite plug placed on top of the silica be about 15 cm thick. The bentonite is also sometimes installed as a slurry, being allowed to hydrate before emplacement. Mixing the bentonite with fine sand at a 1 to 9 ratio, respectively, reduces the potential for shrinking and swelling inherent with pure bentonite **(1)**. The excavated soil should be backfilled above the bentonite in the order in which it was withdrawn. An effort to compact the soil to its original bulk density should be made. When more than one suction lysimeter is installed in one borehole, these procedures are repeated at the various desired sampling depths (see Fig. 17). Care must be taken with these installations to ensure that lines from lower samplers do not interfere with the hydraulic contact of shallower samplers. Designs are available to avert these problems **(25**[7]**)**.

7.4.2.7 U.S. EPA **(12)** recommended removal of the water within the sampler and silica slurry after installation. Litaor **(58)** recommended installation of samplers a year before sampling is to begin, in order to allow them to equilibrate with the surrounding soil. The lines at the surface should be labeled, clamped and housed in locked containers such as valve boxes or casing **(1)**. Methods for cutting and splicing tubing may be found in Timco **(25**[7]**)**. The user should be careful when using clamps and tubing provided by different manufacturers, inappropriate clamps may damage tubing. Clamps must be restricted to permanently flexible tubing

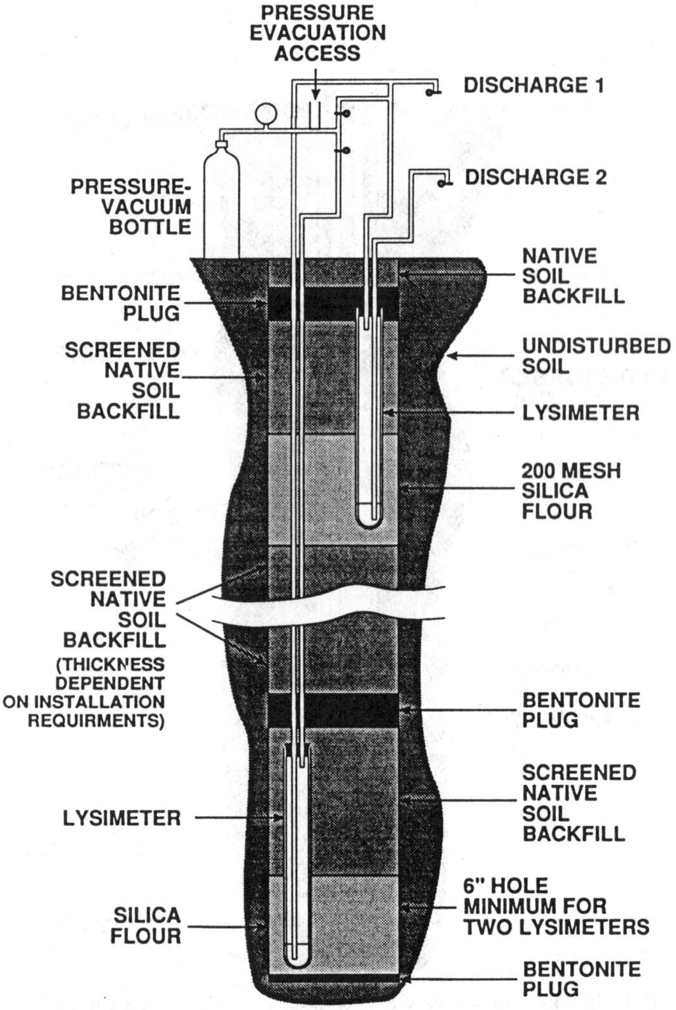

FIG. 17 Multiple Pressure-Vacuum Lysimeter Installations in a Borehole

otherwise stopcocks should be used.

7.4.3 *Experimental Suction Sampler Installation:*

7.4.3.1 Cellulose-acetate hollow-fiber sampler installation procedures were described by Everett et al **(9)**. Membrane filter sampler installation procedures were described by Stevenson **(35)**, Everett et al **(9)**, and Morrison **(1)**. Vacuum plate sampler installation procedures were described by Everett et al **(9)** and Morrison **(1)**.

7.4.3.2 Cellulose-acetate hollow-fiber samplers have been used almost exclusively in laboratory studies **(34)**. Because the samplers operate on the same principles as vacuum lysimeters, the goals and concerns of installation are similar. Good hydraulic contact between the hollow-fiber and the soil is critical. However, the fibers are too thin and fragile to be pushed into place. Therefore, the fibers must be placed in a predrilled hole (vertical or horizontal). Silkworth and Grigal **(59)** installed these samplers within a length of perforated, protective PVC tubing filled with soil slurry.

7.4.3.3 Membrane filter samplers are placed in a hole dug to the top of the selected sampling depth. First, sheets of the glass fiber "collectors" are placed at the bottom of the hole. These develop the necessary hydraulic contact between the sampler and the soil. In addition, the "collectors" extend the area of sampling as they cover a larger area than the filter holder alone. Second, two or three smaller glass fiber "wick" discs that fit within the filter holder are placed on the "collectors." Third, the filter holder fitted with a glass fiber prefilter and the membrane filter is placed on top of the "wick" disks. The suction line leads to the surface. Finally, the hole is backfilled **(1, 9)**.

7.4.3.4 Vacuum plate lysimeters are normally installed on the ceiling of a cavity cut into the side of a trench. In order to obtain the necessary contact between the porous plate and the soil, pneumatic bladders, inner tubes, or similar devices are placed beneath the sampler and are used to force it against the cavity ceiling **(1)**. The cavity ceiling is not a smooth surface. Therefore, a layer of silica flour between the plate and the soil is sometimes used to enhance hydraulic contact.

7.4.4 *Maintenance:*

7.4.4.1 The major causes of sampler failure are line damage and leaks (caused by freezing, installation, rodents, etc.), connection leaks, and clogging of the porous material. Freeze damage to the lines can be minimized if the lines are emptied of sample prior to applying a vacuum. Care needs to be taken that the tubing line closure devices are freeze proof.

7.4.4.2 The possibility of line and connection leaks is minimized by rigorously sealing and pressure testing all connections and lines before installation. A common precaution to assist in repairing surface damage to lines is to store excess line below the surface (within the riser pipe when used) when backfilling the borehole. In the event of severed lines, an excavation to this buried length allows restoration of an operational system **(1)**. Lines should be clamped shut when not in use to prevent foreign objects or insects from entering them. The lines should be protected from weather, sunlight exposure, and vandalism with a locked housing. The use of riser pipe around the sampler lines prevents punctures by backfill materials and prevents rodents from damaging the lines.

7.4.4.3 When shallow samplers are used, the ground surface above the sampler should be maintained in a fairly representative state. Large line housings and excessive traffic around the sampler (causing compaction of the soil) will reduce the amount of infiltration in that area. This will affect the representativeness of the pore-liquid samples. Methods to avoid these effects include angled installations, and remote operation of sampler lines.

7.4.4.4 Porous sections may clog as a function of soil composition, type of porous section material, biofilm growth, suction application, and pore-liquid content **(1, 17, 20, 50)**. However, porous section clogging appears to be less severe than once thought **(12, 17)**. Soils and the 200 mesh silica flour filter out fine materials before they reach the porous section **(60, 61, 62)**. Clogging can be further reduced by periodically filling the sampler with distilled water and allowing it to drain out of the sampler. Debyle et al **(50)** suggested removing shallow samplers on a seasonal basis for flushing with HCl and distilled water. This process restores samplers to their original operational and chemical states. A "clogged" lysimeter may be cleaned out by filling the lysimeter with pure water and applying a pressure of 5 psi for 30 min. However, reinstallation at the same location and depth does not guarantee resumption of sampling from the same soil volume.

7.4.4.5 Often no sample is retrieved during a sampling attempt. The first check should be a continuity test of the lines and connections. This test can be done by applying a gentle pressure or vacuum to the V/P line and detecting air movement from the open sample line. This could be due to sampler failure or high pore-liquid tensions. Because of this, it is prudent to install a tensiometer near the sampler at a similar depth. The tensiometer that measures pore-liquid tensions allows the operator to determine if failure to obtain a sample is due to high pore-liquid tensions or due to sampler damage. The tensiometer can also be used to gage the effect of sampling on local pore-liquid flow regimes. Pore-liquid tension should be determined as an initial condition during lysimeter installation.

7.4.4.6 If a tensiometer is not available to measure pore-liquid tensions, the lysimeter can be tested to help determine reasons for failure to recover a sample. The sampler is tested by applying a suction of 80 cbar, and monitoring the decay of suction with time. Figure 18 depicts the various types of suction decay that might be found in a suction lysimeter with a 200 cbar bubbling pressure ceramic section. An almost instantaneous decay of suction is associated with lysimeter leakage. A suction decay over a period of minutes is associated with pore-liquid tensions greater than 200 cbar. Under these conditions, the porous section is desaturated and air enters the sampler. A suction decay over a period of hours reflects normal sample collection. This suggests that failure to retrieve a sample is related to damage of the sample retrieval system (for example, discharge line damage). When suction does not decay, or does so over a period of days, the pressure-vacuum line may be clogged or pore-liquid tensions may be greater than 60 cbar (but less than 200 cbar) causing liquid inflow rates that are too low for sample collection.

7.4.4.7 Morrison and Szecsody **(63)** described devices that could be used as tensiometers and then converted to pressure-vacuum lysimeters. However, they found that gases entering the devices prevented accurate measurement of pore-liquid tensions. Baier et al **(64)** described methods for converting tensiometers to pressure-vacuum lysimeters. It would also appear reasonable to convert suction lysimeters to tensiometers. However, Taylor and Ashcroft **(65)** found that the volume of water drawn from a converted lysimeter into the surrounding soil would significantly affect natural pore-liquid tensions. In addition, they found that the larger porous section of a lysimeter would cause more diffusion of dissolved air into the device, and that the time constant for measurement would be significantly increased over that of a tensiometer. Filter tip samplers can be converted to tensiometers with pressure transducers **(32)**.

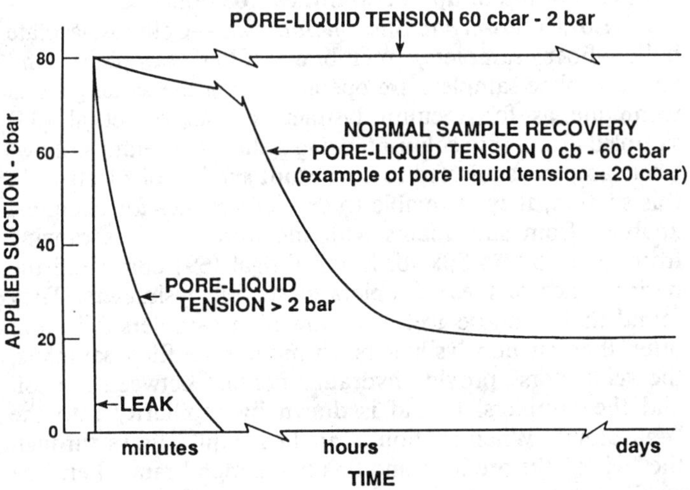

NOTE—Also shown is the almost instantaneous decay associated with an appreciable leak in the instrument.

FIG. 18 Decay Characteristics of Suction Applied to a Two Bar (Bubbling Pressure) Ceramic Cup Lysimeter in Equilibrium With Soils in Varying Ranges of Pore-Liquid Tension

7.4.4.8 Operational lifetimes of suction samplers are dependent on installation, subsurface conditions, maintenance, and sampling frequency. Some samplers have been reported to be operational for as long as 25 years **(64)**.

7.4.5 *Comments:*

7.4.5.1 Vacuum lysimeters and experimental samplers use suction to retrieve samples. Therefore, the maximum sampling depth is limited by the maximum suction lift of water (about 7.5 m) **(12)**. In practice, these samplers are generally used to about 2 m below the surface **(12)**. They are primarily used to monitor near-surface movement of pollutants such as those from land disposal facilities and those from irrigation return flow.

7.4.5.2 Pressure-vacuum lysimeters are generally not used at depths below about 15 m. At greater depths, sample loss and overpressurization problems are considered significant enough to warrant the use of high pressure-vacuum lysimeters that do not have these limitations. High pressure-vacuum lysimeters are not preferred at the shallower depths because they are more expensive. In addition, high pressure-vacuum units have more moving parts than pressure-vacuum units, and as a result, the possibility of failure for the former is higher.

7.4.5.3 As discussed in 7.6, two problems with suction samplers are that they may not sample from macropores (under unsaturated conditions; unless the macropores are directly intercepted) and that their results cannot be used in quantitative mass balance studies. Hornby et al **(66)** described an installation that could be used to surmount these problems. A barrel-sized casing (for example, 57 cm outside diameter by 85.7 cm high) is placed in a support device and gently pushed into the soil with a backhoe. As the casing is pushed, soil is excavated around it to help with insertion. The process results in an encased monolith of undisturbed soil. The monolith is then rotated and lifted, pressure-vacuum lysimeters are placed in its base, and the bottom is sealed. Subsequently the assembly is placed back into the ground at the monitoring site (see Fig. 19). All fluid draining through the monolith is collected by the samplers. Inasmuch as the boundaries of the system are sealed, the flux of liquid through the system requires maintaining a vertical hydraulic gradient by applying continual suction to the samplers.

7.5 *Operation:*

7.5.1 *Methods:*

7.5.1.1 *Vacuum Lysimeters*—Sampling methods are described by the U.S. EPA **(12)**, by Soilmoisture **(18[7])** and by Timco **(25[7])**. To collect a sample, suction is applied to the sampler, and the suction line is clamped shut. After sample has collected in the body tube, it is retrieved through a

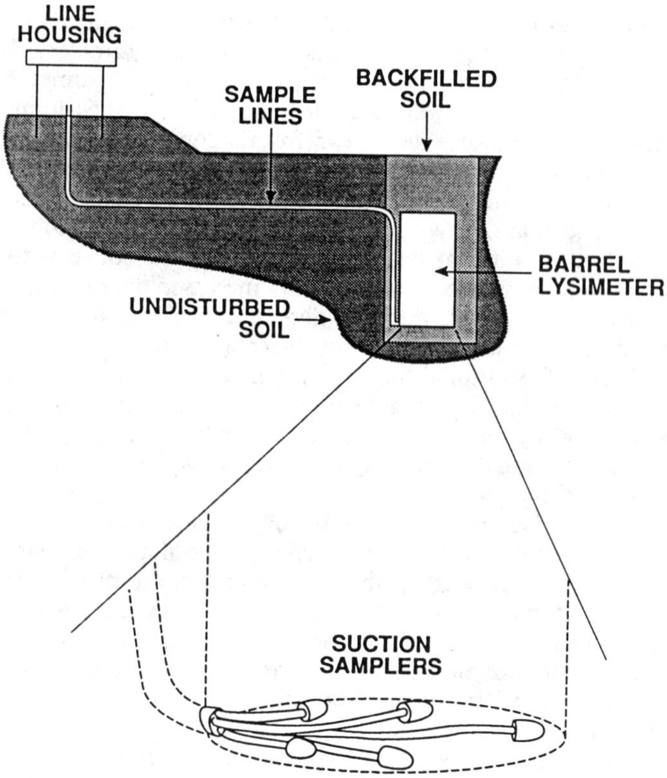

FIG. 19 Barrel Lysimeter

discharge line extending to the base of the porous cup. In shallow installations, with the body tube extending above the soil surface, the discharge line is sometimes inserted and removed as needed. For deeper installations, the discharge line is permanently installed. At the surface, the line is connected to a sample collection flask. Suction is applied to the flask, and liquid is pulled from the sampler, up the discharge line, and into the collection flask. Cole **(42)** constructed an array of samplers that were attached to a vacuum tank connected to an electric power source. This system allowed remote operation at a constant suction. Wengel and Griffen **(67)** described methods by which samplers can be connected to a central control board and operated remotely. Brown et al **(68)** employed a solar panel to power a similar setup. Chow **(44)** described a sampler that shuts off automatically when the desired sample volume has been collected.

7.5.1.2 *Pressure-Vacuum Lysimeters*—Sampling methods are described in U.S. EPA **(12)**, by Soilmoisture **(18**[7]**)** and by Timco **(25**[7]**)**. To sample, suction is applied to the system via the pressure-vacuum line. The discharge line to the sample bottle is clamped shut during this time. When sufficient time has been allowed for the unit to fill with pore-liquid, suction is released and the clamp on the discharge line is opened. Gas pressure (for example, air or nitrogen; see 7.6.2) is then applied through the pressure-vacuum line. This forces the sample through the discharge line and into the collection flask at the surface **(12)**. A variety of systems have been developed by which the pressure, suction, and sample volume can be controlled remotely or manually **(44, 49, 67, 69)**.

7.5.1.3 *High Pressure-Vacuum Lysimeters (Lysimeters With a Transfer Vessel)*—Sampling methods may be found in U.S. EPA **(12)**, in Soilmoisture **(18**[7]**)** and in Timco **(25**[7]**)**. When suction is applied to the system, it extends to the porous section through an open, one-way check valve at the bottom of the transfer vessel or chamber. A second one-way check valve in the discharge line is closed during this time. As soil solution enters the sampler it is pulled by the suction into the transfer vessel or chamber through a line attached to the open valve at its base. The sample is brought to the surface by releasing the suction and applying pressure (for example, air or nitrogen) through the pressure-vacuum line. This shuts the one-way valve to the porous segment and opens the one-way valve in the discharge line. The sample is then pushed to the surface **(12)**. A variety of systems have been developed to control pressure, suction and sample volume remotely or manually **(44, 49, 67, 69)**.

7.5.1.4 *Suction Lysimeters With Low Bubbling Pressures (Samplers With PTFE Porous Sections)*—Sampling methods for this group of samplers are a bit different than those for the three designs described in 7.5.1.1, 7.5.1.2, and 7.5.1.3. This system is designed to allow the soil pore water extraction process to occur separately from the movement of the collected pore water to the transfer vessel. The only difference is that maximum sampling suctions for these units are much lower (see 7.6.1.3).

7.5.1.5 *Filter Tip Samplers*—Sampling methods may be found in Torstensson and Petsonk **(32)**. Samples are collected by first evacuating the sample vial. The vial is then inserted in the sampling adaptor that contains a two way hypodermic needle. The adaptor is then lowered down the extension pipe. When the adaptor connects with the nozzle of the filter tip, the needle penetrates the septa in the vial and in the filter tip. Sample then flows through the porous segment and into the sample vial due to the negative pressure in the vial. As sample is collected, the negative pressure in the vial falls towards that of the pore-liquid tension. When these negative pressures are equal, sampling ends and the sample vial is retrieved. The standard sample volume is about 35 mL. However, by connecting several vials in series, sample volumes of up to 500 mL can be obtained.

7.5.1.6 *Experimental Suction Samplers*—Cellulose-acetate hollow-fiber samplers, membrane filter samplers, and vacuum plate samplers are operated using the same general technique as for vacuum lysimeters. Jackson et al **(33)** sampled from soil columns using cellulose-acetate hollow-fiber samplers subjected to a constant suction of 81 cbar. At this suction, they were able to extract samples for chemical analyses from silty loams with moisture contents ranging from 20 to 50 %. Silkworth and Grigal **(59)** compared the performance of these samplers to suction lysimeters. They found that cellulose-acetate hollow-fiber samplers fail more often than suction lysimeters. In membrane filter samplers, the "collectors" provide hydraulic contact between the soil and the samplers. Liquid is drawn by capillarity into the "collectors." When suction is applied, liquid flows through the "wick," the prefilter, and finally the membrane filter. The prefilter reduces clogging of the membrane filter by fine soil materials **(9)**. Stevenson **(35)** recommended using a suction of between 50 and 60 cbar when sampling with membrane filter samplers. A variety of constant suction methods for sampling with vacuum plates are described by Morrison **(1)**.

An advantage of the larger plates is that they have large contact areas with the soil. Therefore, larger sample volumes can be collected in shorter times than with vacuum lysimeters which have porous sections with smaller surface areas.

7.5.2 *Comments:*

7.5.2.1 Nagpal **(70)** recommended several consecutive extractions of liquids during a sampling event and use of only the last one for chemical analyses. The purpose of this is to flush out cross contaminants from previous sampling periods, and to ensure that any porous segment/soil solution interactions have reached equilibrium. Debyle et al **(50)** also suggested discarding the first one or two sample volumes when sampling dilute solutions with newly flushed (HCL method) and installed samplers. The purpose of this is to allow cation exchange between the porous segment and the pore-liquid (caused by the HCL flushing) to equilibrate.

7.5.2.2 Factors which affect the volume and source of a pore-liquid sample include the amount of suction applied, the schedule of suction application, the pore-liquid content, the distribution of pore-liquid, the soil grain size distribution, the soil structure, the porous section design, and the porous section age.

7.5.2.3 Samples collected with lower suctions (about 10 cbar or less) tend to come from liquids migrating through soil macropores **(1)**. Samples collected with higher suctions (greater than about 10 cbar) also include fluids held at higher tensions in micropores. The sampler may disrupt normal flow patterns due to the applied suctions. The effects may extend several meters from the sampler although the area nearest the sampler is most disturbed **(71, 72, 73)**. This disturbance causes samples to be averages of the affected flow area rather than point samples **(1)**. Warrick and Amoozegar-Fard **(72)** developed an approximate analytical equation which can be used to estimate the maximum radius of influence on the flow regime by a suction sampler. Narasimhan and Driess **(74)** developed a numerical technique to simulate the effects of suction samplers on the pore-liquid regime.

7.5.2.4 Sampling with falling suction produces samples with compositions that are "averages" of the liquids held at the range of tensions applied. Because suctions and therefore inflow rates decrease with time, these "averages" are weighted toward those portions of the samples obtained in early times. Samples collected over prolonged periods (due to slow inflow rates) are "averages" of the liquids fluxing past the sampling region during those times.

7.5.2.5 During wet periods, samplers affect a small volume of soil and pull liquids from a sequence of pores that may include macropores. During dry periods samplers affect a larger volume of soil, draw from micropores because the macropores have been drained, and collect less liquid **(75, 76)**. The net result of this is that sampled soil solutions are "averaged" over different volumes and derived from different pores as a function of the soil moisture content and distribution.

7.5.2.6 Soil textures and pore-liquid tensions control the amount of liquid that can be removed by a sampler and its radius of influence. The slope on the pore-liquid release curve for a sand is greater than that for a clay at low pore-liquid tensions (see Fig. 20). This indicates that there

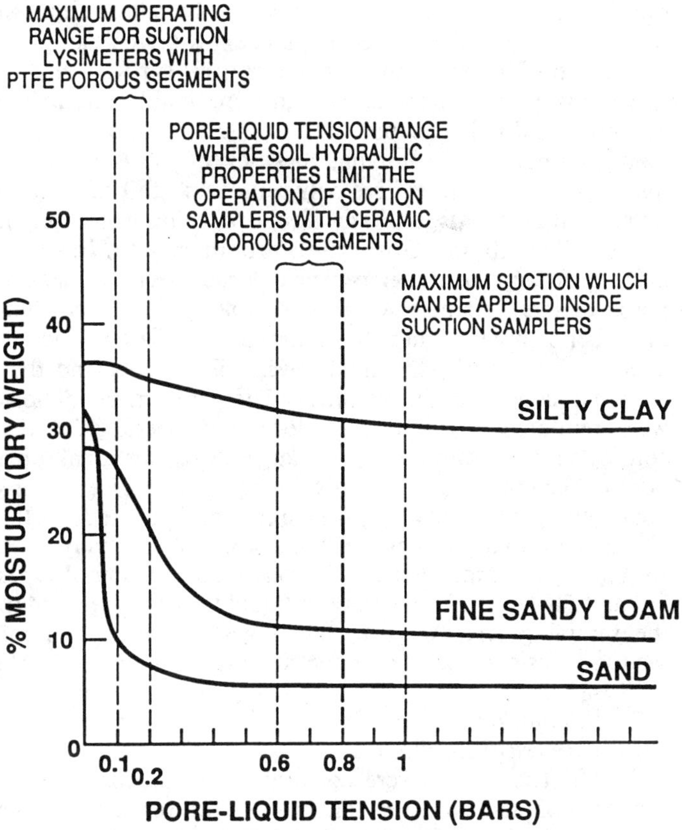

FIG. 20 Water Release Curves for Three Soils, Showing Operating Conditions for Suction Samplers

will be a larger quantity of pore-liquid released from a sand than from a clay for an equal change of pore-liquid tension at these low tensions. At higher tensions, the slope of a clay pore-liquid release curve is greater than that for a sand (see Fig. 20). This indicates that more pore-liquid will be released from a clay than from a sand for an equal change in pore-liquid tension at the higher tensions. A consequence of this is that suction samplers may not obtain samples from coarse grained soils at higher pore-liquid tensions. Morrison and Szecsody **(30)** found that (under the conditions of their study) radii of influence for suction lysimeters ranged from 10 cm in coarse soils up to 92 cm in fine grained soils.

7.5.2.7 Hansen and Harris **(20)**, demonstrated that intake rates may vary substantially due to variability in the ceramic sections from one manufacturer's batch to another. As discussed in 7.4, the intake rate of a sampler is also a function of the degree of clogging. As discussed in 7.6, the range of pore-liquid tensions over which a sampler can operate is a direct function of the maximum pore size of the porous section and the surrounding silica flour pack. Finally, Morrison and Szecsody **(30)** found that the radius of influence of a sampler increases with the diameter of the porous section.

7.5.2.8 Because of these factors the following recommendations have been made for sampling with suction lysimeters. Hansen and Harris **(19)**, suggested using uniform initial suctions, short sampling intervals, and uniform sampling times for different sampling events and locations to increase the uniformity of samples. Debyle et al **(50)** also recommended sampling with uniform suctions that do not significantly exceed the tension at which the percolating soil solution is being held. U.S. EPA **(12)** suggested sampling after infiltration events such as rain storms, spring melts, or irrigations as these periods of high pore-liquid content are accompanied by higher pore-liquid flow rates and contaminant transport. For sampling these events, it is useful to install samplers at interfaces between coarse and fine materials to take advantage of any liquid perching which might occur. Silkworth and Grigal **(59)** recommended using samplers with large diameter ceramic porous sections (as opposed to small diameter ceramic samplers, or hollow cellulose fiber samplers) since they showed less of a tendency to alter the pore-liquid, they had lower failure rates, and they collected larger sample volumes. These recommendations were reinforced by van der Ploeg and Beese **(73)** who concluded that samplers with large cross sectional area porous sections used with low extraction rates (suctions approaching those of the pore-liquid tensions) reduce the effects of sampling on compositions of samples. Finally, U.S. EPA **(12)** recommended that porous section material types be carefully chosen based on pore-liquid tensions expected in the sampling area. Operational ranges of various porous section types are discussed in 7.6 and are presented in Table 1.

7.6 *Limitations:*

7.6.1 *Physical Limitations:*

7.6.1.1 The most severe constraint on the operation of suction samplers involves soil around the porous sections becoming so dry (and pore-liquid tensions so high) that samples cannot be collected. The limiting factors in these conditions will be the porous segment or the soil hydraulic properties. For porous segments with bubbling pressures less than 60 cbar (for example, PTFE), the porous segment will be the limiting factor because the high suctions required to move liquids into the samplers will cause meniscuses in the porous segments to break down and air to enter. Soil hydraulic properties will be the limiting factors for porous segments with bubbling pressures greater than 60 cbar (for example, ceramics) because unsaturated hydraulic conductivity of the soil and pressures gradients across the porous segments will be so low (due to high pore-liquid tensions) that flow into the samplers will be negligible.

7.6.1.2 The maximum suction that the saturated porous section of a sampler can withstand before air enters is a function of the pore configuration and size, and its hydrophilicity/hydrophobicity (see Appendix X1 and **(65)**). The following variation of the capillary rise equation **(8, 12, 18, 77)** combines these factors:

$$P_b = \frac{-2\delta \cos\alpha}{r}$$

where:

P_b = bubbling pressure (gage), units – FLT^{-2},
δ = surface tension between pore-liquid and air, units – FT^{-1},
α = contact angle between the liquid and the material of the porous segment, *D*, and
r = maximum pore radius of the pore segment, units – L.

This equation shows that the bubbling pressure decreases with increasing contact angle and with increasing maximum pore radius. The maximum sampling suction that can be applied is 100 cbar (1 atmosphere). For a hydrophilic material (that has an acute contact angle) the maximum pore size that will allow the application of 100 cbar of suction is 2.8 μm. For a hydrophobic material (that has an obtuse contact angle) a smaller pore size will be required **(65)**. The maximum pore sizes of presently available ceramics (that are hydrophilic) used for suction lysimeters and filter tip samplers vary from 1.2 to 3 μm (as measured by the bubbling pressures) **(18[7], 45, 78[7])**. The maximum pore sizes of cellulose-acetate hollow-fibers and membrane filters range from less than 2.8 μm and 0.4 to 5.0 μm respectively **(1, 35, 36)**. These pore sizes result in maximum sampling suctions near 100 cbar. Therefore, these materials will not allow air to enter during sampling, and the limiting factors will be the soil hydraulic properties. The combination of soil limiting effects result in negligible sampling rates when pore-liquid tensions are above 60 cbar (for coarse grained soils) to 80 cbar (for finer grained soils) **(45)**. At tensions above these levels, inflow rates are too low to allow sampling.

7.6.1.3 The maximum pore sizes of presently available porous PTFE segments for suction lysimeters range from about 15 to 30 μm (calculated from bubbling pressures) **(25[7])**. These pore sizes allow maximum sampling suctions of about 10 to 21 cbar **(25)**. The hydrophobicity of PTFE will further reduce the magnitude of the maximum sampling suction. Applied suctions of greater than 10 to 21 cbar (or less) will cause air to enter, and sampling to cease. Because a suction greater than 10 to 21 cbar cannot be applied to these samplers, pore-liquids held at tensions greater than 10 to 21 cbar cannot be sampled with these devices. Because of this, PTFE will be the limiting factor when it is used for the porous segment. A consequence of the small suction range available to PTFE porous sections is that only very moist soils approaching saturation may be sampled **(17)**.

7.6.1.4 The silica flour pack, that has smaller pore sizes than PTFE, can act as an extension of the porous segment, and may extend the range of suctions that can be applied to the sampler. Everett and McMillion **(45)** found that the pack extended the suction range of earlier, larger pore size PTFE (70 to 90 μm) from less than 4 to 7 cbar. Timco **(25[7])** suggested that the operational range of the presently available PTFE samplers (15 to 30 μm) can be extended from 10 to 21 cbar to between 61 to 71 cbar when "properly" installed within a silica flour pack (this has not been verified in peer reviewed literature). For this to be true, the silica flour pack must be able to remain saturated over the range of applied suctions. However, the results of Everett and McMillion **(45)** suggest that the bubbling pressure of the silica flour pack is only 7 cbar. Trainor **(27)** found that even if these samplers are "properly" installed, air may still enter if applied suctions exceed pore-liquid tensions by more than 30 %. Pore-liquid tensions are not always known, and technicians may not carefully control applied suctions. In addition, pressurization of pressure-vacuum lysimeters for sample retrieval appears to damage the silica flour pack **(27, 28)**. Thus, dependency on the silica flour pack to provide the needed suction range is an extremely limited option. Because of this, suction lysimeters with PTFE porous sections are limited to near saturated sampling and have been classified separately (see Fig. 20).

7.6.1.5 Samples can be collected (using ceramic porous sections) from clays with high pore-liquid tensions (approaching 60 to 80 cbar). However, because liquid inflow rates are low at higher tensions, the amount of time required to collect sufficient sample volumes may exceed the maximum allowable holding time for many chemical analyses. Law **(76)** pointed out that when soils have liquid contents that allow little or no sample collection (high pore-liquid tensions), there is little or no liquid movement in the soil. Consequently, there will be little or no contaminant migration. If samples of pore-liquids held at tensions above 60 to 80 cbar are desired, soil core sampling with subsequent laboratory liquid extraction may be used **(76)**. However, Law **(76)**, and Brown **(79)** concluded that results from the two sampling methods are not comparable. Liquid from soil core samples will include constituents that are held at tensions greater than 60 to 80 cbar and that would not be picked up by suction samplers. Because of this and because samples removed from soil cores may include some of the constituents from the soil itself (for example, cations preferentially adsorbed in electrical double layers) or sorbed organics, Law **(76)** concluded that soil cores are more conservative estimators of cation contaminant presence in soil. Brown **(79)** concluded that organic contaminant concentrations derived from soil cores and pore-liquid samplers are not comparable because of preferential sorption of some compounds. Amter **(80)** developed an alternative to extraction of samples from soil cores. The method involves injecting a chemically blank fluid through an existing lysimeter. After a time, the fluid (now containing dilute pore-liquid) is recovered through the sampler and analyzed. The results, although qualitative, were shown to correlate well with known relative pore-liquid constituent concentrations.

7.6.1.6 Suction samplers may not intercept macropores because of the small size of their porous sections. Because of this, they may miss the majority of flow at high moisture contents in structured soils **(81)**. The ability to intercept this flow can be enhanced by installing the samplers in large diameter silica flour packs. However, this involves drilling larger holes. Because suction samplers only sample when suction is applied, they may miss infiltration events unless a constant suction is applied. Therefore, under conditions of high moisture content in structured soils, free drainage samplers are recommended (see Section 8) **(81)**. Pore-liquid composition changes with time. Because suction samplers sample over an extended period (especially in drier soils), the resulting sample should be considered an average of the total flux past the sampler during the sampling interval.

7.6.1.7 A major factor limiting the operation of shallow suction samplers in cold climates is that pore-liquid may freeze near the porous segments. In addition, liquid may freeze within porous segments and lines, preventing sample retrieval and perhaps fracturing the sampler during ice expansion. Because of this, lines should be emptied before the onset of cold weather. Additionally, some soils tend to heave during freezing and thawing. Consequently, the samplers may be displaced in the soil profile, resulting in a break of hydraulic contact **(12)**.

7.6.2 *Chemical Limitations:*

7.6.2.1 The inherent heterogeneities of unsaturated pore-liquid movement and chemistry limit the degree to which samples collected with suction samplers can be considered representative. This is because the small cross sectional areas of suction samplers may not adequately integrate for spatial variability in liquid movement rates and chemistry **(51, 82, 83)**. Biggar and Nielsen **(84)** suggested that results of analyses from suction lysimeter samples are good for qualitative comparisons, but that they cannot be used for quantitative analysis unless the variabilities of the parameters involved are established. Law **(76)** came to similar conclusions, stating that results from suction lysimeter sampling could not be used for quantitative mass balance studies.

7.6.2.2 Well structured soils have two distinct flow regions including macropores (for example, cracks, burrows, and root traces) and micropores. Under saturated conditions, liquids move more rapidly through macropores than through micropores. Because of this, the movement of liquid-borne pollutants into the finer pores may be limited. Consequently, pore-liquids in macropores may have different chemistries than those in micropores **(85)**. This is enhanced by the fact that oxygen content in macropores can change in a matter of hours during an infiltration event, whereas micropores may remain suboxic regardless of flow conditions **(75)**. In addition, micropores are less susceptible to leaching than macropores **(1, 86, 87, 88)**. Because of these differences, sample chemistry can vary widely from location to location and from time to time depending on the amount of liquid drawn from these two flow systems.

7.6.2.3 Suction samplers may affect pore-liquid chemistry as it is being sampled. The major sampler related factors that can affect the sample chemistry are the porous segment material and sample storage time. The degree of chemical interaction may also be affected by the amount of porous section clogging **(1)**. Clogging slows the rate of flow through the porous section so that contact time and chances for chemical interaction are increased **(50)**. In addition, the types of adhesives used to attach porous segments (for example, epoxy) may alter the pore-liquid chemistry.

7.6.2.4 Interactions between porous materials and liquid can include sorption, desorption, cation exchange, precipitation (for example, ferric precipitation), and screening **(20)**. These interactions can also occur with all other parts of the samplers that the samples contact. However, the much higher surface area of pores within porous segments makes them the most critical element chemically. Table 2 presents the results of a literature review for porous section/pore-liquid interactions. An attempt has been made to document the pertinent features of the listed studies. However, the reader should refer to the original papers to determine if experimental techniques are applicable to the situation of interest. The absence of entries for a constituent relative to a material does not infer absence of interactions. Although studies for membrane filter interactions have been performed, the results have not been included in Table 2. This is because membrane filters are made from a variety of materials that have differing chemical characteristics **(36)**.

7.6.2.5 Suarez **(89)** showed that the pH of a sample may be affected by 0.28 to 0.44 pH units due to CO_2 degassing during sampling. He reduced this error by reducing the gas-liquid ratio in the sampler, and by flushing several sampler volumes of soil solution through the sampler before collecting a sample. Alternately, Suarez **(89)** developed a

TABLE 2 Porous Material Interactions[A,B]

	Material Sorbs Species	Material Desorbs Species[C,D]	Material Screens Species[C,D]	No Significant Interaction[C,D]	No Interaction
Al	. . .	C(2)	. . .	C(16)	. . .
		A(14)	. . .	A(14)	. . .
Alkalinity	. . .	. . .	. . .	SF(11)	. . .
Ca	. . .	C(1,2,18)	. . .	C(3,6,10,11,25)	PTFE(13)
		CAF(18)	. . .	PTFE (3)	. . .
		A(14)	. . .	A(3)	. . .
				FG(18,22)	. . .
				CAF(10)	. . .
C	. . .	FG(22)	. . .	. . .	. . .
CO3	. . .	C(2)	. . .	. . .	. . .
HCO3	. . .	C(2)	. . .	. . .	. . .
Cd	C(11)	. . .	. . .	C(3)	. . .
			. . .	PTFE(3)	. . .
			. . .	A(3)	. . .
Cl	. . .	. . .	. . .	C(11,25)	PTFE(13)
			. . .	SF(11)	. . .
Cr	C(19)	C(3)	. . .	. . .	. . .
		PTFE(3)	. . .		. . .
		A(3)	. . .		. . .
Cu	C(11)	C(3)	. . .	A(3)	. . .
		PTFE(3)	. . .		. . .
Fe	C(11)	PTFE(3)	. . .	C(3,25)	PTFE(13)
		A(3)	. . .	A(14)	. . .
H	. . .		. . .	SF(11)	. . .
K	C(5,6,15)	C(18)	. . .	C(1,25)	. . .
		A(14)	. . .	CAF(18)[E]	
			. . .	FG(18,22)	. . .
Mg	C(6)	C(2,3,11,18)	. . .	C(10,25)	PTFE(13)
		A(3,14)	. . .	PTFE(3)	. . .
		CAF(18)	. . .	CAF(10)	. . .
			. . .	FG(18,22)	. . .
Mn	C(11)	A(3)	. . .	C(3)	PTFE(13)
			. . .	PTFE(3)	. . .
			. . .	A(14)	. . .
Na	C(6)	C(2,18)	. . .	C(1,11,25)	PTFE(13)
		A(14)			
		CAF(18)			
		FG(18,22)			
NH4	C(4,12)		. . .	PTFE(4)	. . .
N	. . .	FG(22)	. . .		. . .
NO2	. . .	. . .	. . .	C(4,5)	. . .
	. . .	. . .		PTFE(4)	. . .
NO3	. . .	. . .	. . .	C(4,8)	. . .
	. . .	. . .	CAF(10)	PTFE(4)	. . .
NO3-N	. . .	. . .	C(10)	. . .	. . .
	. . .	. . .	CAF(10)		. . .
(NO2+NO3)–N	. . .	. . .	. . .	C(5)[F]	
P	C(1,5,8,15,18)	. . .	. . .	CAF(18)	. . .
		. . .	. . .	FG(18)	. . .
PO4	C(4,5,7)	. . .	. . .	PTFE(4)	. . .
		. . .	. . .	CAF(10)	. . .
PO4-P	. . .	. . .	. . .	C(10)	. . .
	. . .	. . .	. . .	CAF(10)	. . .
Pb	. . .	. . .	. . .	. . .	PTFE(13)
SiO2	. . .	C(2)	. . .	. . .	. . .
Si	. . .	. . .	. . .	C(4)	. . .
	. . .		. . .	PTFE(4)	. . .
SO4	. . .	. . .	. . .	C(11)	. . .
Sr	. . .	C(11)	. . .	. . .	. . .
Zn	. . .	C(11)	. . .	. . .	PTFE(13)
High Molecular Weight Compounds	. . .		C(17,21)	. . .	. . .
			CAF(10)		
4-nitrophenol	PTFE(23)	. . .	. . .	. . .	. . .
Chlorinated hydrocarbons	PTFE(23,24)	. . .	. . .	. . .	. . .
Diethyl phthalate		. . .	. . .	PTFE(23)	. . .
Naphthalene	PTFE(23)	. . .	. . .	. . .	. . .
Acenaphthene	PTFE(23)	. . .	. . .	. . .	. . .

model by which pH values could be corrected. He noted that multichamber samplers had minimal pH errors and that pH corrections due to CO_2 degassing were not necessary. Peters and Healy **(53)** found that there was no significant change in

References and Notes on Experimental Techniques[G]

Reference Number in Table	Cited Reference	Porous Section was Washed	Results are a Function of Several Factors	Dilute Solutions were Tested	Experiments were Performed on Nonporous Materials
1	Ref **(137)**	X	...	...	...
2	Ref **(48)**	X	...	...	...
3	Ref **(93)**	X	...	...	...
4	Ref **(138)**	X	...	...	...
5	Ref **(139)**	...	X	...	...
6	Ref **(50)**	X	X	...	...
7	Ref **(52)**	X	...	...	...
8	Ref **(19)**	X	...	...	...
9	Ref **(56)**	...	X	...	...
10	Ref **(34)**	...	...	X	...
11	Ref **(53)**	X	...	...	...
12	Ref **(69)**	...	...	...	...
13	Ref **(1)**	X	...	...	...
14	Ref **(47)**	X	X	X	...
15	Ref **(139)**	X	...	...	...
16	Ref **(58)**	...	X	...	...
17	Ref **(76)**	...	...	...	...
18	Ref **(59)**	X	X	...	...
19	Ref **(75)**	...	X	...	...
20	Ref **(140)**	X	...	...	...
21	Ref **(57)**	...	...	...	...
22	Ref **(36)**	X	X	...	...
23	Ref **(142)**	...	X	X	X
24	Ref **(143)**	...	X	X	X
25	Ref **(17)**	X	X	...	...

[A] Abbreviations:
C = porous ceramic,
PTFE = porous PTFE,
A = porous alundum,
CAF = cellulose-acetate fibers,
FG = fritted glass or glass fibers, and
SF = silica flour.

[B] Comparisons of materials based on this table should be made cautiously. Differing experimental techniques should be considered as a source of differing conclusions. Undocumented factors often include material age and sampling history.

[C] Numbers in parenthesis refer to references cited in the second part of this table. This is indicated by (2).

[D] Valence states are often not reported in studies. This is indicated by (3).

[E] Reference **(59)** found that there is no significant interaction of cellulose-acetate fibers with potassium in solution. The porous section was washed prior to testing and results were found to be a function of several factors.

[F] Reference **(139)** found that there was no significant interaction of porous ceramic with nitrate plus nitrite nitrogen in solution. The results are a function of several factors.

[G] Absence of information on experimental techniques means that the techniques were not specified in the citation.

pH due to CO_2 degassing during long sampling times, although they recognized that pH changes could occur when the solution is originally more acidic than that which they tested. Ransom and Smeck **(90)** and Anderson **(75)** suggested purging the sampler with N_2 to preserve the subsurface redox states when sampling for redox dependent ions. Filter tip samplers do not use a purging gas, therefore, pore-liquid redox states are preserved in the samples.

7.6.2.6 Nightingale et al **(31)** indicated that normal suction sampling techniques are not suitable for sampling volatile organic compounds due to potential loss. Wood et al **(91)** devised a body tube connected to a purging chamber that is in turn connected to a trap packed with resin. Compounds that volatilize during sampling are captured in the trap. Pettyjohn et al **(92)** described a suitable system for sampling highly volatile organics. However, the reported system was limited to a maximum sampling depth of 6 m and a small sample volume (5 to 10 mL). Torstensson and Petsonk **(32)** described methods that can be used to collect samples with filter tip samplers that result in no head space in the sample vial and consequently no loss of volatile compounds.

7.6.2.7 A newly forming consensus is that the effects of suction samplers (when properly pre-treated) on sample chemistry of non-dilute solutions are generally less significant than the inherent uncertainties of sampling discussed in 7.6.2.1 and 7.6.2.2 **(17, 53, 93)**.

7.6.3 *Microbial Limitations*—Viruses or bacteria are sometimes monitored in areas where there are livestock lots, leach fields, septic tanks, or sewage sludge spreading plots. However, it has been found that although porous ceramics will allow viruses to pass, they will screen out bacteria (for example, escherichia coli and fecal coliform) **(12, 26, 94, 95)**.

8. Free Drainage Samplers

8.1 Free drainage samplers are classified differently by various authors, depending on the installation methods. Many free drainage samplers are installed in the side walls of trenches and are referred to as trench lysimeters. However, free drainage samplers are also installed in the walls of vertical caissons. The principle behind each of the samplers is essentially the same. However, the materials and construction differ. The general types of free drainage samplers include:

8.1.1 *Pan lysimeters*,

8.1.2 *Glass block lysimeters*,

8.1.3 *Trough lysimeters,*
8.1.4 *Vacuum trough lysimeters,*
8.1.5 *Caisson lysimeters,*
8.1.6 *Wicking soil pore-liquid samplers,* and
8.1.7 *Sand filled funnel samplers.*

8.2 *Operating Principles:*

8.2.1 A free drainage sampler consists of some sort of collection chamber that is placed at depth in the soil. Pore-liquid in excess of field capacity is free to drain through soil (usually through macropores) under the influence of gravity. This gravity drainage creates a slightly positive pressure at the soil-sampler interface causing fluid to drip into the sampler. Hence, these samplers collect liquid from those portions of the vadose zone that are intermittently saturated due to events such as rainfall, flooding, or irrigation. Some free drainage samplers apply a small suction in order to break the initial surface tension at the soil-sampler interface. Samples are retrieved either by accessing the samplers at depth or by drawing samples to the surface through a suction line.

8.2.2 As described in 4.2, suction samplers can also be used to sample free drainage flow. However, the small area of those samplers compared to the spacing of macropores limits their usefulness for this application. As described in 7.4.5.3, Hornby et al **(66)** developed an installation that includes pressure-vacuum lysimeters within an encased monolith. This enhances collection of macropore flow with these samplers.

8.3 *Description:*

8.3.1 *Pan Lysimeters:*

8.3.1.1 A pan lysimeter generally consists of a galvanized, metal pan of varying dimensions. A copper tube is soldered to a raised edge of the pan. Plastic or vinyl tubing connects the copper tube to a collection vessel. Any liquid that accumulates on the pan drains through the tubing into the vessel (see Fig. 21) **(26)**.

8.3.2 *Glass Block Lysimeters*—Barbee and Brown **(81)** developed free drainage samplers made from hollow glass bricks. These glass bricks, that are produced for ornamental masonry work, have dimensions of 30 by 30 by 10 cm and have a capacity of 5.5 L. To build a sampler, nine holes, 0.47 cm in diameter, are drilled along the perimeter of one of the square surfaces of a brick. Nylon tubing is inserted into one of the holes to allow for sample removal. The collecting surface is fitted with a fiberglass sheet to improve contact with the soil. Pore-liquid collection is enhanced by a raised lip along the edge of the surface (see Fig. 22).

8.3.3 *Trough Lysimeters:*

8.3.3.1 Trough lysimeters, also known as Ebermayer lysimeters, rely on a trough or pail to collect pore-liquid. In order for the edges of the sampler to maintain a firm contact with the soil, a fiberglass screen is suspended inside the trough. The screen is lined with glass wool and covered with soil until the soil is even with the top of the trough **(96)**.

8.3.3.2 Morrison **(1)** reported a trough lysimeter in which two parallel metal rods are inside the trough, in contact with the bottom side of the screen, and bent toward the collection tube. Liquid that enters the trough migrates along these rods towards the collection tube in response to capillary forces (see Fig. 23). A modification of this design consists of a metal trough with a length of perforated PVC pipe mounted inside.

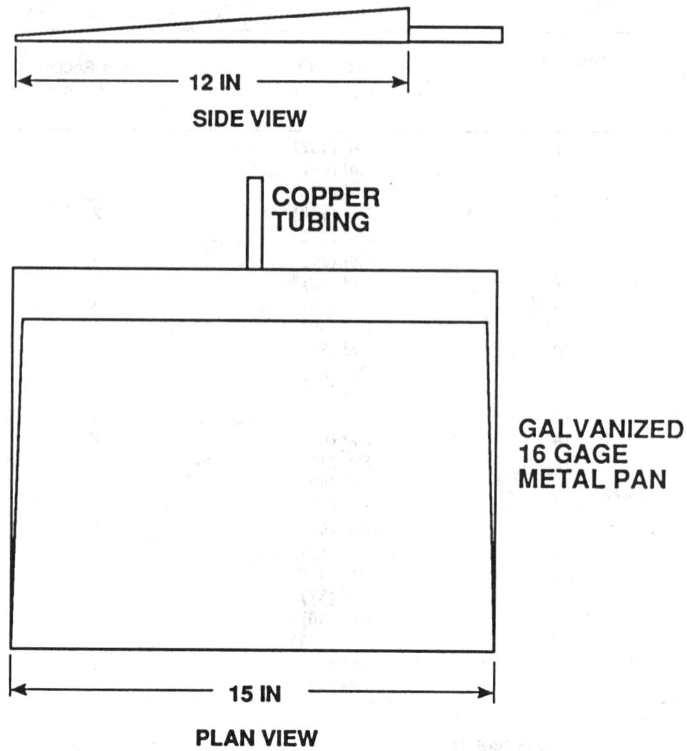

FIG. 21 Example of a Pan Lysimeter

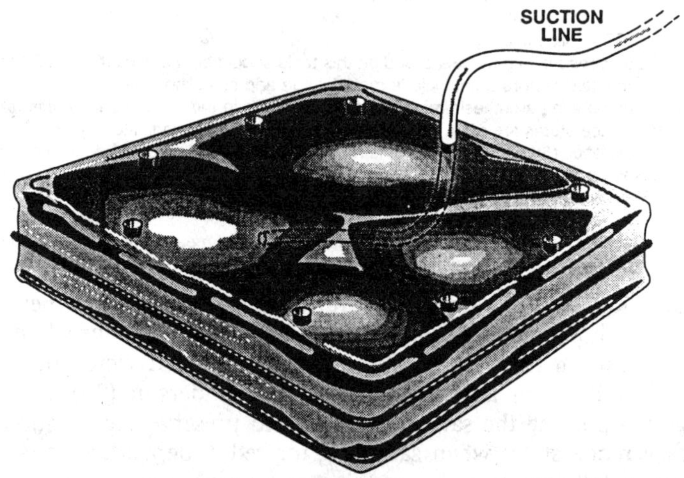

FIG. 22 Glass Block Lysimeter

The trough is filled with graded gravel so that coarse material is immediately adjacent to the PVC pipe and fine sand is at the edges and the top of the trough. The pipe is capped at one end while the other end is connected to a sample container via a drainage tube **(1)**.

8.3.4 *Vacuum Trough Lysimeters*—Montgomery et al **(97)** described a vacuum trough lysimeter consisting of a metal trough equipped with two independent strings of ceramic pipe, each 13 mm in diameter. The design, otherwise similar to trough lysimeters, allowed extraction of samples under applied suctions of up to 50 cbar. The ceramic pipes act as a vacuum system, and samples are extracted through a suction line.

8.3.5 *Caisson Lysimeters*—A caisson lysimeter consists of

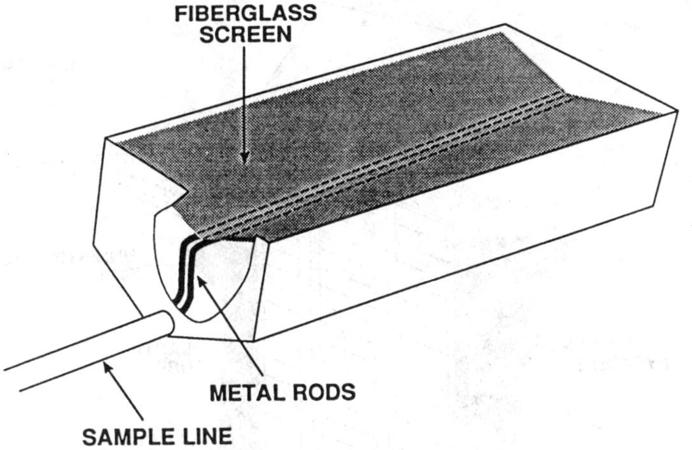

FIG. 23 Trough Lysimeter

collector pipes, radiating from a vertical chamber **(1)**. A design used by Schmidt and Clements **(98)** consists of a nearly horizontal, half-screened PVC casing (see Fig. 24). Schneider et al **(99)** designed a similar system consisting of: a 15.2 cm diameter stainless steel tube extending diagonally upward through the caisson wall into the native soil; a screened plate assembly within the tube to retain the soil; a purging system that can be used to redevelop the sampler when it becomes clogged; and an airtight cap that prevents exchange between the air in the caisson and the soil air.

8.3.6 *Wicking Soil Pore-Liquid Samplers*—Hornby et al **(66)** described a wicking sampler, alleged to combine the attributes of free drainage samplers and pressure-vacuum lysimeters. The sampler collects both free drainage liquid and liquid held at tensions to about 4 cbar. A hanging "Hurculon" fibrous column acts as a wick to exert a tension on the soil pores in contact with a geotextile fiber which serves as a plate covering a 30.5 by 30.5 by 1.3 cm pan. The terminus of the fibrous column is sealed into the cap of a tubular sample collector. The collection tube also contains an inlet pressure-vacuum line and a sample collection tube. Materials for the sample collector depend on the constituents being sampled. Glass and PTFE were recommended materials when sampling for organics (see Fig. 25) **(66)**.

8.3.7 *Sand Filled Funnel Samplers*—K. W. Brown and Associates **(100)** discussed a sand-filled funnel for collecting freely draining liquid. The funnel is filled with clean sand and inserted into the sidewall of a trench. The funnel is connected through tubing to a collection bottle. Application of suction to a separate collection tube pulls the sample to land surface (see Fig. 26).

8.3.8 *Comments*—The dimensions of the free drainage samplers discussed are purposely left vague. Because the samplers collect fluid flowing primarily through macropores, the dimensions are often dictated on a site-by-site basis by the configurations and spacings of the macropores.

8.4 *Installation Methods:*

8.4.1 *Installation:*

8.4.1.1 Free drainage samplers are commonly installed into the side walls of trenches or caissons. The trenches can be dug either by hand or with backhoes. But they should be stabilized with timbers and siding if deeper than 1.5 m, in order to allow safe access **(26)** (see Occupational Health and Safety Administration Regulations). All wood used in the construction of permanent trenches should be treated with preservatives to protect it against degradation due to semi-saturated conditions. This may pose a problem when monitoring for organics. Any spaces between the bare trench side walls and the siding are filled with soil and peagravel to allow for free drainage. The excavations should be covered to provide positive surface drainage away from the area. Some free drainage samplers require only temporary excavations for installation. After the samplers have been installed, the excavations are backfilled with native soil.

8.4.1.2 Caissons for housing free drainage samplers are constructed with corrugated culverts or concrete drainage pipes. Schneider and Oaksford **(101)** installed caissons by excavating soil from within a concrete pipe using a crane operated shovel and manual labor. Each concrete pipe section, weighing 222.5 kN (25 tons), was set in place with a

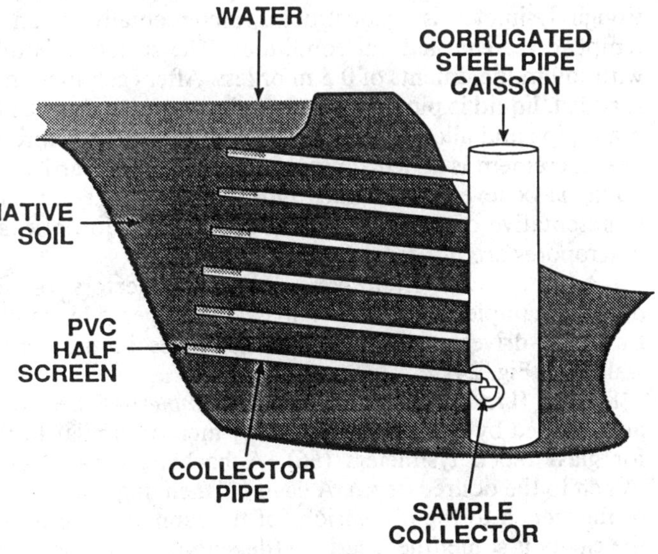

FIG. 24 Example of a Caisson Lysimeter

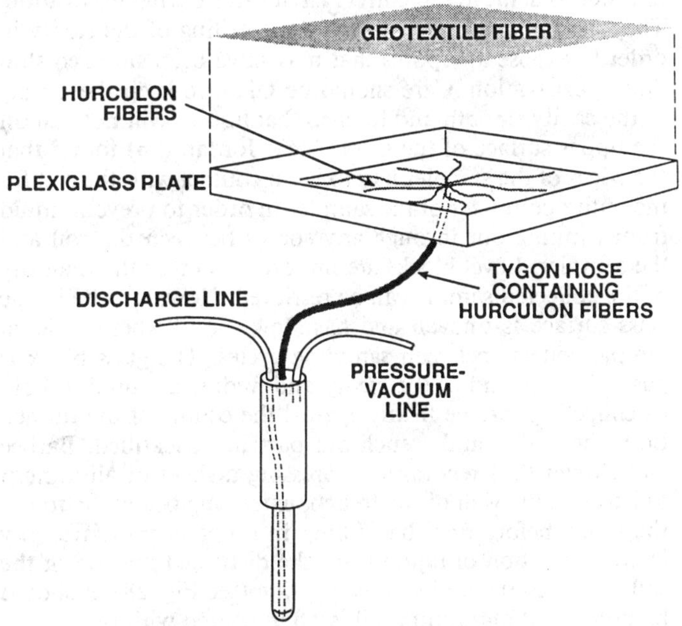

FIG. 25 Wicking Type Soil Pore-Liquid Sampler

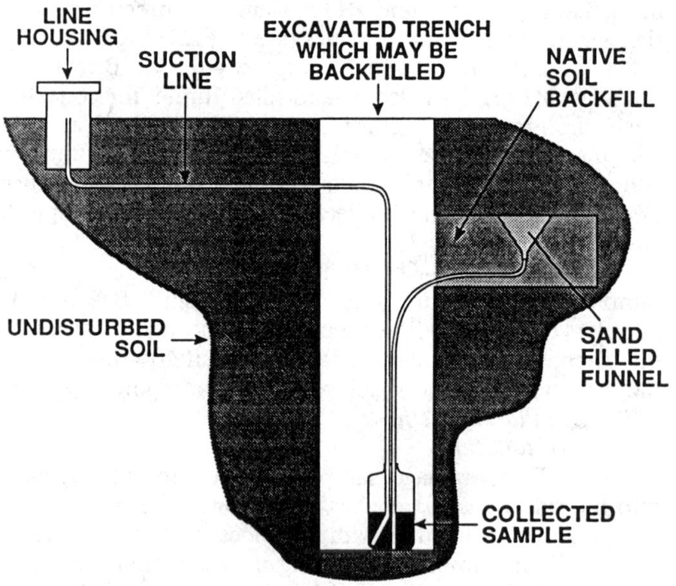

FIG. 26 Sand Filled Funnel Sampler Installation

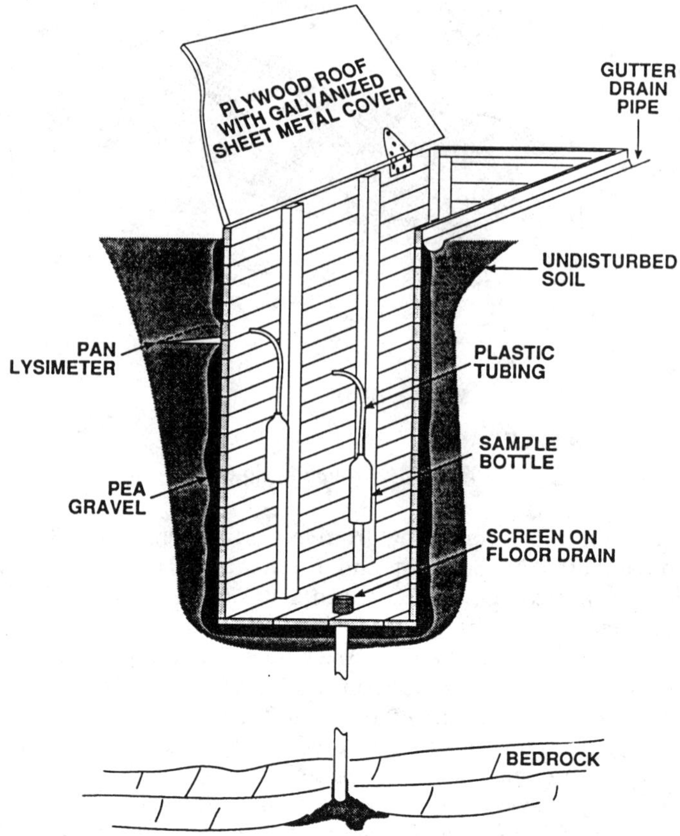

FIG. 27 Example of Pan Lysimeter Installation

crane. As excavation inside the pipe progressed, the pipe advanced downward under its own weight.

8.4.1.3 *Pan Lysimeters*—A pan lysimeter can often be pushed or driven directly into the side wall of a trench. However, if the soil is resistant, an opening for the sampler can be created by hammering a sheet metal blade into the soil profile with a sledge hammer. The pan is placed in the side wall so that it slopes gently toward the trench. Any voids above or below the pan are filled with soil **(26)**. The end of the copper tubing is allowed to project through the trench siding and is connected to plastic tubing and a sample bottle (see Fig. 27).

8.4.1.4 *Glass Block Lysimeters*—A glass block lysimeter is installed in a cavity that is excavated in the side of a trench. Barbee and Brown **(81)** used a wooden model of the sampler in order to achieve the correct cavity size during excavation. They used a small knife to score the ceiling of the cavity in order to expose any pores that may have been smeared shut during excavation. Care should be taken to keep the ceiling of the cavity smooth and level so that liquids will not run off the upper surface of the glass block. Jordan **(96)** found that the edges of the sampler had to be in contact with the soil for the entire perimeter of the sampler in order to prevent liquid from running out through any spaces between the soil and the sampler. Level blocks are important so that the majority of the collected sample can be retrieved. However, the inside glass surface is uneven and has "low spots" where residual sample collects between sampling cycles. The glass block is pushed to the end of the cavity and wedges are used to hold its collecting surface firmly against the ceiling of the tunnel. Both the cavity and trench are partially backfilled. Barbee and Brown **(81)** recommend pressing a sheet of aluminum foil against the wall of the trench, extending below the top of the brick, before final backfilling in order to minimize any lateral migration of liquid from the disturbed portion of the soil profile to the undisturbed portion (see Fig. 28). It should be noted that aluminum foil is often coated with oil.

8.4.1.5 *Trough Lysimeters*—Trough lysimeters are installed in the same manner as glass block lysimeters (see Fig. 28).

8.4.1.6 *Vacuum Trough Lysimeters*—The vacuum trough lysimeter described by Montgomery et al **(97)** is housed in a box-like structure, with four walls and a floor, but no ceiling. The floor of the structure and the lower portions of the walls are made of steel. The upper portions of the walls are composed of fiberglass coated plywood. A slotted, plastic drain pipe is set 20 mm above the floor of the structure and is surrounded by gravel. The soil profile surrounding the trough lysimeter is reconstructed incrementally in an attempt to recreate natural conditions. The structure is filled with soil in increments of 0.5 m or less. After each increment is added, liquid is piped slowly into the structure through the drain pipe and allowed to drain back out for 24 h before the next increment is added. This working of the soil particles by liquid is believed to produce bulk densities that are fairly representative of the undisturbed conditions. However, soil macropores are not reproduced.

8.4.1.7 *Caisson Lysimeters*—Lateral collectors or free drainage samplers are installed in cavities augered by hand or by power-driven equipment through holes in the caisson walls (see Fig. 24).

8.4.1.8 *Wicking Soil Pore-Liquid Samplers*—These units are installed by the trench and cavity method similar to that for glass block lysimeters **(66)**. A backhoe excavates the trench to the desired depth. A cavity is then dug into the wall of the trench to the dimensions of the sampler. The roof of the cavity is sometimes scarified (depending on the soil type) to remove smearing caused during excavation. The sampler

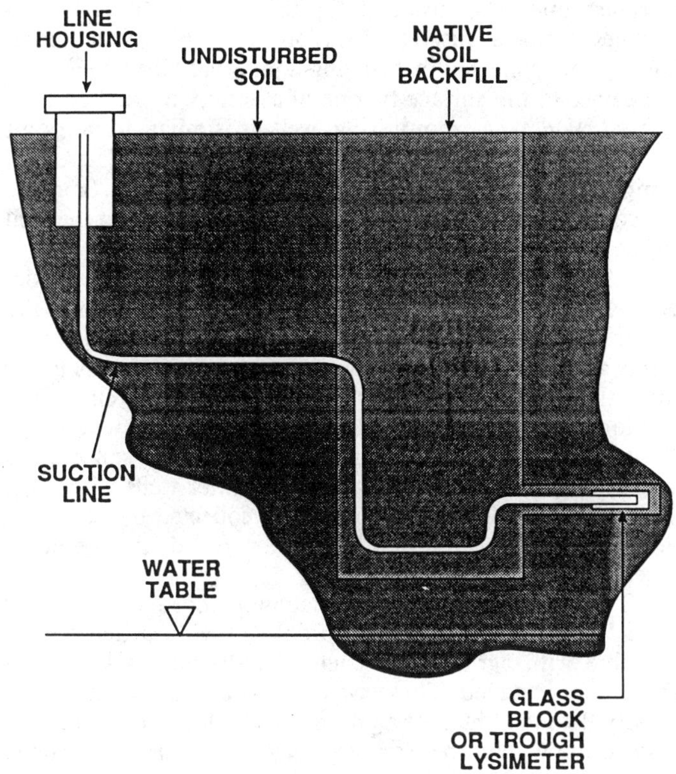

FIG. 28 Example of Glass Block or Trough Lysimeter Installation

is then forced tightly into place to ensure good contact with the roof of the cavity. The cavity is large enough to accommodate the sampler, the hanging wick and the collection tube. The pressure-vacuum line and the sample collection line are extended to the surface. During backfilling of the tunnel and trench, the bulk density of the fill should be equal to or greater than the native soil.

8.4.1.9 *Sand Filled Funnel Samplers*—The installation procedure for these samplers is similar to that used for glass block lysimeters (see Fig. 26).

8.4.2 *Maintenance:*

8.4.2.1 Where samplers are accessed through permanent trenches and caissons, the sampling station must be protected against flooding due to excessive infiltration. Parizek and Lane **(26)** drilled a floor drain about 27 m through underlying soil into the unsaturated bedrock. This allowed drainage of excess liquid from the floor of the sampling trench, and also decreased the chances of contamination of soil surrounding the structure. Alternately, a sump pump can be used if a drain is not feasible. Parizek and Lane **(26)** also found that stratified soils intensified lateral flow of pore-liquid, thus aggravating any flooding problems. They concluded that flooding may be a problem in humid areas where more than about 5 cm of liquid per week is applied to the land surface.

8.4.2.2 The ground surface above the sampler should be maintained in a fairly representative state. Large housings and excessive traffic around the sampler (causing compaction of the soil) will reduce the amount of infiltration in that area. This will affect the representativeness of the pore-liquid samples.

8.4.3 *Comments*—A significant advantage of samplers installed in the sidewalls of trenches is that in sufficiently cohesive soils, installation produces no disturbance in the overlying soil. In cohesionless, sandy soils, stable cavities may not be possible. As a result, backfilling of the fallen material may be required. This disturbs the soil profile and macropores are not preserved **(97)**.

8.5 *Operation:*

8.5.1 *Methods:*

8.5.1.1 Since pore-liquid flows into free drainage samplers under the influence of gravity, sampling is a relatively simple procedure. Liquid accumulates in the collection device and then drains through tubing into a sample bottle. The sample can be retrieved either through access to the sampling trench or by pulling it to the surface by a suction pump. The wicking pore-liquid sampler allows the application of a slight suction (4 cbar) to improve sampling. However, this design also has a tendency to clog.

8.5.1.2 Jordan **(96)** found that surface tension develops in trough lysimeters at the soil-air interface and prevents some of the liquid from entering the collector. Cole **(41)** addressed this problem by inserting an aluminum oxide disc between the soil and the collection surface, and then applying suction to break the surface tension and draw liquid out of the soil. The problem with this approach is that it requires the soil adjacent to the aluminum oxide disc to be free of roots, cracks, and channels **(96)**. The two parallel rods included in the trough lysimeter design overcome this problem. If one end of the metal rod touches the fiberglass screen, then the surface film of liquid surrounds the rod and the liquid moves down the rod toward the sample container. Two rods, barely touching, facilitate this migration by allowing the liquid to move in response to the capillary forces between them **(96)**.

8.5.2 *Comments*—Under near saturated conditions with macropore flow, free drainage samplers tend to collect larger and more consistent samples than suction samplers. Since free drainage samplers are continuous samplers, they need to be emptied after each infiltration event in order to ensure sample integrity, to prevent sample container overflows, and to prevent cross contamination between hydrologic events **(12)**.

8.6 *Advantages and Limitations:*

8.6.1 *Physical Advantages and Limitations*—A major advantage of free drainage samplers is that they are essentially passive, thus they do not alter pore-liquid flow paths. The major disadvantage of free drainage samplers is that samples can only be obtained when soil moisture conditions are in excess of field capacity. Such saturated conditions usually require constant application of surface liquid, as in the case of agricultural irrigation or at land treatment sites. Under drier conditions, free drainage samplers fail to yield any liquid and suction samplers are required.

8.6.2 *Chemical Advantages and Limitations:*

8.6.2.1 There are both advantages and disadvantages to using free drainage samplers to collect pore-liquid for chemical analysis. A major advantage is that the samplers do not distort natural flow patterns as do suction samplers. Because samples are collected over known areas, quantitative mass balance estimates are possible. Because the samplers are continuous collectors, infiltration events can be sampled without having to go to the field. The major limitation of free drainage samplers is that they cannot sample pore-liquids held at tensions greater than the field capacity. As with all samplers, analytical parameters/sampler material compatibilities should be considered. As an example, samples collected from pan lysimeters should not be analyzed for copper or zinc, particularly if the pH of the collected fluid is below seven.

8.6.2.2 Free drainage samplers tend to collect pore-liquids that drain rapidly through macropores. Since the residence time of this liquid is less than that of liquid moving under tension, the major ion chemistry appears more dilute than the fluid sampled from unsaturated pores with suction samplers. In some cases, this decrease in residence time in combination with other factors can result in an actual change in the chemical signature rather than just an overall dilution. This is because insufficient time may be available for reactions to occur with soil components that act as chemical sources or sinks **(102)**. However, free drainage samplers have large cross-sectional areas and they are cumulative collectors. As a result, they collect samples which average soil heterogeneities and therefore give a more representative picture than suction samplers of chemical movement through wet soil, particularly through well-structured soils **(81)**. In addition, the samplers use suction only to retrieve samples. As a result there is less potential for loss of volatile compounds than with suction samplers **(12)**.

8.6.3 *Microbial Advantages and Limitations*—Since free drainage samplers do not have the minute openings that porous ceramic suction samplers have, they do not screen out colloidal-sized particles and soil bacteria. Consequently, they yield more representative values for suspended solids or BOD measurements **(26)**.

9. Perched Ground Water Samplers

9.1 Perched water occurs where varying permeability layers in the vadose zone retard downward movement of liquid. Over time, liquid collects above lower permeability layers and moisture contents rise until the soil becomes saturated with liquid **(9, 103)**. Once soil becomes saturated, wells and other devices normally installed below the water table can be used to collect samples. Separate guides are available for ground water sampling, therefore, the topics are covered briefly, with reference to appropriate documents.

9.2 Sampling perched liquid is attractive because the perching layer collects liquid over a large area. This integration allows samples to be more representative of areal conditions than suction samples **(103)**. This also allows the sampler to potentially detect contaminants that may not be moving downward immediately adjacent to the sampler. In addition, larger sample volumes can be collected than those that can be obtained by suction samplers. Everett et al **(6)** and Everett et al **(9)** discussed the incorporation of perched ground water sampling into monitoring programs. There are a variety of systems that can be used. These include the following:

9.2.1 *Point samplers,*

9.2.2 *Wells,*

9.2.3 *Cascading water samplers,* and

9.2.4 *Drainage samplers.*

9.3 *Operating Principles:*

9.3.1 *Point Samplers*—Point samplers are open ended pipes or tubes, such as piezometers or wells with short screened intervals, installed for the purpose of collecting samples from a discrete location in saturated material. Samples are collected by bringing liquid that flows freely into the device to the surface by one of a variety of methods.

9.3.2 *Wells*—A monitoring well is similar to a point sampler except the screened interval is longer. Therefore, samples are averaged over the screened length **(104)**. Samples are collected by bringing liquid that flows freely into the well to the surface by one of a variety of methods.

9.3.3 *Cascading Water Samplers*—Cascading water occurs when a well is screened across a perched layer and the underlying water table or when water leaks through casing joints at the perched layer. Because the water table is lower than the perched layer, water flows into the well in the portion open to the perched layer, and cascades downward to the water table. This situation is common in some areas where the practice has been to install water wells with large screened intervals **(105)**. Samples are collected by capturing liquid flowing into the well from the perched layer before it cascades down to the water table.

9.3.4 *Drainage Samplers*—Shallow perched systems may spread contamination, cause problems with structures, or interfere with agriculture. Therefore, drainage systems are sometimes installed. These systems usually funnel liquid via gravity flow to a ditch or sump from which it is pumped out. This outflow can be sampled. Typical drainage systems include tile lines or manifold collectors. Depending on the design of the system, it may be possible to sample outflows that drain different areas.

9.4 *Description:*

9.4.1 *Point Samplers:*

9.4.1.1 Point samplers can be installed in separate boreholes or clustered together in one borehole at different depths. Fig. 29 presents different configurations that have been used. Piezometers, that are often used as point samplers, are similar to wells, in that they consist of a small diameter casing open at one end or connected to a short screened interval **(106)**. Reeve **(107)**, Patton and Smith **(108)** and Morrison **(1)** discussed different designs.

9.4.1.2 Point samplers can be made from a variety of materials including steel, PVC, PTFE, ABS, fiberglass, and additional materials for joints, seals and other components **(106, 107)**.

9.4.2 *Wells:*

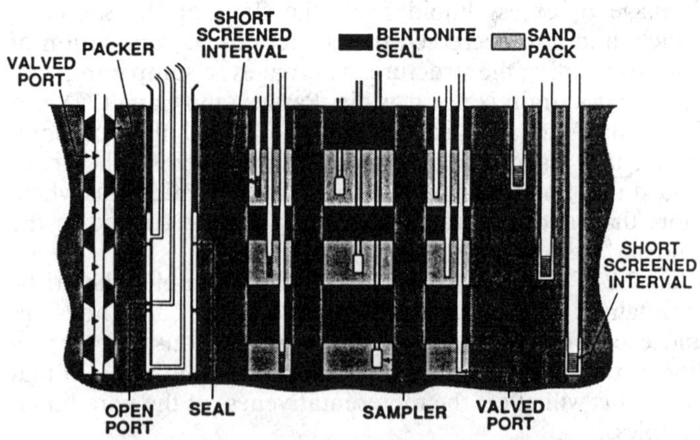

FIG. 29 Examples of Point Sampling Systems

9.4.2.1 Monitoring wells (as depth averaged samplers) are normally installed with one well in each borehole. Components of a well generally include a bottom plug, a length of screen, a length of blank casing, a cap, and a protective cover. Different monitoring well designs are presented in Figs. 30, 31, and 32. Authors who described methods for designing and installing monitoring wells include U.S. EPA **(109)**, Driscoll **(110)**, Gass **(111)**, Keely **(112)**, Minning **(113)**, Richter and Collentine **(114)**, Riggs **(115)**, Riggs and Hatheway **(116)**, Scalf et al **(117)**, Morrison **(1)**, Everett et al **(9)**, Campbell and Lehr **(118)**. Hackett **(119, 120)** summarized methods for designing and installing monitoring wells with hollow stem augers. Screened hollow stem augers can also be used as temporary wells for sampling (Taylor and Serafini, **(121)**).

9.4.2.2 Monitoring wells can be made from a variety of materials including steel, PVC, PTFE, ABS, fiberglass, and additional materials for joints, seals, and other components. Details are provided by Barcelona et al **(122)**, U.S. EPA **(109)**, Morrison **(1)** and many of the references listed above.

9.4.3 *Cascading Water Samplers:*

9.4.3.1 Cascading water is most often seen in production wells in areas with extensive ground water pumpage. Samplers simply consist of a bucket or bailer lowered to a point below the inflow of cascading water. Wilson and Schmidt **(103)** described methods for developing cascading water samplers (see Fig. 33).

9.4.3.2 Cascading wells differ from other wells only by the way in which water flows into them. Otherwise, the materials used for these wells are identical to those used for other types of wells. Bailers or buckets used to collect samples are also available in steel, PVC, PTFE, acrylic and other materials.

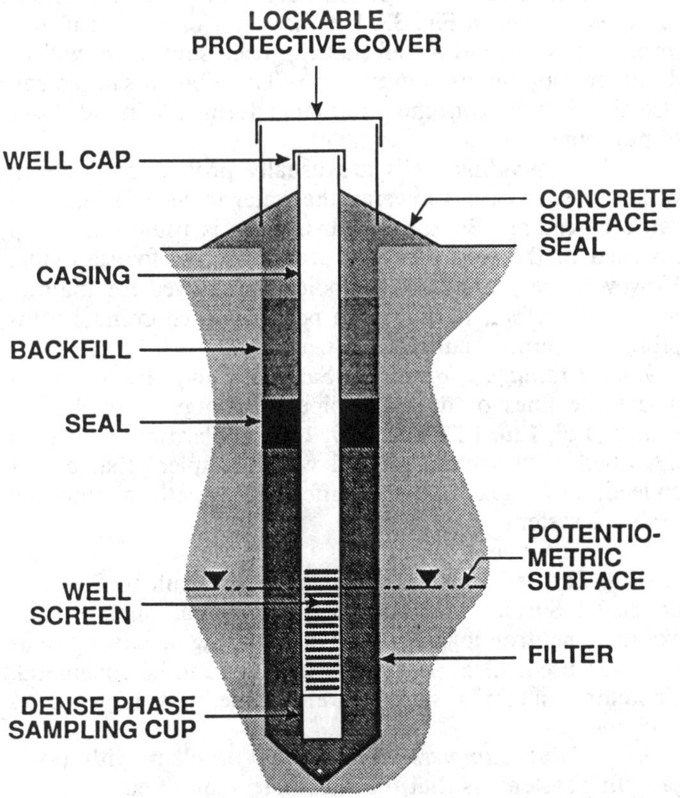

FIG. 30 A Monitoring Well With the Uppermost Ground-Water Level Intersecting the Slotted Well Screen

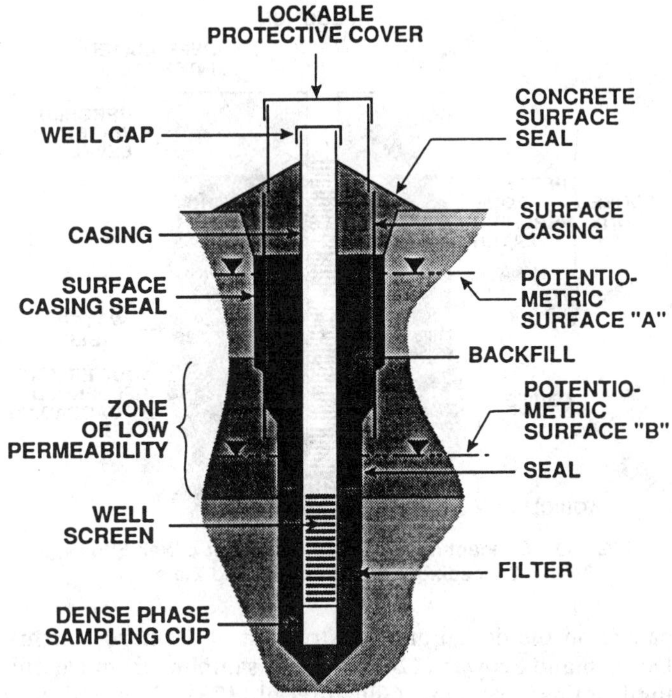

FIG. 31 A Monitoring Well Installed to Sample From the Lower of Two Ground-Water Zones

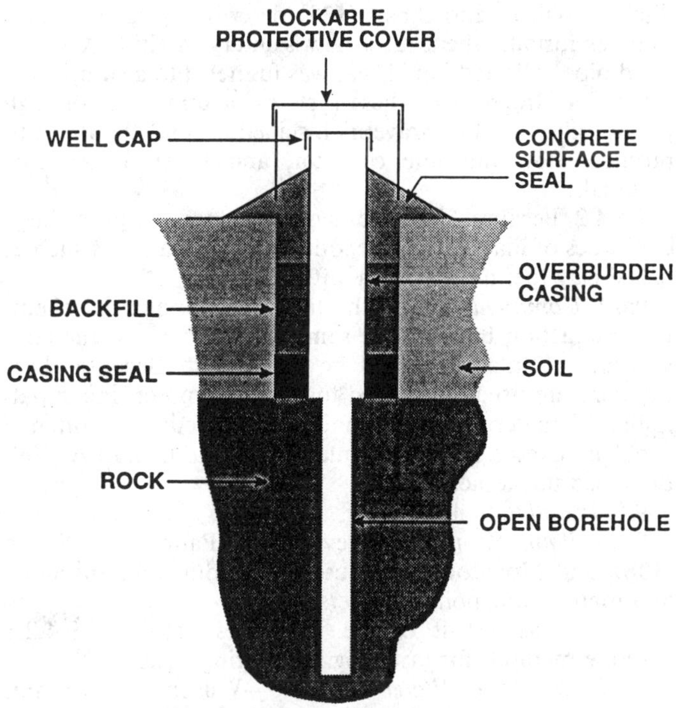

FIG. 32 An Open-Hole Ground-Water Monitoring Well in Rock

9.4.4 *Drainage Samplers:*

9.4.4.1 Drainage systems consist of conduits installed within the perched zone at sufficient slopes for water to flow to a central ditch or drain. The conduits can be tile drains, half perforated pipes, synthetic sheeting, or even layers of gravel and sand. Schilfgaarde ed. **(123)**, contains numerous

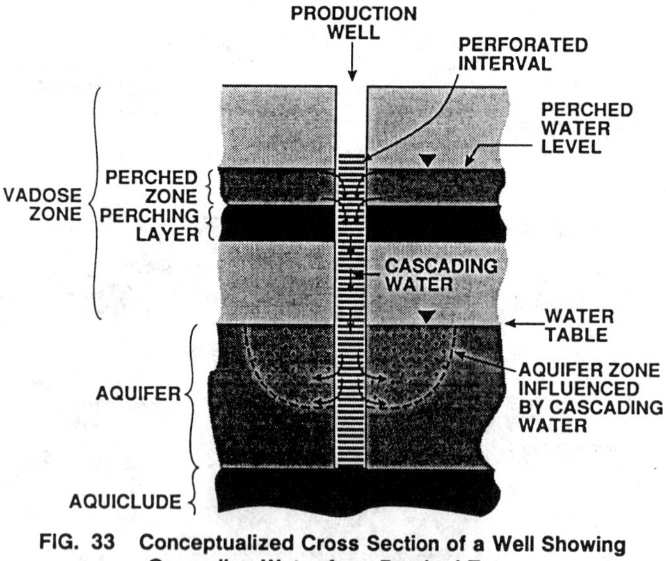

FIG. 33 Conceptualized Cross Section of a Well Showing Cascading Water from Perched Zone

papers on the design and construction of drainage systems. Donnan and Schwab **(124)** described sampling from agricultural drainage systems. Gilliam et al **(125)**, Gambrell et al **(126)**, Eccles and Gruenberg **(127)** and Gilliam et al **(128)** described sampling from tile drains. Gilliam et al **(125)** and Jacobs and Gilliam **(129)** described sampling from drainage ditches. Wilson and Small **(130)** described a lateral drain sampler installed beneath a new sanitary landfill. A perforated pipe collected liquid that was funneled to a sump via a drain line. In most of these systems, a thin layer of high permeability sand or gravel is installed around the drain to promote flow into the collector, and to sieve out fine materials.

9.4.4.2 Because drainage systems often require large quantities of materials, less exotic, cheaper materials such as baked clay tiles and PVC are often used.

9.4.5 *Comments*—As with all samplers, potential chemical interaction between the sampler material and the constituents of interest should be considered. Because these samplers are usually installed for other purposes, incompatibility of materials with monitoring objectives is often a problem. Everett et al **(9)**, Dunlap **(131)**, and U.S. EPA **(109)** discussed this topic.

9.5 *Installation Methods:*

9.5.1 *Point Samplers*—Reeve **(107)**, Patton and Smith **(108)**, and Morrison **(1)** discussed procedures for installing and maintaining point samplers.

9.5.2 *Wells*—Most of the references listed in 9.4.2.1 describe methods for installing monitoring wells.

9.5.3 *Cascading Water Samplers*—Wilson and Schmidt **(103)** discussed methods for installing cascading water samplers.

9.5.4 *Drainage Samplers*—Schilfgaarde ed. **(123)** contains articles that discussed installation of agricultural drainage systems. Associated hazards and costs often prohibit the installation of these systems at existing landfills. As a result, inclusion of these systems, as leachate collectors, in new landfills is more common. Everett et al **(9)** discussed methods for installing drainage sampling systems at hazardous waste sites.

9.6 *Operation:*

9.6.1 *Point Samplers*—Point samplers usually have diameters that are too small to allow the use of submersible pumps. As a result, suction methods are usually required **(1)**. Sampling techniques are described in Pickens et al **(106)**, Reeve **(107)**, Patton and Smith **(108)** and Morrison **(1)**.

9.6.2 *Wells*—Samples may be retrieved from wells in the same manner as from piezometers. However, because wells are designed for sampling or pumpage, diameters are usually large enough to accommodate most pumps. Samples can be brought to the surface by a variety of systems including bailers, suction pumps (for example, peristaltic pumps), air lift pumps, piston pumps, submersible pumps and swabbing. Each of these methods have advantages and disadvantages relating to considerations such as depth to water, required sample volume, sampling speed, alteration of the sample chemistry, equipment requirements, manpower requirements, and cost. These considerations were discussed by Everett et al **(9)**, Fenn et al **(132)**, Gibb et al **(133)**, U.S. EPA **(109)**, Dunlap et al **(131)**, Driscoll **(110)**, and Anderson **(75)**. Sampling methods were described in most of the references of 9.4.2.1. As described in 9.3.2, samples from wells are averaged over the screened interval. However, samples from discrete depths along the screened interval can also be obtained using packer-pump setups such as those described by Fenn et al **(132)**.

9.6.3 *Cascading Water Samplers:*

9.6.3.1 Wilson and Schmidt **(103)** described techniques for sampling from cascading wells. A bailer or bucket is decontaminated and then lowered to a position in the well below the cascading water but above the water table. When the sampler is full, it is pulled back to the surface. Alternately, as shown in Fig. 33, the chemistry of the water table immediately around a well that has been shut down will be dominated by the cascading water. Therefore, a sample can also be collected from the water table during the initial stages of pumping.

9.6.3.2 Cascading wells are usually production wells in which drawdown has lowered the water table sufficiently to cause cascading. Because of this, there is usually a pump installed in the well that will prevent access for sampling. However, the pumps are periodically removed for maintenance. Therefore, it should be possible to coordinate sampling with pump maintenance personnel.

9.6.4 *Drainage Samplers*—Samples may be collected where tile lines or drainage pipes discharge to ditches or sumps **(125, 126, 127, 128, 129, 130)**. Willardson et al **(134)** described a "flow-path ground water sampler" that allows collection of water following different flowpaths along a tile drainage system.

9.7 *Limitations:*

9.7.1 Perched water systems can be difficult to find and delineate. Surface and borehole geophysical methods (for example, neutron logging) and video logging of existing wells are often used. Also, perched systems tend to be ephemeral. Therefore, suctions samplers are sometimes required as backups.

9.7.2 *Point Samplers*—The major problem with point sampling systems is that their diameters are often too small to allow adequate development after installation or to allow sampling by any method other than suction. Because the maximum suction lift of water is about 7.5 m, this is the maximum sampling depth for many of the small diameters systems. Systems such as those depicted in Fig. 29 require tight contact with the surrounding material to prevent side leakage of liquid. Depending on the material, this tight contact may not be achievable **(9)**.

9.7.3 *Wells*—Wells provide samples that are averaged over the screened interval. As a result, when contaminants are detected, packer-pump arrangements must be used if zonation of the contaminants is to be delineated. When separate phases of water-immiscible fluid (for example, oil) are found floating in the well, it is difficult to obtain samples of the underlying water without contamination from the overlying fluid. As with all samplers, care should be taken to ensure that materials used to construct a well are compatible with the chemical analyses to be performed.

9.7.4 *Cascading Water Samplers*—Cascading water may enter a well from several distinct perched systems **(103)**. As a result, the sample may be a mixture of water from several depths. Cascading water is most often sampled from pre-existing wells used for other purposes. As a result, materials used in the well construction may alter those chemical constituents of interest. Wells used for irrigation and water supply often have lubricant oils from the pump floating in them. With fluctuating water levels, these oils become smeared along the casing, and may even move out into the surrounding soils. Therefore, traces of these oils may appear in samples.

9.7.5 *Drainage Samplers*—Because of the limitations of excavation equipment, drainage samplers are limited to shallow depths. In addition, the systems are difficult to install in rocky or steep terrains. In areas that experience freeze-thaw cycles, they may be damaged by soil heaving **(9)**. Drainage systems are often susceptible to clogging over time as fine particles and chemically precipitated material accumulate on the drain openings. Collected samples may or may not be representative of average conditions, depending on the distribution of soil types and contaminants in the drained area. If the area of contamination is small compared to the drained area, dilution may prevent the detection of contaminants. In addition, pollutants that are heavier than water may move below the drain if it is not located at the bottom of the perched zone. As with all samplers, there is the possibility of chemical interaction between the sampling system and the chemical constituents of interest. In the case of drainage sampling systems, this effect is amplified as contaminants may have to travel considerable distances through drains before being sampled. In addition, normally non-aerated solutions may be aerated and chemically altered as they travel through drains.

10. Experimental Absorption Samplers

10.1 *Operating Principles*—Absorbent samplers depend on the ability of the material to absorb pore-liquid **(1)**. Samples are collected by placing the sampler in contact with soil. Liquid is allowed to absorb into the sampler material over time. The sampler is then removed, and the sample liquid is extracted for analyses.

10.2 *Description:*

10.2.1 Two designs have been described. The first design includes a cellulose-nylon sponge (0.5 by 4.8 by 30 cm) seated in a galvanized iron trough. The trough is pressed against a soil surface with a series of lever hinges **(135)**.

10.2.2 The second absorbent sampler design consists of tapered ceramic rods that are driven into soil **(136)**. The rods are made from unglazed ceramic similar to that used as the porous segments of suction samplers.

10.3 *Installation:*

10.3.1 *Pre-Installation:*

10.3.1.1 Sponge samplers are prepared by soaking them for 24 h in a 1 to 5 % NaOH solution containing a washing powder, Tadros and McGarity **(135)**. Sponges are then pressed dry using rollers, stored in a moisture tight container, and taken to the field.

10.3.1.2 Ceramic rod samplers are weighed, boiled in distilled water, oven dried, and stored in a desiccator. The rods are weighed again and then taken to the field.

10.3.2 *Installation:*

10.3.2.1 A sponge is placed in the sampling trough. The trough is then placed in a horizontal cavity cut into the side of a trench. The trough is then pressed against the cavity ceiling with the lever hinges.

10.3.2.2 A ceramic rod sampler is installed by simply driving it into the soil.

10.3.3 *Maintenance:*

10.3.3.1 The only field maintenance required for sponge samplers is the preservation of the sampling trench if future sampling is desired at that location.

10.3.3.2 There is no field maintenance for ceramic rod samplers as they are completely removed to retrieve the samples.

10.3.4 *Comments*—Theoretically, there is no maximum installation depth for sponge samplers. However, because access trenches are required for operation, installations are restricted to shallow depths dictated by excavation equipment and safety considerations. Given the NaOH treatment (without rinsing) of the sponge, measurements of pH, conductivity, TDS, metals and major cations and anions might be affected by the residual NaOH. Depending on the composition of the "washing powder", phosphate, BOD, and MBAS might also be affected.

10.3.4.1 Ceramic rod samplers will have maximum installation depths if pushed or driven from the surface. This maximum depth will generally decrease with increasing soil grain size. However, deeper installations can be achieved by drilling to the top of the interval to be sampled, lowering a rod down the hole and pushing or driving the rod into the sampling interval **(136)**.

10.4 *Operation:*

10.4.1 *Methods:*

10.4.1.1 Sponge samplers are pressed against the soil until a sufficient volume of liquid for planned analyses has been absorbed. The sampler is then removed and the sponge is placed in a moisture proof container. The sample is extracted from the sponge with rollers.

10.4.1.2 Ceramic rod samplers are pushed or driven into the soil and left in place for a period of time. The rods are then withdrawn, and weighed. The rods are leached by boiling in a known volume of distilled water. This solution is then analyzed. The concentrations of constituents in the original pore-liquid are estimated by using the ratio of the volume of absorbed water to the volume of the boiling water.

10.4.2 *Comments*—The amount of liquid that can be sampled is dependent on time, soil type, moisture content, absorbency of the sampler material, volume of the absorbent material, and surface area of the absorbent material in contact with the soil. Generally, sponge samplers function only at higher moisture contents approaching saturation (**1**). Shimshi (**136**) used ceramic rod samplers to sample from a sandy loam with moisture contents varying from 7 to 20 %.

10.5 *Limitations:*

10.5.1 Physically, absorbent methods are limited to soils approaching saturation. Sampling requires removal of the absorbent material. Because of this, repeat sampling at the same location is difficult. Although the sampler may be placed back at its original location, identical hydraulic contact with the soil cannot be guaranteed.

10.5.2 Chemically, as with other samplers, there are problems with absorption, desorption, precipitation, cation exchange, and screening of various pore-liquid components as a function of the sampler materials. Tadros and McGarity (**135**) discussed these concerns in relation to sponge samplers. Shimshi (**136**) provided a good discussion of the limitations of sampling for nitrate with ceramic rod samplers. Specifically, he found that at lower moisture contents, sampled solutions became less representative due to vapor transfer and chromatographic separation. However, he suggested that these effects could be reduced by increasing the length of the rod insertion period. Clearly, boiling will affect analyses for organics, BOD, COD, NH_3, and some other species of nitrogen, among others.

11. Keywords

11.1 pore fluids, pressure vacuum lysimeters; soil moisture; soil water; suction lysimeters; unsaturated; vadose zone sampling

APPENDIX

(Nonmandatory Information)

X1. DESCRIPTIONS OF TERMS SPECIFIC TO THIS STANDARD

X1.1 *air entry value*—the applied suction at which water menisci of the porous segment of a suction sampler break down, and air enters.

X1.2 *bubbling pressure*—the applied air pressure at which water menisci of the porous segment of a suction sampler break down, and air exits.

X1.3 *cascading water*—perched ground water that enters a well casing via cracks or uncovered perforations, trickling, or pouring down the inside of the casing.

X1.4 *cation exchange capacity (CEC)*—the total capacity of a porous system to adsorb cations from a solution.

X1.5 *hydraulic gradient*—the change in total hydraulic head of water per unit distance of flow.

X1.6 *hydrophobicity*—the property that defines a material as being water repellent. Water exhibits an obtuse contact angle with hydrophobic materials.

X1.7 *hydrophelicity*—the property that defines a material as attracting water. Water exhibits an acute contact angle with hydrophilic materials.

X1.8 *lysimeter*—a device to measure the quantity or rate of water movement through a block of soil, usually undisturbed or in-situ; or to collect such percolated water for analyses.

X1.9 *macropore*—interaggregate cavities that serve as the principal avenues for the infiltration and drainage of water and for aeration.

X1.10 *matric potential*—the energy required to extract water from a soil against the capillary and adsorptive forces of the soil matrix.

X1.11 *matric suction*—for isothermal soil systems, matric suction is the pressure difference across a membrane separating soil solution, in-place, from the same bulk (see soil-water pressure).

X1.12 *micropore*—intraaggregate capillaries responsible for the retention of water and solutes.

X1.13 *percolation*—the movement of water through the vadose zone, in contrast to infiltration at the land surface and recharge across a water table.

X1.14 *pore-liquid*—Liquid that occupies an open space between solid soil particles. Within this guide, pore-liquid is limited to aqueous pore-liquid; that includes water and its solutes.

X1.15 *pore-liquid tension*—see *matric-suction* or *soil-water pressure.*

X1.16 *pressure head*—the head of water at a point in a porous system; negative for unsaturated systems, positive for saturated systems. Quantitatively, it is the water pressure divided by the specific weight of water.

X1.17 *Richard's outflow principle*—the principle that states that pore-liquid will not generally flow into an air-filled cavity (at atmospheric pressure) in unsaturated soil.

X1.18 *soil-water pressure*—the pressure on the water in a soil-water system, as measured by a piezometer for a saturated soil, or by a tensiometer for an unsaturated soil.

X1.19 *tensiometer*—a device for measuring soil-water matric potential (or tension or suction) of water in soil in-situ; a porous, permeable ceramic cup connected through a water filled tube to a pressure measuring device.

X1.20 *total soil-water potential*—the sum of the energy-related components of a soil-water system; for example, the sum of the gravitational, matric and osmotic potentials.

X1.21 *tremie method*—the method whereby materials are emplaced in the bottom of a borehole with a small diameter pipe.

X1.22 *vacuum*—a degree of rarefaction below atmospheric pressure: negative pressure.

X1.23 *Vadose zone*—the hydrogeological region extending from the soil surface to the top of the principle water table; commonly referred to as the "unsaturated zone" or "zone of aeration". These alternate names are inadequate as they do not take into account locally saturated regions above the principle water table (for example, perched water zones).

X1.24 *water content*—the amount of water stored within a porous matrix, expressed as either a volume (volume per unit volume) or a mass (mass per unit mass) of a given solid.

REFERENCES

(1) Morrison, Robert D., *Ground Water Monitoring Technology*, Timco Mfg., Inc., Prairie Du Sac, Wisconsin, 1983.

(2) Wilson, L. G., *Monitoring in the Vadose Zone: A Review of Technical Elements and Methods*, U.S. Environmental Protection Agency, EPA-600/7-80-134, 1980.

(3) Wilson, L. G., "The Fate of Pollutants in the Vadose Zone, Monitoring Methods and Case Studies", *Thirteenth Biennial Conference on Ground Water*, September 1981.

(4) Everett, L. G., "Monitoring in the Vadose Zone", *Ground Water Monitoring Review*, Summer 1981, pp. 44–51.

(5) Wilson, L. G., "Monitoring in the Vadose Zone: Part II", *Ground Water Monitoring Review*, Winter 1982, pp. 31–42.

(6) Everett, L. G., Wilson, L. G., and McMillion, L. G., "Vadose Zone Monitoring Concepts for Hazardous Waste Sites", *Ground Water*, Vol 20, May/June 1982, pp. 312–324.

(7) Wilson, L. G., "Monitoring in the Vadose Zone: Part III", *Ground Water Monitoring Review*, Winter 1983, pp. 155–165.

(8) Everett, Lorne G., Hoylman, Edward W., Wilson, L. Graham, and McMillion, Leslie G., "Constraints and Categories of Vadose Zone Monitoring Devices", *Ground Water Monitoring Review*, Winter 1984, pp. 26–32.

(9) Everett, L. G., Wilson, L. G., and Hoylman, E. W., *Vadose Zone Monitoring Concepts for Hazardous Waste Sites*, Noyes Data Corporation, New Jersey, 1984.

(10) Robbins, Gary A., and Gemmell, Michael M., "Factors Requiring Resolution in Installing Vadose Zone Monitoring Systems", *Ground Water Monitoring Review*, Summer 1985, pp. 75–80.

(11) Merry, W. M., and Palmer, C. M., "Installation and Performance of a Vadose Monitoring System", *Proceedings of the NWWA Conference on Characterization and Monitoring of the Vadose Zone*, NWWA, 1986, pp. 107–125.

(12) Environmental Monitoring Systems Laboratory, *Permit Guidance Manual on Unsaturated Zone Monitoring for Hazardous Waste Land Treatment Units*, U.S. Environmental Protection Agency, Office of Solid Waste and Emergency Response, EPA/530-SW-86-040, October 1986.

(13) Ball, John, and Coley, David M., "A Comparison of Vadose Monitoring Procedures", *Proceedings of the Sixth National Symposium and Exposition on Aquifer Restoration and Ground Water Monitoring*, NWWA/EPA, 1986, pp. 52–61.

(14) Wilson, L. G., "Methods for Sampling Fluids in the Vadose Zone", *Ground Water and Vadose Zone Monitoring, ASTM STP 1053*, ASTM, 1990, pp. 7–24.

(15) Bond, William R., and Rouse, Jim V., "Lysimeters Allow Quicker Monitoring of Heap Leaching and Tailing Sites", *Mining Engineering*, April 1985, pp. 314–319.

(16) Doanhue, Miller, and Shickluna, *Soils—An Introduction to Soils and Plant Growth*, 1977.

(17) Johnson, Thomas, M., and Cartwright, Keros, *Monitoring of Leachate Migration in the Unsaturated Zone in the Vicinity of Sanitary Landfills*, State Geological Survey Circular 514, Urbana, Illinois, 1980.

(18) Sales Division, *Catalog of Products*, Soilmoisture Equipment Corp., Santa Barbara, California, 1988.

(19) Hansen, Edward A., and Harris, Alfred Ray, "Validity of Soil-Water Samples Collected with Porous Ceramic Cups", *Soil Science Society of America Proceedings*, Vol 39, 1975, pp. 528–536.

(20) Quin, B. F., and Forsythe, L. J., "All-Plastic Suction Lysimeters for the Rapid Sampling of Percolating Soil Water", *New Zealand Journal of Science*, Vol 19, 1976, pp. 145–148.

(21) Long, F. Leslie, "A Glass Filter Soil Solution Sampler", *Soil Science Society of America Journal*, Vol 42, 1978, pp. 834–835.

(22) Starr, Michael R., "Variation in the Quality of Tension Lysimeter Soil Water Samples from a Finnish Forest Soil", *Soil Science*, Vol 140, December 1985, pp. 453–461.

(23) Sales Division, *Catalog of Products*, Mott Metallurgical Corp., Farmington, Conn., 1988.

(24) Smith, C. N., and Carsel, R. F., "A Stainless-Steel Soil Solution Sampler for Monitoring Pesticides in the Vadose Zone", *Soil Science Society of America Journal*, Vol 50, 1986, pp. 263–265.

(25) Caster, A., and Timmons, R., "Pourous Teflon®: Its Application in Groundwater Sampling", Timco Mfg. Inc., Prairie Du Sac, Wisc., 1988.

(26) Parizek, Richard R., and Lane, "Soil-Water Sampling Using Pan and Deep Pressure-Vacuum Lysimeters", *Journal of Hydrology*, Vol 11, 1970, pp. 1–21.

(27) Trainor, David P., *The Relationship Between Two Laboratory Leaching Procedures and Leachate Quality at Foundry Waste Landfills*, M. S. Thesis/Independent Report, Dept. of Civil and Environmental Engineering, The University of Wisconsin, Madison, Wisconsin, February, 1983.

(28) Young, Mark, "Use of Suction Lysimeters for Monitoring in the Landfill Linear Zone", *Proceedings of Monitoring Hazardous Waste Sites*, Geotechnical Engineering Division, American Society of Civil Engineers, Detroit, Mich., October, 1985.

(29) Morrison, R. D., and Tsai, T. C., *Modified Vacuum-Pressure Lysimeter for Vadose Zone Sampling*, Calscience Research Inc., Huntington Beach, California, 1981.

(30) Morrison, Robert, and Szecsody, James, "Sleeve and Casing Lysimeters for Soil Pore Water Sampling", *Soil Science*, Vol 139, May 1985, pp. 446–451.

(31) Nightingale, H. I., Harrison, Doug, and Salo, John E., "An Evaluation Technique for Ground Water Quality Beneath Urban Runoff Retention and Percolation Basins", *Ground Water Monitoring Review*, Vol 5, Winter 1985, pp. 43–50.

(32) Torstensson, B. A., and Petsonk, A. M., "A Hermetically Isolated Sampling Method for Ground Water Investigations", *Ground Water Contamination: Field Methods, ASTM STP 963*, ASTM, 1988, pp. 274–289, and Knighton, M. Dean, and Streblow, Dwight E., "A More Versatile Soil Water Sampler", *Soil Science Society of America Journal*, Vol 45, 1981, pp. 158–159.

(33) Jackson, D. R., Brinkley, F. S., and Bondietti, E. A., "Extraction of Soil Water Using Cellulose-Acetate Hollow Fibers", *Soil Science Society of America Journal*, Vol 40, 1976, pp. 327–329.

(34) Levin, M. J., and Jackson, D. R., "A Comparison of In Situ Extractors for Sampling Soil Water", *Soil Science Society of America Journal*, Vol 41, 1977, pp. 535–536.

(35) Stevenson, Craig D., "Simple Apparatus for Monitoring Land Disposal Systems by Sampling Percolating Soil Waters", *Environmental Science and Technology*, Vol 12, March 1978, pp. 329–331.

(36) Wagemann, R., and Graham, B., "Membrane and Glass Fibre Filter Contamination in Chemical Analysis of Fresh Water", *Water Research*, Vol 8, 1974, pp. 407–412.

(37) Sales Division, *Catalog of Products*, Cole-Parmer Instrument Company, 1988.

(38) Sales Division, *Catalog of Products*, Corning Glass Works, New York, 1988.

(39) Duke, H., Kruse, E., and Hutchinson, G., "An Automatic Vacuum Lysimeter for Monitoring Percolation Rates", *USDA Agricultural Research Service*, ARS, 1970, pp. 41–165.

(40) Tanner, C. B., Bourget, S. J., and Holmes, W. E., "Moisture Tension Plates Constructed from Alundum Filter Discs", *Soil Science Society of America Proceedings*, Vol 18, 1954, pp. 222–223.

(41) Cole, D. W., "Alundum Tension Lysimeter", *Soil Science*, Vol 85, June 1958, pp. 293–296.
(42) Cole, D., Gessell, S., and Held, E., "Tension Lysimeter Studies of Ion and Moisture Movement in Glacial Till and Coral Atoll Sites", *Soil Science Society of America Proceedings*, Vol 25, 1968, pp. 321–325.
(43) Nielson, D., and Phillips, R., "Small Fritted Glass Bead Plates For Determination of Moisture Retention", *Soil Science Society of America Proceedings*, Vol 22, 1958, pp. 574–575.
(44) Chow, T. L., "Fritted Glass Bead Materials as Tensiometers and Tension Plates", *Soil Science Society of America Journal*, Vol 41, 1977, pp. 19–22.
(45) Everett, Lorne, and McMillion, Leslie, G., "Operational Ranges for Suction Lysimeters", *Ground Water Monitoring Review*, Summer 1985, pp. 51–60.
(46) Haldorsen, Sylvi, Petsonk, Andrew M., and Tortensson, Bengt-Arne, "An Instrument for In-Situ Monitoring of Water Quality and Movement in the Vadose Zone", *Proceedings of the NWWA Conference on Characterization and Monitoring of the Vadose Zone*, NWWA, 1986, pp. 158–172.
(47) Neary, A. J., and Tomassini, F., "Preparation of Alundum/Ceramic Plate Tension Lysimeters for Soil Water Collection", *Canadian Journal of Soil Science*, Vol 65, February 1985, pp. 169–177.
(48) Wolff, R. G., "Weathering Woodstock Granite, near Baltimore, Maryland", *American Journal of Science*, Vol 265, 1967, pp. 106–117.
(49) Wood, Warren W., "A Technique Using Porous Cups for Water Sampling at Any Depth in the Unsaturated Zone", *Water Resources Research*, Vol 9, April 1973, pp. 486–488.
(50) Debyle, Norbert V., Hennes, Robert W., and Hart, George E., "Evaluation of Ceramic Cups for Determining Soil Solution Chemistry", *Soil Science*, Vol 146, July 1988, pp. 30–36.
(51) England, C. B., "Comments on A Technique Using Porous Cups for Water Sampling at Any Depth in the Unsaturated Zone by Warren W. Wood", *Water Resources Research*, Vol 10, October 1974, pp. 1049.
(52) Bottcher, A. B., Miller, L. W., and Campbell, K. L., "Phosphorous Adsorption in Various Soil-Water Extraction Cup Materials: Effect of Acid Wash", *Soil Science*, Vol 137, 1984, pp 239–244.
(53) Peters, Charles A., and Healy, Richard W., "The Representativeness of Pore Water Samples Collected from the Unsaturated Zone Using Pressure-Vacuum Lysimeters", *Ground Water Monitoring Review*, Spring 1988, pp. 96–101.
(54) Linden, D. R., "Design, Installation and Use of Porous Ceramic Samplers for Monitoring Soil-Water Quality", *U.S. Dep. Agric. Technical Bull.*, 1562, 1977.
(55) Rhoades, J. D., and Oster, J. D., "Solute content," Methods of Soil Analysis, Part I, Agronomy Series Monograph No 9, Second Edition, American Society of Agronomy, Inc. Madison, Wisconsin, 1986, pp. 985–1006.
(56) Klute, A., ed., *Methods of Soil Analysis, Part I*, Agronomy Series Monograph No. 9, Second Edition, American Society of Agronomy, Inc., Madison, Wisconsin, 1986.
(57) Brose, Richard J., Shatz, Richard W., and Regan, Thomas M., "An Alternate Method of Lysimeter and Flour Pack Placement in Deep Boreholes", *Proceedings of the Sixth National Symposium and Exposition on Aquifer Restoration and Ground Water Monitoring*, NWWA, 1986, pp. 88–95.
(58) Litaor, M. Iggy, "Review of Soil Solution Samplers", *Water Resources Research*, Vol 24, May 1988, pp. 727–733.
(59) Silkworth, D. R., and Grigal, D. F., "Field Comparison of Soil Solution Samplers", *Soil Science Society of America Journal*, Vol 45, 1981, pp. 440–442.
(60) Smith, J. L., and McWhorter, D. M., "Continuous Subsurface Injection of Liquid Organic Wastes", *Land as a Waste Management Alternative*, Ann Arbor Science, 1977, pp. 646–656.
(61) Grier, H. E., Burton, W., and Tiwari, "Overland Cycling of Animal Waste", *Land as a Waste Management Alternative*, Ann Arbor Science, 1977, pp. 693–702.
(62) Smith, J. H., Robbins, C. W., Bondurant, J. A., and Hayden, C. W., "Treatment of Potato Processing Wastewater on Agricultural Land: Water and Organic Loading, and the Fate of Applied Plant Nutrients", *Land as a Waste Management Alternative*, Ann Arbor Science, 1977, pp. 769–781.
(63) Morrison, Robert D., and Szecsody, Jim E., "A tensiometer and Pore Water Sampler for Vadose Zone Monitoring", *Soil Science*, Vol 144, November 1987, pp. 367–372.
(64) Baier, Dwight C., Aljibury, Falih K., Meyer, Jewell K., Wolfenden, Allen K., *Vadose Zone Monitoring is Effective for Evaluating the Integrity of Hazardous Waste Pond Liners*, Baier Agronomy, Inc., Woodland, California, November 1983.
(65) Taylor, S. A., and Ashcroft, G. L., *Physical Edaphology, The Physics of Irrigated and Non Irrigated Soils*, W. H. Freeman and Company, San Francisco, 1972.
(66) Hornby, W. J., Zabick, J. D., and Crawley, W., "Factors Which Affect Soil-Pore Liquid: A Comparison of Currently Available Samplers with Two New Designs", *Ground Water Monitoring Review*, Vol 6, Spring 1986, pp. 61–66.
(67) Wengel, R. W., and Griffin, G. F., "Remote Soil-Water Sampling Technique", *Soil Science Society of America Journal*, Vol 35, 1971, pp. 661–664.
(68) Brown, K. W., Thomas, J. C., and Aurelius, M. W., "Collecting and Testing Barrel Sized Undisturbed Soil Monoliths", *Soil Science Society of America Journal*, Vol 49, 1985, pp. 1067–1069.
(69) Wagner, George H., "Use of Porous Ceramic Cups to Sample Soil Water Within the Profile", Soil Science, Vol 94, 1962, pp. 379–386.
(70) Nagpal, N. K., "Comparison Among and Evaluation of Ceramic Porous Cup Soil Water Samplers for Nutrient Transport Studies", *Canadian Journal of Soil Science*, Vol 62, 1982, pp. 685–694.
(71) Bouma, J., Jongerius A., and Schoondebeek, D., "Calculation of Hydraulic Conductivity of Some Saturated Clay Soils Using Micromorpho-Meteric Data", *Soil Science Society of America Journal*, Vol 43, 1979, pp. 261–265.
(72) Warrick, A. W., and Amoozegar-Fard, A., "Soil Water Regimes Near Porous Cup Water Samplers", *Water Resources Research*, Vol 13, February 1977, pp. 203–207.
(73) Van der Ploeg, R. R., and Beese, F., "Model Calculations for the Extraction of Soil Water by Ceramic Cups and Plates", *Soil Science Society of America Journal*, Vol 41, 1977, pp. 466–470.
(74) Narasimhan, T. N., and Dreiss, Shirley, J., "A Numerical Technique for Modeling Transient Flow of Water to a Soil Water Sampler", *Soil Science*, Vol 141, March 1986, pp. 230–236.
(75) Anderson, Linda Davis, "Problems Interpreting Samples Taken with Large-Volume, Falling Suction Soil-Water Samplers", *Ground Water*, 1986.
(76) Law Engineering Testing Company, *Lysimeter Evaluation Study*, American Petroleum Institute, May 1982.
(77) Hillel, D., *Applications of Soil Physics*, Academic Press, NY, 1980.
(78) Sales Division, *Catalog of Products*, BAT Envitech Inc., Long Beach California, 1988.
(79) Brown, K. W., *Efficiency of Soil Core and Soil Pore-Liquid Sampling Systems*, U.S. EPA/600/52-86/083, Virginia, February, 1987.
(80) Amter, S., *Injection/Recovery Lysimeter Technique for Soil-Water Extraction*, M. S. Thesis, Dept. of Hydrology and Water Resources, The University of Arizona, Tucson, Arizona, 1987.
(81) Barbee, G. C., and Brown, K. W., "Comparison Between Suction and Free-Drainage Soil Solution Samplers", *Soil Science*, Vol 141, February 1986, pp. 149–154.
(82) Amoozegar-Fard, A. D., Nielsen, D. R., and Warrick, A. W., "Soil Solute Concentration Distributions for Spatially Varying Pore Water Velocities and Apparent Diffusion Coefficients" *Soil Science Society of America Journal*, Vol 46, 1982, pp. 3–9.
(83) Haines, B. L., Waide, J. B., and Todd, R. L., "Soil Solution Nutrient Concentrations Sampled with Tension and Zero-Tension Lysimeters: Report of Discrepancies", *Soil Science Society of America Journal*, Vol 46, 1982, pp. 658–660.
(84) Biggar, J. W., and Nielsen, D. R., "Spatial Variability of the

Leaching Characteristics of a Field Soil", *Water Resources Research*, Vol 12, 1976, pp. 78–84.

(85) Thomas, G. W., and Phillips, R. E., "Consequences of Water Movement in Macropores", *Journal of Environmental Quality*, Vol 8, 1979, pp. 149–152.

(86) Severson, R. C., and Grigal, D. F., "Soil Solution Concentrations: Effect of Extraction Time Using Porous Ceramic Cups Under Constant Tension", *Water Resources Bulletin*, Vol 12, December 1976, pp. 1161–1169.

(87) Shuford, J. W., Fritton, D. D., and Baker, D. E., "Nitrate-Nitrogen and Chloride Movement Through Undisturbed Field Soil", *Journal of Environmental Quality*, Vol 6, 1977, pp. 736–739.

(88) Tyler, Donald D., and Thomas, Grant W., "Lysimeter Measurements of Nitrate and Chloride Losses from Soil Under Conventional and No-Tillage Corn", *Journal of Environmental Quality*, Vol 6, 1977, pp. 63–66.

(89) Suarez, D. L., "Prediction of pH Errors in Soil Water Extractors Due to Degassing", *Soil Science Society of America Journal*, Vol 51, 1987, pp. 64–68.

(90) Ransom, M. D., and Smeck, N. E., "Water Table Characteristics and Water Chemistry of Seasonally Wet Soils of Southwestern Ohio", *Soil Science Society of America Journal*, Vol 50, 1986, pp. 1281–1290.

(91) Wood, A. L., Wilson, J. T., Cosby, R. L., Hornsby, A. G., and Baskin, L. B., "Apparatus and Procedure for Sampling Soil Profiles for Volatile Organic Compounds", *Soil Science Society of America Journal*, Vol 45, 1981, pp. 442–444.

(92) Pettyjohn, Wayne A., Dunlap, W. J., Cosby, Roger, and Keeley, Jack W., "Sampling Ground Water for Organic Contaminants", *Ground Water*, Vol 19, March-April 1981, pp. 180–189.

(93) Creasey, Carol L., and Dreiss, Shirley J., "Soil Water Sampler: Do They Significantly Bias Concentrations in Water Samples?", *Proceedings of the NWWA Conference on Characterization and Monitoring of the Vadose Zone*, NWWA, 1986, pp. 173–181.

(94) Bell, R. G., "Porous Ceramic Soil Moisture Samplers, An Application in Lysimeter Studies on Effluent Spray Irrigation", *New Zealand Journal of Experimental Agriculture*, Vol 2, June 1974, pp. 173–175.

(95) Dazzo, F., and Rothwell, D., "Evaluation of Porcelain Cup Soil Water Samplers For Bacteriological Sampling", *Applied Microbiology*, Vol 27, 1974, pp. 1172–1174.

(96) Jordan, Carl F., "A Simple, Tension-Free Lysimeter", *Soil Science*, Vol 105, 1968, pp. 81–86.

(97) Montgomery, B. R., Prunty, Lyle, and Bauder, J. W., "Vacuum Trough Extractors for Measuring Drainage and Nitrate Flux Through Sandy Soils", *Soil Science Society of America Journal*, Vol 51, 1987, pp. 271–276.

(98) Schmidt, C., and Clements, E., *Reuse of Municipal Wastewater For Groundwater Recharge*, U.S. Environmental Protection Agency, 68-03-2140, 1978, Ohio, pp. 110–125.

(99) Schneider, B. J., Oliva, J., Ku, H. F. H., and Oaksford, E. T., "Monitoring the Movement and Chemical Quality of Artificial-Recharge Water in the Unsaturated Zone on Long Island, New York", *Proceedings of the Characterization and Monitoring of the Vadose (Unsaturated) Zone*, National Water Well Association, Las Vegas, Nevada, 1984, pp. 383–410.

(100) K. W. Brown and Associates, *Hazardous Waste Land Treatment*, U.S. EPA, Office of Research and Development, SW-874, Cincinnati, Ohio, September, 1980.

(101) Schneider, B. J., and Oaksford, E. T., *Design and Monitoring Capability of an Experimental Artificial-Recharge Facility at East Meadow, Long Island, New York*, U.S. Geological Survey Open-File Report 84-070, 1986.

(102) Joslin, J. D., Mays, P. A., Wolfe, M. H., Kelly, J. M., Garber, R. W., Brewer, P. F., "Chemistry of Tension Lysimeter Water and Lateral Flow in Spruce and Hardwood Stands", *Journal of Environmental Quality*, Vol 16, 1987, pp. 152–160.

(103) Wilson, L. G., and Schmidt, K. D., "Monitoring Perched Water in the Vadose Zone", *Establishment of Water Quality Monitoring Programs, Proceedings of a Symposium*, American Water Resources Association, 1978, pp. 134–149.

(104) Pickens, J. F., and Grisak, G. E., "Reply to the Preceding Discussion of Vanhof et al. at a Multilevel Device for Groundwater Sampling and Pieziometric Monitoring", *Ground Water*, Vol 17(4), 1979, pp. 393–397.

(105) Smith, S. A., Small, G. S., Phillips, T. S., and Clester, M., *Water Quality in the Salt River Project, A Preliminary Report*, Salt River Project Water Resource Operations, Ground Water Planning Division, Phoenix, Arizona, August, 1982.

(106) Pickens, J. F., Cherry, J. A., Coupland, R. M., Grisak, G. E., Merritt, W. F., and Risto, G. A., "A Multi-Level Device for Ground Water Sampling", *Ground Water Monitoring Review*, Vol 1(1), 1981.

(107) Reeve, R. C., and Doering, E. J., "Sampling the Soil Solution for Salinity Appraisal", *Soil Science*, Vol 99, 1965, pp. 339–344.

(108) Patton, F. D., and Smith, H. B., "Design Considerations and the Quality of Data from Multiple-Level Groundwater Monitoring Wells", *Ground-Water Contamination: Field Methods, ASTM STP 963*, ASTM, 1988, pp. 206–217.

(109) U.S. Environmental Protection Agency, *RCRA Ground-Water Monitoring Technical Enforcement Guidance Document*, Office of Waste Programs Enforcement, Office of Solid Waste and Emergency Response, OSWER-9950.1, 1986.

(110) Driscoll, F. G., *Groundwater and Wells*, Johnson Division, St. Paul, Minnesota, 1986.

(111) Gass, T. E., "Methodology for Monitoring Wells", *Water Well Journal*, Vol 38(6), 1984, pp. 30–31.

(112) Keely, J. F., and Boateng K., "Monitoring Well Installation, Purging and Sampling Techniques", *Ground Water*, 1987, Vol 25 (3, 4).

(113) Minning, R. C., "Monitoring Well Design and Installation", *Proceedings of the Second National Symposium on Aquifer Restoration and Ground Water Monitoring*, Columbus, Ohio, 1982, pp. 194–197.

(114) Richter, H. R., and Collentine, M. G., "Will My Monitoring Well Survive Down There?: Design and Installation Techniques for Hazardous Waste Studies", *Proceedings of the Third National Symposium on Aquifer Restoration and Ground Water Monitoring*, Columbus, Ohio, 1983, pp. 223–229.

(115) Riggs, C. O., "Monitoring Well Drilling Methods", *Proceedings of the Workshop on Resource Conservation Recovery Act Ground Water Monitoring Enforcement: Use of the Technical Enforcement Guidance Document and Compliance Order Guide*, Philadelphia, Pennsylvania, 1987.

(116) Riggs, C. O., and Hatheway, A. W., "Ground Water Monitoring Field Practice—An Overview", *Ground Water Contamination: Field Methods, ASTM STP 963*, ASTM, pp. 121–136, 1988.

(117) Scalf, M. R., McNabb, J. F., Dunlap, W. J., Cosby, R. L., and Fryberger, J., *Manual of Ground Water Sampling Procedures*, National Water Well Association, Ohio, 1981.

(118) Campbell, M., and Lehr, J., *Water Well Technology*, McGraw-Hill Book Co., New York, New York, 1973.

(119) Hackett, G., "Drilling and Constructing Monitoring Wells with Hollow Stem Augers", *Ground Water Monitoring Review*, Fall 1987, pp. 51–62.

(120) Hackett, G., "Drilling and Constructing Monitoring Wells with Hollow Stem Augers", *Ground Water Monitoring Review*, Winter 1988, pp. 60–68.

(121) Taylor, T. W., and Serafini, M. C., "Screened Auger Sampling: The Technique and Two Case Studies", *Ground Water Monitoring Review*, Summer 1988, pp. 145–152.

(122) Barcelona, M. J., Gibb, J. P., and Miller, R. A., *A Guide to the Selection of Materials For Monitoring Well Construction and Ground Water Sampling*, Illinois State Water Survey, Champaign, Illinois, 1983.

(123) Schilfgaarde, J. V., ed., *Drainage for Agriculture*, Number 17 in Agronomy Series, American Society of Agronomy, Madison, Wisconsin, 1974.

(124) Donnan, W. W. and G. O. Schwab, "Current Drainage Methods in the USA," in *Drainage for Agriculture*, Agronomy Monograph Number 17, American Society of Agronomy, Madison,

Wisconsin, 1974, pp 93–114.

(125) Gilliam, J. W., Daniels, R. B., and Lutz, J. F., "Nitrogen Content of Shallow Ground Water in the North Carolina Coastal Plain", *Journal of Environmental Quality*, Vol 2, 1974, pp. 147–151.

(126) Gambrell, R. P., Gilliam, J. W., and Weed, S. B., "Denitrification in Subsoils of the North Carolina Coastal Plain as Affected by Soil Drainage", *Journal of Environmental Quality*, Vol 4, 1975, pp. 311–316.

(127) Eccles, L. A., and Gruenberg, P. A., "Monitoring Agricultural Waste Water Discharges For Pesticide Residues", *Proceedings: Establishment of Water Quality Monitoring Programs*, American Water Resources Association, 1978, pp. 319–327.

(128) Gilliam, J. W., Skaggs, R. W., and Weed, S. B., "Drainage Control to Diminish Nitrate Loss from Agricultural Fields", *Journal of Environmental Quality*, Vol 8, 1979, pp. 137–142.

(129) Jacobs, T. C., and Gilliam, J. W., "Riparian Losses of Nitrate from Agricultural Drainage Waters", *Journal of Environmental Quality*, Vol 14, 1985, pp. 472–478.

(130) Wilson, L. G., and Small, G. G., "Pollution Potential of a Sanitary Landfill Near Tucson, Hydraulic Engineering and the Environment", *Proceedings 21st Annual Hydraulics Division Specialty Conference*, ASCE, 1973.

(131) Dunlap, William J., *Some Concepts Pertaining to Investigative Methodology for Subsurface Process Research*, U.S. Environmental Protection Agency, 1977, pp. 167–172.

(132) Fenn, D. E., Hanley, K. J., Isbister, J., Briads, O., Yare, B., and Roux, P., *Procedures Manual for Ground Water Monitoring at Solid Waste Disposal Facilities*, U.S. Environmental Protection Agency, EPA/530/SW-611, 1977.

(133) Gibb, J. P., Schuller, R. M., and Griffin, R. A., *Procedures for the Collection of Representative Water Quality Data from Monitoring Wells*, Cooperative Groundwater Report 7, Illinois State Water Survey, 1981.

(134) Willardson, L. S., Meek, B. D., and Huber, M. J., "A Flow Path Ground Water Sampler", *Soil Science Society of America Proceedings*, Vol 36, 1973, pp. 965–966.

(135) Tadros, V. T., and McGarity, J. W., "A Method for Collecting Soil Percolate and Soil Solution in the Field", *Plant and Soil*, Vol 44, June 1976, pp. 655–667.

(136) Shimshi, Daniel, "Use of Ceramic Points for the Sampling of Soil Solution", *Soil Science*, Vol 101, 1966, pp. 98–103.

(137) Grover, B. L., and Lamborn, R. E., "Preparation of Porous Seramic Cups to be Used for Extraction of Soil Water Having Low Solute Concentrations", *Soil Science Society of America Proceedings*, Vol 34, 1970, pp. 706–708.

(138) Zimmermann, C. F., Price, M. T., and Montgomery, J. R., "A Comparison of Ceramic and Teflon In-Situ Samplers for Nutrient Pore Water Determinations", *Estuarine and Coastal Marine Science*, Vol 7, 1978, pp. 93–97.

(139) Faber, W. R., and Nelson, P. V., "Evaluation of Methods for Solution from Container Root Media", *Communications in Soil Science Plant Analysis*, Vol 15, No. 9, 1984, pp. 1029–1040.

(140) Barbarick, K. A., Sabey, B. R., and Klute, A., "Comparison of Various Methods of Sampling Soil Water for Determining Ionic Salts, Sodium, and Calcium Content in Soil Columns", *Soil Science Society of America Journal*, Vol 43, 1979, pp. 1053–1055.

(141) Wagemann, R., and Graham, B., "Membrane and Glass Fibre Filter Contamination in Chemical Analysis of Fresh Water", *Water Research*, Vol 8, 1974, pp. 407–412.

(142) Jones, J. N., and Miller, G. D., "Adsorption of Selected Organic Contaminants onto Possible Well Casing Materials", *Ground-Water Contamination Field Methods, ASTM STP 963*, ASTM, 1988, pp 185–198.

(143) Barcelona, M. J., Helfrich, J. A., and Garske, E. E., "Verification of Sampling Methods and Selection of Materials for Ground Water Contamination Studies", *Ground-Water Contamination Field Methods, ASTM STP 963*, ASTM, 1988.

Standard Guide for Soil Sampling from the Vadose Zone[1]

This standard is issued under the fixed designation D 4700; the number immediately following the designation indicates the year of original adoption or, in the case of revision, the year of last revision. A number in parentheses indicates the year of last reapproval. A superscript epsilon (ε) indicates an editorial change since the last revision or reapproval.

1. Scope

1.1 This guide addresses procedures that may be used for obtaining soil samples from the vadose zone (unsaturated zone). Samples can be collected for a variety of reasons including the following:

1.1.1 Stratigraphic description,

1.1.2 Hydraulic conductivity testing,

1.1.3 Moisture content measurement,

1.1.4 Moisture release curve construction,

1.1.5 Geotechnical testing,

1.1.6 Soil gas analyses,

1.1.7 Microorganism extraction, or

1.1.8 Pore liquid and soils chemical analyses.

1.2 This guide focuses on methods that provide soil samples for chemical analyses of the soil or contained liquids or contaminants. However, comments on how methods may be modified for other objectives are included.

1.3 This guide does not describe sampling methods for lithified deposits and rocks (for example, sandstone, shale, tuff, granite).

1.4 In general, it is prudent to perform all field work with at least two people present. This increases safety and facilitates efficient data collection.

1.5 The values stated in inch-pound units are to be regarded as the standard. The SI units given in parentheses are for information only.

1.6 *This standard does not purport to address all of the safety problems, if any, associated with its use. It is the responsibility of the user of this standard to establish appropriate safety and health practices and determine the applicability of regulatory limitations prior to use.*

2. Referenced Documents

2.1 *ASTM Standards:*

D 420 Practice for Investigating and Sampling Soil and Rock for Engineering Purposes[2]

D 653 Terminology Relating to Soil, Rock, and Contained Fluids[2]

D 1452 Practice for Soil Investigation and Sampling by Auger Borings[2]

D 1586 Method for Penetration Test and Split-Barrel Sampling of Soils[2]

D 1587 Method for Thin-Walled Tube Sampling of Soils[2]

D 2488 Practice for Description and Identification of Soils (Visual-Manual Procedure)[2]

D 2607 Classification of Peats, Mosses, Humus, and Related Products[2]

D 3550 Method for Ring-Lined Barrel Sampling of Soils[2]

D 4083 Practice for Description of Frozen Soils (Visual-Manual Procedure)[2]

D 4220 Practice for Preserving and Transporting Soil Samples[2]

3. Terminology

3.1 *Definitions:*

3.1.1 Except where noted, all terms and symbols in this guide are in accordance with the following publications. In order of consideration they are:

3.1.1.1 Terminology D 653.

3.1.1.2 *Compilation of ASTM Standard Terminology,*[3] and

3.1.1.3 *Webster's New Collegiate Dictionary.*[4]

3.1.2 For definitions and classifications of soil related terms used, refer to Practice D 2488 and Terminology D 653. Additional terms that require clarification are defined in 3.2.

3.2 *Descriptions of Terms Specific to This Standard:*

3.2.1 *cascading water*—perched ground water that enters a well casing via cracks or uncovered perforations, trickling, or pouring down the inside of the casing.

3.2.2 *sludge*—a water charged sedimentary deposit.

3.2.2.1 *Discussion*—The water-formed sedimentary deposit may include all suspended solids carried by the water and trace elements that were in solution in the water. Sludge usually does not cohere sufficiently to retain its physical shape when mechanical means are used to remove it from the surface on which it deposits, but it may be baked in place and be adherent.

4. Summary of Guide

4.1 Sampling vadose zone soil involves inserting into the ground a device that retains and recovers a sample. Devices and systems for vadose zone sampling are divided into two general groups, namely the following: samplers used in conjunction with hand operated devices; and samplers used

[1] This guide is under the jurisdiction of ASTM Committee D-18 on Soil and Rock and is the direct responsibility of Subcommittee D18.21 on Ground Water and Vadose Zone Investigations.

Current edition approved July 15, 1991. Published September 1991.

[2] *Annual Book of ASTM Standards,* Vol 04.08.

[3] *Compilation of ASTM Standard Terminology,* Sixth edition, ASTM, 1916 Race St., Phila., PA 19103, 1986.

[4] *Webster's New Collegiate Dictionary,* Fifth edition, ______ 1977.

Type of Sampler	Obtains Core Sample		Most Suitable Core Types		Operation in Stoney Soils		Most Suitable Soil Moisture Conditions			Access to Sample Sites During Poor Soil Conditions		Relative Sample Size		Labor Req'mts	
A. Drill Rig Samplers	Yes	No	Coh.	Coh'less	Fav.	Unfav.	Wet	Dry	Inter.	Yes	No	Sm.	Lg.	Sngl.	2/More
1. Multipurpose Drill Rig	◆		◆	◆	◆		◆	◆	◆	◆		◆	◆		
2. Split-barrel Drive Sampler	◆		◆		◆			◆			◆		◆		◆
3. Thin-Walled Tube Sampler	◆		◆			◆			◆	◆			◆		◆
4. Piston Sampler	◆		◆			◆	◆			◆			◆		◆
5. Continuous Sample Tube system	◆		◆		◆		◆	◆	◆	◆		◆	◆		◆
6. Hand-Held Power Auger		◆			◆					◆			◆		◆
B. Hand Operated Samplers															
1. Screw-Type Auger		◆				◆	◆			◆		◆		◆	
2. Barrel Auger															
a. Post-Hole Auger		◆	◆		◆		◆			◆			◆	◆	
b. Dutch Auger		◆	◆		◆		◆								
c. Regular Barrel Auger		◆	◆		◆				◆	◆		◆		◆	
d. Sand Auger		◆		◆	◆				◆	◆			◆	◆	
e. Mud Auger		◆	◆		◆		◆			◆			◆	◆	
3. Tube-Type Sampler															
a. Soil Sampling Tube															
(1) Wet Tip	◆					◆	◆			◆				◆	
(2) Dry Tip	◆					◆	◆			◆		◆		◆	
b. Veihmeyer Tube	◆								◆			◆		◆	

FIG. 1 Criteria for Selecting Soil Sampling Equipment

in conjunction with multipurpose or auger drill rigs. This guide discusses these groups and their associated practices.

4.2 The discussion of each device is organized into three sections, describing the device, describing sampling methods, and limitations and advantages of its use.

4.3 This guide identifies and describes a number of sampling methods and samplers. It is advisable to consult available site-specific geological and hydrological data to assist in determining the sampling method and sampler best suited for a specific project. It is also advisable to contact a local firm providing the services required as not all sampling and drilling methods described in this guide are available nationwide.

5. Significance and Use

5.1 Chemical analyses of liquids, solids, and gases from the vadose zone can provide information on the presence, possible source, migration route, and physical-chemical behavior of contaminants. Remedial or mitigating measures can be formulated based on this information. This guide describes devices and procedures that can be used to obtain vadose zone soil samples.

5.2 Soil sampling is useful for the reasons presented in Section 1. However, it should be recognized that the general method is destructive, and that resampling at an exact location is not possible. Therefore, if a long term monitoring program is being designed, other methods for obtaining samples should be considered.

6. Criteria for Selecting Soil Samplers

6.1 Important criteria to consider when selecting devices for vadose zone soil sampling include the following:

6.1.1 Type of sample: An encased core sample, an uncased core sample, a depth-specific representative sample, or a sample according to requirements of the analyses,

6.1.2 Sample size requirements,

6.1.3 Suitability for sampling various soil types,

6.1.4 Maximum sampling depth,

6.1.5 Suitability for sampling soils under various moisture conditions,

6.1.6 Ability to minimize cross contamination,

6.1.7 Accessibility to the sampling site, and

6.1.8 Personnel requirements.

6.2 The sampling devices described in this guide have been evaluated for these criteria. The results are summarized in Fig. 1.

7. Sampling with Hand Operated Devices

7.1 These devices, that have mostly been developed for agricultural purposes, include:

7.1.1 Screw-type augers,

7.1.2 Barrel augers,

7.1.3 Tube-type samplers,

7.1.4 Hand held power augers, and

7.1.5 Trench sampling with shovels in conjunction with machine excavations.

7.2 The advantages of using hand operated devices over drill rigs are the ease of equipment transport to locations with poor vehicle access, and the lower costs of setup and decontamination. However, a major disadvantage is that these devices are limited to shallower depths than drill rigs.

7.3 *Screw-Type Augers:*

7.3.1 *Description*—The screw or ship auger is essentially a small diameter (for example, 1.5 in. (3.81 cm)) wood auger from which the cutting side flanges and tip have been removed **(1)**[5] (see Fig. 2(*a*)). According to the Soil Survey Staff **(1)**, the spiral part of the auger should be about 7 in. (18 cm) long, with the distances between flights about the same as the diameter (for example, 1.5 in.) of the auger. This facilitates measuring the depth of penetration of the tool. Variations on this design include the closed spiral auger and the Jamaica open spiral auger **(2)** (see Fig. 2(*b*) and 1(*c*)). The

[5] The boldface numbers in parentheses refer to the list of references at the end of the text.

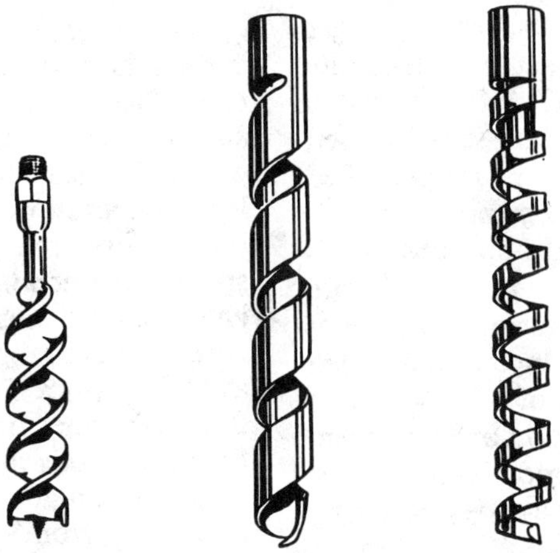

FIG. 2 Screw Type Augers

auger is welded onto a length of solid or tubular rod. The upper end of this rod is threaded, to accept a handle or extension rods. As many extensions are used as are required to reach the target sampling depth. The rod and the extensions are marked in even increments (for example, in 6-in. (15.24-cm) increments) above the base of the auger to aid in determining drilling depth. A wooden or metal handle fits into a tee-type coupling, screwed into the uppermost extension rod.

7.3.2 *Sampling Method*—For drilling, the auger is rotated manually. The operator may have to apply downward pressure to start and embed the auger; afterwards, the auger screws itself into the soil. The auger is advanced to its full length, and then pulled up and removed. Soil from the deepest interval penetrated by the auger is retained on the auger flights. A sample can be collected from the flights using a spatula. A foot pump operated hydraulic system has been developed to advance augers up to 4.5 in. (11.43 cm) in diameter. This larger diameter allows insertion of other sampling devices into the drill hole, once the auger is removed, if desired **(3)**.[6]

7.3.3 *Comments*—Samples obtained with screw-type samplers are disturbed and are not truly core samples. Therefore, the samples are not suitable for tests requiring undisturbed samples, such as hydraulic conductivity tests. In addition, soil structures are disrupted and small scale lithologic features cannot be examined. Nevertheless, screw-type samplers are still suitable for use in collecting samples for the purpose of detecting contaminants. However, it is difficult to avoid transporting shallow soils downward when reentering a drill hole. When representative samples are desired from a discrete interval, the borehole must be made large enough to insert a sampler and extend it to the bottom of the borehole without touching the sides of the borehole. It is suggested that a larger diameter auger be used to advance and clear the borehole, then a smaller diameter auger sampler be used to obtain the sample. Screw-type augers work better in wet, cohesive soils than in dry, loose soils. Sampling in very dry (for example, powdery) soils may not be possible with these augers as soils will not be retained on the auger flights. Also, if the soil contains gravel or rock fragments larger than about one tenth of the hole diameter, drilling may not be possible **(4)**.

7.4 *Barrel Augers:*

7.4.1 *Description*—The barrel auger consists of a bit with cutting edges welded to a short tube or barrel within which the soil sample is retained, welded in turn to shanks. The shanks are welded to a threaded rod at the other end. Extension rods are attached as required to reach the target sampling depth. Extensions are marked in increments above the base of the tool. The uppermost extension rod contains a tee-type coupling for a handle. The auger is available in carbon steel and stainless steel with hardened steel cutting edges **(5, 6)**.

7.4.2 *Sampling Method*—The auger is rotated to advance the barrel into the ground. The operator may have to apply downward pressure to keep the auger advancing. When the barrel is filled, the unit is withdrawn from the soil cavity and a sample may be collected from the barrel.

[6] This reference is manufacturer's literature, and it has not been subjected to technical review.

7.4.3 *Comments*—Barrel augers generally provide larger samples than screw-type augers. The augers can penetrate shallow clays, silts, and fine grained sands **(7)**.[6] The augers do not work well in gravelly soils, caliche, or semi-lithified deposits. Samples obtained with barrel augers are disturbed and are not core samples. Therefore, the samples are not suitable for tests requiring undisturbed samples, such as hydraulic conductivity tests. Nevertheless, the samplers are still suitable for use in collecting samples for the purpose of detecting contaminants. Because the sample is retained inside the barrel, there is less of a chance of mixing it with soil from a shallower interval during insertion or withdrawal of the sampler. The following are five common barrel augers:

7.4.3.1 Post-hole augers (also called Iwan-type augers),

7.4.3.2 Dutch-type augers,

7.4.3.3 Regular or general purpose barrel augers,

7.4.3.4 Sand augers, and

7.4.3.5 Mud augers.

7.4.4 *Post-Hole Augers*—The most readily available barrel auger is the post-hole auger (also called the Iwan-type auger) **(8)**. As shown in Fig. 3, the barrel consists of two-part cylindrical leaves rather than a complete cylinder and is slightly tapered toward the cutting bit. The taper and the cupped bit help to retain soils within the barrel. The barrel is available with a 3 to 12-in. (7.62 to 30.48-cm) diameter. There are two types of drilling systems, one has a single rod and handle, and the other has two handles. In stable, cohesive soils, the auger can be advanced up to 25 ft (7.62 m) **(8)**.

7.4.5 *Dutch-Type Augers*—The Dutch-type auger (commercially developed by Eijkelkamp) is a smaller variation of the post-hole auger design. As shown in Fig. 4, the pointed bit is continuous with two, narrow part-cylindrical barrel segments, welded onto the shanks. The barrel generally has a 3 in. (7.62 cm) outside diameter. This tool is best suited for sampling wet, clayey soils.

7.4.6 *Regular or General Purpose Barrel Augers*—A version of the barrel auger commonly used by soil scientists and county agricultural agents is depicted in Fig. 5(*a*) and (*b*). As shown, the barrel is a complete cylinder. As with the post-hole auger, the cutting blades are cupped so that soil is loosened and forced into the barrel as the unit is rotated and pushed into the ground. Each filling of the barrel corresponds to a depth of penetration of 3 to 5 in. (7.62 to 12.70 cm) **(1)**. The most popular barrel diameter is 3.5 in. (8.89 cm), but sizes ranging from 1.5 to 7 in. (3.81 to 17.78 cm) are available **(6)**.[6] Plastic, stainless steel, PTFE (polytetrafluoroethylene) or aluminum liners can also be used **(6)**.[6] Extension rods are available in 4 ft (1.22 m) lengths. The rods can be made from standard black pipe, from lightweight conduit or from seamless steel tubing. The extensions have evenly spaced marks to facilitate determining sample depth. The regular barrel auger is suitable for use in loam type soils.

7.4.7 *Sand Augers*—For dry, sandy soils it may be necessary to use a variation of the regular barrel auger that includes a specially-formed bit to retain the sample in the barrel (see Fig. 5(*c*)). Sand augers with 2, 3, or 4-in. (5.08, 7.62, or 10.16-cm) diameters are available **(5)**.[6]

7.4.8 *Mud Augers*—Another variation on the regular barrel auger design is available for sampling wet, clayey soils. As shown in Fig. 5(*d*), the barrel is designed with open sides to facilitate extraction of samples. The bits are the same as those used on the regular barrel auger **(6)**.[6] Mud augers with 2, 3, or 4-in. (5.08, 7.62, or 10.16-cm) diameters are available **(5)**.[6]

7.5 *Tube-Type Samplers:*

7.5.1 Tube-type samplers generally have proportionally smaller diameters and greater body lengths than those of barrel augers.

7.5.2 For sampling, these units are perched into the soil causing the tube to fill with material from the interval penetrated. The assembly is then pulled to the surface and a sample can be collected from the tube. Since the device is not

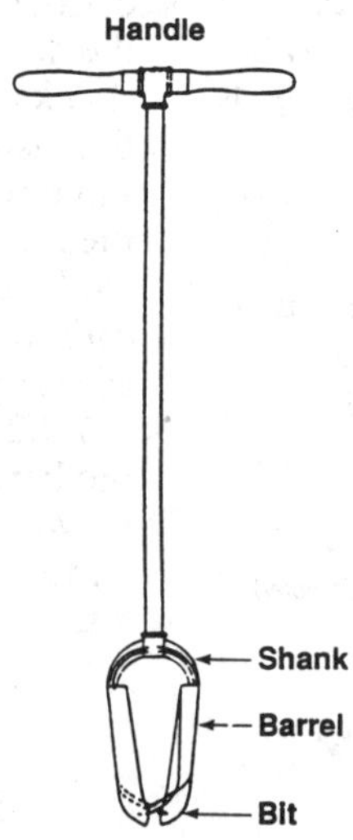

FIG. 3 Post-Hole Type Barrel Auger

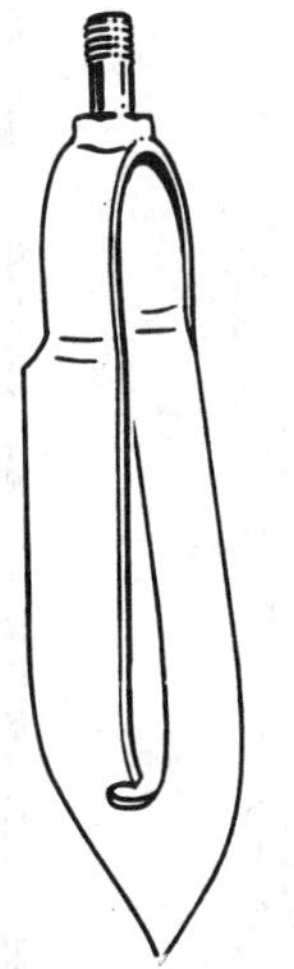

FIG. 4 Dutch Type Auger

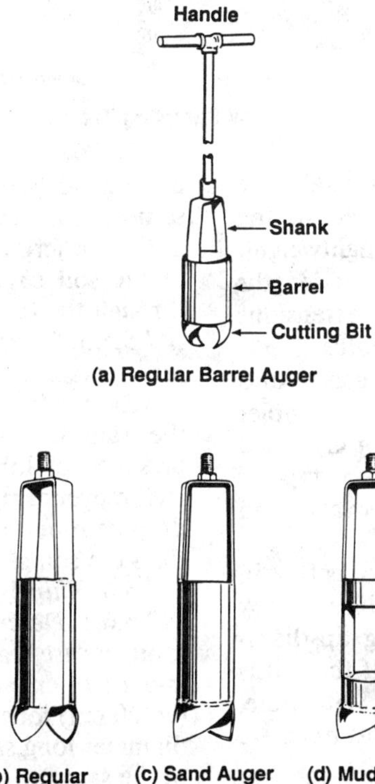

(a) Regular Barrel Auger

(b) Regular Barrel Auger (c) Sand Auger (d) Mud Auger

FIG. 5 Barrel Auger Variations and Soil Moisture

rotated, a nearly undisturbed sample can be obtained. Commercial units are available with foot lever attachments, a hydraulic apparatus, or drop-hammers to aid in driving the sampler into the ground **(5)**[6]. Vibratory heads have also been developed to advance tube-type samplers **(9)**.[6]

7.5.3 These units are not as suitable for sampling in compacted, gravelly soils as are the barrel augers. They are preferred if an undisturbed sample is required. Commonly used varieties of the tube type samplers include:

7.5.3.1 Soil sampling tubes (also called Lord samplers),

7.5.3.2 Veihmeyer tubes (also called King tubes),

7.5.3.3 Thin-walled tube samplers (also called Shelby tubes),

7.5.3.4 Ring-lined barrel samplers, and

7.5.3.5 Piston samplers.

7.5.4 *Soil Sampling Tubes:*

7.5.4.1 *Description*—As depicted in Fig. 6, the soil sampling tube consists of a hardened cutting tip, a cut-away barrel, and an uppermost threaded segment. The cut-away barrel allows textural examination and easy removal of soil samples. Generally, the tube is constructed from high strength alloy steel **(10)**.[6] The samplers are available with 6, 12, 15, 18, and 24-in. (15.24, 30.48, 38.10, 45.72, 60.96-cm) lengths **(5, 6)**. The tubes are available with 1.13 or 0.88-in. (2.87 or 2.22-cm) outside diameter. Two modified versions of the tip are available, for sampling in wet or dry soils. The

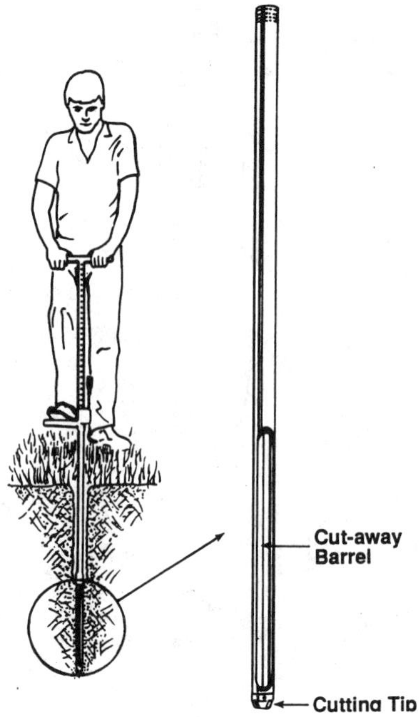

FIG. 6 Soil Sampling Tube

sampling tube is attached to extension rods to attain the target sampling depth. A cross-handle is attached to the uppermost rod. Extension rods are made of lightweight, durable metal. They are available in a variety of lengths depending on the manufacturer. Markings on the extensions and the sampler facilitate determining sample depths.

7.5.4.2 *Sampling Method*—The sampler is pushed into the ground by leaning on the unit's handle. Once the sampler has reached the bottom of the sampling interval, it is twisted to break soil continuity at the tip. Depending on the type of cutting edge, the tube sampler may obtain samples varying in diameter from 0.69 to 0.75 in. (1.75 to 1.91 cm).

7.5.4.3 *Comments*—The soil sampling tube works best in soft, clayey, cohesive soils. If the soil contains cobbles or rock fragments larger than about one-half the cutting tip diameter, satisfactory sampling may not be possible. If the soil is cohesionless, it will not be retained in the tube. With time, the cutting tip will be damaged and worn dull. Most units are designed so that this part can be replaced.

7.5.5 *Veihmeyer Tubes:*

7.5.5.1 *Description*—The Veihmeyer tube is a long, complete cylinder. As shown in Fig. 6, this unit consists of a bevelled tip, that is threaded into the lower end of the tube, and a drive head threaded onto the upper end of the tube. The sampler is constructed of hardened steel. The tube is generally marked in even increments (for example, 1 ft or 0.30 m). These samplers are available in 4 to 16-ft (1.22 to 4.88-m) lengths with a 0.75-in. (1.91-cm) inside diameter.

7.5.5.2 *Sampling Method*—The lower guide rod of the drop hammer is slipped into the upper tube, through the drive head (see Fig. 7). The hammer is used to pound the sampler into the ground. The sampler is then retrieved by pulling or jerking up on the hammer to force the sampler out of the soil cavity. Samples are extruded by forcing a rod through the tube.

7.5.5.3 *Comments*—Prior to sampling, the inside of the tube is sometimes coated with a lubricant to facilitate extrusion. However, the types of analyses to be performed on the samples should be considered to determine if the presence of lubricant will cause interference. Because the Veihmeyer sampler is a solid-walled tube and is fitted with a drop hammer, it can generally be used in more resistant soils than the soil sampling tube.

7.5.6 *Thin-Walled Tube Samplers:*

7.5.6.1 *Description*—Thin-walled tube (Shelby Tube) samplers are readily available with 2, 3, and 5-in. (5.08, 7.62, and 12.70-cm) outside diameters and are commonly 30 in. (76.20 cm) long. The 3 by 30-in. (7.62 by 76.20-cm) outside diameter long sampler is most common. The advancing end of the sampler is rolled inwardly and has a cutting edge with a smaller diameter than the tube inside diameter. The cutting edge inside diameter reduction, defined as a "clearance ratio," is usually in the range of 0.0050 to 0.0150 or 0.50 to 1.50 % (Refer to Practice D 1587). The sampler tube is usually connected with set screws to a sampler head that in turn is threaded to connect with extension rods. Plastic and PTFE sealing caps for use after sampling are readily available for the 2, 3, and 5-in. (5.08, 7.62, and 12.70-cm) diameter tubes (refer to Practice D 4220). Shelby tubes are commonly available in carbon steel but can be manufactured from other metal (see Fig. 8).

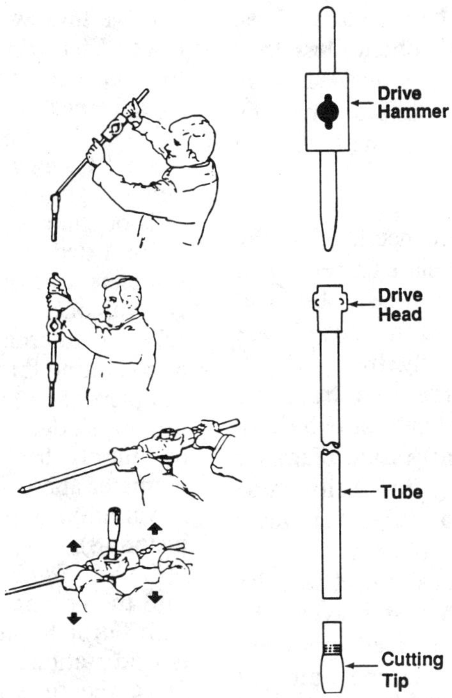

FIG. 7 Veihmeyer Tube

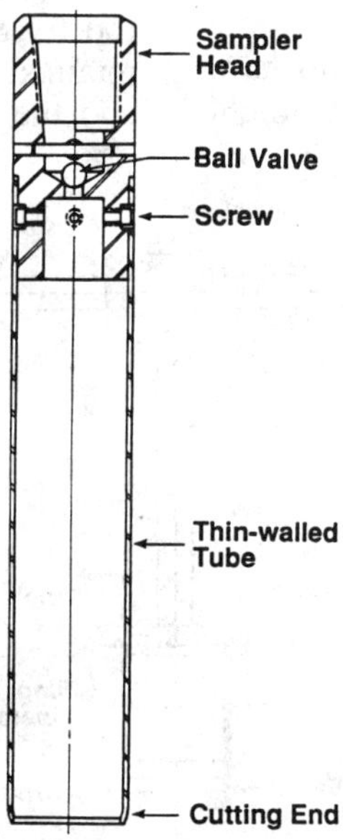

FIG. 8 Thin-Walled Tube Sampler

7.5.6.2 *Sampling Method*—The Shelby tube is pushed into soil by hand, with a jack-like system or with a hydraulic piston. The sample recovered is often less than the distance pushed, that is, the recovery ratio is less than 1.0. The recovery ratio is less than 1.0 because of soil compaction during sampling, and because friction between soil and the inner tube walls becomes greater than the shear strength of the soil in front of the tube. Consequently, soil in front of the advancing end of the tube is displaced laterally rather than entering the tube **(11)**. In general, shorter tubes provide less-disturbed samples than longer tubes. Samples are extruded from the Shelby tube with a hydraulic ram. As with all sampling devices, the most disturbed portion of the sample in contact with the tube is considered unrepresentative. Wilson et al. **(12)** developed a paring device to remove this outer layer of the core during extrusion.

7.5.6.3 *Comments*—Shelby tubes are best used in clays, silts, and fine-grained sands. If the soils are cohesionless, they may not be retained in the tube. If firm to very hard soils are encountered, driving (hammering) the sampler may be required. However, this should be avoided as the tube may buckle under the drive stress.

7.5.7 *Ring-Lined Barrel Samplers:*

7.5.7.1 *Description*—As described in Practice D 3550, the ring-lined barrel sampler consists of a one piece barrel or two split barrel halves, a drive shoe, rings, and a sampler head (see Fig. 9). The rings, that are usually brass, fit snugly inside the barrel and are designed to be directly inserted into geotechnical testing apparatuses when removed from the barrel. Most samplers are designed to hold at least two rings. The barrel is commonly 3.5 in. (8.89 cm) inside diameter and 3.94 to 5.91 in. (10 to 15 cm) long **(5)**.[6] With these lengths, the barrel can be fitted with a variety of liners ranging in length from 1 to 2.36 in. (2.54 to 6 cm).

7.5.7.2 *Sampling Method*—The ring-lined barrel sampler can be driven or pushed into soil. Once retrieved, the sampler is disassembled, and the sample-filled rings are removed. The rings are usually removed as one unit and placed into a capped container. Alternately, the individual soil-filled rings can be capped with plastic or PTFE and then sealed with wax or adhesive tape (refer to Practice D 4220).

7.5.7.3 *Comments*—Because barrel samplers are more rigid than thin-walled tubes, they can be driven into hard soils and soils containing sands and gravels that might damage thin-walled tubes. The sampler provides samples in rings which can be handled without further disturbance of the soil. Because of this, these devices are most often used when geotechnical or chemical analyses are to be performed.

7.5.8 *Piston Samplers:*

7.5.8.1 *Description*—Locally saturated (for example, by perched ground water), or cohesionless soils, and very soft soils or sludges may not be retained in most samplers, even when fitted with retainer baskets or flap valves. Piston samplers can be used in these situations. The sampler consists of a sampling tube, extension pipe attached to the tube, an internal piston, and rods connected to the piston and running through the extension pipe (see Fig. 10). These samplers are often built, as needed, out of common PVC (for use in sludge) or steel pipe fittings. The sampling tube commonly has a 0.75 to 3-in. (1.91 to 7.62-cm) inside diameter and is 8 in. to 9 ft (20.32 cm to 2.74 m) long **(13)**. A variation designed for sampling peat has a cone shaped piston **(8)**.

7.5.8.2 *Sampling Method*—The sampler can be pushed into the ground with the handle or driven into the ground with a drop hammer **(13)**. As the tube is advanced, the piston is held stationary or pulled upward with the attached rods. Once the tube has been advanced through the sampling interval, it is rotated to break suction that might have developed between the soil and the outside wall of the tube. The sampler is then pulled to the surface keeping the piston rod fixed with respect to the extension pipe. The sample is retained because of suction that develops between the piston and the sample. Upon retrieval, the sample is extruded by

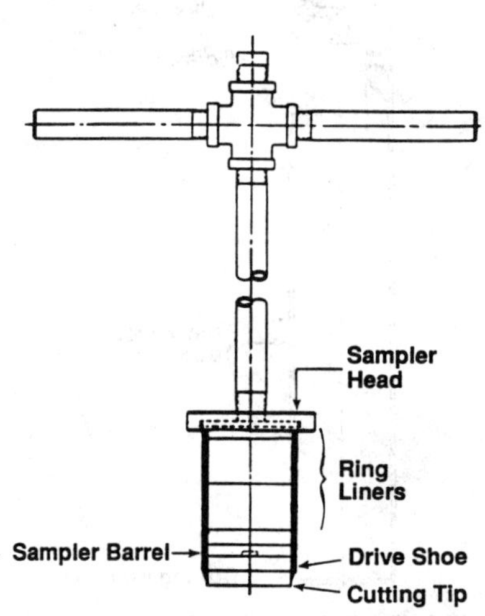

FIG. 9 Hand Operated Ring-Lined Barrel Sampler

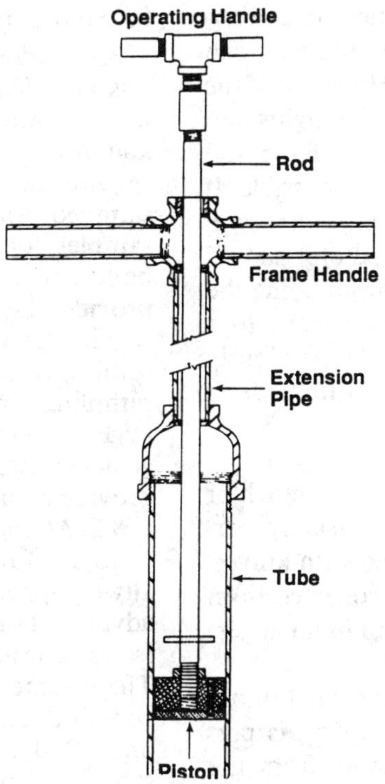

FIG. 10 Hand Operated Piston Sampler

using the piston to force the sample out of the tube. Sharma and De Dalta **(14)** described a cylindrical sampler for use in puddled soils that would flow back out of most samplers. The design includes a basal shutter that retains the sample while the sampler is withdrawn from the soil.

7.5.8.3 *Comments*—Because the sampler depends on development of suction between the sample and the piston, it may not work in unsaturated, coarse-grained sands and gravels. This is due to the high air permeability of such material that prevents the creation of high suction.

7.6 *Hand Held Power Augers:*

7.6.1 *Description*—A very simple, commercially available auger consists of a solid flight auger attached to and driven by a small air-cooled engine (see Fig. 11). Two handles on the head assembly allow two operators to guide the auger into the soil. Throttle and clutch controls are integrated into grips on the handles. Augers are available with diameters ranging from 2 to 16 in. (5.08 to 40.64 cm). The auger sections are commonly 3 ft (0.91 m) long.

7.6.2 *Sampling Method*—As the auger rotates into soil, cuttings advance up the flights and are discharged at the surface. Soil samples can be collected from the surface discharge, or from the auger flights after pulling the auger out of the ground. Alternatively, samples can be collected with other samplers (for example, a thin-walled tube) after auger removal.

7.6.3 *Comments*—As discussed in 7.3, if samples are collected from surface discharge or from the flights, they are disturbed and are not suitable for some uses. In addition, if samples are collected from surface discharge, it is difficult to

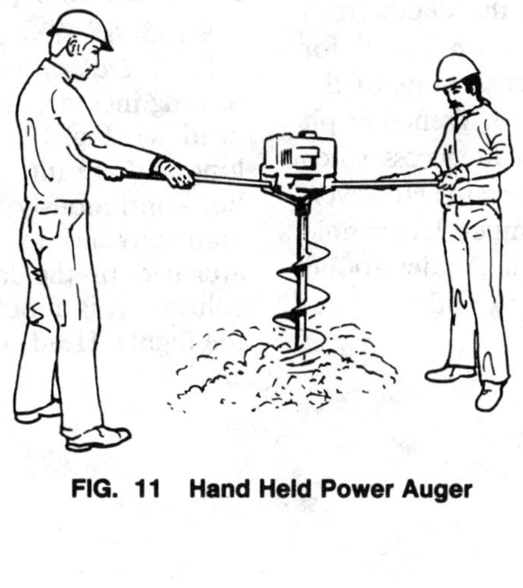

FIG. 11 Hand Held Power Auger

determine the depth from which the soil came and uncontrolled mixing of soil from different depth intervals can occur. The auger operates well in most soils. However, if the soil is cohesionless, it may not be retained on the flights and sampling in that fashion may not be possible. If the soil contains cobbles or boulders, drilling may not be possible. If the auger "hangs up" on an obstruction, the machine will start to rotate at the surface. Otherwise, the operator should not attempt to stop rotation of the machine by grabbing the handles. An alternate design that transfers the torque to a separate engine prevents this problem **(15)**.[6] As previously stated, it is prudent to perform the field work with at least two people present.

7.7 *Trench Sampling:*

7.7.1 *Description*—Soils may be sampled from a trench or pit excavated for that purpose. Excavation is usually performed by a backhoe, and samples are collected with knives, trowels, or shovels. Occasionally, samples are collected from the sides or the bottom of the trench or pit with hand augers or tube-type samplers.

7.7.2 *Sampling Method*—Excavation is performed under the guidance of the sampling technician. Sampling is performed only after the backhoe has moved away from the trench or pit. When the trench or pit is in unstable material or is more than a few feet deep, the sampling technician should only enter the trench or pit after it has been shored up or the sidewalls have been cut back to within the angle of repose (see Occupational Safety and Health Administration regulations). Otherwise, samples are more commonly collected at the surface from the bucket of the backhoe as excavation occurs.

7.7.3 *Comments*—The maximum sampling depth for the trench or pit method is dictated by the reach of the backhoe, the soil type and the moisture content of the soil. Maximum depths of up to 20 ft (6.10 m) can be obtained in moist clays. Maximum depths of less than 10 ft (3.05 m) are common in dry sands. Samples collected from the backhoe bucket should be taken from the center of the material to prevent collecting soil contaminated by the bucket surface, and to prevent inclusion of materials that may have fallen from above the desired sampling interval. However, when this is done, it is difficult to accurately estimate the depth from which the sample was obtained. Trenches are useful for obtaining lithologic information since cross sections of the vadose zone can be studied and photographed. Trench or pit sampling is often used in areas with difficult access since backhoes are designed to travel on rough terrain. However, because the process involves excavating a much larger hole than drilling methods, chances of encountering underground utilities are increased, and proper backfilling and compaction of the trench is often very difficult.

8. Multipurpose and Auger Drill Rigs

8.1 Vadose zone samplers used in conjunction with drill rigs are identical to those used to sample below the water table. However, commonly used drill rigs such as cable tool and rotary units are not recommended as they generally require the introduction of drilling fluids to the soils to be sampled. Air rotary drilling is also undesirable for obtaining samples for pore liquid or gas extraction. In most cases, hollow-stem augers with some type of cylindrical sampler provide the greatest level of assurance that soil sampled within the vadose zone was not carried downward by the drilling or sampling process. For some situations, such as sampling firm to very hard ground, using multipurpose auger-core-rotary drill rigs will be necessary. For some geologic circumstances the use of solid stem augers will provide an adequate drilling method.

8.2 *Multipurpose Auger-Core-Rotary Drill Rigs:*

8.2.1 Multipurpose auger-core-rotary drill rigs are generally equipped with rotary power and vertical feed control to advance both hollow-stem augers and continuous flight (solid stem) augers to depths greater than 100 ft (30.48 m). These same drills have secondary capability for rotary and core drilling. The larger of these drills are typically mounted on 20 000 to 30 000-lb (9070 to 13605-kg) GVW trucks. The same multipurpose drill rigs are available on both rubber-tired and track-driven all-terrain carriers. The smaller of the multipurpose drills are typically mounted on trailers or one-ton, 4 by 4 trucks.

8.2.2 When equipped with augers, the sampling process is identical to that for auger drill rigs. When multipurpose auger-core-rotary drill rigs or auger drill rigs are used, the speed of drilling and sampling is much greater than with hand operated equipment. Therefore it is useful to have a larger crew to efficiently handle, log, identify, and preserve the samples.

8.3 *Auger Drill Rigs*—Auger drill rigs are similar to multipurpose auger-core-rotary drill rigs. They are manufactured specifically for efficient auger drilling but do not have the pumps and hoists that are required for efficient core or rotary drilling. The rigs can be equipped with either solid stem or hollow stem augers. There are relatively few auger drills available in comparison to multipurpose auger-core-rotary drills.

9. Auger Drilling and Sampling

9.1 *Solid Stem Auger Drilling and Sampling:*

9.1.1 *Description*—The tools used for solid-stem auger drilling include: auger sections, the drive cap, and the cutter head (see Fig. 12). Auger sections are typically 5 ft (1.52 m) long and are interchangeable for assembly in an articulated but continuously flighted column. Augers are available in diameters up to 24 in. (60.96 cm). The cutter head is attached to the lowermost or leading flight of the auger column. It is about 0.5 in. (1.25 cm) larger in diameter than the flights. Head types include fish tail or drag bits for use in

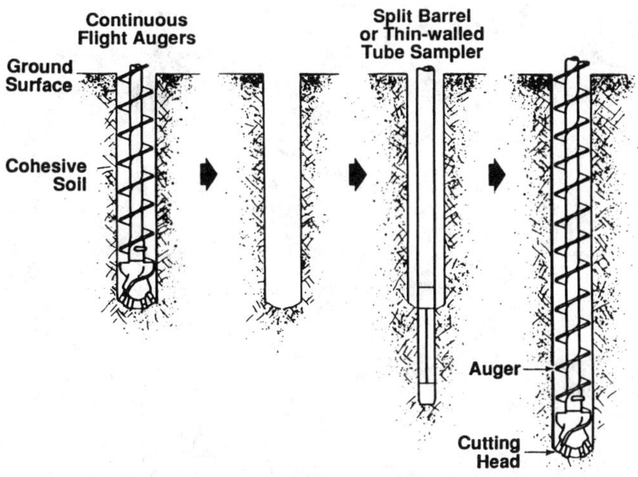

FIG. 12 Solid Stem Auger Sampling

cohesionless materials, and clay or stinger bits for use in more consolidated material **(16)**.

9.1.2 *Sampling Method*—As the auger column is rotated into soil, cuttings are retained on the flights. The augers are then removed from the hole and samples are taken from the retained soil. Samples obtained with solid stem augers are disturbed and are not core samples. Therefore, the samples are not suitable for analyses requiring undisturbed samples, such as hydraulic conductivity tests. This sampling method can provide an adequately clean borehole in some clayey and silty soils. However, when using the method in caving or squeezing ground, the quality and the origin of the recovered samples are questionable because soils from different intervals may have mixed. Therefore, when representative samples from discrete depths are desired, the borehole should be made large enough to insert a smaller diameter auger or another sampler (for example, a thin-walled tube) to the bottom of the borehole, without touching the sides of the borehole (see Fig. 11), to collect a discrete sample from the interval ahead.

9.1.3 *Comments*—Typical drilling depths with solid stem augers range from 50 to 120 ft (15.24 to 36.58 m). The greater drilling depths are attained in firm, silty and clayey soils. However, the depth to which the hole will remain open for sampling once the auger column has been removed is usually less than the maximum drilling depth. If cascading water or cohesionless soils are encountered, it can be expected that the hole will cave at that depth. The sample depth measurement, as taken from its location on an auger, is not precise. This is because soil may move up the flights in an uneven fashion as the auger column is advanced. As with hollow-stem augers, solid stem augers are often painted by the driller or manufacturer. It is prudent to remove this paint before drilling. The majority of the paint can be removed by drilling through sandy soils or by sand blasting. As with all sampling devices, decontamination (for example, steam cleaning) should be performed between holes when chemical analyses are to be performed on the samples. This is especially important with the solid stem auger as it doubles as the drilling and sampling tool.

9.2 *Bucket Auger Drilling and Sampling:*

9.2.1 *Description*—The bucket auger is a large diameter cylindrical bucket with auger-type cutting blades on the bottom. The bucket can have a diameter ranging from 12 in. (30.48 cm) up to 6 ft (1.83 m) with lengths varying from 24 to 48 in. (60.96 to 121.92 cm) **(17)**. The bottom is hinged to allow cuttings to be emptied out (see Fig. 13).

9.2.2 *Sampling Method*—The bucket is rotated to depth in the vadose zone until the bucket is full. Therefore, depending on the bucket length, sampling intervals can range from 24 to 48 in. (60.96 to 121.92 cm). Sampling consists of extracting small diameter core samples from the interior of the bucket after lowering the full bucket to the ground (see Section 7). This approach minimizes problems with undiscrete mixing of discrete portions to be sampled.

9.2.3 *Comments*—The bucket auger is best suited for sampling from relatively stable clays as the caving problems discussed in 9.1.3 are amplified by the larger hole diameter. Boulders can impede drilling and may have to be individually removed from the hole before sampling can continue **(15)**[6]. Generally, boulders up to one-third or one-fourth the bucket diameter can be picked up by the bucket. Common sampling depths are less than 50 ft (15.24 m) but holes up to 250 ft (76.20 m) deep have been drilled **(16, 17)**.

9.3 *Hollow Stem Auger Drilling and Sampling:*

9.3.1 *Description*—Outer components of the hollow stem auger system include: hollow auger sections, the hollow auger head, and the drive cap. Inner components include: the pilot assembly, the center rod column, and the rod-to-cap adaptor (see Fig. 14). The auger head contains replaceable carbide teeth that pulverize the formation during flight

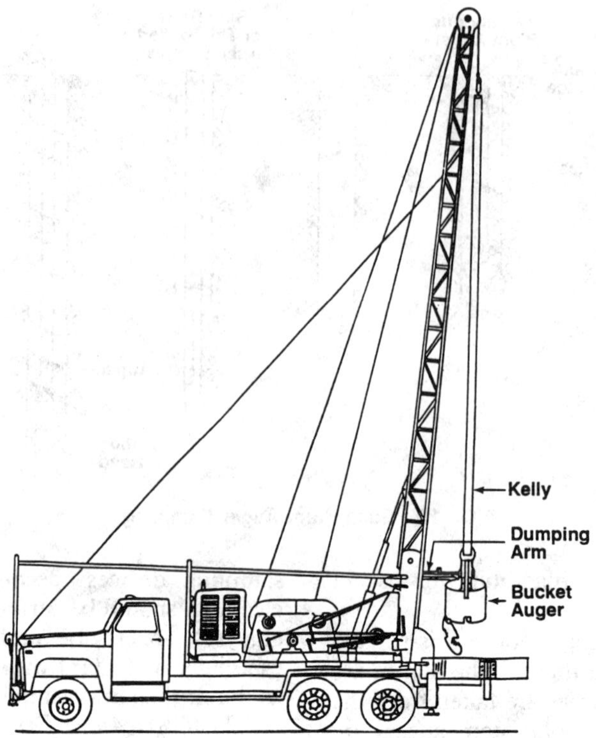

FIG. 13 Bucket Auger and Drilling Rig

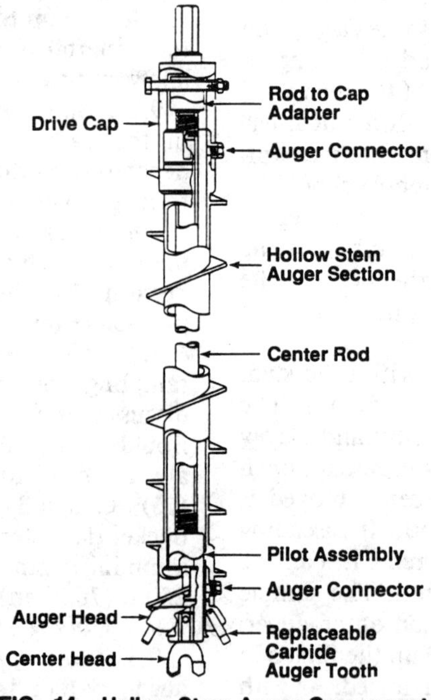

FIG. 14 Hollow-Stem Auger Components

column rotation. The cutting diameter is somewhat greater than the flighting diameter because of the protruding teeth. Auger sections are typically 5 ft (1.52 m) long and are interchangeable for assembly in an articulated but continuously flighted column. Drilling progresses in 5 ft (1.52 m) or shorter increments and sampling can be accomplished at any depth within that increment. Upon advancement of a 5 ft (1.52 m) increment, another 5 ft (1.52 m) section of hollow-stem auger and center rod is added. Hollow-stem augers are readily available with 2.25, 2.75, 3.25, 3.75, 4.25, 6.25, and 8.25-in. (5.72, 6.99, 8.26, 9.53, 10.80, 15.88, and 20.96-cm) inside diameters.

9.3.2 *Sampling Method*—The auger column and pilot assembly are advanced to the top of the desired sampling interval. Sampling is accomplished by removing the pilot assembly and center rod, if they are used, and inserting the sampler through the hollow stem of the auger column (see Fig. 15). The sampler may be lowered to the sampling depth by attaching it to center rods or by using a wireline assembly **(12)**. When the sampler is attached to center rods, a sample is collected by pushing or driving the sampler into undisturbed soil with the rig hydraulic system or with a drop hammer. When a wireline is used, the sampler is locked into place ahead of the lower-most auger and advanced into the sampling interval by rotating the auger column **(18)**.[6] Hollow stem augers with a 6.25-in. (15.88-cm) inside diameter allow the use of 5-in. (12.70-cm) outside diameter Shelby tubes and 4.5-in. (11.43-cm) outside diameter split barrel samplers (see 9.4).

9.3.3 *Comments*—The purpose of the center head (pilot) assembly is to prevent soils from entering the auger column as it is advanced **(19)**. Driscoll **(17)** suggests that the assembly may be omitted when drilling through hard, silty and clayey soils as these materials will usually form a 2 to 4 in. (5.08 to 10.16 cm) long plug at the auger opening. However, Hackett **(19)** recommends that the pilot assembly be used when detailed samples are required. When perched water is encountered, "heaving sands" that move up into the auger column upon pilot assembly removal during sampling, may be a concern. Various one-way plugs that allow sampling, but that prevent sand from moving into the auger column, are described in Hackett **(19)**. The important capability of being able to obtain samples that do not contain mixed material from shallow sources in the hole is enhanced by using the hollow-stem auger method. However, because the sections are hollow, decontamination of the auger interiors between holes to prevent cross contamination is difficult. High pressure steam cleaners are usually necessary to remove caked-on soils and contaminants. Hollow stem augers may advance rapidly through unconsolidated materials.

9.4 *Sampling Devices:*

9.4.1 Sampling devices used in conjunction with hollow stem augers and occasionally in holes advanced by solid stem augers include:

9.4.1.1 Thin-walled tube samplers (also called Shelby tubes),

9.4.1.2 Split-barrel drive samplers (also called Split spoons),

9.4.1.3 Ring-lined barrel samplers,

9.4.1.4 Continuous sample tube systems, and

9.4.1.5 Piston samplers.

9.4.2 These samplers are either pushed or driven in sequence with an increment of drilling or advanced simultaneously with the advance of a hollow stem auger column.

9.4.3 *Thin-Walled Tube Samplers:*

9.4.3.1 *Description*—The thin-walled tube sampler consists of a tube connected to a head with screws. The head is threaded to connect with standard drill rods. The head contains a ball check valve. Thin-walled tube (Shelby tube) samplers are readily available with 2, 3, and 5-in. (5.08, 7.62, and 12.70-cm) outside diameter and are commonly 30 in. (76.20 cm) long. The 3 by 30 in. (7.62 by 76.20 cm) outside diameter long sampler is most common. The advancing end of the sampler is constructed with an inward lip, machined to a cutting edge, that has a smaller diameter than the tube inside diameter. The cutting edge inside diameter reduction, defined as a "clearance ratio," is usually in the range of 0.0050 to 0.0150 or 0.50 to 1.50 % (refer to Practice D 1587). PTFE or plastic sealing caps and other sealing devices for use after sampling are readily available for the 2, 3, and 5-in. (5.08, 7.62 and 12.70-cm) diameter tubes (refer to Practice D 4220). Shelby tubes are commonly available in carbon steel but can be manufactured from other metal (see Fig. 8).

9.4.3.2 *Sampling Methods*—When a Shelby tube is pushed into soil, the length of the sample recovered is often less than the distance pushed, that is, the recovery ratio is less than 1.0 (see 7.5.6.2). In addition, a portion of the sample frequently remains in the borehole after retrieval of the sampler. This is due to suction that develops at the sampler-soil interface. This suction may be broken by twisting the

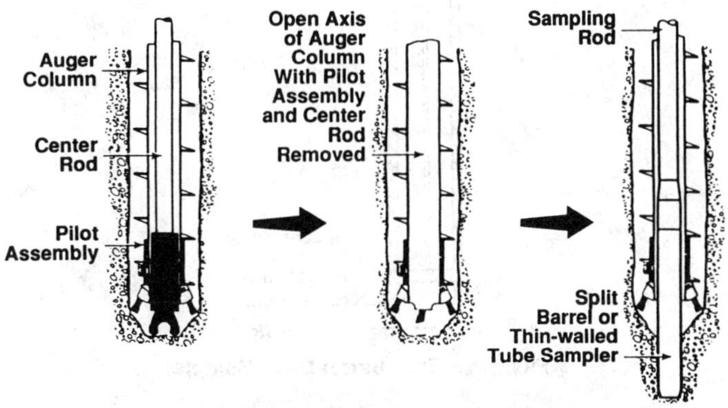

FIG. 15 Hollow-Stem Auger Sampling

sampler prior to retrieval or by advancing the auger column below the base of the sampler before retrieval **(20)**. Samples are extruded from the Shelby tube with a hydraulic ram. As with all sampling devices, the portion of the sample in contact with the tube is considered disturbed and unrepresentative. Wilson et al. **(12)** developed a paring device to remove this outer layer of the core during extrusion.

9.4.3.3 *Comments*—The ball check valve was originally intended to provide a vent for drilling fluids when pushing the tube into soil, and also to prevent the column of fluid within the drill stem from forcing the sample out of the tube during retrieval. Since drilling fluids are not used when sampling in the vadose zone, these considerations are not important. However, the valve does provide a vent for air displaced as the sampler is pushed into soil. Shelby tubes are best used in clays, silts, and fine grained sands. They can be pushed with the hydraulic system of most drill rigs in fine grained sands that are loose to moderately consolidated or in clays and silts that are soft to firm. If the soils are cohesionless, they may not be retained in the tube. If consolidated or hard soils are encountered, driving the sampler may be required. However, some tubes may buckle under the drive stress. A spring-loaded barrel has been developed to protect the Shelby tube from buckling when sampling these soils **(21)**.[6]

9.4.4 *Split-Barrel Drive Samplers:*

9.4.4.1 *Description*—The split-barrel drive sampler consists of two split-barrel halves, a drive shoe, and a sampler head containing a ball check valve, all of which are threaded together (see Fig. 15). The most common size has a 2-in. (5.08-cm) outside diameter and a 1.5-in. (3.81-cm) inside diameter split barrel with a 1.375-in. (3.49-cm) inside diameter drive shoe. This sampler is used extensively in geotechnical exploration (Refer to Method D 1586). When fitted with a 16 gage liner for encased cores, the sampler has a 1.375-in. (3.49-cm) inside diameter throughout. A 3-in. (7.62-cm) outside diameter by 2.5-in. (6.35-cm) inside diameter split-barrel sampler with a 2.375-in. (6.03-cm) inside diameter drive shoe is also available **(22)**.[6] Other split-barrel samplers in the size range of 2.5-in. (6.35-cm) to 4.5-in. (11.43-cm) outside diameter are manufactured but are less common. A plastic or metal retainer basket, or a flap valve is often fitted into the drive shoe to prevent samples from falling out during retrieval.

9.4.4.2 *Sampling Method*—As described in Method D 1586 the sampler is threaded onto drilling rods and is lowered to the bottom of the boring. The sampler is then driven into the soil with blows from a drop hammer attached to the drill rig. The hammer usually weighs 140 lb and is operated by the driller. The sampler is extracted from the soil in a manner that will ensure maximum sample recovery. A

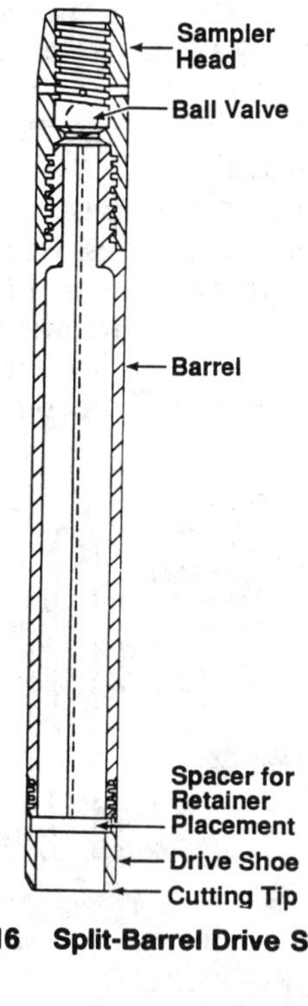

FIG. 16 Split-Barrel Drive Sampler

sample is obtained by disassembling the drive shoe and head, and splitting the barrel to expose the core of soil. Material disturbed by contact with the barrel can be scraped away, or a less disturbed interior portion collected with a spatula.

9.4.4.3 *Comments*—Split barrel drive samplers can be used in all soil types if the larger grain sizes can enter through the opening of the drive shoe. Because the sampler can be fitted with a retainer basket, it is typically used in place of thin-walled tubes when cohesionless soils are to be sampled.

9.4.5 *Ring-Lined Barrel Samplers:*

9.4.5.1 *Description*—As described in Practice D 3550, the ring-lined barrel sampler consists of a one piece barrel or two split-barrel halves, a drive shoe, rings, a waste barrel and a sampler head containing a ball check valve (see Fig. 17). The rings fit snugly inside the barrel and are designed to be directly inserted into geotechnical testing apparatus when removed from the barrel. Most samplers are designed to hold at least six rings. The waste barrel provides a space above the rings into which disturbed soil, originally at the bottom of the hole, can move. The samplers are commonly available with 2, 3, and 4-in. (5.08, 7.62, and 10.16-cm) outside diameter.

9.4.5.2 *Sampling Method*—The ring-lined barrel sampler can be driven or pushed into soil. It is important to insert the sampler deep enough to allow all disturbed soil to move through the rings and into the waste barrel. Once retrieved, the sampler is disassembled, and the sample filled rings are carefully removed. The rings are usually removed as one unit and placed into a capped container. Alternately, the individual soil filled rings can be capped with plastic or PTFE and even sealed with wax or adhesive tape (refer to Practice D 4220).

9.4.5.3 *Comments*—Because ring-lined barrel samplers are more rigid than thin-walled tubes, they can be driven into soils containing sands and gravels that might damage thin-walled tubes. The sampler provides samples in rings that can be handled without further disturbance of the soil. Because of this, these devices are most often used when geotechnical or chemical analyses are to be performed.

9.4.6 *Continuous Sample Tube System:*

9.4.6.1 *Description*—Continuous sample tube systems that fit within a hollow-stem auger column are readily available in North America. The barrel is typically 5 ft (1.52 m) long, and fits within the lead auger of the hollow auger column. The sampler is prevented from rotating as the auger column is turned **(20)**. For many conditions the sampler provides continuous, 5-ft (1.52-m) samples (see Fig. 18). The assembly can be split- or solid-barrel and can be used with or without liners of various metallic and nonmetallic materials **(20)**. Two clear, plastic, 30 in. (76.20 cm) long liners are often used. The sampler may also be fitted with a plastic or metal retainer basket, or a falp valve to prevent cohesionless soils from falling out of the sampler during retrieval **(20)**.

9.4.6.2 *Sampling Method*—The sampler is locked in place inside the auger column with its open end protruding a short distance beyond the end of the column. While advancing the column, soil enters the non-rotating sampling barrel. After a 5-ft (1.52-m) advance, the sampler is withdrawn, and the liner (if used) is removed and capped.

9.4.6.3 *Comments*—The continuous sample tube system replaces the pilot head assembly in the hollow-stem auger column. Because of this, sampling speed is greatly increased since the pilot assembly does not have to be removed before taking a sample. The continuous sample tube system is best used in clays, silts, and in fine grained sands. It can be used to sample soils that are much more consolidated or harder than can be sampled with Shelby tubes.

9.4.7 *Piston Samplers:*

9.4.7.1 *Description*—Locally saturated (for example, perched ground water), or cohesionless soils, and very soft soils or sludges may not be retained in most samplers, even when they have been fitted with retainer baskets or flap valves. Piston samplers are often used under these conditions. The sampler consists of a sampling tube, an internal piston, and a drive head. The piston fits snugly inside the tube. The piston is attached to a rod assembly or a cable that leads to the surface. Tubes made of steel are available in 5.5 and 30-in. (13.97 and 76.20-cm) and 5-ft (1.5-m) lengths with 0.75, 2, 3, 4, and 5-in. (1.91, 5.08, 7.62, 10.16, and 12.70-cm) inside diameter **(22, 23)**. When equipped with a hardened steel drive shoe, the tube can be fitted with a liner made of aluminum clear PVC, or another material (see Fig. 19) **(24)**. A version of the sampler designed for peat sampling has a cone shaped piston **(8)**.

9.4.7.2 *Sampling Method*—Prior to sampling, the piston is placed at the base (advancing end) of the tube. The sampler is then attached to drill rods and lowered down the borehole or hollow-stem auger column to the bottom of the hole (top of the sampling interval). The sampler is then pushed or driven into the sampling interval. As the tube

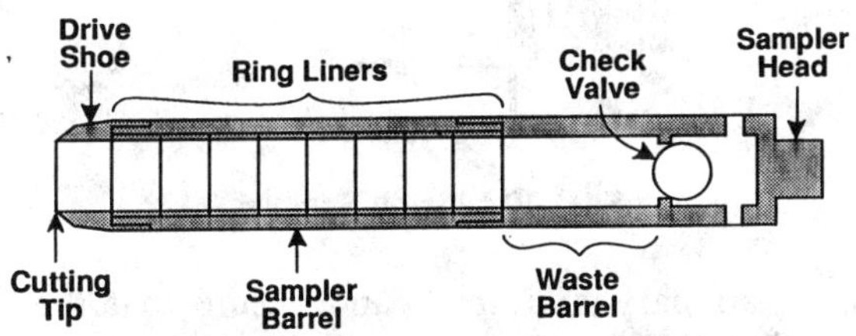

FIG. 17 Ring-Lined Barrel Sampler

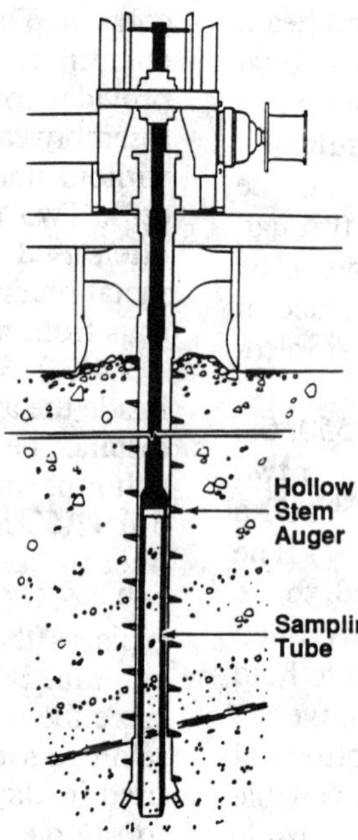

FIG. 18 Continuous Sample Tube System

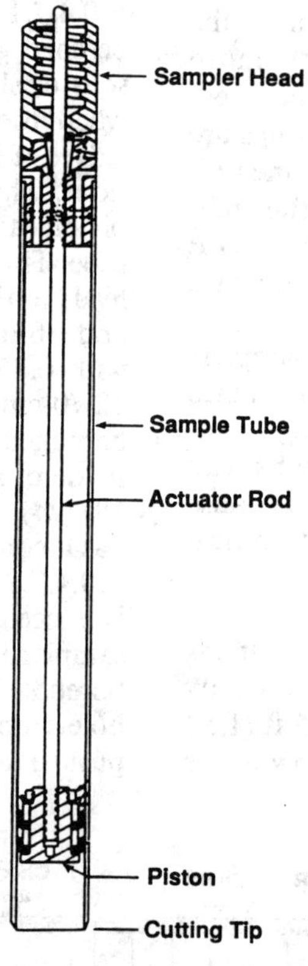

FIG. 19 Piston Sampler

moves downward, the piston remains stationary and in contact with the top of the soil sample. When the sampler is withdrawn, soil is retained because of suction that develops between the piston and the soil core within the sampler. This suction is stronger than the suction at the bottom of the sampler that would tend to extract soil from the sampler. Even so, it is often useful to twist the sampler with the drill rods prior to retrieval, to break suction at the bottom end and ensure that the sample will not be pulled out of the sampler.

9.4.7.3 *Comments*—Average recovery ratios greater than 0.9 can be attained with this sampling tool **(24, 25)**. However, because the sampler depends on development of suction between the sample and the piston, it may not work in unsaturated, coarse grained sands and gravels. This is due to the high air permeability of such material that prevents the creation of suction with the sampler. Samples collected with piston samplers are relatively undisturbed. Zapico et al. **(24)** described techniques for extracting fluid samples directly from liners, and for converting liners into permeameters.

REFERENCES

(1) Soil Survey Staff, *Soil Survey Manual*, U.S. Dept. of Agriculture, Superintendent of Documents, Washington, D.C., 1951.

(2) Acker, W. L., *Basic Procedures for Soil Sampling and Core Drilling*, Acker Drill Co. Inc., Scranton, Pa, 1974.

(3) Materials Testing Division, *Catalog of Products*, Soiltest Inc., Evanston, Illinois, 1983.

(4) Bureau of Reclamation, *Earth Manual*, U.S. Dept. of the Interior, United States Government Printing Office, Washington, D.C., 1974.

(5) Sales Division, *Catalog of Products*, Soilmoisture Equipment Corp., Santa Barbara, California, 1988.

(6) Sales Division, *Catalog of Products*, Art's Manufacturing and Supply, American Falls, Idaho, 1988.

(7) Sales Division, *Catalog of Products*, Brainard Kilman, Stone Mountain, Georgia, 1988.

(8) Everett, L. G., and Wilson, L. G., *Permit Guidance Manual on Unsaturated Zone Monitoring For Hazardous Waste Land Treatment Units*, EPA/530-SW-86-040, 1986.

(9) Sales Division, *Catalog of Products*, VI-COR Technologies Inc., Bellevue, Washington, 1988.

(10) Sales Division, *J.M.C. Soil Investigation Equipment, Catalog No. 6*, Clements Associates Inc.

(11) Hvorslev, M. J., *Subsurface Exploration and Sampling of Soils for Civil Engineering Purposes*, U.S. Army Corp of Engineers, Waterways Experiment Station, Vicksburg, Mississippi, 1949.

(12) Wilson, J. T., and McNabb, J. F., "Biological Transformation of Organic Pollutants in Ground Water," *EOS-Transactions*, AGU, Vol 64, pp. 505–507.

(13) Brakensiek, D. L., Osborn, H. B., and Rawls, W. L., *Field Manual for Research in Agricultural Hydrology, Agriculture Handbook No. 224*, Science and Education Administration, United States Dept. of Agriculture, Washington, D.C., 1979 (revised), pp. 258–275.

(14) Sharma, P. K., and DeDalta, S. K., "A Core Sampler for Puddled Soils," *Soil Science Society of America Journal*, Vol 49, 1985, pp. 1069–1070.

(15) Sales Division, *Catalog of Products*, Little Beaver Inc., Livingston, Texas, 1988.

(16) Scalf, M. R., McNabb, J. F., Dunlap, W. J., Cosby, R. L., and Fryberger, J., *Manual of Groundwater Sampling Procedures*, National Water Well Association, Dublin, Ohio, 1981.

(17) Driscoll, F. G., *Groundwater and Wells*, Johnson Division, St. Paul, Minnesota, (2nd ed.), 1986.

(18) Sales Division, *Catalog of Products*, Mobile Drilling Co. Inc., Indianapolis, Indiana, 1988.

(19) Hackett, G., "Drilling and Constructing Monitoring Wells with Hollow-Stem Augers Part 1: Drilling Considerations," *Ground Water Monitoring Review*, NWWA, Fall 1987, pp. 51–62.

(20) Riggs, C. O., "Soil Sampling in the Vadose Zone," *Proceedings of the NWWA/U.S. EPA Conference on Characterization and Monitoring of the Vadose Zone*, NWWA, Las Vegas, Nevada, 1983, pp. 611–622.

(21) Sales Division, *Product Literature*, Pitcher Drilling Co., Palo Alto, California, 1986.

(22) Sales Division, *Catalog of Products*, Diedrich Drilling Equipment, LaPorte, Indiana, 1988.

(23) Sales Division, *Instrumentation for Soil and Rocks, Catalog of Products*, Solinst, Burlington, Ontario, Canada, 1988.

(24) Zapico, M. M., Vales, S., and Cherry, J. A., "A Wireline Core Barrel for Sampling Cohesionless Sand and Gravel Below the Water Table," *Ground Water Monitoring Review*, NWWA, Spring 1987, pp. 75–82.

(25) Shuter, E., Teasdale, W. E., "Application of Drilling, Coring, and Sampling Techniques to Test Holes and Wells," *U.S. Geol. Survey Techniques of Water Resource Investigations*, Book 2, Chapter F-1, 1989.

Designation: D 4750 – 87

Standard Test Method for Determining Subsurface Liquid Levels in a Borehole or Monitoring Well (Observation Well)[1]

This standard is issued under the fixed designation D 4750; the number immediately following the designation indicates the year of original adoption or, in the case of revision, the year of last revision. A number in parentheses indicates the year of last reapproval. A superscript epsilon (ε) indicates an editorial change since the last revision or reapproval.

1. Scope

1.1 This test method describes the procedures for measuring the level of liquid in a borehole or well and determining the stabilized level of liquid in a borehole.

1.2 The test method applies to boreholes (cased or uncased) and monitoring wells (observation wells) that are vertical or sufficiently vertical so a flexible measuring device can be lowered into the hole.

1.3 Borehole liquid-level measurements obtained using this test method will not necessarily correspond to the level of the liquid in the vicinity of the borehole unless sufficient time has been allowed for the level to reach equilibrium position.

1.4 This test method generally is not applicable for the determination of pore-pressure changes due to changes in stress conditions of the earth material.

1.5 This test method is not applicable for the concurrent determination of multiple liquid levels in a borehole.

1.6 The values stated in inch-pound units are to be regarded as the standard.

1.7 *This standard may involve hazardous materials, operations, and equipment. This standard does not purport to address all of the safety problems associated with its use. It is the responsibility of the user of this standard to establish appropriate safety and health practices and determine the applicability of regulatory limitations prior to use.*

2. Referenced Document

2.1 *ASTM Standard:*

D 653 Terminology Relating to Soil, Rock, and Contained Fluids[2]

3. Terminology

3.1 *Descriptions of Terms Specific to This Standard:*

3.1.1 *borehole*—a hole of circular cross-section made in soil or rock to ascertain the nature of the subsurface materials. Normally, a borehole is advanced using an auger, a drill, or casing with or without drilling fluid.

3.1.2 *earth material*—soil, bedrock, or fill.

3.1.3 *ground-water level*—the level of the water table surrounding a borehole or well. The ground-water level can be represented as an elevation or as a depth below the ground surface.

3.1.4 *liquid level*—the level of liquid in a borehole or well at a particular time. The liquid level can be reported as an elevation or as a depth below the top of the land surface. If the liquid is ground water it is known as water level.

3.1.5 *monitoring well* (*observation well*)—a special well drilled in a selected location for observing parameters such as liquid level or pressure changes or for collecting liquid samples. The well may be cased or uncased, but if cased the casing should have openings to allow flow of borehole liquid into or out of the casing.

3.1.6 *stabilized borehole liquid level*—the borehole liquid level which remains essentially constant with time, that is, liquid does not flow into or out of the borehole.

3.1.7 *top of borehole*—the surface of the ground surrounding the borehole.

3.1.8 *water table* (*ground-water table*)—the surface of a ground-water body at which the water pressure equals atmospheric pressure. Earth material below the ground-water table is saturated with water.

3.2 *Definitions:*

3.2.1 For definitions of other terms used in this test method, see Terminology D 653.

4. Significance and Use

4.1 In geotechnical, hydrologic, and waste-management investigations, it is frequently desirable, or required, to obtain information concerning the presence of ground water or other liquids and the depths to the ground-water table or other liquid surface. Such investigations typically include drilling of exploratory boreholes, performing aquifer tests, and possibly completion as a monitoring or observation well. The opportunity exists to record the level of liquid in such boreholes or wells, as the boreholes are being advanced and after their completion.

4.2 Conceptually, a stabilized borehole liquid level reflects the pressure of ground water or other liquid in the earth material exposed along the sides of the borehole or well. Under suitable conditions, the borehole liquid level and the ground-water, or other liquid, level will be the same, and the former can be used to determine the latter. However, when earth materials are not exposed to a borehole, such as material which is sealed off with casing or drilling mud, the borehole water levels may not accurately reflect the ground-water level. Consequently, the user is cautioned that the liquid level in a borehole does not necessarily bear a relationship to the ground-water level at the site.

4.3 The user is cautioned that there are many factors which can influence borehole liquid levels and the interpretation of borehole liquid-level measurements. These factors are not described or discussed in this test method. The

[1] This test method is under the jurisdiction of ASTM Committee D-18 on Soil and Rock and is the direct responsibility of Subcommittee D18.02 on Sampling and Related Field Testing for Soil Investigations.

Current edition approved Nov. 27, 1987. Published January 1988.

[2] *Annual Book of ASTM Standards*, Vol 04.08.

interpretation and application of borehole liquid-level information should be done by a trained specialist.

4.4 Installation of piezometers should be considered where complex ground-water conditions prevail or where changes in intergranular stress, other than those associated with fluctuation in water level, have occurred or are anticipated.

5. Apparatus

5.1 Apparatus conforming to one of the following shall be used for measuring borehole liquid levels:

5.1.1 *Weighted Measuring Tape*—A measuring tape with a weight attached to the end. The tape shall have graduations that can be read to the nearest 0.01 ft. The tape shall not stretch more than 0.05 % under normal use. Steel surveying tapes in lengths of 50, 100, 200, 300, and 500 ft (20, 30, 50 or 100 m) and widths of ¼ in. (6 mm) are commonly used. A black metal tape is better than a chromium-plated tape. Tapes are mounted on hand-cranked reels up to 500 ft (100 m) lengths. Mount a slender weight, made of lead, to the end of the tape to ensure plumbness and to permit some feel for obstructions. Attach the weight to the tape with wire strong enough to hold the weight but not as strong as the tape. This permits saving the tape in the event the weight becomes lodged in the well or borehole. The size of the weight shall be such that its displacement of water causes less than a 0.05-ft (15-mm) rise in the borehole water level, or a correction shall be made for the displacement. If the weight extends beyond the end of the tape, a length correction will be needed in measurement Procedure C (see 7.2.3).

5.1.2 *Electrical Measuring Device*—A cable or tape with electrical wire encased, equipped with a weighted sensing tip on one end and an electric meter at the other end. An electric circuit is completed when the tip contacts water; this is registered on the meter. The cable may be marked with graduations similar to a measuring tape (as described in 5.1.1).

5.1.3 *Other Measuring Devices*—A number of other recording and non-recording devices may be used. See Ref. **(1)** for more details.[3]

6. Calibration and Standardization

6.1 Calibrate measuring apparatus in accordance with the manufacturers' directions.

7. Procedure

7.1 Liquid-level measurements are made relative to a reference point. Establish and identify a reference point at or near the top of the borehole or a well casing. Determine and record the distance from the reference point to the top of the borehole (land surface). If the borehole liquid level is to be reported as an elevation, determine the elevation of the reference point or the top of borehole (land surface). Three alternative measurement procedures (A, B, and C) are described.

NOTE 1—In general, Procedure A allows for greater accuracy than B or C, and B allows for greater accuracy than C; other procedures have a variety of accuracies that must be determined from the referenced literature **(2–5)**.

7.2 *Procedure A—Measuring Tape:*

7.2.1 Chalk the lower few feet of tape by drawing the tape across a piece of colored carpenter's chalk.

7.2.2 Lower a weighted measuring tape slowly into the borehole or well until the liquid surface is penetrated. Observe and record the reading on the tape at the reference point. Withdraw the tape from the borehole and observe the lower end of the tape. The demarcation between the wetted and unwetted portions of the chalked tape should be apparent. Observe and record the reading on the tape at that point. The difference between the two readings is the depth from the reference point to the liquid level.

NOTE 2—Submergence of the weight and tape may temporarily cause a liquid-level rise in wells or boreholes having very small diameters. This effect can be significant if the well is in materials of very low hydraulic conductivity.

NOTE 3—Under dry surface conditions, it may be desirable to pull the tape from the well or borehole by hand, being careful not to allow it to become kinked, and reading the liquid mark before rewinding the tape onto the reel. In this way, the liquid mark on the chalked part of the tape is rapidly brought to the surface before the wetted part of the tape dries. In cold regions, rapid withdrawal of the tape from the well is necessary before the wet part freezes and becomes difficult to read. The tape must be protected if rain is falling during measurements.

NOTE 4—In some pumped wells, or in contaminated wells, a layer of oil may float on the water. If the oil layer is only a foot or less thick, read the tape at the top of the oil mark and use this reading for the water-level measurement. The measurement will not be greatly in error because the level of the oil surface in this case will differ only slightly from the level of the water surface that would be measured if no oil was present. If several feet of oil are present in the well, or if it is necessary to know the thickness of the oil layer, a water-detector paste for detecting water in oil and gasoline storage tanks is available commercially. The paste is applied to the lower end of the tape that is submerged in the well. It will show the top of the oil as a wet line and the top of the water as a distinct color change.

7.2.3 As a standard of good practice, the observer should make two measurements. If two measurements of static liquid level made within a few minutes do not agree within about 0.01 or 0.02 ft (generally regarded as the practical limit of precision) in boreholes or wells having a depth to liquid of less than a couple of hundred feet, continue to measure until the reason for the lack of agreement is determined or until the results are shown to be reliable. Where water is dripping into the hole or covering its wall, it may be impossible to get a good water mark on the chalked tape.

7.2.4 After each well measurement, in areas where polluted liquids or ground water is suspected, decontaminate that part of the tape measure that was wetted to avoid contamination of other wells.

7.3 *Procedure B—Electrical Measuring Device:*

7.3.1 Check proper operation of the instrument by inserting the tip into water and noting if the contact between the tip and the water surface is registered clearly.

NOTE 5—In pumped wells having a layer of oil floating on the water, the electric tape will not respond to the oil surface and, thus, the liquid level determined will be different than would be determined by a steel tape. The difference depends on how much oil is floating on the water. A miniature float-driven switch can be put on a two-conductor electric tape that permits detection of the surface of the uppermost fluid.

[3] The boldface numbers in parentheses refer to the list of references at the end of this standard.

BOREHOLE OR WELL SCHEDULE FORM

Date ______, 19___ Field No. ______
Record by ______ Office No. ______
Source of data ______

1. *Location:* State ______ County ______
   Map ______
   ______ ¼ ______ ¼ sec. ______ T ______ N/S R ______ E/W
2. *Owner:* ______ Address ______
   Tenant ______ Address ______
   Driller ______ Address ______
3. *Topography* ______
4. *Elevation* ______ ft. above/below ______
5. *Type:* Dug, drilled, driven, bored, jetted ______ 19___
6. *Depth:* Rept. ______ ft. Meas. ______ ft.
7. *Casing:* Diam. ______ in., to ______ in., Type ______
   Depth ______ ft., Finish ______
8. *Chief Aquifer* ______ From ______ ft. to ______ ft.
   Others ______
9. *Water level* ______ ft. rept./meas. ______ 19___ above/below ______
   ______ which is ______ ft. above/below surface
10. *Pump:* Type ______ Capacity ______ G. M. ______
    Power: Kind ______ Horsepower ______
11. *Yield:* Flow ______ G. M., Pump ______ G. M., Meas., Rept. Est. ______
    Drawdown ______ ft. after ______ hours pumping ______ G. M.
12. *Use:* Dom., Stock, PS., RR., Ind., Irr., Obs. ______
    Adequacy, permanence ______
13. *Quality* ______ Temp ______ °F.
    Taste, odor, color ______ Sample Yes/No ______
    Unfit for ______
14. *Remarks:* (Log, Analyses, etc.) ______

FIG. 1 Example of a Borehole or Well Schedule Form

LIQUID LEVEL MEASUREMENT FORM

Field No. ______
Owner ______ Office No. ______
Location ______ Project ______
Measuring Point ______
Elevation of Measuring Point ______

Date	Hour	Depth to Water	Elev. of Water Surface	Meas. by	Remarks (Nearby wells pumping, etc.)

FIG. 2 Example of a Liquid Level Measurement Form

7.3.2 Dry the tip. Slowly lower the tip into the borehole or well until the meter indicates that the tip has contacted the surface of the liquid.

7.3.3 For devices with measurement graduations on the cable, note the reading at the reference point. This is the liquid-level depth below the reference point of the borehole or well.

7.3.4 For measuring devices without graduations on the cable, mark the cable at the reference point. Withdraw the cable from the borehole or well. Stretch out the cable and measure and record the distance between the tip and the mark on the cable by use of a tape. This distance is the liquid-level depth below the reference point.

7.3.5 A second or third check reading should be taken before withdrawing the electric tape from the borehole or well.

7.3.6 Decontaminate the submerged end of the electric tape or cable after measurements in each well.

NOTE 6—The length of the electric line should be checked by measuring with a steel tape after the line has been used for a long time or after it has been pulled hard in attempting to free the line. Some electric lines, especially the single line wire, are subject to considerable permanent stretch. In addition, because the probe is usually larger in diameter than the wire, the probe can become lodged in a well. Sometimes the probe can be attached by twisting the wires together by hand and using only enough electrical tape to support the weight of the probe. In this manner, the point of probe attachment is the weakest point of the entire line. Should the probe become "hung in the hole," the line may be pulled and breakage will occur at the probe attachment point, allowing the line to be withdrawn.

7.4 *Procedure C—Measuring Tape and Sounding Weight:*

7.4.1 Lower a weighted measuring tape into the borehole or well until the liquid surface is reached. This is indicated by an audible splash and a noticeable decrease in the downward force on the tape. Observe and note the reading on the tape at the reference point. Repeat this process until the readings are consistent to the accuracy desired. Record the result as the liquid-level depth below the reference point.

NOTE 7—The splash can be made more audible by using a "plopper," a lead weight with a concave bottom surface.

7.4.2 If the liquid level is deep, or if the measuring tape adheres to the side of the borehole, or for other reasons, it may not be possible to detect the liquid surface using this method. If so, use Procedure A or Procedure B.

8. Determination of a Stabilized Liquid Level

8.1 As liquid flows into or out of the borehole or well, the liquid level will approach, and may reach, a stabilized level. The liquid level then will remain essentially constant with time.

SITE NO.

BOREHOLE OR WELL SCHEDULE FORM

Recorded by ____________ Date ____________

Check One ____ English ____ Metric Units

GENERAL SITE DATA (0)

Site Ident No [5 … 19] — RG Number [R = 0 *] — Transaction [T = A D M V *] add delete modify verified

Site-Type [2 = C D E H I M Ø P S T W X *] collector, drain, excava-tion, sink-hole, connector well, multiple, outcrop, pond, spring, tunnel or shaft, well, test hole — Data Reliability [3 = C U *] field checked, unchecked — Reporting Agency [4 = *]

Project No. [5 = *] — District [6 = *] — State [7 = *] — County (or town) ____ [8 = *]

Latitude [9 = *] deg min sec — Longitude [10 = *] deg min sec — Lat-Long Accuracy [11 = S F T M *] sec, 5 sec, 10 sec, Min

Local Number [12 = *] — Land Net Loc [13 = S T R *] 1/4 1/4 1/4 section, township, range, merid

Location Map [14 = *] — Scale [15 = *]

Altitude [16 = *] — Method of Measurement [17 = A L M *] altimeter, level, map — Accuracy [18 = *]

Topo Setting [19 = A B C D E F G H K L M Ø P S T U V W *] alluvial fan, playa, stream channel, depression, dunes, flat, flood plain, hill top, sink, swamp, mangrove swamp, offshore, pediment, hill side, terrace, undulating, valley flat, upland draw — Hydrologic Unit (OWDC) [20 = *]

Use of Site [23 = A C D E G H M Ø P R S T U W X Z *] anode, standby emergency supply, drain, geothermal, seismic, heat reservoir, mine, observation, oil or gas, recharge, repressure, test, unused, withdrawal, waste, destroyed — Secondary Site Use [301 = *] — Tertiary Site Use [302 *]

Use of Water [24 = A B C D E F H I J K M N P Q R S T U Y Z *] air cond, bottling, commercial, dewater, power, fire, domestic, irrigation, industrial (cooling), mining, medicinal, industrial, public supply, aquaculture, recreation, stock, institution, unused, desal, other

Secondary Water Use [25 = *] — Tertiary Use of Water [26 = *] — Depth of Hole [27 = *] — Depth of Well [28 = *] — Source of Depth Data① [29 = *]

Water Level [30 = *] — Date Measured [31 = / / *] month day year — Source① [33 = *]

Method of Measurement [34 = A B C E G H L M N R S T V Z *] airline, analog, calibrated airline, estimated, pressure gage, calibrated pressure gage, geophysical logs, manometer, non-rec. gage, reported, steel tape, electric tape, calibrated electric tape, other

Site Status [37 = D E F G H I J N Ø P R S T V W X Z *] dry, recently flowing, flowing, nearby flowing, nearby recently flowing, injector site, inj site monitor, meas. discon., obstruction, pumping, recently pumped, nearby pumping, nearby recently pumped, foreign substance, well destroyed, surface water effects, other

Source of Geohydrologic Data① [36 = *] — Pump Used [35 = *] no — Date of First Construction Completion [21 = / / *] month day year

OWNER IDENTIFICATION (1)

[R = 158 *] [T = A D M *] add, delete, modify — Date of Ownership [159 = / / *] month day year

Name: Last [161 = *] — First [162 = *] — Middle Initial [163 = *]

OTHER SITE IDENTIFICATION NUMBERS (1)

[R = 189 *] [T = A D M *] add, delete, modify — Ident [190 = *] — Assigner [191 = *]

[New Card Same R & T] — Ident [190 = *] — Assigner [191 = *]

SITE VISIT DATA (1)

[R = 186 *] [T = A D M *] add, delete, modify — Date of Visit [187 = / / *] month day year — Name of Person [188 = *]

FIELD WATER QUALITY MEASUREMENTS (1)

[R = 192 *] [T = A D M *] add, delete, modify — Date [193 = / / *] month day year — Geohydrologic Unit [195 = *]

[New Card Same R thru 195]

Temperature [196 = 0 0 0 1 0 *] — Degrees C [197 = . *]

Conductance [196 = 0 0 0 9 5 *] — μ Mhos [197 = . *]

Other (STORET) Parameter [196 = *] — Value [197 = . *]

Other (STORET) Parameter [196 = *] — Value [197 = . *]

FOOT NOTES

① Source of Data Codes

A	D	G	L	M	O	R	S	Z
other gov't	driller	geologist	logs	memory	owner	other reported	reporting agency	other

FIG. 3 Example of a Borehole or Well Schedule Form

NOTE 8—The time required to reach equilibrium can be reduced by removing or adding liquid until the liquid level is close to the estimated stabilized level.

8.2 Use one of the following two procedures to determine the stabilized liquid level.

8.2.1 *Procedure 1*—Take a series of liquid-level measurements until the liquid level remains constant with time. As a minimum, two such constant readings are needed (more readings are preferred). The constant reading is the stabilized liquid level for the borehole or well.

NOTE 9—If desired, the time and level data could be plotted on graph paper in order to show when equilibrium is reached.

8.2.2 *Procedure 2*—Take at least three liquid-level measurements at approximately equal time intervals as the liquid level changes during the approach to a stabilized liquid level.

8.2.2.1 The approximate position of the stabilized liquid level in the well or borehole is calculated using the following equation:

$$h_o = \frac{y_1^2}{y_1 - y_2}$$

where:

h_o = distance the liquid level must change to reach the stabilized liquid level,

y_1 = distance the liquid level changed during the time interval between the first two liquid-level readings, and

y_2 = distance the liquid level changed during the time interval between the second and the third liquid level readings.

8.2.2.2 Repeat the above process using successive sets of three measurements until the h_o computed is consistent to the accuracy desired. Compute the stabilized liquid level in the well or borehole.

NOTE 10—The time span required between readings for Procedures 1 and 2 depends on the permeability of the earth material. In material with comparatively high permeability (such as sand), a few minutes may be sufficient. In materials with comparatively low permeability (such as clay), many hours or days may be needed. The user is cautioned that in clayey soils the liquid in the borehole or well may never reach a stabilized level equivalent to the liquid level in the earth materials surrounding the borehole or well.

9. Report

9.1 For borehole or well liquid-level measurements, report, as a minimum, the following information:

9.1.1 Borehole or well identification.

9.1.2 Description of reference point.

9.1.3 Distance between reference point and top of borehole or land surface.

9.1.4 Elevation of top of borehole or reference point (if the borehole or well liquid level is reported as an elevation).

9.1.5 Description of measuring device used, and graduation.

9.1.6 Procedure of measurement.

9.1.7 Date and time of reading.

9.1.8 Borehole or well liquid level.

9.1.9 Description of liquid in borehole or well.

9.1.10 State whether borehole is cased, uncased, or contains a monitoring (observation) well standpipe and give description of, and length below top of borehole of, casing or standpipe.

9.1.11 Drilled depth of borehole, if known.

9.2 For determination of stabilized liquid level, report:

9.2.1 All pertinent data and computations.

9.2.2 Procedure of determination.

9.2.3 The stabilized liquid level.

9.3 *Report Forms*—An example of a borehole or well-schedule form is shown in Fig. 1. An example of a liquid-level measurement form, for recording continuing measurements for a borehole or well, is shown in Fig. 2. An example of a borehole or well schedule form designed to facilitate computer data storage is shown in Fig. 3.

10. Precision and Bias

10.1 Borehole liquid levels shall be measured and recorded to the accuracy desired and consistent with the accuracy of the measuring device and procedures used. Procedure A multiple measurements by wetted tape should agree within 0.02 ft (6 mm). Procedure B multiple measurements by electrical tape should agree within 0.04 ft (12 mm). Procedure C multiple measurements by tape and sounding weight should agree within 0.04 ft (12 mm). Garber and Koopman **(2)** describe corrections that can be made for effects of thermal expansion of tapes or cables and of stretch due to the suspended weight of tape or cable and plumb weight when measuring liquid levels at depths greater than 500 ft (150 m).

REFERENCES

(1) "*National Handbook of Recommended Methods for Water Data Acquisition—Chapter 2—Ground Water*", Office of Water Data Coordination, Washington, DC, 1980.

(2) Garber, M. S., and Koopman, F. C., "Methods of Measuring Water Levels in Deep Wells," *U.S. Geologic Survey Techniques for Water Resources Investigations*, Book 8, Chapter A-1, 1968.

(3) Hvorslev, M. J., "Ground Water Observations," in *Subsurface Exploration and Sampling of Soils for Civil Engineering Purposes*, American Society Civil Engineers, New York, NY, 1949.

(4) Zegarra, E. J., "Suggested Method for Measuring Water Level in Boreholes," *Special Procedures for Testing Soil and Rock for Engineering Purposes, ASTM STP 479*, ASTM, 1970.

(5) "Determination of Water Level in a Borehole," CSA Standard A 119.6 – 1971, Canadian Standards Association, 1971.

Standard Practice for Decontamination of Field Equipment Used at Nonradioactive Waste Sites[1]

This standard is issued under the fixed designation D 5088; the number immediately following the designation indicates the year of original adoption or, in the case of revision, the year of last revision. A number in parentheses indicates the year of last reapproval. A superscript epsilon (ϵ) indicates an editorial change since the last revision or reapproval.

1. Scope

1.1 This practice covers the decontamination of field equipment used in the sampling of soils, soil gas, sludges, surface water, and ground water at waste sites which are to undergo both physical and chemical analyses.

1.2 This practice is applicable only at sites where chemical (organic and inorganic) wastes are a concern and is not intended for use at radioactive or mixed (chemical and radioactive) waste sites.

1.3 Procedures are included for the decontamination of equipment which comes into contact with the sample matrix (sample contacting equipment) and for ancillary equipment that has not contacted the portion of sample to be analyzed (non-sample contacting equipment).

1.4 This practice is based on recognized methods by which equipment may be decontaminated. When collecting environmental matrix samples, one should become familiar with the site specific conditions. Based on these conditions and the purpose of the sampling effort, the most suitable method of decontamination can be selected to maximize the integrity of analytical and physical testing results.

1.5 This practice is applicable to most conventional sampling equipment constructed of metallic and synthetic materials. The manufacturer of a specific sampling apparatus should be contacted if there is concern regarding the reactivity of a decontamination rinsing agent with the equipment.

1.6 *This standard does not purport to address the safety problems associated with its use. It is the responsibility of the user of this standard to establish appropriate safety and health practices and determine the applicability of regulatory limitations prior to use.*

2. Referenced Document

2.1 *ASTM Standard:*

D 653 Terminology Relating to Soil, Rock, and Contained Fluids[2]

3. Terminology

3.1 *Definitions:*

3.1.1 *contaminant*—an undesirable substance not normally present or an unusually high concentration of a naturally occurring substance in water or soil.

3.1.2 *control rinse water*—water used for equipment washing and rinsing having a known chemistry.

3.1.3 *decontamination*—the process of removing or reducing to a known level undesirable physical or chemical constituents, or both, from a sampling apparatus to maximize the representativeness of physical or chemical analyses proposed for a given sample.

3.1.4 *non-sample contacting equipment*—related equipment associated with the sampling effort, but that does not directly contact the sample (for example, augers, drilling rods, excavations machinery).

3.1.5 *quality assurance/quality control (QA/QC)*—the efforts completed to evaluate the accuracy and precision of a sampling or testing procedure, or both.

3.1.6 *sample contacting equipment*—equipment that comes in direct contact with the sample or portion of sample that will undergo chemical analyses or physical testing (for example, ground water well bailer, split-spoon sampler, soil gas sampling probe).

3.1.7 For definitions of other terms used in this practice, see Terminology D 653.

4. Summary of Practice

4.1 Two different procedures are presented for the decontamination of sample-contacting and non-sample contacting equipment. The procedures have been developed based on a review of current state and federal guidelines, as well as a summary of commonly employed procedures. In general, sample contacting equipment should be washed with a detergent solution followed by a series of control water, desorbing agents and deionized water rinses. Nonsample contacting equipment should be washed with a detergent solution and rinsed with control water. Although such techniques may be difficult to perform in the field, they may be necessary to most accurately evaluate low concentrations of the chemical constituent(s) of interest.

4.2 Prior to initiating a field program that will involve equipment decontamination, a site specific equipment decontamination protocol should be prepared for distribution to the individuals involved with the particular sampling program. Information to be presented in the protocol should include:

4.2.1 Site location and description,

4.2.2 Statement of the sampling program objective and desired precision and accuracy, that is, is sampling effort for gross qualitative evaluation or for trace concentration, parameter specific evaluations,

4.2.3 Summary of available information regarding soil types, hydrogeology and anticipated chemistry of the materials to be sampled,

[1] This practice is under the jurisdiction of ASTM Committee D-18 on Soil and Rock and is the direct responsibility of Subcommittee D18.14 on Geotechnics of Waste Management.

Current edition approved June 29, 1990. Published September 1990.

[2] *Annual Book of ASTM Standards*, Vol 04.08.

4.2.4 Listing of equipment to be used for sampling and materials needed for decontamination,

4.2.5 Detailed step by step procedure for equipment decontamination for each piece or type of equipment to be utilized and procedures for rinse fluids containment and disposal as appropriate,

4.2.6 Summary of QA/QC procedures and QA/QC samples to be collected to document decontamination completeness including specific type of chemical analyses and their associated detection limit, and

4.2.7 Outline of equipment decontamination verification report.

5. Significance and Use

5.1 An appropriately developed, executed and documented equipment decontamination procedure is an integral and essential part of waste site investigations. The benefits of its use include:

5.1.1 Minimizing the spread of contaminants within a study area and from site to site,

5.1.2 Reducing the potential for worker exposure by means of contact with contaminated sampling equipment, and

5.1.3 Improved data quality and reliability.

5.2 This practice is not a substitute for a well-documented Quality Assurance/Quality Control (QA/QC) program. Because the ultimate test of a decontamination procedure is its ability to minimize erroneous data, a reasonable QA/QC program must be implemented.

5.3 This practice may not be applicable to all waste sites. When a sampling effort is completed to determine only the general range of chemical concentrations of interest less rigorous decontamination procedures can be adequate. Investigators should have the flexibility to modify the decontamination procedures with due consideration for the sampling objective or if QA/QC documentation supports alternative decontamination methods.

5.4 At sites where the reactivity of sampling equipment to decontamination washes creates concern for the generation of undesirable chemical by-products, the use of dedicated sampling equipment should be considered.

5.5 This practice, where applicable, should be used before, between, and after the completion of sampling events.

6. Reagents

6.1 *Detergent*, non-phosphate detergent solution.[3]

6.2 *Acid rinse (inorganic desorbing agent)*, 10 % nitric or hydrochloric acid solution-made from reagent grade nitric or hydrochloric acid and deionized water (1 % is to be applied to low-carbon steel equipment).

6.3 *Solvent rinse (organic desorbing agent)*, isopropanol, acetone, or methanol; pesticide grade.

6.4 *Control rinse water*, preferably from a water system of known chemical composition.

6.5 *Deionized water*, organic-free reagent grade.

7. Procedure for Sample Contacting Equipment

7.1 At a minimum, sample contacting equipment should be washed with a detergent solution and rinsed with control water.

7.2 For programs requiring more rigorous decontamination to meet the sampling or QA/QC objectives, the following procedures are indicated:

7.2.1 Wash with detergent solution, using a brush made of inert material to remove any particles or surface film.

7.2.1.1 For equipment that, because of internal mechanism or tubing cannot be adequately cleaned with a brush, the decontamination solutions should be circulated through the equipment.

7.2.2 Rinse thoroughly with control water.

7.2.3 Rinse with an inorganic desorbing agent (may be deleted if samples will not undergo inorganic chemical analysis).

7.2.4 Rinse with control water.

7.2.5 Rinse with organic desorbing agent (may be deleted if samples will not undergo organic chemical analyses).

7.2.6 Rinse with deionized water.

7.2.7 Allow equipment to air dry prior to next use.

7.2.8 Wrap equipment for transport with inert material (aluminum foil or plastic wrap) to direct contact with potentially contaminated material.

7.3 *Nonsample Contact Equipment:*

7.3.1 Clean the equipment with portable power washer or steam cleaning machine. Alternatively, hand wash with brush using detergent solution.

7.3.2 Rinse with control water.

7.3.3 The more rigorous decontamination procedures may be employed if necessary to meet sampling or QA/QC objectives.

7.4 Depending on site conditions, it may be appropriate to contain spent decontamination rinse fluids. If this is the case the appropriate vessel[4] for fluid containment should be used depending on the ultimate disposition of the material.

7.5 Depending on site conditions, it may be desirable to perform all equipment decontamination at a centralized location as opposed to the location where the equipment was used. If this is the case, care must be taken to transport the equipment to the decontamination area such that the spread of contaminants is minimized.

8. Quality Assurance/Quality Control

8.1 It is important to document the effectiveness of the decontamination procedure. To that end the projects QA/QC program should include provisions for the collection of samples to evaluate the completeness of a specific decontamination procedure. This could include:

8.1.1 Collection of rinse or wipe samples before the initial equipment decontamination prior to its use for sampling to establish a base line level of contaminants residing on or in the equipment,

8.1.2 Collection of final rinse or wipe samples after equipment decontamination following its use, and

8.1.3 The frequency of sampling to demonstrate the completeness of equipment decontamination is dependent upon objectives of the project as they relate to QA/QC. At a

[3] Alquinox or Liquinox or similar solution has been found suitable for this purpose.

[4] A drum approved by the Department of Transportation or similar container has been found suitable for this purpose.

minimum it is recommended after every ten decontamination washings.

9. Report

9.1 The activities completed for each equipment decontamination should be documented in writing. Included in this report should be the following information:

9.1.1 Site location, date, time, and weather,

9.1.2 Sample location where equipment was employed,

9.1.3 Location where decontamination was performed,

9.1.4 Individuals performing the decontamination,

9.1.5 Decontamination procedures,

9.1.6 Source of materials (solutions) used for decontamination,

9.1.7 Handling of rinse fluids and accumulates solids, if any, and

9.1.8 QA/QC sampling performed and analytical results of QA/QC samples whether completed in the field or laboratory subsequent to sampling event.

10. Keywords

10.1 contaminant; decontamination; sampling; waste

Standard Practice for Design and Installation of Ground Water Monitoring Wells in Aquifers[1]

This standard is issued under the fixed designation D 5092; the number immediately following the designation indicates the year of original adoption or, in the case of revision, the year of last revision. A number in parentheses indicates the year of last reapproval. A superscript epsilon (ε) indicates an editorial change since the last revision or reapproval.

INTRODUCTION

This practice for the design and installation of ground water monitoring wells in aquifers will promote (*1*) durable and reliable construction, (*2*) extraction of representative ground water quality samples, and (*3*) efficient and site hydrogeological characterizations. The guidelines established herein are affected by governmental regulations and by site specific geological, hydrogeological, climatological, topographical, and subsurface chemistry conditions. To meet these geoenvironmental challenges, this guidance promotes the development of a conceptual hydrogeologic model prior to monitoring well design and installation.

1. Scope

1.1 This practice considers the selection and characterization (that is, defining soil, rock types, and hydraulic gradients) of the target monitoring zone as an integral component of monitoring well design and installation. Hence, the development of a conceptual hydrogeologic model for the intended monitoring zone(s) is recommended prior to the design and installation of a monitoring well.

1.2 These guidelines are based on recognized methods by which monitoring wells may be designed and installed for the purpose of detecting the presence or absence of a contaminant, and collecting representative ground water quality data. The design standards and installation procedures herein are applicable to both detection and assessment monitoring programs for facilities.

1.3 The recommended monitoring well design, as presented in this practice, is based on the assumption that the objective of the program is to obtain representative ground water information and water quality samples from aquifers. Monitoring wells constructed following this practice should produce relatively turbidity-free samples for granular aquifer materials ranging from gravels to silty sand and sufficiently permeable consolidated and fractured strata. Strata having grain sizes smaller than the recommended design for the smallest diameter filter pack materials should be monitored by alternative monitoring well designs which are not addressed in this practice.

1.4 The values stated in inch-pound units are to be regarded as standard. The values in parentheses are for information only.

1.5 *This standard does not purport to address the safety problems associated with its use. It is the responsibility of the user of this standard to establish appropriate safety and health practices and determine the applicability of regulatory limitations prior to use.*

2. Referenced Documents

2.1 *ASTM Standards:*

C 150 Specification for Portland Cement[2]

C 294 Descriptive Nomenclature of Constituents of Natural Mineral Aggregates[3]

D 653 Terminology Relating to Soil, Rock, and Contained Fluids[4]

D 1452 Practice for Soil Investigation and Sampling by Auger Borings[4]

D 1586 Method for Penetration Test and Split-Barrel Sampling of Soils[4]

D 1587 Practice for Thin-Walled Tube Sampling of Soils[4]

D 2113 Practice for Diamond Core Drilling for Site Investigation[4]

D 2487 Test Method for Classification of Soils for Engineering Purposes[4]

D 2488 Practice for Description and Identification of Soils (Visual-Manual Procedure)[4]

D 3282 Practice for Classification of Soils and Soil Aggregate Mixtures for Highway Construction Purposes[4]

D 3550 Practice for Ring Lined Barrel Sampling of Soils[4]

D 4220 Practice for Preserving and Transporting Soil Samples[4]

3. Significance and Use

3.1 An adequately designed and installed ground water monitoring well system for aqueousphase liquids provides essential information for decisions pertaining to one or more of the following subjects:

3.1.1 Aquifer and aquitard properties, both geologic and hydraulic;

[1] This practice is under the jurisdiction of ASTM Committee D-18 on Soil and Rock and is the direct responsibility of Subcommittee D18.21.05 on Design and Installation of Ground-Water Monitoring Wells.

Current edition approved June 29, 1990. Published October 1990.

[2] *Annual Book of ASTM Standards*, Vol 04.01.

[3] *Annual Book of ASTM Standards*, Vol 04.02.

[4] *Annual Book of ASTM Standards*, Vol 04.08.

3.1.2 Potentiometric surface of a particular hydrologic unit(s);

3.1.3 Water quality with respect to various indicator parameters;

3.1.4 Migration characteristics of a contaminant release;

3.1.5 Additional installations or decommissioning of installations, or both, no longer needed.

4. Terminology

4.1 *Definitions:*

4.1.1 *annular space; annulus*—the space between two concentric tubes or casings, or between the casing and the borehole wall. This would include the space(s) between multiple strings of tubing/casings in a borehole installed either concentrically or multi-cased adjacent to each other.

4.1.2 *assessment monitoring*—an investigative monitoring program that is initiated after the presence of a contaminant in ground water has been detected. The objective of this program is to determine the concentration of constituents that have contaminated the ground water and to quantify the rate and extent of migration of these constituents.

4.1.3 *ASTM cement types*—Portland cements meeting the requirements of Specifications C 150. Cement types have slightly different formulations that result in various characteristics which address different construction conditions and different physical and chemical environments. They are as follows:

4.1.3.1 *Type I (Portland)*—a general-purpose construction cement with no special properties.

4.1.3.2 *Type II (Portland)*—a construction cement that is moderately resistant to sulfates and generates a lower head of hydration at a slower rate than Type I.

4.1.3.3 *Type III (Portland; high early strength)*—a construction cement that produces a high early strength. This cement reduces the curing time required when used in cold environments, and produces a higher heat of hydration than Type I.

4.1.3.4 *Type IV (Portland)*—a construction cement that produces a low head of hydration (lower than Types I and II) and develops strength at a slower rate.

4.1.3.5 *Type V (Portland)*—a construction cement that is a high sulfate resistant formulation. Used when there is severe sulfate action from soils and ground water.

4.1.4 *bailer*—a hollow tubular receptacle used to facilitate withdrawal of fluid from a well or borehole.

4.1.5 *ballast*—materials used to provide stability to a buoyant object (such as casing within a borehole filled with water).

4.1.6 *blow-in*—the inflow of ground water and unconsolidated material into a borehole or casing caused by differential hydraulic heads; that is, caused by the presence of a greater hydraulic head outside of a borehole/casing than inside.

4.1.7 *borehole* a circular open or uncased subsurface hole created by drilling.

4.1.8 *borehole log*—the record of geologic units penetrated, drilling progress, depth, water level, sample recovery, volumes, and types of materials used, and other significant facts regarding the drilling of an exploratory borehole or well.

DISCUSSION—The definition of aquifer as currently included in Terminology D 653 varies from the definition as prescribed by US federal regulations. Since this federal definition is associated with the installation of many monitoring wells it is provided herein as a technical note:

aquifer—a geologic formation, group of formation, or part of a formation that is saturated, and is capable of providing a significant quantity of water.

4.1.9 *bridge*—an obstruction within the annulus which may prevent circulation or proper emplacement of annular materials.

4.1.10 *casing*—pipe, finished in sections with either threaded connections or bevelled edges to be field welded, which is installed temporarily or permanently to counteract caving, to advance the borehole, or to isolate the zone being monitored, or combination thereof.

4.1.11 *casing, protective*—a section of larger diameter pipe that is emplaced over the upper end of a smaller diameter monitoring well riser or casing to provide structural protection to the well and restrict unauthorized access into the well.

4.1.12 *casing, surface*—pipe used to stabilize a borehole near the surface during the drilling of a borehole that may be left in place or removed once drilling is completed.

4.1.13 *caving; sloughing*—the inflow of unconsolidated material into a borehole which occurs when the borehole walls lose their cohesive strength.

4.1.14 *cement; Portland cement*—commonly known as Portland cement. A mixture that consists of a calcareous, argillaceous, or other silica-, alumina-, and iron-oxide-bearing materials that is manufactured and formulated to produce various types which are defined in Specification C 150. Portland cement is also considered a hydraulic cement because it must be mixed with water to form a cement-water paste that has the ability to harden and develop strength even if cured under water (see *ASTM cement types*).

4.1.15 *centralizer*—a device that assists in the centering of a casing or riser within a borehole or another casing.

4.1.16 *circulation*—applies to the fluid rotary drilling method; drilling fluid movement from the mud pit, through the pump, hose and swivel, drill pipe, annular space in the hole and returning to the mud pit.

4.1.17 *conductance (specific)*—a measure of the ability of the water to conduct an electric current at 77°F (25°C). It is related to the total concentration of ionizable solids in the water. It is inversely proportional to electrical resistance.

4.1.18 *confining unit*—a term that is synonymous with "aquiclude," "aquitard," and "aquifuge;" defined as a body of relatively low permeable material stratigraphically adjacent to one or more aquifers.

4.1.19 *contaminant*—an undesirable substance not normally present in water or soil.

4.1.20 *detection monitoring*—a program of monitoring for the express purpose of determining whether or not there has been a contaminant release to ground water.

4.1.21 *drill cuttings*—fragments or particles of soil or rock, with or without free water, created by the drilling process.

4.1.22 *drilling fluid*—a fluid (liquid or gas) that may be used in drilling operations to remove cuttings from the borehole, to clean and cool the drill bit, and to maintain the integrity of the borehole during drilling.

4.1.23 *d-10*—the diameter of a soil particle (preferably in millimetres) at which 10 % by weight (dry) of the particles of a particular sample are finer. Synonymous with the effective size or effective grain size.

4.1.24 *d-60*—the diameter of a soil particle (preferably in millimetres) at which 60 % by weight (dry) of the particles of a particular sample are finer.

4.1.25 *flow path*—represents the area between two flow lines along which ground water can flow.

4.1.26 *flush joint or flush coupled*—casing or riser with ends threaded such that a consistent inside and outside diameter is maintained across the threaded joints or couplings.

4.1.27 *gravel pack*—common nomenclature for the terminology, primary filter of a well (see *primary filter pack*).

4.1.28 *grout (monitoring wells)*—a low permeability material placed in the annulus between the well casing or riser pipe and the borehole wall (that is, in a single-cased monitoring well), or between the riser and casing (that is, in a multi-cased monitoring well), to maintain the alignment of the casing and riser and to prevent movement of ground water or surface water within the annular space.

4.1.29 *grout shoe*—a plug fabricated of relatively inert materials that is positioned within the lowermost section of a permanent casing and fitted with a passageway, often with a flow check device, through which grout is injected under pressure to fill the annular space. After the grout has set, the grout shoe is usually drilled out.

4.1.30 *head (static)*—the height above a standard datum of the surface of a column of water (or other liquid) that can be supported by the static pressure at a given point. The static head is the sum of the elevation head and the pressure head.

4.1.31 *head (total)*—the sum of three components at a point: (*1*) elevation head, h_e, which is equal to the elevation of the point above a datum; (*2*) pressure head, hp, which is the height of a column of static water than can be supported by the static pressure at the point; and (*3*) velocity head, hv, which is the height the kinetic energy of the liquid is capable of lifting the liquid.

4.1.32 *hydrologic unit*—geologic strata that can be distinguished on the basis of capacity to yield and transmit fluids. Aquifers and confining units are types of hydrologic units. Boundaries of a hydrologic unit may not necessarily correspond either laterally or vertically to lithostratigraphic formations.

4.1.33 *jetting*—when applied as a drilling method, water is forced down through the drill rods or casings and out through the end aperture. The jetting water then transports the generated cuttings to the ground surface in the annulus of the drill rods or casing and the borehole. The term jetting may also refer to a development technique (see well screen jetting).

4.1.34 *loss of circulation*—the loss of drilling fluid into strata to the extent that circulation does not return to the surface.

4.1.35 *mud pit*—usually a shallow, rectangular, open, portable container with baffles into which drilling fluid and cuttings are discharged from a borehole and that serves as a reservoir and settling tank during recirculation of the drilling fluids. Under some circumstances, an excavated pit with a lining material may be used.

4.1.36 *multi-cased well*—a well constructed by using successively smaller diameter casings with depth.

4.1.37 *neat cement*—a mixture of Portland cement (Specification 150) and water.

4.1.38 *observation well*—typically, a small diameter well used to measure changes in hydraulic heads, usually in response to a nearby pumping well.

4.1.39 *oil air filter*—a filter or series of filters placed in the air flow line from an air compressor to reduce the oil content of the air.

4.1.40 *oil trap*—a device used to remove oil from the compressed air discharged from an air compressor.

4.1.41 *packer (monitoring wells)*—a transient or dedicated device placed in a well that isolates or seals a portion of the well, well annulus, or borehole at a specific level.

4.1.42 *potentiometric surface*—an imaginary surface representing the static head of ground water. The water table is a particular potentiometric surface.

DISCUSSION—Where the head varies with depth in the aquifer, a potentiometric surface is meaningful only if it describes the static head along a particular specified surface or stratum in that aquifer. More than one potentiometric surface is required to describe the distribution of head in this case.

4.1.43 *primary filter pack*—a clean silica sand or sand and gravel mixture of selected grain size and gradation that is installed in the annular space between the borehole wall and the well screen, extending an appropriate distance above the screen, for the purpose of retaining and stabilizing the particles from the adjacent strata. The term is used in place of *gravel pack.*

4.1.44 *PTFE tape*—joint sealing tape composed of polytetrafluoroethylene.

4.1.45 *riser*—the pipe extending from the well screen to or above the ground surface.

4.1.46 *secondary filter pack*—a clean, uniformly graded sand that is placed in the annulus between the primary filter pack and the over-lying seal, or between the seal and overlying grout backfill, or both, to prevent movement of seal or grout, or both, into the primary filter pack.

4.1.47 *sediment sump*—a blank extension beneath the well screen used to collect fine-grained material from the filter pack and adjacent strata. The term is synonymous with rat trap or tail pipe.

4.1.48 *shear strength (monitoring wells)*—a measure of the shear or gel properties of a drilling fluid or grout.

4.1.49 *single-cased well*—a monitoring well constructed with a riser but without an exterior casing.

4.1.50 *static water level*—the elevation of the top of a column of water in a monitoring well or piezometer that is not influenced by pumping or conditions related to well installation, hydrologic testing, or nearby pumpage.

4.1.51 *tamper*—a heavy cylindrical metal section of tubing that is operated on a wire rope or cable. It slips over the riser and fits inside the casing or borehole annulus. It is generally used to tamp annular sealants or filter pack materials into place and prevent bridging.

4.1.52 *target monitoring zone*—the ground water flow path from a particular area or facility in which monitoring wells will be screened. The target monitoring zone should be a stratum (strata) in which there is a reasonable expectation that a vertically placed well will intercept migrating contaminants.

4.1.53 *test pit*—a shallow excavation made to characterize the subsurface.

4.1.54 *transmissivity*—the rate at which water of the prevailing kinematic viscosity is transmitted through a unit width of the aquifer under a unit hydraulic gradient.

DISCUSSION—It is equal to an integration of the hydraulic conductivities across the saturated part of the aquifer perpendicular to the flow paths.

4.1.55 *tremie pipe*—a pipe or tube that is used to transport filter pack materials and annular sealant materials from the ground surface into the borehole annulus or between casings and casings or riser pipe of a monitoring well.

4.1.56 *uniformly graded*—a quantitative definition of the particle size distribution of a soil which consists of a majority of particles being of the same approximate diameter. A granular material is considered uniformly graded when the uniformity coefficient is less than about five (Test Method D 2487). Comparable to the geologic term *well sorted.*

4.1.57 *vented cap*—a cap with a small hole that is installed on top of the riser.

4.1.58 *washout nozzle*—a tubular extension with a check valve utilized at the end of a string of casing through which water can be injected to displace drilling fluids and cuttings from the annular space of a borehole.

4.1.59 *weep hole*—a small diameter hole (usually 1/4 in.) drilled into the protective casing above the ground surface that serves as a drain hole for water that may enter the protective casing annulus.

4.1.60 *well completion diagram*—a record that illustrates the details of a well installation.

4.1.61 *well screen*—a filtering device used to retain the primary or natural filter pack; usually a cylindrical pipe with openings of a uniform width, orientation, and spacing.

4.1.62 *well screen jetting (hydraulic jetting)*—when jetting is used for development, a jetting tool with nozzles and a high-pressure pump is used to force water outwardly through the screen, the filter pack, and sometimes into the adjacent geologic unit.

4.1.63 *zone of saturation*—a hydrologic zone in which all the interstices between particles of geologic material or all of the joints, fractures, or solution channels in a consolidated rock unit are filled with water under pressure greater than that of the atmosphere.

5. Site Characterization

5.1 *General*—Soil mechanics, geomorphological concepts, geologic structure, stratigraphy, and sedimentary concepts, as well as the nature and behavior of the solutes of interest, must be combined with a knowledge of ground water movement to make a complete application of the results of the monitoring well design and installation guidance. Therefore, development of a conceptual hydrogeologic model that identifies potential flow paths and the target monitoring zone(s) is recommended prior to monitoring well design and installation. Development of the conceptual model is accomplished in two phases—an initial reconnaissance and a field investigation. When the hydrogeology of a project area is relatively uncomplicated and well documented in the literature, the initial reconnaissance may provide sufficient information to identify flow paths and the target monitoring zone(s). However, where little background data is available or the geology is complicated, a field investigation will generally be necessary to completely develop a conceptual hydrogeologic model.

5.2 *Initial Reconnaissance of Project Area*—The goal of the initial reconnaissance of the project area is to identify and locate those zones with the greatest potential to transmit a fluid from the project area. Identifying these flow paths is the first step in selecting the target ground water monitoring zone(s).

5.2.1 *Literature Search*—Every effort should be made to collect and review all applicable field and laboratory data from previous investigations of the project area. Data such as, but not limited to, topographic maps, aerial imagery, site ownership and utilization records, geologic and hydrogeologic maps and reports, mineral resource surveys, water well logs, personal information from local well drillers, agricultural soil reports, geotechnical engineering reports, and other engineering maps and report related to the project area should be reviewed.

5.2.2 *Field Reconnaissance*—Early in the investigation, the soil and rocks in open cut areas in the vicinity of the project should be studied, and various soil and rock profiles noted. Special consideration should be given to soil color and textural changes, landslides, seeps, and springs within or near the project area.

5.2.3 *Preliminary Conceptual Model*—The distribution of the predominant soil and rock units likely to be found during subsurface exploration may be hypothesized at this time in a preliminary hydrogeologic conceptual model using data obtained in the literature search and field reconnaissance. In areas where the geology is relatively uniform, well documented in the literature, and substantiated by the field reconnaissance, further refinement of the conceptual model may not be necessary unless anomalies are discovered in the well drilling stage.

5.3 *Field Investigation*—The goal of the field investigation is to refine the preliminary conceptual hydrogeologic model so that the target monitoring zone(s) is selected prior to monitoring well installation.

5.3.1 *Exploratory Borings and Test Pits*—Characterization of the flow paths conceptualized in the initial reconnaissance involves defining the porosity, hydraulic conductivity, gradation, stratigraphy, lithology, and structure of each hydrologic unit. The characteristics are defined by conducting an exploratory boring program which may include test pits. Exploratory borings and test pits should be deep enough to develop the required engineering and hydrogeologic data for determining the flow path(s), target monitoring zone, or both.

5.3.1.1 *Sampling*—Soil and rock properties should not be predicted wholly on field identification or classification, but should be checked by laboratory and field tests made on samples. Representative soil or rock samples, or both, of each material that is significant to the analysis and design of the monitoring system should be obtained and evaluated by a geologist, hydrogeologist, or engineer trained and experienced in soil and rock analysis. Soil sample extraction should be conducted according to Practice D 1452, Method D 1586,

Practice D 3550, or Practice D 1587, whichever is appropriate given the anticipated characteristics of the soil samples. Rock samples should be extracted according to Practice D 2113. Soil samples obtained for evaluation of hydraulic properties should be containerized and identified for shipment to a laboratory. Special measures to preserve either the continuity of thc sample or the natural moisture are not usually required. However, soil and rock samples obtained for evaluation of chemical properties often require special field preparation and preservation to prevent significant alteration of the chemical constituents during transportation to a laboratory (see Practice D 4220). Rock samples for evaluation of hydraulic properties are usually obtained using a split-inner-tube core barrel. Evaluation and logging of the core samples is usually made in the field before the core is removed from half of the split inner tube core barrel.

5.3.1.2 *Boring Logs*—Care should be taken to prepare and retain a complete boring log and sampling record for each exploratory borehole and test pit.

NOTE 1—Site investigations for the installation of ground-water monitoring wells can vary greatly due to the availability of reliable site data or the lack thereof. The general procedure would however be as follows: (*1*) gather factual data regarding the surficial and subsurface conditions, (*2*) analyze the data, (*3*) develop a conceptual model of the site conditions, (*4*) locate the monitoring wells based on the first three steps. Monitoring wells should only be installed with sufficient understanding of the geologic and hydrogeologic conditions present on site. Monitoring wells often serve as part of an overall site investigation for a specific purpose, such as determining the extent of contamination present, or for prediction of the effectiveness of aquifer remediations. In these cases extensive additional geotechnical and hydrogeologic information may be required that would go beyond the Section 5 Site Characterization description.

Boring logs should include the location, geotechnical (that is, penetration rates or blow counts), and sampling information for each material identified in the borehole either by symbol or word description, or both. Identification of all soils should be in accordance with Practice D 2488 or Practice D 3282. Identification of rock material should be based on Nomenclature C 294 or by an appropriate geologic classification system. Observations of seepage, free water, and water levels should also be noted. The boring logs should be accompanied by a report that includes a description of the area investigated; a map illustrating the vertical and horizontal location (with reference to nearest National Geodetic Vertical Datum [NGVD] and to a standardized survey grid, respectively) of each exploratory borehole or test pit, or both; and color photographs of rock cores, soil samples, and exposed strata labeled with a date and identification.

5.3.2 *Geophysical Exploration*—Geophysical surveys may be used to supplement borehole and outcrop data and to aid in interpretation between boreholes. Surface geophysical methods such as seismic surveys, and electrical-resistivity and electromagnetic conductance surveys can be particularly valuable when distinct differences in the properties of contiguous subsurface materials are indicated. Borehole methods such as resistivity, gamma, gamma-gamma, neutron, and caliper logs can be useful to confirm specific subsurface geologic conditions. Gamma logs are particularly useful in existing cased wells.

5.3.3 *Ground Water Flow Direction*—Ground water flow direction is generally determined by measuring the vertical and horizontal hydraulic gradient within each conceptualized flow path. However, because water will flow along the path of least resistance, flow direction may be oblique to the hydraulic gradient (buried stream channels or glacial valleys, for example). Flow direction is determined by first installing piezometers in the exploratory boreholes. The depth and location of the piezometers will depend upon anticipated hydraulic connections between conceptualized flow paths and their respective lateral direction of flow. Following careful evaluation, it may be possible to utilize existing private or public wells to obtain water level data. The construction integrity of such wells should be verified to ensure that the water levels obtained from the wells are representative only of the zones of interest. Following water level data acquisition, a potentiometric surface map should be prepared. Flow paths are ordinarily determined to be at right angles, or nearly so, to the equipotential lines.

5.4 *Completing the Conceptual Model*—A series of hydrogeologic cross sections should be developed to refine the conceptual model. This is accomplished by first plotting logs of soil and rock observed in the exploratory borings or test pits, and interpreting between these logs using the geologic and engineering interrelationships between other soil and rock data observed in the initial reconnaissance or with geophysical techniques. Extrapolation of data into adjacent areas should be done only where geologically uniform subsurface conditions are known to exist. The next step is to integrate the profile data with the piezometer data for both vertical and horizontal hydraulic gradients. Plan view and cross-sectional flow nets may need to be constructed. Following the analysis of these data, conclusions can be made as to which flow path(s) is the appropriate target monitoring zone(s).

NOTE 2—Ground water monitoring is difficult and may not be a reliable technology in fine-grain, low hydraulic conductivity, primary porosity strata because of (*1*) the disproportionate influence that microstratigraphy has on ground water flow in fine-grain strata; (*2*) flow lines proportionally higher for the vertical flow component in low hydraulic conductivity strata; and (*3*) the presence of indigenous metallic and inorganic constituents that make water quality data evaluation difficult.

6. Monitoring Well Construction Materials

6.1 *General*—The materials that are used in the construction of a monitoring well and that come in contact with the water sample should not measurably alter the chemical quality of the sample for the constituents being examined using the appropriate sampling protocols. Furthermore, the riser, well screen, and annular sealant injection equipment should be steam cleaned or high-pressure water cleaned (if appropriate for the selected riser material) immediately prior to well installation or certified clean from the manufacturer and delivered to site in a protective wrapping. Samples of the cleaning water, filter pack, annular seal, and mixed grout should be retained to serve as quality control until the completion of at least one round of ground water quality sampling and analysis.

6.2 *Water*—Water used in the drilling process, to prepare grout mixtures and to decontaminate the well screen, riser, and annular sealant injection equipment, should be obtained from a source of known chemistry that does not contain constituents that could compromise the integrity of the well installation.

6.3 *Primary Filter Pack:*

6.3.1 *Materials*—The primary filter pack (gravel pack) consists of a granular material of known chemistry and selected grain size and gradation that is installed in the annulus between the screen and the borehole wall. The filter pack is usually selected to have a 30 % finer (d-30) grain size that is about 4 to 10 times greater than the 30 % finer (d-30) grain size of the hydrologic unit being filtered (see Fig. 1). Usually, the filter is selected to have a low (that is, less than 2.5) uniformity coefficient. The grain size and gradation of the filter are selected to stabilize the hydrologic unit adjacent to the screen and permit only the finest soil grains to enter the screen during development. Thus, after development, a correctly filtered monitoring well is relatively turbid-free.

NOTE 3—When installing a monitoring well in Karst or highly fractured bedrock, the borehole configuration of void spaces within the formation surrounding the borehole is often unknown. Therefore, the installation of a filter pack becomes difficult and may not be possible.

6.3.2 *Gradation*—The filter pack should be uniformly graded and comprised of hard durable siliceous particles washed and screened with a particle size distribution derived by multiplying the d-30 size of the finest-grained screened stratum by a factor between 4 and 10. Use a number between four and six as the multiplier if the stratum is fine and uniform; use a factor between six and ten where the material has highly nonuniform gradation and includes silt-sized particles. The grain-size distribution of the filter pack is then plotted using the d-30 size as the control point on the graph. The selected filter pack should have a uniformity coefficient of approximately 2.5 or less.

NOTE 4—This practice presents a design for monitoring wells that will be effective in the majority of aquifers. Applicable state guidance may differ from the designs contained in this practice.

NOTE 5—Because the well screen slots have uniform openings, the filter pack should be composed of particles that are as uniform in size as is practical. Ideally, the uniformity coefficient (the quotient of the 60 % passing, D-60 size divided by the 10 % passing D-10 size [effective size]) of the filter pack should be 1.0 (that is, the D-60 % and the D-10 % sizes should be identical). However, a more practical and consistently achievable uniformity coefficient for all ranges of filter pack sizes is 2.5. This value of 2.5 should represent a maximum value, not an ideal.

NOTE 6—Although not recommended as standard practice, often a project requires drilling and installing the well in one phase of work.

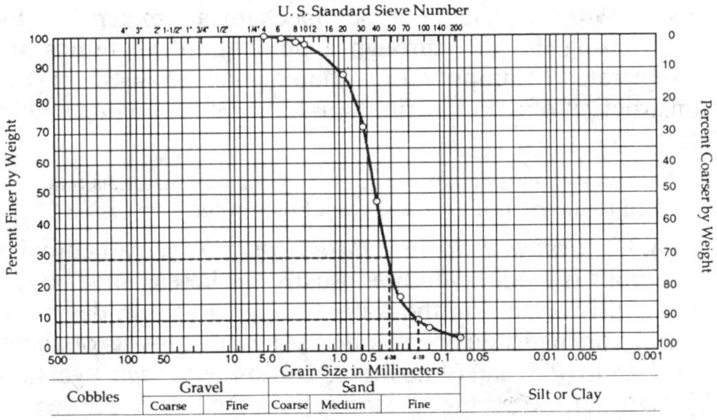

FIG. 1 Example Grading Curve for Design of Monitoring Well Screens

Therefore, the filter pack materials must be ordered and delivered to the drill site before soil samples can be collected. In these cases, the suggested well screen slot size and filter pack materials are presented in Table 1.

6.4 *Well Screen:*

6.4.1 *Materials*—The well screen should be new, machine-slotted or continuous wrapped wire-wound and composed of materials most suited for the monitoring environment and site characterization findings. The screen should be plugged at the bottom. The plug should be of the same material as the well screen. This assembly must have the capability to withstand installation and development stresses without becoming dislodged or damaged. The length of the slotted area should reflect the interval to be monitored. Immediately prior to installation, the well screen should be steam cleaned or high-pressure water cleaned (if appropriate for the selected well screen materials) with water from a source of known chemistry if not certified by the manufacturer, delivered, and maintained clean at the site.

NOTE 7—Well screens are most commonly composed of PVC, stainless steel, fiberglass, or fluoropolymer materials.

6.4.2 *Diameter*—The minimum nominal internal diameter of the well screen should be chosen based on the particular application. However, in most instances, a minimum of 2 in. (50 mm) is needed to allow for the introduction and withdrawal of sampling devices.

6.4.3 *Slot Size*—The slot size of the well screen should be determined relative to the grain size analysis of the stratum interval to be monitored and the gradation of the filter pack material. In granular non-cohesive strata that will fall in easily around the screen, filter packs are not necessary. In these cases of natural development, the slot size of the well screen is to be determined using the grain size of the materials in the surrounding strata. The slot size and arrangement should retain at least 90 % and preferably 99 % of the filter pack. The method for determining the correct gradation of filter pack material is described in 6.3.2.

6.5 *Riser:*

6.5.1 *Materials*—The riser should be new and composed of materials that will not alter the quality of water samples for the constituents of concern and that are appropriate for the monitoring environment. The riser should have adequate wall thickness and coupling strength to withstand installation and development stresses. Each section of riser should be steam cleaned or high-pressure water cleaned (if appropriate for the selected material) using water from a source of known chemistry immediately prior to installation.

NOTE 8—Risers are generally constructed of PVC, stainless steel, fiberglass, or fluoropolymer materials.

6.5.2 *Diameter*—The minimum nominal internal diameter of the riser should be chosen based on the particular application. However, in most instances, a minimum of 2 in. (50 mm) is needed to accommodate sampling devices.

6.5.3 *Joints (Couplings)*—Threaded joints are recommended. Glued or solvent welded joints of any type are *not* recommended since glues and solvents may alter the chemistry of the water samples. In most cases, square profile flush joint threads do not require PTFE taping, however, tapered thread joints should be PTFE taped to prevent leakage of water into the riser. Alternatively, O-rings composed of

TABLE 1 Recommended (Achievable) Filter Pack Characteristics for Common Screen Slot Sizes

Size of Screen Opening, mm (in.)	Slot No.	Sand Pack Mesh Size Name(s)	1 % Passing Size (D-1), mm	Effective Size, (D-10), mm	30 % Passing Size (D-30), mm	Range of Uniformity Coefficient	Roundness (Powers Scale)
0.125 (0.005)	5[A]	100	0.09 to 0.12	0.14 to 0.17	0.17 to 0.21	1.3 to 2.0	2 to 5
0.25 (0.010)	10	20 to 40	0.25 to 0.35	0.4 to 0.5	0.5 to 0.6	1.1 to 1.6	3 to 5
0.50 (0.020)	20	10 to 20	0.7 to 0.9	1.0 to 1.2	1.2 to 1.5	1.1 to 1.6	3 to 6
0.75 (0.030)	30	10 to 20	0.7 to 0.9	1.0 to 1.2	1.2 to 1.5	1.1 to 1.6	3 to 6
1.0 (0.040)	40	8 to 12	1.2 to 1.4	1.6 to 1.8	1.7 to 2.0	1.1 to 1.6	4 to 6
1.5 (0.060)	60	6 to 9	1.5 to 1.8	2.3 to 2.8	2.5 to 3.0	1.1 to 1.7	4 to 6
2.0 (0.080)	80	4 to 8	2.0 to 2.4	2.4 to 3.0	2.6 to 3.1	1.1 to 1.7	4 to 6

[A] A 5-slot (0.152-mm) opening is not currently available in slotted PVC but is available in Vee wire PVC and Stainless; 6-slot opening may be substituted in these cases.

materials that would not impact the water sample for the constituents of concern may be selected for use on flush joint threads.

6.6 *Casing*—Where conditions warrant, the use of permanent casing installed to prevent communication between water-bearing zones is encouraged. The following subsections address both temporary and permanent casings.

6.6.1 *Materials*—The material type and minimum wall thickness of the casing should be adequate to withstand the forces of installation. All casing that is to remain as a permanent part of the installation (that is, multi-cased wells) should be new and cleaned to be free of interior and exterior protective coatings.

NOTE 9—The exterior casing (temporary or permanent multi-cased) is generally composed of steel, although other appropriate materials may be used.

6.6.2 *Diameter*—Several different casing sizes may be required depending on the subsurface geologic conditions penetrated. The diameter of the casing for filter packed wells should be selected so that a minimum annular space of 2 in. (50 mm) is maintained between the inside diameter of the casing and outside diameter of the riser. In addition, the diameter of the casings in multi-cased wells should be selected so that a minimum annular space of 2 in. is maintained between the casing and the borehole (that is, a 2-in. diameter screen will require first setting a 6-in. (152-mm) diameter casing in a 10-in. (254-mm) diameter boring).

NOTE 10—Under difficult drilling conditions (collapsing soils, rock, or cobbles), it may be necessary to advance temporary casing, under these conditions a smaller annular space may be maintained.

6.6.3 *Joints (Couplings)*—The ends of each casing section should be either flush-threaded or bevelled for welding.

6.7 *Protective Casing:*

6.7.1 *Materials*—Protective casings may be made of aluminum, steel, stainless steel, cast iron, or a structural plastic. The protective casing should have a lid capable of being locked shut by a locking device.

6.7.2 *Diameter*—The inside dimensions of the protective casing should be a minimum of 2 in. (50 mm) and preferably 4 in. (101 mm) larger than the nominal diameter of the riser to facilitate the installation and operation of sampling equipment.

6.8 *Annular Sealants*—The materials used to seal the annulus may be prepared as a slurry or used un-mixed in a dry pellet, granular, or chip form. Sealants should be selected to be compatible with ambient geologic, hydrogeologic, and climatic conditions and any man-induced conditions anticipated to occur during the life of the well.

6.8.1 *Bentonite*—Bentonite should be powdered, granular, pelletized, or chipped sodium montmorillonite furnished in sacks or buckets from a commercial source and free of impurities which adversely impact the water quality in the well. Pellets consist of roughly spherical or disk shaped units of compressed bentonite powder. Chips are large, irregularly shaped, and coarse granular units of bentonite free of additives. The diameter of pellets or chips selected for monitoring well construction should be less than one fifth the width of the annular space into which they are placed to reduce the potential for bridging. Granules consist of coarse particles of unaltered bentonite, typically smaller than 0.2 in. (50 mm).

6.8.2 *Cement*—Each type of cement has slightly different characteristics that may be appropriate under various physical and chemical conditions. Cement should be one of the five Portland cement types that are specified in Specification C 150. The use of quick-setting cements containing additives is not recommended for use in monitoring well installation. Additives may leach from the cement and influence the chemistry of the water samples.

6.8.3 *Grout*—The grout backfill that is placed above the bentonite annular seal and secondary filters (see Fig. 2) is ordinarily a liquid slurry consisting of either a bentonite (powder or granules, or both) base and water, or a Portland cement base and water. Often, bentonite-based grouts are used when it is desired that the grout remain flexible (that is, to accommodate freeze-thaw) during the life of the installation. Cement or bentonite-based grouts are often used when the filling in of cracks in the surrounding geologic material, adherence to rock units, or a rigid setting is desired.

6.8.3.1 *Mixing*—The mixing (and placing) of a grout backfill should be performed with precisely recorded weights and volumes of materials, and according to procedures stipulated by the manufacturer that often include the order of component mixing. The grout should be thoroughly mixed with a paddle type mechanical mixer or by recirculating the mix through a pump until all lumps are disintegrated. Lumpy grout should not be used in the construction of a monitoring well to prevent bridging within the tremie.

NOTE 11—Lumps do not include lost circulation materials that may be added to the grout if excessive grout losses occur.

6.8.3.2 *Typical Bentonite Base Grout*—When a bentonite base grout is used, bentonite, usually unaltered, *must* be the first additive placed in the water through a venturi device. A typical unbeneficiated bentonite base grout consists of about 1 to 1.25 lb (0.57 kg) of unaltered bentonite to each 1 gal (3.8 L) of water. After the bentonite is mixed and allowed to "yield or hydrate," up to 2 lb (0.9 kg) of Type I Portland cement (per gallon of water) is often added to stiffen the mix.

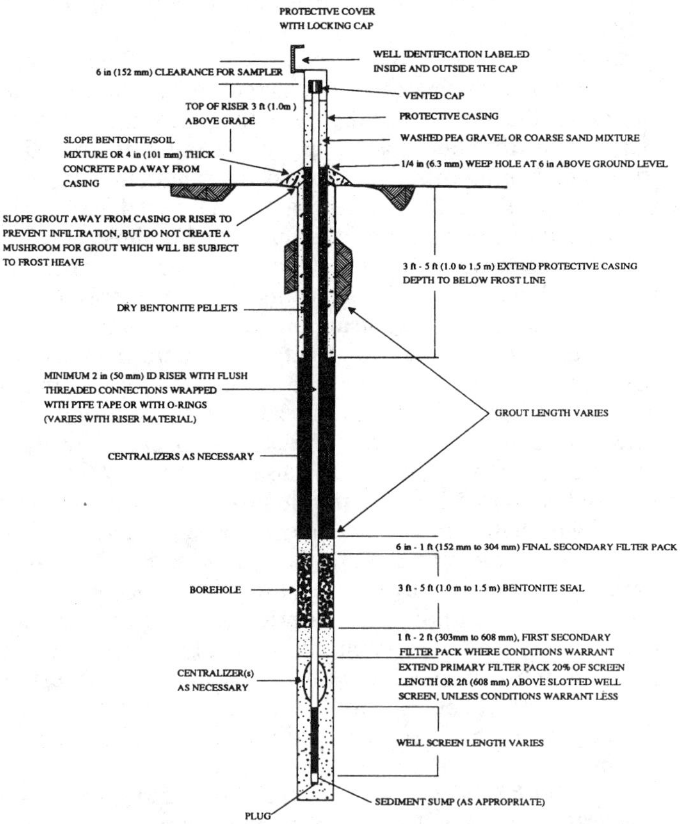

FIG. 2 Monitoring Well Design—Single-Cased Well

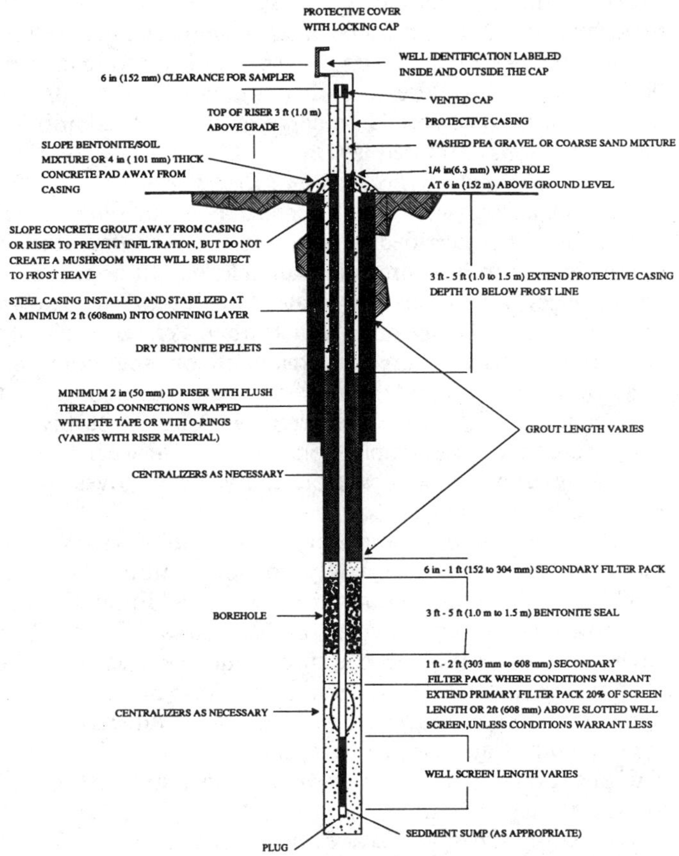

FIG. 3 Monitoring Well Design—Multi-Cased Well

100 % Bentonite grouts should not be used solely for monitoring well annular sealants in the vadose zone of arid regions because of their propensity to desiccate. This could result in non-representative waters affecting the target monitoring zone.

NOTE 12—High solids bentonite grouts (minimum 20 % by weight with water) and other bentonite-based grouts may contain granular bentonite to increase the solids content and other components added under manufacturer's directions to either stiffen or retard stiffening of the mix.

All additives to grouts should be evaluated for their effects on subsequent water samples.

6.8.3.3 *Typical Cement Base Grout*—When a cement-based grout is used, cement is usually the first additive placed in the water. A typical cement-based grout consists of about 6 to 7 gal (23 to 26 L) of water per 94-lb (43-kg) bag of Type I Portland cement. From 0 to 10 % (by dry weight of cement) of unaltered bentonite powder is often added after the initial mixing of cement and water to retard shrinkage and provide plasticity. The bentonite is added *dry* to the cement-water slurry without first mixing it with water.

6.9 *Secondary Filter Packs:*

6.9.1 *Materials*—A secondary filter pack is a layer of material placed in the annulus between the primary filter pack and the bentonite seal, and between the bentonite seal and the grout backfill (see Figs. 2 and 3).

6.9.2 *Gradation*—The secondary filter pack should be uniformly graded fine sand with a 100 % by weight passing the No. 30 U.S. Standard sieve, and less than 2 % by weight passing the 200 U.S. Standard sieve.

6.10 *Annular Seal Equipment*—The equipment used to inject the annular seals and filter pack should be steam cleaned or high-pressure water cleaned (if appropriate for the selected material) using water from a source or known quality prior to use. This procedure is performed to prevent the introduction of materials that may ultimately alter the water sample quality.

7. Drilling Methods

7.1 The type of equipment required to create a stable, open, vertical borehole for installation of a monitoring well depends upon the site geology, hydrology, and the intended use of the data. Engineering and geological judgment is required for the selection of the drilling methods utilized for drilling the exploratory boreholes and monitoring wells. Whenever feasible, drilling procedures should be utilized that do not require the introduction of water or liquid fluids into the borehole, and that optimize cuttings control at ground surface. Where the use of drilling fluid is unavoidable, the selected fluid should have as little impact as possible on the water samples for the constituents of interest. In addition, care should be taken to remove as much drilling fluid as possible from the well and the aquifer during the well development process. It is recommended that if an air compressor is used, it is equipped with an oil air filter or oil trap.

8. Monitoring Well Installation

8.1 *Stable Borehole*—A stable borehole must be constructed prior to attempting to install the monitoring well screen and riser. Steps must be taken to stabilize the borehole before attempting installation if the borehole tends to cave or blow-in, or both. Boreholes that are not straight or are partially obstructed should be corrected prior to attempting the installations described herein.

8.2 *Assembly of Well Screen and Riser:*

8.2.1 *Handling*—The well screen, bottom plug, riser, should be either certified clean from the manufacturer or steam cleaned or high-pressure water cleaned (if appropriate for the selected material) using water from a source of known chemistry immediately prior to assembly. Personnel should take precautions to assure that grease, oil, or other contaminants that may ultimately alter the water sample do not contact any portion of the well screen and riser assembly. As one precaution, for example, personnel should wear a clean pair of cotton or surgical (or equivalent) gloves while handling the assembly.

8.2.2 *Riser Joints (Couplings)*—Flush joint risers with square profile threads normally do not require additional PTFE taping to obtain a water tight seal. In addition, O-rings of known chemistry, selected on the basis of prevailing environmental or physical conditions, may be used to assure a tight seal of flush-joint couplings. Couplings are often tightened by hand; however, if necessary, steam cleaned or high-pressure water cleaned wrenches may be utilized. Precautions should be taken to prevent damage to the threaded joints during installation.

8.3 *Setting the Well Screen and Riser Assembly*—When the well screen and riser assembly is lowered to the predetermined level and held into position, the assembly may require ballast to counteract the tendency to float in the borehole. Ballasting may be accomplished by continuously filling the riser with water from a source of known chemistry or, preferably, water which was previously removed from the borehole. Alternatively, the riser may be slowly pushed into the fluid in the borehole with the aid of hydraulic rams on the drill rig and held in place as additional sections of riser are added to the column. Care must be taken to secure the riser assembly so that personnel safety is assured during the installation. The assembly must be installed straight with the appropriate centralizers to allow for the introduction and withdrawal of sampling devices. Difficulty in maintaining a straight installation may be encountered where the weight of the well screen and riser assembly is significantly less than the buoyant force of the fluid in the borehole. The riser should extend above grade and be capped temporarily to deter entrance of foreign materials during completion operations.

8.4 *Installation of the Primary Filter Pack:*

8.4.1 *Volume of Filter Pack*—The volume of filter pack required to fill the annular space between the well screen and borehole should be computed, measured, and recorded on the well completion diagram during installation. To be effective, the filter pack should extend above the screen for a distance of about 20 % of the length of the well screen but not less than 2 ft (600 mm) (see Figs. 2 and 3). Where there is hydraulic connection between the zone to be monitored and the overlying strata, this upward extension should be gauged to prevent seepage from overlying hydrologic units into the filter pack. Seepage from other units may alter the water sample.

8.4.2 *Placement of Primary Filter Pack*—Placement of the well screen is preceded by placing no less than 2 % and no more than 10 % of the primary filter pack into the bottom of the borehole using a decontaminated, flush threaded, 1-in. (25-mm) minimum internal diameter tremie pipe. Alternatively, the filter pack may be added directly between the riser pipe and the auger or borehole or casing and the top of the filter pack located using a tamper or a weighted line. The well screen and riser assembly is then centered in the borehole using one or more centralizer(s) or alternative centering device located not more than 10 ft (3 m) above the bottom of the well screen (see Figs. 2 and 3). The centralizer should not be located in the bentonite seal. The remaining primary filter pack is then placed in increments as the tremie is gradually raised. As primary filter pack material is poured into the tremie pipe, water from a source of known chemistry may be added to help move the filter pack. The tremie pipe or a weighed line inserted through the tremie pipe can be used to measure the top of the primary filter pack as work progresses. If bridging of the primary filter pack occurs, the bridged material should be broken mechanically prior to proceeding with the addition of more filter pack material. The elevation, volume, and gradation of primary filter pack is recorded on the well completion diagram.

8.4.3 *Withdrawal of the Temporary Casing/Augers*—If used, the temporary casing or hollow stem auger is withdrawn, usually in stipulated increments. Care should be taken to minimize lifting the riser with the withdrawal of the temporary casing/augers. To limit borehole collapse, the temporary casing or hollow stem auger is usually withdrawn until the lower most point on the temporary casing or hollow stem auger is at least 2 ft (608 mm), but no more than 5 ft (1.5 m), above the filter pack for unconsolidated materials; or at least 5 ft, but no more than 10 ft (3.0 m), for consolidated materials. In highly unstable formations, withdrawal intervals may be much less. After each increment, it should be ascertained that the primary filter pack has not been displaced during the withdrawal operation (that is, a weighed measuring device).

8.5 *Placement of First Secondary Filter*—A secondary filter pack may be installed above the primary filter pack to prevent the intrusion of the bentonite grout seal into the primary filter pack (see Figs. 2 and 3). To be effective, measured and recorded volume of secondary filter material should be added to extend 1 to 2 ft (304 to 608 mm) above the primary filter pack. As with the primary filter, a secondary filter must not extend into an overlying hydrologic unit (see 8.4.1). The well designer should evaluate the need for this filter pack by considering the gradation of the primary filter pack, the hydraulic heads between adjacent units, and the potential for grout intrusion into the primary filter pack. The secondary filter material is poured into the annular space through a decontaminated, flush threaded, 1-in. (25-mm) minimum internal diameter tremie pipe lowered to within 3 ft (1.0 m) of the placement interval. Water from a source of known chemistry may be added to help move the filter pack into its proper location. The tremie pipe or weighed line inserted through the tremie pipe can be used to measure the top of the secondary filter pack as work progresses. The elevation, volume, and gradation of the secondary filter pack is recorded on the well completion diagram.

8.6 *Installation of the Bentonite Seal*—A bentonite pellet or a slurry seal is placed in the annulus between the borehole and the riser pipe on top of the secondary or primary filter pack (see Figs. 2 and 3). This seal retards the movement of cement-based grout backfill into the primary or secondary filter packs. To be effective, the bentonite seal should extend above the filter packs approximately 3 to 5 ft (1.0 to 1.5 m)—depending on local conditions. The bentonite seal should be installed using a tremie pipe lowered to the top of the filter packs and slowly raised as the bentonite pellets or the slurry fill the annular space. Bentonite pellets may bridge and block the tremie pipe in deep wells. In these cases, pellets may be allowed to free-fall into the borehole. As a bentonite pellet seal is poured into the tremie pipe or allowed to free-fall into the borehole, a tamper or weighed line may be necessary to tamp pellets into place. If the seal is installed above the water level, water from a source of known chemistry would be added to allow proper hydration of the annular seal. The tremie pipe or a weighed line inserted through the tremie pipe can be used to measure the top of the bentonite seal as the work progresses. If a bentonite pellet seal is being constructed above the water level, approximately 5 gal (20 L) of water from a source of known chemistry can be poured into the annulus to ensure that the pellets hydrate. Sufficient time should be allowed for the bentonite pellet seal to hydrate or the slurry annular seal to expand prior to grouting the remaining annulus. The volume and elevation of the bentonite seal material should be measured and recorded on the well completion diagram.

8.7 *Final Secondary Filter Pack*—A 6-in. to 1-ft (152 to 304-mm) secondary filter may be placed above the bentonite seal in the same manner described in 8.5 (see Figs. 2 and 3). This secondary filter pack will provide a confining layer over the bentonite seal to limit the downward movement of cement-based grout backfill into the bentonite seal. The volume, elevation, and gradation of this final secondary filter pack should be documented on the well completion diagram.

8.8 *Grouting the Annular Space:*

8.8.1 *General*—Grouting procedures vary with the type of well design. The following procedures will apply to both single- and multi-cased monitoring wells. Paragraphs 8.8.2 and 8.8.3 detail those procedures unique to single- and multi-cased installations, respectively.

8.8.1.1 *Volume of Grout*—The volume and location of grout used to backfill the remaining annular space is recorded on the well completion diagram. An ample volume of grout should be premixed on site to compensate for unexpected losses. The use of alternate grout materials, including grouts containing gravel, may be necessary to control zones of high grout loss.

8.8.1.2 *Injection Procedures*—The grout backfill should be injected under pressure to reduce the chance of leaving voids in the grout, and to displace any liquids and drill cuttings that may remain in the annulus. Depending upon the well design, grouting may be accomplished using a pressure grouting technique or by gravity feed through a tremie pipe. With either method, grout is introduced in one continuous operation until full strength grout flows out at the ground surface without evidence of drill cuttings or fluid. The grout should slope away from the riser or casing at the surface, but care should be taken not to create a grout mushroom that would be subjected to frost heave.

8.8.1.3 *Grout Setting and Curing*—The riser or casing or both should not be disturbed until the grout sets and cures for the amount of time necessary to prevent a break in the seal between the grout and riser or grout and casing or both. The amount of time required will vary with grout content and climatic conditions and should be documented on the well completion diagram.

8.8.2 *Specific Procedures for Single-Cased Wells*—Grouting should begin at a level directly above the final secondary filter pack (see Fig. 2). Grout should be injected using a tremie pipe equipped with a side discharge; this dissipates the fluid-pumping energy against the borehole wall and riser, reducing the potential for infiltration of grout into the primary filter pack. The tremie pipe should be kept full of grout from start to finish with the discharge end of the pipe completely submerged as it is slowly and continuously lifted. Approximately 5 to 10 ft (1.5 to 3.0 m) of tremie pipe should remained submerged until grouting is complete. For deep installations or where the joints or couplings of the selected riser cannot withstand the shear or collapse stress exerted by a full column of grout as it sets, a staged grouting procedure may be considered. If used, the temporary casing or hollow stem auger should be removed in increments immediately following each increment of grout installation and in advance of the time when the grout begins to set. If casing removal does not commence until grout injection is completed, then, after the casing is removed, additional grout may be periodically injected into the annular space to maintain a continuous column of grout up to the ground surface.

8.8.3 *Specific Procedures for Multi-Cased Wells*—If the outer casing of a multi-cased well cannot be driven to form a tight seal between the surrounding stratum (strata) and the casing, it should be installed in a predrilled borehole. After the borehole has penetrated not less than 2 ft (608 mm) of the first targeted confining stratum, the outer casing is lowered to the bottom of the boring and the annular space is filled with grout. Grouting may be accomplished using a pressure grouting method or gravity feed through a tremie pipe. Pressure grouting will require the use of a grout shoe or packer installed at the end of the outer casing to prevent grout from moving up into the casing. If a tremie pipe is used to inject grout into the annular space, it should be equipped with a side discharge. With each alternative, the grout must be allowed to cure and form a seal between the casing and the grout prior to advancing the hole to the next hydrologic unit. This procedure is repeated as necessary to advance the borehole to the desired depth. Upon reaching the final target depth, the riser and screen is set through the inner casing. Subsequent to the placement of the filter packs and bentonite seal, the remaining annular space is grouted as described in 8.8.2 (see Fig. 3).

NOTE 13—When using a packer, pressure may build up during grout injection and force grout up the sides of the packer and into the casing.

8.9 *Well Protection*—Well protection refers specifically to installations made at the ground surface to deter unauthorized entry to the monitoring well and to prevent surface water from entering the annulus.

8.9.1 *Protective Casing*—The protective casing should extend from below the frost line (3 to 5 ft [1.0 to 1.5 m]) below the grade depending on local conditions to slightly above the well casing tip. The protective casing should be initially placed before final set of the grout backfill. The protective casing should be sealed and immobilized in concrete placed around the outside of the protective casing above the set grout backfill. The casing should be positioned and stabilized in a position concentric with the riser (see Figs. 1 and 2). Sufficient clearance, usually 6 in. (152 mm) should be maintained between the lid of the protective casing and the top of the riser to accommodate sampling equipment. A ¼-in. (6.3-mm) diameter weep hole should be drilled in the casing 6 in. above the ground surface to permit water to drain out of the annular space. In cold climates, this hole will also prevent water freezing between the well protector and the well casing. Dry bentonite pellets, granules, or chips should then be placed in the annular space below ground level within the protective casing. Coarse sand or pea gravel or both is placed in the annular space above the dry bentonite pellets and above the weep hole to prevent entry of insects. All materials chosen should be documented on the well completion diagram. The monitoring well identification number should be clearly visible on the inside and outside of the lid of the protective casing.

8.9.2 *Completion of Surface Installation*—The well protection installation may be completed in one of three ways:

8.9.2.1 In areas subject to frost heave, place a soil or bentonite/sand layer adjacent to the protective casing sloped to direct water drainage away from the well.

8.9.2.2 In regions *not* subject to frost heave, a 4-in. (101-mm) thick concrete pad sloped to provide water drainage away from the well may be placed around the installation. Care must be taken not to lock the concrete pad onto the protective casing if subsidence of the surface may occur in the future.

8.9.2.3 Where monitoring well protection must be flushed with the ground, an internal cap should be fitted on top of the riser within the manhole or vault. This cap should be leak-proof so that if the vault or manhole should fill with water, the water will not enter the well casing. Ideally, the manhole cover cap should also be leak-proof.

8.9.3 *Additional Protection*—In areas where there is a high probability of damaging the well (high traffic, heavy equipment, poor visibility), it may be necessary to enhance the normal protection of the monitoring well through the use of posts, markers, signs, etc. The level of protection should meet the damage threat posed by the location of the well.

9. Well Development

9.1 *General*—The development serves to remove the finer grained material from the well screen and filter pack that may otherwise interfere with water quality analyses, restore the ground-water properties disturbed during the drilling process and to improve the hydraulic characteristics of the filter pack and hydraulic communication between the well and the hydrologic unit adjacent to the well screen. Methods of well development vary with the physical characteristics of hydrologic units in which the monitoring well is screened and with the drilling method used.

9.2 *Development Methods*—Methods of development most often used include mechanical surging and bailing or pumping, over-pumping, air-lift pumping, and jetting. An important factor in any method is that the development work be stated slowly and gently and be increased in vigor as the well is developed. Most methods of well development require the application of sufficient energy to disturb the filter pack, thereby freeing the fines and allowing them to be drawn into the well. The coarser fractions then settle around and stabilize the screen. The well development method chosen should be documented on the well completion diagram.

NOTE 14—Any time an air compressor is used, it should be equipped with an oil air filter or oil trap to minimize the introduction of oil into the screen area. The presence of oil would impact the organic constituent concentrations of the water samples.

NOTE 15—Development procedures for wells completed in fine sand and silt strata should involve methods that are relatively gentle so that the strain material will not be incorporated into the filter pack. Vigorous surging for development can produce mixing of the fine strata and filter pack and produce turbid samples from the installation. Also, development methods should be carefully selected based upon the potential contaminant(s) present, quality of waste water generated, and requirements for containerization or treatment of waste water.

9.2.1 *Mechanical Surging*—In this method, water is forced to flow into and out of the well screen by operating a plunger (or surge block) or bailer up and down in the riser. A pump or bailer should then be used to remove the dislodged sediments following surging.

9.2.2 *Over Pumping*—With this method, the monitoring well is pumped at a rate considerably higher than it would be during normal operation. The fine-grain materials would be dislodged from the filter pack and surrounding strata influenced by the higher pumping rate. This method is usually conducted in conjunction with mechanical surging.

9.2.3 *Air Lift Pumping*—In this method, an air lift pump is operated by cycling the air pressure on and off for short periods of time. This operation will provide a surging action that will dislodge fine-grained particles. Applying a steady, low pressure will remove the fines that have been drawn into the well by the surging action. Efforts should be made (that is, through the use of a foot valve) to avoid pumping air into the filter pack and adjacent hydrologic unit because the air may lodge there and inhibit future sampling efforts and may alter ambient water chemistry. Furthermore, application of high air pressures should be avoided to prevent damage to small diameter PVC risers, screens, and filter packs.

9.2.4 *Well Jetting*—Another method of development involves jetting the well screen area with water while simultaneously air-lift pumping the well. However, the water added during this development procedure will alter the natural, ambient water quality and may be difficult to remove. Therefore, the water added should be obtained from a source of known chemistry. Water from the monitoring well being developed may also be used if the suspended sediments are first removed.

9.3 *Duration of Well Development*—Well development should begin after the monitoring well is completely installed and prior to water sampling. Development should be continued until representative water, free of the drilling fluids, cuttings, or other materials introduced during well construction is obtained. Representative water is assumed to have been obtained when pH, temperature, and specific conductivity readings stabilize and the water is visually clear of suspended solids. The minimum duration of well development should vary in accordance with the method used to develop the well. For example, surging and pumping the well may provide a stable, sediment-free sample in a matter of minutes; whereas, bailing the well may require several hours of continuous effort to obtain a clear sample. The duration of well development and the pH, temperature, and specific conductivity readings should be recorded on the well completion diagram.

9.4 *Well Recovery Test*—A well recovery test should be performed immediately after and in conjunction with well development. The well recovery test not only provides an indication of well performance but also provides data for determining the transmissivity of the screened hydrologic unit. Estimates of the hydraulic conductivity of the unit can then be determined. Readings should be taken at intervals suggested in the table below until the well has recovered to 90 % of its static water level.

NOTE 16—If a monitoring well does not recover sufficiently for sampling within a 24-h period and the well has been properly developed, the installation should not generally be used as a monitoring well for detecting or assessing low level organic constituents. The installation may, however, be used for long-term water level monitoring if measurements of shorter frequency water level changes are not required.

10. Installation Survey

10.1 *General*—The vertical and horizontal position of each monitoring well in the monitoring system should be surveyed and subsequently mapped by a licensed surveyor. The well location map should include the location of all monitoring wells in the system and their respective identification numbers, elevations of the top of riser position to be used as the reference point for water level measurements, and the elevations of the ground surface protective installations. The locations and elevations of all permanent benchmark(s) and pertinent boundary marker(s) located on-site or used in the survey should also be noted on the map.

TABLE 2 Suggested Recording Intervals for Well Recovery Tests

Time Since Starting Test	Time Interval
0 to 15 min	1 min
15 to 50 min	5 min
50 to 100 min	10 min
100 to 300 min (5 h)	30 min
300 to 1440 min (24 h)	60 min

10.2 *Water Level Measurement Reference*—The water level measurement reference point should be permanently marked, for instance, by cutting a V-notch into the top edge of the riser pipe. This reference point should be surveyed in reference to the nearest NGVD reference point.

10.3 *Location Coordinates*—The horizontal location of all monitoring wells (active or decommissioned) should be surveyed by reference to a standardized survey grid or by metes and bounds.

11. Monitoring Well Network Report

11.1 To demonstrate that the goals as set forth in Section 1, the Scope, have been met, a monitoring well network report should be prepared. This report should:

11.1.1 Locate the area investigated in terms pertinent to the project. This should include sketch maps or aerial photos on which the exploratory borings, piezometers, sample areas, and monitoring wells are located, as well as topographic items relevant to the determination of the various soil and rock types, such as contours, streambeds, etc. Where feasible, include a geologic map and geologic cross sections of the area being investigated.

11.1.2 Include copies of all well boring test pits and exploratory borehole logs, initial and post-completion water levels, all laboratory test results, and all well completion diagrams.

11.1.3 Include the well installation survey.

11.1.4 Describe and relate the findings obtained in the initial reconnaissance and field investigation (Section 5) to the design and installation procedures selected (Sections 7 to 9) and the surveyed locations (Section 10).

11.1.5 This report should include a recommended decommission procedure that is consistent with the well construction and local regulatory requirements.

12. Keywords

12.1 aquifer; borehole drilling; geophysical exploration; ground water; monitoring well; site investigation

Designation: D 5126 – 90

Standard Guide for
Comparison of Field Methods for Determining Hydraulic Conductivity in the Vadose Zone[1]

This standard is issued under the fixed designation D 5126; the number immediately following the designation indicates the year of original adoption or, in the case of revision, the year of last revision. A number in parentheses indicates the year of last reapproval. A superscript epsilon (ε) indicates an editorial change since the last revision or reapproval.

1. Scope

1.1 This guide provides a review of the test methods for determining hydraulic conductivity in unsaturated soils and sediments. Test methods for determining both field-saturated and unsaturated hydraulic conductivity are described.

1.2 Measurement of hydraulic conductivity in the field is used for estimating the rate of water movement through clay liners to determine if they are a barrier to water flux, for characterizing water movement below waste disposal sites to predict contaminant movement, and to measure infiltration and drainage in soils and sediment for a variety of applications. Test methods are needed for measuring hydraulic conductivity ranging from 1×10^{-2} to 1×10^{-8} cm/s, for both surface and subsurface layers, and for both field-saturated and unsaturated flow.

1.3 For these field test methods a distinction must be made between "saturated" (K_s) and "field-saturated" (K_{fs}) hydraulic conductivity. True saturated conditions seldom occur in the vadose zone except where impermeable layers result in the presence of perched water tables. During infiltration events or in the event of a leak from a lined pond, a "field-saturated" condition develops. True saturation does not occur due to entrapped air **(1)**.[2] The entrapped air prevents water from moving in air-filled pores that, in turn, may reduce the hydraulic conductivity measured in the field by as much as a factor of two compared to conditions when trapped air is not present **(2)**. Field test methods should simulate the "field-saturated" condition.

1.4 Field test methods commonly used to determine field-saturated hydraulic conductivity include various double-ring infiltrometer test methods, air-entry permeameter test methods, and borehole permeameter tests. Many empirical test methods are used for calculating hydraulic conductivity from data obtained with each test method. A general description of each test method, and special characteristics affecting applicability is provided.

1.5 Field test methods used to determine unsaturated hydraulic conductivity in the field include direct measurement techniques and various estimation methods. Direct measurement techniques for determining unsaturated hydraulic conductivity include the instantaneous profile (IP) test method, and the gypsum crust method. Estimation techniques have been developed using borehole permeameter data, and using data obtained from desorption curves (a curve relating water content to matric potential).

1.6 The values stated in SI units are to be regarded as standard.

1.7 *This standard does not purport to address the safety problems associated with its use. It is the responsibility of the user of this standard to establish appropriate safety and health practices and determine the applicability of regulatory limitations prior to use.*

2. Referenced Documents

2.1 *ASTM Standards:*

D 653 Terms and Symbols Relating to Soil and Rock[3]

D 2434 Test Method for Permeability of Granular Soils (Constant Head)[3]

D 3385 Test Method for Infiltration Rate of Soils in the Field Using Double-Ring Infiltrometers[3]

D 4643 Test Method for Determination of Water (Moisture) Content of Soil by the Microwave Oven Method[3]

3. Terminology

3.1 *Definitions:*

3.1.1 Definitions shall be in accordance with Terms and Symbols D 653.

3.2 *Descriptions of Terms Specific to This Standard:*

3.2.1 Descriptions of terms shall be in accordance with Ref **(2)**.

4. Summary of Guide

4.1 *Test Methods for Measuring Saturated Hydraulic Conductivity Above the Water Table*—There are several test methods available for determining the field saturated hydraulic conductivity of unsaturated materials above the water table. Most of these methods involve measurement of the infiltration rate of water into the soil from an infiltrometer or permeameter device. Infiltrometers typically measure conductivity at the soil surface, whereas permeameters may be used to determine conductivity at different depths within the soil profile. A representative list of the most commonly used equipment includes the following: infiltrometers, (single and double ring infiltrometers); double tube method; air-entry permeameter; borehole permeameter methods, (constant and multiple head methods).

4.1.1 *Infiltrometer Test Method:*

4.1.1.1 Infiltrometer test methods measure the rate of infiltration at the soil surface, (see Test Method D 2434), that is influenced both by saturated hydraulic conductivity as well

---

[1] This guide is under the jurisdiction of ASTM Committee D-18 on Soil and Rock and is the direct responsibility of Subcommittee D18.21.02 on Vadose Zone Monitoring.

Current edition approved Oct. 26, 1990. Published December 1990.

[2] The boldface numbers in parentheses refer to a list of references at the end of the text.

[3] *Annual Book of ASTM Standards*, Vol 04.08.

as capillary effects of soil **(4)**. Capillary effect refers to the ability of dry soil to pull or wick water away from a zone of saturation faster than would occur if soil were uniformly saturated. The magnitude of the capillary effect is determined by initial moisture content at the time of testing, the pore size, soil physical characteristics (texture, structure), and a number of other factors. By waiting until steady-state infiltration is reached the capillary effects are minimized.

4.1.1.2 Most infiltrometers generally employ the use of a metal cylinder placed at shallow depths into the soil, and include the single ring infiltrometer, the double ring infiltrometer, and the infiltration gradient method. Various adaptations to the design and implementation of these methods have been employed to determine the field-saturated hydraulic conductivity of material within the unsaturated zone **(5)**. The principles of operation of these methods are similar in that the steady volumetric flux of water infiltrating into the soil enclosed within the infiltrometer ring is measured. Saturated hydraulic conductivity is derived directly from solution of Darcy's Equation for saturated flow. Primary assumptions are that the volume of soil being tested is field-saturated and that the saturated hydraulic conductivity is a function of the flow rate and the applied hydraulic gradient across the soil volume.

4.1.1.3 Additional assumptions common to infiltrometer tests are as follows:

(*a*) The movement of water into the soil profile is one-dimensional downward.

(*b*) Equipment compliance effects are minimal and may be disregarded or easily accounted for.

(*c*) The pressure of soil gas does not offer any impedance to the downward movement of the wetting front.

(*d*) The wetting front is distinct and easily determined.

(*e*) Dispersion of clays in the surface layer of finer soils is insignificant.

(*f*) The soil is non-swelling, or the effects of swelling can easily be accounted for.

4.1.2 *Single Ring Infiltrometer:*

4.1.2.1 The single ring infiltrometer typically consists of a cylindrical ring 30 cm or larger in diameter that is driven several centimetres into the soil. Water is ponded within the ring above the soil surface. The upper surface of the ring is often covered to prevent evaporation. The volumetric rate of water added to the ring sufficient to maintain a constant head within the ring is measured. Alternatively, if the head of water within the ring is relatively large, a falling head type test may be used wherein the flow rate, as measured by the rate of decline of the water level within the ring, and the head for the later portion of the test are used in the calculations. Infiltration is terminated after the flow rate has approximately stabilized. The infiltrometer is removed immediately after termination of infiltration, and the depth to the wetting front is determined either visually, with a penetrometer-type probe, or by moisture content determination for soil samples (see Test Method D 4643).

4.1.2.2 A special type of single ring infiltrometer is the ponded infiltration basin. This type of test is conducted by ponding water within a generally rectangular basin that may be as large as several metres on a side. The flow rate required to maintain a constant head of water within the pond is measured. If the depth of ponding is negligible compared to the depth of the wetting front, the steady state flux of water across the soil surface within the basin is presumed to be equal to the saturated hydraulic conductivity of the soil.

4.1.2.3 Another variant of the single ring infiltrometer is the air-entry permeameter. The air-entry permeameter is discussed in 4.1.4.

4.1.3 *Double Ring Infiltrometer:*

4.1.3.1 The underlying principles and method of operation of the double ring infiltrometer are similar to the single ring infiltrometer, with the exception that an outer ring is included to ensure that one-dimensional downward flow exists within the tested horizon of the inner ring. Water that infiltrated through the outer ring acts as a barrier to lateral movement of water from the inner ring. Double ring infiltrometers may be either open to the atmosphere, or most commonly, the inner ring may be covered to prevent evaporation. For open double ring infiltrometers the flow rate is measured directly from the rate of decline of the water level within the inner ring for falling head tests, or from the rate of water input necessary to maintain a stable head within the inner ring for the constant head case; for sealed double ring infiltrometers, the flow rate is measured by weighing a sealed flexible bag that is used as the supple reservoir for the inner ring **(6)**.

4.1.3.2 Refer to Test Method D 3385 for measuring infiltration rates in the range of 10^{-2} to 10^{-5} cm/s. A modified double-ring infiltrometer test method for infiltration rates from 10^{-5} to 10^{-8} cm/s is also being developed.

4.1.4 *Double Tube Test Method:*

4.1.4.1 The double tube test method proposed by Bouwer **(6, 7, 8)** has been described by Boersma **(9)** as a means of measuring the horizontal, as well as the vertical, field-saturated hydraulic conductivity of material in the vadose zone.

4.1.4.2 This test method as proposed by Bouwer **(6, 7, 8)** utilizes two coaxial cylinders positioned in an auger hole. The difference between the rate of flow in the inner cylinder and the simultaneous rate of combined flow from in the inner and outer cylinders is used to calculate K_{fs}.

4.1.4.3 A borehole is augured to the desired depth and a hole conditioning device is used to square the bottom of the hole. The hole is then cleaned and a 1 to 2 cm layer of coarse protective sand is placed in the bottom of the hole. An outer tube is then placed in the hole and sunken about 5 cm into the soil. The outer tube is then filled with water and a smaller inner tube is placed at the center of the outer tube. It is then driven into the soil. A top plate assembly (see Fig. 2) consisting of water supply valves and standpipes for the inner and outer cylinders is installed. Water is then supplied to both cylinders. The standpipe for the outer cylinder is allowed to overflow and the standpipe gage for the inner cylinder is set at 0 by adjusting the appropriate water supply values. After an equilibrium period of approximately 1 h, the hole is saturated.

4.1.4.4 After saturation is achieved, the level of fall of water in the inner standpipe, H, is recorded at given time intervals, t. H is recorded at least every 5 cm, for a total of at least 30 cm (Test 2). During this test, water in the outer standpipe remains at a constant head.

4.1.4.5 After the data is recorded, the inner reservoir is again filled and the inner standpipe water level is set to 0.

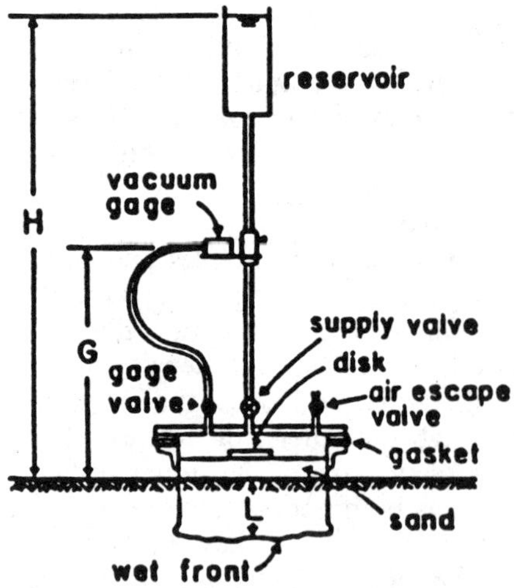

FIG. 1 Diagram of the Equipment for the Air-Entry Permeameter Technique (from Klute, 1986)

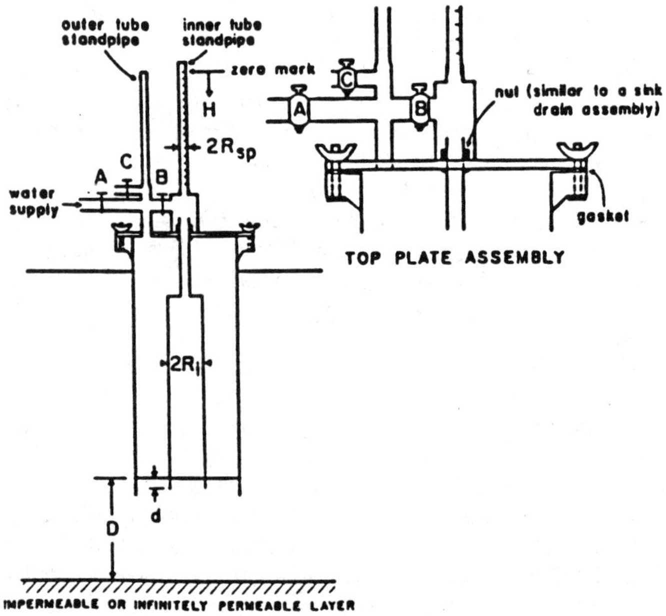

FIG. 2 Diagram of the Equipment Used for Double-Tube Test Method (from Klute, 1986)

The system is allowed to re-equilibrate for a period of time at least ten times as long as the time required to collect the first data set.

4.1.4.6 After waiting, Test 2 is performed. The levels in the outer standpipe and inner standpipe are both brought to 0. Once again the drop in the inner standpipe in cm, H, is recorded as a function of time, t. During the second test, however, water levels in both tubes drop simultaneously. Both tests are then performed a second time or until the results of two consecutive runs are consistent.

4.1.5 *Air-Entry Permeameter:* (Fig 1.)

4.1.5.1 The air-entry permeameter is similar to a single ring infiltrometer in design and operation in that the volumetric flux of water into the soil within a single permeameter ring is used to calculate field-saturated hydraulic conductivity. The primary differences between the two test methods are that the air-entry permeameter typically penetrates deeper into the soil profile and measures the air-entry pressure of the soil. Air-entry pressure is used as an approximation of the wetting front pressure head for determination of the hydraulic gradient, and consequently field-saturated hydraulic conductivity.

4.1.5.2 The air-entry permeameter consists of a single ring, typically 30 cm in diameter, sealed at the top, that is driven into the soil approximately 15 to 25 cm. Water is introduced into the permeameter through a standpipe, to the top of which is attached a water supply reservoir. Water is allowed to infiltrate into the soil within the permeameter ring, and the flow rate is measured by observing the decline of the water level within the reservoir. After a predetermined amount of water has infiltrated (based upon the estimated available storage of the soil interval contained within the ring), and the flow rate is relatively stable, infiltration is terminated and the wetted profile is allowed to drain. The air-entry value is the minimum pressure measured over the standing water inside of the permeameter ring attained during drainage. Once the minimum pressure is achieved, the permeameter is removed, and the depth to the wetting front is determined **(10)**.

4.1.6 *Borehole Permeameter:*

4.1.6.1 Borehole permeameter test methods encompass a wide range of test designs, methods of operation, and methods of solution. The common feature among the different types of borehole tests is that the rate of water infiltration into a cylindrical borehole is used to determine field-saturated hydraulic conductivity. One of the most popular borehole infiltration tests is the constant-head borehole infiltration test, wherein the flow rate necessary to maintain a constant water level within a borehole is measured. The steady state flow rate, borehole geometry, borehole radius (r), and depth of ponding within the borehole (h), and along with certain capillary parameters are typically used in the solution. Hence, by accounting for capillary effects, borehole test methods attempt to measure field-saturated hydraulic conductivity rather than infiltration rate. Another variation of this test consists of conducting multiple constant head borehole infiltration tests within with the same borehole. Different water levels are established within the borehole for each individual test. Results from one or more tests at different ponded heights are solved simultaneously to independently find hydraulic conductivity and capillarity.

4.1.6.2 Borehole infiltration tests are the only currently available tests which can measure field-saturated hydraulic conductivity at depth within the unsaturated zone. Borehole tests may be conducted at great depth within the unsaturated zone, and are frequently used to measure the variability of conductivity with depth by conducting tests at selected horizons within an advancing borehole.

4.1.6.3 During constant head borehole tests water is introduced into a cylindrical borehole and maintained at a predetermined level. This may be accomplished by use of a float valve connected to an external water supply reservoir, or with a Mariotte-siphon device **(2, 10)**. The flow rate into the borehole necessary to maintain the water at the prescribed level is measured at various times. The flow rate at steady state is used in the solution of field-saturated hydraulic conductivity. The dimensions and geometry of the borehole and the depth to the water table are also required for the solution.

4.1.7 *Empirical Methods—Saturated Hydraulic Conductivity:*

4.1.7.1 A number of empirical methods have been developed for estimation of hydraulic conductivity from grain size data (Shepard **(11)**). Shepard suggested that hydraulic conductivity could be predicted from the following:

$$K = cd^a$$

where:

c = a dimensionless constant found through regression analysis,
d = the mean pore throat or particle diameter, and
a = an exponent generally ranging from 1.65 to 1.85.

4.1.7.2 Values for c and a were found to vary substantially depending on the degree of sorting of particles and the amount of induration. Both c and a decreased as the degree of sorting became poorer and as the induration increased. The amount of secondary porosity ("structure" in soils, or "fractures" in rock and sediment) is also expected to affect the values for c and a. Estimates of K for a particular value of d varied by nearly three orders of magnitude depending on the choice of values for c and a **(11)**.

4.2 *Test Methods for Measuring Unsaturated Hydraulic Conductivity:*

4.2.1 *Instantaneous Profile Test Method (IP):*

4.2.1.1 Several references describe the IP test method including Watson **(12)**. The relationship between water potential and hydraulic conductivity can be determined by measuring the rate of drainage and water potential and then solving a form of the Richards equation. The Richards equation solves for the change in water content through time for non-steady, uniform unsaturated flow by relating water potential and unsaturated hydraulic conductivity.

4.2.1.2 To conduct an IP test a small basin is constructed in which water is ponded. Neutron access tubing and a nest of tensiometers at varying depths are installed in the center of the basin. Water is ponded in the basin until the wetting front passes the bottom of the horizon being investigated. Movement of the wetting front is detected with a neutron probe. The soil basin is then covered to reduce evaporation and water content and water potential are measured periodically as water drains downward under the influence of gravity.

4.2.2 *Gypsum Crust Test Method:*

4.2.2.1 The gypsum crust test method is similar to infiltrometer methods in that the rate of water flux across an infiltrative surface is measured. A crust composed of varying mixtures of gypsum and coarse sand is poured over the surface of an exposed excavated cylinder of soil. After the crust cures water is ponded on the crust. The presence of the crust causes unsaturated conditions to form in the soil beneath the crust.

4.2.2.2 The cylinder of soil is instrumented with a nest of tensiometers to measure water potential below the gypsum crust. The rate of flux of water necessary to maintain a constant head over the gypsum crust and the diameter of the cylinder is also recorded **(13, 14)**.

4.2.3 *Empirical Test Methods—Unsaturated Hydraulic Conductivity:*

4.2.3.1 A number of empirical test methods have been developed to estimate unsaturated hydraulic conductivity from other hydraulic parameters. Van Genuchten **(15)** and Mualem **(16)** developed methods for predicting unsaturated hydraulic conductivity from the desorption curve (that relates water content to water potential) and from K_s measurements. Reynolds and Elrick **(2)** developed a borehole permeameter method for measuring a fitting parameter used for estimating unsaturated hydraulic conductivity according to a model proposed by Gardner. The fitting parameter is found by solving simultaneous equations developed from borehole water flux data for two ponded heights. The two ponded height test method is discussed further in 6.4. Infiltration data can be used to estimate hydraulic conductivities by solving the Green-Ampt or Philips Eq. **(4)**.

5. Significance and Use

5.1 Saturated hydraulic conductivity measurements are made for a variety of purposes varying from design of landfills, construction of clay liners, to assessment of irrigation systems. Infiltrometers are commonly used where infiltration or percolation rates through a surface or subsurface layer are desired. Evaluation of the rate of water movement through a pond liner is one example of this kind of measurement. Penetration of the liner by a borehole would invalidate the measurement of liner permeability. It has been noted that small-ring infiltrometers are subject to error due to lateral divergence of flow. Therefore, techniques using very large (1 to 2 m diameter) infiltration basins have been recommended for measuring the very slow percolation rates typically required for clay liners. The air-entry permeameter can be used instead of infiltrometer tests to avoid lateral divergence of flow. However, because a cylinder must be driven into the media tested, the actual soil column tested may be disrupted by introduction of the cylinder, especially in structured soils.

5.2 Borehole tests for determining saturated hydraulic conductivity are applicable for evaluating the rate of water movement through subsurface layers. For slowly permeable layers, an accurate method of measuring the rate of water movement into the borehole must be developed. Use of a flexible bag as a reservoir that can be periodically weighed is advisable for these conditions. A number of mathematical solutions for borehole outflow data are available (Stephens et al. **(17)**, Reynolds et al. **(18)**, and Philip **(19)**).

5.3 Information on unsaturated flow rates is needed to design hazardous waste landfills and impoundments where prevention of flow of contaminants into ground water is required. Of the test methods available, the primary differences are cost and resultant bias and precision. The instantaneous profile test method appears to provide very reliable data because it uses a large volume of soil (several cubic metres) and is performed on undisturbed soils in the field. However, a single test can cost several thousand dollars. The gypsum crust test method, although more rapid than the instantaneous profile test method, sacrifices precision of results due to the smaller spatial extent of the tested area. Methods for estimating unsaturated hydraulic conductivity from fundamental soil hydraulic functions like the desorption curve may readily deviate from true values by an order of magnitude, but may be of use where relative differences in permeability between materials or across water content ranges is of interest.

6. Report

6.1 The reporting requirements for each test vary substantially. However, the variability of hydraulic conductivity in soils, and the sensitivity of some test methods to factors such as textural stratifications, anisotropic conditions, changes in temperature or barometric pressure, initial and final water contents, and depth to groundwater, suggest that a detailed description of each test site be recorded. Record the following:

6.1.1 Soil series (for comparison to existing data).

6.1.2 Soil horizon characteristics above and below layer tested (to help interpret deviations from theoretical response).

6.1.3 Initial and final water content (measure or describe subjectively depending upon method and to identify which numerical solution is most applicable).

6.1.4 General climatic conditions (for example, barometric pressure, temperature, precipitation, cloud cover to estimate possible evaporation, pressure responses, accumulation of prescription that might bias results).

6.1.5 Diameter of borehole, or infiltration ring (parameter used in solution).

6.1.6 Rate of outflow, infiltration, or drainage (parameter used in solution).

6.1.7 Water potential (tensiometer) readings as required (parameter used in solution).

6.1.8 Temperature of water used.

6.1.9 Chemical composition of water used.

6.2 *Infiltrometer Tests:*

6.2.1 Infiltrometer tests are useful for measuring the rate of infiltration but do not provide a direct measure of field-saturated hydraulic conductivity. Since entrapped air exists within the wetting front, true saturated conditions do not form during infiltration tests. Experience indicates that field saturated K_{fs} is approximately 50 to 75 % less than K_s **(1, 2)**.

6.2.2 Infiltration data can be fitted to empirical models such as those developed by Green and Ampt and Philip (described by Bouwer **(4)**).

$$I = S_i t^{1/2} + At$$

where:

I = cumulative infiltration (cm of H_2O),
S_i = sorbtivity of soil (determined from plot of cumulative infiltration against $t^{1/2}$),
t = time increment in seconds, and
A = approximates ½ K_{fs}.

6.3 *Air-Entry Permeameter:*

6.3.1 As soon as minimum pressure is reached, air begins to bubble up through the wetting front. Field-saturated K_{fs} can be calculated from the critical "air-entry value" or minimum pressure. Field-saturated K_{fs} is approximately equal to ½ of K_s in most soils or ¼ of K_s in fine-textured (clayey) soils.

6.3.2 Field saturated K_{fs} is calculated (from Amoozegar and Warrick **(12)**) as follows:

$$K_{fs} = L(dH/dt)(R/Rc)^2/\ (H + L - (P/2\ pg))$$

where:

K_{fs} = field-saturated hydraulic conductivity (cm/s),
L = depth of wetting front (cm),
H = ponded height of water above the soil (cm),
dH/dt = rate of fall just before water supply was shut off (cm/s),
R/Rc = Radius of the reservoir divided by the cylinder radius, and
$P/2\ pg$ = air entry value (minimum pressure divided by the unit weight of liquid (cm)).

6.4 *Double-Tube Test Method:*

6.4.1 Data from both tests are plotted on a graph of H versus t (H is on the y axis). Due to the decrease in head in the inner tube and the greater head in the outer tube, in Test 1, H decreases more rapidly through time than in Test 2. A curve of H verses t data for Test 2 will lie above the curve for Test 1 because in Test 2 the head is the same in both the inner and outer tubes.

6.4.2 Saturated hydraulic conductivity (K) is calculated using the H versus t graphs and the following equation (Amoozegar and Warrick, **(11)**):

$$K = R^2{}_{sp}\ dHt_1/(FR_i \int_{t0}^{t1} Hdt)$$

where:

R_{sp} = radius inner tube standpipe,
R_i = radius inner tube,
dHt_1 = vertical distance between the two curves (Fig. 2) at $t = t_1$,
$\int_{t0}^{t1} Hdt$ = areas under the lower curve between $t = 0$ and $t = t_1$, and
F = a dimensionless quantity dependent on the geometry of the flow system. See Fig. 3 (nomograph).

6.5 *Borehole Permeameter Test Methods:*

6.5.1 Unlike the previous described infiltrometer and permeameter test methods, borehole permeameters account for three-dimensional flow as a result of lateral, as well as downward, flow components. The actual configuration of the flow field around the borehole is highly dependent on the geometry of the borehole, the hydraulic properties of the soil and the capillary suction of the soil. Many of the earlier solutions for falling-head and constant-head type borehole tests ignore the effects of unsaturated flow away from the borehole. Several authors (Glover **(20)**); U.S. Bureau of

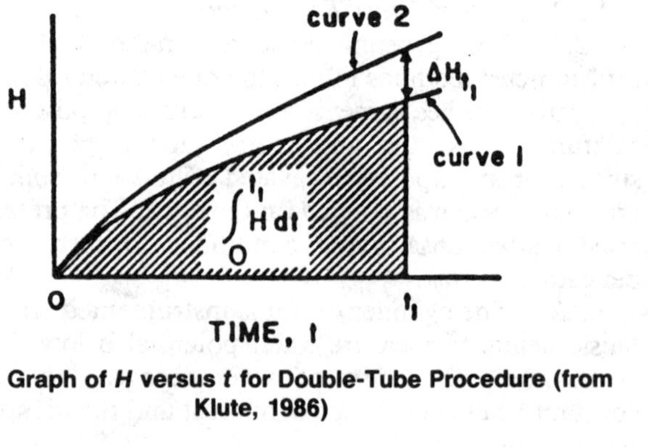

FIG. 3 Graph of *H* versus *t* for Double-Tube Procedure (from Klute, 1986)

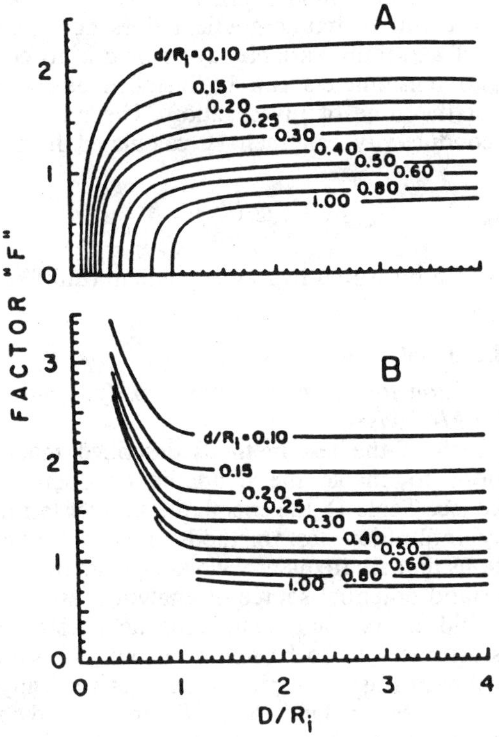

FIG. 4 Values of *F* for the Double-Tube Test Method, (*A*) An Impermeable Layer Below the Hole; and (*B*) An Infinitely Permeable Layer Below the Hole (from Klute, 1986)

Reclamation (**21**); have proposed borehole test methods that are entirely dependent on "free surface" solutions that ignore capillarity. More recently, Stephens et al. (**17**); Philip (**19**); and Reynolds and Elrick (**18**), have shown that unsaturated flow can greatly affect the infiltration rate from a borehole—especially in fine-textured soils, and must be considered in the solution for hydraulic conductivity. Each of these workers has proposed testing methods and/or solutions which account for unsaturated flow away from a wetted bulb around the borehole.

6.5.2 The solution methods of Stephens et al. (**17**) and Philip (**19**) require that certain capillary parameters be either determined separately or be estimated based on soil texture.

6.5.3 The methods of solution proposed by Stephens et al. (**17**) account for capillary effects and are based on multivariate regression equations developed from numerical simulations. Capillary parameters are determined from a catalog of soil hydraulic properties based on soil texture (for example, Mualem (**16**)), or by a fit to moisture retention curves using a model developed by Van Genuchten (**16**).

6.5.4 The Philip (**19**) method is an approximate quasi-analytical solution that accounts for unsaturated flow from a borehole. The solution is based on an approximation of the borehole geometry as an elongate half-spheroid. The capillary parameter must be either known *a priori*; or estimated from a catalog of soil hydraulic properties based on soil texture.

6.5.6 Reynolds and Elrick (**18**) described an analytical solution for borehole permeameter data that involves a simultaneous solution for data collected at two different ponded heights. This approach was found to be sensitive to

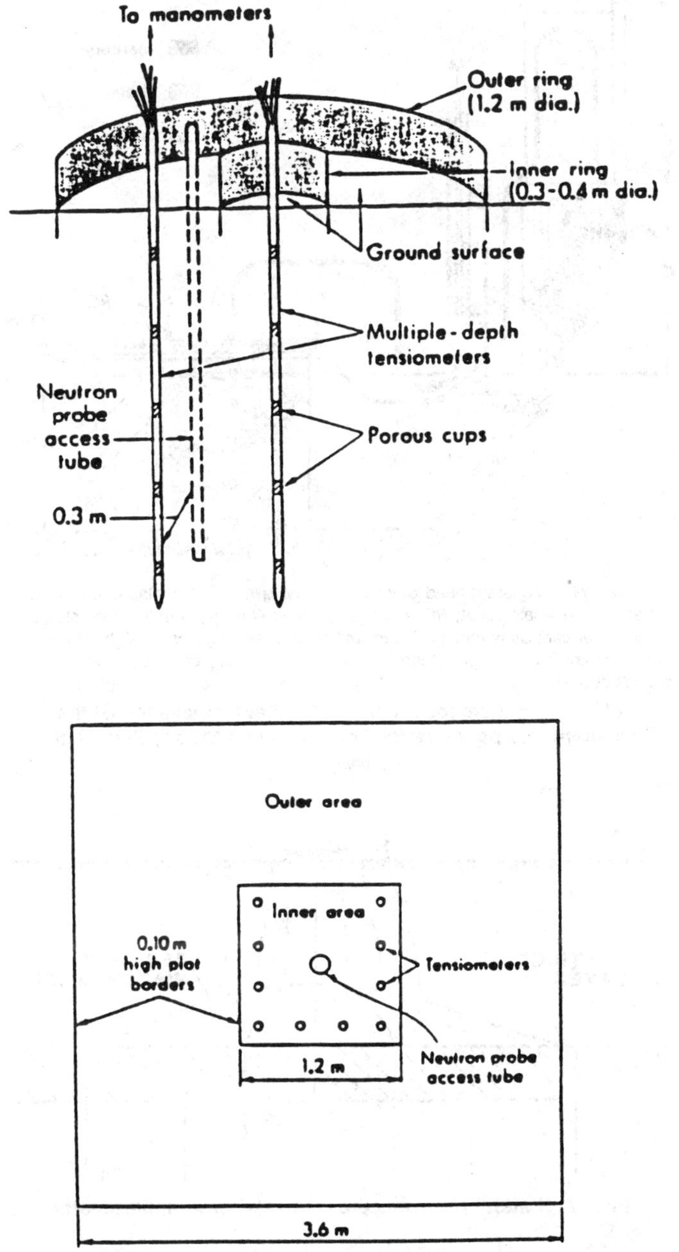

FIG. 5 Sectional and Plan View of Double-Ring Infiltrometer With Instruments Installed for Unsteady Drainage-Flux Method (from Klute, 1986)

slight field measurement error and to texturally stratified systems with the result that negative values for K_{fs} are frequently obtained. Reynolds and Elrick (**18**) suggested an alternative analytical solution where capillary effects are estimated based on soil texture and structure.

6.6 *Instantaneous Profile (IP) Test Method:*

6.6.1 A detailed description of calculating unsaturated hydraulic conductivity (or diffusivity) for different depth increments is provided in Green and others (**22**). Graphical plots of tensiometric data, and soil water content data through time are used to estimate instantaneous water flux at known levels of water content and water potential. An alternative analytical solution was described by Hillel (**23**). Unsaturated hydraulic conductivity data are subject to

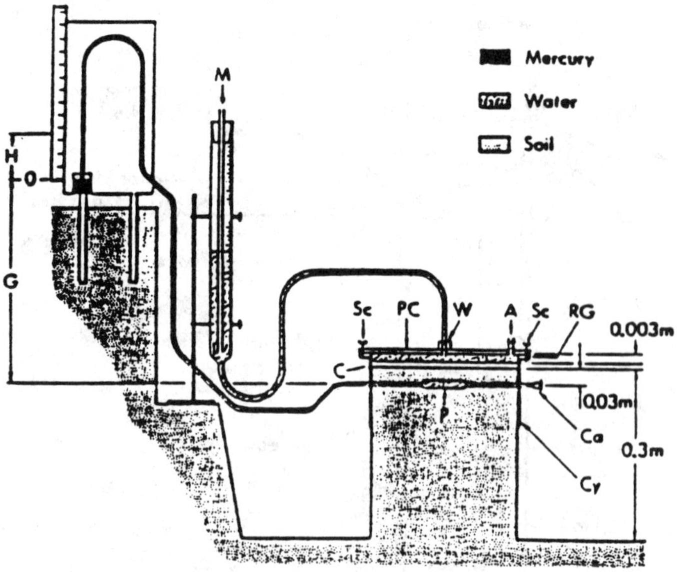

NOTE—M = constant-head device; Sc = wing nut; PC = plastic cover; W = water inlet; A = air outlet; RG = rubber gasket; C = gypsum-sand crust; Ca = tensiometer cap; Cy = metal cylinder with sharpened edge; H = height of mercury column above mercury pool; and H = height of mercury pool above tensiometer porous cup, P.

FIG. 6 Schematic Diagram of a Field Installation of the Measurement Apparatus for Crust-Imposed Steady Flux Method

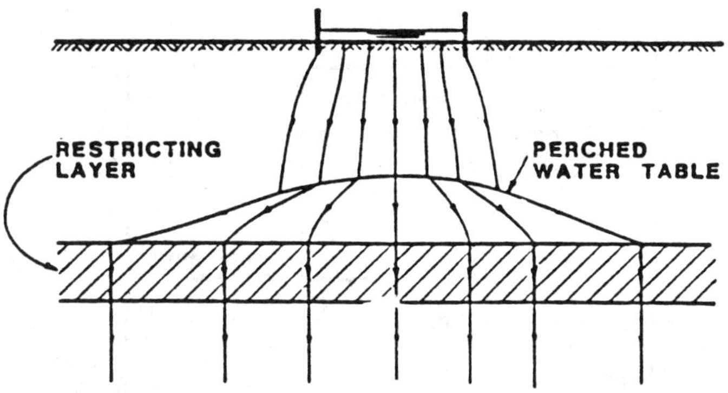

FIG. 7 Lateral Divergence of Flow Below an Infiltrometer

hysteresis. The IP test method provides data from the desorption loop.

6.7 *Gypsum Crust Test Method:*

6.7.1 The gypsum crust test method yields a single measurement of unsaturated hydraulic conductivity as a function of measured water potential for each crust constructed. The unsaturated hydraulic conductivity values are associated with the absorption loop rather than the desorption loop obtained with drainage methods.

6.7.2 In the crust test method a steady unsaturated flux of water is attained with a unit hydraulic gradient (influenced only by gravity). Under these conditions the measured water flux is equal to the hydraulic conductivity:

$$K(h) = Q/A$$

where:

$K(h)$ = unsaturated hydraulic conductivity at head = h,
Q = outflow of water (cm^3/s), and
A = area of soil cylinder (cm^2).

6.7.3 If a unit hydraulic gradient does not form due to presence of texturally variable layers or due to compacted zones, two tensiometers can be installed at the top and bottom of the zone of investigation. The unsaturated hydraulic conductivity can then be calculated from the following:

$$K(h) = Q/(dH/dz)A$$

where:

dH/dz = the measured hydraulic gradient (unitless).

7. Precision and Bias

7.1 *Precision and Bias of Saturated Hydraulic Conductivity Test Methods:*

7.1.1 Each of the test methods described make certain assumptions to enable the hydraulic conductivity to be calculated (see Table 1). In general, the simpler test methods, especially infiltration test methods, rely on many more assumptions that are frequently violated. Care must be taken to understand potential source of analytical error and bias, and to avoid errors while conducting field tests. Most test methods assume soils to be homogeneous (the same in all directions) or at least isotropic (no changes vertically) that is seldom the case in field soils. Errors caused by non-homogeneous or anisotropic conditions vary from test to test. Appropriate test methods for a particular situation should be chosen based on the cost of testing, the bias and precision required, the depth at which a layer is to be tested, the characteristics of the soil profile (uniform or layered), and the approximate hydraulic conductivity range expected.

7.1.1.1 The accuracy of hydraulic conductivity tests is highly dependent on the spatial variability of soils or sediments to be tested. Studies indicate that hydraulic properties of field soils are highly variable (Neilsen et al. **(24)**) and that numerous readings (at closely spaced locations) would typically be needed to characterize a "site" or field-sized area. Field-saturated hydraulic conductivity values tend to be log-normally distributed rather than normally distributed meaning that a majority of the net flux of water may occur in a few permeable spots **(24)**.

7.1.1.2 Infiltrometer test methods should be used cautiously if the saturated hydraulic conductivity is to be determined. Infiltration is affected by both hydraulic conductivity as well as by capillary effects. Infiltration measurements are sensitive to: disruption of the infiltration surface (for example, compaction, sealing by rain splash); presence of textural stratification; chemistry of the water used; and water temperature. Water that is low in salts or high in sodium is dispersive and may result in lower calculated values of K_{fs}.

7.1.2 *Single-Ring Infiltrometer:*

7.1.2.1 The single ring infiltrometer is subject to divergent flow due to the effects of unsaturated flow heterogeneities, and anisotropy. Because with most applications the wetting front is allowed to propagate below the bottom of the ring. These effects may lead to inaccuracies in the determination of saturated hydraulic conductivity.

7.1.3 *Double-Ring Infiltrometers:*

7.1.3.1 As with single ring infiltrometers the wetting front is allowed to advance below the bottom of the ring, but it's

TABLE 1 Review and Comparison of Test Methods for Measuring Hydraulic Conductivity in the Vadose Zone

Characteristics	Single Ring Infiltrometer (4)	Double Ring Infiltrometer (4)	Double-Tube Test Method (8)	Air-Entry Permeameter (10)	Borehole Permeameter Methods: Free Surface (20)	Borehole Permeameter Methods: Capillarity Fixed (2)	Borehole Permeameter Methods: Capillarity Predicted (16)	Borehole Permeameter Methods: 2-Head Simultaneous Solution (2)	IP (21)	Crust (21)	Empirical (14, 15)
Relative accuracy	Low	Fair	Fair	Good	Poor	Good	Good	Variable	Good	Good	Low
Relative cost	Low	Low-Moderate	Moderate	Moderate	.... Low to Moderate				High	High	Low
Time required (at K_{fs} = 10^{-5} cm/s)	<4 h	<4 h	4 h to 1 day	<4 h	.... <4 h				1 day	4 h to 1 day	4 h
Depth of Testing Possible	Surface	Surface	0 to 1 ft	0 to 1 ft	.... Any				0 to 5 ft	0 to 1 ft	Any
Range of K_{fs} (cm/s) for which test is suited	10^{-2} to 10^{-6}	10^{-2} to 10^{-6} 10^{-6} to 10^{-8} (with flexible bag for inner reservoir)	$<10^{-6}$	$<10^{-8}$	.... $<10^{-6}$ .... $<10^{-8}$ (with precautions for temperature effects on reservoir volume				—	—	—
Advantages	Simple apparatus, rapid, can estimate K_{is} from infiltration data, can increase diameter to reduce scale effects and edge effect	Similar to single ring	....	Measures vertical K_{fs} only, accounts for capillary effects	Simple numerical solution, good approximation for sands	Accounts for capillarity effects	—	Simple solution, accounts for capillarity	Excellent method for deriving K (unset) curve	Good for values of Y near zero	Simple, rapid
Drawbacks	Lateral flow affects accuracy, measures infiltration not K_{fs}, surface crust reduces infiltration, measured on surface of soil only	Similar to single ring	Cumbersome apparatus, time-consuming numerical solution	Sometimes difficult to drive tube, difficult to identify wetting front in wet soil	Does not account for capillary effects, high error for medium to fine unstructured soil	Must assume ratio of capillary to flux effects, difficult to predict	Requires description data	Occasionally gives negative values	Time-consuming, affected by barometric changes	Time-economical, only one value of K(unset) and water potential per test	Low accuracy

assumed that infiltration through the outer ring functions as an effective barrier to lateral flow beneath the ring. However, the accuracy of this assumption may be limited.

7.1.3.2 Bouwer **(4)** discussed the ratio of the inner ring to the outer ring necessary to maintain vertical flow in the inner ring. He suggested that an error of several hundred percent can occur unless cylinders of very large diameter are used because of edge effects. For large diameter (1 m or more) cylinders the "edge" effects become small enough that a "double-ring" system is not necessary. Edge effects are not corrected through use of a double-ring infiltrometer though they are somewhat reduced. Bouwer **(4)** mentioned that true vertical infiltration below a ring infiltrometer only occurs after a surface soil crust forms that limits the rate of water intake. Hence, the rate of infiltration thus measured does not represent fully saturated flow.

7.1.4 *Air-Entry Permeameter:*

7.1.4.1 The same restrictions and assumptions apply for the air-entry permeameter as for the infiltrometer test methods. However, since the wetting front is not allowed to advance below the bottom of the permeameter ring, one-dimensional vertically downward flow is ensured. In addition, since the hydraulic gradient is measured during the test, the infiltration rate need not necessarily reach steady state during the first portion of the test. One potential problem with the air-entry permeameter test method is determining the depth of the wetting front after completion of the test. Visual determination is especially difficult in soils with higher initial moisture content.

7.1.5 *Double-Tube Test Method:*

7.1.5.1 Depending on the permeability of the soil, the double-tube method requires over 200 L of water and 2 to 6 h for completion.

7.1.5.2 The test method is not suitable for rocky soils because of the difficulty in driving tubes into the ground. Due to soil disturbance around the inner tube, the diameter of the outer tube should be at least two times that of the inner tube (Bouwer and Riel, 1967). An inner tube with a diameter <10 cm is not recommended. The *K* value obtained utilizing this method is affected by both the horizontal and vertical conductivities of soil.

7.1.6 *Borehole Permeameter Test Methods:*

7.1.6.1 Permeameter test methods rely on an accurate measure of steady-state flow. The length of time required to establish steady flow can range from minutes, for small-diameter borehole tests in coarse-textured soils, to months, for large-diameter borehole tests in clays. The analytical solution used will also affect the accuracy of the results. Stephens and others **(25)** compared large and small-diameter borehole test methods to the air-entry permeameter method for several geologic materials. Various analytical solutions were used to find K_{fs} from the borehole data. It was found that accuracy of permeameter test methods is sensitive to the spatial variability of soils both vertically and laterally. Test results are sensitive to the condition of the sidewall of the test borehole. Care should be taken to avoid smearing of the borehole that creates a hydraulic barrier.

7.2 *Precision and Bias of Unsaturated Hydraulic Conductivity Measurements:*

7.2.1 Test methods for unsaturated hydraulic conductivity are subject to the same limitations as methods for saturated hydraulic conductivity. Little comparative information is available concerning precision and bias of hydraulic conductivity tests, hence no one method can be clearly judged to be superior to another.

7.2.2 *Instantaneous Profile Test Method (IP):*

7.2.2.1 The IP test method is thought to be an accurate test of unsaturated hydraulic conductivity. However, it is costly and time-consuming to perform. Errors in measurement of water content or water potential will affect the accuracy of calculated unsaturated hydraulic conductivity. Mercury manometer-type tensiometers are suggested for measuring water potential. Rapid changes in barometric pressure may affect soil water potential readings. As with all methods, stratification within the soil profile being measured will also affect accuracy. Presence of a water table within about three to four feet of the base of the zone of measurement should be avoided **(12)**.

7.2.3 *Crust Test Method:*

7.2.3.1 The crust test method provides a single measurement of unsaturated hydraulic conductivity at a specific water potential which is read off of a tensiometer installed below the crust. The water potentials that evolve below the crust are a function of the crust material used (for example, specific gypsum/sand mixture). A steady-state flow rate must be measured. This may take hours. Tensiometer readings must be made accurately. The geometry of the excavated block of soil is critical to the solution, hence soil cylinders with a consistent diameter must be accurately excavated. This is difficult in highly-structured or rocky soils **(13)**.

8. Keywords

8.1 Air-entry permeameter, air-entry value, borehole permeameter, hydraulic conductivity, infiltrometer, vadose zone monitoring

REFERENCES

(1) Bouwer, H. "Rapid Field Measurement of Air Entry Value and Hydraulic Conductivity of Soil as Significant Parameters in Flow System Analysis" Water Resources Research, Vol 2, No. 4, 1966, pp. 729–738.

(2) Reynolds, D. and Elrick, D. E. "A Method for Simultaneous *In-Situ* Measurement in the Vadose Zone of Field-Sa Hydraulic Conductivity, Sorptivity and the Conductivity-Pressure Head Relationship" Groundwater Monitoring Review, Vol 6, No. 4, 1986, pp. 84.

(3) *Glossary of Science Terms*, Soil Science Society of America, Madison, WI, 1987.

(4) Bouwer, H. "Intake Rate: Cylinder Infiltrometer." Methods of Soil Analysis, Part 1: Physical and Mineralogical Methods, Agronomy Monograph No. 9, American Society of Agronomy, Madison, WI, 1986, pp. 825–844.

(5) Philip, J. R. "The Theory of Infiltration: The Infiltration Equation and its Solution" Soil Science, 1957, 83:345–357.

(6) Bouwer, H. "A Double Tube Method for Measuring Hydraulic Conductivity of Soil in Sites Above a Water Table" Soil Science Soc. Amer. Proc. 25:334–342, 1961.

(7) Bouwer, H. "Field Determination of Hydraulic Conductivity Above a Water Table with a Double-Tube Method" Soil Science

Soc. Amer. Proc. 26:330–335, 1962.

(8) Bouwer, H. "Measuring Horizontal and Vertical Hydraulic Conductivity of Soil with the Double-Tube Method" Soil Science Soc. Amer. Proc. 28:19–23, 1964.

(9) Boersma, L. "Field Measurement of Hydraulic Conductivity Below a Water Table" Methods of Soil Analysis Part 1: Physical and Mineralogical Methods. Agronomy Monograph No. 9. American Society of Agronomy, Madison, WI, 1965.

(10) Amoozegar, A. and Warrick A. W. "Hydraulic Conductivity of Saturated Soils—Field Methods" Methods of Soil Analysis Part 1: Physical and Mineralogical Methods, Agronomy Monograph 9, American Society of Agronomy, Madison, WI, 1986.

(11) Shepard, R. G. Correlations of Permeability and Grain Size, Ground Water. 27(5):633–638, 1989.

(12) Watson, K. K. "An Instantaneous Profile Method for Determining the Hydraulic Conductivity of Unsaturated Porous Materials" Water Resources Res. 2:709–715, 1966.

(13) Bouma, J., Hillel, D. I., Hole, F. D., and Amerman, C. R. "Field Measurement of Hydraulic Conductivity by Infiltration through Artificial Crusts" Soil Science Soc. Amer. Proc. 33:362–344, 1971.

(14) Bouma, J. and Denning, J. C. "Field Measurement of Unsaturated Hydraulic Conductivity by Infiltration through Gypsum Crusts" Soil Science Soc. Amer. Proc. 36:846–847, 1972.

(15) Van Genuchten, M. T. "A Closed Form Equation for Predicting the Hydraulic Conductivity of Unsaturated Soil" Soil Sci. Soc. Amer. J. 44:892–898, 1980.

(16) Mualem, Y. "A New Model for Predicting the Hydraulic Conductivity of Unsaturated Porous Media" Water Resources Res. 12:513–522, 1976.

(17) Stephens, D. B., Lambet, K., and Watson, D. "Regression Models for Hydraulic Conductivity and Field Test of the Borehole Permeameter" Water Resources Res. 23:2207–2214, 1987.

(18) Reynolds, W. D. and Elrick, D. E. "A Laboratory and Numerical Assessment of the Guelph Permeameter Method" Soil Science 144:282–299, 1987.

(19) Philip, J. R. "Approximate Analysis of the Borehole Permeameter in Unsaturated Soil" Water Resour. Res. 21(7):1025–1033, 1985.

(20) Glover, R. E. "Flow from a Test-Hole Located Above Groundwater Level" U.S. Bur. Rec. Eng. Meng. 8:69–71, 1953.

(21) Bureau of Reclamation. Drainage Manual U.S. Govt. Print Offc. Wash. DC, 1978, pp. 74–97.

(22) Green, R. E., Ahuja, L. R., and Chong, S. K. "Hydraulic Conductivity, Diffusivity, and Sorptivity of Unsaturated Soils: Field Methods." Methods of Soil Analysis Part 1: Physical and Mineralogical Methods. Agron. Mono. No. 9. American Soc. of Agron. Madison, WI, 1986, pp. 771–798.

(23) Hillel, D. Fundamentals of Soil Physics. Academic Press. New York, 1980.

(24) Neilsen, D. R., Biggar, J. W., and Erh, K. T. "Spatial Variability of Field-Measured Soil Water Properties: Hilgardia 42:215–260, 1973.

(25) Stephens, D. B., Unruh, M., Havlena, J., Knowlton, R. G., Mattson, E., and Cox, W. "Vadose Zone Characterization of Low-Permeability Sediments Using Field Permeameters" Ground Water Monitoring Res. Vol 8, No. 2, 1988, pp. 59–66.

Designation: D 5254 – 92

Standard Practice for Minimum Set of Data Elements to Identify a Ground-Water Site[1]

This standard is issued under the fixed designation D 5254; the number immediately following the designation indicates the year of original adoption or, in the case of revision, the year of last revision. A number in parentheses indicates the year of last reapproval. A superscript epsilon (ε) indicates an editorial change since the last revision or reapproval.

1. Scope

1.1 This practice specifies what information should be obtained for any individual ground-water site, also known as monitoring location or sampling station. As used in this practice, a site is meant to be a single point, not a geographic area or property. A ground-water site is defined as any source, location, or sampling station capable of producing water or hydrologic data from a natural stratum from below the surface of the earth. A source or facility can include a well, spring or seep, and drain or tunnel (nearly horizontal in orientation). Other sources, such as excavations, driven devices, bore holes, ponds, lakes, and sinkholes, that can be shown to be hydraulically connected to the ground water, are appropriate for the use intended (see 6.4.2.3).

NOTE 1—There are many additional data elements that may be necessary to identify a site, but are not included in the minimum set of data elements. An agency or company may require additional data elements as a part of their minimum set.

1.2 This practice includes those data elements that will distinguish a site as to its geographical location on the surface of the earth, political regimes, source identifiers, and individual site characteristics. These elements apply to all ground-water sites. Each category of site, such as a well or spring, may individually require additional data elements to be complete. Many of the suggested components and representative codes for coded data elements are those established by the Water Resources Division of the U.S. Geological Survey and used in the National Water Information Systems computerized data base **(1)**.[2]

NOTE 2—The data elements presented in this practice do not uniquely imply a computer data base, but rather the minimum set of ground-water data elements that should be collected for entry into any type of permanent file.

1.3 The values stated in SI units are to be regarded as the standard. The inch-pound units given in parentheses are for information only.

1.4 *This standard does not purport to address all of the safety problems, if any, associated with its use. It is the responsibility of the user of this standard to establish appropriate safety and health practices and determine the applicability of regulatory limitations prior to use.*

2. Referenced Documents

2.1 *ASTM Standards:*

D 653 Terminology Relating to Soil, Rock, and Contained Fluids[3]

3. Terminology

3.1 *Definitions:*

3.1.1 For definitions of terms applicable to this practice refer to Terminology D 653.

4. Summary of Practice

4.1 This practice includes the following data elements to identify a ground-water site:

4.1.1 *Geographic Location*—Including latitude, longitude, latitude-longitude coordinate accuracy, altitude, and altitude accuracy.

4.1.2 *Political Regimes*—Including state or country identification, and county or county equivalent.

4.1.3 *Source Identifiers*—Including owner's name, source agency or company and address, unique identification, and date of first record for the ground-water site.

4.1.4 *Individual Site Characteristics*—Including hydrologic unit, setting, type of ground-water site, use of site, use of water from site, reason for data collection.

5. Significance and Use

5.1 Normally, the basic ground-water data are gathered by trained personnel during the field investigation phase of a study. Each agency or company has its own methods of obtaining, recording, and storing the information. Usually, these data are recorded onto forms that serve both in organizing the information in the field and the office, and many times as entry forms for a computer data base. For ground-water data to be of maximum value to the current project and any future studies, it is essential that a minimum set of key data elements be recorded for each site. The data elements presented in this practice do not uniquely imply a computer data base, but rather the minimum set of ground-water data elements that should be collected for entry into any type of permanent file.

5.2 When obtaining basic data concerning a ground-water site, it is necessary to identify thoroughly that site so that it may be readily field located again with minimal uncertainty and that it may be accurately plotted and interpreted for data parameters in relationship to other sites. For example,

[1] This practice is under the jurisdiction of ASTM Committee D-18 on Soil and Rock and is the direct responsibility of Subcommittee D18.21 on Ground Water and Vadose Zone Investigations.

Current edition approved July 15, 1992. Published November 1992.

[2] The boldface numbers given in parentheses refer to a list of references at the end of the text.

[3] *Annual Book of ASTM Standards*, Vol 04.08.

information can be presented on scientific maps and in summary tables.

6. Documentation

6.1 *Geographic Location:*

6.1.1 *Introduction*—The universally accepted coordinates defining the absolute two-dimensional location of a site on the Earth's surface are latitude and longitude. The coordinates are determined by careful measurement from an accurate map or by survey. The third-dimension of the location is established by determining the altitude at the site, usually from topographic maps or by surveying techniques **(2)**.[4]

NOTE 3—If sites are located by plane coordinates, plant location grids, or referenced to recoverable benchmarks, they may be recorded if the position is converted to absolute location coordinates by an acceptable method.

6.1.2 *Documentation Procedures:*

6.1.2.1 *Latitude*—Latitude is a coordinate representation that indicates locations on the surface of the earth using the earth's equator as the respective latitudinal origin. Record the best available value for the latitude of the site in degrees, minutes, seconds and fractions of a second (DDMMSSss). If latitude of the site is south of the Equator, precede the numbers with a minus sign (−). The use of *N* or *S* is also appropriate **(1, 2, 3, 4, 5, 6, 7, 8)**.

6.1.2.2 *Longitude*—Longitude is a coordinate representation that indicates locations on the surface of the Earth using the prime meridian (Greenwich, England) as the longitudinal origin. Record the best available value for the longitude of the site, in degrees, minutes, seconds, and fractions of a second (DDDMMSSss). If longitude of the site is measured east of the Greenwich Meridian, precede the numbers with a minus sign (−). The use of *E* or *W* is also appropriate **(1, 2, 3, 4, 5, 6, 7, 8)**.

6.1.2.3 *Latitude-Longitude Coordinate Accuracy*—Record the accuracy of the latitude and longitude values. Suggested coordinate accuracy components and representative codes are as follows **(1, 6, 7, 8)**:

H — The measurement is accurate to ±0.01 s.
U — The measurement is accurate to ±0.1 s.
S — The measurement is accurate to ±1 s.
F — The measurement is accurate to ±5 s.
T — The measurement is accurate to ±10 s.
M — The measurement is accurate to ±1 min.

NOTE 4—Components and corresponding codes listed under data elements, such as latitude-longitude coordinate accuracy and setting, are only suggestions. An agency or company may require additional components to fully describe their ground-water sites. Also, having the data element components written out, for example, "accurate to within 1 s" for the latitude-longitude accuracy, may be preferred to the use of codes. The important factor is that each data element in the "minimum set of data elements" be included with every ground-water site.

6.1.2.4 *Altitude*—Record the altitude of land surface or measuring point. Altitude of the land surface is the vertical distance in feet (or metres) either above or below a reference datum surface. The reference datum surface must be noted.

NOTE 5—In the United States, this reference surface should be the North American Vertical Datum (NAVD) of 1988 or National Geodetic Vertical Datum (NGVD) of 1929. If another vertical reference datum is used to determine the altitude, describe the system. Altitudes below the reference datum must be preceded by a minus sign (−) **(1, 2, 4, 7, 8)**.

NOTE 6—The measuring point is usually a clearly defined mark or permanently fixed object at a ground-water site that is used for conducting repeated evaluations, such as water levels in a monitoring well.

6.1.2.5 *Altitude Accuracy*—Record the accuracy of the altitude. As an example, record 1.0 for an accuracy of ± 1 m or 0.1 for ± 0.1"th" m to denote the judged error of the measurement **(1, 3)**.

6.2 *Political Regimes:*

6.2.1 *Introduction*—The placement of the ground-water site into a political jurisdiction assists in the proper identification of the site.

6.2.2 *Documentation Procedures:*

6.2.2.1 *State or Country Identification*—Record the state or country in which the site is physically located. The common systems for identifying states and countries are the Federal Information Processing Standard code (FIPS), a two-digit numeric code or the American National Standard abbreviation two-letter code. The country codes are a two-character and a set of three-character alphabetic codes **(1, 3, 9, 10, 11)**.

NOTE 7—The publications **(9, 10, 12)** containing the codes for countries, states, and counties are available from the address given in Footnote 4.

6.2.2.2 *County and County Equivalent*—Record the county or county equivalent in which the site is physically located. The common code system for identifying counties is the FIPS code, a three-digit numeric code. The documentation of political subdivisions will depend on the system used in each individual country **(1, 3, 7, 11)**.

NOTE 8—In many cases it is necessary to record a subdivision of the local government to further identify the area where the ground-water site is located. Some of the local subdivisions are a city, town, village, municipality, township, or borough. Identify the local subdivision, for example "City of Rockville", to clearly denote the unit.

6.3 *Source Identifiers:*

6.3.1 *Introduction*—The ground-water site must be identified as to the owner, the agency or company that recorded data, and its distinctive classification.

6.3.2 *Documentation Procedures:*

6.3.2.1 *Owner's Name*—Record the name of the property owner of the ground-water site. The recommended format for an individual's name is: last name, first name, middle initial. If a company's name is lengthy, use meaningful abbreviations **(1, 8)**.

6.3.2.2 *Source Agency or Company and Address*—Record the name and address of the agency or company that collected the data for the ground-water site. This data element is necessary to determine the original source of the data for the site. A coded list of agency and company names is available through National Water Data Exchange (NAWDEX);[5] the list has over 1200 organizations that actively collect and store water data throughout the United States **(1, 3, 4, 5, 6, 12)**.

6.3.2.3 *Unique Identification*—Record the unique naming that the agency or company uses to identify the ground-water

[4] Available from National Technical Information Service, U.S. Department of Commerce, 5285 Port Royal Road, Springfield, VA 22161.

[5] Available from National Water Data Exchange, U.S. Geological Survey, 421 National Center, Reston, VA 22092.

site. This identification is called by several terms such as "local site number", "site identification", "well number", etc. The description is commonly a combination of letters and numbers that could represent a land-net location or a sequential assignment for a site in a county, city, or company. This identification is very important to precisely differentiate a site in the records of an agency or company **(1, 5, 6, 7, 8)**.

6.3.2.4 *Date of First Record for the Ground-Water Site*—Record the date that the first valid transaction occurred for any element of the specified site. This could be the date of permit application, start of construction, or first used as a monitoring site. This element is important to facilitate in the proper identification of the record **(1, 3, 12)**.

6.4 *Individual Characteristics of the Site:*

6.4.1 *Introduction*—Each ground-water site has very specific features that, in combination, uniquely identify that site, that is, water from a ground-water sustained pond used for aquaculture. These characteristics should be recorded as a means of further defining the site.

6.4.2 *Documentation Procedures:*

6.4.2.1 *Hydrologic Unit*—Record the hydrologic unit code for the Office of Water Data Coordination (OWDC) cataloging unit in which the site is located. This eight-digit code consists of four 2-digit parts **(1, 4, 5, 6, 13, 14)**: hydrographic region code, subregion code designated by the Water Resources Council, accounting unit within the National Water Data Network, and cataloging unit of the USGS's "Catalog of Information on Water Data".

NOTE 9—An explanation of a hydrologic unit code, for example Code 07080107, is the following; Region Code "07" is the Upper Mississippi River Basin above the confluence with the Ohio River, Subregion Code "08" is the Mississippi River Basin below Lock and Dam 13 to the confluence with the Des Moines River Basin, excluding the Rock River Basins, Accounting Unit Code "01" is the Mississippi River Basin below Lock and Dam 13 to the confluence with the Des Moines River Basin, excluding the Iowa and Rock River Basins, and Catalog Unit Code "07" is the Skunk River Basin of Iowa.

NOTE 10—State hydrologic unit maps delineating the hydrographic boundaries of these units are available[6] (see Ref **(13)**).[4]

6.4.2.2 *Setting*—Record the information that best describes the setting in which the site is located. Setting refers to the topographic or geomorphic features in the vicinity of the site. Suggested setting components and representative codes are as follows **(1, 8)**:

A — Alluvial fan
B — Playa
C — Stream channel
D — Local depression
E — Dunes
F — Flat surface
G — Flood plain
H — Hilltop
I — Inland wetlands
J — River delta
K — Sinkhole
L — Lake
M — Mangrove swamp or coastal wetlands
O — Offshore (estuary)
P — Pediment
S — Hillside (slope)
T — Alluvial or marine terrace
U — Undulating
V — Valley flat (valleys of all sizes)
W — Upland draw
X — Unknown
Y — Wetlands
Z — Other (describe)

6.4.2.3 *Type of Ground-Water Site*—This data element helps to identify the physical type of ground-water site. Record the type of site to which these data apply. Suggested site type components and representative codes are as follows **(1, 8)**:

C — Collector (radial-collector) well
D — Drain dug to intercept the water table or potentiometric surface to either lower the ground-water level or serve as a water supply
E — Excavation
H — Sinkhole
I — Interconnected wells, also called connector or drainage wells; that is, a well interconnected via an underground lateral
M — Multiple wells—Use only for well field consisting of a group of wells that are pumped through a single header and for which little or no data about the individual wells are available
O — Outcrop
P — Pond that intercepts the water table or potentiometric surface
S — Spring
T — Tunnel, shaft, or mine from which ground water is obtained
W — Well, for single wells other than wells of the collector (radial collector) type
X — Test hole, not completed as a well
Z — Other (describe)

6.4.2.4 *Use of Site*—Record the use of the site or the purpose for which the site was constructed (the former always holds precedence over the latter). If site is used for more than one purpose, also record the subordinate uses. Suggested site use components and representative codes are as follows **(1, 4, 7, 8)**:

A — Electrical anode
C — Standby emergency supply
D — Drain or dry well
E — Geothermal—for geothermal extraction or injection
G — Seismic exploration
H — Heat reservoir—fluid circulated in closed system
M — Mine—primary use for extraction of minerals
O — Observation/monitoring
P — Oil or gas well
R — Recharge
S — Repressurize—to increase pressure in aquifer
T — Test—for hydrologic testing
U — Unused
W — Withdrawal of water
X — Waste disposal
Y — Other (describe)
Z — Destroyed

6.4.2.5 *Use of Water from Site*—Record the use of the water from the site. If water from the site is used for more than one purpose, also record the subordinate uses. Suggested water use components and representative codes are as follows **(1)**:

A — Air conditioning
B — Bottling
C — Commercial
D — Dewater
E — Power
F — Fire
G — Hydrogeologic interpretation
H — Domestic
I — Irrigation
J — Industrial (cooling)
K — Mining
L — Chemical screening for contaminants
M — Medicinal
N — Industrial (manufacturing)

[6] Available from USGS Books and Reports Sales Federal Center, P.O. Box 25425, Denver, CO, 80225.

Ground–Water Site Minimum Set of Data Elements	Date prepared:		
GEOGRAPHIC LOCATION			
Latitude:		Accuracy:	
Longitude:		Accuracy:	
ALTITUDE			
Land Surface:	(meters/feet)	Accuracy:	
Other (Specify):	(meters/feet)	Accuracy:	
Altitude Reference Datum:			
POLITICAL REGIMES			
State (or Country) Identification:			
County (or County Equivalent) Identification:			
SOURCE IDENTIFIERS			
Owner's Name:			
Source Agency (or Company):			
Address of Source Agency:			
Unique Identification of Site:			
Date of First Record for Site:			
INDIVIDUAL CHARACTERISTICS OF THE SITE			
Hydrologic Unit:			
Setting:			
Type of Ground–Water Site:			
Use of Site:			
Use of Water from Site:			
Reason for Data Collection:			

FIG. 1 Example of Minimum Set of Data Elements Form

Ground–Water Site Minimum Set of Data Elements	Date prepared: SEPT 4, 1991		
GEOGRAPHIC LOCATION (5.1)			
Latitude:	39° 18' 26"	Accuracy:	± 1 SEC
Longitude:	74° 37' 09"	Accuracy:	± 1 SEC
ALTITUDE (5.1.2.4)			
Land Surface:	10.00 FT (meters/feet)	Accuracy:	0.01 FT
Other (Specify):	9.34 FT RECORDER SHELF (meters/feet)	Accuracy:	0.01 FT
Altitude Reference Datum:	NGVD (1929)		
POLITICAL REGIMES (5.2)			
State (or Country) Identification:	NEW JERSEY		
County (or County Equivalent) Identification:	ATLANTIC COUNTY		
SOURCE IDENTIFIERS (5.3)			
Owner's Name:	U.S. GEOLOGICAL SURVEY		
Source Agency (or Company):	U.S. GEOLOGICAL SURVEY		
Address of Source Agency:	810 BEAR TAVERN ROAD, SUITE 206 WEST TRENTON, N.J. 08628		
Unique Identification of Site:	JOBS POINT OBS. NJ-WRD WELL #01-0578		
Date of First Record for Site:	OCTOBER, 1959		
INDIVIDUAL CHARACTERISTICS OF THE SITE (5.4)			
Hydrologic Unit:	02040302		
Setting:	UNKNOWN; NEAR EXIT 29 - GARDEN STATE PRKWY		
Type of Ground–Water Site:	WELL		
Use of Site:	OBSERVATION/MONITORING		
Use of Water from Site:	HYDROGEOLOGIC INTERPRETATION		
Reason for Data Collection:	HYDROLOGIC BENCHMARK/RESEARCH		

FIG. 2 Example of Filled-Out Minimum Set of Data Elements Form

P — Public supply
Q — Aquaculture
R — Recreation
S — Stock
T — Institutional
U — Unused
Y — Desalination
Z — Other (describe)

6.4.2.6 *Reason for Data Collection*—Record the reason for which data were collected from the site. If the data were collected for more than one purpose, record the subordinate reasons. Suggested data collection components and representative codes are as follows:

A — Construction/dewatering
B — Research
C — CERCLA
R — RCRA
D — Drinking water regulations
E — Exploration (water)
L — Local ordinance
S — State regulations, other than CERCLA or RCRA
F — Federal regulations, other than CERCLA or RCRA
G — Geothermal
H — Hydrologic benchmark
I — Environmental issues
J — Judicial/litigation
M — Mining regulations
N — Natural resources exploration
U — Unknown
Z — Other (describe)

7. Sample Form

7.1 An example of a generalized form for recording a minimum set of data elements for a ground-water site is shown in Fig. 1. An example of a filled-out form is shown in Fig. 2.

8. Keywords

8.1 ground water; ground-water sampling site; hydrologic unit; key data elements; site identification; site location; setting; monitoring location; site coordinates

REFERENCES

(1) Mathey, S. B., ed., *National Water Information System User's Manual*, Vol 2, Chapter 4, "Ground-Water Site Inventory System," U.S. Geological Survey, Open-File Report 89-587, 1990.

(2) U.S. Department of Commerce, Representation of Geographic Point Locations for Information Interchange; *Federal Information Standards (FIPS) Publication 70-1*, National Institute for Standards and Technology, Washington, DC, June 23, 1986.

(3) Perry, R. A., and Williams, O. O., *Data Index Maintained by the National Water Data Exchange*, U.S. Geological Survey, Open-File Report 82-327.

(4) Texas Natural Resources Information System, *Ground-Water Data INTERFACE, Users Reference Manual*, Texas Natural Resources Information System, November 20, 1986.

(5) U.S. Environmental Protection Agency, *STORET Users Handbook*, Vols 1 and 2; U.S. EPA, Washington, DC, February 1982.

(6) U.S. Environmental Protection Agency, *Ground-Water Data Management With STORET*, Office of Ground-Water Protection, U.S. EPA, Washington, DC, March 1986.

(7) U.S. Environmental Protection Agency, *Definitions for the Minimum Set of Data Elements for Ground Water Quality*, U.S. Environmental Protection Agency (Draft), July 22, 1991.

(8) U.S. Geological Survey, *National Handbook of Recommended Methods for Water-Data Acquisition*, Chapter 2, "Ground Water," Office of Data Coordination, Reston, VA, 1980.

(9) U.S. Department of Commerce, "American National Standard Codes for the Representation of Names of Countries, Dependencies, and Areas of Special Sovereignty for Information Interchange," *Federal Information Standards (FIPS) Publication 104-1*, National Institute for Standards and Technology, Washington, DC, May 12, 1986.

(10) U.S. Department of Commerce, "Codes for the Identification of the States, the District of Columbia and Outlying Areas of the United States, and Associated Areas," *Federal Information Standards (FIPS) Publication 5-2*, National Institute for Standards and Technology, Washington, DC, May 28, 1987.

(11) U.S. Department of Commerce, "Counties and Equivalent Entities the United States, Its Possessions, and Associated Areas," *Federal Information Standards (FIPS) Publication 6-4*, National Institute for Standards and Technology, Washington, DC, August 31, 1990.

(12) Edwards, M. D., and Josefson, B. M., "Identification Codes for Organizations listed in Computerized Data Systems of the U.S. Geological Survey," U.S. Geological Survey, Open-File Report 82-921, 1982.

(13) U.S. Geological Survey, "Codes for the Identification of Hydrologic Units in the United States and the Caribbean Outlying Areas," U.S. Geological Survey, Circular 878-A, Reston, VA, (also *FIPS PUB 103*), 1982.

(14) Seaber, P. R., Kapinos, F. P., and Knapp, G. L., State Hydrologic Unit Maps, U.S. Geological Survey, Open-File Report 84-708, Reston, VA, 1984.

Standard Test Method for Determining Transmissivity of Nonleaky Confined Aquifers by the Theis Recovery Method[1]

This standard is issued under the fixed designation D 5269; the number immediately following the designation indicates the year of original adoption or, in the case of revision, the year of last revision. A number in parentheses indicates the year of last reapproval. A superscript epsilon (ϵ) indicates an editorial change since the last revision or reapproval.

1. Scope

1.1 This test method covers an analytical procedure for determining the transmissivity of a confined aquifer. This test method is used to analyze data from the recovery of water levels following pumping or injection of water to or from a control well at a constant rate.

1.2 The analytical procedure given in this test method is used in conjunction with the field procedure in Test Method D 4050.

1.3 *Limitations*—The valid use of the Theis recovery method is limited to determination of transmissivities for aquifers in hydrogeologic settings with reasonable correspondence to the assumptions of the Theis theory (see 5.1).

1.4 The values stated in SI units are to be regarded as standard.

1.5 *This standard does not purport to address all of the safety problems, if any, associated with its use. It is the responsibility of the user of this standard to establish appropriate safety and health practices and determine the applicability of regulatory limitations prior to use.*

2. Referenced Documents

2.1 *ASTM Standards:*

D 653 Terminology Relating to Soil, Rock and Contained Fluids[2]

D 4043 Guide for Selection of Aquifer-Test Methods in Determining Hydraulic Properties by Well Techniques[2]

D 4050 Test Method (Field Procedure) for Withdrawal and Injection Well Tests for Determining Hydraulic Properties of Aquifer Systems[2]

D 4105 Test Method (Analytical Procedure) for Determining Transmissivity and Storage Coefficient of Nonleaky Confined Aquifers by the Modified Theis Nonequilibrium Method[2]

D 4106 Test Method (Analytical Procedure) for Determining Transmissivity and Storage Coefficient of Nonleaky Confined Aquifers by the Theis Nonequilibrium Method[2]

D 4750 Test Method for Determining Subsurface Liquid Levels in a Borehole or Monitoring Well (Observation Well)[2]

3. Terminology

3.1 *Definitions:*

3.1.1 *aquifer, confined*—an aquifer bounded above and below by confining beds and in which the static head is above the top of the aquifer.

3.1.2 *confining bed*—a hydrogeologic unit of less permeable material bounding one or more aquifers.

3.1.3 *control well*—a well by which the aquifer is stressed, for example, by pumping, injection, or change of head.

3.1.4 *drawdown*—vertical distance the static head is lowered due to the removal of water.

3.1.5 *hydraulic conductivity (field aquifer tests)*—the volume of water at the existing kinematic viscosity that will move in a unit time under unit hydraulic gradient through a unit area measured at right angles to the direction of flow.

3.1.6 *observation well*—a well open to all or part of an aquifer.

3.1.7 *piezometer*—a device used to measure head at a point in the subsurface.

3.1.8 *residual drawdown*—The difference between the projected prepumping water-level trend and the water level in a well or piezometer after pumping or injection has stopped.

3.1.9 *specific storage*—the volume of water released from or taken into storage per unit volume of the porous medium per unit change in head.

3.1.10 *step-drawdown test*—a test in which a control well is pumped at constant rates in "steps" of increasing discharge. Each step is approximately equal in duration, although the last step may be prolonged.

3.1.11 *storage coefficient*—the volume of water an aquifer releases from or takes into storage per unit surface area of the aquifer per unit change in head. For a confined aquifer it is equal to the product of specific storage and aquifer thickness. For an unconfined aquifer, the storage coefficient is approximately equal to the specific yield.

3.1.12 *transmissivity*—the volume of water of the prevailing kinematic viscosity transmitted in a unit time through a unit width of the aquifer under a unit hydraulic gradient.

3.2 *Symbols and Dimensions:*

3.2.1 b [L]—aquifer thickness.

3.2.2 K [LT^{-1}]—hydraulic conductivity.

3.2.2.1 *Discussion*—The use of the symbol K for the term hydraulic conductivity is the predominant usage in groundwater literature by hydrogeologists, whereas the symbol k is commonly used for this term in rock mechanics and soil science.

3.2.3 K_r—hydraulic conductivity in the plane of the aquifer, radially from the control well.

3.2.4 K_z—hydraulic conductivity in the vertical direction.

3.2.5 ln—natural logarithm.

---

[1] This test method is under the jurisdiction of ASTM Committee D-18 on Soil and Rock and is the direct responsibility of Subcommittee D18.21 on Ground Water and Vadose Zone Investigations.

Current edition approved July 15, 1992. Published December 1992.

[2] *Annual Book of ASTM Standards*, Vol 04.08.

3.2.6 $\log_{10}$—logarithm to the base 10.
3.2.7 Q [L^3T^{-1}]—discharge.
3.2.8 r [L]—radial distance from control well.
3.2.9 r_c [L]—equivalent inside radius of control well.
3.2.10 S [nd]—storage coefficient.
3.2.11 s [L]—drawdown.
3.2.12 s_c [L]—drawdown corrected for the effects of reduction in saturated thickness.
3.2.13 S_y [nd]—specific yield.
3.2.14 s' [L]—residual drawdown.
3.2.15 $\Delta s'$ [L]—change in residual drawdown over one log cycle of t/t'.
3.2.16 T [L^2T^{-1}]—transmissivity.
3.2.17 t [T]—time since pumping or injection began.
3.2.18 t' [T]—time since pumping or injection stopped.
3.2.19 u—dimensionless parameter, equal to $r^2/4Tt$.
3.2.20 u'—dimensionless parameter, equal to $r^2/4Tt'$.

4. Summary of Test Method

4.1 This test method describes an analytical procedure for determining transmissivity using data collected during the recovery phase of a withdrawal or injection well test. The field test (see Test Method D 4050) requires pumping or injecting a control well that is open to the entire thickness of a confined aquifer at a constant rate for a specified period. The water-levels in the control well, observation wells, or piezometers are measured after pumping is stopped and used to calculate the transmissivity of the aquifer using the procedures in this test method. Alternatively, this test method can be performed by injecting water into the control well at a constant rate. With some modification, this test method can also be used to analyze the residual drawdown following a step test. This test method is used by plotting residual drawdown against either a function of time or a function of time and discharge and determining the slope of a straight line fitted to the points.

4.2 *Solution*—The solution given by Theis **(1)**[3] can be expressed as follows:

$$s = \frac{Q}{4\pi T}\int_u^\infty \frac{e^{-y}}{y}\,dy \tag{1}$$

and:

$$u = \frac{r^2S}{4Tt} \tag{2}$$

4.3 At a control well, observation well, or piezometer, for large values of time, t, and small values of radius, r, the Theis equation reduces, as shown by Cooper and Jacob **(2)** and Jacob **(3)** to the following:

$$s' = \frac{Q}{4\pi T}\,ln(t/t') \tag{3}$$

where:
t = the time after pumping began and
t' = the time after pumping ceases. From which it can be shown that:

$$T = \frac{2.3Q}{4\pi\Delta s'} \tag{4}$$

where:
$\Delta s'$= the measured or projected residual drawdown over one $\log_{10}$ cycle of t/t'.

4.4 A similar analysis (see 4.3) may also be used for a step-drawdown test in which a well is pumped at a constant rate for an initial period, and then the pumping rate is increased through several new constant rates in a series of steps. Harrill **(4)** shows that:

$$s' = \frac{2.3\Delta Q_1}{4\pi T}\left(\log_{10}\frac{t_1}{t'}\right) + \frac{2.3\Delta Q_2}{4\pi T}\left(\log_{10}\frac{t_2}{t'}\right) + \ldots \frac{2.3\Delta Q_n}{4\pi T}\left(\log_{10}\frac{t_n}{t'}\right) \tag{5}$$

where:
$t_1, t_2, \ldots t_n$ = the elapsed times since either pumping was begun or the discharge rate was increased,
$Q_1, Q_2, \ldots Q_n$ = the well discharge rates, and
$\Delta Q_1, \Delta Q_2 \ldots \Delta Q_n$ = the incremental increases in discharge.
Equation 5 can be rewritten as follows:

$$T = \frac{2.3Q_n}{4\pi s'}\log_{10}f(t, Q) \tag{6}$$

where:

$$f(t, Q) = \frac{t_1^{\Delta Q_1/Q_n}t_2^{\Delta Q_2/Q_n}t_3^{\Delta Q_3/Q_n}\ldots t_n^{\Delta Q_n/Q_n}}{t'} \tag{7}$$

and:

$$T = \frac{2.3Q_n}{4\pi\Delta s'_h} \tag{8}$$

where:
$\Delta s'_h$ = the residual drawdown over one log cycle of the expression $f(t, Q)$ in Eq 6.
Equation 8 can also be used to analyze the residual drawdown following a test in which discharge varies significantly, so long as the discharge can be generalized as a series of constant-discharge steps.

5. Significance and Use

5.1 *Assumptions:*

5.1.1 The well discharges at a constant rate, Q, or at steps of constant rate $Q_1, Q_2 \ldots Q_n$.

5.1.2 Well is of infinitesimal diameter and is open through the full thickness of the aquifer.

5.1.3 The nonleaky aquifer is homogeneous, isotropic, and areally extensive.

5.1.4 Discharge from the well is derived exclusively from storage in the aquifer.

5.1.5 The geometry of the assumed aquifer and well are shown in Fig. 1.

5.2 *Implications of Assumptions:*

5.2.1 Implicit in the assumptions are the conditions of radial flow. Vertical flow components are induced by a control well that partially penetrates the aquifer, that is, not open to the aquifer through the full thickness of the aquifer.

[3] The boldface numbers given in parentheses refer to a list of references at the end of the text.

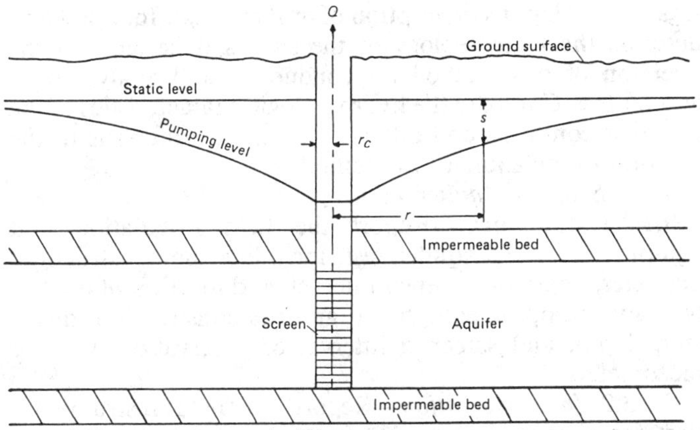

FIG. 1 Cross Section Through a Discharging Well in a Nonleaky Aquifer

If vertical flow components are significant, the nearest partially penetrating observation well should be located at a distance, r, beyond which vertical flow components are negligible. See 5.2.1 of Test Method D 4106 for assistance in determining the minimum distance to partially penetrating observation wells and piezometers.

5.2.2 The Theis method assumes the control well is of infinitesimal diameter. The storage in the control well may adversely affect drawdown measurements obtained in the early part of the test. See 5.2.2 of Test Method D 4106 for assistance in determining the duration of the effects of well-bore storage on drawdown.

5.2.3 *Application of Theis Recovery Method for Unconfined Aquifers:*

5.2.3.1 Although the assumptions are applicable to artesian or confined conditions, the Theis solution may be applied to unconfined aquifers if (*A*) drawdown is small compared with the saturated thickness of the aquifer or if the drawdown is corrected for reduction in thickness of the aquifer and (*B*) the effects of delayed gravity yield are small. See 5.2.3 of Test Method D 4106 for guidance in treating reduction in saturated thickness and delayed gravity drainage in unconfined aquifers.

6. Apparatus

6.1 Analysis of data by this test method from the field procedure given in Test Method D 4050 requires that the control well and observation wells meet the requirements specified in the following subsections.

6.2 *Construction of Control Well*—Install the control well in the aquifer and equip with a pump capable of discharging water from the well at a constant rate, or several steps at constant rate, for the duration of the test. Preferably, the control well should be open throughout the full thickness of the aquifer. If the control well partially penetrates the aquifer, take special precautions in the placement or design of observation wells (see 5.2.1).

6.3 *Construction of Observation Wells and Piezometers*—Construct one or more observation wells or piezometers at a distance from the control well. Observation wells may be open through all or part of the thickness of the aquifer.

6.4 *Location of Observation Wells and Piezometers*—Wells may be located at any distance from the control well within the area of influence of pumping. However, if vertical flow components are significant and if piezometers or partially penetrating observation wells are used, locate them at a distance beyond the effect of vertical flow components. If the aquifer is unconfined, constraints are imposed on the distance to partially penetrating observation wells and the validity of early time measurements (see 5.2.1).

7. Procedure

7.1 The overall procedure consists of conducting the field procedure for withdrawal or injection well tests (described in Test Method D 4050) and analysis of the field data, which is addressed in Section 8.

8. Calculation and Interpretation of Results

8.1 The Theis recovery method gives satisfactory results when properly used. However, the method is valid only for values of $u' < 0.01$, that is:

for confined aquifers:

$$u' = \frac{r^2S}{4Tt'} < 0.01 \tag{9}$$

or for unconfined aquifers:

$$u' = \frac{r^2S_y}{4Tt'} < 0.01 \tag{10}$$

NOTE—The limiting value for u of less than 0.01 may be excessively restrictive in some applications. The errors for small values of u, from Kruseman and De Ridder (**5**) are:

Error less than, %	1	2	5	10
For u smaller than	0.03	0.05	0.1	0.15

8.1.1 This test method allows only the calculation of transmissivity, T, not storage coefficient, S, or specific yield, S_y. Therefore, to determine whether the assumption in Eqs 9 or 10 has been violated it is necessary to estimate a value for storage coefficient for confined aquifers or specific yield for unconfined aquifers. If data are available during the pumping period, the storage may be computed using the procedures in Test Method D 4105. Storage coefficients can be estimated as about $10^{-6}b$, where b is the aquifer thickness, whereas the specific yield of unconfined aquifers averages about 0.2 according to Lohman (**6**). After calculating T, substitute the appropriate values into Eqs 9 or 10 and solve for u'. It is not adequate to simply note that the data described a straight line on semi-log graph paper.

8.2 Plot either residual drawdown, s', or water level, on the arithmetic axis of semilogarithmic graph paper versus either t/t' (for recovery from a constant-discharge test) (see Fig. 2) or $f(t, Q)$ (for recovery from a step-drawdown test) (see Fig. 3) on the logarithmic axis. Fit a straight line to the linear part of the data plot, usually at smaller values of t/t'. Extend the straight line to intercept the $t/t' = 1$ axis. At $t/t' = 1$, residual drawdown should be approximately equal to zero, or if water levels were plotted, the intercept should be equal to the prepumping water levels corrected for prepumping water-level trends. Substitute the values for $\Delta s'$ or $\Delta s'_h$ in Eqs 4 or 8 and solve for transmissivity. Check that all values of t' for the points used in defining the straight line meet the criterion that $u' < 0.01$ (Eqs 9 and 10), as described in 8.1.

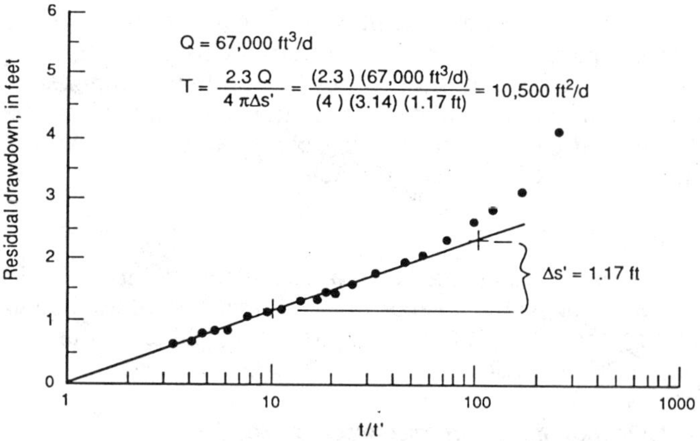

FIG. 2 Example Analysis Using the Theis Recovery Method

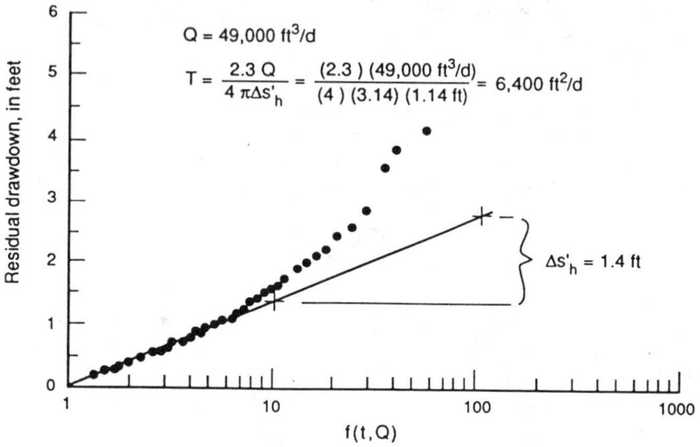

FIG. 3 Example Analysis Using Harrill's Method

9. Report

9.1 Prepare a report including the information described below. The report of the analysis will include information from the field testing procedure.

9.1.1 *Introduction*—The introductory section is intended to present the scope and purpose of the Theis recovery method for determining transmissivity in a confined nonleaky aquifer. Summarize the field hydrogeologic conditions and the field equipment and instrumentation including the construction of the control well and observation wells and piezometers, the method of measurement of discharge and water levels, and the duration of the test and pumping rates. Discuss rationale for selecting the Theis recovery method.

9.1.2 *Hydrogeologic Setting*—Review the information available on the hydrogeology of the site. Include driller's logs and geologist's description of drill cuttings. Interpret and describe the hydrogeology of the site as it pertains to the selection of this method for conducting and analyzing an aquifer test. Compare the hydrogeologic characteristics of the site as it conforms and differs from the assumptions in the solution to the aquifer test method.

9.1.3 *Scope of Aquifer Test:*

9.1.3.1 *Equipment*—Report the field installation and equipment for the aquifer test, including the construction, diameter, depth of screened interval, and location of control well and pumping equipment, and the construction, diameter, depth, and screened interval of observation wells or piezometers.

9.1.3.2 *Instrumentation*—Report the field instrumentation for observing water levels, pumping rate, barometric changes, and other environmental conditions pertinent to the test. Include a list of measuring devices used during the test, the manufacturer's name, model number, and basic specifications for each major item, and the name and date of the last calibration, if applicable.

9.1.3.3 *Testing Procedures*—State the steps taken in conducting pretest, drawdown, and recovery phases of the test. Include the frequency of measurements of discharge rate, water level in observation wells, and other environmental data recorded during the testing procedure.

9.1.4 *Presentation of Interpretation of Test Results:*

9.1.4.1 *Data*—Present tables of data collected during the test. Show methods of adjusting water levels for barometric changes and calculation of drawdown and residual drawdown.

9.1.4.2 *Data Plots*—Present data plots used in analysis of the data. Show data plots with straight line segments and intercepts of the $t/t' = 1$ axis.

9.1.4.3 Evaluate qualitatively the overall accuracy of the test on the basis of the adequacy of instrumentation and observations of stress end response, and the conformance of the hydrogeologic conditions and the performance of the test to the assumptions of the test method (see 5.1) and the implications of the assumptions (see 5.2).

10. Precision and Bias

10.1 It is not practicable to specify the precision of this test method because the response of aquifer systems during aquifer tests is dependent upon ambient system stresses. No statement can be made about bias because no true reference values exist.

11. Keywords

11.1 aquifers; aquifer tests; confined aquifers; control wells; ground water; hydraulic properties; observation wells; step tests; transmissivity; unconfined aquifers

REFERENCES

(1) Theis, C. V., "The Relation Between the Lowering of the Piezometric Surface and the Rate and Duration of Discharge of a Well Using Ground-Water Storage," *American Geophysical Union Transactions*, Vol 16, Part 2, 1935, p. 519–524.

(2) Cooper, H. H., and Jacob, C. E., "A Generalized Graphical Method for Evaluating Formation Constants and Summarizing Well-Field History," *Trans. Amer. Geophys. Union*, Vol 27, 1946, pp. 526–534.

(3) Jacob, C. E., "The Recovery Method for Determining the Coefficient of Transmissibility," *in* Bentall, Ray, Compiler, "Methods of Determining Permeability, Transmissibility, and Drawdown," *U.S. Geological Survey Water-Supply Paper 1536-I*, 1963, pp. 288–292.

(4) Harrill, J. R., "Determining Transmissivity from Water-Level Recovery of a Step-Drawdown Test," *U.S. Geological Survey Professional Paper 700-C*, 1970.

(5) Kruseman, G. P., and DeRidder, N. S., "Analysis and Evaluation of Pumping Test Data," Intern. Inst. for Land and Reclamation and Improvement, Bull. 11 Wageningen, The Netherlands, 1979.

(6) Lohman, S. W., "Ground-Water Hydraulics," *U.S. Geological Survey Professional Paper 708*, 1970.

Designation: D 5270 – 92

Standard Test Method for Determining Transmissivity and Storage Coefficient of Bounded, Nonleaky, Confined Aquifers[1]

This standard is issued under the fixed designation D 5270; the number immediately following the designation indicates the year of original adoption or, in the case of revision, the year of last revision. A number in parentheses indicates the year of last reapproval. A superscript epsilon (ϵ) indicates an editorial change since the last revision or reapproval.

1. Scope

1.1 This test method covers an analytical procedure for determining the transmissivity, storage coefficient, and possible location of boundaries for a bounded confined aquifer. This test method is used to analyze water-level or head data from one or more observation wells or piezometers during the pumping of water from a control well at a constant rate. This test method also applies to flowing artesian wells discharging at a constant rate. With appropriate changes in sign, this test method also can be used to analyze the effects of injecting water into a control well at a constant rate.

1.2 The analytical procedure in this test method is used in conjunction with the field procedure in Test Method D 4050.

1.3 *Limitations*—The valid use of this test method is limited to determination of transmissivities and storage coefficients for aquifers in hydrogeologic settings with reasonable correspondence to the assumptions of the Theis nonequilibrium method (see Test Method D 4106) (see 5.1), except that the aquifer is limited in areal extent by a linear boundary that fully penetrates the aquifer. The boundary is assumed to be either a constant-head boundary (equivalent to a stream or lake that hydraulically fully penetrates the aquifer) or a no-flow (impermeable) boundary (equivalent to a contact with a significantly less permeable rock unit). The Theis nonequilibrium method is described in Test Methods D 4105 and D 4106.

1.4 The values stated in SI units are to be regarded as standard.

1.5 *This standard does not purport to address all of the safety problems, if any, associated with its use. It is the responsibility of the user of this standard to establish appropriate safety and health practices and determine the applicability of regulatory limitations prior to use.*

2. Referenced Documents

2.1 *ASTM Standards:*

D 653 Terminology Relating to Soil, Rock, and Contained Fluids[2]

D 4043 Guide for Selection of Aquifer-Test Method in Determining Hydraulic Properties by Well Techniques[2]

D 4050 Test Method (Field Procedure) for Withdrawal and Injection Well Tests for Determining Hydraulic Properties of Aquifer Systems[2]

D 4106 Test Method (Analytical Procedure) for Determining Transmissivity and Storage Coefficient of Nonleaky Confined Aquifers by the Theis Nonequilibrium Method[2]

D 4105 Test Method (Analytical Procedure) for Determining Transmissivity and Storage Coefficient of Nonleaky Confined Aquifers by the Modified Theis Nonequilibrium Method[2]

D 4750 Test Method for Determining Subsurface Liquid Levels in a Borehole or Monitoring Well (Observation Well)[2]

3. Terminology

3.1 *Definitions:*

3.1.1 *constant-head boundary*—the conceptual representation of a natural feature such as a lake or river that effectively fully penetrates the aquifer and prevents water-level change in the aquifer at that location.

3.1.2 *equipotential line*—a line connecting points of equal hydraulic head. A set of such lines provides a contour map of a potentiometric surface.

3.1.3 *image well*—an imaginary well located opposite a control well such that a boundary is the perpendicular bisector of a straight line connecting the control and image wells; used to simulate the effect of a boundary on water-level changes.

3.1.4 *impermeable boundary*—the conceptual representation of a natural feature such as a fault or depositional contact that places a boundary of significantly less-permeable material laterally adjacent to an aquifer.

3.1.5 See Terminology D 653 for other terms.

3.2 *Symbols and Dimensions:*

3.2.1 K_l [nd]—constant of proportionality, r_i/r_r.

3.2.2 Q [L^3T^{-1}]—discharge.

3.2.3 r [L]—radial distance from control well.

3.2.4 r_i [L]—distance from observation well to image well.

3.2.5 r_r [L]—distance from observation well to control well.

3.2.6 S [nd]—storage coefficient.

3.2.7 s [L]—drawdown.

3.2.8 s_i [L]—component of drawdown due to image well.

3.2.9 s_o [L]—drawdown at an observation well.

3.2.10 s_r [L]—component of drawdown due to control well.

3.2.11 T [L^2T^{-1}]—transmissivity.

3.2.12 t [T]—time since pumping or injection began.

3.2.13 t_o [T]—time at projection of zero drawdown.

[1] This test method is under the jurisdiction of ASTM Committee D-18 on Soil and Rock and is the direct responsibility of Subcommittee D18.21 on Ground Water and Vadose Zone Investigations.

Current edition approved July 15, 1992. Published December 1992.

[2] *Annual Book of ASTM Standards*, Vol 04.08.

4. Summary of Test Method

4.1 This test method prescribes two analytical procedures for analysis of a field test. This test method requires pumping water from, or injecting water into, a control well that is open to the entire thickness of a confined bounded aquifer at a constant rate and measuring the water-level response in one or more observation wells or piezometers. The water-level response in the aquifer is a function of the transmissivity and storage coefficient of the aquifer, and the location and nature of the aquifer boundary or boundaries. Drawdown or build up of the water level is analyzed as a departure from the type curve defined by the Theis nonequilibrium method (see Test Method D 4106) or from straight-line segments defined by the modified Theis nonequilibrium method (see Test Method D 4105).

4.2 A constant-head boundary such as a lake or stream that fully penetrates the aquifer prevents drawdown or build up of head at the boundary, as shown in Fig. 1. Likewise, an impermeable boundary provides increased drawdown or build up of head, as shown in Fig. 2. These effects are simulated by treating the aquifer as if it were infinite in extent and introducing an imaginary well or "image well" on the opposite side of the boundary a distance equal to the distance of the control well from the boundary. A line between the control well and the image well is perpendicular to the boundary. If the boundary is a constant-head boundary, the flux from the image well is opposite in sign from that of the control well; for example, the image of a discharging control well is an injection well, whereas the image of an injecting well is a discharging well. If the boundary is an impermeable boundary, the flux from the image well has the same sign as that from the control well. Therefore, the image of a discharging well across an impermeable boundary is a discharging well. Because the effects are symmetrical, only discharging control wells will be described in the remainder of this test method, but this test method is equally applicable, with the appropriate change in sign, to control wells into which water is injected.

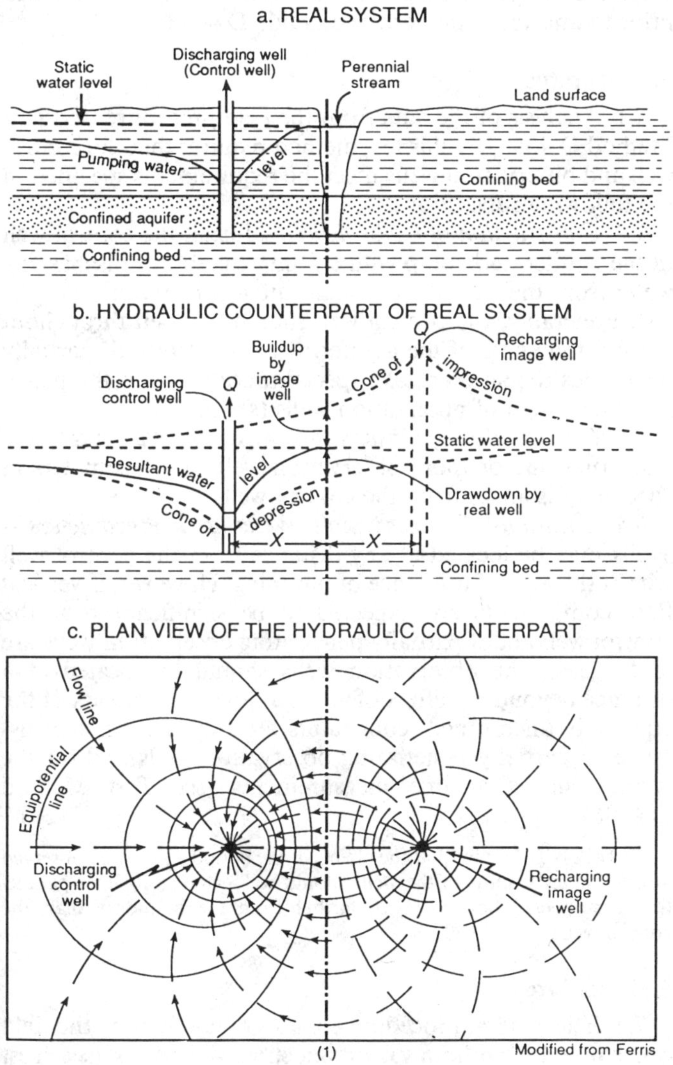

NOTE—Modified from Ferris and others **(6)** and Heath **(7)**.

FIG. 1 Diagram Showing Constant-Head Boundary

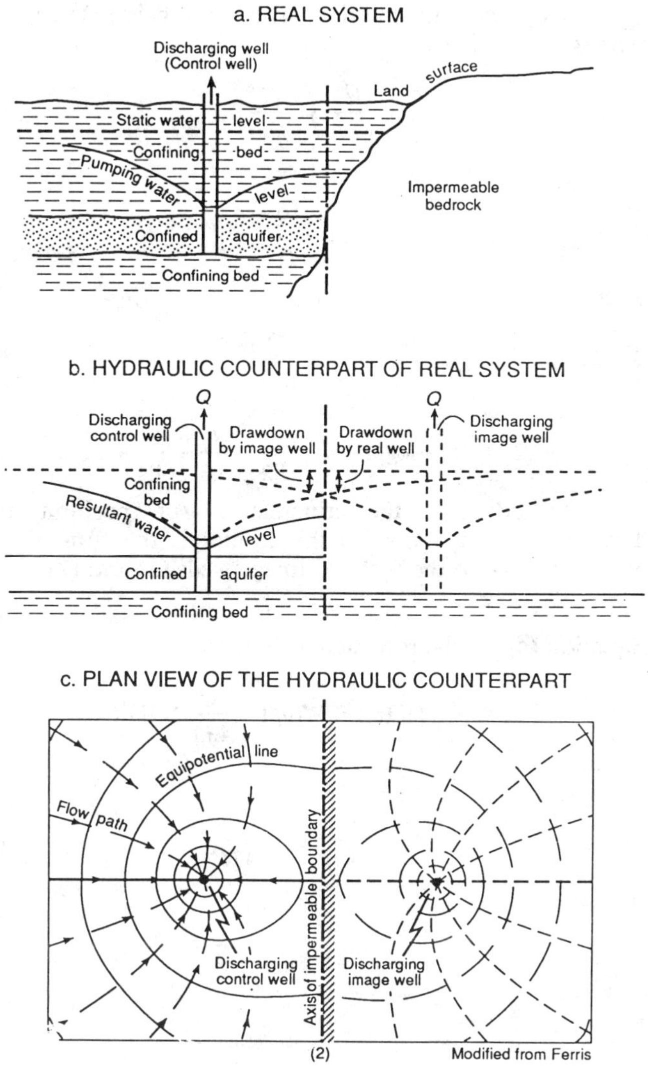

NOTE—Modified from Ferris and others **(6)** and Heath **(7)**.

FIG. 2 Diagram Showing Impermeable Boundary

4.3 *Solution*—The solution given by Theis (**1**)[3] can be expressed as follows:

$$s = \frac{Q}{4\pi T}\int_u^\infty \frac{e^{-y}}{y}\,dy \tag{1}$$

and:

$$u = \frac{r^2S}{4Tt} \tag{2}$$

where:

$$\int_u^\infty \frac{e^{-y}}{y}\,dy = W(u) =$$

$$-0.577216 - \log_e u + u - \frac{u^2}{2!2} + \frac{u^3}{3!3} - \frac{u^4}{4!4} + \ldots \tag{3}$$

4.4 According to the principle of superposition, the drawdown at any point in the aquifer is the sum of the drawdown due to the real and image wells (**1**) and (**2**):

$$s_o = s_r \pm s_i \tag{4}$$

Equation (**4**) can be rewritten as follows:

$$s_o = \frac{Q}{4\pi T}[W(u_r) \pm W(u_i)] = \frac{Q}{4\pi T}\Sigma\, W(u) \tag{5}$$

where:

$$u_r = \frac{r_r^2S}{4Tt},\ u_i = \frac{r_i^2S}{4Tt} \tag{6}$$

so that:

$$u_i = \left(\frac{r_i}{r_r}\right)^2 u_r,\ u_i = K_l^2u_r \tag{7}$$

where:

$$K_l = \frac{r_i}{r_r} \tag{8}$$

NOTE 1—K_l is a constant of proportionality between the radii, not to be confused with hydraulic conductivity.

5. Significance and Use

5.1 *Assumptions:*

5.1.1 The well discharges at a constant rate.

5.1.2 Well is of infinitesimal diameter and is open through the full thickness of the aquifer.

5.1.3 The nonleaky confined aquifer is homogeneous, isotropic, and areally extensive except where limited by linear boundaries.

5.1.4 Discharge from the well is derived initially from storage in the aquifer; later, movement of water may be induced from a constant-head boundary into the aquifer.

5.1.5 The geometry of the assumed aquifer and well are shown in Figs. 1 or 2.

5.1.6 Boundaries are vertical planes, infinite in length that fully penetrate the aquifer. No water is yielded to the aquifer by impermeable boundaries, whereas recharging boundaries are in perfect hydraulic connection with the aquifer.

5.1.7 Observation wells represent the head in the aquifer; that is, the effects of wellbore storage in the observation wells are negligible.

5.2 *Implications of Assumptions:*

5.2.1 Implicit in the assumptions are the conditions of a fully-penetrating control well and observation wells of infinitesimal diameter in a confined aquifer. Under certain conditions, aquifer tests can be successfully analyzed when the control well is open to only part of the aquifer or contains a significant volume of water or when the test is conducted in an unconfined aquifer. These conditions are discussed in more detail in Test Method D 4105.

5.2.2 In cases in which this test method is used to locate an unknown boundary, a minimum of three observation wells is needed. If only two observation wells are available, two possible locations of the boundary are defined, and if only one observation well is used, a circle describing all possible locations of the image well is defined.

5.2.3 The effects of a constant-head boundary are often indistinguishable from the effects of a leaky, confined aquifer. Therefore, care must be taken to ensure that a correct conceptual model of the system has been created prior to analyzing the test. See Guide D 4043.

6. Apparatus

6.1 Analysis of the data from the field procedure (see Test Method D 4050) by this test method requires that the control well and observation wells meet the requirements specified in the following subsections.

6.2 *Construction of Control Well*—Install the control well in the aquifer and equip with a pump capable of discharging water from the well at a constant rate for the duration of the test. Preferably, the control well should be open throughout the full thickness of the aquifer. If the control well partially penetrates the aquifer, take special precautions in the placement or design of observation wells (see 5.2.1).

6.3 *Construction of Observation Wells and Piezometers*—Construct one or more observation wells or piezometers at specified distances from the control well.

6.4 *Location of Observation Wells and Piezometers*—Wells may be located at any distance from the control well within the area of influence of pumping. However, if vertical flow components are expected to be significant near the control well and if partially penetrating observation wells are to be used, the observation wells should be located at a distance beyond the effect of vertical flow components. If the aquifer is unconfined, constraints are imposed on the distance to partially penetrating observation wells and on the validity of early time measurements (see Test Method D 4106).

NOTE 2—To ensure that the effects of the boundary may be observed during the tests, some of the wells should be located along lines parallel to the suspected boundary, no farther from the boundary than the control well.

7. Procedure

7.1 The general procedure consists of conducting the field procedure for withdrawal or injection wells tests (see Test Method D 4050) and analyzing the field data, as addressed in this test method.

[3] The boldface numbers given in parentheses refer to a list of references at the end of the text.

7.2 Analysis of the field data consists of two steps: determination of the properties of the aquifer and the nature and distance to the image well from each observation well, and determination of the location of the boundary.

7.3 Two methods of analysis can be used to determine the aquifer properties and the nature and distance to the image well. One method is based on the Theis nonequilibrium method; the other method is based on the modified Theis nonequilibrium method.

7.3.1 *Theis Nonequilibrium Method*—Expressions in Eqs 5 through 8 are used to generate a family of curves of $1/u_r$ versus $\Sigma\ W(u)$ for values of K_l for recharging and discharging image wells as shown in Fig. 3 **(2)**. Table 1 gives values of $W(u)$ versus $1/u$. This table may be used to create a table of $\Sigma W(u)$ versus $1/u$ for each value of K_l by picking values for $W(u_r)$ and $W(u_i)$, and computing the $\Sigma\ W(u)$ for the each value of $1/u$.

7.3.1.1 Transmissivity, storage coefficient, and the possible location of one or more boundaries are calculated from parameters determined from the match point and a curve selected from a family of type curves.

7.3.2 *Modified Theis Nonequilibrium Method*—The sum of the terms to the right of $\log_e u$ in Eq 3 is not significant when u becomes small, that is, equal to or less than 0.01.

NOTE 3—The limiting value for u of less than 0.01 may be excessively restrictive in some applications. The errors for small values of u, from Kruseman and DeRidder **(3)** are as follows:

Error less than, %:	1	2	5	10
For u smaller than:	0.03	0.05	0.1	0.15

7.3.2.1 The value of u decreases as time, t, increases and decreases as radial distance, r, decreases. Therefore, for large values of t and small values of r, the terms to the right of $\log_e u$ in Eq 3 may be neglected, as recognized by Theis **(1)**. The modified Theis equation can then be written as follows:

$$s = \frac{Q}{4\pi T}\left(-0.577216 - \log_e\left(\frac{r^2S}{4Tt}\right)\right) \tag{9}$$

from which it has been shown by Lohman **(4)** that:

$$T = \frac{2.3Q}{4\pi\Delta s} \tag{10}$$

where:

Δs = the drawdown (measured or projected) over one log cycle of time.

8. Calculation and Interpretation of Results

8.1 Determine the aquifer properties and the nature and distance to the image well by either the Theis nonequilibrium method or the modified Theis method.

8.1.1 *Theis Nonequilibrium Method*—The graphical procedure for solution by the Theis nonequilibrium method is based on the relationship between $\Sigma W(u)$ and s, and between $1/u$ and t/r^2.

8.1.1.1 Plot values of $\Sigma W(u)$ on the vertical coordinate and $1/u$ on the horizontal coordinate on logarithmic-scale paper. Plot a family of curves for several values of K_l, for both recharging and discharging images. This plot (see Fig. 3) is referred to as a family of type curves. Plots of the family of type curves are contained in **(2)** and **(4)**.

8.1.1.2 On logarithmic tracing paper of the same scale and size as the type curve, plot values of drawdown, s, on the vertical coordinate versus t/r^2 on the horizontal coordinate. Use a different symbol for data from each observation well.

8.1.1.3 Overlay the data plot on the type curve plot and, while the coordinate axes are held parallel, shift the plot to align the data with the type curve. The data points for small values of t/r^2 should fall on or near the central (standard) type curve, and larger values of t/r^2 should fall on curves representing different values of K_l, ordinarily a different

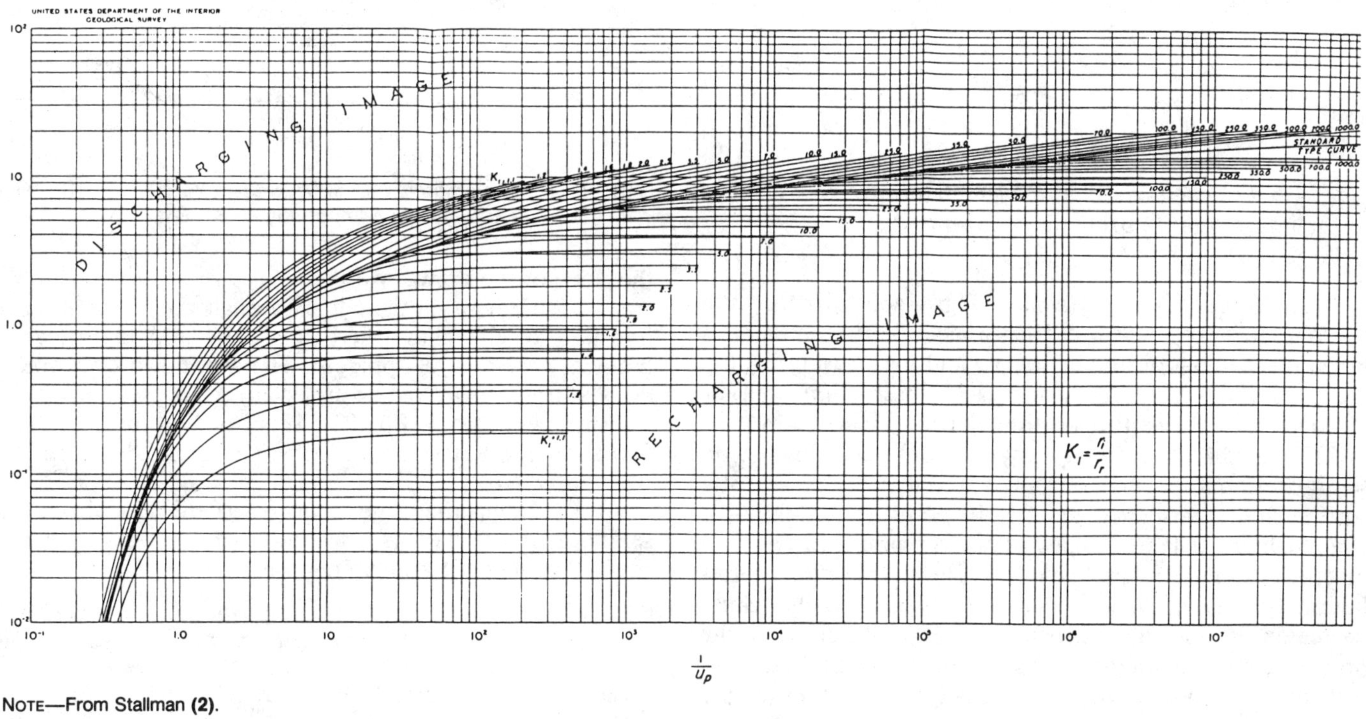

NOTE—From Stallman **(2)**.

FIG. 3 Family of Type Curves for the Solution of the Modified Theis Formula

TABLE 1 Values of Theis equation $W(u)$ for values of $1/u$ (8)

$1/u$	$1/u \times 10^{-1}$	1	10	10^2	10^3	10^4	10^4	10^4
1.0	0.00000[A]	0.21938	1.82292	4.03793	6.33154	8.63322	10.93572	13.23830
1.2	0.00003	0.29255	1.98932	4.21859	6.51369	8.81553	11.11804	13.42062
1.5	0.00017	0.39841	2.19641	4.44007	6.73667	9.03866	11.34118	13.64376
2.0	0.00115	0.55977	2.46790	4.72610	7.02419	9.32632	11.62886	13.93144
2.5	0.00378	0.70238	2.68126	4.94824	7.24723	9.54945	11.85201	14.15459
3.0	0.00857	0.82889	2.85704	5.12990	7.42949	9.73177	12.03433	14.33691
3.5	0.01566	0.94208	3.00650	5.28357	7.58359	9.88592	12.18847	14.49106
4.0	0.02491	1.04428	3.13651	5.41675	7.71708	10.01944	12.32201	14.62459
5.0	0.04890	1.22265	3.35471	5.63939	7.94018	10.24258	12.54515	14.84773
6.0	0.07833	1.37451	3.53372	5.82138	8.12247	10.42490	12.72747	15.03006
7.0	0.11131	1.50661	3.68551	5.97529	8.27659	10.57905	12.88162	15.18421
8.0	0.14641	1.62342	3.81727	6.10865	8.41011	10.71258	13.01515	15.31774
9.0	0.18266	1.72811	3.93367	6.22629	8.52787	10.83036	13.13294	13.43551
$1/u$	$1/u \times 10^1$	10^1	10^9	10^{10}	10^{11}	10^{12}	10^{13}	10^{14}
1.0	15.54087	17.84344	20.14604	22.44862	24.75121	27.05379	29.35638	31.65897
1.2	15.72320	18.02577	20.32835	22.63094	24.93353	27.23611	29.53870	31.84128
1.5	15.94634	18.24892	20.55150	22.85408	25.15668	27.45926	29.76184	32.06442
2.0	16.23401	18.53659	20.83919	23.14177	25.44435	27.74693	30.04953	32.35211
2.5	16.45715	18.75974	21.06233	23.36491	25.66750	27.97008	30.27267	32.57526
3.0	16.63948	18.94206	21.24464	23.54723	25.84982	28.15240	30.45499	32.75757
3.5	16.79362	19.09621	21.39880	23.70139	26.00397	28.30655	30.60915	32.91173
4.0	16.92715	19.22975	21.53233	23.83492	26.13750	28.44008	30.74268	33.04526
5.0	17.15030	19.45288	21.75548	24.05806	26.36054	23.66322	30.96582	33.26840
6.0	17.33263	19.63521	21.93779	24.24039	26.54297	28.84555	31.14813	33.45071
7.0	17.48677	19.78937	22.09195	24.39453	26.69711	28.99969	31.30229	33.60487
8.0	17.62030	19.92290	22.22548	24.52806	26.83064	29.13324	31.43582	33.73840
9.0	17.73808	20.04068	22.34326	24.64584	29.94843	29.25102	31.55360	33.85619

[A] Value shown as 0.00000 is nonzero but less than 0.000005.

value of K_l for each observation well.

8.1.1.4 Select and record the values of $\Sigma W(u)$, $1/u$, s, and t/r^2 for a point (called the match point) common to both the type curve and the data plot. For convenience, the point may be selected where $\Sigma W(u)$ and $1/u$ are major axes, that is, 0.1, 1.0, 10.0, etc. Record a value of K_l for each observation well.

8.1.1.5 Using the match point coordinates, determine the transmissivity and storage coefficient from the following equations:

$$T = \frac{Q}{4\pi s} \Sigma W(u) \tag{11}$$

and:

$$S = 4T(t/r^2)u \tag{12}$$

8.1.1.6 For each observation well, determine the distance to the image well, r_i, using the following:

$$r_i = K_l r_r \tag{13}$$

8.1.2 *Modified Theis Method*—The graphical procedure for solution by the modified Theis nonequilibrium method is based on the relationship between s and $\log_{10} t$ using Eq 10.

8.1.2.1 Plot values of s for each observation well or piezometer on the vertical (arithmetic) coordinate and values of t on the horizontal (logarithmic) coordinate on semilogarithmic paper. For values of t that are sufficiently large such that u is less than 0.01, the points should fall on a straight line. At larger values of t, the points will begin to diverge from the straight line due to the effects of the nearest boundary (see Fig. 4). A constant-head boundary will cause decreased drawdown, and measurements will fall above the projected straight line, whereas an impermeable boundary will cause increased drawdown and points will fall below the projected line. Note that an impermeable boundary doubles the slope of the drawdown plot.

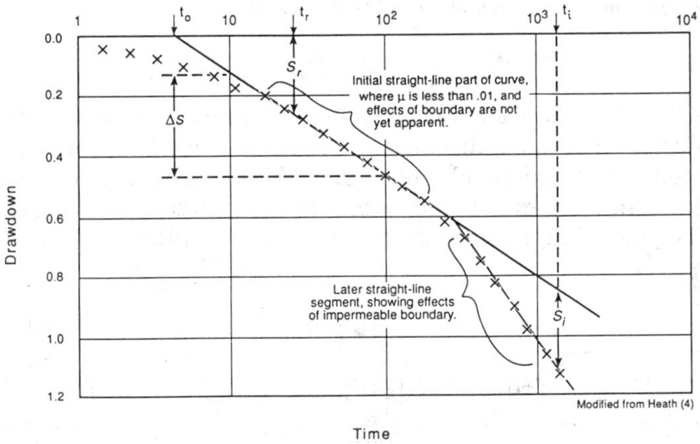

FIG. 4 Semilogarithmic Plot of Drawdown Versus Time Showing Effects of an Impermeable Boundary

8.1.2.2 Draw a straight line through the initial straight-line part of the data where $u < 0.01$ and the effects of boundary are not yet apparent. The drawdown over one log cycle of time (measured or projected) Δs, is used to calculate transmissivity from Eq 10. This method of calculating hydraulic properties is prescribed in more detail in Test Method D 4105.

8.1.2.3 Determine the storage coefficient from the semilogarithmic plot of drawdown versus $\log_{10}$ time by a method proposed by Jacob (**5**), where:

$$s = \frac{2.3Q}{4\pi T} \log_{10} \left(\frac{2.25Tt}{r^2S}\right) \tag{14}$$

Project the initial straight-line part of the curve to the left until it intercepts the line of zero drawdown. Taking $s = 0$ at the zero-drawdown intercept of the straight-line plot of drawdown versus $\log_{10}$ time:

$$S = \frac{2.25\ Tt_0}{r^2} \qquad (15)$$

where:

t_o = the value of time at the projection of zero.

Additional discussion of the limits of the modified Theis nonequilibrium method is found in Test Method D 4105.

8.1.2.4 Select a convenient value of s within the initial straight-line part of the plot. Because the drawdown has not yet been affected by the boundary, $s = s_r$. Note the value of t_r that corresponds to this value of s_r.

8.1.2.5 Graphically extend the initial straight-line part of the curve to the right. The departure of the measured drawdown from the extended straight line is the drawdown due to the presence of the boundary, the image-well drawdown, s_i. Select a point within the second straight-line part of the curve such that $s_i = s_r$ and note the value of time, t_i, at which s_i is found.

8.1.2.6 Because t_r and t_i were selected such that $s_r = s_i$, u_r is equal to u_i and $r_r^2S/4Tt_r = r_i^2S/4Tt_i$, so that:

$$K_l = \frac{r_i}{r_r} = \sqrt{\frac{t_i}{t_r}} \qquad (16)$$

Determine the radius to the image well using Eq 13.

8.2 *Determine Location of Boundary:*

8.2.1 On a map showing the locations of the control and observation wells, with a compass describe a circle around each observation well. The radius of the circle should be the radius to the image well, r_i, from that observation well.

8.2.2 The image well is located at the intersection of the circles. If the circles do not intersect exactly, the most probable location is the centroid of the intersections.

8.2.3 Draw a straight line between the control well and the image well. The boundary is represented by the perpendicular bisector of this line.

9. Report

9.1 Prepare a report including the following information:

9.1.1 *Introduction*—The introductory section is intended to present the scope and purpose of this test method for determining transmissivity, storage coefficient, and boundary location in a confined nonleaky aquifer. Summarize the field hydrogeologic conditions and the field equipment and instrumentation including the construction of the control well and observation wells, the method of measurement of discharge and water levels, and the duration of the test and pumping rates. Discuss the rationale for selecting a method that incorporates the effects of boundaries.

9.1.2 *Hydrogeologic Setting*—Review the information available on the hydrogeology of the site. Include driller's logs and geologist's description of drill cuttings. Interpret and describe the hydrogeology of the site as it pertains to the selection of this test method for conducting and analyzing an aquifer test. Compare the hydrogeologic characteristics of the site as they conform and differ from those assumed in the solution to the aquifer test method. In particular, locate all possible boundaries and describe their characteristics.

9.1.3 *Scope of Aquifer Test:*

9.1.3.1 *Equipment*—Report the field installation and equipment for the aquifer test, including the construction, diameter, depth of screened interval, and location of control well and pumping equipment, and the construction, diameter, depth, and screened interval of observation wells or piezometers.

9.1.3.2 *Instrumentation*—Report the field instrumentation for observing water levels, pumping rate, barometric changes, and other environmental conditions pertinent to the test. Include a list of measuring devices used during the test, the manufacturer's name, model number, and basic specifications for each major item, and the name and date of the last calibration, if applicable.

9.1.3.3 *Testing Procedures*—State the steps taken in conducting pretest, drawdown, and recovery phases of the test. Include the frequency of measurements of discharge rate, water level in observation wells, and other environmental data recorded during the testing procedure.

9.1.4 *Presentation of Interpretation of Test Results:*

9.1.4.1 *Data*—Present tables of data collected during the test. Show methods of adjusting water levels for barometric changes or other background water level changes and calculation of drawdown.

9.1.4.2 *Data Plots*—Present data plots used in analysis of the data. Show overlays of data plots and type curves with match points and corresponding values of parameters at match points. Show values of K_l, selected for each observation well.

9.1.4.3 *Calculation*—Show calculation of transmissivity, storage coefficient, radius to image well and radius to boundary.

9.1.5 Evaluate qualitatively the overall accuracy of the test on the basis of the adequacy of instrumentation and observations of stress and response, and the conformance of the hydrogeologic conditions and the performance of the test to the model assumptions.

10. Precision and Bias

10.1 It is not practicable to specify the precision of this test method because the response of aquifer systems during aquifer tests is dependent upon ambient system stresses. No statement can be made about bias because no true reference values exist.

11. Keywords

11.1 aquifers; aquifer boundaries; aquifer tests; confined aquifers; control wells; ground water; hydraulic properties; image wells; observation wells; storage coefficient; transmissivity

REFERENCES

(1) Theis, C. V., "The Relation Between the Lowering of the Piezometric Surface and the Rate and Duration of Discharge of a Well Using Ground-Water Storage," *American Geophysical Union Transactions*, Vol 16, part 2, 1935, pp. 519–524.

(2) Stallman, Robert W., "Type Curves for the Solution of Single-Boundary Problems," *in* "Shortcuts and Special Problems in Aquifer Tests," Ray, Bentall, Compiler, *U.S. Geological Survey Water-Supply Paper 1545-C*, 1963, pp. 45–47.

(3) Kruseman, G. P., and DeRidder, N. A., "Analysis and Evaluation of Pumping Test Data," Inter. Inst. for Land Reclamation and Improvement, Bull. 11, Wageningen, The Netherlands, 1979.

(4) Lohman, S. W., "Ground-Water Hydraulics," *U.S. Geological Survey Professional Paper 708*, 1972.

(5) Jacob, C. E., "The Recovery Method for Determining the Coefficient of Transmissibility," *in* Bentall, Ray, Compiler, "Determining Permeability of Water-Table Aquifers," *U.S. Geological Survey Water-Supply Paper 1536-I*, 1963, pp. 283–292.

(6) Ferris, J. G., Knowles, D. B., Brown, R. H., and Stalman, R. W., "Theory of Aquifer Tests," *U.S. Geological Survey Water-Supply Paper 1536-E*, 1962, pp. 69–174.

(7) Heath, R. W., "Basic Ground-Water Hydrology," *U.S. Geological Survey Water Supply-Paper 2220*, 1983.

(8) Reed, J. E., "Type Curves for Selected Problems of Flow to Wells in Confined Aquifers," *U.S. Geological Survey Techniques of Water Resources Investigations*, Book 3, Ch. B3, 1980.

Designation: D 5299 – 92

Standard Guide for Decommissioning of Ground Water Wells, Vadose Zone Monitoring Devices, Boreholes, and Other Devices for Environmental Activities[1]

This standard is issued under the fixed designation D 5299; the number immediately following the designation indicates the year of original adoption or, in the case of revision, the year of last revision. A number in parentheses indicates the year of last reapproval. A superscript epsilon (ε) indicates an editorial change since the last revision or reapproval.

1. Scope

1.1 This guide covers procedures that are specifically related to permanent decommissioning (closure) of the following as applied to environmental activities. It is intended for use where solid or hazardous materials or wastes are found, or where conditions occur requiring the need for decommissioning. The following devices are considered in this guide:

1.1.1 A borehole used for geoenvironmental purposes (see Note 1),

1.1.2 Monitoring wells,

1.1.3 Observation wells,

1.1.4 Injection wells (see Note 2),

1.1.5 Piezometers,

1.1.6 Wells used for the extraction of contaminated ground water, the removal of floating or submerged materials other than water such as gasoline or tetrachloroethylene, or other devices used for the extraction of soil gas,

1.1.7 A borehole used to construct a monitoring well, and

1.1.8 Any other vadose zone monitoring device.

1.2 Temporary decommissioning of the above is not covered in this guide.

NOTE 1—This guide may be used to decommission boreholes where no contamination is observed at a site (see Practice D 420 for details); however, the primary use of the guide is to decommission boreholes and wells where solid or hazardous waste have been identified. Methods identified in this guide can also be used in other situations such as the decommissioning of water supply wells and boreholes where water contaminated with nonhazardous pollutants (such as nitrates or sulfates) are present. This guide should be consulted in the event that a routine geotechnical investigation indicates the presence of contamination at a site.

NOTE 2—The term "well" is used in this guide to denote monitoring wells, piezometers, or other devices constructed in a manner similar to a well. Some of the devices listed such as injection and extraction wells can be decommissioned using this guide for information, but are not specifically covered in the text.

NOTE 3—Details on the decommissioning of multiple-screened wells are not provided in this guide due to the many methods used to construct these types of wells and the numerous types of commercially available multiple-screened well systems. However, in some instances, the methods presented in this guide may be used with few changes. An example of how this guide may be used is the complete removal of the multiple-screened wells by overdrilling.

1.3 Most monitoring wells and piezometers are intended primarily for water quality sampling, water level observation, or soil gas sampling, or combination thereof, to determine quality. Many wells are relatively small in diameter (<4-in. (10.1 cm) inside diameter) and are used to monitor for hazardous chemicals in ground water. Decommissioning of monitoring wells is necessary to:

1.3.1 Eliminate the possibility that the well is used for purposes other than intended,

1.3.2 Prevent migration of contaminants into an aquifer or between aquifers,

1.3.3 Prevent migration of contaminants in the vadose zone,

1.3.4 Reduce the potential for vertical or horizontal migration of fluids in the well or adjacent to the well, and

1.3.5 Remove the well from active use when the well is no longer capable of rehabilitation, or has failed structurally; no longer required for monitoring; no longer capable of providing representative samples or is providing unreliable samples; or required to be decommissioned; or to meet regulatory requirements.

NOTE 4—The determination of whether a well is providing a representative water quality sample is not defined in this guide. Examples of when a representative water quality sample may not be collected include the biological or chemical clogging of well screens, a drop in water level to below the base of the well screen, or complete silting of a tail pipe. These conditions may indicate that a well is not functioning properly.

1.4 This guide is intended to provide information for effective permanent closure of wells so that the physical structure of the well does not provide a means of hydraulic communication between aquifers or react chemically in a detrimental way with the environment.

1.5 The intent of this guide is to provide procedures that when followed result in a reasonable level of confidence in the integrity of the decommissioning activity. However, it may not be possible to verify the integrity of the decommissioning procedure. At this time, methods are not available to substantially determine the integrity of the decommissioning activity.

1.6 The values stated in inch-pound units are to be regarded as the standard. The SI units given in parentheses are for information only.

1.7 *This standard does not purport to address all of the safety problems, if any, associated with its use. It is the responsibility of the user of this standard to establish appropriate safety and health practices and determine the applicability of regulatory limitations prior to use.*

---

[1] This guide is under the jurisdiction of ASTM Committee D-18 on Soil and Rock and is the direct responsibility of Subcommittee D18.21 on Ground Water and Vadose Zone Investigations.

Current edition approved Sept. 15, 1992. Published January 1993.

NOTE 5—If state and local regulations are in effect where the decommissioning is to occur, the regulations take precedence over this guide.

2. Referenced Documents

2.1 *ASTM Standards:*

C 150 Specification for Portland Cement[2]

C 618 Specification for Fly Ash and Raw or Calcined Pozzolan for Use as a Mineral Admixture in Portland Concrete Cement[2]

D 420 Practice for Investigating and Sampling Soil and Rock for Engineering Purposes[3]

D 4380 Test Method for Density of Bentonitic Slurries[3]

D 5088 Practice or the Decontamination of Field Equipment Used at Non-Radioactive Waste Sites[3]

D 5092 Practice for Design and Installation of Ground Water Monitoring Wells in Aquifers[3]

3. Terminology

3.1 *Descriptions of Terms Specific to This Standard:*

3.1.1 **abandonment**—see *decommissioning.*

3.1.2 *attapulgite clay*—a chain-lattice clay mineral. The term also applies to a group of clay minerals that are lightweight, tough, matted, and fibrous.

3.1.3 *borehole television log*—a borehole or well video record produced by lowering a television camera into the borehole or well. This record is useful in visually observing downhole conditions such as collapsed casing or a blocked screen.

3.1.4 *blowout*—a sudden or violent uncontrolled escape of fluids or gas, or both, from a borehole.

3.1.5 *caliper log*—a geophysical borehole log that shows to scale the variations with depth in the mean diameter of a cased or uncased borehole.

3.1.6 *cement, API, Class A*—a cement intended for use from the surface to a depth of 6000 ft (1828 m). This cement is similar to ASTM Type I cement.

3.1.7 *cement, API, Class B*—a cement intended for use from the surface to a depth of 6000 ft (1828 m) when conditions require moderate- to high-sulfate resistance. This cement is similar to ASTM Type II cement.

3.1.8 *cement, API, Class C*—this cement is intended for use from the surface to a depth of 6000 ft (1828 m) when conditions require high early strength. This cement is similar to ASTM Type III cement. Also available as a high sulfate resistant type.

3.1.9 *cement, API, Class G*—this cement is intended for use from the surface to a depth of 8000 ft (2438 m). It can be used with accelerators or retarders to cover a wide range of well depths and temperatures. No additions other than calcium sulfate or water, or both, can be interground or blended with the clinker during manufacture of the cement. Also available as several sulfate-resistant types.

3.1.10 *cement, API, Class H*—this cement is intended for use from the surface to a depth of 8000 ft (2438 m). It can be used with accelerators or retarders to cover a wide range of well depths and temperatures. No additions other than calcium sulfate or water, or both, can be interground or blended with the clinker during manufacture of the cement. Also available as a sulfate-resistant type.

3.1.11 *cement, API, Class J*—this cement is intended for use from depths of 12 000 to 16 000 ft (3658 to 4877 m) under conditions of extremely high temperatures and pressures. It can be used with accelerators and retarders to cover a range of well depths and temperatures. No additions of retarders other than calcium sulfate, or water, or both, can be interground or blended with the clinker during manufacture of the cement.

3.1.12 *cement bond (sonic) log*—a borehole geophysical log that can be used to determine the effectiveness of a cement seal of the annular space of a well.

3.1.13 *channeling*—the process of forming a vertical cavity resulting from a faulty cement job in the annular space.

3.1.14 *curing accelerator*—a material added to cement to decrease the time for curing. Examples are sodium chloride, calcium sulfate (gypsum), and aluminum powder.

3.1.15 *curing retarder*—a material added to cement to increase the time for curing. Sodium chloride in high concentrations is an example.

3.1.16 *decommissioning (closure)*—the engineered closure of a well, borehole, or other subsurface monitoring device sealed with plugging materials. Decommissioning also includes the planning and documenting of all associated activities. A synonym is abandonment.

3.1.17 *decontamination*—the process of removing undesirable physical or chemical constituents, or both, from equipment to reduce the potential for cross-contamination.

3.1.18 *fallback*—shrinkage, settlement, or loss of plugging material placed in a borehole or well.

3.1.19 *fire clay*—a silicious clay rich in hydrous aluminum silicates.

3.1.20 *flow log*—a borehole geophysical log used to record vertical movement of ground water and movement of water into or out of a well or borehole and between formations within a well.

3.1.21 *geophysical borehole log*—a log obtained by lowering an instrument into a borehole and continuously recording a physical property of native or backfill material and contained fluids. Examples include resistivity, induction, caliper, sonic, and natural gamma logs.

3.1.22 *grout*—material consisting of bentonite, cement, or a cement-bentonite mixture.

3.1.23 *grout pipe*—a pipe or tube that is used to transport cement, bentonite, or other plugging materials from the ground surface to a specified depth in a well or borehole. The material may be allowed to flow freely or it may be injected under pressure. The term tremie pipe is frequently used interchangeably.

3.1.24 *hydraulic communication*—the migration of fluids from one zone to another, with reference to this guide; especially along a casing, grout plug, or through backfill materials.

3.1.25 *multiple-screened wells*—two or more monitoring wells situated in the same borehole. These devices can be either individual casing strings and screen set at a specific depth, a well with screens in more than one zone, or can consist of devices with screens with tubing or other collecting devices attached that can collect a discrete sample.

[2] *Annual Book of ASTM Standards*, Vol 04.02.

[3] *Annual Book of ASTM Standards*, Vol 04.08.

3.1.26 *native material*—in place geologic (or soil) materials encountered at a site.

3.1.27 *overdrilling*—the process of drilling out a well casing and any material placed in the annular space.

3.1.28 *perforation*—a slot or hole made in well casing to allow for communication of fluids between the well and the annular space.

3.1.29 *permanent plugging*—a seal that has a hydraulic conductivity that is equivalent or less than the hydraulic conductivity of the geologic formation. This term is often used with uncased boreholes.

3.1.30 *plow layer*—the depth typically reached by a plow or other commonly used earth turning device used in agriculture. This depth is commonly one to two feet (.3 m to .61 m) below land surface.

3.1.31 *plugging material*—a material that has a hydraulic conductivity equal to or less than that of the geologic formation(s) to be sealed. Typical materials include portland cement and bentonite.

3.1.32 *pre-conditioning*—an activity conducted prior to placing plugging material into a borehole in order to stabilize the hole.

3.1.33 *temporary decommissioning*—the engineered closure of a well intended to be returned to service at some later date (generally no more than six months). Temporary plugging should not damage the structural integrity of the well. Plugging materials consist of sand, bentonite, or other easily removed materials.

4. Summary of Guide

4.1 Information is provided on the significance of properly decommissioning boreholes and wells at sites containing or formerly containing solid or hazardous waste or hazardous materials or their byproducts, or that may be affected by solid or hazardous waste materials or their byproducts in the future. This guide may be used in situations where water quality in one aquifer may be detrimental to another aquifer either above or below the aquifer. The primary purpose of decommissioning activities is to permanently decommission the borehole or monitoring device so that the natural migration of ground water or soil vapor is not significantly influenced. Decommissioned boreholes and wells should have no adverse influence on the local environment than the original geologic setting.

4.2 It is important to have a good understanding of the geology, hydrogeology, well construction, historic and future land use, chemicals encountered, and the regulatory environment for successful decommissioning to occur.

4.3 Various materials suitable for decommissioning boreholes and wells are discussed, including their positive and negative attributes for decommissioning. A generalized procedure is provided that discusses the process from planning through implementation and documentation. Examples of typical practices are provided in the appendix.

5. Significance and Use

5.1 Decommissioning of boreholes and monitoring wells, and other devices requires that the specific characteristics of each site be considered. The wide variety of geological, biological, and physical conditions, construction practices, and chemical composition of the surrounding soil, rock, waste, and ground water precludes the use of a single decommissioning practice. The procedures discussed in this guide are intended to aid the geologist or engineer in selecting the tasks required to plan, choose materials for, and carry out an effective permanent decommissioning operation. Each individual situation should be evaluated separately and the appropriate technology applied to best meet site conditions. Considerations for selection of appropriate procedures are presented in this guide, but other considerations based on site specific conditions should also be taken into account.

NOTE 6—Ideally, decommissioning should be considered as an integral part of the design of the monitoring well. Planning at this early stage can make the decommissioning activity easier to accomplish. See Practice D 5092 for details on monitoring well construction.

5.2 This guide is intended to provide technical information and is not intended to supplant statutes or regulations. Approval of the appropriate regulatory authorities should be an important consideration during the decommissioning process.

6. Materials

6.1 The materials used for construction of a monitoring well or other monitoring device to be decommissioned in part determines how it is decommissioned. Various materials are available for use in plugging boreholes and monitoring wells. This section provides information on these materials.

6.2 *Casing and Screen Materials:*

6.2.1 Various materials are used for well casing and screen. The most common materials used are: PVC, PTFE, fiberglass, carbon steel, stainless steel, and aluminum. Typically, the same material is used for casing and screen in a well, however, in some instances different materials may be used in a well to achieve a particular purpose such as corrosion protection, reduction of material costs, or improving the integrity of ground water or soil vapor samples. This guide does not specifically address the use of more than one type of casing or screen material used in a well, however, the same decommissioning methods can frequently be used when more than one material is used (for example, PVC and PTFE, or stainless steel and carbon steel) in a well.

6.2.2 In selecting a well decommissioning method, PVC, PTFE, and fiberglass wells can be decommissioned using similar methods as all three types of materials tend to be low in tensile strength and easy to drill out or perforate. Appendix X1 provides a discussion on various procedures that can be used for the decommissioning of PVC wells and by reference PTFE and fiberglass wells.

6.2.3 Wells constructed of carbon steel, stainless steel, and heavy walled aluminum can be decommissioned using similar methods as these materials tend to have a higher tensile strength that allows for the casing to be removed. Appendix X1 provides a discussion on various procedures that can be used for the decommissioning of steel wells and by reference stainless steel and aluminum wells.

6.3 *Plugging Materials:*

6.3.1 Plugging materials should be carefully chosen for well closure to be permanent. Basic material characteristics are listed as follows:

6.3.1.1 Plugging materials should not react with contaminants or adversely react with ground water or geologic materials.

6.3.1.2 Plugging materials used in decommissioning wells, borings, etc. should have hydraulic conductivity (saturated condition) that is comparable to or lower than that of the lowest hydraulic conductivity of the geologic material being sealed.

6.3.1.3 Plugging materials must have sufficient structural strength to withstand pressures expected from native conditions.

6.3.1.4 Plugging materials must maintain sealing capabilities and not degrade due to chemical interaction, corrosion, dehydration, or other physical or chemical processes. Materials should maintain their design characteristics for the length of time contamination is present at the site.

6.3.1.5 Plugging materials should not be readily susceptible to cracking or shrinkage, or both.

6.3.1.6 Plugging materials must be capable of being placed at the position in the well or borehole in which they are needed and must have properties that reduce their unintended movement vertically and horizontally.

6.3.1.7 Plugging materials must be capable of forming a tight bond and seal with well casing and the formation.

6.3.18 Plugging materials must have properties that eliminate leaching or erosion of the material, under the conditions the material will be subjected. These include vertical or horizontal movement, or both, or contact with ground water or other existing conditions.

NOTE 7—The grain size of plugging material used in decommissioning operations conducted in areas where thick vadose zones occur should be coarser than materials used in areas where thin vadose zones or shallow saturated conditions occur. This is necessary as water is not transported effectively in coarse-grained materials under negative pore pressures. Coarse-grained materials should not be used where saturated conditions are likely to exist during the period of time that hazardous materials can be expected to occur at the site. It is important to determine the lithology and grain size distribution of materials adjacent to the borehole or well prior to selection of plugging materials.

NOTE 8—If coarse-grained materials are used to decommission the borehole or well, a layer of fine-grained material (such as cement or bentonite, or both) 1 or 2 ft (.3 or .61 m) thick should be placed at 10 ft (3 m) intervals in the borehole in the saturated zone. This layer should extend 2 to 3 ft (.61 to .91 m) above the highest expected level saturation is expected based on historical information on the water table for unconfined aquifers. A similar thickness of these materials should be used for confined aquifers. A similar 5-ft (1.5-m) seal of a low-permeability material should be placed near the ground surface to reduce the potential for entrance of fluids at the ground surface.

6.4 *Commonly Used Materials*—Subsections 6.2 and 6.3 introduced the general criteria that must be evaluated during the process of selecting the appropriate procedure and material for plugging a specific well. Because well construction and local geological conditions are site specific, a wide variety of materials and procedures may be used to complete the closure.

6.4.1 Section 6.4 presents a review of the plugging materials most commonly used to decommission monitoring wells. Table 1 summarizes these materials and lists the most important considerations (positive and negative) for their use. A detailed discussion of each material is presented in the following subsections.

6.4.2 *Portland Cement*—Portland cement may be used in any of its various forms to meet placement, strength, and durability criteria listed in 6.1. The amount of shrinkage or settling of neat cement is dependent on the amount of water used. Higher water to cement ratios tend to increase shrinkage **(1)**.[4]

6.5 *Specification C 150:*

6.5.1 *Type I*—Type 1 cement, a general-purpose material, is the most commonly used cement. This material has a tendency to develop a relatively high heat of hydration when used in confined situations and has relatively low-sulfate resistance.

6.5.2 *Type II*—Somewhat slower strength development than Type I; however, Type II cement has moderate heat of hydration and moderate sulfate resistance.

6.5.3 *Type III*—Type III cement is used when high early strength is desired. This material is not commonly used in decommissioning activities because of its ability to quickly set. Care must be used in working with this material.

6.5.4 *Type IV*—Type IV cement is used where a low heat of hydration is desired. It is not commonly used in decommissioning activities.

6.5.5 *Type V*—Type V cement has high resistance to sulfate, and brine solutions. This material has ultimate strength development somewhat less than either Types I or II.

6.5.6 Type K cement is expansive and can be used to compensate for shrinkage. This cement is essentially Type I or more commonly Type II Portland Cement with additives to produce expansion. It can be of use in plugging situations where water-tightness is important. Type K cement contains calcium sulfoaluminate. When mixed with water, the hydration causes an expansion ranging from approximately 0.05 to 0.20 % **(2)**.

6.6 *API Cements* **(3)**:

6.6.1 *Class A*—Class A cement corresponds closely to ASTM Type 1. This cement is intended to be used from the surface to a depth of 6000 ft (1828 m).

6.6.2 *Class B*—Class B cement corresponds closely to ASTM Type II. It is intended for use from the surface to a depth of 6000 ft (1828 m) and is also available as a high-sulfate resistant variety.

6.6.3 *Class C*—Class C cement corresponds closely to ASTM Type III. It is intended for use from the surface to a depth of 6000 ft (1828 m). It is also available as a high-sulfate resistant variety.

6.6.4 *Class G*—Class G cement is intended for use from the surface to a depth of 8000 ft (2438 m) and can be used with accelerators or retarders to cover a wide range of depths and temperatures. The cement is also available as a high-sulfate resistant variety.

6.6.5 *Class H*—Class H cement is intended for use from the surface to a depth of 8000 ft (2438 m). It can be used with a wide variety of accelerators and retarders to cover a wide range of depths and temperatures. It is available only as a moderate-sulfate resistant type.

6.6.6 *Class J*—This cement is intended for use from a depth of 12 000 to 16 000 ft (3658 to 4877 m) where

[4] The boldface numbers in parentheses refer to a list of references at the end of the text.

TABLE 1 Properties of Common Plugging Materials

Plugging Material	Description	Positive Attributes	Negative Attributes
ASTM C-50 Portland Cement			
Type I	Most commonly used type of cement for plugging	Forms a good seal when used with bentonite in 3 to 5 % concentration. Commonly available and can be purchased premixed on-site.	High heats of hydration may be a problem in PVC-cased wells. Can shrink and crack; low-sulfate resistance. Should not be used in the presence of strong acids or in low-pH environments.
Type II	Similar to Type I, but with a moderate heat of hydration.	Moderate heat of hydration. Moderate resistance to sulfate.	Somewhat slower strength development than Type I; expensive. Can shrink and crack. Can be difficult to use. Should not be used in the presence of strong acids or in low-pH environments.
Type III	High early strength.	May prove useful in situations where high early strength is needed, such as borehole walls that have a tendency to collapse.	Not a common cement. Can set very quickly before decommissioning is completed. Should not be used in the presence of strong acids or in low-pH environments.
Type IV	Low heat of hydration.	May prove useful in situations where a low heat of hydration is required	Not a common cement. Should not be used in the presence of strong acids or in low-pH environments.
Type V	Similar to Type I, with high resistance to sulfate and brine.	High resistance to sulfate and brine. Low heat of hydration.	Ultimate strength is less than Types I and III. Expensive; should not be used in the presence of strong acids or in low-pH environments. Can be difficult to use. Can shrink or crack.
K	Expansive cement.	Basically Type I or Type II Portland Cement with additions (tricalcium sulfo aluminate for example) to provide for expansion. Expansion is generally in the range from 0.05 to 0.20 % Good resistance to sulfate attack.	
API 10			
Class A	Similar to ASTM Type I.	Can be used to a depth of 6000 ft (1828 m). Forms a good seal when used with bentonite in 3 to 5 % concentration. Commonly available and can be purchased premixed on-site.	High heats of hydration may be a problem in PVC-cased wells. Can shrink and crack; low-sulfate resistance. Should not be used in the presence of strong acids or in low-pH environments.
Class B	Similar to ASTM Type II.	Can be used to depth of 6000 ft (1828 m). Moderate heat of hydration. Moderate resistance to sulfate. Available as a high-sulfate resistant variety.	Somewhat slower strength development than Type I; expensive. Can shrink and crack. Can be difficult to use. Should not be used in the presence of strong acids or in low-pH environments.
Class C	Similar to ASTM Type III.	Can be used to a depth of 6000 ft (1828 m).	Can set very quickly before decommissioning is completed. Should not be used in the presence of strong acids or in low-pH environments. Can shrink and crack.
Class G	Useful in a wide range of temperatures and depths through the use of accelerators or retarders.	Can be used to a depth of 8000 ft (2438 m). Available as a sulfate-resistant variety.	Should not be used in the presence of strong acids or in low-pH environments. Can shrink and crack.
Class H	Useful in a wide range of depths and temperatures through the use of accelerators or retarders.	Can be used to a depth of 8000 ft (2438 m). Available only as a moderate sulfate type.	Should not be used in the presence of strong acids or in low-pH environments. Can shrink and crack.
Class J	Intended for use from a depth 12 000 to 16 000 ft (3658 to 4877 m).	Has use where extremely high temperatures and pressures occur.	Should not be used in the presence of strong acids or in low-pH environments. Can shrink and crack.
Pozzolanic cement	Addition of silicious materials to ASTM Type V or API Class A cement.	Good resistance to corrosive conditions and in reducing the permeability of cement.	Many types of materials can be used that can result in variable results.
Epoxy cements	Vinyl ester resins.	Good chemical resistance to acids and bases. Can use available equipment to place cement.	Very expensive. Poor chemical resistance to chlorinated hydrocarbons and acetic aid. Should be used only by experienced personnel. Water accelerates curing, must use diesel oil to precondition hole (diesel may increase contamination of site if hydrocarbons are a concern).
Bentonite			
Pellets	Granular bentonite compressed into a tablet	Uniform in size. Easy to use. Produces a low permeability seal.	Must be hydrated after placement. Shrinkage may occur when desiccated or when in contact with high concentrations of organic compounds (greater than 2 %) or materials that are strongly acidic or alkaline. Expensive.
Chips	Raw mined montmorillonite in the form of chunks 1/4 to 3/4 in. (.64 to 1.91 m) in size.	Inexpensive. No mixing equipment required. Forms a low-permeability seal.	Difficult to place. Must be hydrated after placement. Less swelling than beneficiated bentonite. Shrinkage may occur when desiccated when in contact with high concentrations of organic compounds (greater than 2 %) or materials that are strongly acidic or alkaline.
Granular	Raw mined montmorillonite crushed and seared to an 8 to 20-mesh size.	Can be placed at depth in dry holes. Forms a low- permeability seal.	Difficult to place in holes containing water as it quickly hydrates. Can bridge in hole. May desiccate when in contact with high concentrations of organic compounds (greater than 2 %) or materials that are strongly acidic or alkaline causing shrinkage.

TABLE 1 *Continued*

Plugging Material	Description	Positive Attributes	Negative Attributes
Powdered	Pulverized and seared bentonite that passes a 200- mesh screen. Used as drilling mud or as an additive to cement.	Used with cement to compensate for shrinkage (under saturated conditions). Other additives can be used to inhibit swelling, etc. Retards cement set; lowers heat of hydration.	May not be a desirable plugging material in deep vadose zones due to the drying out of the material, resulting in cracking. Difficult to place in holes containing water, as it quickly hydrates. Can bridge in hole. May desiccate when in contact with high concentrations of organic compounds (greater than 2 %) or materials that are strongly acidic or alkaline causing shrinkage.
High solids clay grout	Powdered bentonite (200 mesh) mixed with fresh water to form a slurry with a minimum of 20 % solids and a density of 9.4 lb/gal (1126 Kg/m^3 g/L).	Does not shrink during curing. Low density reduces formation losses. Forms a low-permeability seal that stays flexible as long as it is hydrated.	May not be a desirable plugging material in deep vadose zones due to the drying out of the material, resulting in cracking. May desiccate when in contact with high concentrations of organic compounds (greater than 2 %) or materials that are strongly acidic or alkaline causing shrinkage. A low-strength material subject to expansion under low-pressure differentials such as artesian conditions.

extremely high temperatures and pressures can be expected to occur.

6.6.7 *Other Cements*—Other cements have been developed that may have applicability in decommissioning activities. These include the following:

6.6.7.1 Ultralight cements with a slurry density that can be as low as 6 lb/gal (719 cm Kg/L). This material can be made by foaming the cement with nitrogen or through the addition of hollow glass microspheres between 60 and 315 μm in diameter. The latter forms a slurry of between 9 and 12 lb/gal (1078 and 1438 Kg/L). Ultralight cements and microspheres have been reported **(4)** for cement unconsolidated sands and for plugging cavernous formations and lost circulation zones. Reference **(5)** provided similar information on microspheres. Microspheres can also be used in high-pressure applications when it may be desirable to limit density increases. Another advantage is the low water/cement ratio due to the low water absorbency and low density **(5)**.

6.6.7.2 *Pozzolanic-Portland Cements*—These cements consist of silicious materials that develop into a cement in the presence of lime and water. Both natural materials of volcanic origin such as perlites (volcanic ashes), heat-treated clays, shales, tuffs, opaline cherts, diatomaceous earth and artificial materials consisting of byproducts from glass factories, furnace slag, and fly ash have been used **(2, 4, 6)**. The large variety of materials that can be used as a source for pozzolans may result in variable results.

6.6.7.3 Pozzolans act to extend cement and decrease density. The specific gravity of fly ash ranges from 2.3 to 2.7 (depending upon the source) while portland cement is 3.1 to 3.2 **(2)**. These materials can also provide improved resistance to corrosive fluids. Table 2 provides a comparison of sulfate resistance between ASTM Type V cement with and without pozzolans.

6.6.7.4 The improved resistance to corrosive materials is accomplished in part as many pozzolans contain zeolites which have the ability for ion exchange between the corrosive material and the alkaline component in the cement **(7)**. Secondly, the use of pozzolans also decreases cement permeability over time. This occurs as a result of the increased percentage in hydrated cement containing materials resulting from the release of calcium hydroxide and the silica combining with lime from the cement to form a stable material **(8)**.

6.6.7.5 Pozzolans are added to portland cement by adding 74 lb (33.6 Kg) (as fly ash) per sack of cement. If perlites are used, 2 to 6 % of bentonite by weight is needed to keep the perlite from floating **(9)**.

6.6.8 Gypsum cements can be used for high early strength development and their ability to set rapidly. These cements expand approximately 0.3 %. This cement may have use in plugging highly permeable formations. Care should be taken in using gypsum cements due to the solubility of gypsum.

6.6.9 *Epoxy Resin Cements*—Epoxy (vinyl ester resins) cements may have applicability in decommissioning wells and boreholes where corrosive materials may be present **(5)**.

6.6.9.1 Epoxy cement consists of an epoxy base, hardener, accelerator, and inert filler. Resin viscosity is reduced by the addition of a nonreactive liquid diluent that also controls exothermic heat during polymerization. An inert solid filler such as very fine silica or barite is added to further reduce reaction heat and increase strength **(10)**.

6.6.9.2 There are several advantages of using epoxy cements. Reference **(11)** reports that a stronger bond occurred between the casing and the formation. Cole also reported resistance to various chemicals (see Table 3). The cement is highly resistant to high concentrations of hydrochloric and sulfuric acid, but is not suitable for use in environments where acetic acid, chlorinated hydrocarbons, or toluene are present.

TABLE 2 Comparison of ASTM Type V Cement With and Without Pozzolan Materials[A]

Cement Type	Relative Degree of Sulfate Attack	Percentage of Water Soluble Sulfate (as SO_4) in Soil, ppm	Sulfate (as SO_4) in Water Samples, ppm
V	Severe	0.20 to 2.00	1 500 to 10 000
V (plus pozzolan)	Very severe	2.00 or more	10 000 or more

[A] See Ref **(4)**.

TABLE 3 List of Chemicals Reported Not to Affect Epoxy Cement[A]

Chemical	Concentration, %
Hydrochloric acid	30
Sulfuric acid	25
Chromic acid	up to 10
Nitric acid	5
Hydroxide	up to 20
Hypochlorite	up to 6

[A] See Ref **(11)**.

6.6.9.3 This cement is expensive and requires removal of water (that reduces settling time) through the use of gelled and weighted diesel oil to precondition the hole. The use of this material may increase contamination at the site, if diesel oil (hydrocarbons) are a concern at the site.

6.6.10 *Cement Additives*—A number of materials can be added to cement to modify properties to meet a specific need. Cement additives can be used to extend, accelerate, retard, increase density, control fluid losses, control circulation losses, or reduce friction **(4)**. Several of these materials have more than one use; for example, sodium chloride can be used to accelerate or retard cement. The most common cement additives are discussed in the following subsections. Figure 1 lists these additives and presents the relative impact of their use on selected performance criteria.

6.7 *Extenders:*

6.7.1 Bentonite is the most commonly used material in modifying cement properties. It can be added to most ASTM and API cements. In decommissioning activities, the percentage of bentonite added is generally no more than 4 %. It has the following effects when added to cement **(4)**:

6.7.1.1 Lowers the hydraulic conductivity of the cement;

6.7.1.2 Increases slurry viscosity;

6.7.1.3 Reduces fluid loss to the formation;

6.7.1.4 Provides for a longer pumpability at normal pressures as a result of delaying strength development;

6.7.1.5 Reduces compressive strength; and

6.7.1.6 Lowers resistance to chemical attack.

6.7.2 Bentonite increases shrinkage as it ties up large volumes of water that would normally be in the cement. A second common extender are pozzolans which have already been addressed in 6.6.7.2.

6.8 *Accelerators:*

6.8.1 Accelerators hasten the settling of cement and are useful when voids occur, or when cement plugs are to be used in the first pour. Two common materials are used; calcium chloride and sodium chloride.

6.8.2 Calcium chloride is available as a powder or flake. Flakes are the most commonly used form, as it is easy to store and can absorb some moisture without becoming lumpy **(2)**. Two to four % of calcium chloride by weight is used to achieve maximum acceleration. The use of calcium chloride should be considered when a rapid set, a decrease in viscosity, and early strength are desired.

6.8.3 Sodium chloride can be added between 1.5 to 5 % by weight of cement to reduce setting time. Maximum acceleration occurs at a concentration of 2 to 2.5 % except when higher water ratios are used **(2)**.

6.9 *Retarders:*

6.9.1 Sodium chloride can be used to retard the settling of cement as well as accelerate the setting of cement. Fifteen to seventeen % salt (14 to 16 lb (6.35 to 7.26 Kg) of salt per sack) by weight is added to retard cement **(2)**.

6.9.2 Other chemicals (cellulose, lignosulfates) have been used as retarders, but are not appropriate for decommissioning activities without additional information on their compatibility with waste and their effect on water quality. Reference **(12)** indicates that sugar-derived retarders such as cellulose lignosulfates are destructive to cement strength and should not be used where strength is important. Organic

		Bentonite	Diatomaceous Earth	Pozzolan	Sand	Barite	Hematite	Calcium Chloride	Sodium Chloride*	Lignosulfonates	CMHEC+	Low-Water-Loss Materials	Lost Circulation Materials
DENSITY	Decrease	⊗	⊗	⊗									
	Increase				⊗	⊗	⊗		x				
WATER REQUIRED	Less												
	More	⊗	⊗	x	x	x							x
VISCOSITY	Decrease							x		⊗			
	Increase	x	x	x	x	x	x						
THICKENING TIME	Accelerated	x						⊗	⊗				
	Retarded		x						x	⊗	⊗	x	
EARLY STRENGTH	Decreased	x	x	x		x				⊗	⊗	x	x
	Increased								x				
FINAL STRENGTH	Decreased	x	⊗	x		x	x				⊗	x	x
	Increased									x			
DURABILITY	Decreased	x	x										x
	Increased			⊗									
WATER LOSS	Decreased	⊗								x	⊗	⊗	x
	Increased		x										

x Denotes minor effect.
⊗ Denotes major effect and/or principal purpose for which use.
+ Carboxymethyl hydroxyethyl cellulose.
* Small percentages of sodium chloride accelerate thickening.

NOTE—See Ref **(3)**.

FIG. 1 Effects of Some Additives on the Physical Properties of Cement

retarders should not be used for decommissioning activities.

6.10 *Density Improvers*—The density of cement can be improved to increase hydrostatic pressure. Sand can be used to increase density without affecting the cement chemically although additional water is required. Barite has been used, but may interact with waste and should not be used.

6.11 *Fluid Loss Controllers:*

6.11.1 Various organic materials such as cellulose can be used to produce a constant water to solids ratio that may have applicability when a grout is placed under pressure and water loss can occur. However, these materials may not be suitable for decommissioning activities, as they may contribute to contamination.

6.11.2 These organic materials (fibrous materials, cellophane flakes) act to block the movement of the grout into the formation. It is not desirable to use these materials in decommissioning activities due to their organic content that may adversely affect water quality and also may not result in a good plug.

6.12 *Friction Reducers (Dispersants):*

6.12.1 These materials reduce friction to improve flow and can be effective when the water cement ratio is reduced. Reduction of the water cement ratio is a method to decrease cement friction. (It is possible to reduce the amount of water added by using a dispersant **(5)**. These materials (sodium chloride, polymers, and calcium lignosulfonate) also help to reduce the energy required to pump the grout. Polymers and calcium lignosulfonate may not be appropriate materials for decommissioning activity as they may affect water quality.

6.13 *Bentonite*—Bentonite is predominantly composed of the clay mineral sodium montmorillonite. It has the ability to absorb large quantities of water and swell to many times its original size when hydrated, and the material remains flexible. Bentonite clay may be used in any of its various forms to meet placement, strength, and sealing criteria listed in 6.3. The amount of shrinkage or settling of a bentonite seal is dependent on the percent solids of bentonite, composition of surrounding formation and its soil moisture. Higher water to bentonite ratios increase the likelihood of dehydration.

6.13.1 The permeability of bentonite is very low; hydraulic conductivities of 1×10^{-6} cm/s or less can be achieved. However, bentonite may desiccate in the presence of high concentrations of some organic chemicals, strong acids or bases, saline ground water, or when allowed to dry, thereby increasing its hydraulic conductivity. Bentonite is commercially available in the following forms:

6.13.1.1 *Pellets*—Pellets are made from granular powdered bentonite that has been compressed into tablets, commonly ¼ to ¾ in. (.64 cm to 1.91 cm) in diameter. Pellets have a low-moisture content, high density, and uniform size. Pellets should be composed of additive-free, high-swelling granular sodium bentonite. Properly placed in a well or borehole, pellets hydrate and expand creating a low permeability (1×10^{-6} cm/s) plug. Pellets can be used in the saturated zone provided the length of the water column is short. The rate of pour into the hole should not be more than 50 lb (22.7 Kg) of bentonite in 5 min **(1)**.

6.13.1.2 *Preformed Donuts*—Commercial preformed donuts consist of compressed bentonite and may have use in decommissioning activities.

6.13.2 *Chips*—Raw mined sodium montmorillonite in the form of chunks that are ¼ to ¾ in. (6.4 to 1.91 cm) in diameter. Their angular shape can make it difficult to place chips to the desired depth in a small-diameter well or borehole without bridging.

6.13.2.1 Fine-grained material resulting from the mechanical breakdown during shipping may cause a problem in the placement of chips due to clumping. Fines should be screened through a ¼-in. (6.4-mm) mesh screen before use.

6.13.2.2 The lower affinity for water that chips have allow them to fall through a water column without rapid hydration.

6.13.2.3 Chips have applicability in large-diameter boreholes and when carefully dropped into the hole to reduce bridging.

6.13.3 *Granular*—Raw-mined sodium montmorillonite without any additives that has been crushed and seared to an 8 to 20-mesh size. This material can be placed at depth in dry holes but hydrates quickly when placed into water. It often sticks to wet borehole walls and bridges when placed through water. Granular material is best suited for use in the unsaturated zone with enough water added to provide adequate hydration.

6.13.3.1 Fines can clump when in contact with water **(1)**. Fines result from mechanical breakdown of the material during shipping. Granular bentonite should be poured slowly to reduce the potential for bridging. In some situations, a pour rate not exceeding 50 lb (227 Kg) in 5 min has been used successfully **(1)**.

6.13.4 *Powdered*—Untreated, seared, and ground bentonite that passes through a 200-mesh screen. It is designed to be used in drilling fluids (muds) and as an additive to other plugging materials such as cement. Bentonite powder slurry can become an effective grout material when combined with density-increasing additives and swelling inhibitors. Powdered bentonite should not be placed in dry form through water as it can bridge and stick to the borehole walls.

6.13.5 *High Solids Clay Grout*—This material is a blend of powdered polymer-free bentonite clays mixed with fresh water that forms a slurry with a minimum 20 % solids by weight and a density of 9.4 lb/gal (1126m^3 Kg/L). The slurry sets to a low-permeable plastic grout that generates no heat of hydration and does not shrink during curing in the presence of moisture. High solids clay grouts are commonly used for borehole plugging.

6.14 *Other Materials:*

6.14.1 A number of other materials have been used for plugging:

6.14.1.1 Attapulgite clay (may have applicability when used with a salt cement grout)[5]

6.14.1.2 Fire clay

6.14.1.3 Commercial packing materials

6.14.1.4 Packers

6.14.2 These materials are either inappropriate for use in decommissioning wells or boreholes where hazardous waste are encountered, or are not well studied for decommissioning wells in hazardous waste situations. Therefore, they

[5] Sutton, Fred, Personal Communication, 1990.

are not discussed in this guide.

6.14.3 Other materials have been used in the past for plugging wells and boreholes. Such materials as wooden or lead plugs should not be used because wood plugs may decay and lead is a potentially hazardous material. Mechanical packers composed of steel, plastic, or other materials can be used to assist in plugging.

7. Procedure

7.1 The primary purpose of most boreholes and monitoring wells is to monitor chemical compounds in the soil and ground water; however, there are other uses for subsurface monitoring including the measurement of temperature, soil gas sampling, or measurement of geophysical parameters. Use significant care in planning and implementing the decommissioning activity. It is important to obtain any required approvals from regulatory agencies, land owners, responsible parties, and other parties involved with the site. The following subsections present a recommended list of tasks in order that the decommissioning activity is successfully completed. Several of the steps outlined below do not pertain to boreholes and may be omitted.

7.2 *Planning:*

7.2.1 *Records Review*—Carefully review all available records and information relating to use of the monitoring well, borehole, etc. This review may include the following information:

7.2.1.1 Review applicable Federal, state, and local regulations relating to decommissioning activities. This may include contacting the applicable state or local agency having jurisdiction over drilling activities and preparation of the necessary documentation to drill (start card),

7.2.1.2 Collection of drillers' logs, geophysical logs, well construction, or geologic logs, including stratigraphy, structural geology, subsurface information, construction materials, screened interval, depth, hydraulic gradients (if water levels are available from other wells for its determination), legal location, date of installation, and photographs of the well,

7.2.1.3 Review of analytical chemical data for soil and ground water over the life of the well, and variations in water levels over time,

7.2.1.4 Review of records of the repairs, modifications, or other changes made to the well during the lifetime of the well,

7.2.1.5 Evaluation of historic, current, and planned land use,

7.2.1.6 Interviews with local workers and collection of other pertinent data such as discussing site conditions with local drillers, and

7.2.1.7 While not directly part of the decommissioning activity, proper disposal of displaced fluids and other materials (such as pulled or drilled out casing and cement seals) should be considered. Some of these materials may be classified as a hazardous waste under Federal, state, or local regulations. Conduct a review of these regulations and appropriate analytical documentation prior to classifying a material as a hazardous waste.

NOTE 9—This information may be summarized in a work plan. A work plan would also include a description of the site geology and hydrogeology and the decommissioning method to be used.

7.2.2 *Verification of Field Data*—The variety and quality of field practices and reporting require that the well be inspected to verify the actual field situation prior to decommissioning the well. The following list of procedures is recommended so that the actual condition of the well is known. Some of the borehole geophysical logs may not be applicable or may not be available for small diameter (2 in. (5.08 cm) or less) holes or wells.

7.2.2.1 Inspection of well head installation for integrity,

7.2.2.2 Current depth measurement of the casing and well. (The original depth of the borehole may be different than the well.),

7.2.2.3 *Water Quality Sampling and Analysis*—A final water quality sample taken from the well may be required for regulatory purposes;

7.2.2.4 *Downhole Inspections*—Including caliper logs to measure inside diameter; television logs to determine in-well conditions such as casing breaks, screen size, etc.; gamma logs to verify geologic information, if not already available; cement bond logs (sonic) to determine if the casing is firmly attached to grout (presently available for holes 2½ in. (6.35 cm) or larger in diameter); flow logs (flow meter or spinners) to determine if vertical flow occurs within the casing; and hydraulic integrity test to determine if the well casing is intact.

NOTE 10—Care should be taken in running any of these tools in a well with a collapsed or broken casing, or in boreholes that may collapse on the tool. Tools with active radioactive sources should not be used under these circumstances. Conduct downhole inspections only after obstructions are removed from the well casing or borehole.

7.2.2.5 Contact local owner, resident operator/observer to verify operations at the site.

7.2.2.6 Verification of field data is an ongoing responsibility. Use verified information to modify plans in order that the decommissioning activity is correctly conducted. Continue this activity during the field phase and change specifications as needed.

7.2.3 *Review of Decommissioning Options*—After the records have been thoroughly reviewed and verified in the field, select an appropriate decommissioning procedure. Evaluate each possible option to determine the most appropriate method for the selection. The following list of evaluation criteria is recommended:

7.2.3.1 The potential for fluid movement from one aquifer into another by means of the borehole or well should be eliminated.

7.2.3.2 Materials to be used in plugging must be compatible with well casing and screen (if left in place), and with subsurface formation and ground water, etc. over the period of time hazardous materials are found at these sites.

7.2.3.3 Future land use (as is known at the time of decommissioning) should be compatible with decommissioning plans.

7.2.3.4 Closure options should be compatible with applicable federal, state, and local requirements.

7.3 *Implementation:*

7.3.1 *Field Procedure:*

7.3.1.1 Satisfactory completion of decommissioning is the primary purpose of this guide. All work performed on the borehole or well should be completed by competently trained drillers, equipped with appropriate tools, under the direction of a geological or engineering professional who is qualified to certify that the decommissioning is completed according to the planned procedures and is consistent with applicable regulations.

7.3.1.2 Approve any modifications to the proposed work plan and record in writing by the on-site geologist or engineer (or their representatives) prior to implementation.

7.3.1.3 The geologist or engineer should be on-site during the field activities to verify that the activities are completed as planned. Decommissioning operations can be successfully accomplished by careful planning and documentation (see 7.5). Maintain documentation of decommissioning activities for the post-closure period or period required by regulations (if specified). While regulations may require documentation for a period of 30 years, it is advisable to continue this activity for a period lasting as long as hazardous materials occur at the site.

7.3.1.4 Remove casing from the ground by either pulling or overdrilling (see 7.3.7). Depending upon construction, it may be necessary to leave the casing in place and produce suitable perforations in the screen and blank casing to allow for the plugging material to penetrate the annular space and formation (see 7.3.7). If grout in the annular space can be verified to be in good condition, the well can be decommissioned by cutting the blank casing and filling the screened interval with grout. Verifying the integrity of grout may be difficult to impossible. If a filter pack is present, it may be necessary to remove the filter pack after perforating the casing by washing or overdrilling. Several of the methods identified in this subsection are briefly discussed in the Appendix X1.

7.3.1.5 If well construction conditions are not adequately known and the well site contains hazardous materials, it may not be appropriate to remove the casing and screen, as this may increase the mobility of hazardous materials.

7.3.1.6 The borehole or well, or both, may require preconditioning for decommissioning to be successful. Preconditioning can reduce the potential for sloughing of the borehole wall if for example, a sodium montmorillonite clay occurs naturally in the formation and cement is used as the plugging material. The calcium contained in Portland cement exchanges with the sodium cation in bentonite clay decreasing the water contained in the clay and inducing sloughing. This problem can be significant in sediments or rocks that are under considerable pressure, causing a loss in part of the hole, thereby not completely plugging the borehole and possibly causing loss of all downhole equipment. These conditions are usually known by local drilling and well servicing contractors who should be contacted prior to the start of field operations.

7.3.1.7 Preconditioning consists of removing mud from borehole walls (when mud is used for drilling the borehole), or stabilizing a borehole prior to placement of the plugging material. If a drilling mud has been used to drill a borehole, preconditioning can involve the circulation of a high-quality, low-solids drilling fluid to remove gelled mud from the borehole and borehole walls prior to plugging. For the above situation, a high-quality bentonite or drilling fluid can be used. If a drilling mud is used as the plugging material, prepare fresh mud. The mud used in drilling contains cuttings and may also not have suitable properties **(13)**. The selection and use of material(s) should meet the requirements specified in Section 6.

7.3.2 *Volume of Plugging Material Required:*

7.3.2.1 The geologist or engineer should calculate the volume of plugging material required for the borehole or well after first taking into consideration applicable loss of material to the formation, voids intersecting the borehole, changes in borehole diameter, washout zones, and swelling or shrinkage of material. An approximation of the volume of material that may be required can frequently be provided by contacting local drillers or professionals.

7.3.2.2 Loss of plugging material into the formation may occur rapidly (within minutes) or after several hours or days. The volume of plugging material required to be on-site should be at a minimum, enough to fill one borehole volume, however, it is advisable to have available a minimum of 25 to 50 % in excess of the calculated borehole volume **(1, 2)**. Additional plugging material should be readily available to site personnel under short notice, especially if it is common to lose plugging material into the formation or the material is required for clearing a hole (see 7.3.8.2).

7.3.2.3 A caliper log is helpful in boreholes or overdrilled holes to define hole diameter. If this information is not available, use an estimate of borehole diameter from the available well construction specifications. Calculate the volume of the borehole using the following:

$$v = \pi r^2 L$$

where:

v = volume,
L = length of borehole or well to be plugged, and
r = radius of hole.

7.3.2.4 Manuals listing the volume of a hole per linear foot (metre) such as Ref **(3)** can also be used. These manuals are available through major deep well cementing service companies. Assume in the calculations or table used that the derived volume is the minimum required for actual conditions due to possible loss of plugging material into the formation. Verify grout emplaced through estimating the volume of material leaving the hole for holes filled with mud or water. For all boreholes, measure the volume of material emplaced and check for the appearance of the material at the top of the interval grouted or land's surface, whichever is applicable.

7.3.2.5 Location of the grout at depth can be difficult because the top of the grout may not be able to be distinguished from water or other fluids in the borehole. One field procedure that has been applied to measure the level of the grout is to use a wooden sounding block with a weight attached to the block. The weight should be slightly denser than the grout.

NOTE 11—Location of the depth of plugging material in a borehole or well does not confirm that the plug has performed its intended purpose. Plugging materials may not have adhered to the walls of the borehole or well or may have been bypassed from the target zone. Therefore, placement of plugging materials is highly dependent upon proper preparation of the hole.

7.3.2.6 In areas where coarse materials are encountered, or considerable fracturing or solution openings occur, two or three times the calculated volume of material may be required to fill the hole. In extreme cases such as in some karst terrains where large conduits may exist, it may be difficult to plug a well or borehole. In these situations, packers can be used to isolate critical intervals for filling with plugged materials above or below the karstic formations. Packers may also be used to isolate critical intervals within a well or borehole throughout the entire depth.

7.3.3 *Quality of Water Used for Grout*—Grout must be carefully mixed using water of a known chemical quality. The quality of the water must be compatible with the grouting material and not introduce contamination into the subsurface. For cement, water used should be free of silt, organic matter, alkali compounds, and have a total solids content of 500 mg/L or less **(14)**. The grout should be weighed using a mud balance.

NOTE 12—Thicker grout mixtures may have greater effectiveness in plugging highly permeable materials or materials with large voids, but may be very difficult to pump through a grout pipe. Less viscous mixtures are easy to pump, but may be too mobile, have greater shrinkage, and take longer to set if additives to compensate these problems are not added.

7.3.4 *Mixing of Cement and Bentonite On-Site*—Both cement and bentonite can be mixed on site. Cement can also be purchased from a contractor and delivered premixed on site, provided the water used is of known and acceptable quality.

7.3.5 *Use of Curing Accelerators*—Curing accelerators can be used to hasten the initial set time for cement from 6 to 8 h to perhaps 2 or 3 h, provided the accelerator (typically calcium chloride, sodium chloride, aluminum powder, or gypsum) does not degrade the cement or react with the environment. The use of these materials may also require approval by regulatory authorities.

7.3.6 *Decommissioning of Boreholes:*

7.3.6.1 To achieve an effective seal, the borehole should be free of debris and foreign matter that may restrict the adhesion of the plugging materials to the borehole wall. Clear boreholes of excessive mud filtercake, or gelled mud (if used) and any bridges resulting from the removal of temporary casing, or when noncohesive materials (such as sand and gravel) are encountered that can lead to a collapsed borehole during decommissioning activities.

7.3.6.2 One method commonly used for effective removal of these materials is to advance a small grout pipe to the bottom of the borehole by use of water or a high-quality bentonite slurry and flush the hole. Flushing is continued until the blockage is removed, or noncohesive formations are stabilized. Use at least two borehole volume of materials. If the flushing fluid returning to the surface contains potentially hazardous materials, it may be necessary to place this material into containers for proper disposal.

7.3.6.3 As soon as the borehole is prepared, pump plugging material slowly through a grout pipe to displace the flushing fluid. Inject grout starting from the bottom of the hole, forcing other fluids upward. Complete grouting slowly to prevent channeling of the grout around any undesirable material remaining in the hole. Complete this operation in one continuous operation. Raise the grout pipe when pumping pressure increases significantly, or when undiluted grout reaches the surface. Regularly sample and evaluate overflowing grout for weight, presence of foreign material, or other changes. When the overflowing grout is similar to that being pumped down the hole, the plugging is considered complete.

7.3.6.4 Grout pipes should be the largest diameter that is practical for field operations to reduce the required pumping pressure. Cut the lower end of the pipe at an angle to allow for the side discharge of the grout. Side discharge reduces the potential for erosion of the borehole wall. Typically, heavy-walled PVC (Schedule 40 or greater) or thin-walled steel pipe is used for grouting purposes. Use caution while grouting deeper holes in order that downhole pressure does not exceed the rupture strength of the grout pipe.

7.3.6.5 Grouting of shallow auger holes or other boreholes where grout pumping equipment is not readily available may be accomplished by placing grout through a side discharge grout pipe that has a funnel attached to the top. As the grouting progresses, slowly raise the pipe. Take care when using this procedure, as the low placement pressure may not completely fill or flush undesirable materials from the borehole. This procedure is only recommended when the entire borehole depth is 10 to 15 ft (3 to 4.6 m) or less.

7.3.6.6 Small diameter (<2-in. (5.08 cm-mm)) holes are difficult to plug. A small diameter (¾-in. (1.9 cm-mm)) grout pipe can be used, however, high-pumping pressures may be required or less viscous plugging materials may be necessary.

NOTE 13—It is important to avoid hydrofracturing or blowout in shallow holes in unconsolidated materials when pressure grouting. A general rule to avoid hydrofracturing is to restrict pumping pressure to about 0.6 lb/in.^{2} (0.42 Kg/cm^2) of hydraulic pressure for each foot (metre) of overburden. In some instances, the pressure used must be reduced further.

7.3.6.7 A conservative method **(15)** of calculating maximum pressure is as follows:

$$Pm = (0.733 - 0.433\ Sg)d$$

where:

Pm = pressure of fluid injected at wellhead, lb/in.^{2} (Kg/cm^2),

Sg = density (specific gravity) of the plugging material (grout) (unitless), and

d = depth measured for the surface to the opening of the grout pipe (metres) feet.

Hydrofracturing typically occurs during start-up or restart of grouting.

7.3.6.8 Inject grout at the bottom of the hole, displacing grout, loose formation materials, and borehole fluids upward. Slowly pump grout to avoid channeling. Do this in one continuous operation. Raise the grout pipe when grout can no longer be easily forced from the pipe into the hole or when undiluted grout reaches the top of the hole and flows out. Sample and weigh this material. The weight of the grout returning to the surface should be the same as the grout pumped into the hole.

7.3.7 *Control of Elevated Formation Pressures*—Occasionally, a borehole or well may penetrate a formation that is under confined conditions (artesian head), or from which a gas is being released, under pressure. (Gas bubbling upward through the grout may result in open channels, or increase the permeability of the grout.) When this condition is encountered, sealing of the borehole requires that the grout pressure be maintained greater than the formation pressure until initial grout set occurs. The "shut-in-pressure" of the formation can usually be determined by use of pressure gages attached to a casing that has been pushed or driven into the confining unit. This casing must be tightly sealed to prevent leakage around its annulus. (Separate packers or casing grout, or both, may be required to prevent leakage around the casing.)

7.3.7.1 Several procedures may be used to balance the formation pressure until the initial set of the grout has occurred. (See Fig. 2.) The procedure most often used to contain the pressure is to use a sufficient column of grout. If additional head pressure is required, increase the unit weight of the grout by the addition of sand, densifying additives (barite, or hematite) or decreasing the water cement ratio. Another alternative commonly used is to pump the grout through a check valve into the casing until the top of grout reaches the desired elevation. Maintain pressure above the grout by use of compressed air, or by use of a simple standpipe filled with water. If the standpipe procedure is used, pump grout through a tremie pipe extending through the water to prevent dilution.

7.3.7.2 After the initial grout set has occurred (minimum 24 h), the sustaining pressure (above atmospheric pressure) may be released. However, record the air pressure (or fluid level) in the air filled casing above the grout frequently for several days to verify that no leakage is occurring. If gas (of any variety) is present in the formation, analyze a sample of the air from the casing above the grout. Care should be taken under these conditions, as explosive or hazardous conditions may be present. Methane, the most commonly produced gas, is odorless, colorless, toxic, and extremely flammable. Safety precautions are advised.

7.3.7.3 Only after testing results confirm that the seal is effective, is it advisable to complete the permanent closure or removal of the near (or above) surface casing.

7.3.7.4 Measure any fallback or settlement of the grout surface (whether by gravity or under pressure) that occurs after grout emplacement to the nearest foot and record. Correct any fallback with native or imported materials to grade, or to a specified depth below grade, such as below the plow layer. The depth is based on existing and proposed land use and regulatory requirements.

7.3.8 *Decommissioning of Wells and Other Monitoring Devices:*

7.3.8.1 This discussion has applicability in decommissioning ground water monitoring and injection wells and soil gas monitoring wells, neutron probe access tubes, lysimeter and tensiometer and installations, and similar devices. Appendix X1 should be consulted for additional information.

7.3.8.2 It is desirable to remove all existing well construction materials such as screen, casing, filter pack, seal, and grout from the hole to reduce the potential for the formation of a vertical conduit to occur at the contact between casing and annular seal; reduce the potential of these materials interfering with the decommissioning operation; decrease the potential of a reaction between the materials used; or to minimize the reaction with the native materials or ground water. In situations where well materials are removed and borehole collapse occurs, redrill the borehole following the guidance provided in 7.3.6.

7.3.8.3 Steel casing may be removed using jacks to free casing from the hole followed by lifting the casing out by using a drill rig, backhoe, cranes, etc. of sufficient capacity. If

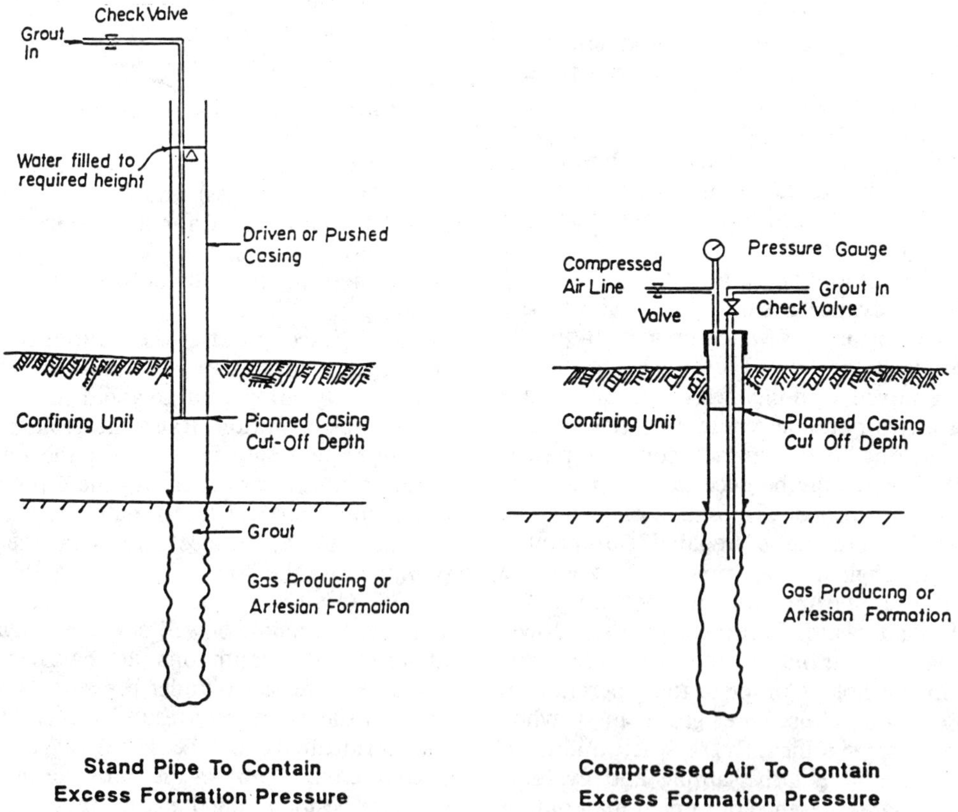

FIG. 2 Two Procedures That Can Be Used to Overcome Elevated Formation Pressure

the annular space has been cemented over a long distance, this method may not be readily used unless a poor contact occurs between the casing and borehole. Small lengths of cement (typically less than 10 ft (3 m)) can be removed along with the casing if the drill rig has sufficient lifting capacity. When the casing cannot be removed, perforate the casing and screen using a suitable tool. This is necessary as encrustation or corrosion of the screen can block or completely close openings. A wide variety of commercial equipment is available for perforating casings and screens. Due to the diversity of application, consult an experienced contractor prior to selection of the technique. A minimum of four rows of perforations several inches (millimetres) long and a minimum of five perforations per linear foot (metre) of casing or screen is recommended.

7.3.8.4 Remove steel casing by overdrilling (overreaming) the casing. Overreaming can be done using an overreaming tool (see Fig. 3). Select a pilot bit that is nearly the same size as the inside diameter of the casing. The reaming bit should be slightly larger than the borehole diameter to remove all well construction materials and a small amount of native material. As drilling proceeds, the casing, grout, bentonite seal, filter pack, and other well materials are destroyed and returned to the surface. In situations when the grout in the annular space can be verified to be in good condition, it may be very difficult to remove casing and grout and grout can be left in place by pressure grouting the screened interval and casing. Verifying the integrity of the grout can be difficult.

7.3.8.5 PVC and other low tensile strength materials generally cannot be removed by pulling if they have been properly cemented into place. Overdrilling is necessary to remove these casing and screen materials.

7.3.8.6 A hollow stem auger equipped with outward facing carbide cutting teeth with a diameter 2 to 4 in. (50.8 to 1016 cm) larger than the casing may be used for overdrilling. Place the lead auger over the casing and rotate downward (see Fig. 4). The casing guides the cutting head and remains inside the auger. When the full diameter and length of the well has been penetrated, the casing and screen can be retrieved from the center of the auger. It is important to use outward facing cutting teeth in order that the cutting tool does not sever the casing and drift off center. An alternative is to install a steel guide pipe inside the well casing so that the augers can be centered.[6] Firmly attach this temporary working pipe to the inside of the casing by use of a packer, or other type of expansion or friction device. When the auger reaches full depth and the well materials have been removed, pump plugging materials through the hollow stem as the augers are withdrawn.

NOTE 14—Local regulations may allow leaving the PVC casing in place, if a good annular seal exists, or just removing blank casing if the integrity of the annular seal can be documented. It is generally difficult to document a good annular seal (tight fit, no fractures or channels) without the use of a cement bond log or similar method. A cement bond log cannot be run in small diameter wells (less than 2.5 in. (6.35 cm) in diameter) because a small diameter tool is not currently available. Also, this method is not reliable for PVC casings.

7.3.8.7 After removal of the casing, decommissioning can be completed in accordance with 7.3.6.

[6] Baker, Robert, Personal Communication, Layne Environmental, 1990.

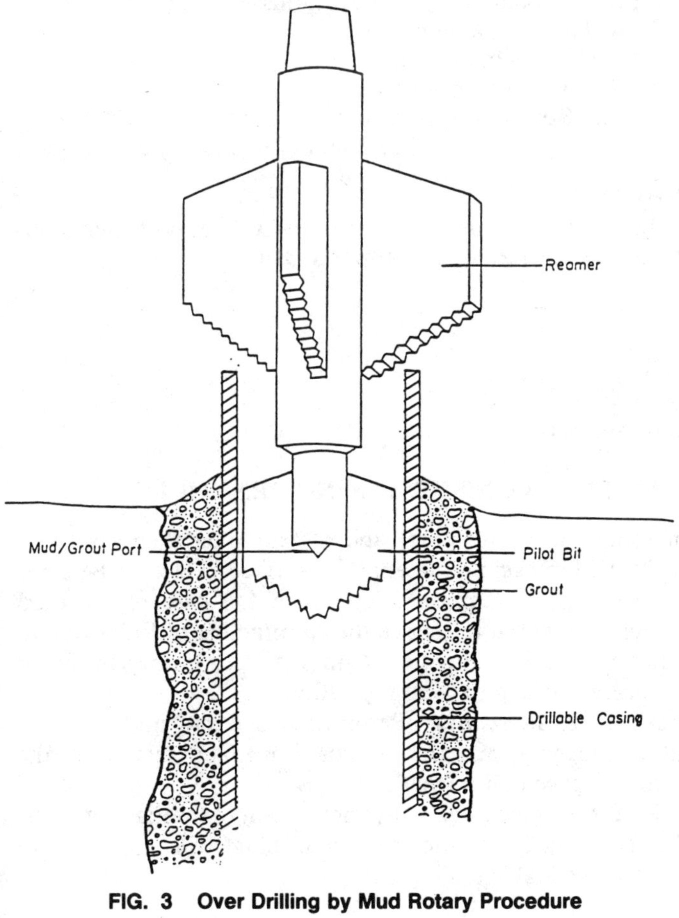

FIG. 3 Over Drilling by Mud Rotary Procedure

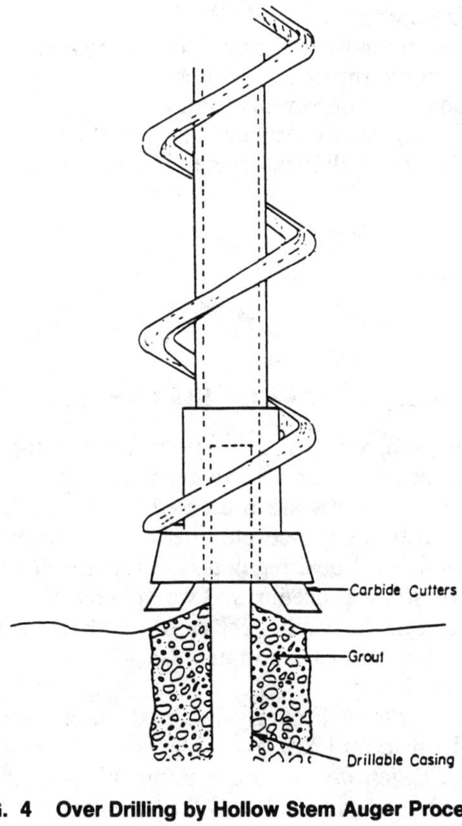

FIG. 4 Over Drilling by Hollow Stem Auger Procedure

NOTE 15—Wells with a maximum depth of 10 to 15 ft (3 to 4.6 m) can be removed by overdrilling, or when possible by pulling the casing out of the ground. The open hole is then filled with grout or other suitable plugging materials. A tamping rod can be used to tamp the material into place.

7.4 *Decontamination of Downhole Equipment:*

7.4.1 If hazardous waste or other contaminants occur at the site, decontaminate all tools placed into the borehole or well prior to entry (see Practice D 5088). Follow state and local regulations, if applicable. A hot water pressure wash or steam cleaning are examples of two typical decontamination methods. Other methods such as the use of solvents followed by rinsing with clean tap water and allowing equipment to air dry may also be acceptable. Also contain water and other materials used for decontamination for chemical analysis to determine proper disposal methods.

7.4.2 Do not flush grout pipes out with water while in or above the hole. Conduct all cleaning operations aboveground and manage appropriately as a hazardous or nonhazardous situation depending upon chemical analytical data on the site.

7.5 *Documentation:*

7.5.1 *Inspection Records*—The primary purpose of records of field work is to provide that appropriate measures have been taken so that the borehole or monitoring well is permanently decommissioned in a manner that minimizes it from being a conduit for fluid, water, or vapor migration. Properly decommissioned boreholes and wells should have no adverse influence on the local environment than the original geological setting.

7.5.2 *Narrative Report*—Maintain the narrative report of activities as a permanent record of the decommissioning activities and include the full set of field activities including:

7.5.2.1 Decommissioning date,

7.5.2.2 Personnel (with listing of company representation, including phone number and address),

7.5.2.3 Source of decommissioning method,

7.5.2.4 Step-by-step procedures used in the field,

7.5.2.5 Record of all measurements made, depths encountered, types and volume of fluids pumped and photographs before and after, and

7.5.2.6 All other pertinent information that is required based on site conditions or regulatory requirements. Any problems encountered should also be documented in detail.

8. Report

8.1 At the completion of field work, present a formal report to document the entire procedure used in the decommissioning process for the borehole or well. Additional information or requirements may be specified in government regulations and should also be included in this report. Report in detail the following items as a minimum:

8.1.1 *Background Information:*

8.1.1.1 Location of the borehole/well,

8.1.1.2 Purpose of the borehole/well,

8.1.1.3 History of the borehole/well,

8.1.1.4 Chemical parameters and concentrations present during the active life of the well and at the time of plugging; and

8.1.1.5 Information collected prior to the inspection record.

8.1.2 *Testing Prior to Decommissioning:*

8.1.2.1 Site characterization (if applicable),

8.1.2.2 Physical testing (depth, structural condition, etc.),

8.1.2.3 Chemical testing results (last chemical analysis for record), and

8.1.2.4 Geophysical logging results (if any).

8.1.3 *Decommissioning Design Procedure:*

8.1.3.1 Rationale for selection of method used, and

8.1.3.2 Presentation of a decommissioning plan.

8.1.4 *Field Implementation:*

8.1.4.1 Description of activities,

8.1.4.2 Variance from plan, and

8.1.4.3 Result of testing and measurements.

8.1.5 *Confirmation*—Statement of regulatory compliance.

9. Keywords

9.1 abandonment; decommissioning; ground water monitoring wells; hazardous waste; plugging

APPENDIX

(Nonmandatory Information)

X1. EXAMPLES OF SUCCESSFUL MONITORING WELL DECOMMISSIONING PRACTICES

X1.1 This appendix covers a general discussion of field procedures that have been used to decommission monitoring wells. These procedures are presented to provide technical guidance to assist in the development of decommissioning wells or boreholes. Local regulations may not allow implementation of some procedures. Two subsets of procedures are outlined, one for rigid steel casing and the second for lower tensile strength casing materials (plastic).

X1.2 *Steel Casing:*

X1.2.1 *Condition*—Properly grouted steel casing and screen, difficult to pull casing:

X1.2.1.1 *Suggestion*—Pressure pump plugging materials into well to above aquifer, place 3-ft (.91-mm) plug of plastic material (bentonite plug) above rigid plug then complete with rigid plugging material to surface. Place a PVC or neoprene plug or packer, located at the base of the rigid (cement) material to reduce the co-mingling of fluids due to different specific gravities. Complete decommissioning at surface according to local regulations.

X1.2.1.2 *Alternative*—Perforate steel casing and screen to allow plugging material to come into contact with the annular space and formation.

X1.2.1.3 *Advantage*—If (when) casing material corrodes, the continually plastic bentonite plug will continue to provide a suitable sealing zone.

X1.2.2 *Condition*—Steel well materials, that are removable:

X1.2.2.1 *Suggestion*—Pull screen and casing, then redrill borehole to original depth using slightly larger diameter drill bit than the borehole. Complete closure by pressure sealing the borehole by grout pipe from bottom to top.

X1.2.2.2 *Advantage*—Casing and annular materials are completely removed and replaced with sealing material of equal or lower permeability (saturated zone) than the native geologic materials.

X1.2.3 *Condition*—Shallow screened well completed into permeable sand or gravel aquifer:

X1.2.3.1 *Suggestion*—Pull the casing and screen (grout also, if possible) then redrill the borehole to original depth. Completion should be by pressure sealing to the surface.

X1.2.3.2 *Advantage*—The possibility of hydraulic connections with the surface is greatly reduced.

X1.2.4 *Condition*—Steel casing and screen set into highly permeable formation (such as fractured limestone) that results in lost fluid circulation during drilling or decommissioning procedures:

X1.2.4.1 *Suggestion*—Fill highly permeable section with quick setting cement grout (addition of calcium chloride), or coarse gravel to base of casing, or hang "cement basket" at base of casing. Complete decommissioning with plugging materials pumped into place, including at least one section of permanent plastic seal.

X1.2.4.2 *Advantage*—Extremely permeable or highly cavernous materials may not be able to be filled sufficiently to allow closure similar to other cases. In such situations, prevention of vertical migration of fluids to or from that zone is the primary purpose of decommissioning.

X1.3 *Low-Tensile Strength Materials (that is, PVC, ABS, PTFE):*

X1.3.1 *Condition*—Properly constructed well that has a sealed annulus from the surface to the top of screen:

X1.3.1.1 *Suggestions*—Pressure grout in place after perforating the entire casing so that the screen and gravel pack are filled with grout; or over-ream the well using a drill bit diameter that is a minimum or slightly larger than the diameter of the annular space.

X1.3.2 *Condition*—Annulus is not permanently sealed and casing/screen cannot be pulled:

X1.3.2.1 *Suggestion*—Overdrill casing, screen and filter pack with a hollow stemmed auger or other suitable drilling method. Remove casing from bottom to top as augers are removed.

X1.3.2.2 *Advantage*—Complete well structure is removed. Grouting materials provide no avenue of fluid or gas migration after decommissioning.

X1.3.3 *Condition*—Annulus seal questionable; casing screen cannot be pulled:

X1.3.3.1 *Suggestion*—Use rotary tools equipped with pilot bit. Large portion of bit must be equal to or larger than diameter of original borehole. Drill out casing, annular material, to depth of well. Pressure seal after tools are removed.

X1.3.3.2 *Advantage*—Complete well structure is removed. Grouting materials do not provide pathways for fluid or gas migration after decommissioning.

X1.3.4 *Condition*—Annulus seal questionable, casing and screen may be pulled from firm clay soils:

X1.3.4.1 *Suggestion*—Use pulling rig equipped with down hole casing latch tool. Anchor tool at bottom of well and lift casing and screen. Pressure grout open hole from bottom to top. (This method may have applicability for shallow wells.)

X1.3.4.2 *Advantage*—This method is often possible in firm clay soils. Requires less expensive equipment than other procedures.

X1.3.5 *Condition*—Vadose zone monitoring devices such as lysimeters, tensiometers, neutron access holes, and similar devices installed vertically:

X1.3.5.1 *Suggestion*—Drill out the device and backfill as suggested in 7.3.8.5 and 7.3.8.6. (See also Note 15 in the text.)

X1.3.5.2 *Advantage*—The device and materials used to construct the device are removed reducing the potential for fluid movement from the surface to underlying materials and ground water.

REFERENCES

(1) Gaber, M. S., and Fisher, B., *Michigan Water Well Grouting Manual*, Michigan Department of Public Health, Lansing, MI, 1988, pp. 83.

(2) Roscoe Moss Company, *Handbook of Ground Water Development, Chapter 12, Water Well Cementing*, John Wiley, New York, NY 1990, pp. 232–252.

(3) Dowell Division of Dow Chemical, U.S.A., *Engineer's Handbook*, Houston, 1983, p. 77.027.

(4) Williams, Camila, and Evans, L., "Guide to the Selection of Cement, Bentonite, and Other Additives for Use in Monitor Well Construction," *First National Outdoor Action Conference on Aquifer Restoration, Monitoring and Geophysical Methods*, NWWA, Las Vegas, NV, 1987, pp. 325–343.

(5) George, Charles, and Bruce, T., "Cementing to Achieve Zone Isolation in Disposal Wells," in *International Symposium Subsurface Inspection of Liquid Wastes*, NWWA, New Orleans, LA, March 3–6, 1986, pp. 77–89.

(6) Anonymous, "Special Cements and Their Uses," *Concrete Construction*, Vol 18, No. 1, January, 1973, pp. 3–7.

(7) Berry, E., and Malhortra, V., "Fly Ash for Use in Concrete—A Critical Review," *Journal of the American Concrete Institute*, Vol 77, No. 2, 1980, pp. 57–73.

(8) Coleman, R. J., and Corrigan, G., "Fineness and Water-to-Cement Ratio in Relation to Volume and Permeability of Cement," *Petroleum Technology, Technical Publication No. 1266*, 1941, pp. 1–11.

(9) Dowell Schumberger, *Cementing Technology*, Nova Communications Ltd., New York, 1984.

(10) Creech, John, 1986, "Class I Injection Well Design Considerations Using Fiberglass Tubular and Epoxy Cement," *Proceedings of International Symposium Subsurface Injections of Liquid Wastes*, New Orleans, LA, March 3–5, 1986.

(11) Cole, Robert C., "Epoxy Sealant for Combating Well Corrosion," *International Symposium on Oilfield and Geothermal Chemistry*, SPE of AIME, Houston, 1979.

(12) Jeffries, S. A., "Bentonite-Cement Slurries for Hydraulic Cut-Offs," *Soil Mechanics and Foundation Engineering*, Tenth International Congress, Vol 1, Stockholm, 1981, pp. 435–440.

(13) Herndon, Joe, and Smith, Dwight K., "Settling Downhole Plugs: A State-of-the-Art," *Petroleum Engineer International*, Vol 50, No. 4, April 1978, p. 56.

(14) Smith, D. K., Cementing, "Society of Petroleum Engineers Monograp Series," Vol 4, *Society of Petroleum Engineers of AIME*, 1976.

(15) U.S. EPA, "Construction Requirements for Wells Authorized by Rule," 40 CRF 147.2912b(1), (*Subpart GGG, Osage Mineral Reserve, Class II Wells*), 1988.

(16) Portland Cement Association, *Design and Control of Concrete Mixtures*, Skokie, IL, 1979.

Designation: D 5314 – 93

Standard Guide for Soil Gas Monitoring in the Vadose Zone[1]

This standard is issued under the fixed designation D 5314; the number immediately following the designation indicates the year of original adoption or, in the case of revision, the year of last revision. A number in parentheses indicates the year of last reapproval. A superscript epsilon (ϵ) indicates an editorial change since the last revision or reapproval.

1. Scope

1.1 This guide covers information pertaining to a broad spectrum of practices and applications of soil atmosphere sampling, including sample recovery and handling, sample analysis, data interpretation, and data reporting. This guide can increase the awareness of soil gas monitoring practitioners concerning important aspects of the behavior of the soil-water-gas-contaminant system in which this monitoring is performed, as well as inform them of the variety of available techniques of each aspect of the practice. Appropriate applications of soil gas monitoring are identified, as are the purposes of the various applications. Emphasis is placed on soil gas contaminant determinations in certain application examples.

1.2 This guide suggests a variety of approaches useful to successfully monitor vadose zone contaminants with instructions that offer direction to those who generate and use soil gas data.

1.3 This guide does not recommend a standard practice to follow in all cases nor does it recommend definite courses of action. The success of any one soil gas monitoring methodology is strongly dependent upon the environment in which it is applied.

1.4 Concerns of practitioner liability or protection from or release from such liability, or both, are not addressed by this guide.

1.5 This guide is organized into the following sections and subsections that address specific segments of the practice of monitoring soil gas:

Section	
4	Summary of Practice
4.1	Basic principles, including partitioning theory, migration and emplacement processes, and contaminant degradation
4.7	Summary Procedure
5	Significance and Use
6	Approach and Procedure
6.1	Sampling Methodology
6.5	Sample Handling and Transport
6.6	Analysis of Soil Gas Samples
6.7	Data Interpretation
7	Reporting

1.6 *This guide does not purport to set standard levels of acceptable risk. Use of this guide for purposes of risk assessment is wholly the responsibility of the user.*

1.7 The values stated in either inch-pound or SI units are to be regarded separately as the standard. The values given in parentheses are for information only.

1.8 *This standard does not purport to address all of the safety problems, if any, associated with its use. It is the responsibility of the user of this standard to establish appropriate safety and health practices and determine the applicability of regulatory limitations prior to use.*

2. Referenced Documents

2.1 *ASTM Standards:*

D 653 Terminology Related to Soil, Rock, and Contained Fluids[2]

D 1356 Terminology Relating to Atmospheric Sampling and Analysis[3]

D 1357 Practice for Planning the Sampling of the Ambient Atmosphere[3]

D 1452 Practice for Soil Investigation and Sampling by Auger Borings[2]

D 1605 Practices for Sampling Atmospheres for Analysis of Gases and Vapors[3]

D 1914 Practice for Conversion Units and Factors Relating to Atmospheric Analysis[3]

D 2652 Definitions of Terms Relating to Activated Carbon[4]

D 2820 Test Method for C_1 Through C_5 Hydrocarbons in the Atmosphere by Gas Chromatography[3]

D 3249 Practice for General Ambient Air Analyzer Procedures[3]

D 3416 Test Method for Total Hydrocarbons, Methane, and Carbon Monoxide (Gas Chromatographic Method) in the Atmosphere[3]

D 3584 Practice for Indexing Papers and Reports on Soil and Rock for Engineering Purposes[2]

D 3614 Guide for Laboratories Engaged in Sampling and Analysis of Atmospheres and Emissions[3]

D 3670 Guide for Determination of Precision and Bias of Methods of Committee D-22[3]

D 3686 Practice for Sampling Atmospheres to Collect Organic Compound Vapors (Activated Charcoal Tube Adsorption Method)[3]

D 3687 Practice for Analysis of Organic Compound Vapors Collected by the Activated Charcoal Tube Adsorption Method[3]

D 4220 Practices for Preserving and Transporting Soil Samples[2]

D 4490 Practice for Measuring the Concentration of Toxic Gases or Vapors Using Detector Tubes[3]

---

[1] This guide is under the jurisdiction of ASTM Committee D-18 on Soil and Rock and is the direct responsibility of Subcommittee D18.21 on Ground Water and Vadose Zone Investigations.

Current edition approved Nov. 15, 1992. Published January 1993.

[2] *Annual Book of ASTM Standards*, Vol 04.08.

[3] *Annual Book of ASTM Standards*, Vol 11.03.

[4] *Annual Book of ASTM Standards*, Vol 15.01.

D 4597 Practice for Sampling Workplace Atmospheres to Collect Organic Gases or Vapors with Activated Charcoal Diffusional Samplers[3]
D 4696 Guide for Pore-Liquid Sampling from the Vadose Zone[2]
D 4700 Guide for Soil Core Sampling from the Vadose Zone[2]
D 5088 Practice for the Decontamination of Field Equipment Used at Non Radioactive Waste Sites
E 177 Practice for Use of the Terms Precision and Bias in ASTM Test Methods[5]
E 260 Practice for Packed Column Gas Chromatography[6]
E 355 Practice for Gas Chromatogaphy Terms and Relationships[6]
E 594 Practice for Testing Flame Ionization Detectors Used in Gas Chromatography[6]
E 697 Practice for Use of Electron-Capture Detectors in Gas Chromatography[6]

3. Terminology

3.1 *Descriptions of Terms Specific to This Standard:*

3.1.1 *capillary fringe*—the basal region of the vadose zone comprising sediments that are saturated, or nearly saturated, near the water table, gradually decreasing in water content with increasing elevation above the water table. Also see Terminology D 653.

3.1.2 *contaminant*—substances not normally found in an environment at the observed concentration.

3.1.3 *emplacement*—the establishment of contaminant residence in the vadose zone in a particular phase.

3.1.4 *free product*—liquid phase contaminants released into the environment.

3.1.5 *free vapor phase*—a condition of contaminant residence in which volatilized contaminants occur in porosity that is effective to free and open gaseous flow and exchange, such porosity generally being macroporosity.

3.1.6 *liquid phase*—contaminant residing as a liquid in vadose zone pore space, often referred to as "free product."

3.1.7 *macroporosity*—large intergranular porosity with large pore throats, including soil cracks, moldic porosity, animal burrows and other significant void space.

3.1.8 *microporosity*—intragranular porosity and microscopic intergranular porosity with submicroscopic pore throats.

3.1.9 *occluded vapor phase*—condition of contaminant residence in which volatilized contaminants occur in porosity that is ineffective to free and open gaseous flow and exchange, such porosity generally being microporosity; frequently termed dead-end pore space.

3.1.10 *partitioning*—the act of movement of contaminants from one soil residence phase to another.

3.1.11 *soil gas*—vadose zone atmosphere.

3.1.12 *solute phase*—a condition of contaminant residence in which contaminants are dissolved in ground water in either the saturated or the vadose zone.

3.1.13 *sorbed phase*—a condition of contaminant residence in which contaminants are adsorbed onto the surface of soil particles or absorbed by soil organic matter.

3.1.14 *vadose zone*—the hydrogeological region extending from the soil surface to the top of the principal water table.

4. Summary of Guide

4.1 Soil gas monitoring in the vadose zone is a method used to directly measure characteristics of the soil atmosphere that are frequently utilized as an indirect indicator of processes occurring in and below a sampling horizon. Soil gas monitoring is used as a method to suggest the presence, composition, and origin of contaminants in and below the vadose zone. Among other applications, this method is also employed in the exploration for natural resources, including petroleum, natural gas and precious metals. Soil gas monitoring is a valuable screening method for detection of volatile organic contaminants, the most abundant analytical group of ground-water contaminant compounds (**1**).[7]

4.2 *Basic Theoretical Principles*—The processes indicated by the soil gas monitoring method are partitioning, migration, emplacement and degradation. Partitioning represents a group of processes that control contaminant movement from one physical phase to another, these phases being liquid, free vapor (that is, through-flowing air (**2**)), occluded vapor (that is, locally accessible air and trapped air (**2**)), solute and sorbed. Migration refers to contaminant movement over distance with any vertical, horizontal or temporal component. Emplacement refers to establishment of contaminant residence in any phase within any residence opportunity. Degradation is the process whereby contaminants are attenuated by oxidation or reduction in the vadose zone, either through biogenic or abiogenic processes. Soil gas monitoring measures the result of the interaction of these processes in a dynamic equilibrium. Measurement of these processes in static equilibrium is unrealistic.

4.3 The following subsections provide detailed information on partitioning, migration, emplacement and degradation. Subsection 4.4 provides a summary procedure for soil gas sampling. Users of this guide who do not wish to study details of partitioning, migration, emplacement and degradation at this time may skip to 4.4.

4.3.1 Partitioning is the initial step by which contaminants begin to move away from their source. Partitioning occurs in water saturated and unsaturated environments. This group of processes is complex and difficult to quantify when considered in the vadose zone due to the unique makeup of the vadose matrix, i.e. air-filled porosity (microporous and macroporous), pore water, free product, solid-phase soil organic matter, clay and discrete inorganic soil particles. Important individual processes of partitioning are dissolution, volatilization, air-water partitioning, soil-water partitioning and soil-air partitioning (**3**).

4.3.2 Dissolution is the process whereby volatile contaminants move between the liquid phase (free product) and the solute phase (dissolved in water). At equilibrium, the product of the mole fraction of a particular compound in the liquid phase and the activity coefficient of that compound in the liquid phase is equal to the product of the mole fraction of

[5] *Annual Book of ASTM Standards*, Vol 14.02.

[6] *Annual Book of ASTM Standards*, Vol 14.01.

[7] The boldface numbers given in parentheses refer to a list of references at the end of the text.

that compound in the solute phase and the activity coefficient of that compound in the solute phase. This process is more clearly described by the following expression:

$$X^L{}_I\Gamma^L{}_I = X^W{}_I\Gamma^W{}_I \quad (1)$$

where:

$X^L{}_I$ = the mole fraction of compound (I) in the liquid (L) phase (free product),
$X^W{}_I$ = the mole fraction of compound (I) in the solute (W) phase (dissolved in water),
$\Gamma^L{}_I$ = the activity coefficient of compound (I) in the liquid (L) phase (free product), and
$\Gamma^W{}_I$ = the activity coefficient of compound (I) in the solute (W) phase (dissolved in water).

Dissolution equilibrium is therefore influenced by concentration of the subject compound in both the free product contaminant mixture and water. The most common practical application of expression Eq (1) in soil gas monitoring is in hydrocarbon detection. Simplification of Eq (1) is achieved by the following:

assume:

$$\Gamma^W{}_I = 1/S_I,$$

where:

S = the solubility of compound (I) in water

and:

$$\Gamma^L{}_I = 1, \text{ acceptable for hydrocarbons } \mathbf{(3)},$$

then:

$$X^W{}_I = X^L{}_I S_I \quad (2)$$

4.3.2.1 Dissolution equilibrium is impacted by the presence of liquid phase cosolvents, such as gasoline additives, at low concentrations in liquid phase mixtures. This change in dissolution equilibrium can enhance the solubility of certain liquid phase components in water beyond what is indicated by partitioning coefficient data generated in the laboratory. This can have significant impact on downstream concentrations of the contaminant(s) in the soil atmosphere.

4.3.2.2 The effects of temperature upon dissolution equilibrium are generally insignificant for aliphatic hydrocarbons between 15 and 50°C **(4)**, the temperature range from which most soil gas samples are recovered. However, temperature effects upon dissolution equilibrium can be significant for other common families of contaminant compounds within similar temperature ranges **(5)**. These effects must be considered when planning or interpreting the results of a soil gas survey.

4.3.2.3 Dissolution equilibrium is altered by changes in water salinity. Modest decreases in the solubility of contaminants in water are to be expected with increases in salinity of the solution.

4.3.2.4 The rate of dissolution is strongly dependent upon the partitioning coefficient of the particular contaminant of interest and the amount of mixing of the liquid phase and water **(3)**. For example, partitioning of a particular contaminant into ground water is accelerated by frequent water level fluctuations within a contaminated capillary fringe. The downstream implications for subsequent partitioning of the contaminant from the solute to the vapor phase for eventual soil gas recovery are obvious.

4.3.3 Volatilization is the process during which volatile contaminants move between the liquid phase (free product) or solute phase and a vapor phase, either the free vapor phase or the occluded vapor phase or both. Contaminant mixtures can contain compounds with a considerable range of vapor pressures that can contribute contaminants to the soil atmosphere by volatilization. This atmosphere will exhibit a composition similar to that of the parent contaminant but lacking in those constituents with the lowest vapor pressures. The likelihood of the presence of a particular contaminant introduced into the soil atmosphere by volatilization can be estimated by considering the partial pressure of that contaminant in a vapor phase. This partial pressure is equal to the product of the mole fraction concentration of the subject component in the liquid contaminant solution, the activity coefficient of the subject component and the vapor pressure of the pure component. This concept is more clearly expressed as follows:

$$P = X_I\Gamma_I P^o \quad (3)$$

where:

P = the partial pressure of the subject contaminant compound in the vapor phase,
X_I = the mole fraction concentration of contaminant (I) in the liquid contaminant solution,
Γ_I = the activity coefficient of the subject contaminant in the liquid contaminant solution, and
P^o = the vapor pressure of the pure component.

4.3.3.1 The quantity of contaminant volatilized into a vapor phase and the rate of that process is strongly dependent upon temperature. Rate of volatilization is also controlled by the rate of transport of contaminant vapors from the liquid phase-vapor phase interface **(3)**. This rate is probably higher when macroporous flow paths are available for vapor phase transport, and is promoted by a number of driving forces. These are concentration gradient, density gradient between soil atmosphere and contaminant-saturated soil atmosphere, convection currents related to temperature gradient, barometric pressure pumping and introduction of water onto the liquid phase-vapor phase interface.

4.3.4 Air-water partitioning is the process by which volatile contaminants move between the solute phase and a vapor phase, either the free vapor phase or the occluded vapor phase or both. For dilute solutions, air-water partitioning is controlled by Henry's Law, which states that the vapor pressure of a volatile compound above a dilute aqueous solution of that compound is equal to the product of the Henry's Law constant and the mole fraction of that compound in the aqueous solution. Henry's Law may be represented as:

$$P_I = kX_{I(aq)} \quad (4)$$

where:

P_I = vapor pressure of compound (I) above a dilute aqueous solution of (I),
k = the Henry's Law constant for compound (I) at a given temperature, and
$X_{I(aq)}$ = the mole fraction of the subject contaminant compound in the aqueous solution.

Care must be exercised in using Henry's Law to approximate contaminant vapor pressures because of unknowns related to the concentration of contaminants in solution and the contribution of other partitioning processes. Some available literature pertaining to soil gas surveying places emphasis on

Henry's Law constant at 25°C and atmospheric pressure as a primary controlling factor in determining the suitability of a particular volatile contaminant to the soil gas monitoring method. Such emphasis may be inappropriate when, for example, free product is the source of contaminant vapors or when contaminants have not reached ground water. Care must also be exercised in noting the units in which Henry's Law constants are expressed, as these vary from source to source. Volatile but very highly water soluble compounds behaving according to Henry's Law may not be detectable in soil gas because of their persistence for residence in the solute phase **(6)**.

4.3.5 Soil-water partitioning is the process by which volatile contaminants move between the sorbed phase and the solute phase. This process is generally underestimated in its importance to the success or failure of contaminant recovery by soil gas sampling, especially when utilizing the majority of active soil gas sampling techniques generally available to field personnel.[8] There is uncertainty with respect to factors controlling soil-water partitioning, creating doubt as to the reliability of soil sorption data in most applications. Problems with soil sorption data include variability in measurement protocols, the variable nature of organic matter in soils, the effect of dissolved organic matter, unusual pH effects and the effect of salinity, among others **(3)**.

4.3.5.1 The contribution of soil-water partitioning to contaminant phase residence equilibria is strongly controlled by sorbed contaminant concentration in soil, soil makeup, vadose zone pore water content, and soil porosity configuration. Important variables in soil makeup are the quantity, type and distribution of clay in soil and the quantity, type and distribution of soil organic matter. These variables impact the surface area available to sorptive processes, that is, the storage capacity of the soil for contaminants in the sorbed phase, and the pH of the sorption environment. Variations in vadose zone pore water content directly affect the storage capacity of the soil for contaminants in the solute phase. Soil porosity configuration, principally microporosity versus macroporosity, is critical to the rate of soil-water partitioning due to the contrast in surface area between micropores and macropores and the related storage capacity of this porosity for both pore water and sorbed contaminants.

4.3.6 Soil-air partitioning is the process by which volatile contaminants move between the sorbed phase and a vapor phase, either the free vapor phase or the occluded vapor phase or both. Like soil-water partitioning, this process is underestimated in its importance to the recoverability of contaminants by many soil gas sampling techniques. In vadose zone horizons with very low pore water contents, soil-air partitioning can yield vapor phase contaminant composition that differs from free product composition. In vadose zone horizons with higher pore water content, the responsibility for this compositional inconsistency is shared, largely with soil-water partitioning. In wet soil conditions, threshold soil water content values exist for trapped soil atmosphere content to become significant **(7)**, suggesting that responsibility for this compositional inconsistency can be largely attributed to occluded phase residence. Additional important variables are soil clay content, type and distribution, and soil organic matter content, type and distribution. Studies have demonstrated significant impact of soil organic matter and clay content on volatile organic compound emissions from soils **(8)**. Due to the strong control on vapor phase contaminant content by the soil-air partitioning process, it is unreasonable to expect soil contaminants with high affinity for sorption to be efficiently recovered by most soil gas sampling techniques.

4.4 Migration of contaminants in the vadose zone, that is, unsaturated flow, is highly complex and is controlled by soil characteristics, contaminant composition and contaminant phase **(9)**. Migration through unsaturated matrix can occur through a variety of diffusion, dispersion and mass transport mechanisms which behave in a manner unique to saturated flow.

4.4.1 A major division in migratory behavior of contaminants is defined by their solubility or immiscibility in water. Contaminants are often introduced into the soil as liquid mixtures, the components of which immediately begin to partition into other phases upon soil entry. Contaminants that establish soil residence behind a migratory front change in composition with distance from their point of entry. As contaminant migration continues, pathways for individual components can become divergent, such that the composition of the liquid mixture continues to change as migration proceeds. Eventually, migration of liquid mixtures may reach ground water. This can be retarded if the contaminants partition into other phases before reaching ground water and if contaminant vapor is less dense than the uncontaminated soil atmosphere. Transport of contaminants by downward percolation of meteoric waters and upward movement of ground water accelerate the contact of contaminants with ground water. When these contaminants do reach ground water, a radically different set of migration mechanisms begins to govern contaminant transport via saturated flow. Further divergence of contaminant pathways is dependent upon the tendency of each component of the contaminant mixture to float on ground water, become dissolved in ground water or sink to an impermeable layer within the aquifer. Detailed descriptions of these phenomena are available in the literature **(10)**.

4.4.2 The impact of migration processes on soil gas measurement is significant. Although it is impractical to estimate actual migration mechanisms by modelling prior to most soil gas monitoring efforts, a rudimentary knowledge of site characteristics can guide investigators to realistic interpretations of soil gas data expressing unusual or highly variable compositions. More thorough knowledge of relevant site characteristics, such as the presence or absence of barriers to vertical or horizontal migration, that is, foundations, buried pavement, or perched ground water, as well as preferential pathways for contaminant migration, that is, backfill rubble, utility vaults, storm sewers or soil cracks, can assist investigators to assess the migration impact on soil gas survey design.

4.5 The vadose zone is a highly complex soil-air-water-hydrocarbon system with abundant opportunity to store contaminants in all phases. Contaminants partition ac-

[8] See 6.2 for a discussion of active soil gas sampling techniques.

cording to their physical properties and the residence opportunity presented to them along their migratory path. This process has been described as an in-situ chromatographic-like separation of contaminants (**11**). Emplacement, or the establishment of contaminant residence, is a highly dynamic process. Contaminants move from one phase to another as changes occur in both chemical and physical equilibria. Important changes impacting phase residence change include temporal variations in moisture content, soil temperature and level of microbial activity.

4.5.1 One interesting example of disruption in equilibrium conditions is the act of sampling soil gas. Many soil gas sampling systems rely on large volume recovery of soil gas to provide a sample that is believed to be representative of the soil atmosphere in situ. Movement of this soil gas by convective flow through unsaturated soils can cause upward changes in vapor phase contaminant concentration at the expense of other phases.

4.5.2 In natural systems, temporal increases in soil moisture cause gradual increases in solute phase emplacement at the expense of other phases. It is unrealistic to attempt to characterize a static soil gas equilibrium in the vadose zone because this equilibrium is never achieved. For this reason, soil gas data sets based on specific contaminant concentrations and generated at different times are usually not comparable for the absolute values generated by each temporal sampling event. Qualitative comparison of data generated by the same soil gas method and performed at different times is permissible. Generation of a single data set by reconnaissance soil gas sampling and subsequent infilling of data to form a single data set is strongly discouraged.

4.5.3 Attempts to compensate for temporal variations in phase equilibria have been attempted by collecting samples that approximate replicates at known locations and adjusting succeeding data up or down to compensate for observed changes. This procedure is also strongly discouraged, because the number of variables affecting observed changes are too great. Moreover, the ability or willingness of most investigators to determine the most significant effects upon phase equilibria is insufficient to be of use.

4.5.4 Data sets generated by different soil gas sampling techniques may not be comparable as a direct result of differences in efficiency of recovery of contaminants from specific phases. Not only can these data sets differ in measured contaminant concentration, but they can vary substantially in composition as well.

4.6 Degradation of contaminants occurs in the vadose zone through oxidation or reduction reactions that can be biogenic or abiogenic in nature. This process can occur both aerobically and anaerobically to mitigate contaminant levels. Degradation is most often recognized in shallow, permeable soils where favorable conditions exist for oxidation of labile compounds, however other vadose environments can be conducive to degradation. Specific environmental conditions are required for degradation processes to occur. For abiogenic degradation, redox potential and soil pH can be rate controlling factors. For biodegradation, necessary environmental conditions include the presence of microorganisms capable of adaptation to the contaminant as substrate, conditions favorable to population increases of these microorganisms and migration pathways for contaminants to come in contact with these microorganisms. Most soils contain naturally occurring populations of various microorganisms that can degrade petroleum products (**12**). Contaminant biodegradation is known to occur in groundwater (**13**) and in soils (**14**) prior to contaminant partitioning into a vapor phase. Contaminant biodegradation rates for some compounds are highly variable and are controlled by a number of kinetic factors influencing the distribution of microorganisms responsible for degradation. These include aerobic versus anaerobic environments, contaminant type and temperature (**15, 16**).

4.6.1 Degradation rate can approach, equal or periodically exceed the rate of contaminant emplacement into the vadose zone, such that contaminants are not detectable by soil gas monitoring. This mechanism can result in soil gas data which are not representative of an underlying contaminated condition (**17**).

4.6.2 Labile contaminants can be degraded to compounds that may or may not be detectable in soil gas. Aerobic degradation can produce carbon dioxide which can be monitored as an indirect indicator of the presence of contaminants (**18**), or organic acids and phenols (**13**) that are not routinely detectable in active whole air soil gas samples. In alternative to whole air methods, use of an appropriate adsorption medium may facilitate recovery of such compounds for analysis by desorption and gas chromatography-mass spectroscopy. Anaerobic degradation can produce compounds including methane, ethylene, propylene, acetylene, and vinyl chloride which also can be monitored as an indirect indicator of the presence of contaminants. Caution must be used in attributing elevated levels of these compounds to biodegradation, because competitive processes can confuse the interpretation of absolute concentration values and potential sources.

4.6.3 Biodegradation of contaminants in the vadose zone can proceed naturally by adaptation of indigenous microbial populations to metabolize contaminants as primary substrate, or by introduction of foreign populations which have been preconditioned to metabolize contaminants of interest. Case histories demonstrate the absence of certain compounds in soil gas contaminant suites for which biodegradation has been named as the responsible process (**17, 19, 20**). Such cases address the attenuation or complete absence of simple aromatic hydrocarbons, some of which are halogenated, in soil gas. This phenomenon may be controlled by the availability of oxygen as has been demonstrated in the laboratory (**13**). Other compound classes can exhibit similar effects.

4.6.4 Other processes may share responsibility for the actual or apparent absence or attenuation of some contaminants in soil gas sample sets. In some cases where attenuation of contaminant concentration is attributed to degradation, combinations of high soil clay, organic matter and pore water content can reduce the recovery efficiency of certain soil gas sampling techniques for certain contaminants such that contaminant concentrations fall below detection limits. Care must be exercised in attributing a lack of contaminants in soil gas samples to degradation.

4.7 *Summary Procedure for Soil Gas Sampling*—Vadose zone monitoring methods have a set of procedures, both general and specific, that must be consistently followed in order to provide maximum data quality and usefulness. Soil gas monitoring is no exception, with six primary procedures common to all soil gas monitoring techniques. The procedures are a planning and preparation step including definition of data quality objectives, the act of sampling soil gas in the field, handling and transporting the sample, sample analysis, interpretation of the results of analysis, and preparation of a report of findings.

4.7.1 The planning and preparation step begins with the formulation of project objectives, including purpose of the survey, appropriate application of the data to be collected and data quality objectives.

4.7.2 Data can vary in quality due to sampling methodology, sample preparation, analytical procedures, laboratory quality control, and available documentation. Quality assurance programs include all of the activities necessary to provide measurement data at a requisite precision and bias (see Practice 1357). Quality assurance objectives for soil gas monitoring are similar to those for atmospheric air monitoring. The overall quality assurance objective for measurement data is to ensure that data of known and acceptable quality are provided. In order to meet these objectives, data quality objectives should be defined for data measurements in support of the soil gas data interpretation. These are comparability, completeness, representativeness, bias and precision. The comparability of the data collected refers to the ability to interpret the results in light of previous data collection efforts. Completeness refers to the number of samples collected and analyzed compared to the planned number of samples. Representativeness is a measure of the degree to which analytical results reflect true field conditions. Field contamination and sampling intensity are two factors affecting representativeness. Bias is a generic concept of exactness related to the closeness of agreement between the average of one or more test results and an accepted reference value (see Practice E 177). The precision of a measurement process is a generic concept related to the closeness of agreement between test results obtained under prescribed like conditions from the measurement process being evaluated. Overall precision and bias targets for chemical contaminant measurements can be set at 10 % allowable deviation with 90 % confidence limits. In all of these quality assurance activities one must take into consideration that factors including geophysical conditions and definition of sampling volume in the vadose zone often have higher variability than analytical equipment calibration procedures.

4.7.3 Table 1 provides suggested quantitative limits for data quality objectives.

4.7.4 The planning and preparation step continues with the evaluation of available information already gathered for the project area. These efforts culminate in the selection of an appropriate soil gas monitoring method and a survey design which best fits the project objectives within budgetary constraints. Prior to actual field work, investigators must obtain the necessary permits and landowner permission for property access. When a survey area is pending sale, investigators should obtain written permission to conduct the survey from both the buyer and the seller. Moreover, when a soil gas survey is being performed as a service, no work should proceed on the survey without a fully executed consulting agreement between the investigator and the client

TABLE 1 Suggested Quantitative Limits for Data Quality Objectives

QA/QC Objective	Measure	Formula	Limit
Accuracy	Laboratory standard	Standard recovery	90 to 110 %
Precision	Field replicate	Relative standard deviation	< 20 %
	Laboratory replicate	Relative standard deviation	< 20 %
Representativeness	Air blank	Bias	< 10 %
	Cross contam. blank	Bias	< 10 %
Completeness	Completion (%)	Relative compl.	> 90 %
Comparability	Prof. judgment	NA	NA

for whom the survey is being conducted.

4.7.5 Actual field work consists of recovery of soil gas samples. The method selected should be based upon site specific factors and dictated by the project objectives. A detailed discussion of soil gas sampling methods is provided in 6.1.

4.7.6 As samples are being recovered, they must be handled and transported in such a way as to assure preservation prior to analysis. A detailed discussion of sampling and transport is located in 6.5.

4.7.7 The presence of contaminants is determined through analysis of the soil gas samples. This step is controlled to a large degree by the QA/QC objectives of the survey. A discussion of sample analysis is provided as 6.6.

4.7.8 Data interpretation is largely an iterative process of review of the raw soil gas data out of context, a review of the soil gas data in context of other site characteristics and the formulation of conclusions based upon all known information. A discussion of soil gas data interpretation is located in 6.7.

4.7.9 Finally, a report of findings is generated in a format that is selected to be appropriate to the requirements of the end users. Section 7 provides options that can be addressed in reporting as well as recommendations of topics that should be included in all soil gas summations.

5. Significance and Use

5.1 *Application of Soil Gas Monitoring*—Soil gas monitoring is an extremely versatile method in that it can be adapted to conform to the requirements of dissimilar industries for a wide variety of applications. A number of soil gas techniques have been utilized in the agricultural **(21)**, petroleum **(22, 23)** and minerals **(24)** industries. Certain applications have been exercised for well over 50 years. Soil gas monitoring has been utilized in research efforts, including the monitoring of underground coal gasification retorts **(25)**. Application to the environmental industry is comparably recent but very effective as a rapid and relatively inexpensive method of detecting volatile contaminants in the vadose zone. Field screening, of which soil gas monitoring is a basic component, has been demonstrated to be effective for selection of suitable and representative samples for other more costly and definitive monitoring methods **(26)**. Soil gas monitoring is useful to assess the extent of ground water contamination for certain contaminants and field environments **(27)**. Soil gas monitoring is also a viable method of monitoring subsurface contaminant discharges from underground storage tanks **(28)**. New applications of the soil gas monitoring are periodically developed and published in the referenced literature. The method may be useful in the study of unsaturated flow. In most instances, the method can make use of very light-weight, portable and inexpensive tools made from commonly available materials. Soil gas monitoring has become a widely accepted method for locating subsequent environmental monitoring and remediation activities such as ground water monitoring wells, contaminant product recovery wells or excavations to recover contaminated soil. Soil gas monitoring has made a significant contribution to ground water monitoring and remedial planning on sites that fall under the Comprehensive Environmental Response, Compensation and Liability Act (CERCLA) **(29)**. This method is highly useful at the initiation of Phase II environmental assessment action in determining the presence of volatile organic contamination of real property in a pending sale.

5.1.1 In any application, soil gas monitoring can be performed over a wide range of both spatial and temporal designs. Spatial designs include soil gas sampling in profiles or grid patterns at a single depth or multiple depths. Multiple depth sampling is particularly useful for contaminant determinations in cases with complex soil type distribution and multiple sources. Depth profiling can also be useful in the determination of the most appropriate depth(s) at which to monitor soil gas, as well as the demonstration of migration and degradation processes in the vadose zone. Temporal designs include the long-term monitoring of the vadose zone for the appearance of volatile organic contaminants from known potential sources such as underground storage tanks and solid waste landfills. Temporal designs are especially useful in monitoring the effectiveness of contaminant remediation efforts.

5.1.2 Soil gas monitoring in the vadose zone is an ideal reconnaissance tool and screening technique in most applications. However, site specific and contaminant specific limitations can cause this technique to be unsuccessful in meeting project objectives. Caveats exist in all soil gas monitoring procedures that can frustrate efforts to successfully apply the method to any application.

5.2 *Limitations*—The most significant limitation on soil gas monitoring is the inability to utilize the method as a stand alone technique. Soil gas monitoring does not provide repeatable quantitative information over time due primarily to the dynamic nature of phase equilibria in the vadose zone and secondarily to unavoidable inconsistencies in sampling practice. As a result of geologic variability in the vadose zone and the multitude of unique sampling devices currently being used in the field, quality assurance and quality control protocol, discussed in 6.4, cannot provide the rigor required as in a test method. For these reasons, soil gas data in itself cannot be used to provide definitive answers about the location or absence of buried contaminants. Moreover, the success of any soil gas monitoring method is strongly dependent upon effects related to geologic variation and moisture content in the sampling horizon as well as the physical properties of the target contaminants.

5.2.1 False negative results can occur as a direct result of the incompatibility of a specific procedure with the properties of the sampling horizon or the target contaminants, or both. Soil gas data cannot be used to establish bulk volume or the commerciality of buried petroleum, natural gas, or ore bodies.

5.2.2 With the necessary analytical procedures, soil gas can be examined for compositional anomalies, a very useful technique for multiple source problems. In some instances, contaminant occurrences are limited to single species (compounds, mercury, etc.), however more often than not the contaminant source is a mixture of organic chemicals that have a unique chemical compositional character consisting of both normally evaluated priority pollutants and non-priority pollutant chemicals that may be overlooked. By identifying and using compositional information, many problematic site situations such as degradation can be minimized by targeting the more refractory compounds associated with the contaminant occurrence. This interpretive method is impossible to model for an industry wide application due to variation in methods and technique.

5.2.3 A basic limitation of the technique is that due to the ease of procurement and use of soil gas sampling devices, there is a tendency for inexperienced personnel to oversimplify any and all aspects of the method. Investigators must consider the experience level and technical ability of personnel who acquire soil gas samples and attempt to interpret the results. Certain procedural facets are not trivial, as discussed in Section 6. The results of certain techniques tend to be affected by minor variations in procedure despite apparent adherence to a "Standard Operating Procedure."

5.2.4 Atmospheric air contamination is not a trivial problem corrected by simple device-oriented field practice. Many sampling systems recover very large volumes of "soil gas" that may actually represent a mixture of soil gas and atmospheric air. This mixing occurs through the introduction of ambient air adjacent to the sampling device and through macroporous pathways in the soil which are far from the sampling device. Some environmental investigators avoid the impact of this problem by reasoning that contaminant quantities in the soil are so great that they are detected despite atmospheric mixing. For qualitative approaches with non-rigorous quality assurance/quality control (QA/QC) objectives this mixing problem can be insignificant. For detection of compounds that exhibit only marginal partitioning preference for the free vapor phase, the mixing problem can be a fatal flaw in procedure. Moreover, contaminant concentration and composition investigations can be rendered useless by variations in the magnitude of mixing at various sample locations and depths in a survey area.

5.3 *Comments on Limitations of Soil Gas Monitoring*—Many investigators believe that soil gas monitoring is not an effective vadose zone monitoring method for certain volatile organic applications, in certain geographic regions or during certain seasons of the year, or both. The applicability of soil gas monitoring is controlled by physical and chemical properties and processes in the subsurface and not by factors that are obvious at or above the surface. For example, one common misconception is that soil gas monitoring is not effective during the winter season. The impacts upon soil gas measurement of elevated soil pore water content, reduced vadose zone temperature and the presence of frost, typical of numerous regions in winter, are obvious for many facets of most soil gas monitoring methods. Modification of standard operating procedure, such as an increase in sampling depth, or selection of another soil gas monitoring method altogether can minimize the negative impacts of seasonal field conditions. It is important to understand that the responsibility for success or failure in soil gas monitoring can reside as much in the planning phase of a survey, including the method chosen, as in factors controlling the chemical and physical processes at work in the subsurface. Even with apparently ideal field conditions and with a carefully planned survey, soil gas monitoring can succeed or fail due to unknown factors controlling contaminant migration and emplacement. Soil gas monitoring is no different than any other measurement method, in that investigators must maximize effort in planning and implementation of procedure to maximize the likelihood of success.

6. Approach

6.1 *Sampling Methodology*—Soil gas sampling methodology has evolved over time and through practice in several industries. The equipment with which to perform this monitoring technique is highly varied, although it may be categorized into basic types (see 6.2.2). The literature provides numerous discussions about the design of some of this equipment (**10, 30, 31, 32, 33**). The selection of a soil gas sampling method involves consideration of three primary issues. These are the type of sampling system, the methodology of application of that sampling system and the rigor of the field QA/QC protocol. Each of these issues is discussed in this guide, however, no single method or procedure is recommended to the reader due to the variation in site specific factors. As many as one hundred unique soil gas sampling systems exist that arise from variations or combinations, or both, of the many facets described in this guide. Some systems are highly versatile for numerous applications. Others are functional for more limited or specific applications. Informed investigators must assume the responsibility of selecting the technique most appropriate to the subject application, whether that technique is commercially available from contractors or equipment suppliers, or reliant upon the ingenuity of the investigator in the field utilizing commonly available materials. Success in choosing an appropriate sampling device or an entire sampling system is dependent upon the investigator's level of understanding of vadose zone processes, contaminant properties and appropriate applicability of the soil gas method.

6.1.1 The application of any of these methods must be controlled by strict adherence to a standard operating procedure. Occasional deviations as dictated by unusual field conditions should be recorded in the project field notebook. Inadvertent minor deviations in field procedure can result in misinterpretation of the data acquired.

6.2 *Sampling Systems*—Six basic sampling systems exist. These are based upon the collection of soil gas by a whole-air or sorbent method in an active or passive approach, or upon the principle of collection of a soil or water sample for subsequent sampling of a contained headspace atmosphere. Contained atmosphere methods do not yield samples representative of in situ vadose zone atmospheres.

6.2.1 Whole-air methods sample the soil atmosphere as a mixture of gases, including contaminant and non-contaminant vapors. Sorbent methods sample contaminants adsorbed onto a collection medium exposed to a whole-air sample stream. Active methods are those that obtain a soil gas sample by positioning a sampling device in the subsurface and the withdrawal of soil atmosphere through the device from the sampling horizon. Passive methods are those that obtain a soil gas sample by placing a collection device in the soil or on the soil surface, and allowing the atmosphere within the device to come into compositional equilibrium with the soil atmosphere. Four of the six basic sampling systems arise from these approaches, namely the whole air-active approach, the sorbed contaminants-active approach, the whole air-passive approach, and the sorbed contaminants-passive approach. Two additional systems exist that are based respectively upon the collection of a soil or water sample for subsequent sampling of a small volume headspace atmosphere.

6.2.2 *Whole Air-Active Approach*—This method of soil gas sample collection involves the forced movement of bulk soil atmosphere from the sampling horizon to a collection or contaminant device through a probe or other similar apparatus (**10, 34**). Contained samples of soil atmosphere are then transported to a laboratory for analysis, or the sampling device is directly coupled to an analytical system. Whole air-active sampling is best suited to soil gas monitoring efforts where contaminant concentrations are expected to be high and the vadose zone is highly permeable to vapor. Probes exist that must utilize pre-existing holes or that can penetrate the vadose zone by driven means. These devices can be very simple and light-weight for low cost mobilization (**35**), or they can be affixed to vehicle mounted drills or hammers useful for larger, more complex surveys at a higher cost of mobilization. The whole air-active technique can be combined with other monitoring methods such as soil monitoring for engineering purposes (**36**) in some survey environments. The success of this practice can be highly site-specific.

6.2.2.1 Ground probes can be of small to large internal volume. The development of sampling devices with smaller internal volumes equating to smaller purge volumes is a significant improvement, providing samples which are more representative of soil atmosphere, and a greater ease of equipment decontamination between usages. Sample size can vary from a few millilitres to many tens of litres depending upon the sample rate through the probe, the vapor storage capacity of the soil and the ability of the soil to deliver vapor to a probe under vacuum.

6.2.2.2 The success of the active approach is strongly dependent upon soil clay, organic matter and moisture content. Driven probes tend to destroy natural soil permeability around the body of the probe due to soil compaction concurrent with insertion. This can be a severe limitation in moist, heavy clay soils. In very dry, cemented soils, driven probes can create radial fractures that can enhance soil permeability to vapor concurrent with insertion. These fractures can communicate atmospheric air with soil atmosphere, a limiting factor for obtaining representative, large-volume soil gas samples. The effect can be so severe as to lower recovered contaminant concentrations in the soil gas sample below the limits of analytical detection. This is especially true for highly sorptive or water soluble compounds, or both. Some investigators have attributed the poor recoveries of these compounds exclusively to other processes, that is, degradation (**21, 37**).

6.2.2.3 Methods requiring a pre-existing hole for probe insertion (**38**) made with a commercially available "slam bar" can provide supportable contaminant data where contaminant concentrations and soil permeability to vapor are high, however the act of making a hole with a "slam bar" and subsequent removal of the "slam bar" can encourage soil contaminant venting and lower sample representativeness. Insertion of the sampling probe into this hole further degrades representativeness by additional venting of contaminants as the probe displaces the atmosphere in the hole upon insertion. Purging of the probe prior to sampling under conditions of low soil permeability and low contaminant concentration may lower contaminant levels below the limits of analytical detection. Methods requiring a pre-existing hole for probe insertion are not recommended for soil gas sampling from soils with high clay and moisture contents.

6.2.2.4 Excellent discussions of numerous whole air-active sampling systems may be found in the literature (**10, 21, 37, 39**). Investigators must consider the caveats and limitations of the whole air-active approach when selecting a certain method for a specific application.

6.2.3 *Sorbed Contaminants-Active Approach*—The sorbed contaminants-active method of soil gas sample collection also involves the forced movement of bulk soil atmosphere from the sampling horizon through a probe or other similar apparatus, but to a collection device designed to extract and trap sample stream contaminants by adsorption (**40, 41**). This system is well suited to sites where the soil may be highly permeable to vapor and where the contaminant concentration may be lower than required for successful whole-air surveys. Sorbent devices are designed to concentrate the components of interest and remove some of the soil gas components known to interfere with sample analysis.

6.2.3.1 Contaminant trapping is accomplished by use of an adsorbent collection medium such as charcoal or a carbonized molecular sieve adsorbent (**43, 44**), as well as porous polymers, silica gel and activated alumina (**10**). This approach is especially amenable to the detection of nonpolar volatile organic compounds. Organic compounds that are reactive, oxygenated or are gaseous at room temperature are either not adsorbed by or are not efficiently desorbed (**42**) from charcoal. Sorbent collection devices are commercially available or can be specially prepared with an appropriate sorbent material that concentrates desired compounds for future analysis. Colorimetric detector tubes are available which will provide an indication of the presence of target compounds at the time of sampling. These devices are limited in application by the high concentration requirements for many compounds and the compound-specific nature of these tubes.

6.2.3.2 The effectiveness of the sorbed contaminants-active approach can be limited by high vadose zone clay and water content, reducing the ability of the soil to transmit vapor through the sorbent trap. Commercially available sorbent traps come with information suggesting maximum, minimum and optimum sampling rate through the trap. Soil characteristics can limit flow rate to a point below the minimum recommended rate, affecting the performance of the trap and the reproducibility of adjacent samples. Interaction of the sorption media with target compounds during desorption in the laboratory can form artifacts, restricting the interpretive value of the data. Some sorption media are prone to irreversible adsorption (see Definitions D 2652). Some may be affected by high soil gas relative humidity. Humidity greater than 60 % (very common for soil gas) can reduce the adsorptive capacity of activated charcoal to 50 % for some chemicals. Presence of condensed water in the sample tube will indicate a suspect sample (see Practice D 3686). Anticipation of these problems is recommended for all sorbent techniques, and a thorough quality control plan should be designed and implemented as is discussed in 6.4 of this guide.

6.2.3.3 Special sample preparation is required for samples adsorbed onto a trapping medium. This preparation step consists of the thermal or solvent desorption of the contaminants from the trapping medium. Proper practice will promote needed accuracy and precision in the determination of contaminant concentrations above specified values (see Practice D 3687).

6.2.4 *Whole Air-Passive Approach*—This method of soil gas sample collection involves the entry of bulk soil atmosphere or soil atmosphere components from a near-surface sampling horizon to a collection or containment device through a flux chamber or other similar apparatus (**30**). Enclosure devices sample vaporous emissions from a known soil surface area capped by a chamber. The volume of the chamber is continuously swept by injection of a gas of known composition, and the resultant carrier gas-contaminant mixture is collected for analysis. The rate of emission or "flux" of contaminants can be calculated if flow rate of injected gas and contaminant concentration in the sample are determined.

6.2.4.1 The whole air-passive approach is useful to some very specific applications. This method may be used, for example, to monitor contaminant emissions from soil or water to assess the health hazard risk of such emissions to the general public. Determination of the extent of contamination by volatile organic compounds has been performed with whole air-passive devices, however the application of other types of systems is far more common.

6.2.4.2 A key to successful operation of a whole air-passive system is that the system is able to recover volatile compounds as they are emitted from the vadose zone. The effects of changes in barometric pressure, soil temperature and soil moisture content are not quantifiable from site to site due to site specific variables controlling vapor phase contaminant migration and the rate of contaminant partitioning into the vapor phase. The presence of contaminants or naturally occurring organic matter floating on surface water may impact the rate of entry of certain vapor phase contaminants into the chamber.

6.2.4.3 The whole air-passive method is limited in application primarily due to the great degree of dilution of contaminants in the sample stream by injected gas. This can decrease method sensitivity by lowering contaminant concentrations to levels below the detection limits of the analysis method chosen. Further decrease in method sensitivity results from the fact that soil gas contaminant concentrations are generally lower at the surface than even at nominal depths. Soil characteristics such as high water saturation, soil cements, clay content and organic matter content will negatively impact results of these systems by restricting the rate of contaminant flux to the chamber.

6.2.4.4 Additional limitations exist. Certain devices limit flux rates into the chamber due to aspects of design. Soil macroporosity such as desiccation cracks extending beyond the collecting device will vent soil vapors to the atmosphere that will not be collected by flux chambers unless monitoring locations are biased to include these features.

6.2.5 *Sorbed Contaminants-Passive Approach*—This method of soil gas sample collection involves the passive movement of contaminants in soil to a sorbent collection device over time. Passive samplers that have been applied to sampling soil gases of environmental concern include occupational health volatile organic compound monitors **(44)** and a sampler originally developed for detecting the presence of hydrocarbons in petroleum exploration **(33, 46)**. Both devices use charcoal as a sorbent; the former as a flat film and the latter coated on a wire. Passive samplers are housed in containers up to several inches in diameter, depending upon the design. They are placed open end down in holes that are usually less than 5 ft (1.5 m) deep, that are then backfilled **(32)**. These monitors are generally left in place from two to ten days, although certain passive collectors can be left in place for a period of 30 days or more for certain applications. For at least one device, exposure efficiency can be determined.

6.2.5.1 The sorbed contaminants-passive approach can be employed in a wide range of geological conditions. Frozen ground and high water saturation may not limit the ability of the monitors to collect contaminants **(46)**, although the composition of the contaminant suite may be impacted by related alterations in partitioning equilibria.

6.2.5.2 The sorbed contaminants-passive approach depends upon the ability of contaminants to move through the vadose zone to the passive collection device. Numerous adsorption media can be used to collect contaminants (see 6.2.4). The principle of passive-sorbent monitors relies on adsorbent reduction of the equilibrium concentration of contaminants around the monitor over time, therefore creating a concentration sink, that is, a continuous state of disequilibrium, in the vicinity of the monitor. This can encourage continued migration of contaminants toward the monitor when conditions for contaminant partitioning into the vapor phase are favorable. Migration of contaminants in the vadose zone toward a passive-sorbent device is strongly controlled by vadose zone character and the chemical and physical properties of the subject contaminants. Contaminants may move from a few feet to thousands of feet, or not at all.

6.2.5.3 Many investigators attribute the principle mechanism of contaminant migration to a passive-sorbent device to diffusion, that is, the movement of organic vapor or gas molecules from a region of high concentration to a region of low concentration as described by Fick's law (see Practice D 4597). Fick's law of diffusion states that for a constant concentration gradient, the mass of material transferred to the sampling layer can be expressed as:

$$M = \{DA(C - C_o)t\}/L \tag{5}$$

where:

M = mass of the material, ng,
D = diffusion coefficient, cm^2/min,
A = cross sectional area of diffusion cavitie(s), cm^2,
L = length of diffusion path, cm,
C = concentration at face of sampler, ng/cm^3,
C_o = concentration at adsorbing layer surface, ng/cm^3, and
t = exposure time, min.

6.2.5.4 The cross sectional area of a diffusion cavity, the length of the diffusion path and the quantity $(C - C_o)$ are impossible to accurately measure for soil gas contaminants interacting with a passive-sorbent sampler. There is some debate as to whether passive samplers measure flux or total contaminant concentration **(32)** in the vicinity of the trap. Due to the fact that the mass of the material transferred to the sampler by diffusion, a key measurement, cannot be determined, the debate will no doubt continue. It is reasonable to assume that a combination of processes is responsible for contaminant migration to sorbent traps, including diffusion, dispersion and mass transfer. All migration processes are impacted by partitioning equilibria.

6.2.5.5 Ambient air represents an atmospheric contaminant concentration sink that encourages a strong vertical vector of contaminant migration. This prevailing upward movement of contaminants from sources at depth results in contaminant concentration gradients throughout the vadose zone. The sorbed contaminants-passive method makes use of this contaminant flux (see 6.2.4) to collect long-term, nondisruptive samples of volatile contaminants. The method can collect contaminants which are compositionally representative of the contaminant mixture favoring the vapor phase. The quantity of volatile organic compounds trapped by these devices is proportional to the concentration gradients of contaminants present near the collection device and the affinity of the contaminant(s) for the collection medium.

6.2.5.6 As with active sampling protocols, specific issues exist affecting the function and calibration of passive monitors. Soil gas, even in the drier climates, will be at a relatively high humidity condition. This humidity can affect the collection efficiency of the adsorbent media. In soils of low permeability, contaminants commonly move very slowly. This can create a condition of near-zero contaminant concentration in the soils immediately adjacent to the monitor if the sorptive potential of the monitor is higher than that of the soil. When soil contaminant concentrations are rapidly depleted, that is, as the result of invasion of the sampling horizon by meteoric water, the passive monitor can source contaminants back to the soil.

6.2.5.7 The sorbed contaminant-passive approach to soil gas monitoring is not immune to the migration, emplacement and degradation factors affecting all soil gas monitoring techniques. It is not possible to measure the efficiency of passive-sorbent monitoring devices because the bulk volume of soil gas affected by the sorbent trap cannot be measured. Care must be taken not to contaminate the sorbent samples during installation or by backfilling with contaminated soil. Such care is comparable to potential problems for any measurement method in which a contaminated layer is penetrated.

6.2.6 *Soil Sampling for Subsequent Headspace Atmosphere or Extraction Sampling*—This method examines contaminants that are present in a headspace atmosphere above a contained soil sample. Note well that this headspace atmosphere is not true soil gas (see 3.1.11), but is an artificial atmosphere formed above a potential contaminant source, that is, the soil sample. Contained atmosphere methods do not yield samples representative of in situ vadose zone atmospheres. Headspace atmospheres differ from in situ vadose zone atmospheres in that large percentages of vapor phase and moderate percentages of solute and sorbed phase contaminants can be lost in the act of soil sampling. This method is not generally recommended for a broad spectrum of cases due to numerous limitations and caveats. In comparison to other methods described in this guide, soil sampling for subsequent headspace atmosphere or extraction sampling can be a relatively poor method for determining many of the more volatile contaminants. Headspace atmospheres contain residual sorbed and solute phase contaminants that have partitioned to the vapor phase in the contained environment; most headspace approaches are reasonably efficient in recovery of some fraction of sorbed and solute phase contaminants. Contaminants in these phases in situ are recovered from a headspace after they have partitioned into the vapor phase. Recovery efficiency of contaminants in the vapor phase in situ ranges from moderate to poor.

6.2.6.1 Important criteria exist to consider when selecting a device that will provide suitable samples (see Guide D 4700). The equipment required is simple and readily available. Some commonly used augers are not suitable for soil sampling in support of subsequent headspace atmosphere sampling due to soil disturbance. Depths of auger investigations are limited by ground water conditions, soil characteristics and the equipment used (see Practice D 1452). Suitable procedures for some methods are described in the literature **(47, 48)**. Current soil preservation practice may not apply (see Practice D 4220).

6.2.6.2 Limitations and special procedures exist for the application of soil sampling for subsequent headspace gas analysis. Filling head space with solvent can support a subsequent solvent extraction procedure. Some investigators minimize the effects of devolatilization by rapidly recovering small soil core plugs with polypropylene syringes which have been modified to accommodate recovery of soil plugs. Investigators also attempt to maximize partitioning of contaminants into the vapor phase by adding buffering solutions or sodium sulphate and phosphoric acid to the vial prior to sealing, in order to shift the activity coefficients of the subject contaminants to favor the vapor phase. Aqueous suspensions of solvent slurries of soil can be ineffective for the determination of high molecular weight labile compounds. Their persistence in soil is the result of physical entrapment in soil microporosity **(49)**. Recovery efficiency of contaminants in soil headspace can be greatly enhanced by pulverization of the soil **(50)** in a ball mill or other similar apparatus. The method is biased toward recovery of contaminants in the sorbed, solute and occluded phases in situ due to the loss of pore space gas in preference to contaminants adsorbed onto the soil particles or trapped in soil micropores. Contaminant degradation, especially biodegradation, in the container is encouraged by the creation of an aerobic, moist environment during sample handling and transport prior to analysis. However, a simple method to minimize the effects of biodegradation can be achieved by storing samples, when necessary, at approximately 4°C in the dark.

6.2.6.3 Acid extraction of volatile organic compounds is widely used in geochemical exploration for petroleum and natural gas. Soil samples are placed in a closed vessel, heated and evacuated to remove vapor phase contaminants. The addition of acid to the evacuated chamber causes release of hydrocarbons believed to be bound to the soils by carbonates **(22)**. Hydrocarbons are determined by analysis of resulting vessel atmospheres. Refinements to this method have been developed **(48)**, however the method is designed not to determine compounds in the vapor, sorbed, or solute phases. Method sensitivity is therefore greatly reduced.

6.2.7 *Soil Pore Liquid Headspace Gas Approach*—In the vadose zone, soil gas monitoring can be accomplished in combination with soil pore liquid sampling through the use of a suction lysimeter, a pan lysimeter or a free drainage glass block sampler. The suction lysimeter installed in the vadose zone is most commonly employed for this purpose. Temporally designed surveys are ideally suited to this method.

6.2.7.1 After a lysimeter has been installed for some period of time, initial aliquots of vapor sampled from a soil pore liquid sampler will be in compositional equilibrium with solute phase contaminants when pore liquid tensions are within the operating range of the lysimeter and if pore sizes are not so great as to cause loss of hydraulic contact between the soil and the porous segment of the lysimeter. Subsequent aliquots of soil gas may compositionally resemble soil vapor in situ if soil atmosphere enters the porous segment of the sampling device. When the lysimeter cannot recover a pore liquid sample, the soil gas recovered will be compositionally similar to soil vapor in situ.

6.2.7.2 The most common effort to recover soil gas from a suction lysimeter occurs when polytetrafluoroethylene (PTFE) porous segments are employed in sampling environments with high soil moisture tensions (low moisture contents). At tensions above 60 to 80 centibars, soil pore liquid samples cannot be collected (see Guide D 4696). However, soil gas can be recovered through the porous segment and collected at the surface. This alternative sampling effort can monitor soil vapor contaminants utilizing an otherwise unsuccessful procedure until soil moisture contents increase or until an alternative soil pore liquid sampler can be installed.

6.2.7.3 This technique is limited by the relative expense and complexity of installation of the sampling devices as a primary soil gas sampling method. The completeness criterion for quality assurance is difficult to satisfy due to the inability to anticipate the performance of the soil pore liquid sampler with respect to vapor recovery. Moreover, compositional bias toward solute phase contaminants and contaminants volatilized from free product is likely in soil gas samples recovered concurrently with soil pore liquid samples.

6.3 *Methodology in Application of a Sampling Technique*—The likelihood of success of the soil gas sampling technique selected is controlled in part by the methodology in application of that sampling technique. This methodology should be guided by the objectives of the subject project and the perceived spatial and temporal array of the potential sampling targets.

6.3.1 *Grids*—Many problems suitable for soil gas monitoring are best solved by obtaining data distributed over a geographic area. Sampling in grid patterns of variable design and spacing can be a very effective way to provide data coverage over a large area for a very low cost of acquisition. Common applications of soil gas grid sampling are environmental contaminant assessments, exploration for natural resources and the siting of locations for other monitoring or exploratory techniques. Compositional analyses in conjunction with properly designed grid systems are often fundamental to successful evaluation of soil gas monitoring.

6.3.1.1 Grid spacing provides for the location of soil gas samples in grid cells. The selection of grid cell size is strongly dependent upon the relationship between project confidence level requirements and cost budget. Small survey targets and complex vadose zone geology require decreased spacing between soil gas sample locations for grid methodology to be successful. Some applications, for example, defining the boundaries of contaminated soil or ground-water contaminant plumes, may require the grid cell area to be as small as 100 to 400 ft^2 (9 to 37 m^2). Most applications to natural resource exploration monitor naturally occurring volatile compounds in soil atmospheres, requiring closely spaced grids to increase the signal to noise ratio. However, a closely spaced exploratory grid equates to a broadly spaced grid for environmental application in most situations. Common petroleum exploration grid spacing utilizes a grid cell area of approximately 250 000 ft^2 (23 000 m^2), however grid cells can range from 10 000 to 1 000 000 ft^2 (9 to 90 000 m^2) depending upon perceived reservoir target area. Widely spaced grid sample arrays are useful in reconnaissance applications such as the establishment of contaminant baselines or evaluation of the exploration potential of a geologic basin. Grid cells for such purposes can be as large as a square mile or more.

6.3.1.2 The tendency exists for investigators with constrained budgets to utilize overly large grid cell spacings. This action normally results in inadequate, over-interpreted data supporting meaningless conclusions. Care must be taken to avoid this caveat.

6.3.1.3 Grid arrays can be designed as regularly spaced and predetermined locations for soil gas sampling or they can be irregularly spaced and continually field modified. Predetermined and widely spaced grid patterns are most useful for reconnaissance work, while closely spaced, irregularly situated or field modified soil gas grid sample sites, or both, are commonly used when targeting contaminant plume boundaries, contamination from underground storage tanks or other detail work.

6.3.1.4 Multiple depth sampling, discussed in 6.3.3, when coupled with a soil gas grid sampling methodology, can provide useful data in complex geologic settings and sites with multiple contaminant sources. Computer mapping of closely spaced three-dimensional soil gas grids can provide the investigator with horizontal or vertical cross sections through the subject site, making difficult observations possible.

6.3.2 *Profiling*—Profiling is a soil gas sampling methodology useful to test a linear array for the existence of contaminants. Profiling is most often performed by sampling at closely spaced intervals in a linear array and is displayed as contaminant concentration or composition versus distance sampled on an $X - Y$ plot. Concentration data are often displayed logarithmically on the ordinant (Y) axis, while single components or ratios of compositional data are often displayed linearly on the ordinant axis.

6.3.2.1 For environmental applications such as leak detection along the length of a pipeline or monitoring of contaminant encroachment across a property boundary, soil gas samples are recovered along a profile at intervals from 25 to 100 ft (8 to 30 m) **(23)**. Profiling for natural resource exploration can be performed at sample intervals from 50 to 500 ft (15 to 50 m), depending upon the application.

6.3.2.2 Profiling is useful as a corroborative tool for other monitoring or exploration methods. For example, a soil gas sample profile acquired coincident with a seismic profile can suggest primary contaminant migration pathways or the boundaries of confining layers in shallow, complex geologic settings. This technique has been demonstrated as highly effective in reducing exploratory risk prior to drilling for petroleum and natural gas, by suggesting the presence of hydrocarbon seepage coincident with structures with reservoir potential defined by the seismic method **(51)**.

6.3.2.3 Soil gas profiling is also a convenient methodology effective in comparative evaluation of multiple soil gas sampling techniques. Due to variations common to the dynamic equilibrium conditions over small spatial and temporal intervals in the vadose zone (see 4.1), comparisons of multiple soil gas techniques using only one or a few soil gas samples recovered from nearly identical locations will not result in a valid comparison. However, a visual overlay of soil gas profiles resulting from the implementation of the various sampling techniques can provide a rapid and definitive comparison as to the efficiency of recovery of subject contaminants by a particular sampling system in a specific sampling environment. Similarly, comparison of profiles obtained by using the same soil gas sampling system can provide a direct measurement of system accuracy for quality control purposes.

6.3.2.4 Some investigators compare geographically coincident profiles obtained with the same sampling system at times differing by days or even years in order to generate a data correction factor in order to enhance data comparability. This practice is strongly discouraged. Factors not anticipated in this practice such as the effects of the dynamic equilibrium in the vadose zone, unavoidable changes in procedure due to personnel substitutions, contaminant movement or cultural influence on the sampling environment can have impact on results that are far more significant than the apparent correction.

6.3.3 *Multiple Depth Sampling*—Methodologies encompassing multiple depth sampling normally have one of two goals, that is, to monitor changes in soil gas contaminant fractions versus depth, and to closely follow a single sampling horizon for an entire soil gas grid or profile.

6.3.3.1 When the goal of a survey is to monitor contaminants over varying depths, some sampling systems can recover soil gas samples as probes are advanced deeper into the vadose zone. This practice is helpful in determining the optimum sampling depth for a particular site or to demonstrate the presence or absence of soil atmosphere contamination in a certain horizon. Soil gas contaminant concentrations often increase with depth as the sampling horizon approaches contaminated ground water or other source of soil gas contaminants (**52**). Caution must be exercised when soil gas sampling tools are advanced to increasing depths due to the fact that cross contamination of some or all of the sampling system is unavoidable. This situation limits quality control for this type of multiple depth sampling. Attempts to eliminate cross contamination in multiple depth sampling by replacement or decontamination of sampling equipment with each new sample aliquot also result in limited quality control. Tool withdrawal and tool reinsertion result in venting of the sampling environment via an open hole. The open hole behaves as a macroporous pore space, allowing enhanced partitioning into the vapor phase and convective migration to the atmosphere. The end result is a reduction in representativeness for each subsequently recovered soil gas sample.

6.3.3.2 Multiple depth sampling can also be used to focus a sampling program into a single geologic unit or suite of units without regard to depth. This practice is helpful at sites with complex lithologic changes in the vadose zone. Samples can be recovered from lithologies with greater permeability to vapor or greater storage capacity for vapor when bias in sampling depth is necessary to accomplish project goals. This practice involves greater effort and expense than most methodologies due to the necessity to establish the presence, thickness and depth of the target horizons prior to soil gas sampling. The most common application of this methodology is the sampling of soil gas at the top of the capillary fringe.

6.3.4 *Time Variant Methodologies*—Monitoring soil gas in the vadose zone over time can suggest process rates of contaminant partitioning, emplacement, migration and degradation. Practical application of this methodology includes the monitoring of the effectiveness of remedial air-injection systems, the appearance of contaminants sourced from underground storage tanks, the encroachment of contamination onto a subject property from an abutting property and the mitigation of soil and ground-water contamination by microorganisms.

6.3.4.1 Some investigators and regulators with responsibilities at more than one location delegate seemingly simple time variant soil gas monitoring tasks to local personnel. Numerous problems with time variant monitoring can arise in the field as the result of poor system maintenance and record keeping by inexperienced or unmotivated personnel (property owners or parties responsible for contamination).

6.3.4.2 Certain maintenance problems are easily corrected, that is, cleaning bacteria and other foreign matter from detectors or replacing damaged components. Other maintenance problems can be fatal flaws in the methodology. These are principally related to ice formation in the sampling system and destruction of system integrity due to soil frost heaving.

6.3.5 *Combination of Soil Gas Monitoring With Other Vadose Zone Monitoring Techniques*—Soil gas monitoring is not a stand-alone technique. Corroborative support of this reconnaissance and screening tool by other vadose zone monitoring techniques is strongly encouraged. The possible combinations of the various vadose zone techniques with soil gas surveys are numerous. Soil gas can commonly be used as a reconnaissance tool to locate other monitoring devices such as lysimeters, neutron probes or ground water monitoring/sampling wells. Limits upon such combinations are controlled by budgetary constraints and the investigator's imagination.

6.4 *Field QA/QC*—Quality assurance and quality control procedures (QA/QC) are essential to establishing support for any interpretation of measurement data. Soil gas monitoring data requires a thorough QA/QC protocol confirming that data have been generated to satisfy the data quality objectives for the survey. This requirement is well known, however few investigators subject their soil gas data sets to the rigors of such protocol. Conclusions based upon data of unknown quality may be without merit. Justification for interpretations based upon data of unknown quality is not possible.

6.4.1 QA/QC requirements are dependent upon the data quality objectives defined in the planning phase of the survey. For example, simple contaminant audits require a less demanding QA/QC protocol than contaminant source identification. The goals of the QA/QC effort must be understood by field personnel to assure effective implementation of field QA/QC. A document control officer who is a member of the field team can provide this assurance.

6.4.2 Persons collecting descriptive data should not be varied during a soil gas survey. Soil descriptions, for example, can be somewhat subjective when estimations are made as to soil moisture or clay content. Changes in field personnel can translate into apparent changes in soil lithology that are merely functions of this subjectivity. The document control officer can review field records to discover any obvious errors related to descriptive data.

6.4.3 The results of a soil gas survey are highly sensitive to procedure. Field personnel should closely follow a standard operating procedure. This procedure should include the method(s) selected for the survey including the sampling system, means of sample collection, handling and transport of samples and field based equipment decontamination. A standard practice for equipment decontamination is essential to maximize the integrity of samples that may undergo chemical analyses (see Practice D 5088). Any deviations in the standard operating procedure should be recorded by the document control officer in a field notebook, with notes outlining the justification for the deviation. Data comparability can be severely compromised by deviations from the standard operating procedure.

6.4.4 Field based equipment decontamination can have impact on data quality. This results from the potential for cross contamination of samples due to poorly controlled field cleaning procedure or difficulties presented by the inconvenience of field decontamination. Field based equipment decontamination should not be considered a method of choice, but if unavoidable, must be performed with the data quality objectives for the survey as driving forces for procedure.

6.4.5 Bias of soil gas data describes a situation of consistently lower-than-actual or higher-than-actual soil gas contaminant concentration measurements (**32**). The bias of a measurement process is a generic concept related to consistent or systematic difference between a set of test results from the process and an accepted reference value of the property being measured (see Practice E 177). Bias can be imparted to the data through sample site selection, that is, exposure of a sampling device to an environment of enhanced contaminant concentration due to a preferential contaminant migration pathway, or exposure of a sampling device to an environment devoid of contaminants due to barriers to contaminant migration. Bias may also result from malfunction of the sampling system, contaminant degradation or numerous other factors. False positive or false negative values can result, lowering the value of the soil gas data set.

6.4.6 Table 2 summarizes some common problems in soil gas monitoring that can result in biased results.

6.4.7 A sampling program must be conducted during the survey to support evaluation of both the sampling system in the field and the analytical system employed. These samples are known as QA/QC samples. The type and magnitude of QA/QC sampling depends upon the purpose of the soil gas survey and the requirements for data quality attendant to it. It is the responsibility of the investigator to determine the appropriate rigor of field QA/QC protocol. The variation in QA/QC protocol from survey to survey is controlled by the purpose and magnitude of the survey, and can vary to a great degree.

6.4.8 The types of field QA/QC samples are field blanks, travel blanks, sample container blanks, sample probe blanks and sample replicates. Other types of QA/QC samples are analytical in nature and are discussed in 6.6.

6.4.9 Field blanks are samples of ambient air or nitrogen recovered from the sampling system which are recovered to determine contamination of samples by ambient atmospheric air, or, to act as system blanks to test for contamination of the sampling system. Field blanks are used to provide an indication of the probability of leakage in the sampling system or the breakthrough of atmospheric air to the sampling device through macroporous migration pathways in the vadose zone such as soil cracks or moldic porosity. If nitrogen is employed instead of atmospheric air, field blanks can have higher contaminant levels than soil gas. This is especially true for petroleum hydrocarbons in urban environments. At least one field blank should be recovered for each ten soil gas samples, or at least one field blank per sample batch or container type **(53)**.

6.4.10 Travel blanks are the contents of a sample container handled in the same manner as those containers holding samples, except that there has been no sample inserted into the travel blank. The purpose for travel blanks

TABLE 2 Summary of Possible Causes of False Positive and False Negative Values[A]

Result	Causes
False negatives, that is, falsely low values	Barriers to gaseous diffusion, such as perched water, clay lenses, impervious man-made debris, saturation of soil pores with water (as from rain), low subsurface temperatures.
	Biological or chemical degradation.
	Leakage or blockage in the sample train, improper purge procedure, loss of sample from sample container, problem with analytical system.
False positives, that is, falsely high values	Contamination in sampling train, sample container, or analytical system.
	Contribution of volatile organic contaminants from vegetation.
	Significant contamination in overlying soil.

[A] See Ref **(32)**.

is to audit sample integrity for loss due to sample handling and transport. Travel blanks are useful when analysis is performed at an off-site laboratory. The results obtained by analysis of travel blanks can be used to indicate a potential need to modify sample handling and transport procedure. At least one travel blank should be included in each batch of samples.

6.4.11 Sample container blanks are obtained by sampling the contents of a clean sample container to ensure that residual contaminants are not present in the container prior to sample collection. If contamination is detected in the cleaned containers, the decontamination procedure must be modified to remedy the problem. Sample container blanks should be collected and analyzed prior to each use of a sample container.

6.4.12 Sample probe blanks, consisting of carrier gas or atmospheric air contrasted to atmospheric air blanks, are drawn through the sampling device and recovered in the same manner as soil gas. The purpose for sample probe blanks is to check for the presence of sample train contaminants that would impact data quality. If contaminants are detected in sample probe blanks, the decontamination procedure must be modified to remedy this condition. Sample probe blanks should be collected and analyzed prior to each use of a probe and/or other components of the sampling system.

6.4.13 Field replicates are recovered as separate soil gas samples collected from the same sample site into multiple containers. Field replicates can be used to estimate the combined precision of sampling and analysis. The recovery of field replicates is not a common practice. When field replicates are demanded by a client or as dictated by a particular situation, field replicates should be recovered as often as is economically and practically possible, however in no instance should the number of replicates fall below ten percent of the total number of soil gas samples **(53)**.

6.4.14 Sample spiking, or the addition of a known quantity of a known compound or mixture to the soil gas sample, is sometimes performed in the field to provide internal checks of analytical quality. Sample spiking in the field is not recommended due to measurement uncertainties in the field. Moreover, caution must be exercised with this procedure because of the potential for contaminant interaction with the known compound(s).

6.4.15 A paperwork audit is recommended at the end of each working day or at the conclusion of recovery of each batch of samples recovered. The paperwork audit should be conducted by the document control officer and include evidence of an equipment inventory, sample inventory including QA/QC samples, review of field notes and chain-of-custody documentation.

6.4.16 Chain-of-custody documentation is recommended at all times, and is mandatory for soil gas surveys when samples are transmitted to an off-site laboratory. It is recommended for soil gas surveys when sample custody is transferred to someone other than the field team leader for any reason. Chain-of-custody documentation assures that samples have not been altered or mishandled prior to analysis. This procedure is mandatory for sample handling and transport in situations where there is likely to be a cost recovery effort or demonstration of contaminant responsibility in a court of law.

6.5 *Sample Handling and Transport*—Soil gas sampling and analysis usually involve the monitoring of contaminants at very low levels. Consideration of sample handling and transport is not trivial to this exercise.

6.5.1 The period of sample handling and transport represents the greatest opportunity for loss or gain of contaminants from or to sample containers. Loss occurs by contaminant condensation within the sampling train, sorption onto materials within the sampling train, solution into condensed water in the sampling train, chemical changes or leakage to the atmosphere through defects in the sampling apparatus or sample container. Gain of contaminants from sources other than the sampling horizon can occur through related mechanisms working in reverse. Both processes can severely limit the value of data obtained from a survey, and they must be minimized.

6.5.2 In general, the time between sample collection and analysis should be minimized. Investigators should protect samples against light and heat, and exercise precautions against leaks (see Practice D 1605).

6.5.3 *Acceptable Materials*—Investigators are responsible for selecting materials for soil gas sampling, transfer and containment that will not impact sample integrity. Containers that have parts made from porous or synthetic materials such as PTFE, rubber or many plastics are likely to retain or contribute contaminants to soil gas samples. Corrosive metals such as steel or brass become difficult to decontaminate upon corrosion due to the increased surface area of the corroded material and its enhanced sorptive capacity. Septa of any material will be responsible for measurable contaminant loss over time due to leakage. Acceptable materials can be conveniently decontaminated prior to soil gas recovery. Materials that cannot be decontaminated effectively between samples must either be replaced between samples, considered in QA/QC planning as a survey limitation or abandoned in favor of more suitable materials.

6.5.4 *Integral Systems*—Problems of sample handling and transport are minimized by integration of the sampling and analytical system. For example, a whole air-active sampling system can be coupled directly to a portable VOC (volatile organic compound) analyzer. The sample stream is fed directly to the intake port of the analyzer and passed through the detector. If there are no system malfunctions in the sample path, problems of sample degradation become trivial.

6.5.4.1 Care must be exercised with integral systems, however. The dead volume of integral systems is much higher than separate sampling and analytical systems. If the sampling system is not capable of delivering constant sample flow rates at or exceeding the requirements of the analyzer employed, data accuracy and comparability can be seriously affected. Moreover, a large sample volume is required merely to purge the sample system. In soils with moderate moisture contents or even nominal clay contents, it may not be possible to recover the volume of soil gas required to purge the system without serious negative impact to the composition of the soil gas sample recovered. Vapor phase contaminants can be lost to purge volume and atmospheric break-through can occur, leading toward a false negative result. Although this problem may not be apparent in seriously contaminated environments, it can become a fatal flaw at low contaminant levels.

6.5.4.2 Cross-contamination is a concern with integral systems. Many integral systems employ common elements from sample to sample, namely tubing, flow meters and analyzer components. Overcoming persistent contaminants can be difficult in integral systems, especially when high soil humidity and cold weather complicate the field effort.

6.5.5 *Transfer of Samples from Sampler to Container*—The method of transfer of samples from sampling device to containers is largely dependent upon the volume of soil gas recovered.

6.5.5.1 Small volume samples are commonly recovered by syringe for immediate injection into an analyzer or small volume container. Glass gas-tight chromatography syringes are employed when rigorous QA/QC protocol is required and samples are injected into the analyzer immediately upon recovery. These syringes must be decontaminated prior to recovery of each sample aliquot. Disposable syringes are employed when samples are to be transferred to a small volume container for transport. They are inexpensive, commercially available and convenient to use. However, disposable syringes can present a disposal problem. They should be inventoried prior to use and destroyed after use, the number destroyed equalling the number inventoried and used. Destruction includes smashing the syringe cylinder and clipping the needle.

6.5.5.2 Hand pumps are also used to transfer samples into tedlar bags or glass bulbs. Hand pumps are preferably installed behind the analyzer or container in the sample train to avoid contribution from or loss of contaminants to the hand pump. Hand pumps commonly contain petroleum-based lubricants which will contribute to the hydrocarbon content of soil gas. These devices must be placed at the end of the sample train or abandoned in favor of another tool.

6.5.5.3 Large volumes of soil gas are commonly recovered by hand or mechanical pumps installed at the end of the sample train. Large volume systems can be metered for soil gas flow rate, which is controlled by the capacity of the vadose zone sampling horizon to transmit vapor to the sampling device, the volume and configuration of the sampling system and the requirements of the analyzer or sorptive trap employed.

6.5.5.4 Small volume sampling is quite sensitive to variations in sample transfer technique. Septum coring by syringe is a common problem that restricts flow of soil gas through the needle. Coring can be corrected by decreasing the needle size and using a relatively hard septum material. Coring does not occur with side-port needles, a high-cost alternative. Needles of 25 to 27 gage seldom core septa. However, flow rates through these small gage needles are slow enough to require great care in consistency of sampling rate to minimize septum bleeding during sampling. This consistency is highly subjective and must be obtained through experience. Polypropylene disposable needles may provide opportunity for contaminant loss by sorption or gain by contribution to the soil gas sample. This can be minimized by using the polypropylene syringe to purge the sampling device prior to sampling, thereby reducing the potential for loss or gain of contaminants to that of the sampling device. Luer-lock needles should be checked for tightness by twisting prior to each use.

6.5.5.5 Tubing is commonly used in large volume sampling. For low level detection, tubing can present a cross contamination problem if not replaced in the sampling train prior to sampling at a new location. Some particulate matter and condensate may be trapped in tubing prior to entry into the flow meter and analyzer by looping the tubing into three or four small diameter loops at a point near the sampling device. This can eliminate the need for water traps or particulate filters in the system that can contribute to system loss or gain of contaminants.

6.5.5.6 Vacuum can be employed to transfer soil gas from a sampler to a container. Evacuated glass bulbs, some containing adsorbents or absorbing liquids (see Practice D 1605), can be affixed to an in-place and purged sampling device and allowed to come to pressure equilibrium. Care must be exercised in recovery of the gas sample from a vacuum cylinder. Upon recovery, the sample is immediately subjected to negative pressure and atmospheric contamination of the sample is encouraged.

6.5.6 *Sample Collection: Containers*—A wide variety of sample containers is employed by field investigators. Container selection is based upon the physical properties of the contaminants sampled, the volume of the sample recovered, the physical properties of suspected contaminants, the sampling system employed, the anticipated sample holding time prior to analysis and the analytical method chosen. Container type for a soil gas survey should be held constant within the survey. A change in container type can impart bias to a portion of the data due to sorptive or desorptive processes related to container type.

6.5.6.1 Whole air samples can be contained in any device made of suitable materials (see 6.5.3) that conveniently satisfy survey, handling, transport and analytical requirements. Certain containers require special handling practice. The literature provides discourse on atmospheric sampling bags **(54)**.

6.5.6.2 Sorbent traps are commonly self-contained. Care must be exercised to select a trapping device that is compatible with the properties of the target compounds and the technique of desorption chosen. Good practice for use of these devices, including handling and desorption procedure is required for successful implementation of sorbent traps when sampling organic compound vapors (see Practice D 3686).

6.5.6.3 Table 2 provides an inventory of sample containers, their applications, advantages and limitations **(32)**.

6.5.6.4 Containers exist that provide for both whole-air and sorbent fractions as well as removal of sample by displacement (see Practice D 1605). Some are convenient for field use, however most are too complex or fragile to be of effective use for a field screening technique requiring rapid mobility.

6.5.6.5 Detector tubes should not be considered as a primary containment vehicle for the purpose of storage and transport of soil gas. A discussion of detector tube application is provided in 6.6.1.

6.5.6.6 Containers for soil samples to be preserved for a subsequent headspace analysis range from glass sample vials to metal cans. The choice of container for soil headspace determination is dependent upon the method of sampling chosen. For soil samples obtained by backhoe, bucket auger or other destructive technique, that is, a disturbed sample, extrusion into a sample vial is not necessary since most of the highly volatile components have already been lost through the act of soil sampling. Metal cans should be made from a material that does not rust. Coating materials and sealing waxes are likely to react with or adsorb soil contaminants, presenting limitations to the value of the data collected. Glass containers with screw threads or crimped seals are difficult to use for soil headspace methods due to the inability of investigators to consistently, thoroughly and rapidly clean the threads or crimp surfaces of all containers prior to capping.

6.5.6.7 Soil pore liquid headspace samples are whole-air or whole-air plus pore liquid samples. They may be contained in most devices suitable for whole-air containment, however investigators are cautioned to select containers from which a vapor sample can be extracted for analysis independently of the liquid present.

6.5.7 *Sample Processing*—Some investigators process soil vapor samples prior to analysis. Processing is performed in an effort to control sample degradation in containers. Efforts to check this degradation by sample processing include refrigeration, pressurization, and pasteurization. As a general practice, sample processing is strongly discouraged. Refrigeration may be somewhat effective in controlling sample degradation, however the best method is to limit or avoid soil gas sample storage whenever possible. The limited shelf life of soil gas samples is discussed in 6.5.9.

6.5.7.1 Extraction is a sample processing step used to remove soil contaminants from soil cores or other similar samples. This technique can efficiently recover contaminants from all residence phases, not just the vapor phase. As a result, the technique yields samples that are not representative of soil atmosphere contaminant suites.

6.5.8 *Sample Transport*—If samples are to be transported to an off-site laboratory for analysis, they must be properly packaged to avoid damage to sample containers. Care must be taken to keep samples from becoming overly warm or agitated during transport. Overnight air express is highly convenient if samples are properly contained, but air freight is not recommended if samples are held in containers such as gas tight syringes or tedlar bags. These containers have other limitations as discussed in 6.5.6.

6.5.9 *Sample Life*—Soil gas samples have limited shelf life even in the most effective containers. Soil gas sample life is strongly container dependent. Numerous factors limit shelf life; most involve degradation in a container. Exposure to light, heat and agitation during shipping will accelerate sample degradation. Biodegradation may occur in some sample containers if water vapor condenses in a container containing microorganisms capable of metabolizing contaminants as substrate.

6.5.9.1 The safest practice is to minimize sample storage time. This problem is greatest when off-site laboratories are engaged to analyze the samples. Prior to recovering the soil gas samples, arrangements can be made with the selected off-site testing laboratory to schedule the necessary personnel and equipment in anticipation of sample delivery.

6.5.10 *Soil Gas Archiving*—Sample archiving in anticipation of a future analytical or descriptive requirement is a common practice. Minimal effects of degradation or loss may be noted in storing certain sorbed samples. Soil gas archiving is, however, not recommended. Although dependent upon the type of container and the storage environment, the likelihood of degradation of soil gas samples is great enough to raise concern. Insertion of standard gases into an archived sample set and spiking of archived soil gas samples with standards provides a reference to determine the likelihood or extent of sample degradation.

6.6 *Analysis of Soil Gas Samples*—Soil gas analysis procedure is based upon pre-existing protocol established for the analysis of contaminants in ambient air. A common reference practice defining terms, sampling information, calibration techniques and methods for validating results may be applied to all automatic analyzers (see Practice D 3249). Basic laboratory practice common to investigators engaged in sampling and analysis of atmospheres applies to soil gas analysis. Note that air sampling protocols and soil gas sampling protocols are not equivalent; geophysical and geochemical factors as well as definition of air sample volume contribute to this lack of equivalency. This guide includes the criteria, guidelines and recommendations for analytical segments including the mode of operation of the laboratory and data validation (see Practice D 3614).

6.6.1 *Basic Analytical Approach*—Soil gas analysis is performed to identify the presence of contaminants, their type and relative concentrations. Various analytical methods are highly general, satisfying only the most rudimentary requirements of contaminant screening. Others are sophisticated, providing identification and relative concentration information for numerous chemical compounds determined to be present in a soil gas sample. The choice of basic analytical approach in soil gas analysis is driven by the purpose of the soil gas survey, quality assurance objectives and budgetary constraints placed upon investigators.

6.6.1.1 Soil gas surveying as a field screening technique can often be effective without the commitment of expenditure for highly sophisticated techniques. This survey purpose is merely to locate other, more direct, techniques. Caution is suggested when choosing highly sophisticated analytical methods for field screening by soil gas monitoring. This selection is controlled largely by the need for the analytical method chosen to be cost-effective.

6.6.1.2 Other applications of soil gas monitoring require more thorough analytical protocol. It is not possible, for example, to suggest the locations of partitioned miscible and immiscible ground-water contaminant plumes with elementary analytical systems. Moreover, the independent monitoring of multiple classes of contaminants in soil gas normally requires analytical systems with multiple detectors. Successful soil gas monitoring for petroleum exploration requires an analytical system which can separate and identify extremely similar volatile compounds occurring at very low concentration levels.

6.6.1.3 Contaminant concentrations in soil gas can vary from levels below the detection limit of the most sophisticated equipment to percent of a whole-air sample. Ideally, the analytical system chosen has enough flexibility to determine contaminants in a wide range of concentrations. Care should be taken to select an analytical system sensitive enough to avoid false negative results which can lead to invalid conclusions. Many analytical systems are not designed to perform to specifications in very high concentration environments, requiring sample dilution prior to analysis or selection of a less sensitive method.

6.6.1.4 Of primary importance to the successful analysis of soil gas is the familiarity and experience of the analyst with the analytical system chosen. The analyst must be able to independently care for and maintain the equipment as well as recognize symptoms of procedural error. The success of an analytical effort lies wholly with operator ability and experience. Excessive machine capability cannot compensate for operator inexperience.

6.6.1.5 Soil gas may be analyzed by a number of methods, including portable VOC (volatile organic compound) analyzers, gas elution chromatography, gas chromatography-mass spectroscopy, and colorimetric and color-indicating detector tubes. Infrared spectroscopy and fiber optic chemical sensors can be applied to soil gas gas analysis, however their use is currently limited and few investigators have experience with this instrumentation. In practice, gas chromatography (GC) or GC-based handheld detectors are the most widely used analytical instruments **(32)** for soil gas analysis. This guide uses numerous terms relating to various GC methods for soil gas analysis. Most of the terms should apply to other GC methods (see Practice E 355).

6.6.1.6 Portable VOC analyzers used for fugitive emission screening and industrial hygiene monitoring have been adopted for soil gas analytical purposes by numerous investigators. These devices are easily transported to and from the field, require minimal operator skill, provide immediate data and serve to eliminate many sample handling and transport steps which can result in uncertainty. Portable VOC analyzers are limited in application to very low level detection due to the absence of a concentration step. They exhibit limited selectivity and do not have the ability to separate contaminant compounds, leading to potential interference. These devices also are limited in accuracy due to the inability to calibrate for the wide variety of contaminant compounds encountered in soil gas, each compound having its own character of detector response. Portable VOC analyzers contain three types of detectors. These are the flame ionization detector (FID), the photoionization detector (PID) and the infrared (IR) detector. The literature contains a thorough treatment of these devices **(10, 55)**.

6.6.1.7 Soil gas analysis by GC is by far the most versatile and the most costly soil gas analytical method. Instrumentation can be varied to accommodate field mobility, however this is not always required. The technique provides separation of compounds in a chromatographic column, tentative identification of compounds determined to be present and a relative quantitation of compound concentration based upon comparison to a known standard. Soil gas is introduced into the GC and conveyed through a chromatographic column by a carrier gas, separating the contaminants as they pass through the column. The separation is obtained when the sample mixture in the vapor phase passes through a column containing a stationary phase possessing special adsorptive properties. As the gas stream emerges from the column, it passes through a detector, providing for measurement of a specific sample property through the recording of detector electrical response. These responses, or peaks, are recorded as a function of time. Comparison of known standard compound response time with the response time of an unknown represented by a peak results in the tentative identification of the unknown. Comparison of the magnitude of detector response to the newly identified compound versus detector response to the same compound of known concentration, a laboratory standard, results in a relative quantitation of subject compound concentration in the sample.

6.6.1.8 Gas chromatography is essentially a physical separation technique. The degree of separation depends upon the differences in the distribution of volatile compounds, organic or inorganic, between a gaseous mobile phase and a selected stationary phase that is contained in a tube or GC column (see Practice E 260).

6.6.1.9 Numerous factors can impact the ability of the GC to determine contaminants in a soil gas sample. These include column characteristics, sample flow rate, sample temperature, the composition of the carrier gas and the type of detector employed. Instrumentation can be expanded to include multiple columns, multiple detectors, sample loops and temperature programming, all of which make an instrument more versatile, albeit at additional cost.

6.6.1.10 Simple GCs are portable analyzers with GC options. Field GCs are more advanced instruments with temperature programmable ovens and provide opportunity for multiple columns and detectors. They can be carried in mobile laboratories or established in a temporary base laboratory in the field. Research-grade instruments are normally based at off-site laboratories with strictly controlled environments. These are used when positive identification or very low detection limits are specified. The literature contains excellent comparisons of the advantages, limitations and applications of the various configurations of GCs, including instrument specifications **(10, 32, 56, 57)**.

6.6.1.11 Detector tubes have been applied to safety and health atmospheric monitoring, agriculture and the chemical industry. These devices are designed to be compound specific, although this characteristic is dependent upon the contaminant compounds present in the sample drawn through the tube. Detector tubes may be used for short-term sampling (grab sampling; 1 to 10 min) or long-term sampling (dosimeter sampling; 1 to 8 h). Short-term sampling involves the movement of a given volume of gas through the tube by a mechanical pump. If the substance for which the detector tube was designed is present, the indicator chemical in the tube will change color (stain). The concentration of the gas may be estimated by either the length of the stain compared to a calibration chart or by the intensity of the color change compared to a set of standards (see Practice D 4490). Long-term sampling involves the movement of gas at a very slow rate through the tube by means of an electric pump. The use of long-term detector tube sampling for soil gas monitoring is limited to specific temporal survey designs.

6.6.1.12 Detector tubes are relatively inexpensive and provide immediate results. Their use is restricted to applications with few interfering compounds. Depending upon the contaminants present, they may be of low sensitivity and can be affected by humidity, normally high in soil gas, sample flow rate, temperature extremes **(32)**, storage conditions and shelf life.

6.6.1.13 The literature contains excellent discourse on the detector tube apparatus, reagents, procedure accuracy and amenable compounds (see Practice D 4490).

6.6.2 *Specific Analytical Approaches*—This subsection discusses various detectors and methods that may be integrated into soil gas analytical instrumentation. For methods providing detector alternatives, the choice of an appropriate detector should be guided by knowledge of detector properties. Key properties are as follows (after Mayer, 1989 **(32)**):

6.6.2.1 *Selectivity or Specificity*—Selectivity refers to the responsiveness of the detector to the compound of interest. Detectors responding to a wide range of classes of compounds are termed universal or non-selective detectors. Those that respond to only certain classes of compounds are termed selective detectors.

6.6.2.2 *Sensitivity*—Sensitivity refers to the relationship between the detector response and the quantity of the subject compound injected. It is the smallest detectable quantity of a compound; it is usually considered to be the amount that produces a response equal to twice the baseline noise of the detector.

6.6.2.3 *Linear Dynamic Range*—Linear dynamic range is the range over which the detector response to a compound is directly proportional to the amount of compound injected. Detectors vary in the range of component concentrations over which they are linear. Wide linear dynamic range is desirable because it simplifies quantitation of samples having widely varying ranges of concentrations.

6.6.2.4 *Stability*—Stability is a factor referring to detector responsivity over time. Stability is controlled by numerous factors and is seldom quantified. The required frequency of instrument calibration is determined by detector stability.

6.6.3 Specific analytical approaches are as follows:

6.6.3.1 *Flame Ionization Detectors (FID)*—Flame ionization detectors generate electric current when gases containing carbon atoms are oxidized to carbon dioxide in a hydrogen flame and potential is applied across the flame. The magnitude of the electric current generated is termed the detector response. FIDs are responsive to hydrocarbon contaminants in soil gas and are commonly employed for this purpose. These detectors are durable for field application, and have a wide linear range and nearly uniform response to organic gas species. FIDs are generally unresponsive to inorganic gases and water vapor, common constituents in soil gas. FID performance can be evaluated independently of the chromatographic column (see Practice E 594). Although highly versatile, these detectors are not selective for halogenated compounds. They require supplies of fuel gas which require careful safety practices in handling and flame ignition.

6.6.3.2 *Photoionization Detectors (PID)*—Photoionization detectors employ ultraviolet radiation to ionize contaminant molecules. Positive ions and free electrons are formed which migrate to the detector electrode(s), resulting in an electric current that is proportional to contaminant concentration at the detector. PIDs are extremely sensitive to aromatic hydrocarbons due to the great efficiency of ionization of pi bonds under ultraviolet radiation. Efficiency of ionization of sigma bonds is lower, resulting in a higher PID detection limit for aliphatic hydrocarbons. The selectivity of the method can be adjusted by selecting lamps of different energies, causing a change in response of contaminants with fixed ionization potentials to changing lamp energies. Tables exist of ionization potentials of compounds within classes common to soil gas contaminants **(58)**. Methane has an ionization potential higher than the energies of commercially available lamps, limiting the PID to detection of compounds other than methane. PIDs are further limited by their tendency to conceal the presence of low-sensitivity compounds when high-sensitivity compounds (aromatics) are present. PID response can be impacted by condensation of water vapor in the lamp.

6.6.3.3 *Electron Capture Detectors (ECD)*—Electron capture detectors are highly sensitive to and selective for compounds with electronegative functional groups such as CFCs (chloro-fluorocarbons). The sensitivity of the detector is proportional to the number of these groups on a compound, resulting in a unique detector response to each compound. The ECD comprises a source of thermal electrons inside a reaction chamber (a radioactive source emits β radiation which ionizes the carrier gas to produce electrons). The device detects compounds with electronegative functional groups capable of reaction with thermal electrons to form negative ions. Such reactions cause a decrease in the concentration of free electrons. The detector is designed to measure changes in the concentration of these electrons inside the chamber (see Practice E 697). Calibration of the ECD is therefore linked to each compound to be determined by the detector. ECDs are also sensitive to water, oxygen and other common components of soil gas, causing potential problems in method performance. ECDs emit β radiation that should be properly vented. Operation of an ECD requires licensing under Federal regulation.

6.6.3.4 *GC/Mass Spectroscopy*—Combination of gas chromatography and mass spectroscopy results in the GC/MS method of analysis. A mass spectrometer is used to obtain a mass spectrum of each eluting compound. Positive identification of these compounds is sometimes obtained by comparison of the unknown mass spectrum to a library of known spectra. GC/MS can be extremely selective for target compounds. Use of the technique for soil gas monitoring is limited, primarily due to the cost of analyses.

6.6.3.5 *GC/Fourier Transform Infrared Spectroscopy*—This analytical method combines gas chromatography with Fourier transform infrared spectroscopy. GC/FTIR can provide a rapid identification of eluting compounds by comparison of their infrared spectra with a known spectral library. Quantitation is achieved by subsequently passing the sample through an appropriate GC detector such as the FID or ECD. This method, like GC/MS, is limited in application to soil gas monitoring by the high cost of analysis.

6.6.3.6 Other detectors are applied to soil gas analysis by GC, albeit rarely in comparison to FID, PID and ECD. They include the argon ionization detector, a non-destructive device similar in operating design to the ECD, the flame photometric detector (FPD) used to determine organic compounds containing sulfur and phosphorus, and the hot-wire (pyrolyzer) used to determine compounds containing nitrogen.

6.6.4 *Analytical QA/QC*—The validation of the analytical aspects of soil gas monitoring is fundamental to the technique. Analytical equipment and procedure must be evaluated by laboratory QA/QC, just as the sampling system, sampling plan and field procedure are evaluated by field QA/QC methods. Analytical QA/QC defines a confidence limit of performance. The utilization of well tested and uniform analytical practices is essential to the production of reliable and defensible data, the validity of which can be demonstrated at a later date through the use of written field and laboratory records (see Practice D 3614).

6.6.4.1 Most analytical QA/QC plans contain calibration steps, linearity checks, standard analyses, blank analyses, duplicate analyses and audit checks. The various analytical approaches discussed in 6.6.3 require a variety of different protocols which will satisfy the QA/QC requirements for each method. Four types of analytical QA/QC samples are required for determination of quality assurance. These are analytical reagent blanks (used to determine the potential of sample or standard contamination from a reagent), laboratory blanks (used to determine the impact potential of the laboratory atmosphere on analytical results), analytical sample replicates (used to estimate the analytical precision for samples) and analytical standard replicates (used to estimate the analytical precision for standards). Table 3 provides a summary of suggested calibration and quality control requirements for analytical systems **(10)**.

6.6.4.2 The aspects of bias, precision, representativeness, completeness and comparability must be considered to evaluate analytical equipment performance, including the establishment of minimum detectable quantities of contaminant compounds, retention time drift and the linearity of instrument response. Bias and precision must be quantified in order to compare actual survey performance with goals established in the survey plan.

6.6.5 A data validation summary report is a common method of evaluating analytical system performance. A guide for determining parameters key to the data validation summary report is provided as follows.

6.6.5.1 *Bias*—For determination of bias, the percent recovery can be determined using the following formulas:

$$recovery\ reproducibility = (DCS/KCS)*100 \tag{6}$$

where:
DCS = determined concentration of standard, and
KCS = known or certified concentration of standard.

The standard deviation of all standards analyzed can be determined as follows:

$$SD = \{(sum(recovery\text{-}i - recovery\text{-}ave)^2)/(n-1)\}^{0.5} \tag{7}$$

Finally, the range of uncertainty can be determined using the following equation:

$$\pm R = \pm t*(SD)/(n^{0.5}) \tag{8}$$

where:
t = the value of Studentized t at the 90 % confidence level and $(n - 1)$ degrees of freedom.

The bias statements for data collected should be expressed as the average recovery plus or minus the range.

6.6.5.2 *Precision*—For the determination of precision, the relative standard deviation of replicates can be calculated using the following equation:

$$RSD\text{-}pair = SD/Mean \tag{9}$$

TABLE 3 Soil Gas Sampling Containers[A]

Type	Applications	Advantages	Limitations
Stainless steel canisters	Collection of samples for delayed analysis	Durability Ease of sample handling Can be re-used Sample holding time longer than that for other whole-air sample containers Sample volume measurement not required Desorption not required Allows replicate analysis	Expense Requires vacuum pump or gage Can be difficult to decontaminate
Glass bulb	Collection of samples for delayed analysis	Glass is more inert than other sample container materials Septa possible Allows replicate analyses	Easily breakable Leakage through stopcocks or septa possible Adsorption to PTFE or other parts
Bag	Collection of samples for delayed analysis Sampling of very high vapor pressure compounds for which absorption methods are unsuitable	Bulk loss of sample is readily apparent Containers are light-weight and easy to handle Sample volume measurement not required Desorption not required Allows replicate analyses	Expense Some compounds may be lost through or adsorbed to bag walls Some container materials may contaminate samples Containers cannot be easily re-used Leaks in valves
Syringe[B]	Collection of samples for on-site analysis	Ease of sample collection Does not require special equipment to introduce sample into GC Desorption not required	PTFE plungers can adsorb sample Holding time short due to leakage or absorption Sample volume smaller than for other containers
Sorbent sampler	Allows concentration of low level samples If samples are solvent-desorbed, allows analysis of liquid sample	Ease of handling Relatively long holding time	Requires precise sample volume measurements Sorbent type must be tailored to compounds to be measured; adsorption behavior of each compound for solvent used must be accounted for Requires desorption (thermal or solvent) for analysis

[A] See Ref **(32)**.
[B] Syringes may also be used to transfer samples from the sampling device to a container for off-site analysis.

$$RSD\text{-}ave = \{\{(sum(RSD\text{-}pair))^2\}/(n-1)\}^{0.5} \quad (10)$$

where:
RSD-pair = relative standard deviation for each pair of replicates, and
RSD-avg = relative standard deviation overall.

Next, the precision can be determined as follows:

$$precision = \{(t * RSD\text{-}avg)/DF\} * 100 \quad (11)$$

where:
precision = the percent precision,
t = the t value for $n-1$ pairs of replicates, and
DF = the degrees of freedom = $(n-1)$.

Finally, mean value is reported with associated uncertainty:

$$x \pm (x * t * SD\text{-}ave)/(DF)^{0.5} \quad (12)$$

where:
x = reported chemical concentration, and
t = the value of t at the 90 % confidence level for the appropriate degrees of freedom.

6.6.5.3 *Representativeness*—Representativeness is determined by the results of the cross contamination blanks and the air blanks. The results should be presented as a bias estimate, as follows:

$$bias\ (\%) = \{(CCC - CA)/Mean\} * 100 \quad (13)$$

where:
CCC = concentration in cross contamination sample,
CA = concentration in air, and
Mean = mean concentration in sample set (bias may also be expressed for a single sample by substituting sample concentration).

6.6.5.4 *Completeness*—The completeness goal is 90 % or higher. Completeness is the number of samples collected that can be validated through the procedures for bias, precision, and representativeness.

6.6.5.5 *Comparability*—Comparability is based upon professional judgment and is provided through planning steps carried out prior to initiation of field work.

6.7 *Data Interpretation*—Soil gas data interpretation is an iterative process including the examination of the raw data, selection of appropriate and useful data displays, and establishment of correlation of the data set to other vadose zone monitoring data and ground truth. Interpretation of soil gas data is not like other interpretive exercises involving measurement data, in that mathematical expressions relating soil gas contaminant concentrations to underlying soil, rock and ground-water contaminant concentrations cannot be written for most applications at a high confidence level. This is a function of a lack of site characteristics information at even the most comprehensively studied sites. Soil gas data cannot be consistently interpreted in a manner that establishes direct correlation between contaminants in a soil gas horizon and contaminants in other horizons. Processes including migration and degradation can have profound influence on the correlation of soil gas data to ground truth. Interpretive efforts excluding consideration of these influencing processes can be highly misleading. For example, the presence of contamination in an underlying horizon will not necessarily correlate to the detection of contaminants in overlying soil atmospheres, that is, the potential for a false negative result. The converse is also true, that is, the potential for a false positive result. Interpretation of GC results in the laboratory without consideration of pertinent hydrogeological information may lead to incorrect conclusions **(59)**. However, the detection of contaminants in soil gas does suggest the existence of a contaminant source, and increases in contaminant concentration can suggest close proximity to the source or an increased quantity of the subject contaminant in the subsurface. It is the responsibility of the interpreter to examine soil gas data in context of other site characteristics, and provide an interpretation based upon sound judgment and thorough yet practical data treatment.

6.7.1 *Manipulating Data*—Soil gas data is normally interpreted as raw data. The application of correction factors is not recommended, as it is difficult if not impossible to determine if the magnitude of the correction factor is greater than that of the variance between data populations in a survey. Moreover, the need for correction factors can indicate a flaw in survey design, sampling system performance or the objectivity of the interpreter.

6.7.2 *Defining Data Subpopulations*—Soil gas monitoring seeks to define anomalous subpopulations of data that contain measurable quantities of contaminants or unusual compositions. These populations can easily be described by their contrast to normal populations, for example, contrasting populations with and without measurable contaminants. Establishment of contaminant baselines or conditions "at background" make this contrast possible. If all soil gas samples are recovered in a contaminated area, there may be no apparent contrast.

6.7.2.1 Statistical treatment of soil gas monitoring data allows the interpreter to estimate the amount of variation noted in the survey data due to errors. This practice also permits the interpreter to evaluate the data quality objectives suggested for the survey during the planning phase. Statistical treatment of soil gas data can also be of use to define anomalous data subpopulations when the boundaries of a contaminated area are not clearly defined or if the existence of multiple populations of data (that is, contaminated and uncontaminated) within a single data set is in doubt. The literature contains discourse on statistical treatment of soil gas data **(10, 60)**.

6.7.3 *Interpreting Soil Gas Data Profiles*—Soil gas data from survey profiles displayed on an $X - Y$ plot is an effective aid to data interpretation. This display is useful to examine the overall context for soil gas measurement data potentially indicating contamination. If the profile is displayed as a cross section through a grid pattern or as a linear array of sample points, the profile display can illustrate spatially significant groupings of data subpopulations.

6.7.3.1 It is quite common for concentration data to be highly variant within a contaminated area. Soil gas profiles can be used to show variation in spatially related data. This is one method of defining subpopulations of data indicating contamination or other anomalous characteristics.

6.7.3.2 Multiple data sets can be displayed on a single profile. Comparison of one data set to another on a single profile is a simple visual method to screen for suggested data subpopulations. Comparison of concentration data and compositional data (see 6.7.5) on a single profile can further resolve this problem.

6.7.4 *Mapping Soil Gas Data*—Soil gas data obtained by sampling at a single depth is often mapped to suggest the lateral extent of subsurface contamination. Map suites of soil gas data obtained from multiple depths can sometimes aid investigators in determining the depth to the contaminant source.

6.7.4.1 Numerous algorithms can be used to interpolate between data points, including linear, inverse distance squared, inverse distance cubed, splines and kriging. The various interpolation methods will yield similar results, suggesting a general pattern of contaminant distribution in soil gas. Kriging requires a probability model for each survey site mapping application for which it is employed, the derivation of which requires data which are not normally available for a given soil gas survey area.

6.7.4.2 Caveats exist in using computer mapping programs as interpretive aids. Difficulties can arise in treatment of adjacent data points differing in contaminant concentration by an order of magnitude and more due to vapor migration barriers, preferential vapor flow paths or changes in soil moisture or porosity content. It is possible to model these characteristics and input such a model into some computer mapping programs, however this introduces bias into the mapping effort. Single point soil gas contaminant concentration highs may exist due to a sample density which is insufficient to resolve the cause for the single point anomaly. Contour mapping of such data may be meaningless without the complement of other information, especially detailed knowledge of site characteristics.

6.7.5 *Analyzing the Composition of Soil Gas Contaminants*—Certain applications of soil gas monitoring require detailed analyses available from off-site bench laboratories or mobile laboratories. Determination of a number of contaminant compounds in a soil gas sample set with either of these analytical systems enables the interpreter to make a comparative analysis of the changes in soil gas contaminant composition within that sample set.

6.7.5.1 Compositional analyses can range in scope from a simple listing of the various compounds determined in each sample to thorough data treatments. Profiles of soil gas data can be constructed to illustrate the spatial relationship between two potentially different groupings of data (see 6.7.3). Crossplots of contaminant compound concentrations are highly effective in the definition of data subpopulations, and can be used to relate contaminant types to known on-site waste streams and sources in complex settings. Known as fingerprinting, this guide compares vapor composition over a known contaminant product and the known soil atmosphere composition over that product to soil gas contaminant composition in areas being investigated on the subject site. Subtle divisions in data subpopulations can be defined by crossplots of contaminant ratios. In addition to simple ratioing, computerized multivariate pattern recognition techniques such as cluster, factor and discriminant analyses can assist in the evaluation of intra-data set compositional variations and their relationship to the physical contamination issues at a site.

6.7.5.2 Soil gas data can be examined for the appearance of target compounds determined to be present in contaminant mixtures. The success of this practice, used primarily to establish the location and extent of underlying ground water contamination, relies upon selection of appropriate target compounds and the persistence of target compounds in soil vapor.

6.7.5.3 Monitoring specific compounds in soil gas data can be utilized to determine the progress of degradation or migration of contaminants in the vadose zone and in ground water. Biodegradation has been monitored by the appearance of excessive quantities of carbon dioxide in soil gas **(61)**.

6.7.6 *Interpretation in Context of Other Vadose Zone Monitoring*—Soil gas monitoring is not a technique that can consistently support conclusions based upon interpretations of survey results. For this reason it is strongly recommended that other vadose zone monitoring methods be used to corroborate data obtained from a soil gas survey, especially when investigators are attempting to do more than simply audit a subject site for the presence of contaminants. Useful models of contaminant emplacement and transport in the vadose zone can be constructed by combining techniques. Examples of useful combinations are soil pore liquid and soil gas monitoring or neutron probe and soil gas monitoring.

6.7.7 *Correlation With Ground Truth*—Interpretation of soil gas data is difficult without establishing some form of ground truth with which to substantiate survey results. Ground truth can be in the form of monitoring well data, for purposes of determining the extent of contamination by a ground-water contaminant plume. Examples of other forms of ground truth usable in support of soil gas data interpretation are soil cores, the presence of contaminant odors in basements, observed floating contaminants in storm sewers or utility vaults, or other field observations.

7. Data Reporting Requirements

7.1 *Purpose of Reporting*—Of primary concern in a report of findings pertaining to a soil gas survey is that the report includes the information necessary to describe the results of that survey performed for a particular application. In many instances, certain interpretative methods or data reporting formats useful to end users for one particular application are not relevant to the needs of end users applying the information to a different application. Examples of these differing applications that require unique report subject matter are soil gas contaminant determinations for real property environmental assessments, soil gas monitoring of volatile organic contaminants from underground storage tanks and soil gas sampling as a tool useful in the exploration for natural resources. Certain applications require a thorough treatment of a significant number of factors impacting the meaning and usefulness of soil gas data interpretations. Examples of such applications include damage assessments, contaminant source identification or tests of the effectiveness of remediation. Other applications command minimum reporting requirements. An example of such an application is the monitoring of releases from underground storage tanks over time. Included in a discussion of the report objectives should be an identification of the end user category (for example, regulatory agency, land acquisition negotiations).

7.1.1 A decision must be made regarding the units expressed in reporting, that is, qualitative or quantitative. If quantitative, the appropriate expression of units in volume/volume or weight/volume must be determined. SI units are recommended for reporting of atmospheric measurement data (see Practice D 1914).

7.2 *Report Format*—Certain reporting requirements are commanded without regard to data application. In large part they are related to the QA/QC objectives, and include data comparability, representativeness, bias, precision accuracy, completeness and analytical detection limits whenever possible. At a minimum, a general discussion of the reliability of results and analytical detection limits is warranted; soil gas test data may be evaluated in the same manner as is other atmosphere test data (see Practice D 3614).

7.3 *Salient Points to be Addressed in Reporting*—The report of findings of any soil gas monitoring effort can contain discussions within any number of topics that should be selected to best suit the requirements of the end user. Selection of appropriate topics is discretionary, usually based upon a scope of work determined by prior agreement between the data provider and the data end user. Efforts to limit reporting requirements for the sake of short term time and money cost savings usually result in low-confidence level treatment of the report or an ultimate time and money cost gain, or both. Discussions that should be included when appropriate and whenever possible are provided below.

7.3.1 The purpose of the soil gas study should be stated, as well as the rationale for selection of a particular soil gas monitoring technique.

7.3.2 Selection of a particular soil gas monitoring technique is typically controlled by the chemical and physical properties of the chemical compounds of interest which are known to occur or suspected to occur on site. A discussion of the sample array in three dimensions, sampling method employed and the analytical scheme chosen in context of these properties should be provided.

7.3.3 The rationale for selection of a particular soil gas monitoring technique should always be based upon the physical properties of the vadose zone as well as the chemical and physical properties of the compounds of interest. A discussion of the impact of these vadose zone properties on survey design should be included in the report. The regional and local hydrogeologic conditions within the survey area should be described. A discussion of the regional geology should include the physiographic province, a generalized geologic column, geologic structure and general ground water occurrence. The local conditions should be described with regard to soil type(s), moisture content in the vadose zone, soil/bedrock interface, stratigraphy and lithology, ground water bearing zones, flow directions and gradients, potentiometric levels, aquifer characteristics and ground water quality.

7.3.4 If known and appropriate, the characteristics of a contaminant source or spill should be addressed. Examples of such characteristics are contaminant composition, the likelihood of single or multiple contamination events or the reaction potential (above, within and beneath the vadose zone) of multiple contaminant mixtures.

7.3.5 Every subject of every vadose zone monitoring effort has unique characteristics. Those characteristics that could impact the results of the soil gas monitoring effort should be described to provide a meaningful context in which to interpret the soil gas data.

7.3.6 There are a number of topics common to most soil gas data reporting that are useful in the majority of applications. The regional and specific site location should be identified using a site plot plan. The site plot plan could include an insert showing the regional location. A discussion should be included regarding the physical structures at the site that may impact the location of sampling points and the migration of soil gas, for example, asphalt and concrete pads, buried pipelines and surface water impoundments. Site history must be considered, including the types of chemical compounds known or suspected to have been used at the site. These compounds should be listed with their chemical and physical properties as they relate to volatilization, solubility and other migration characteristics or soil gas recovery characteristics.

7.3.7 The site should be evaluated in the report of findings for the impact of the regional and local hydrogeologic conditions within the survey area on the results of the survey.

7.3.8 A detailed description should be given of the type of soil gas survey conducted. Details should include selection of active or passive method, whole air or passive sample collection method, sampling array, background sampling, equipment decontamination procedure employed prior to the survey, field or laboratory analytical methods and QA/QC procedures. Any unusual conditions should be noted, such as rainfall events during the course of the survey (especially when moveable soil gas chiefly originates from vadose zone microporosity), high pressure or low pressure front movement across the survey area during the course of the survey (especially when moveable soil gas chiefly originates from vadose zone macroporosity), or visual observations of contamination at sampling points.

7.3.9 If a subject property is found to be contaminated, a separate discussion of soil gas characterization of uncontaminated or non-anomalous contiguous property should be provided in the report of findings. This can be useful in highlighting naturally occurring petroleum hydrocarbons in soil and in establishing a regional baseline of contamination.

7.3.10 Data collected during the field sampling and field or laboratory analyses should be compiled in table form and be included in a preliminary or final report, preferably as appendices. Such data should include a listing of sampling and analysis dates, soil/rock description at each sampling point, depth and diameter of sampling point, quantity of soil gas purged prior to sampling, quantity of sample extracted, chromatogram and/or mass spectra for each sample and a tabulation of QA/QC samples recovered.

7.3.11 The report of findings should include a discussion of the results of the QA/QC efforts, establishing performance within limits set prior to the survey. Data validation involves review of the data collected for the purpose of isolating spurious values **(32)**. Systematic errors or bias can be detected in this review. Suggestions should be made as to the origin of the errors or bias.

7.3.12 Results of analyses should be displayed on plan maps and should include sampling point locations, physical features, contours of equal concentrations of specific compounds or compound groups (for example, alkanes) and any necessary keys or other notes to guarantee map clarity. Cross-sections showing changes in contaminant concentration with depth and concentration profiles of more than one contaminant through several sample locations can be highly useful displays. The report should include text describing each map, cross-section or profile.

7.3.13 Whenever possible, discussion should be provided that correlates soil gas data to ground truth. The most common and widely accepted form of ground truth is data from ground water monitoring wells.

7.3.14 When appropriate, the report of findings should attempt to identify the source of the contaminants encountered in the soil gas survey.

7.3.15 The report should contain a section which discusses the conclusions drawn from the results of the soil gas study and any recommendations which seem appropriate to enhance the value of conducting such a soil gas study. Conclusions should include identification of the compounds detected, if any, an assessment of the appropriateness of the soil gas study method used, and any circumstances that may have significantly impacted the results of the investigation, such as weather conditions or equipment calibration. Recommendations should address need for establishing ground truth, extension of the study to adjacent areas of interest, the need for a different soil gas study method, actions to resolve questionable QA/QC results, or need for additional chemical analyses for contaminant identification.

7.4 *Disadvantages of Real-Time Reporting*—In actual practice, many end users request real-time reporting of soil gas data obtained from field-based laboratories. Presentation of such data presents the opportunity for misunderstanding by end users who are not familiar with the caveats presented by data not examined in light of the QA/QC program or site specific factors. Real-time reporting of soil gas data is therefore not recommended.

8. Keywords

8.1 contaminant; environmental monitoring; geochemistry; ground water; Henry's law; petroleum hydrocarbon; sampling; soil gas; unsaturated flow; vadose zone; vapor monitoring; volatile organic compound

TABLE 4 Summary of Suggested Calibration and Quality Control Requirements for Analytical Systems[A]

Type of Instrument	Detector Type	Type of Calibration/QC Test	Frequency	Gas Standard(s)	Acceptance Criteria	Corrective Action
Portable VOC (THC) Analyzer	FID	(1) Multipoint calibration (zero plus three upscale concentrations)	At start of program	Methane or other aliphatic compound	Correlation coefficient ≥ 0.995	Repeat multipoint calibration after checking calibration dilution system
		(2) Zero (span) calibration	Daily	UHP Air or N_2/Methane	Response factor agreement within ± 20 % of mean RF for multipoint calibration	(1) Repeat zero span calibration (2) If still unacceptable, repeat multipoint calibration
		(3) Control sample analysis	Daily, prior to testing	Methane	Measured concentration within ± 10 % of certified concentration	(1) Repeat zero span calibration (2) Repeat control sample analysis
		(4) Drift check	Daily, at conclusion of testing	Methane	Drift value ≤ 20 % of the input value	(1) Flag day's data as questionable (2) Repair or discontinue use of analyzer
	PID	(1) Multipoint calibration (zero plus three upscale concentrations)	At start of program	Benzene or other aromatic compound	Correlation coefficient ≥ 0.995	Repeat multipoint calibration after checking calibration dilution system
	PID	(2) Zero/span calibration	Daily	Benzene or other aromatic compound	Response factor agreement within ± 20 % of mean RF for multipoint calibration	(1) Repeat zero/span calibration (2) If still unacceptable, repeat multipoint calibration
		(3) Control sample analysis	Daily, prior to testing	Benzene or other aromatic compound	Measured concentration within ± 10 % of certified concentration	(1) Repeat zero/span calibration (2) Repeat control sample analysis
		(4) Drift check	Daily, at conclusion of testing	Benzene or other aromatic compound	Drift ≤ 20 % of the input value	(1) Flag day's data as questionable (2) Repair or discontinue use of analyzer
Portable Gas Chromato-graph	FID	(1) Multipoint calibration (zero plus three upscale concentrations)	At start of program	Benzene or toluene	Correlation coefficient ≥ 0.995	Repeat multipoint calibration after checking calibration dilution system
		(2) Zero/span calibration	Daily	UHP air or N_2/methane	Response factor agreement within ± 20 % of mean RF for multipoint calibration	(1) Repeat zero/span calibration (2) If still unacceptable, repeat multipoint calibration
	FID	(3) Control sample analysis	Daily, prior to testing	Benzene	Measured concentration within ± 10 % of certified concentration	(1) Repeat zero/span calibration (2) Repeat control sample analysis
		(4) Drift check	Daily, at conclusion of testing	Benzene	Drift ≤ 20 % of the input value	(1) Flag day's data as questionable (2) Repair or discontinue use of analyzer
		(5) Retention time checks	Daily	Benzene or toluene	None	None
		(6) Analytical blanks	Daily	UHP air or N_2	Measured concentration ≤ 5 % of the instrument span value	Clean/replace system components until acceptable blank can be obtained
		(7) Sampling system blanks	Daily, plus after very high samples	Sample gas	Measured concentration ≤ 5 % of the instrument span value	Clean/replace system components until acceptable blank can be obtained
	FID	(8) Duplicate samples	10 % of sampling points, minimum	Sample gas	None; provides a measure of total sampling variability	None
		(9) Control point samples	After every ten samples or once per day, whichever is greater	Sample gas	None; provides a measure of temporal variability	None
		(10) Background samples	One sample per day	Sample gas	None; provides a measure of background concentration	None
	PID	(1) Multipoint calibration (zero plus three upscale concentrations)	At start of program	Benzene or toluene	Correlation coefficient ≥ 0.995	Repeat multipoint calibration after checking calibration dilution system
		(2) Zero span calibration	Daily	UHP air or N_2/methane	Response factor agreement within ± 20 % of mean RF for multipoint calibration	(1) Repeat zero/span calibration (2) If still unacceptable, repeat multipoint calibration

TABLE 4 *Continued*

Type of Instrument	Detector Type	Type of Calibration/QC Test	Frequency	Gas Standard(s)	Acceptance Criteria	Corrective Action
	PID	(3) Control sample analysis	Daily, prior to testing	Benzene	Measured concentration within ± 10 % of certified concentration	(1) Repeat zero/span calibration (2) Repeat control sample analysis
		(4) Drift check	Daily, at conclusion of testing	Benzene	Drift ≤ 20 % of the input value	(1) Flag day's data as questionable (2) Repair or discontinue use of analyzer
		(5) Retention time checks	Daily	Benzene or toluene	None	None
		(6) Analytical blanks	Daily	UHP air or N_2	Measured concentration ≤ 5 % of the instrument span value	Clean/replace system components until acceptable blank can be obtained
		(7) Sampling system blanks	Daily (plus after very high samples)	Sample gas	Measured concentration ≤ 5 % of the instrument span value	Clean/replace system components until acceptable blank can be obtained
		(8) Duplicate samples	10 % of sampling points, minimum	Sample gas	None; provides a measure of total sampling variability	None
	PID	(9) Control point samples	After every ten samples or once per day, whichever is greater	Sample gas	None; provides a measure of temporal variability	None
		(10) Background samples	One sample per day	Sample gas	None; provides a measure of background concentration	None
Off-site Gas Chromato-graph	FID	(1) Multipoint calibration (zero plus three upscale concentrations)	1 per month	Propane/hexane	Correlation coefficient ≥ 0.995	Repeat linearity check
		(2) Single point calibration check	Daily, prior to sample analyses	Propane/hexane	Response factor agreement within ± 20 % of most recent average RFs for multipoint calibration	Repeat single point calibration
		(3) Retention time check	Daily, prior to sample analyses	Multicomponent standard	Agreement with preestablished relative retention times	Adjust GC conditions and repeat RT check
	FID	(4) Control sample analysis	Daily, prior to sample analyses	Sample gas	(1) Correct identification of 90 % of components (2) For 90 % of components, measured concentrations within ± 30 % of actual concentrations	Repeat control sample analysis
		(5) Duplicate analyses	Minimum 10 % of samples (all duplicate samples will be analyzed in duplicate)	Sample gas	CV ≤ 20 % for ten major sample components	Repeat sample analysis
		(6) Blank analysis	Daily, prior to sample analysis	UHP air or N_2	Total ≤ 20 ppbv-C	(1) Clean system (2) Repeat blank analysis
	PID	(1) Multipoint calibration (zero plus three upscale concentrations)	1 per month	Propane/hexane	Correlation coefficient ≥ 0.995	Repeat linearity check
	PID	(2) Single point calibration check	Daily, prior to sample analyses	Propane/hexane	Response factor agreement within ± 20 % of most recent average RFs for multipoint calibrations	Repeat single point calibration
		(3) Retention time check	Daily, prior to sample analyses	Multicomponent standard	Agreement with preestablished relative retention times	Adjust QC conditions and repeat RT check
		(4) Control sample analysis	Daily, prior to control sample analyses	Sample gas	(1) Correct identification of 90 % of components (2) For 90 % of components, measured concentrations within ± 30 % of actual concentrations	Repeat control sample analysis
		(5) Duplicate analyses	Minimum 10 % of samples. (Duplicate samples analyzed in duplicate)	Sample gas	CV ≤ 20 % for ten major sample components	Repeat sample analysis
	PID	(6) Blank analysis	Daily, prior to sample analysis	UHP air or N_2	Total ≤ 20 ppbv-C	(1) Clean system (2) Repeat blank analysis

TABLE 4 *Continued*

Type of Instrument	Detector Type	Type of Calibration/QC Test	Frequency	Gas Standard(s)	Acceptance Criteria	Corrective Action
	ECD	(1) Quantitative standard	Daily, prior to sample analysis	Multicomponent standard	Response factor agreement within ± 30 % of three day rolling mean RFs for all components	Repeat calibration
		(2) Retention time check	Daily, prior to sample analyses	Multicomponent standard	None; will provide basis for comparison of FID/PID results to ECD results	None
		(3) Control sample analysis	Daily, prior to sample analyses	Sample gas	(1) Correct identification of all components (2) For 90 % of components, measured concentrations within ± 30 % of actual concentrations	Repeat control sample analysis
	ECD	(4) Duplicate analyses	Minimum of 10 % of samples (all duplicate samples analyzed in duplicate)	Sample gas	CV ≤ 20 % for ten major sample components	Repeat sample analysis
		(5) Blank analysis	Daily, prior to sample analyses	UHP air or N_2	Total ≤ 20 ppbv-C	(1) Clean system (2) Repeat blank analysis

[A] See Ref **(10)**.

APPENDIX

(Nonmandatory Information)

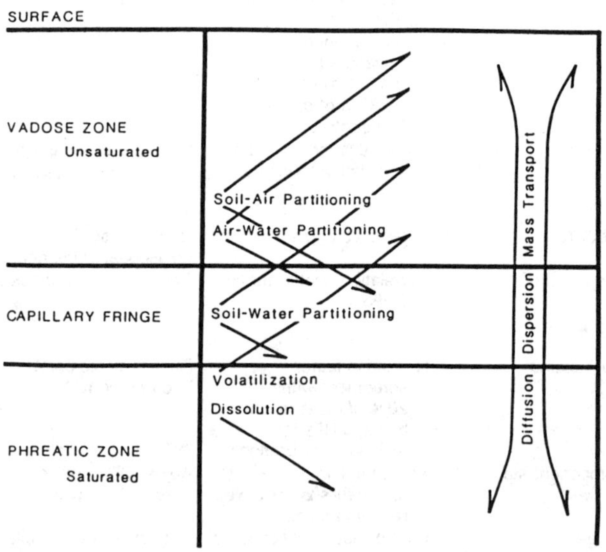

NOTE—The processes indicated by the soil gas monitoring method are partitioning, migration, emplacement and degradation. Partitioning represents a group of processes which control contaminant movement from one physical phase to another, these phases being liquid, free vapor, occluded vapor, solute and sorbed. Migration refers to contaminant movement over distance with any vertical, horizontal or temporal component. Emplacement refers to establishment of contaminant residence in any phase within any residence opportunity. Degradation is the process whereby contaminants are attenuated by oxidation or reduction in the vadose zone, either through biogenic or abiogenic processes. Soil gas monitoring measures the result of the interaction of these processes in a dynamic equilibrium.

FIG. X1.1 Arena of Soil Gas Monitoring

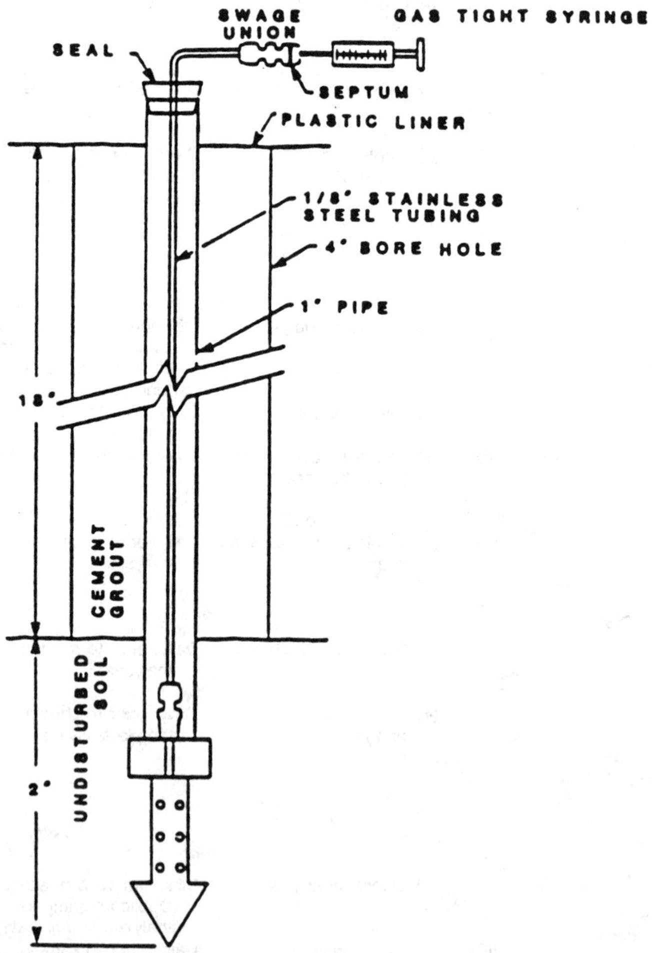

NOTE—Ground probe designed and used by Crow et al., 1985, from Ref **(10)**.

FIG. X1.2 Example of Whole-Air Active Sampling System

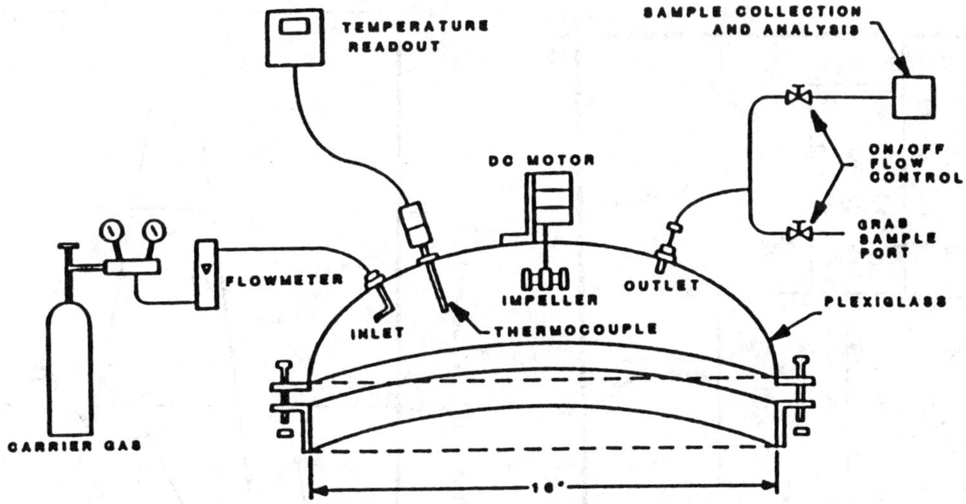

NOTE—Surface flux chamber and peripheral equipment after Eklund et al., 1984, from Ref **(10)**.

FIG. X1.3 Example of Whole-Air Passive Sampling System

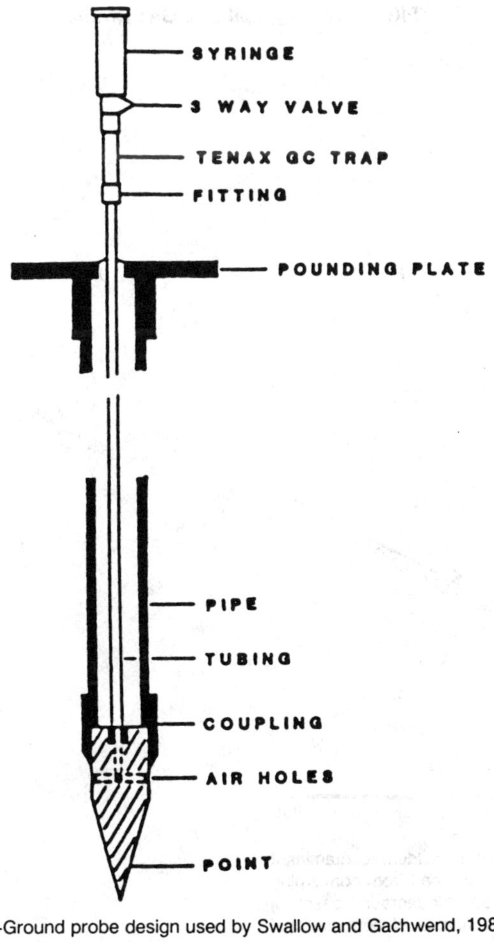

NOTE—Ground probe design used by Swallow and Gachwend, 1983, from Ref **(10)**.

FIG. X1.4 Example of Sorbed Contaminant-Active System

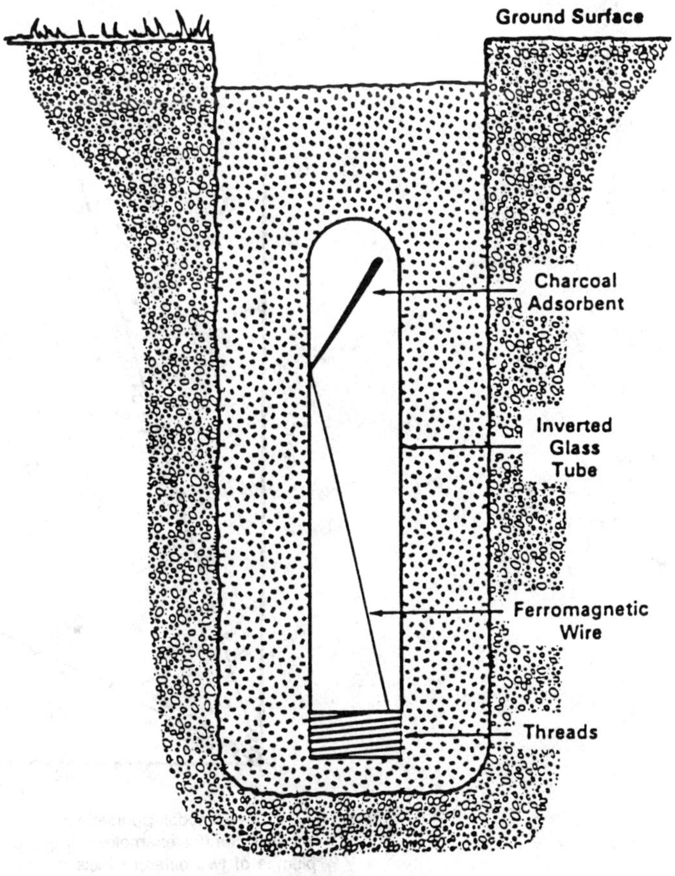

NOTE—Schematic diagram of emplacement of a sorbed contaminant-passive system **(10)**.

FIG. X1.5 Example of Sorbed Contaminant-Passive System

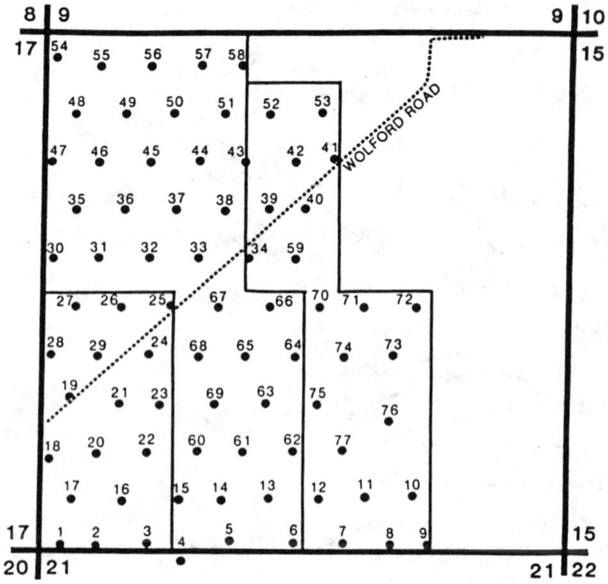

MAP SCALE: 1" = 1,000'

NOTE—In any application, soil gas monitoring can be performed over a wide range of spatial designs, including soil gas sampling in grid patterns at a single depth or multiple depths. This example illustrates a staggered grid pattern of samples recovered at a single depth.

FIG. X1.6 Typical Soil Gas Grid Array and Map Display

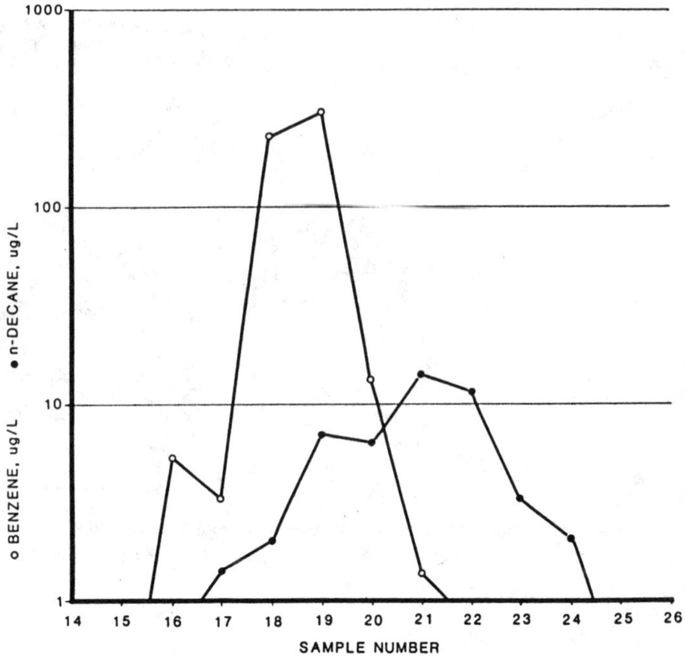

NOTE—Soil gas data from survey profiles displayed on an X – Y plot is an effective aid to data interpretation. This display is useful to examine the overall context for soil gas measurement data potentially indicating contamination. If the profile is displayed as a cross section through a grid pattern or as a linear array of sample points, the profile display can illustrate spatially significant groupings of data subpopulations.

FIG. X1.7 Typical Soil Gas Profile

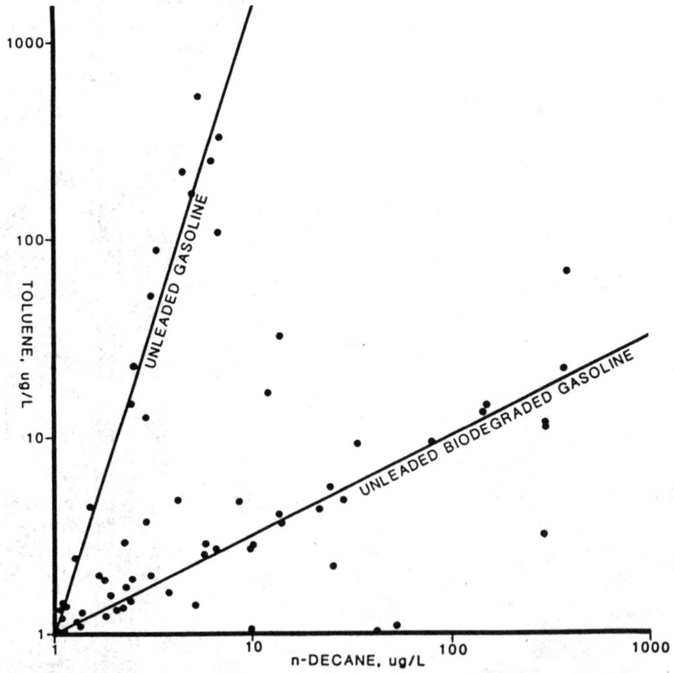

NOTE—Bimodal populations of data that represent coincident contaminant occurrences (for example, soil gas contaminant vapors sourced from converging plumes of two different fuels or mixtures of gasoline and biodegraded gasoline) can be defined using compositional analyses. One technique of compositional analysis is cross-plotting as shown.

FIG. X1.8 Soil Gas Compositional Analysis by Cross Plot

Project # ________ Sample # ____

Sampled by: ____

Date Sampled: ________, 199_____ Time: ________ (AM/PM)

Sampling System (check one):

( ) Whole air-active approach
( ) Whole air-passive approach
( ) Sorbed contaminants-active approach
( ) Sorbed contaminants-passive approach
( ) Headspace or extraction approach
( ) Soil pore liquid headspace approach

Sample Type (check one):

( ) Direct field sample
( ) Field blank
( ) Travel blank
( ) Sample container blank
( ) Sample probe blank
( ) Sample replicate

Spiked? _____ with _____ cc of ____

Potential reaction products due to spiking: ____

System purge volume: _____ Volumes purged: _____ Sample volume: _____

Sorbent Device: Installed _____ (AM/PM), ________, 199_____

Recovered _____ (AM/PM), ________, 199_____

Sample container type: ________ Sample container #_____

Integral analyzer: ________ Detector: _____

Analyzer response: ________ (units) ____

Surface conditions (pavement, wet, frost, etc.) ____

Sample depth: ________ Sampling rate: ____

Sample horizon data-visual estimates:

Vadose zone make-up: ( ) Native soil+rock ( ) Fill ( ) Rock

Soil composition: Clay, ________%

Soil organic matter, ________%

Fine granular material, ________%

Coarse granular material, ________%

100 %

Moisture content of sampling horizon (qualitative):

	( ) Dry
( ) Very	( ) Damp
( ) Slightly	( ) Moist
	( ) Wet

Other characteristics of the sampling horizon:

( ) Free water present	( ) Probable connection to surface macropores
( ) Free product present	
( ) Contaminant odors	( ) Indurated
( ) Poor perm. to vapor	( ) Soil discoloration
( ) Near slope or vent	( ) ________________

____
Investigator Signature/Date

____
Investigator Affiliation

FIG. X1.9 Suggested Soil Gas Sample Data Sheet

REFERENCES

(1) Plumb, R. H., Jr., and Pitchford, A. M., "Volatile Organic Scans: Implications for Ground Water Monitoring," *Proceedings*, NWWA/API Conference on Petroleum Hydrocarbons and Organic Chemicals in Ground Water-Prevention, Detection and Restoration, Houston, Texas, November 13–15, 1985.

(2) Stonestrom, D. A., and Rubin, J., "Air Permeability and Trapped-Air Content in Two Soils," *Water Resources Research*, Vol 25, No. 9, September 1989, pp. 1959–1969.

(3) Lyman, Warren J., "Environmental Partitioning of Gasoline in Soil/Groundwater Compartments," *Seminar on the Subsurface Movement of Gasoline*, Edison, New Jersey, 1987.

(4) Price, L. C., "Aqueous Solubility of Petroleum as Applied to Its Origin and Primary Migration," *Am. Assoc. Petrol. Bull.*, Vol 60, No. 2, February 1976, pp. 213–244.

(5) Owens, J. W., Wasik, S. P., and Devoe, H., "Aqueous Solubilities and Enthalpies of Solution of n-Alkylbenzenes," *J. Chem. Eng. Data*, Vol 31, No. 47, 1986, pp. 47–51.

(6) Reisinger, H. J., Burris, D. R., Cessar, L. R., and McCleary, C. D., "Factors Affecting the Utility of Soil Vapor Assessment Data," *Proceedings*, First National Outdoor Action Conference on Aquifer Restoration, Ground Water Monitoring and Geophysical Methods, May 18–21, 1987, Las Vegas, Nevada, National Water Well Association, Dublin, Ohio, pp. 425–435.

(7) Stonestrom, D. A., and Rubin, J., "Water Content Dependence of Trapped Air in Two Soils," *Water Resources Research*, Vol 25, No. 9, September 1989, pp. 1947–1958.

(8) Manos, C. G. Jr., Williams, K. R., Balfour, W. D., and Williamson, S. J., "Effects of Clay Mineral-Organic Matter Complexes on Gaseous Hydrocarbon Emissions from Soils," *Proceedings*, NWWA/API Conference on Petroleum Hydrocarbons and Organic Chemicals in Ground Water-Prevention, Detection and Restoration, Houston, Texas, November 13–15, 1985.

(9) American Petroleum Institute, "The Migration of Petroleum Products in Soil and Ground Water: Principles and Countermeasures," *API Publication No. 4149*, Washington, D.C., 1972.

(10) Devitt, D. A., Evans, R. B., Jury, W. A., Starks, T. P., Eklund, B., Gnolson, A., and van Ee, J. J., *Soil Gas Sensing for the Detection and Mapping of Volatile Organics*, National Water Well Association, Dublin, Ohio, 1987.

(11) Kerfoot, W. B., and Sanford, W., "Four-Dimensional Perspective of an Underground Fuel Oil Tank Leakage," *Proceedings*, NWWA/API Conference on Petroleum Hydrocarbons and Organic Chemicals in Ground Water, Houston, Texas, November 12–14, 1986, National Water Well Association, Dublin, Ohio, pp. 383–403.

(12) Dragun, J., "Microbial Degradation of Petroleum Products in Soil," *Proceedings*, Conference on the Environmental and Public Health Effects of Soils Contaminated with Petroleum Products, Amherst, Massachusetts, October 30–31, 1985.

(13) Barker, J. F., Patrick, G. C., and Major, D., "Natural Attenuation of Aromatic Hydrocarbons in a Shallow Sand Aquifer," *Ground Water Monitoring Review*, Vol 7, No. 1, pp. 66–71.

(14) Davis, J. B., "Microbiology in Petroleum Exploration," *Unconventional Methods in Exploration for Petroleum and Natural Gas*, W. B. Heroy, ed., Southern Methodist University, Institute for the Study of Earth and Man, SMU Press, 1969, pp. 139–157.

(15) Jensen, B., Arvin, E., and Gundersen, A. T., "The Degradation of Aromatic Hydrocarbons with Bacteria from Oil Contaminated Aquifers," *Proceedings*, NWWA/API Conference on Petroleum Hydrocarbons and Organic Chemicals in Groundwater, Houston, Texas, November 13–15, 1985, pp. 421–435.

(16) White, K. D., Novak, J. T., Goldsmith, C. D., and Bevan, S., "Microbial Degradation Kinetics of Alcohols in Subsurface Systems," *Proceedings*, NWWA/API Conference on Petroleum Hydrocarbons and Organic Chemicals in Ground Water-Prevention, Detection and Restoration, Houston, Texas, November 13–15, 1985.

(17) Kerfoot, H. B., and Barrows, L. J., *Soil Gas Measurement for Detection of Subsurface Organic Contamination*, U.S. Department of Commerce-National Technical Information Service, Springfield, Virginia, 1987.

(18) Diem, D., Kerfoot, H. B., and Ross, B. E., "Field Evaluation of a Soil-Gas Analysis Method for Detection of Subsurface Diesel Fuel Contamination," *Proceedings*, Second National Outdoor Action Conference on Aquifer Restoration, Ground Water Monitoring and Geophysical Methods, National Water Well Association, Dublin, Ohio, 1987.

(19) Chan, D. B., and Ford, E. A., "In-Situ Oil Biodegradation," *Military Engineer*, No. 509, 1986, pp. 447–737.

(20) Thompson, G. M., and Marrin, D. L., "Soil Gas Contaminant Investigations: A Dynamic Approach," *Ground Water Monitoring Review*, Vol 7, No. 3, pp. 88–93.

(21) Boynton, D., and Reuther, W., "A Way of Sampling Soil Gases in Dense Subsoils, and Some of Its Advantages and Limitations," *Proceedings*, Soil Science Society of America, Vol 3, 1938, pp. 37–42.

(22) Horvitz, L., "Hydrocarbon Geochemical Prospecting After Thirty Years," *Unconventional Methods in Exploration for Petroleum and Natural Gas*, W. B. Heroy, ed., Southern Methodist University, Institute for the Study of Earth and Man, SMU Press, 1969, pp. 205–218.

(23) Ullom, W. L., "Ethylene and Propylene in Soil Gas: Occurrence, Sources and Impact on Interpretation of Exploration Geochemical Data," *Bulletin*, Association of Petroleum Geochemical Explorationists, Vol 4, No. 1, December 1988, pp. 62–81.

(24) McCarthy, J. H. Jr., and Reimer, G. M., "Advances in Soil Gas Geochemical Exploration for Natural Resources: Some Current Examples and Practices," *Journal of Geophysical Research*, Vol 91, No. B12, November 1986, pp. 327–338.

(25) Jones, V. T., and Thune, H. W., "Surface Detection of Retort Gases from an Underground Coal Gasification Reactor in Steeply Dipping Beds Near Rawlins, Wyoming," SPE Paper 11050, *57th Annual Fall Technical and Exhibition of the Society of Petroleum Engineers of AIME*, New Orleans, Louisiana, September 26–29, 1982.

(26) Roffman, H. K., Neptune, M. D., Harris, J. W., Carter, A., and Thomas, T., "Field Screening for Organic Contaminants in Samples From Hazardous Waste Sites," *Proceedings*, NWWA/API Conference on Petroleum Hydrocarbons and Organic Chemicals in Ground Water, Houston, Texas, November 13–15, 1985.

(27) Wittmann, S. G., Quinn, K. J., and Lee, R. D., "Use of Soil Gas Sampling Techniques for Assessment of Ground Water Contamination," *Proceedings*, NWWA/API Conference on Petroleum Hydrocarbons and Organic Chemicals in Ground Water-Prevention, Detection and Restoration, Houston, Texas, November 13–15, 1985, pp. 291–309.

(28) Scheinfeld, R. A., and Schwendeman, T. G., "The Monitoring of Underground Storage Tanks, Current Technology," *Proceedings*, NWWA/API Conference on Petroleum Hydrocarbons and Organic Chemicals in Ground Water, Houston, Texas, November 13–15, 1985, pp. 244–264.

(29) Karably, L. S., and Babcock, K. B., "The Effects of Environmental Variables on Soil Gas Surveys," *Hazardous Materials Control*, January/February, 1989, pp. 36–43.

(30) Balfour, W. D., and Schmidt, C. E., *Sampling Approaches for Measuring Emission Rates from Hazardous Waste Disposal Facilities," U.S. EPA,* Contract No. 68-02-3171.

(31) Eklund, B., "Detection of Hydrocarbons in Groundwater by Analysis of Shallow Soil Gas/Vapor," *API Publication No. 4394*, Washington, D.C., 1985.

(32) Mayer, C. L., *Draft Interim Guidance Document for Soil-Gas Surveying*, U.S. EPA Environmental Monitoring Systems Laboratory, Office of Research and Development, Contract No. 68-03-3245, September 1989, 124 pp.

(33) Spittler, T. M., and Clifford, W. S., "A New Method for Detection of Organic Vapors in the Vadose Zone," *Proceedings*, NWWA Conference on Characterization and Monitoring of the Vadose (Unsaturated) Zone, November 19–21, 1985, Denver, Colorado, National Water Well Association, Dublin, Ohio, pp. 236–246.

(34) Kerfoot, H. B., "Shallow-Probe Soil-Gas Sampling for Indication of Ground Water Contamination by Chloroform," *International Journal of Environmental and Analytical Chemistry*, 1987, No. 30, pp. 167–181.

(35) Kerfoot, W. B., "A Portable Well Point Sampler for Plume Tracking," *Ground Water Monitoring Review*, Vol 4, No. 4, pp. 38–42.

(36) Litherland, S. T., Hoskings, T. W., and Boggess, R. T., "A New Ground Water Survey Tool: The Combined Cone Penetrometer/Vadose Zone Vapor Tool," *Proceedings*, NWWA/API Conference on Petroleum Hydrocarbons and Organic Chemicals in Ground Water, Houston, Texas, November 13–15, 1985, pp. 322–330.

(37) Nadeau, R. J., Stone, T. S., and Clinger, G. S., "Sampling Soil Vapors to Detect Subsurface Contamination: A Technique and Case Study," *Proceedings*, NWWA Conference on Characterization and Monitoring of the Vadose (Unsaturated) Zone, November 19–21, 1985, Denver, Colorado, National Water Well Association, Dublin, Ohio, pp. 215–226.

(38) U.S. EPA Environmental Response Team, Standard Operating Procedure 2149: "Soil Gas Sampling," September 30, 1988.

(39) Richers, D. M., et al., "Landsat and Soil Gas Geochemical Study of Patrick Draw Oil Field, Sweetwater County, Wyoming," *Bulletin*, American Association of Petroleum Geologists, Vol 66, No. 7, July 1982, pp. 903–922.

(40) Colenutt, B. A., and Davies, D. N., "The Sampling and Gas Chromatographic Analysis of Organic Vapors in Landfill Sites," *International Journal of Environmental and Analytical Chemistry*, No. 7, 1980, pp. 223–229.

(41) Newman, W., Armstrong, J. M., and Ettenhofer, M., "An Improved Soil Gas Survey Method Using Adsorbent Tubes for Sample Collection," *Proceedings*, Second National Outdoor Action Conference on Aquifer Restoration, Ground Water Monitoring and Geophysical Methods, May 23–26, 1988, Las Vegas, Nevada, National Water Well Association, Dublin, Ohio, pp. 1033–1049.

(42) U.S. EPA Environmental Response Team, Standard Operating Procedure 2051: "Charcoal Tube Sampling," November 7, 1988.

(43) U.S. EPA Environmental Response Team, Standard Operating Procedure 2052: "Tenax Tube Sampling," November 8, 1988.

(44) Kerfoot, H. B., and Mayer, C. L., *The Use of Industrial Hygiene Samplers For Soil-Gas Measurement*, U.S. EPA Environmental Monitoring Systems Laboratory, Advanced Monitoring Division, Contract No. 68-03-3249, September 1987.

(45) Voorhees, K. J., Hickey, J. C., and Klusman, R. W., "Analysis of Ground Water Contamination by a New Surface Static Trapping/Mass Spectrometry Technique," *Analytical Chemistry*, Vol 56, No. 13, 1984, pp. 2602–2604.

(46) Wesson, T. C., and Armstrong, F. E., *The Determination of C_1 – C_4 Hydrocarbons Adsorbed on Soils*, Bartlesville Energy Research Center Report of Investigations BERC/RI-75/13, U.S. Energy Research and Development Administration, Office of Public Affairs, Technology Information Center, Bartlesville, Oklahoma, December 1975.

(47) U.S. EPA Environmental Response Team, Standard Operating Procedure 2012: "Soil Sampling," November 8, 1988.

(48) USEPA, OWSER, SW 846, Method 5030, 1986.

(49) Sawhney, B. L., Pignatello, J. J., and Steinberg, S. M., "Determination of 1,2-Dibromoethane (EDB) in Field Soils: Implications for Volatile Organic Compounds," *Journal of Environmental Quality*, Vol 17, No. 1, January 1988, pp. 149–152.

(50) Steinberg, S. M., Pignatello, J. J., and Sawhney, B. L., "Persistence of 1,2-Dibromoethane in Soils: Entrapment in Intraparticle Micropores," *Environmental Science and Technology*, Vol 21, No. 12, December 1987, pp. 1201–1208.

(51) Rice, Gary K., "Combined Near-Surface Geochemical and Seismic Methods for Petroleum Exploration: Evidence for Vertical Migration," *Bulletin*, Association of Petroleum Geochemical Explorationists, Vol 2, No. 1, December 1986, pp. 46–62.

(52) Evans, O. D., and Thompson, G. M., "Field and Interpretation Techniques for Delineating Subsurface Petroleum Hydrocarbon Spills Using Soil Gas Analysis," *Proceedings*, NWWA/API Conference on Petroleum Hydrocarbons and Organic Chemicals in Ground Water: Prevention, Detection and Restoration, Houston, Texas, November 12–14, 1986, pp. 444–455.

(53) Riggin, R. M., *Technical Assistance Document for Sampling and Analysis of Toxic Organic Compounds in Ambient Air*, U.S. EPA Environmental Monitoring Systems Laboratory, Document No. EPA-600/X-83-025.

(54) U.S. EPA Environmental Response Team, Standard Operating Procedure 2050: "Tedlar Bag Sampling," November 8, 1988.

(55) U.S. EPA Environmental Response Team, Standard Operating Procedure 2056: "Photoionization Detector (PID) HNU," October 18, 1988.

(56) U.S. EPA Environmental Response Team, Standard Operating Procedure 2107: "Photovac 10A10 Portable Gas Chromatograph," December 29, 1988.

(57) U.S. EPA Environmental Response Team, Standard Operating Procedure 2109: "Photovac GC Analysis for Soil, Water and Air/Soil Gas," February 14, 1989.

(58) Brown, G. E., DuBose, D. A., Phillips, W. R., and Harris, G. E., "Project Summary: Response Factors of VOC Analyzers Calibrated with Methane for Selected Organic Chemicals," U.S. EPA, Industrial Environmental Research Laboratory-RTP, EPA Report No. EPA-600/52-81-002, 1981.

(59) Senn, R. B., and Johnson, M. S., "Interpretation of Gas Chromatography Data as a Tool in Subsurface Hydrocarbon Investigations," *Ground Water Monitoring Review*, Vol 7, No. 1, Winter, 1987, pp. 58–63.

(60) Smith, Rosser J. III, "Use of the Empirical Distribution Function in the Estimation of Boundaries of Anomalous Subpopulations," *Bulletin*, Association of Petroleum Geochemical Explorationists, Vol 3, No. 1, December 1987, pp. 64–87.

(61) Marrin, Donald L., "Soil Gas Analysis of Methane and Carbon Dioxide: Delineating and Monitoring Petroleum Hydrocarbons," *Proceedings*, NWWA Conference on Petroleum Hydrocarbons and Organic Chemicals in Ground Water, November 17–19, 1987, Houston, Texas, National Water Well Association, Dublin, Ohio.

Designation: D 5408 – 93

Standard Guide for Set of Data Elements to Describe a Ground-Water Site; Part One—Additional Identification Descriptors[1]

This standard is issued under the fixed designation D 5408; the number immediately following the designation indicates the year of original adoption or, in the case of revision, the year of last revision. A number in parentheses indicates the year of last reapproval. A superscript epsilon (ε) indicates an editorial change since the last revision or reapproval.

1. Scope

1.1 This guide is Part One of three guides to be used in conjunction with Practice D 5254 that delineates the data desirable to describe a ground-water data collection or sampling site. This guide describes additional information beyond the minimum set of data elements that may be needed to identify a ground-water site. Part Two identifies physical descriptors, such as construction, for a site, while Part Three identifies usage descriptors, such as monitoring, for an individual ground-water site.

NOTE 1—A ground-water site is defined as any source, location, or sampling station capable of producing water or hydrologic data from a natural stratum from below the surface of the earth. A source or facility can include a well, spring or seep, and drain or tunnel (nearly horizontal in orientation). Other sources, such as excavations, driven devices, bore holes, ponds, lakes, and sinkholes, that can be shown to be hydraulically connected to the ground water, are appropriate for the use intended.

NOTE 2—Part Two (Guide D 5409) includes individual site characteristic descriptors (7 data elements), construction descriptors (56 data elements), lift descriptors (16 data elements), geologic descriptors (26 data elements), hydraulic descriptors (20 data elements), and spring descriptors (11 data elements). Part Three (Guide D 5410) includes monitoring descriptors (77 data elements), irrigation descriptors (4 data elements), waste site descriptors (9 data elements), and decommissioning descriptors (8 data elements). For a list of descriptors in this guide, see Section 4.

1.2 These data elements are described in terms used by ground-water hydrologists. Standard references, such as the Glossary of Geology and various hydrogeologic professional publications, are used to determine these definitions. Many of the suggested elements and their representative codes are those established by the Water Resources Division of the U.S. Geological Survey and used in the National Water Information Systems computerized data base **(1-9)**.[2]

NOTE 3—The purpose of this guide is to suggest data elements that can be collected for ground-water sites. This does not uniquely imply a computer data base, but rather data elements for entry into any type of permanent file.

NOTE 4—Component and code lists given with some of the data elements, for example "Format of Other Data," are only suggestions. These lists can be modified, expanded, or reduced for the purpose intended by the company or agency maintaining the ground-water data file.

NOTE 5—Use of trade names in this guide is for identification purposes only and does not constitute endorsement by ASTM.

1.3 This guide includes the data elements desirable to identify a ground-water site beyond those given in the "Minimum Set of Data Elements." Some examples of the data elements are map identification, permitting facts, and supporting information. No single site will need every data element, for example, many ground-water sites do not need the data elements described in the legal record group. Each record (group of related data elements) for a site has mandatory data elements, such as the date for the ownership record. However, these elements are considered necessary only when that specific record is gathered for the site.

1.4 The values stated in inch-pound units are to be regarded as the standard. The SI units given in parentheses are for information only.

1.5 *This standard does not purport to address all of the safety problems, if any, associated with its use. It is the responsibility of the user of this standard to establish appropriate safety and health practices and determine the applicability of regulatory limitations prior to use.*

2. Referenced Documents

2.1 *ASTM Standards:*

D 653 Terminology Relating to Soil, Rock, and Contained Fluids[3]

D 5254 Practice for the Minimum Set of Data Elements to Identify a Ground-Water Site[4]

D 5409 Guide for Set of Data Elements to Describe a Ground-Water Site; Part Two—Physical Descriptors[4]

D 5410 Guide for Set of Data Elements to Describe a Ground-Water Site; Part Three—Usage Descriptors[4]

3. Terminology

3.1 *Definitions:*

3.1.1 For definitions of terms applicable to this guide, see Terminology D 653.

3.2 *Descriptions of Terms Specific to This Standard:*

3.2.1 *code*—a suggested abbreviation for a component, for example, "F" is the code suggested for the "Files (Raw Data)" component of data element "Format of Other Data."

3.2.2 *component*—a subdivision of a data element, for example, "Files (Raw Data)" is one of four components suggested for data element "Format of Other Data."

3.2.3 *data element*—an individual segment of information about a ground-water site, for example, "Format of Other Data." The data element is in the "Other Data Record" record.

---

[1] This guide is under the jurisdiction of ASTM Committee D-18 on Soil and Rock and is the direct responsibility of Subcommittee D18.21 on Ground Water and Vadose Zone Investigations.

Current edition approved May 15, 1993. Published November 1993.

[2] The boldface numbers in parentheses refer to a list of references at the end of the text.

[3] *Annual Book of ASTM Standards*, Vol 04.08.

[4] *Annual Book of ASTM Standards*, Vol 04.09.

3.2.4 *record*—a set of related data elements that may need to be repeated to fully describe a ground-water site. For example, a ground-water site that has a series of separate data files will need more than one "Other Data Record" record (the record includes data elements, other data type, other data location, and format of other data) to fully document the history of the site. However, if only a single separate data file exists for the well, the record is utilized once.

3.2.5 *record group*—a set of related records. For example, the "Supporting Information Record Group" includes the owner record, site visits record, other identification record, other data record, and remarks record. Some record groups consist of only one record, for example, the "Legal Record Group" includes only the legal record.

4. Summary of Guide

4.1 This guide includes the following additional identification descriptor data elements to describe a ground-water site. The universal element accompanies any data element requiring a confidence classification. Single elements usually need one entry for a site, while repeated elements commonly require several records to fully describe the conditions and history of the site:

- Universal Element
  - Data Confidence Classification
- Single Elements
  - Geographic Location:
    - Land-Net Location
    - Location Map
    - Map Scale
    - Method Altitude Determined
  - Political Regimes
    - Congressional District
  - Source Identifiers:
    - Mean Greenwich Time Offset
    - Site Reference in Report
    - Site in a Computer Data Base
    - Photography/Sketch Available of Site
- Repeated Elements
  - Legal Record Group:
    - Legal Record:
      - Permitting Agency
      - Priority Date
      - Application Number
      - Application Date
      - Certification Number
      - Certification Date
      - Permit Number
      - Permit Date
      - Water Allocation
- Supporting Information Record Group
  - Owner Record:
    - Date of Ownership
    - Owner's Name
  - Site Visits Record:
    - Date of Visit
    - Person Who Made Visit
    - Purpose of Visit
  - Other Identification Record:
    - Other Name, Number, or Identification
    - Assigner
  - Other Data Record:
    - Other Data Type
    - Other Data Location
    - Format of Other Data
  - Remarks Record:
    - Remark Date
    - Remark
    - Remark Source

5. Significance and Use

5.1 Data at ground-water sites are gathered for many purposes, each of which generally requires a specific set of data elements. For example, when ground-water quality is a concern, not only are the minimum set of data elements required for the site, but information concerning the sample collection depth interval, method of collection, and date and time of collection are needed to fully qualify the data. Another group of elements are recommended for each use of the data, such as aquifer characteristics or water-level records. Normally the more information that is gathered about a site by field personnel, the easier it is to understand the ground-water conditions and to reach valid conclusions and interpretations regarding the site.

5.2 The data elements listed in this guide and Guides D 5409 and D 5410 should assist in planning what information can be gathered for a ground-water site and how to document these data.

NOTE 6—Some important data elements may change during the existence of a site. For example, the elevation of the measuring point used for the measurement of water levels may be modified because of repair or replacement of equipment. This frequently occurs when the measuring point is an opening in the pump and the pump is modified or replaced. Because changes cannot always be anticipated, it is preferable to reference the height of the measuring point to a nearby, permanent altitude datum. The measuring point is referenced by being the same altitude (zero correction) or above (negative correction) or below (plus correction) the altitude datum. All appropriate measurements should be corrected in reference to the altitude datum before entry into the permanent record. Care must be exercised to keep the relationship of these data elements consistent throughout the duration of the site.

5.3 Some data elements have an extensive list of components. For example, the aquifer identification list described in Guide D 5409, has over 5000 components. Lengthy lists of possible components are not included in this guide, however, information on where to obtain these components is included with the specific data element.

NOTE 7—This guide identifies many sources, lists, etc. of information required to completely document information about any ground-water site.

6. Documentation of Universal Element

6.1 For any element that requires a Confidence Classification, document the data confidence classification for that specified critical data element for the ground-water site. Field-measured or laboratory-determined values have varying degrees of accuracy depending upon the methods used to obtain the information. This subjective or judged confidence should be documented for each measured data element by the agency or company that gathered or recorded the information, or both. Suggested components for the data confidence classification and representative codes are as follows:

A—Value is accurate to within the tolerance of the measurement instrument.

I—Value is judged to be inaccurate due to improper instrumentation or bias instrumentation or laboratory methods.

N—Not verified, value was obtained from another source and due to the nature of the data, cannot be verified.

NOTE 8—At a minimum, it is important, and often sufficient, that data be classified subjectively by experienced professionals. It is not always possible, or necessary to objectively quantify the confidence that a data user might have in a data value, but a professional classification can be useful. For the purposes of the three guides, the word confidence refers to a subjective professional judgment on data accuracy as represented by the three data confidence classification components, and does not imply the more rigorous confidence limits or interval as used by statisticians.

NOTE 9—A critical data element is one that the value can be field measured or laboratory determined with an instrument that has a statistically resolved degree of precision. Many data elements gathered for ground-water sites require no accompanying confidence classification, for example, owner's name, location map, type of lift, etc. Each data element that generally requires an accompanying confidence classification will be so noted in these guides.

7. Documentation of Miscellaneous Singular Data Elements

7.1 *Introduction*—A vast number of data elements can be documented about a ground-water site to thoroughly describe its location, physical features, relationship to other features on the earth's surface, and to designate what information is gathered at the site. These data elements typically are transcribed once for a site, in contrast to data elements that may be repetitive, such as water levels. Many of these data are extremely valuable in the characterization of sites that fall into certain categories, for example wells, for which the location map is an essential element to assist in properly positioning the well.

7.2 *Geographic Location:*

7.2.1 *Land-Net Location*—In addition to the locational data required by the minimum set of data elements, land-net location may be a general land office description of the site's position on the surface of the earth. This description is used in many parts of the United states to subdivide the land into sections, townships, and ranges for the purpose of governmental administration and originally was used (beginning in 1786) as a systematic method for the disposal of unoccupied land **(10)**. An abbreviated form of this description is used by many water agencies, in the many parts of the country, as the primary method of systematically cataloging ground-water sites. The method allows for the location of sites to a minimum of a 2½-acre (one hectare) tract (1/256th of a section) within a specified section, township, range, and meridian. The meridian designation must be included to denote where the township and range are located in the National grid system. An example of a 2½-acre (one hectare) location is "Northeast ¼ of the Southeast ¼ of the Northwest ¼ of the Southwest ¼, Section 22, Township 45 South, Range 87 West, Boise meridian." This location is usually abbreviated to a form similar to "NESENWSW Sec. 22, T45S, R87W B." A number of formats comparable to this abbreviation have been established by the various agencies that use the system, however, they basically communicate the same results **(5, 6, 11, 12)**.

NOTE 10—The accuracy of this location method for the minimum 2½-acre (one hectare) area is about 230 ft (70.104 m), that corresponds to between 2 and 3 s of latitude or longitude. Surveying errors are common in the original measurements. See FIPS PUB 70-1.[5]

NOTE 11—To supplement the description of the location of a ground-water site, a common method used is to draw a sketch showing the relationship of the site to other features in the immediate area, such as roads, buildings, etc. In addition, a sketch of the measuring point can assist in defining its exact location at the site. Photographs of the site and measuring point commonly are used as a part of the description.

[5] FIPS PUB 70-1, *Representation of Geographic Point Locations for Information Interchange*, is available from National Technical Information Service, U.S. Department of Commerce, 5285 Port Royal Road, Springfield, VA 22161.

7.2.2 *Location Map*—The location map name that is documented is that or the best available map of the area where the site is located. Much of the United States is covered by U.S. Geological Survey (USGS) topographic quadrangles. However, for those areas without USGS maps, the name of the map that shows the site's location should be documented. In addition, record the map's source, such as county highway or Army Map Service. The availability and identification of the USGS maps are given on individual State topographic map indexes. These indexes and the individual topographic maps can be obtained from USGS Public Inquires Office **(5, 9, 13)**.[6]

NOTE 12—Many mapped areas are available on a computer-stored Geographical Information System (GIS). Document information required to identify and obtain the GIS map of the area where the site is located.

7.2.3 *Map Scale*—Document the scale of the map that is used to locate the site. This value helps to define the accuracy of the site location data **(5)**.

NOTE 13—The map scale is the ratio between the linear distance on a map and the corresponding distance on the surface being mapped. For example, 1 in. = 1 mile (1 mm = 1 m) or the equivalent 1:63 360, are ways of expressing the same ratio.

7.2.4 *Method Altitude Determined*—Document the method used to determine the altitude of the reference datum at the ground-water site. Suggested method altitude determined components and representative codes are as follows **(5)**:

A—Altimeter,
L—Level or other surveying method,
M—Interpolated from topographic map, and
Z—Other, explain (for example, historical local datum).

7.3 *Political Regimes*—Document the political regime (for example, Congressional district) where the site is physically located. The date of documentation should be included because of changes that are commonly made in the boundaries of the districts. This allows for determining the legislative responsibility of the site. A guide to these districts is defined in FIPS PUB 9-1 **(14)**.[7]

NOTE 14—Congressional district boundaries can be modified over time because of population changes. Care must be exercised in using this data element to ensure that the ground-water site is still in the originally assigned District.

7.4 *Source Identifiers:*

7.4.1 *Mean Greenwich Time Offset*—Much of the data collected at a ground-water site is time related, such as water-level measurements or water-quality samples. Document, where applicable, the mean Greenwich time offset or United States time zone of the site, so that the time-dimension can be reduced to a common denominator.

7.4.2 *Site Referenced in Report*—If this site has been used or is referenced in a report, document the data concerning the published or unpublished report(s) and, if available, the identification of the report and the address of where to obtain a copy.

[6] Public Inquires Office, U.S. Geological Survey, 503 National Center, Room 1-C-402, 12201 Sunrise Valley Drive, Reston, VA 22092.

[7] FIPS PUB 9-1, *Congressional Districts of the United States* is available from National Technical Information Service, U.S. Department of Commerce, 5285 Port Royal Road, Springfield, VA 22161.

7.4.3 *Site in a Computer Data Base*—Document whether or not the information concerning the site has been entered into a computer data base and, if in a data base, the location. Show the data base management system (DBMS) used to organize the data base, for example, "INGRES," a relational DBMS. Give the name assigned to the data base containing the site, for example, "WATSTORE," the U.S. Geological Survey water data base.

7.4.4 *Photography/Sketch Available of Site*—Document the existence of a photograph or sketch of the site, or both. Photographs and sketches of the site and associated facility, including the measuring point, are valuable pictorial material to enhance the site description.

NOTE 15—An example of a form (see Fig. 1) for documenting the data elements as described under "Miscellaneous Singular Data Elements" is illustrated here to show a method of design for this tool. These forms are commonly known as field forms or as coding forms (for computer entry). This type of form is routinely used for transcribing field data while at the ground-water site and entering non-field information at the agency's or company's office. It should be noted that each form has the site identification (primary identification as used by the agency or company), date of field visit, and person that recorded the data as the first entries. These three data items are mandatory to ensure correct filing of the information, either in cabinets or in a computer data base, and for quality control.

8. Documentation of Miscellaneous Repetitive Data Elements

8.1 *Introduction:*

8.1.1 Many of the ground-water elements require multiple records to completely describe a site. Time-related elements, such as water levels, discharge measurements, and water chemistry, may need hundreds or thousands of records for a period of many years to document measurements at a single site. These time-related data help to determine historical trends and serve to establish bench-mark standards for the site.

8.1.2 Other data elements that are not time related, such as casing, lengths, spring openings, and an array of geophysical logs, require a sequence of records to thoroughly describe the site. These data are extremely valuable in site characterization, for example, wells for which the construction components are essential to understand the source of the water.

8.2 *Legal Record Group:*

8.2.1 *Legal Record*—The legal record includes information about any regulatory agencies or authorities, such as for establishment of the right-to-use water and the amount of water allocated for use at a ground-water site. This legal record is normally administered by a government agency or government authorized agency (for example, ground-water management district or health department) within the specific state. Some states use the method of permitting to assemble site records.

8.2.1.1 *Permitting Agency*—If applicable, document the name and address of the agency that is responsible for issuing the permit for the legal development of water at the site.

8.2.1.2 *Priority Date*—If applicable, document the date, in year, month, and day (YYYYMMDD), that establishes the legal priority for use of water at the ground-water site. If necessary, show the time of day that the priority was authorized.

8.2.1.3 *Application Number*—If applicable, document the number or identification assigned by the agency to the application for the permit.

8.2.1.4 *Application Date*—If applicable, document the application date, in year, month, and day (YYYYMMDD), for the ground-water site. If necessary, document the time of day.

8.2.1.5 *Certification Number*—If applicable, document the number or identification assigned by the agency to the certification credential.

8.2.1.6 *Certification Date*—If applicable, document the certification date, in year, month, and day (YYYYMMDD), for the ground-water site. If necessary, document the time of day.

8.2.1.7 *Permit Number*—If applicable, document the number or identification assigned by the agency to the permit for the use of water at the site.

8.2.1.8 *Permit Date*—If applicable, document the date, in year, month, and day (YYYYMMDD), the permit was issued for the ground-water site by the responsible agency. If necessary, document the time of day.

8.2.1.9 *Water Allocation*—If applicable, document the amount of water allocated by the permitting agency to the permit holder for the subject permit. Include the measurement unit utilized for the water allocation.

8.3 *Supporting Information Record Group:*

8.3.1 *Owner Record*—The owner's record is used to document a history of ownership of the ground-water site. This record is important to aid in the proper identification of the site and to assign the responsibility for the facility. The following data elements are required to document the history of ownership (**5**).

8.3.1.1 *Date of Ownership*—If applicable, document the date, in year, month, and day (YYYYMMDD), that the owner acquired possession of the ground-water site.

8.3.1.2 *Owner's Name*—Document the name of the owner and owner's address that corresponds with the date of

GROUND-WATER SITE
MISCELLANEOUS SINGULAR DATA ELEMENTS

Site Identification ______ Date Prepared ______
Prepared by ______ (Name)

Geographic Location:
Land-net Location ______ (U.S.General Land Office)
Location Map ______ (Name)
Map Scale ______ (Ratio of Map to Land Surface Distance)
Method Altitude Determined ______ (Altimeter, Surveyed, Interpolated, or Other)

Political Regimes:
Congressional District ______ (Name)
______ (Date)

Source Identifiers:
Mean Greenwich Time Offset ______ (Hour Offset or U.S. Time Zone)
Site Used in Report ______ (Title)
Site in a Computer Data Base ______ (Name)
______ (Location)
Photograph or Sketch Available ______ (Identify)

FIG. 1 Example Form

ownership for the event record.

8.3.2 *Site Visits Record*—The sites visits record is used to document data collection, verification, and quality-control visits to the ground-water site. The following data elements are required to document the history of these site visits **(5)**.

8.3.2.1 *Date of Visit*—If applicable, document the date, in year, month, and day (YYYYMMDD), that the site was visited. If necessary, document the time of day.

8.3.2.2 *Person Who Made Visit*—If applicable, document the name, title, and address of person who made the visit to the ground-water site for the record event.

8.3.2.3 *Purpose of Visit*—If applicable, document a description of the purpose of the visit to the ground-water site.

8.3.3 *Other Identification Record*—Many ground-water sites have more than one identification. These identifiers can be assigned by a company, state agency or federal agency to conform with an internal file system. To aid in the tracking of data for a site the following data elements may be required **(5)**.

8.3.3.1 *Other Name, Number, or Identification*—If applicable, document the ground-water site identification that was assigned by the other company or agency.

8.3.3.2 *Assigner*—Document the name and address of the person, company, or agency that assigned the other identification for this event record.

8.3.4 *Other Data Record*—The other data available record is used to indicate the availability of additional data pertinent to the ground-water site. Many sites have detailed information, such as continuous water-level recorder charts, geophysical logs, detailed geological logs, and extensive water-quality analyses, that may not be filed at a central location. These data are valuable in understanding conditions at the site **(5)**.

8.3.4.1 *Other Data Type*—If applicable, describe the type of other data available for the ground-water site.

8.3.4.2 *Other Data Location*—If applicable, document the location of the other data for the ground-water site. The complete name and address of the holder of the data should be documented.

8.3.4.3 *Format of Other Data*—If applicable, document the format of the other data available. Suggested other data available components and representative codes are as follows:

F—Files (raw data),
M—Machine readable (computer),
P—Published (report or basic-data release), and
Z—Other (describe).

8.3.5 *Remarks Record*—The remarks record is used for documenting meaningful information about the site for which no specific data elements are defined **(5)**.

8.3.5.1 *Remark Date*—If applicable, document the date, in year, month, and day (YYYYMMDD), of the origin of the remark.

8.3.5.2 *Remark*—If applicable, document information concerning the site that does not conform to any of the data elements that are listed in these guides.

8.3.5.3 *Remark Source*—Document the name and address of the person, company, or agency that wrote the remark for this event record.

9. Keywords

9.1 data confidence classification; data element; ground water; monitoring location; sampling site; site identification; site location; water allocation; water quality

REFERENCES

(1) Bates, Robert L., and Jackson, Julia A., *Glossary of Geology*, Third Edition; American Geological Institute, Alexandria, Virginia, 1987.

(2) Bureau of Reclamation, *Ground-Water Manual, A Water Resources Technical Publication*, Revised Reprint; U.S. Department of Interior, Bureau of Reclamation, Washington, DC, 1981.

(3) Campbell, Michael D., and Lehr, Jay H., *Water Well Technology*; McGraw-Hill, New York, NY, 1973.

(4) Heath, Ralph C., *Basic Ground-Water Hydrology; U.S. Geological Survey Water-Supply Paper 2220*, 1983.

(5) Mathey, Sharon B., Editor, *National Water Information System User's Manual*, Vol 2, Chapter 4, Ground-Water Site Inventory System; U.S. Geological Survey, Open-File Report 89-587, 1990.

(6) Texas Natural Resources Information System, *Ground-Water Data INTERFACE, Users Reference Manual*; Texas Natural Resources Information System, Nov. 20, 1986.

(7) U.S. Environmental Protection Agency, *Handbook of Suggested Practices for the Design and Installation of Ground-Water Monitoring Wells*; Office of Research and Development, U.S. EPA, March 1991, Washington, DC, 1991.

(8) U.S. Geological Survey, *National Handbook of Recommended Methods for Water-Data Acquisition*, Chapter 2—Ground Water; Office of Data Coordination, Reston, Virginia, 1980, pp. 2-1 to 2-149.

(9) van der Leedan, Frits, Troise, Fred L., and Todd, David Keith, 1990, *The Water Encyclopedia*, Geraghty and Miller Ground-Water Series, 2nd Edition, Third Printing, Lewis Publishers, Inc., Chelsea, Michigan, 1991.

(10) Stewart, Lowell O., Public Land Survey; Iowa State University Press, Ames, Iowa, 1936.

(11) Morgan, Charles O., and McNellis, Jesse M., *FORTRAN IV Program KANS, for the Conversion of General Land Office Locations to Latitude and Longitude Coordinates*; Kansas State Geological Survey Special Distribution Publication 42, 1969.

(12) U.S. Department of Commerce, *Representation of Geographic Point Locations for Information Interchange*, Federal Information Standards (FIPS) Publication 70-1, National Institute for Standards and Technology, Washington, DC, June 23, 1986.

(13) U.S. Geological Survey, *Guide to Obtaining USGS Information*, U.S. Geological Survey Circular 900, 1989.

(14) U.S. Department of Commerce, *Congressional Districts of the United States*, Federal Information Standards (FIPS) Publication 9-1, National Institute for Standards and Technology, Washington, DC, Nov. 30, 1990.

The American Society for Testing and Materials takes no position respecting the validity of any patent rights asserted in connection with any item mentioned in this standard. Users of this standard are expressly advised that determination of the validity of any such patent rights, and the risk of infringement of such rights, are entirely their own responsibility.

This standard is subject to revision at any time by the responsible technical committee and must be reviewed every five years and if not revised, either reapproved or withdrawn. Your comments are invited either for revision of this standard or for additional standards and should be addressed to ASTM Headquarters. Your comments will receive careful consideration at a meeting of the responsible technical committee, which you may attend. If you feel that your comments have not received a fair hearing you should make your views known to the ASTM Committee on Standards, 1916 Race St., Philadelphia, PA 19103.

Designation: D 5409 – 93

Standard Guide for Set of Data Elements to Describe a Ground-Water Site; Part Two—Physical Descriptors[1]

This standard is issued under the fixed designation D 5409; the number immediately following the designation indicates the year of original adoption or, in the case of revision, the year of last revision. A number in parentheses indicates the year of last reapproval. A superscript epsilon (ε) indicates an editorial change since the last revision or reapproval.

1. Scope

1.1 This guide is Part Two of three guides to be used in conjunction with Practice D 5254 that delineates the data desirable to describe a ground-water data collection or sampling site. This guide identifies physical descriptors, such as construction and geologic elements, for a site. Part One (Guide D 5408) describes additional information beyond the minimum set of data elements that may be specified to identify any individual ground-water site, while Part Three identifies usage descriptors, such as monitoring, for an individual ground-water site.

NOTE 1—A ground-water site is defined as any source, location, or sampling station capable of producing water or hydrologic data from a natural stratum from below the surface of the earth. A source or facility can include a well, spring or seep, and drain or tunnel (nearly horizontal in orientation). Other sources, such as excavations, driven devices, bore holes, ponds, lakes, and sinkholes, that can be shown to be hydraulically connected to the ground water, are appropriate for the use intended.

NOTE 2—Part One (Guide D 5408) includes data confidence classification descriptor (one element), geographic location descriptors (four elements), political regime descriptor (one element), source identifier descriptors (four elements), legal descriptors (nine elements), owner descriptors (two elements), site visit descriptors (three elements), other identification descriptors (two elements), other data descriptors (three elements), and remarks descriptors (three elements). Part Three (Guide D 5410) includes monitoring descriptors (77 data elements), irrigation descriptors (four data elements), waste site descriptors (nine data elements), and decommissioning descriptors (eight data elements). For a list of descriptors in this guide, see Section 3.

1.2 These data elements are described in terms used by ground-water hydrologists. Standard references, such as the Glossary of Geology (**1**)[2] and various hydrogeologic professional publications, are used to determine these definitions. Many of the suggested elements and their representative codes are those established by the Water Resources Division of the U.S. Geological Survey and used in the National Water Information Systems computerized data base (**1–19**).[2]

NOTE 3—The purpose of this guide is to suggest data elements that can be collected for ground-water sites. This does not uniquely imply a computer data base, but rather data elements for entry into any type of permanent file.

NOTE 4—Component and code lists given with some of the data elements, for example "Type of Spring," are only suggestions. These lists can be modified, expanded, or reduced for the purpose intended by the company or agency maintaining the ground-water data file.

NOTE 5—Use of trade names in this guide is for identification purposes only and does not constitute endorsement by ASTM.

1.3 This guide includes the data elements desirable to document a ground-water site beyond those given in the "Minimum Set of Data Elements." Some examples of the data elements are well depth, contributing aquifer, and permanence of spring. No single site will need every data element, for example, springs do not need well depth and well casing data. Each record (group of related data elements) for a site has mandatory data elements, such as the type of lift for the lift record. However, these elements are considered necessary only when that specific record is gathered for the site.

1.4 The values given in either inch-pound units or SI units are to be regarded separately as the standard. The values given in parentheses are for information only.

1.5 *This standard does not purport to address all of the safety problems, if any, associated with its use. It is the responsibility of the user of this standard to establish appropriate safety and health practices and determine the applicability of regulatory limitations prior to use.*

2. Referenced Documents

2.1 *ASTM Standards:*

D 653 Terminology Relating to Soil, Rock, and Contained Fluids[3]

D 2488 Practice for Description and Identification of Soils (Visual—Manual Procedure)[3]

D 5254 Practice for the Minimum Set of Data Elements to Identify a Ground-Water Site[4]

D 5408 Guide for Set of Data Elements to Describe a Ground-Water Site; Part One—Additional Identification Descriptors[4]

D 5410 Guide for Set of Data Elements to Describe a Ground-Water Site; Part Three—Usage Descriptors[4]

3. Terminology

3.1 *Definitions:*

3.1.1 For definitions of terms applicable to this guide, see Terminology D 653.

3.2 *Descriptions of Terms Specific to This Standard:*

3.2.1 *code*—a suggested abbreviation for a component, for example, "G" is the code suggested for the galvanized iron component of data element casing material.

3.2.2 *component*—a subdivision of a data element, for example, galvanized iron is one of 30 components suggested

[1] This guide is under the jurisdiction of ASTM Committee D-18 on Soil and Rock and is the direct responsibility of Subcommittee D18.21 on Ground Water and Vadose Zone Investigations.

Current edition approved May 15, 1993. Published November 1993.

[2] The **boldface** numbers in parentheses refer to a list of references at the end of the text.

[3] *Annual Book of ASTM Standards*, Vol 04.08.

[4] *Annual Book of ASTM Standards*, Vol 04.09.

for data element casing material.

3.2.3 *data element*—an individual segment of information about a ground-water site, for example, casing material. The data element is in the casing record record.

3.2.4 *record*—a set of related data elements that may need to be repeated to fully describe a ground-water site. For example, a well that consists of several diameters of casing from the top end to the bottom will need more than one Casing Record record (the record includes data elements depth to top, depth to bottom, diameter, casing material, and casing thickness) to fully describe the construction of the well. However, if only a single size of casing is used in the well, the record is utilized once.

3.2.5 *record group*—a set of related records. For example, the lift record group includes the lift record, power record, and standby record. Some record groups consist of only one record, for example, the spring record group includes only the spring record.

4. Summary of Guide

4.1 This guide includes the following physical descriptor data elements to describe a ground-water site. Single elements usually need one entry for a site, while repeated elements commonly require several records to fully describe the conditions and history of the site.

- Single Elements:
  - Individual Site Characteristics:
    - Land Use (in vicinity of site)
    - Drainage Basin/Watershed
    - Relationship to Surface Stream/Lake, etc.
    - Hole Depth
    - Well Depth
    - Source of Depth Data
    - Primary Aquifer
- Repeated Elements:
  - Construction Record Group:
    - Construction Record:
      - Date Construction Began
      - Date Construction Ended
      - Name of Contractor
      - Source of Construction Data
      - Method of Construction
      - Type of Drilling Fluid
      - Volume of Drilling Fluid
      - Type of Finish
      - Type of Seal
      - Depth to Bottom of Seal
      - Method of Development
      - Length of Time of Development
      - Volume of Liquid Removed During Development
      - Special Treatment
    - Hole Record:
      - Depth to Top of the Hole Interval
      - Depth to Bottom of the Hole Interval
      - Diameter of the Hole Interval
    - Casing Record:
      - Depth to Top of the Cased Interval
      - Depth to Bottom of the Cased Interval
      - Diameter of the Cased Interval
      - Casing Material
      - Casing Thickness
    - Opening or Screen Record:
      - Depth to Top of the Open Interval
      - Depth to Bottom of the Open Interval
      - Diameter of the Open Interval
      - Type of Material in the Open/Screened Interval
      - Type of Openings in the Open Interval
      - Length of Openings
      - Width of Openings
      - Mesh of Screen
      - Packing Material
      - Size of Packing Material
      - Thickness of Packing Material
      - Depth to Top and Bottom of Packing Material
    - Repairs Record:
      - Date of Repairs
      - Nature of Repairs
      - Name of Contractor Who Made Repairs
      - Percent Change in Performance After Repairs
  - Special Cases Record:
    - Well Clusters:
      - Number of Wells in Cluster
      - Depth of Deepest Well in Cluster
      - Depth of Shallowest Well in Cluster
      - Diameter of Well Cluster
    - Collector Well/Laterals:
      - Number of Laterals in Collector Well
      - Depth of Laterals in Collector Well
      - Length of Laterals in Collector Well
      - Diameter of Laterals in Collector Well
      - Mesh of Screen in Laterals
    - Ponds:
      - Length of Pond
      - Width of Pond
      - Depth of Pond
      - Volume of Pond
    - Tunnel or Drain:
      - Length of Tunnel or Drain
      - Width of Tunnel or Drain
      - Depth of Tunnel or Drain
      - Bearing (Azimuth) Tunnel or Drain
      - Dip of Tunnel or Drain
  - Lift Record Group:
    - Lift Record:
      - Type of Lift
      - Date Permanent Lift was Installed
      - Depth of Intake
      - Manufacturer of Lift Device
      - Serial Number
      - Pump Rating
    - Power Record:
      - Type of Power
      - Horsepower Rating
      - Name of Power Company
      - Power-Company Account Number
      - Power-Meter Number
    - Standby Lift Record:
      - Additional Lift
      - Name of Company that Maintains Lift
      - Rated Pump Capacity
      - Type of Standby Power
      - Horsepower of Standby Power Source
  - Geologic Record Group:
    - Geophysical Log Record:
      - Date of Log
      - Type of Log
      - Depth to Top of Logged Interval
      - Depth to Bottom of Logged Interval
      - Source of Log Data
    - Geohydrologic Unit Record:
      - Aquifer Unit(s)
      - Contributing Unit
      - Depth to Top of Interval
      - Depth to Bottom of Interval
      - Lithology
      - Description of Material
    - Sample/Unconsolidated Material Record:
      - Sample Weight
      - Sample Interval
      - Particle Size
      - Percent of Total Sample
      - Particle Shape
      - Mineralogy
    - Sample/Consolidated Material Record:
      - Drill Cuttings or Core
      - Sample Size (Weight)
      - Sample Interval
      - Mineralogy
      - Core Length
      - Core Diameter

Core Recovery-Percent
Bedding
Structure
Porosity
Hydraulic Record Group:
Hydraulics Record:
Hydraulic/Aquifer Unit
Hydraulic/Aquifer Unit Type
Depth to Top of Unit
Depth to Bottom of Unit
Static Water Level
Measurement Date and Time
Unit Contribution
Aquifer Parameters Record:
Transmissivity
Horizontal Hydraulic Conductivity
Vertical Hydraulic Conductivity
Coefficient of Storage
Leakance
Diffusivity
Specific Storage
Specific Yield
Barometric or Tidal Efficiency
Porosity
Specific Capacity
Method Used to Determine Aquifer Characteristics
Availability of File of Detailed Results
Spring Record Group:
Spring Record:
Name of Spring
Type of Spring
Permanence of Spring
Sphere of Discharge
Discharge
Date of Discharge
Improvements
Number of Spring Openings
Flow Variability
Accuracy of Flow Variability
Magnitude of Spring

5. Significance and Use

5.1 Data at ground-water sites are gathered for many purposes. Each of these purposes generally requires a specific set of data elements. For example, when the ground-water quality is of concern not only are the 'minimum set of data elements' required for the site, but information concerning the sample collection depth interval, method of collection, and date and time of collection are needed to fully qualify the data. Another group of elements are recommended for each use of the data, such as aquifer characteristics or water-level records. Normally the more information that is gathered about a site by field personnel, the easier it is to understand the ground-water conditions and to reach valid conclusions and interpretations regarding the site.

5.2 The data elements listed in this guide and Guides D 5408 and D 5410 should assist in planning what information can be gathered for a ground-water site and how to document these data.

NOTE 6—Some important data elements may change during the existence of a site. For example, the elevation of the measuring point used for the measurement of water levels may be modified because of repair or replacement of equipment. This frequently occurs when the measuring point is an opening in the pump and the pump is modified or replaced. Because changes cannot always be anticipated, it is preferable to reference the height of the measuring point to a permanent nearby altitude datum. The measuring point is referenced by being the same altitude (zero correction) or above (negative correction) or below (plus correction) the altitude datum. All appropriate measurements should be corrected in reference to the altitude datum before entry into the permanent record. Care must be exercised to keep the relationship of these data elements consistent throughout the duration of the site.

5.3 Some data elements have an extensive list of components or possible entries. For example, the aquifer identification list described in 6.1.8 has over 5000 entries. Lengthy lists of possible entries are not included in this guide, however, information on where to obtain these components is included with the specific data element.

NOTE 7—This guide identifies other sources, lists, etc. of information required to completely document information about any ground-water site.

6. Documentation of Individual Site Characteristics

6.1 *Introduction:*

6.1.1 A vast number of data elements can be documented about a ground-water site to thoroughly describe its location, physical features, relationship to other features on the earth's surface, and to designate what information is gathered at the site. These data elements typically are transcribed once for a site, in contrast to data elements that may be repetitive, such as water levels. Many of these data are extremely valuable in the characterization of sites that fall into certain categories, for example wells, for which the primary aquifer is an essential element to assist in the identification of the source of water at the site **(2–5, 7, 8, 10–17, 19)**.

6.1.2 *Land Use (in Vicinity of Site)*—Document the use of the land in the area surrounding the ground-water site. This data element is important if there is a possibility of the use affecting the availability or quality of the water. If more than one significant land use is nearby, such as industrial and farming, document each purpose **(5, 16)**.

6.1.3 *Drainage Basin/Watershed*—Document the name or other identification of the watershed and drainage basin where the site is located. Maps with watersheds delineated are available from the State Conservationists, U.S. Department of Agriculture, Soil Conservation Service located in each of the states, possessions, and associated areas. Information about river basins is available on maps in "Atlas of River Basins of the United States" published by the U.S. Department of Agriculture, Soil Conservation Service **(20)**.[5]

6.1.4 *Relationship to Surface Stream/Lake, etc.*—Document information concerning the influence of any nearby surface-water source upon the ground-water site. For example, the ground-water source for the site could be directly connected to a surface-water body (recharging the aquifer or discharging to the surface-water body) or have no connection and be influenced by a seasonal variation in loading of the surface water body upon the aquifer **(4, 7, 8, 16)**.

NOTE 8—This information is more useful if a quantitative estimate of the amount of connection is given rather than a yes, there is a connection, or no, there is no connection notation. For example, a ground-water body that is only influenced by seasonal loading of a

[5] Regional contacts for obtaining information are as follows: Northeastern United States contact Director, Northeastern National Technical Center, USDA, SCS, 160 East 7th Street, Chester, PA 19013. Southern United States contact Director, Southern National Technical Center, USDA, SCS, Fort Worth Federal Center, Bldg. 23, Room 60, Felix and Hemphill Streets, PO Box 6567, Fort Worth, TX 76115. Midwestern United States contact Director, Midwest National Technical Center, USDA, SCS, Federal Bldg., Room 345, 100 Centennial Mall North, Lincoln, NB 68508-3866. Western United States contact Director, Western National Technical Center, USDA, SCS, Federal Bldg., Room 248, 511 N.W. Broadway, Portland, OR 97209-3489.

surface-water body would have 0 % connection. While a stream or lake that is partially or completely linked to the ground-water body could have from 1 to 100 % connection, however, a quantitative value seldom can be determined. Usually, the range of thickness of the aquifer penetrated by the surface water body or thickness and lithology of the material between the aquifer and surface water body is all that is known about the connection.

6.1.5 *Hole Depth*—If applicable, document the total depth that the hole was drilled, in feet or metres below a datum at or near land surface. Many times the hole is drilled deeper in order to explore stratum below the completed depth of the final well. This number is always equal to or greater than the well depth. The hole depth is important because the information concerning the stratum below the final well can be critical in understanding ground-water conditions at the site. Document the accuracy or confidence classification for this data element **(4, 5, 7, 8, 13, 14, 16)**. If applicable, note orientation and angle of hole if not vertical.

6.1.6 *Well Depth*—If applicable, document the depth of the finished well, in feet or metres below a datum at or near land surface. This depth is important as a means to delineate the maximum depth at which water is entering the well bore. Document the accuracy or confidence classification for this data element **(4, 5, 7, 8, 13, 14, 16)**.

6.1.7 *Source of Depth Data*—If applicable, document the source of the hole and well depth information. Suggested source of depth data components and representative codes are as follows**(13)**:

A—Reported by a government agency
D—From driller's log or report
G—Private geologist-consultant or university associate
L—Depth interpreted from geophysical logs by personnel of source agency
M—Memory (owner, operator, driller)
O—Reported from records by owner of well
R—Reported by person other than owner, driller, or another government agency
S—Measured by personnel of reporting agency
Z—Other source (describe)

6.1.8 *Primary Aquifer*—Document the identification of the primary aquifer unit from which the water is withdrawn or monitoring data are collected. A convenient and systematic method of coding geologic units was described by Cohee **(6)** in the American Association of Petroleum Geologists Bulletin. This method is used by the U.S. Geological Survey to code aquifer and geologic unit names in a National file (Catalog of Aquifer Names and Geologic Unit Codes used by the Water Resources Division) (for example, Edwards Limestone of Texas is coded 218EDRD). Information needed to obtain an ordered list of aquifers and related codes is available from the following **(6, 13)**:[6]

NOTE 9—An example of a form (see Fig. 1) for documenting the data elements as described under "Individual Site Characteristics" is illustrated here to show a method of design for this tool. The forms are commonly known as field forms or as coding forms (for computer entry). This type of form is routinely used for transcribing field data while at the ground-water site and entering non-field information at the agency's or company's office. It should be noted that each form has the site identification (primary identification as used by the agency or company), date of field visit, and person that recorded the data as the first entries. These three data items are mandatory to ensure correct filing of the information, either in cabinets or in a computer data base, and for quality control.

[6] Geologic Names Unit, U.S. Geological Survey, 439 National Center, Reston, VA 22092.

GROUND-WATER SITE
INDIVIDUAL SITE CHARACTERISTICS

Site Identification ____________ Date Prepared ________
Prepared by ______________________________ (Name)

Individual Site Characteristics:
Land Use (In Vicinity of Site) ____________ (Describe)
Drainage Basin/Watershed ____________ (Name)
Relationship to Surface Stream/Lake ____________ (Describe)
Hole Depth ______ (Feet/Meters) Accuracy ________
Well Depth ______ (Feet/Meters) Accuracy ________
Source of Depth Data ____________ (Name)
Primary Aquifer ____________ (Name)

FIG. 1 Example Form

7. Documenting of Miscellaneous Repetitive Data Elements

7.1 *Introduction:*

7.1.1 Many of the ground-water data elements require multiple records or entries to completely describe a site. Time-related elements, such as water levels, discharge measurements, and water chemistry, may present hundreds or thousands of records over a period of many years that answer a specific question about a single site. These time-related data help to determine historical trends and serve to establish bench-mark conditions for the site **(4, 5, 13, 14)**.

7.1.2 Other data elements that are not time related, such as casing lengths, spring openings, and some geophysical logs, require a sequence of records to thoroughly describe the site. These data are extremely valuable in site characterization, for example, wells for which the construction components are required to understand the source of the water **(4, 5, 13, 14)**.

7.2 *Construction Record Group*—The construction record group includes records for documenting data elements relating to any type of structure built for withdrawal of water or monitoring at a ground-water site, including construction, hole, casing, openings or screen, repairs, and special cases, such as well clusters, collector wells, ponds, tunnels, and drains **(2, 3, 5, 7, 8, 13–17)**. If applicable, any construction that may have modified the ambient ground water conditions should be documented. Examples include grouting, blasting, hydrofracturing, and local disruption such as tunnels, underground chambers, or excavations.

7.2.1. *Construction Record*—The construction record includes data elements relating to the date of construction, contractor, construction method, drilling fluids, finish, and development. Data elements that are included in the construction record are the following:

7.2.1.1 *Date Construction Began*—If applicable, document the date (year, month, day in YYYYMMDD format) on which the construction work was initiated at the ground-water site.

NOTE 10—Although this guide is written to be used with any type of data file, date information should be arranged in year, month, day, and time (24-h clock) format (for example, 19910822094158 for 1991, August 22nd, 9 h, 41 min, and 58 s AM), especially for ease of interchanging information among data systems (computerized files). This is the format recommended by the American National Standard for Information Systems (ANSI) and adopted as a Federal Information Processing Standards (FIPS) system **(21, 22)**.

7.2.1.2 *Date Constructed Ended*—If applicable, document

the date (year, month, day in YYYYMMDD format) on which the construction work was completed at the ground-water site.

7.2.1.3 *Name of Contractor*—If applicable, document the name and address of the principal individual or company that did the construction work at the ground-water site (for example, drilled the well).

7.2.1.4 *Source of Construction Data*—If applicable, document the source of the information concerning the construction at the ground-water site (for example, driller's log or geologist's log). Suggested source of construction data components and representative code are as follows:

A—Reported by a government agency
D—From driller's log or report
G—Private geologist-consultant or university associate
L—Depth interpreted from geophysical logs by personnel of source agency
M—Memory (owner, operator, driller)
O—Reported from records by owner of well
R—Reported by person other than owner, driller, or another government agency
S—Measured by personnel of reporting agency
Z—Other source (describe)

7.2.1.5 *Method of Construction*—If applicable, document the method by which the ground-water site was constructed. Suggested method of construction components and representative codes are as follows:

B—Bored or augered, generalized
L—Wash boring
M—Hollow stem auger
N—Solid stem auger
E—Bucket auger
A—Direct air-rotary method, with bit
K—Direct air-rotary method, with downhole hammer
H—Direct mud rotary
R—Reverse circulation rotary (no casing)
F—Dual-wall reverse circulation, generalized
G—Dual-wall reverse rotary
I—Dual-wall reverse percussion
C—Cable-tool
P—Air-percussion drill
Q—Hydraulic percussion
S—Jet percussion
J—Jetted by water
D—Dug or excavated
T—Trenching, dammed pond, or drain
V—Driven pipe
U—Cone penetration
W—Combined driven and jetting
Z—Other (describe)

NOTE 11—Several of the method of construction components are the same or similar methods (jetted by water and wash boring), but with different name identifications. In addition, several of the components that have generalized names, for example, bored or augered also have the specific methods (hollow stem auger, solid stem auger, etc.) included in the list.

7.2.1.6 *Type of Drilling Fluid*—If applicable, document the type and amount of additives (in pounds or kilograms) used in the drilling fluid (water) for the construction of the ground-water site. Suggested additive components and representative codes are as follows:

A—Acrylic polymers
D—Attapulgite
E—Baking soda
B—Barites
F—Biodegradable material
G—Caustic soda
C—Cellulosic polymers
H—Chromelignosulfonates
I—Chrysotile asbestos
J—Complex phosphates
K—Lignitic materials
L—Lime
Q—Lubricants
M—Modified guar gum products
R—Modified polysaccharide
N—Native clay
O—Organic polymers
P—Peptized bentonite
U—Pregelatinized starch
V—Soda ash
W—Sodium carboxymethylcellulose
S—Standard bentonite
X—Surfactants
T—Tannins
Z—Other (describe)

7.2.1.7 *Volume of Drilling Fluid*—If applicable, document the volume (in gallons or litres) of drilling fluid lost in the drilled hole. Specify the unit of measurement. Document the accuracy or confidence classification for this data element. It may be difficult to quantify losses in air drilling. Estimates may be made by comparing output versus compressor capacity.

7.2.1.8 *Type of Finish*—If applicable, document the method of finish or the nature of the openings that allow water to enter the well. Suggested type of finish components and representative codes are as follows:

C—Porous concrete
G—Gravel-packed screen
H—Horizontal gallery or collector
O—Open-ended casing
P—Perforated or slotted casing
S—Screen, commercial
T—Sand point, driven screen
W—Walled or shored
X—Open-hole in aquifer
Z—Other (describe)

7.2.1.9 *Type of Seal*—If applicable, document the type and amount (in pounds or kilograms) of material used to seal the well against the entry of surface water and the leakage of water between aquifers having different hydraulic pressures. Suggested type of seal components and representative codes are as follows:

B—Bentonite
C—Clay or cuttings
G—Cement grout
N—None
Z—Other (describe)

7.2.1.10 *Depth to Bottom of Seal*—If applicable, document the depth to the bottom of the seal in feet or metres below a datum at or near land surface. Document the accuracy or confidence classification for this data element.

7.2.1.11 *Method of Development*—If applicable, document the primary method used to develop the well. Suggested method of development components and representative codes are as follows:

A—Pumped with air lift
B—Bailed
D—Chemical, for example, dry ice
C—Surged, compressed air
J—Jetted, air or water
N—None
P—Overpumped
S—Surge block
Z—Other (describe)

7.2.1.12 *Length of Time of Development*—If applicable, document the number of hours and minutes that the well was bailed, pumped, or surged for development. Document the accuracy or confidence classification for this data element.

7.2.1.13 *Volume of Liquid Removed During Development*—If applicable, document the volume of liquid (in gallons or litres) removed from well during development. Specify the unit of measurement. Document the accuracy or confidence classification for this data element.

7.2.1.14 *Special Treatment*—If applicable, document any special treatment that was applied during development of the well. Suggested special treatment components and representative codes are as follows:

C—Chemical (acid, and so forth)
D—Dry ice
E—Explosives
F—Deflocculent
H—Hydrofracturing
M—Mechanical abrasion
Z—Other (describe)

7.2.2 *Hole Record*—The hole record includes data elements that relate to the description of the opening constructed for emplacement of hardware into the ground for the development of a monitoring or production well at a ground-water site. For many sites, several distinct hole length and size intervals are required for the completion of the well. Data elements that are included in the hole record are the following:

7.2.2.1 *Depth to Top of the Hole Interval*—If applicable, document the depth to the top of the hole interval, in feet or metres below a datum at or near land surface. The first or uppermost section of the hole starts at or near the datum. Document the accuracy or confidence classification for this data element.

7.2.2.2 *Depth to Bottom of the Hole Interval*—If applicable, document the depth to the bottom of the hole interval, in feet or metres below the datum. Document the accuracy or confidence classification for this data element.

7.2.2.3 *Diameter of the Hole Interval*—If applicable, document the nominal diameter of that interval of the hole, in inches or millimetres. Document the accuracy or confidence classification for this data element. Caliper logs may be very useful as documentation.

7.2.3 *Casing Record*—The casing record includes all information that relates to the description of the casing material placed into the ground for the construction of a monitoring or production well at a ground-water site. For many sites, several distinct length and size intervals are required for the completion of the well. Data elements that are included in the casing record are the following:

7.2.3.1 *Depth to Top of the Cased Interval*—If applicable, document the depth to the top of the cased interval, in feet or metres below a datum at or near land surface. The first or uppermost section of the casing starts at or near the datum. Document the accuracy or confidence classification for this data element.

7.2.3.2 *Depth to Bottom of the Cased Interval*—If applicable, document the depth to the bottom of the cased interval, in feet or metres below the datum. Document the accuracy or confidence classification for this data element.

7.2.3.3 *Diameter of the Cased Interval*—If applicable, document the inside diameter of that interval of the casing, in inches or centimetres. Document the accuracy or confidence classification for this data element.

7.2.3.4 *Casing Material*—If applicable, document the type of casing material used for the construction of the well. Note if casing joint or other components are different than casing material. Suggested casing material components and representative codes are as follows:

E—Acrylonitrile butadiene styrene (ABS)
A—Aluminum
H—Asbestos cement
B—Brick
J—Carbon structural steel
L—Chlorotrifluoroethylene (CTFE)
N—Coal tar epoxy coated steel
U—Coated steel
C—Concrete
D—Copper
O—Cupro-nickel
F—Fiberglass-reinforced epoxy
Q—Fluorinated ethylene propylene (FEP)
G—Galvanized iron
K—Kai-well
V—Perfluoroalkoxy (PFA)
X—Polytetrafluoroethylene (PTFE)
Y—Polyvinyl chloride (PVC)
1—Polyvinylidene fluoride (PVDF)
P—PVC, fiberglass, other plastic (general term)
R—Rock or stone
2—Rubber-modified polystyrene
3—Silicon bronze
4—Stainless steel
S—Steel
T—Tile
5—Transite
W—Wood
I—Wrought iron
M—Other metal (describe)
X—Other material, not metal (describe)

7.2.3.5 *Casing Thickness*—If applicable, document the thickness of the casing wall, in inches or centimetres. Document the accuracy or confidence classification for this data element.

7.2.4 *Opening or Screen Record*—The opening or screen record includes all information that relates to the description of the open or screened area that allows for the passage of water into a well at a ground-water site. For some sites, several distinct length and size intervals of open or screened areas are required for the completion of the well. Data elements that are included in the opening or screen record are the following:

7.2.4.1 *Depth to Top of the Open Interval*—If applicable, document the depth to the top of the open interval, in feet or metres below a datum at or near land surface. Document the accuracy or confidence classification for this data element.

7.2.4.2 *Depth to Bottom of the Open Interval*—If applicable, document the depth to the bottom of the open interval, in feet or metres below the datum. Document the accuracy or confidence classification for this data element.

7.2.4.3 *Diameter of the Open Interval*—If applicable, document the diameter of the open interval, in inches or centimetres. The diameter documented normally would be the inside diameter for a screen and the hole diameter for an open hole. Document the accuracy or confidence classification for this data element.

7.2.4.4 *Type of Material in the Open/Screened Interval*—If applicable, document the type of material used for the construction of the open/screened interval. Suggested type of material in the open interval components and representative codes are as follows:

E—Acrylonitrile butadiene styrene (ABS)
A—Aluminum
H—Asbestos cement
B—Brick
J—Carbon structural steel
L—Chlorotrifluoroethylene (CTFE)
N—Coal tar epoxy coated steel
U—Coated steel
C—Concrete
D—Copper
O—Cupro-nickel
F—Fiberglass-reinforced epoxy
Q—Fluorinated ethylene propylene (FEP)
G—Galvanized iron
K—Kai-well
V—Perfluoroalkoxy (PFA)
X—Polytetrafluoroethylene (PTFE)
Y—Polyvinyl chloride (PVC)
1—Polyvinylidene fluoride (PVDF)
P—PVC, fiberglass, other plastic (general term)
R—Rock or stone
2—Rubber-modified polystyrene
3—Silicon bronze
4—Stainless steel
S—Steel
T—Tile
5—Transite
W—Wood
I—Wrought iron
M—Other metal (describe)
X—Other material, not metal (describe)

7.2.4.5 *Type of Openings in the Open Interval*—If applicable, document the type of openings in this interval. Suggested type of openings in the open interval components and representative codes are as follows:

B—Bridge slot
C—Continuous slot wire-wound
F—Open hole in fractured rock
L—Louvered or shutter-type screen
M—Mesh screen
P—Perforated, porous, or slotted casing
R—Wire-wound screen
S—Screen, type not known
T—Sand point
W—Walled or shored
X—Open hole, undefined rock condition
Z—Other (describe)

7.2.4.6 *Length of Openings*—If applicable, document the length or long dimension of the perforations, slots, or mesh of the screen, in inches or centimetres. Document the accuracy or confidence classification for this data element.

7.2.4.7 *Width of Openings*—If applicable, document the short dimension of the perforations, slots, or mesh of the screen, in inches or centimetres. Document the accuracy or confidence classification for this data element.

7.2.4.8 *Mesh of Screen*—If applicable, document the slot or mesh size of the screen, in inches or centimetres. Using the mesh of screen data element may be preferable to using the length and width of openings elements. Document the accuracy or confidence classification for this data element.

7.2.4.9 *Packing Material*—If applicable, document the type of material and supplier (or analysis) used to pack the space or void on the outside of the screened interval. Suggested packing material components and representative codes are as follows:

B—Beads, glass
C—Crushed stone—describe type
G—Gravel, sorted
H—Gravel, unsorted
I—Gravel, graded pack
M—Mixture of sand and gravel
N—Natural formation material
S—Sand, sorted
R—Sand, unsorted
T—Sand, graded pack
Z—Other (describe)

7.2.4.10 *Size of Packing Material*—If applicable, document the grain size of the sorted material or the range of grain size of the unsorted or graded packing material, in inches or millimetres. Document the accuracy or confidence classification for this data element.

7.2.4.11 *Thickness of Packing Material*—If applicable, document the thickness of the material packed between the screen and the natural formation, in inches or centimetres (thickness = (hole size-screen size)/2). Document the accuracy or confidence classification for this data element.

7.2.4.12 *Depth to Top and Bottom of Packing Material*—If applicable, document the depth to the top and to the bottom of the packing material, in feet or metres below the datum. Document the accuracy or confidence classification for this data element.

7.2.5 *Repairs Record*—The repairs record includes all information that relates to repair work done on previously constructed facilities at the ground-water site. For many sites, several occurrences of repairs are normal. Data elements that are included in the repairs record are the following:

7.2.5.1 *Date of Repairs*—If applicable, document the date (year, month, day in YYYYMMDD format) that the repairs were completed.

7.2.5.2 *Nature of Repairs*—If applicable, document the type of repairs that occurred at the ground-water site. Suggested nature of repair components and representative codes are as follows:

B—Blocked off
C—Cleaned
D—Deepened
I—Pump intake lowered
L—Liner installed
O—Slotted or perforated
P—Plugged back
S—Screen replaced
Z—Other (describe)

7.2.5.3 *Name of Contractor Who Made Repairs*—If applicable, document the name and address of the contractor that performed the repairs.

7.2.5.4 *Percent Change in Performance After Repairs*—If applicable, document the percent change in the performance (plus or minus) of the ground-water site. For example, percent change = ((new yield − old yield)/old yield) (100 %). Document the accuracy or confidence classification for this data element.

7.2.6 *Well Cluster*—A well cluster is multiple wells or a gallery of wells that are connected to one pumping source. This type of withdrawal system is used in areas of thin and shallow aquifers where each well in the cluster produces a small amount of water. However, in combination, the wells in the cluster yield the amount of water needed for the use intended. Data elements that are not given below and are needed to document the construction details for the well cluster are found under the construction record, hole record, casing record, and opening or screen record.

NOTE 12—Cluster wells are intended for the production of water (for example, a small public water supply or commercial facility) where potable water sources are limited. This type of withdrawal system is not initially intended for hydrologic investigations or for use in the determination of the water-quality characteristics of an aquifer.

NOTE 13—Well cluster, collector well/laterals, ponds, and tunnel or drain fall in the special cases record for documenting information that is unique to the other sources of ground water (other than wells and springs).

7.2.6.1 *Number of Wells in Cluster*—If applicable, document the number of wells in the cluster that are connected into one pumping system.

7.2.6.2 *Depth of Deepest Well in Cluster*—If applicable, document the depth of the deepest well in the cluster. Document the accuracy or confidence classification for this data element.

7.2.6.3 *Depth of Shallowest Well in Cluster*—If applicable, document the depth of the shallowest well in the cluster. Document the accuracy or confidence classification for this data element.

7.2.6.4 *Diameter of Well Cluster*—If applicable, document the largest diameter or dimension, in feet or metres, of the well cluster field. Document the accuracy or confidence classification for this data element.

7.2.7 *Collector Well/Laterals*—A collector well or radial collector well consists of a large central caisson (for example, 13-ft (3.96-m) inside diameter) with laterals and horizontal screens projecting away (for example, 240 ft (13.15 m)) from the bottom of the central caisson. These radials can be in a radial or linear pattern, depending upon the configuration of the aquifer. This type of water-withdrawal system allows for the optimum development of some low-yielding aquifers and more economical development of large supplies from thin, high-production aquifers (such as under rivers). Data elements that are not given below and are needed to document the construction details for the collector well are found under the construction record, hole record, casing record, and opening or screen record. See Note 12 for an explanation of the purpose of a similar type of withdrawal system.

7.2.7.1 *Number of Laterals in Collector Well*—If applicable, document the number of laterals that are connected to the central caisson.

7.2.7.2 *Depth of Laterals in Collector Well*—If applicable, document the depth or average depth of the laterals connected to the collector well. Document the accuracy or confidence classification for this data element.

7.2.7.3 *Length of Laterals in Collector Well*—If applicable, document the length of the laterals that extend away from the collector well. If there is a large difference in the lengths, document the range in lengths or the length of each individual lateral. Document the accuracy or confidence classification for this data element.

7.2.7.4 *Diameter of Laterals in Collector Well*—If applicable, document the diameter of the laterals, in inches or centimetres. Document the accuracy or confidence classification for this data element.

7.2.7.5 *Mesh of Screen in Laterals*—If applicable, document the slot or mesh size of screens, in inches or centimetres. If there is a large difference in the mesh size of the various screens, document the range in size or the size of each individual screen. Document the accuracy or confidence classification for this data element.

7.2.8 *Ponds*—This category of the ground-water withdrawal system includes natural or constructed ponds that intercept the water table. In areas of shallow water tables, natural ponds occur or ponds can be dug into the water-bearing aquifer and the water pumped from the pond to the area of use.

7.2.8.1 *Length of Pond*—If applicable, document the longest dimension of the pond, in feet or metres. Document the accuracy or confidence classification for this data element.

7.2.8.2 *Width of Pond*—If applicable, document the width of the pond (usually the dimension at right angle to the length), in feet or metres. Document the accuracy or confidence classification for this data element.

7.2.8.3 *Depth of Pond*—If applicable, document the maximum or average depth of the pond (include whether the depth given is the maximum or average), in feet or metres. Document the accuracy or confidence classification for this data element.

7.2.8.4 *Volume of Pond*—If applicable, document the average volume of water contained in the pond, in gallons, litres, or other volume unit. Document the volume unit used. Document the accuracy or confidence classification for this data element.

7.2.9 *Tunnel or Drain*—This category of the ground-water withdrawal system includes tunnels constructed principally to intercept the water table and drains constructed primarily to lower the water table in the vicinity of mines or man-made structures **(23)**.

NOTE 14—Tunnels (called falaj, qanat, karez, and foggara in the Middle East) are used as a water collection and distribution system in many parts of the world, especially in arid regions. This nearly horizontal tunnel system is a very conservative method of skimming the upper surface of the water table. Water from drains, that are used to lower the water table at man-made structures (for example, mines), is commonly used for other purposes (for example, processing of ore). Water from drains that are used for the purpose of lowering the near-surface water table of poorly drained agricultural lands normally are discharged to nearby surface water bodies.

7.2.9.1 *Length of Tunnel or Drain*—If applicable, document the length of the tunnel or drain, in feet or metres. Document the accuracy or confidence classification for this data element.

7.2.9.2 *Width of Tunnel or Drain*—If applicable, document the width of the channel where the water flows, in feet or metres. Document the accuracy or confidence classification for this data element.

7.2.9.3 *Depth of Tunnel or Drain*—If applicable, document the average depth of the tunnel or drain, in feet or metres. Document the accuracy or confidence classification for this data element.

7.2.9.4 *Bearing (Azimuth) Tunnel or Drain*—If applicable, document the orientation in degrees bearing from due north of the tunnel or drain, beginning at the origin and going in direction of the terminus. Document the accuracy or confidence classification for this data element.

7.2.9.5 *Dip of Tunnel or Drain*—If applicable, document the dip in degrees from the horizontal of the tunnel or drain, beginning at the origin and ending at the terminus. Document the accuracy or confidence classification for this data element.

7.3 *Lift Record Group*—The lift record group includes records for documenting data elements relating to any type of equipment or method used for withdrawal of water at a ground-water site, including lift technique, power method, and backup or standby lift and power system **(2, 4, 5, 7, 8, 13, 14, 16, 19)**.

7.3.1 *Lift Record*—The lift record includes all information that relates to the method and equipment that is used to remove the ground water from the aquifer. Commonly, several arrangements and types of lift systems are used over the history of a ground-water site because of maintenance and replacement of worn equipment. In rare cases, several lift systems are used at the site at the same time. Information concerning the pump rating or yield and power consumption may be used to estimate the water usage. Data elements that are included in the lift record are the following:

7.3.1.1 *Type of Lift*—The type of lift is the specific method used to remove the water from the aquifer, either by mechanical or natural means. Suggested type of lift components and representative codes are as follows:

A—Air lift
B—Bucket or bailer
C—Centrifugal pump
G—Natural flow or gravity
J—Jet pump
P—Piston pump
R—Rotary pump
S—Submergible pump
T—Turbine pump
N—None
U—Unknown
Z—Other (describe)

7.3.1.2 *Date Permanent Lift Was Installed*—If applicable, document the date (year, month, day in YYYYMMDD format) that the lift unit was installed. This information is used to identify the age of the lift unit and to further identify the site.

7.3.1.3 *Depth of Intake*—If applicable, document the depth below a datum, in feet or metres, to the bottom of the pump intake. Document the accuracy or confidence classification for this data element.

7.3.1.4 *Manufacturer of Lift Device*—If applicable, document the name and address of the company that manufactured the pump.

7.3.1.5 *Serial Number*—If applicable, document the serial number of the pump. This data element allows for additional identification of the pump and ground-water site.

7.3.1.6 *Pump Rating*—If applicable, document the rating of the pump as the volume of the water lifted per unit of power consumed. Tables are normally available for determining the efficiency of each type of pump according to the amount of lift involved. This pump efficiency table must be used for determining the pump rating. The value should be expressed as million of gallons or litres of water per kilowatt-hour of electricity, cubic foot or metres of natural gas, gallon or litre of liquid fuel or engine hour, depending upon type of power. Document the accuracy or confidence classification for this data element.

7.3.2 *Power Record*—The power record includes all information that relates to the type of power used to drive a lift unit or to remove water from the aquifer. Commonly, several arrangements and types of power are used over the history of a ground-water site because of maintenance and replacement of worn equipment. In rare cases, several types of power are used at the site at the same time. Data elements that are included in the power record are the following:

7.3.2.1 *Type of Power*—Document the type of energy used to power the pump or to remove the water from the aquifer. Suggested type of power components and representative codes are as follows:

A—Animal
C—Compressed air
D—Diesel engine
E—Electric motor
F—Natural flow or gravity
G—Gasoline engine
H—Hand or human
L—LP gas (propane or butane engine)
N—Natural-gas engine
W—Windmill
Z—Other (describe)

7.3.2.2 *Horsepower Rating*—If applicable, document the horsepower rating of the power component given under "Type of Power." For example, 10 hp for the rating of an electric motor used to drive a turbine pump. Document the accuracy or confidence classification for this data element.

7.3.2.3 *Name of Power Company*—If applicable, document the name and address of the company that furnishes the electricity, natural gas, or other fuel for the power source.

7.3.2.4 *Power-Company Account Number*—If applicable, document the account number under which the power company stores information on power consumption at the site.

7.3.2.5 *Power-Meter Number*—If applicable, document the meter number of the electric or gas meter which logs the power consumption of the power source.

7.3.3 *Standby Lift Record*—The standby lift record includes information that relates to the type of lift and power used as a backup to the primary lift and power system. Data elements that are included in the standby lift record are the following:

7.3.3.1 *Additional Lift*—If applicable, document the additional head (above land-surface datum) against which the pump work, in feet or metres of water.

7.3.3.2 *Name of Company that Maintains Lift*—If applicable, document the name and address of the company that is responsible for the maintenance of the pump.

7.3.3.3 *Rated Pump Capacity*—If applicable, document the manufacturer's pump capacity rating. Document the accuracy or confidence classification for this data element.

7.3.3.4 *Type of Standby Power*—If applicable, document the type of standby power available. Suggested type of power components and representative codes are as follows:

A—Animal
C—Compressed air
D—Diesel engine
E—Electric motor
F—Natural flow or gravity
G—Gasoline engine
H—Hand or human
L—LP gas (propane or butane engine)
N—Natural-gas engine
W—Windmill
Z—Other (describe)

7.3.3.5 *Horsepower of Standby Power Source*—If applicable, document the horsepower rating of the standby power source. Document the accuracy or confidence classification for this data element.

7.4 *Geologic Record Group*—The geologic record group includes records for documenting data elements relating to geophysical logs, geologic units, and geologic samples of both unconsolidated and consolidated materials **(4, 5, 7, 8, 12–14, 16)**.

7.4.1 *Geophysical Log Record*—The log record is used to enter information about types of geophysical or other logs available for the site. Data elements that are included in the geophysical log record are the following:

7.4.1.1 *Date of Log*—If applicable, document the date (year, month, day in YYYYMMDD format) that the geophysical log was completed at the ground-water site.

7.4.1.2 *Type of Log*—If applicable, document the type of log available for the hole. If more than one type of log was run on the well, document those with the corresponding depth intervals. Suggested type of log components and representative codes are as follows **(12)**:

A—Drilling time
B—Casing collar
C—Caliper
D—Drillers
E—Electric
R—Single-point resistance
W—Spontaneous potential
Y—Multi-electrode
1—Acoustic velocity
2—Acoustic televiewer
F—Conductivity, fluid
G—Geologists or sample
H—Magnetic
I—Induction
J—Gamma ray
K—Dipmeter survey
L—Lateral log
M—Microlog
N—Neutron
O—Microlateral log
P—Photographic
Q—Radioactive-tracer
S—Sonic
T—Temperature
U—Gamma-gamma
V—Fluid velocity (flow)
X—Core
Z—Other (describe)

7.4.1.3 *Depth to Top of Logged Interval*—Enter the depth to the top of the logged interval, in feet or metres below a datum at or near land surface. Document the accuracy or confidence classification for this data element.

7.4.1.4 *Depth to Bottom of Logged Interval*—Enter the depth to the bottom of the logged interval, in feet or metres below a datum at or near land surface. Document the accuracy or confidence classification for this data element.

7.4.1.5 *Source of Log Data*—If applicable, document the source of the log information. Suggested source of depth data components and representative codes are as follows:

A—Reported by a government agency
D—From driller's log or report
G—Private geologist-consultant or university associate
L—Depth interpreted from geophysical logs by personnel of source agency
M—Memory (owner, operator, driller)
O—Reported from records by owner of well
R—Reported by person other than owner, driller, or another government agency
S—Measured by personnel of reporting agency
Z—Other source (describe)

7.4.2 *Geohydrologic Units Record*—The geohydrologic units record is used to document information about the rock material that yields water or is monitored at the ground-water site. Normally, information is gathered for all rock material encountered in drilling the well, both above and below the water-bearing aquifer unit. Geophysical log data (see 7.4.1) are commonly used to assist in the interpretation of rock material, especially in accurately defining depth intervals and fluid characteristics. The more common data elements that are included in the geohydrologic units record are the following:

NOTE 15—Usually, data describing rock material are documented sequentially (top to bottom) on strip charts. These strip charts can be simple field-compiled logs generated from visual examination of the rock material by a hydrogeologist or drilling engineer. Detailed strip charts are laboratory compiled from the examination of all the properties of the field-collected rock (drilling) samples. These properties include such characteristics as color, mineralogy, luster, structure, induration, inclusions, cementation, sorting, grain size, grain shape, porosity, hardness of material, solubility, etc.

7.4.2.1 *Aquifer Unit(s)*—Document the identification of the water-bearing aquifer unit or units from which the water is withdrawn or monitoring data are collected and the non-water-bearing units above and below the aquifer. See 6.1.8 for additional information on aquifer unit identification.

7.4.2.2 *Contributing Unit*—In combination with the aquifer identification, indicate how the unit is categorized as an aquifer. Suggested contributing unit components and representative codes are as follows:

P—Principal contributing aquifer
S—Secondary contributing aquifer
N—Contributes no water
U—Unknown contribution

7.4.2.3 *Depth to Top of Interval*—If applicable, document the depth, in feet or metres below a datum at or near land surface, to the top of this aquifer or non-water-bearing unit. Document the accuracy or confidence classification for this data element.

7.4.2.4 *Depth to Bottom of Interval*—If applicable, document the depth, in feet or metres below a datum or non-water-bearing unit. Document the accuracy or confidence classification for this data element.

7.4.2.5 *Lithology*—If applicable, document the lithology of the aquifer or non-water-bearing unit. Suggested lithology components and representative codes are as follows:

Rock Term	Abbreviation
Alluvium	ALVM
Anhydrite	ANDR
Anorthosite	ANRS
Arkose	ARKS
Basalt	BSLT
Bentonite	BNTN
Boulders	BLDR
Boulders and sand	BLSD
Boulders, silt, and clay	BLSC
Breccia	BRCC
Calcite	CLCT
Caliche (hard pan)	CLCH
Chalk	CHLK
Chert	CHRT
Clay	CLAY
Clay, some sand	CLSD
Claystone	CLSN
Coal	COAL
Cobbles	COBB
Cobbles and sand	COSD
Cobbles, silt, and clay	COSC
Colluvium	CLVM
Conglomerate	CGLM
Coquina	CQUN
Diabase	DIBS
Diorite	DORT

Rock Term	Abbreviation
Dolomite	DLMT
Drift	DRFT
Evaporite	EVPR
Gabbro	GBBR
Glacial (undifferentiated)	GLCL
Gneiss	GNSS
Granite	GRNT
Granite, gneiss	GRGN
Gravel	GRVL
Gravel and clay	GRCL
Gravel, cemented	GRCM
Gravel, sand, and silt	GRDS
Gravel, silt, and clay	GRSC
Graywacke	GRCK
Greenstone	GNST
Gypsum	GPSM
Hard pan	HRDP
Igneous (undifferentiated)	IGNS
Lignite	LGNT
Limestone	LMSN
Limestone and Dolomite	LMDM
Loam	LOAM
Loess	LOSS
Marble	MRBL
Marl	MARL
Marlstone	MRLS
Metamorphic (undifferentiated)	MMPC
Muck	MUCK
Mud	MUD
Mudstone	MDSN
Other	OTHR
Outwash	OTSH
Overburden	OBDN
Peat	PEAT
Quartzite	QRTZ
Residium	RSDM
Rhyolite	RYLT
Rock	ROCK
Rubble	RBBL
Sand	SAND
Sand and clay	SDCL
Sand and gravel	SDGL
Sand and silt	SDST
Sand, gravel, and clay	SGVC
Sand, some clay	SNCL
Sandstone	SNDS
Sandstone and shale	SDSL
Saprolite	SPRL
Schist	SCST
Sedimentary (undifferentiated)	SDMN
Serpentine	SRPN
Shale	SHLE
Silt	SILT
Silt and clay	STCL
Siltstone	SLSN
Slate	SLTE
Soil	SOIL
Syenite	SYNT
Till	TILL
Travertine	TRVR
Tuff	TUFF
Volcanic (undifferentiated)	VLCC

7.4.2.6 *Description of Material*—If applicable, in combination with the lithology described in 7.4.2.5, document the adjective modifiers needed to describe the rock type of the aquifer or non-water-bearing unit. Use of meaningful abbreviations assists in condensing the description. Standard color guides are valuable in dictating consistency in describing the rock and soil material **(9, 24)**. Various guides are available as aid to standardizing the descriptions of rock and soil materials **(3)**. The following are examples of lithologic descriptions:

Example 1—For soft, chalky grey limestone, suggested description: LMSN, GREY, SOFT, CHALKY.

Example 2—For hard red sandstone, iron stained, suggested description: SNDS, HARD, RED, FE STND.

7.4.3 *Sample/Unconsolidated Material Record*—Samples of geologic materials commonly are collected when drilling holes are to be completed as monitoring or water wells. These samples are used to assist in the determination of the aquifer, vadose zone, and underlying material characteristics for evaluating the movement of water through these materials. Undisturbed unconsolidated samples are collected from driven or cored holes, while disturbed unconsolidated samples are collected from bored, rotary, and cable tool holes. This record is used to describe the geologic material in combination with data given in the sections on construction (see 7.2) and geohydrologic units (see 7.4.2).

NOTE 16—If samples are collected of soil materials that are located in the weathered zone (1.5 to 2.0 metres or 6 ft in depth) and below the weathered zone (an additional 1.5 to 2.0 metres or about 12 ft in depth) a description that is more comprehensive than the following may be required for the features that are found in these zones. Those additional features, when present, are: (*1*) texture (USDA and Unified estimated textures, coarse fragments), (*2*) sorting and roundness, (*3*) moisture condition (moist, wet, dry, presence of water table), (*4*) color and mottling, (*5*) consistency (rupture resistance, cementation), (*6*) secondary porosity features, (*7*) sedimentary structure, (*8*) presence of organic matter, and (*9*) effervescence in dilute HCl. The field pocket guide by Boulding **(3)** and Practice D 2488 presents an excellent summary of these features.

7.4.3.1 *Sample Weight*—If applicable, document the weight, in ounces or grams, of the sample of geologic material. Indicate whether this is wet or dry weight. The volume of the sample may be included so that bulk density can be determined. Document the accuracy or confidence classification for this data element.

7.4.3.2 *Sample Interval*—If applicable, document the depth interval, in feet or metres below a datum at or near land surface, of the sample of geologic material. Document the accuracy or confidence classification for this data element.

7.4.3.3 *Particle Size*—If applicable, document the particle or grain sizes of the unconsolidated geologic material. Indicate whether this is a visual determination using a hand lens or microscope, or a mechanical analysis using calibrated sieves. The grain sizes of sand or larger-sized materials are normally recorded in millimetres. While silt and clay-sized materials are recorded in micrometres (μm). Document the accuracy or confidence classification for this data element.

7.4.3.4 *Percent of Total Sample*—If applicable, document the percentage, by weight, of each particle size (see 7.4.3.3) contained in the total sample (see 7.4.3.1). Document the accuracy or confidence classification for this data element.

7.4.3.5 *Particle Shape*—If applicable, document the particle shape or roundness of the sampled particles. Common shape descriptions are rounded, sub-rounded, subangular, and angular. The shape is usually determined visually by use of a hand lens or microscope.

7.4.3.6 *Mineralogy*—If applicable, document the mineralogy of the particles in the sample. The mineralogy is usually determined visually by use of a hand lens or microscope, however, the exact composition must be determined in the laboratory by powder X-ray diffraction. The most common visually recognizable mineral is silica (SiO_2).

7.4.4 *Sample/Consolidated Material Record*—Samples of

geologic materials commonly are collected when drilling holes to be completed as monitoring or water wells. These samples are used to assist in the determination of the aquifer or vadose zone characteristics for evaluating the movement of water through these materials. Undisturbed consolidated samples are collected from cored holes, while disturbed consolidated samples are collected from bored, rotary, and cable tool holes. This record is used to describe the geologic material in combination with data given in the sections on construction (see 7.2) and geohydrologic units (see 7.4.2).

7.4.4.1 *Drill Cuttings or Core*—If applicable, document whether the samples of consolidated geologic material is undisturbed (cored) or disturbed (for example, rotary tool).

7.4.4.2 *Sample Size (Weight)*—If applicable, document the weight, in ounces or grams, of the sample of geologic material. Indicate whether this is wet or dry weight. The volume of the sample may be included so that bulk density can be determined. Document the accuracy or confidence classification for this data element.

7.4.4.3 *Sample Interval*—If applicable, document the depth interval, in feet or metres below a datum at or near land surface, of the sample of geologic material. Document the accuracy or confidence classification for this data element.

7.4.4.4 *Mineralogy*—If applicable, document the mineralogy of the sampled interval. Indicate whether the mineralogy was field (visually with hand lens or microscope) or laboratory determined by powder X-ray diffraction. This information supplements data given in 7.4.2.

7.4.4.5 *Core Length*—If applicable, document the length of core, if feet and inches or metres and centimetres, recovered from the sample interval (see 7.4.4.3). Document the accuracy or confidence classification for this data element.

7.4.4.6 *Core Diameter*—If applicable, document the core diameter, in inches or centimetres, recovered in this interval. Document the accuracy or confidence classification for this data element.

7.4.4.7 *Core Recovery-Percent*—If applicable, document the percentage of core recovery (core length/distance of sample interval). Document the accuracy or confidence classification for this data element.

7.4.4.8 *Bedding*—If applicable, document the type of bedding or plane of stratification of the geologic materials of the recovered core. This characteristic of the rock material can be critical in the interpretation of the movement of water.

7.4.4.9 *Structure*—If applicable, document any structure or microstructure apparent in the recovered core. This characteristic of the rock material can be critical in the interpretation of the movement of water.

7.4.4.10 *Porosity*—If applicable, document the apparent porosity of the recovered geologic material. Indicate whether the porosity was field (estimated visually with a hand lens or microscope) or laboratory determined. Document the accuracy or confidence classification for this data element.

7.5 *Hydraulic Record Group*—The hydraulic record group includes records for documenting data elements relating to hydraulic characteristics of the aquifer, including both the basic description and hydraulic parameters of the aquifer unit (**4, 5, 7, 8, 10–13, 16, 25**).

NOTE 17—Information given in 7.4 is commonly used to assist in the interpretation of hydraulic characteristics, especially in accuracy defining depth intervals and aquifer properties.

7.5.1 *Hydraulics Record*—The hydraulics record is used to document information about the rock unit that yields water or is monitored at the ground-water site. The more common data elements that are included in the hydraulics record are the following:

7.5.1.1 *Hydraulic/Aquifer Unit*—If applicable, document the identification of the aquifer unit or units from which the hydraulic data were determined. See 6.1.7 for additional information on aquifer unit identification.

7.5.1.2 *Hydraulic/Aquifer Unit Type*—If applicable, document the type of aquifer(s) tested for hydraulic characteristics at the ground-water site. These types normally are confined, unconfined, or a combination of confined and unconfined.

7.5.1.3 *Depth to Top of Unit*—If applicable, enter the depth to the top of the tested interval, in feet or metres below a datum at or near land surface. Document the accuracy or confidence classification for this data element.

7.5.1.4 *Depth to Bottom of Unit*—If applicable, enter the depth to the bottom of the tested interval, in feet or metres below a datum at or near land surface. Document the accuracy or confidence classification for this data element.

7.5.1.5 *Static Water Level*—If applicable, document the water level, in feet or metres below a datum at or near land surface, prior to the hydraulic test. For those water levels that are above the measuring-point datum (normally artesian wells), precede the value with a minus (−) sign to distinguish those water levels from ones at or below the measuring point. Document the accuracy or confidence classification for this data element.

7.5.1.6 *Measurement Date and Time*—If applicable, document the date (year, month, day, and time of day in YYYYMMDDHHMM format) of the measurement of the static water level.

7.5.1.7 *Unit Contribution*—If applicable, enter the percentage of the total yield of the site that is contributed by this hydraulic/aquifer unit, if known. If part of the water that the site would otherwise produce is lost to this unit, enter the percentage of the water lost preceded by a minus sign (−). Document the accuracy or confidence classification for this data element.

7.5.2 *Aquifer Parameters Record*—The aquifers parameters record is for the documentation of hydraulic characteristics as determined by testing of the aquifer or estimation by using data facts about the aquifer.

NOTE 18—Data gathered during aquifer testing and generated as a result of the interpretation process are sometimes extensive. The information discussed here are the final results.

7.5.2.1 *Transmissivity*—If applicable, document the transmissivity of the aquifer. Indicate whether this data element is estimated or is determined by an aquifer test. Document the aquifer test method and the accuracy or confidence classification for this data element.

NOTE 19—Transmissivity is the capability of an aquifer to transmit water of the prevailing kinematic viscosity in a unit time through a unit width of the aquifer under a unit hydraulic gradient, expressed in metres squared per day.

7.5.2.2 *Horizontal Hydraulic Conductivity*—If applicable,

document the hydraulic conductivity parallel to bedding of the aquifer. Indicate whether this data element is estimated or is determined by an aquifer test. Document the aquifer test method and the accuracy or confidence classification for this data element.

NOTE 20—Hydraulic conductivity is the capacity of the rock to transmit water, expressed as the volume of water at the existing kinematic viscosity that will move in unit time under a unit hydraulic gradient through a unit area measured at right angles to the direction of flow.

7.5.2.3 *Vertical Hydraulic Conductivity*—If applicable, document the vertical hydraulic conductivity of the aquifer. Indicate whether this data element is estimated or is determined by an aquifer test. Document the aquifer test method and the accuracy or confidence classification for this data element.

7.5.2.4 *Coefficient of Storage*—If applicable, document the coefficient of storage of the aquifer. Indicate whether this data element is estimated or is determined by an aquifer test. Document the aquifer test method and the accuracy or confidence classification for this data element.

NOTE 21—Coefficient of storage is the volume of water an aquifer releases from or takes into storage per unit surface area of the aquifer per unit change in head. For a confined aquifer, it is equal to the product of specific storage and aquifer thickness. For an unconfined aquifer, the storage coefficient is equal to the specific yield.

7.5.2.5 *Leakance*—If applicable, document the leakance of the confining unit in 1/day. Indicate whether this data element is from estimated hydraulic characteristics or from those determined by an aquifer test. Document the aquifer test method and the accuracy or confidence classification for this data element.

NOTE 22—Leakance (K′/b′) is the vertical hydraulic conductivity of the confining unit (K′) divided by the thickness of the confining unit (b′).

7.5.2.6 *Diffusivity*—If applicable, document the diffusivity of the aquifer. Indicate whether this data element is from estimated hydraulic characteristics or from those determined by an aquifer test. Document the aquifer test method and the accuracy or confidence classification for this data element.

NOTE 23—Diffusivity is the transmissivity divided by the storage coefficient (T/S in feet (metres) squared per day).

7.5.2.7 *Specific Storage*—If applicable, document the specific storage of the aquifer. Indicate whether this data element is from estimated hydraulic characteristics or from those determined by an aquifer test. Document the aquifer test method and the accuracy or confidence classification for this data element.

NOTE 24—Specific storage is the storage coefficient divided by the thickness of the aquifer. It is the volume of water the aquifer released from or taken into storage per unit volume of porous medium per unit change in head.

7.5.2.8 *Specific Yield*—If applicable, document the specific yield of the aquifer. Indicate whether this data element is from estimated hydraulic characteristics or from those determined by an aquifer test. Document the aquifer test method and the accuracy or confidence classification for this data element.

NOTE 25—Specific yield is the ratio of the volume of water that the saturated rock or soil will yield by gravity to the volume of the rock or soil. In the field, specific yield is generally determined by tests of unconfined aquifers and represents the change that occurs in the volume of water in storage per unit area of unconfined aquifer as the result of a unit change in head. Such a change in storage is produced by the draining or filling of pore space and is, therefore, mainly dependent on particle size, rate of change of the water table, and time of drainage.

7.5.2.9 *Barometric or Tidal Efficiency*—If applicable, document the barometric or tidal efficiency of the aquifer. Document the accuracy or confidence classification for this data element.

NOTE 26—Barometric efficiency is the ratio of the change in depth to water in a well to the inverse of water-level change in a water barometer.

7.5.2.10 *Porosity*—If applicable, document the porosity of the aquifer. Indicate whether this data element is estimated or determined by an aquifer test. Document the aquifer test method and the accuracy or confidence classification for this data element.

NOTE 27—Porosity of the aquifer is its property of containing interstices or voids and may be expressed quantitatively as the ratio of the volume of the interstices to the total volume.

7.5.2.11 *Specific Capacity*—If applicable, document the specific capacity of the ground-water site. Indicate whether this data element is estimated or determined by a test. Document the aquifer test method and the accuracy or confidence classification for this data element.

NOTE 28—Specific capacity is the rate of discharge from a well divided by the drawdown of the water level within the well at a specific time since pumping started.

7.5.2.12 *Method Used to Determine Aquifer Characteristics*—If applicable, document the method used to determine the hydraulic characteristics of the aquifer. The following are some generalized descriptions for the method used to determine aquifer characteristics components and their corresponding representative codes. If the specific aquifer test method is known, that information should be documented.

B—Bail test
C—Controlled single-well test methods
D—Controlled multiple-well test methods
N—Natural ground-water fluctuation
P—Cyclic pumping—single well
W—Cyclic pumping—multiple wells
S—Slug test
E—Estimated (describe)

7.5.2.13 *Availability of File of Detailed Results*—If applicable, document the availability and format of a file of detailed aquifer test results. Suggested availability and format of a file of aquifer test components and representative codes are as follows:

F—Files (raw data)
M—Machine readable
P—Published (report or basic-data release)
Z—Other (describe)

7.6 *Spring Record Group*—The spring record group is a single record for documenting data elements relating to properties of a spring, including both the basic description and flow characteristics of the ground-water source (**4, 5, 7, 8, 13, 19, 26**).

7.6.1 *Spring Record*—The spring record includes all information that relates to the description of a ground-water site that is determined to be a spring. A spring is defined as a place where ground water flows naturally from a rock or the soil onto the land surface or into a body of surface water. Data elements that are included in the spring record are the following:

NOTE 29—The spring record includes the data elements that relate directly to the properties of a spring, however, to completely document a spring site additional data elements need to be selected from those described in the three "Standard Guides." For example, the aquifer identification and lithology of the unit(s), along with the location map and photographs or sketches of the site, contribute additional information in interpreting the hydrology of the area. Most of the data elements narrated in Guide D 5408 can apply to springs, as well as applicable portions of this guide and Guide D 5410.

7.6.1.1 *Name of Spring*—If applicable, document the name by which the spring is known locally or, preferably, displayed on a published map.

7.6.1.2 *Type of Spring*—If applicable, document the type of spring at the ground-water site. Suggested type of spring components and representative codes are as follows **(13)**:

A—Artesian
J—Artesian and depression
K—Artesian and seepage or filtration
C—Contact
D—Depression
F—Fracture or fault
L—Fracture fault and depression
G—Geyser
B—Perched and contact
E—Perched and depression
H—Perched and tubular
O—Perched and fracture
P—Perched
R—Perched and seepage or filtration
S—Seepage or filtration
T—Conduit or tubular (cave)
Z—Other (describe)

7.6.1.3 *Permanence of Spring*—If applicable, document the permanence of the spring at the ground-water site. Suggested permanence of spring components and representative codes are as follows:

E—Periodic—Ebb and flow, normally have periods of relatively greater discharge at regular and frequent intervals.
G—Geyser—Discharge at more or less regular intervals. Nature of discharge is caused by expansive force of highly heated steam.
P—Perennial—Springs that discharge continuously.
I—Intermittent—Springs that discharge only during certain periods but at other times are dry.
R—Response to precipitation—Exist only after periods of rainfall.
S—Seasonal—Exist only during periods of high-water levels.
T—Estavelle—A cave that is a spring during some periods and a sinking stream (swallet) during other periods.
Z—Other (describe).

7.6.1.4 *Sphere of Discharge*—If applicable, document the sphere of discharge of the spring. Suggested sphere of discharge components and representative codes are as follows:

A—Subaerial—Discharges at lands surface
W—Subaqueous—Discharges under water

7.6.1.5 *Discharge*—If applicable, document the discharge value for the spring in gallons per minute, cubic feet (or metres) per second, litres per second or any other standard volume/time unit. Document the volume/time unit used. Note whether the discharge is clear or sometimes turbid. Document the accuracy or confidence classification for this data element.

7.6.1.6 *Date of Discharge*—If applicable, document the date (year/month/day/time of day in YYYYMMDD-HHMM format) that the discharge was measured. If the discharge value is averaged over a period of time, explain.

7.6.1.7 *Improvements*—If applicable, document the type of improvements at the spring of the ground-water site. Suggested improvement components and representative codes are as follows:

B—Boxed or small covered basin
C—Concrete basin
G—Gallery
H—Spring house
L—Lined
N—None
P—Pond
R—Pipe (not for conduction of water from spring)
T—Trough
Z—Other (describe)

7.6.1.8 *Number of Spring Openings*—If applicable, document the number of openings through which water discharges from the spring. Document the accuracy or confidence classification for this data element.

7.6.1.9 *Flow Variability*—If known or applicable, document the discharge variability of the spring, in percent, as expressed by the formula:

$$V = 100 \times [(a - b)/c]$$

where:
V = variability, %,
a = maximum discharge,
b = minimum discharge, and
c = average discharge.

Document the accuracy or confidence classification for this data element.

7.6.1.10 *Accuracy of Flow Variability*—If applicable, document the basis on which the variability of the spring was determined. Suggested accuracy of flow variability components and representative codes are as follows:

A—Calculated from less than one year of continuous discharge record.
B—Calculated from one to five years of continuous discharge record.
C—Calculated from more than five years of continuous discharge record.
D—Calculated from intermittent measurements made over a period of more than one year.
E—Calculated from less than one year of record, or estimated.
Z—Determined by other method (describe).

7.6.1.11 *Magnitude of Spring*—If applicable, document the magnitude of the spring. Document the accuracy or confidence classification for this data element. The following is a magnitude classification that has been used in the United States and was published by Meinzer **(19, 26)**.

Magnitude	Inch-Pound Units	SI (Metric) Units
First	Greater than 100 ft³/s	>10 m³/s
Second	10 to 100 ft³/s	1 to 10 m³/s
Third	1 to 10 ft³/s	0.1 to 1 m³/s
Fourth	100 gal/min to 1 ft³/s	10 to 100 l/s
Fifth	10 to 100 gal/min	1 to 10 l/s
Sixth	1 to 10 gal/min	0.1 to 1 l/s
Seventh	1 pint to 1 gal/min	10 to 100 ml/s
Eighth	Less than 1 pint/min	<10 ml/s

NOTE 30—Spring magnitudes were commonly used in the older hydrogeologic literature to classify the approximate discharge from a spring. This terminology is rarely used in the current hydrology reports. The common unit for measurement in recent literature is a cubic-foot per second, that translates to 646 317 gal (2 446 310 L) per day. Meinzer used the second-foot measurement that is equal to 646 000 gal (2 445 110 L) of water per day.

8. Keywords

8.1 aquifer test; confidence classification; drainage basin;

geohydrologic unit; geological sample; geophysical log; ground water; ground-water discharge; ground-water site; ground-water site construction; monitoring site; source agency; spring site; surface-water site; well site

REFERENCES

(1) Bates, Robert L., and Jackson, Julia A., *Glossary of Geology*, Third Edition; American Geological Institute, Alexandria, Virginia, 1987.

(2) Anderson, Keith E., *Water Well Handbook*; Fourth Edition: Missouri Water Well and Pump Contractors Association, Inc., Rolla, Missouri, 1971.

(3) Boulding, J. R., *Description and Sampling of Contaminated Soils, A Field Pocket Guide:* Center for Environmental Research Information, U.S. EPA, EPA/625/12-91/002, Cincinnati, Ohio, November 1991.

(4) Bureau of Reclamation, *Ground-Water Manual, A Water Resources Technical Publication*, Revised Reprint: U.S. Department of Interior, Bureau of Reclamation, Washington, DC, 1981.

(5) Campbell, Michael D., and Lehr, Jay H., *Water Well Technology*: McGraw-Hill, New York, NY, 1973.

(6) Cohee, George V., *Standard Stratigraphic Code Adopted by AAPG, Committee on Standard Stratigraphic Coding: American Association of Petroleum Geologists Bulletin*, Vol 51, No. 10, October 1967, p. 2146–2151.

(7) Fetter, C. W., *Applied Hydrogeology*, Second Edition: Macmillan Publishing Company, 866 Third Avenue, New York, NY, 1988.

(8) Freeze, R. A., and Cherry, J. A., *Groundwater*, Prentice Hall, Englewood Cliffs, NJ, 1975.

(9) Geological Society of America, *Rock Color Chart*, 7th Printing: Geological Society of America, 3300 Penrose Place, PO Box 9140, Boulder, Colorado 80301-9140, 1991.

(10) *Handbook of Ground-Water Development:* Roscoe Moss Co., Los Angeles, CA, Published by John Wiley and Sons, Inc., New York, NY, 1990.

(11) Heath, Ralph C., *Basic Ground-Water Hydrology: U.S. Geological Survey Water-Supply Paper 2220*, 1983.

(12) Keys, W. S., *Borehole Geophysics Applied to Ground-Water Investigations: U.S. Geological Survey Techniques of Water-Resource Investigations*, TWRI 2-E2, 1990.

(13) Mathey, Sharon B. (Editor), *National Water Information System User's Manual*, Vol 2, Chapt. 4. Ground-Water Site Inventory System: U.S. Geological Survey, Open-File Report, 1990, pp. 89–587.

(14) Texas Natural Resources Information System, *Ground-Water Data INTERFACE, Users Reference Manual*, Texas Natural Resources Information System, Nov. 20, 1986.

(15) U.S. Environmental Protection Agency, *Ground Water Monitoring in SW-846, Field Manual Physical/Chemical Methods, Test Methods for Evaluating Soil Wastes*, Vol II, Chapt. 11, Third Edition; Office of Solid Wastes and Emergency Responses, U.S. EPA, Washington, DC, 1991.

(16) U.S. Environmental Protection Agency, *Handbook of Suggested Practices for the Design and Installation of Ground-Water Monitoring Wells*: Office of Research and Development, U.S. EPA, EPA/600/4-89/034 Washington, DC, March 1991.

(17) U.S. Geological Survey, *National Handbook of Recommended Methods for Water-Data Acquisition*, Chapter 2—Ground Water: Office of Data Coordination, Reston, Virginia, 1980.

(18) U.S. Geological Survey, *Guide to Obtaining USGS Information: U.S. Geological Survey Circular 900*, 1989.

(19) van der Leedan, Frits, Troise, Fred L., and Todd, David Keith, *The Water Encyclopedia*, Geraghty and Miller Ground-Water Series, 2nd Edition, Third Printing, Lewis Publishers, Inc., Chelsea, Michigan, 1991.

(20) U.S. Department of Agriculture, *Atlas of River Basins of the United States: Soil Conservation Service*, 82 Maps, 1970.

(21) American National Standards Institute, Inc., American National Standard for Information Systems—*Representations of Local Time of Day for Information Interchange*; American National Standards Institute, Inc. Publication ANSI X3.43, 1430 Broadway, New York, NY 10018, 1986.

(22) U.S. Department of Commerce, *Representation of Local Time of Day for Information Interchange*, Federal Information Standards (FIPS) Publication 58-1, National Bureau of Standards, Washington, DC, Jan. 27, 1988.

(23) Morgan, Charles O., *Transition from the Ancient Underground Falaj to the Modern Pumped Well in Oman: In Minimizing Risk to the Hydrologic Environment*, Proceedings of the American Institute of Hydrology Conference held in Las Vegas, NV, March 13–15, 1990, pp. 155–160.

(24) MacBeth Division of Collmorgen Instrument Corp., *Munsell Soil Color Charts*, revised edition: MacBeth Division of Collmorgen Instrument Corp., PO Box 230, Newburgh, NY 12551-0230, 1990.

(25) Lohman, S. W., Ground-Water Hydraulics: U.S. Geological Survey Professional Paper 708, 1972.

(26) Meinzer, O. E., *Large Springs in the United States: U.S. Geological Survey Water-Supply Paper 557*, 1927.

Standard Guide for Set of Data Elements to Describe a Ground-Water Site; Part Three—Usage Descriptors[1]

This standard is issued under the fixed designation D 5410 ; the number immediately following the designation indicates the year of original adoption or, in the case of revision, the year of last revision. A number in parentheses indicates the year of last reapproval. A superscript epsilon (ε) indicates an editorial change since the last revision or reapproval.

1. Scope

1.1 This guide is Part Three of three guides to be used in conjunction with Practice D 5254 that delineates the data desirable to describe a ground-water data collection or sampling site. This guide identifies usage descriptors, such as monitoring, for an individual ground-water site. Guide D 5408 describes additional information beyond the minimum set of data elements that may be specified to identify a ground-water site, while Guide D 5409 identifies physical descriptors, such as construction, for a site.

NOTE 1—A ground-water site is defined as any source, location, or sampling station capable of producing water or hydrologic data from a natural stratum from below the surface of the earth. A source or facility can include a well, spring or seep, and drain or tunnel (nearly horizontal in orientation). Other sources, such as excavations, driven devices, boreholes, ponds, lakes, and sinkholes, that can be shown to be hydraulically connected to the ground water, are appropriate for the use intended.

NOTE 2—Guide D 5408 includes data confidence classification descriptor (1 element), geographic location descriptors (4 elements), political regime descriptor (1 element), source identifier descriptors (4 elements), legal descriptors (9 elements), owner descriptors (2 elements), site visit descriptors (3 elements), other identification descriptors (2 elements), other data descriptors (3 elements), and remarks descriptors (3 elements). Guide D 5409 includes individual site characteristics (7 data elements), construction descriptors (56 data elements), lift descriptors (16 data elements), geologic descriptors (26 data elements), hydraulic descriptors (20 data elements), and spring descriptors (11 data elements). For a list of descriptors in this guide, see Section 3.

1.2 These data elements are described in terms used by ground-water hydrologists. Standard references, such as Ref **(1)**[2] and various hydrogeologic professional publications, are used to determine these definitions. Many of the suggested elements and their representative codes are those established by the Water Resources Division of the U.S. Geological Survey and used in the National Water Information Systems computerized data base[3] **(1-21)**.

NOTE 3—The purpose of this guide is to suggest data elements that can be collected for ground-water sites. This does not uniquely imply a computer data base, but rather data elements for entry into any type of permanent file.

NOTE 4—Component and code lists given with some of the data elements, for example "method of discharge measurement," are only suggestions. These lists can be modified, expanded, or reduced for the purpose intended by the company or agency maintaining the ground-water data file.

NOTE 5—Use of trade names in this guide is for identification purposes only and does not constitute endorsement by ASTM.

1.3 This guide includes the data elements desirable to document a ground-water site beyond those given in the minimum set of data elements. Some examples of the data elements are water level, discharge, and water-quality sample collection date. No single site will need every data element, for example, a monitoring site may not need a long-term water use record. Each record (group of data elements) for a site has mandatory data elements, such as the date for the water level record. However, these elements are considered necessary only when that specific record is gathered for the site.

1.4 The values stated in both inch-pound and SI units are to be regarded separately as the standard. The values given in parentheses are for information only.

1.5 *This standard does not purport to address all of the safety problems, if any, associated with its use. It is the responsibility of the user of this standard to establish appropriate safety and health practices and determine the applicability of regulatory limitations prior to use.*

2. Referenced Documents

2.1 *ASTM Standards:*

D 653 Terminology Relating to Soil, Rock, and Contained Fluids[4]

D 5254 Practice for the Minimum Set of Data Elements to Identify a Ground-Water Site[5]

D 5408 Guide for Set of Data Elements to Describe a Ground-Water Site, Part One—Additional Identification Descriptors[5]

D 5409 Guide for Set of Data Elements to Describe a Ground-Water Site, Part Two—Physical Descriptors[5]

3. Terminology

3.1 *Definitions:*

3.1.1 For definitions of terms applicable to this guide, see Terminology D 653.

3.2 *Descriptions of Terms Specific to This Standard:*

3.2.1 *code*—a suggested abbreviation for a component, for example, "T" is the code suggested for the "electric tape" component of data element method of measurement.

3.2.2 *component*—a subdivision of a data element, for example, "electric tape" is one of 14 components suggested for data element method of measurement.

3.2.3 *data element*—an individual segment of informa-

---

[1] This guide is under the jurisdiction of ASTM Committee D-18 on Soil and Rock and is the direct responsibility of Subcommittee D18.21 on Ground Water and Vadose Zone Investigations.

Current edition approved May 15, 1993. Published November 1993.

[2] The boldface numbers given in parentheses refer to the list of references at the end of the text.

[3] Guide for the Decommissioning of Ground-Water Wells, Vadose Zone Monitoring Devices, Boreholes, and Other Devices for Environmental Activities (draft); ASTM Subcommittee D18.21.06 on Well Maintenance, Rehabilitation, and Abandonment Section, August 1991.

[4] *Annual Book of ASTM Standards*, Vol 04.08.

[5] *Annual Book of ASTM Standards*, Vol 04.09.

tion about a ground-water site, for example, "method of measurement." The data element is in the water-level record record.

3.2.4 *record*—denotes a set of related data elements that may need to be repeated to fully describe a ground-water site. For example, a ground-water monitoring site where water levels are measured periodically will need more than one water-level record record (the record includes data elements date of measurement, date accuracy, water level, water-level accuracy, status, method of measurement, instrumentation, and statistics method) to fully document the water-level history of the site. However, if only a single water level was measured for the site, the record is utilized once.

3.2.5 *record group*—a set of related records. For example the "monitoring record group" includes the measuring point record, network record, water level record, discharge record, water use record, water quality record, and field water quality record. Some record groups consist of only one record, for example, the "irrigation record group" includes only the irrigation record.

4. Summary of Guide

4.1 This guide includes the following usage descriptor data elements to describe a ground-water site. This guide includes only repeated elements that commonly require several records to fully describe the conditions and history of the site:

Monitoring Record Group:
- *Network Record:*
  - Data Type
  - Date Interval of Network Utilization
  - Source Agency for Network Data
  - Frequency of Data Collection
  - Method of Data Acquisition
  - Power Type of Instruments
  - Network
- *Measuring-Point Record:*
  - Date Interval of Measuring-Point Utilization
  - Height in Reference to Datum
  - Description
- *Water-Level Record:*
  - Measurement Date and Time
  - Water Level
  - Water-Level Accuracy
  - Status
  - Method of Measurement and Instrumentation
  - Instrumentation
  - Source of Data
  - Statistics Method
- *Discharge Record:*
  - Measurement Date and Time
  - Discharge
  - Type of Discharge
  - Source of Data
  - Method of Discharge Measurement
  - Instrumentation
  - Production or Pumping Level
  - Static Level
  - Method of Water-Level Measurement
  - Pumping Period
  - Specific Capacity
  - Drawdown
  - Source of Water-Level Data
- *Water-Use Record:*
  - Date Range of Water-Use Record
  - Data Collection Interval
  - Long-Term Water Use
  - Method Used to Determine Long-Term Water Use
- *Water-Quality Record:*
  - Sample Date and Time
  - Agency That Analyzes Samples
  - Type of Analyses
  - Parameters Requested for Analysis
  - Sample Depth/Interval
  - Water-Quality File Containing Analysis
  - Laboratory Number
  - Laboratory Name
  - Replicate Sequence Number
  - Collecting Agency or Company
  - Agency or Company Code
  - Chain of Custody
  - Sampling Purpose
  - Site Condition
  - Sample Appearance
  - Sample Odor
  - Presence of Immiscible Stratum
  - Thickness of Immiscible Stratum
  - Sensors
  - Sample Preservation Method
  - Sample Filtration Material
  - Pumped Period
  - Casing Volume
  - Amount of Water Purged
  - Sampling Method or Sampler Type
  - Sampler Material
  - Aquifer Sampled
  - Regulating Agency
- *Field Water-Quality Record:*
  - Field Sample Date and Time
  - Parameter Code
  - Value of Parameter
  - Unit for Parameter
  - Instrumentation or Method of Determination
- *Monitoring Site at Waste-Facility Record:*
  - Date Interval in Service
  - State Regulatory Agency
  - State Registration Identification
  - EPA Registration Identification
  - Responsible Company
  - Company's Site Identification
  - Site Location in Relationship to Waste Facility
  - Status of Site
  - Sampled Interval

Irrigation Record Group:
- *Irrigation Record:*
  - Irrigated Land Area
  - Allowance for Irrigating
  - Date Legal Irrigation Begins
  - Date Legal Irrigation Ends

Decommissioning Record Group:
- *Decommissioning Record:*
  - Date Decommissioned
  - Method Used for Decommissioning
  - Reason for Decommissioning
  - Plugging Material
  - Name and Address of Decommissioner
  - Step-by-Step Procedures
  - Availability of Decommissioning Report
  - Regulations Followed, Federal, State, Local

5. Significance and Use

5.1 Data at ground-water sites are gathered for many purposes. Each of these purposes generally requires a specific set of data elements. For example, when the ground-water quality is of concern not only are the minimum set of data elements required for the site, but information concerning the sample collection depth interval, method of collection, and date and time of collection are needed to fully qualify the data. Another group of elements are recommended for each use of the data, such as aquifer characteristics or water-level records. Normally the more information that is gathered about a site by field personnel, the easier it is to understand the ground-water conditions and to reach valid conclusions and interpretations regarding the site.

5.2 The data elements listed in this guide and Guides

D 5408 and D 5409 should assist in planning what information can be gathered for a ground-water site and how to document these data.

NOTE 6—Some important data elements may change during the existence of a site. For example, the elevation of the measuring point used for the measurement of water levels may be modified because of repair or replacement of equipment. This frequently occurs when the measuring point is an opening in the pump and the pump is modified or replaced. Because changes cannot always be anticipated. It is preferable to reference the height of the measuring point to a nearby, permanent altitude datum. The measuring point is referenced by being the same altitude (zero correction) or above (negative correction) or below (plus correction) the altitude datum. All appropriate measurements should be corrected in reference to the altitude datum before entry into the permanent file. Care must be exercised to keep the relationship of these data elements consistent throughout the duration of the site.

5.3 Some data elements have an extensive list of components or possible entries. For example the aquifer identification list described in Guide D 5409 has over 5000 entries. Lengthy lists of possible entries are not included in this guide, however, information on where to obtain these components is included with the specific data element.

NOTE 7—This guide identifies sources, lists, etc. of information required to completely document information about any ground-water site.

6. Documentation

6.1 *Introduction:*

6.1.1 Many of the ground-water data elements require multiple records or entries to completely describe a site. Time-related elements, such as water levels, discharge measurements, and water chemistry, may present hundreds or thousands of records for a period of many years that document measurements at a single site. These time-related data help to determine historical trends and serve to establish bench-mark facts for the site.

6.1.2 Other data elements that are not time related, such as casing lengths, spring openings, and an array of geophysical logs, require a sequence of records to thoroughly describe the site. These data are extremely valuable in site characterization, for example, wells for which the construction components are required to understand the source of the water.

6.2 *Monitoring Record Group:*

6.2.1 *Introduction*—The monitoring record group includes records for documenting data elements relating to any type of information gathered at a ground-water site for the purpose of monitoring hydrologic, usage, and water-quality trends.

6.2.2 *Network Record*—The network record includes the data elements that describes a ground-water site as it relates to a node in a hydrologic, usage, or water-quality network **(10)**.

NOTE 8—An example of a form (see Fig. 1) for documenting the data elements as described for two records of the monitoring record group is illustrated to show a method of design for this tool. These forms are commonly known as "field forms" or as "coding forms" (for computer entry). This type of form is routinely used for transcribing field data while at the ground-water site and entering non-field information at the agency's or company's office. It should be noted that each form has the site identification (primary identification as used by the agency or company), date of field visit, and person that recorded the data as the first entries. These three data items are mandatory to ensure correct filing of the information, either in cabinets or in a computer data base, and for quality control.

6.2.2.1 *Data Type*—If applicable, document the type of monitoring conducted or information collected at the ground-water network site. The information identified as continuous are from automatic recording devices, such as water-level recorders. Suggested network data type components and representative codes are as follows:

A—Water quality, analyzed in the laboratory
B—Water quality, analyzed in the field
C—Water levels, continuous
D—Water levels, intermittent
E—Water discharge, continuous
F—Water discharge, intermittent
G—Prediction and detection of earthquakes
H—Land compaction for subsidence
I—Vadose-zone water pressure
Z—Other (describe)

NOTE 9—A ground-water site can be used as a node in a network for monitoring phenomenon other than the common water-quality, water-level, and discharge parameters. For example, in some areas of the

GROUND-WATER SITE

MONITORING RECORD

Site Identification ______ Date Prepared ______
Prepared by ______ (Name)

Network Record:
Data Type ______ (See List in Manual)
Date Interval of Network Utilization ______ (Begin) ______ (End)
Source Agency to Network Data ______ (Name)
______ (Address)
Frequency of Data Collection ______ (See List in Manual)
Method of Data Acquisition ______ (See List in Manual)
Power Type of Instruments ______ (See List in Manual)
Network ______ (See List in Manual)

Measuring-Point Record:
Date Interval of Measuring-Point Utilization ______ (Begin) ______ (End)
Height in Reference to Datum ______ (feet/metre)
Description ______

FIG. 1 Sample Form

country, wells are used indirectly as part of programs to monitor the subsidence of the land surface as a result of collapse or compaction resulting from extraction of liquid from the underground material. Wells and springs have been used in past research as monitoring sites to predict and detect earthquakes by chemical changes in the water and by water-level fluctuations.

6.2.2.2 *Date Interval of Network Utilization*—If applicable, document the date interval, in year, month, and day (YYYYMMDD), that the ground-water site was used for network monitoring. If still in use, document the date that the site was put into service.

6.2.2.3 *Source Agency for Network Data*—If applicable, document the name and address of the agency responsible for the collection of the ground-water data at the network monitoring site.

6.2.2.4 *Frequency of Data Collection*—If applicable, document the frequency of data collection at the network monitoring site. Suggested frequency of data collection components and representative codes are as follows:

C—Continuously (analog recorder)
J—Fixed interval (digital recorder) give interval
K—Variable interval (digital recorder) give interval
H—Hourly
D—Daily
W—Weekly
F—Semimonthly (twice a month)
M—Monthly
B—Bimonthly (every two months)
Q—Quarterly
S—Semiannually
A—Annually
2—Biennially (every two years)
3—Every three years
4—Every four years
5—Every five years
X—Every ten years
O—One time only
I—Intermittently or variable time scale
Z—Other (describe)

6.2.2.5 *Method of Data Acquisition*—If applicable, document the method of data collection at the network monitoring site. Suggested method of data collection components and representative codes are as follows:

A—Automated instruments accessed by field personnel
B—Automated instruments accessed by direct line
D—Automated instruments accessed by radio
G—Automated instruments accessed by remote data transmission
F—Periodic field visits by agency or company personnel
C—Calculated from records of owner
E—Estimated from other records
U—Unknown
Z—Other (describe)

6.2.2.6 *Power Type of Instruments*—If applicable, document the power type of permanently mounted data collection instruments at the ground-water monitoring site in the network. A detailed description of the instruments should be included with each specific type of monitoring (for example, digital recorder with water levels). Suggested power type of instrument components and representative codes are as follows:

M—Mechanical
B—Battery operated
S—Spring driven
E—Electrical
R—Solar batteries
Z—Other (describe)

6.2.2.7 *Network*—If applicable, document the areal extent or management level of the network that includes the ground-water monitoring site. Document any additional networks that may include the site. Suggested network data components and representative codes are as follows:

N—National
S—State or Province
R—Regional, multiple state or county
C—County
D—Drainage basin
P—Project
Z—Other (describe)

6.2.3 *Measuring-Point Record*—The measuring point represents a convenient position at a ground-water facility to reference repeated measurements, such as water levels. For some ground-water sites the measuring point is at the same location as the altitude datum **(10)**.

NOTE 10—The altitude of the datum is described as one of the geographic locational data elements in the "Standard Practice for the Minimum Set of Data Elements to Identify a Ground-Water Site." This record describes the relationship between the measuring point and the datum, thereby linking measurements made at the site to the third dimension, the altitude. The datum normally remains the same for the life of the site; the measuring point can change as the water-withdrawal facilities are modified.

6.2.3.1 *Date Interval of Measuring-Point Utilization*—If applicable, document the date interval, in year, month, and day (YYYYMMDD), that the measuring point was used for conducting measurements at the ground-water site. If still in use, document the date that the measuring point was first used.

NOTE 11—The measuring point can be modified because of changing conditions at the ground-water site, therefore, several different measuring points may be used over the life of the site. These changes must be dated in order to relate to time-significant information, such as ground-water levels, that are collected over the history of the site.

6.2.3.2 *Height in Reference to Datum*—Document the height of the measuring point above or below the datum, in feet or metres. If the position of the measuring point is the same as the datum, the value is 0.0. Document the accuracy or confidence classification for this data element.

6.2.3.3 *Description*—Document a detailed description of the measuring point in relationship to the ground-water withdrawal facilities. A sketch or photograph of the facility and measuring point is valuable as an aid to future identification of the location.

6.2.4 *Water-Level Record*—This record is used to document water-level measurements of ground-water sites. Each water-level record requires most of the following data elements to thoroughly document the event. A single site may have thousands of records **(3–6, 9, 10, 12, 18, 19)**.

6.2.4.1 *Measurement Date and Time*—Document the date and time of day (standard time and 2400 clock) of the water-level measurement. Many historical measurements do not have the time of day information and are accurate only to the nearest day. Some measurements are only accurate to the nearest year, however are extremely valuable in documenting long-term water-level trends. Unless obvious, document the exactness of the date of measurement. An example is "193800000000" for the year 1938 without the month, day, and time **(2, 13)**.

6.2.4.2 *Water Level*—If applicable, document the water level, in feet or metres, in reference to the measuring point, for the ground-water site. For water levels that are above the measuring point (normally artisan wells), precede the value with a minus (−) sign to distinguish those water levels from ones at or below the measuring point. For those events where the condition at the site is dry, plugged, discontinued, certain flowing situations, or the site is destroyed, the water level is not transcribed (see 6.2.4.4). Document the accuracy or confidence classification for this data element (see 6.2.4.3 for suggestions).

6.2.4.3 *Water-Level Accuracy*—Document the accuracy of the water level as an aid for interpretation. Suggested water-level accuracy components and representative codes are as follows:

0—Accurate to nearest foot or metre
1—Accurate to nearest 1/10th ft or 1 cm
2—Accurate to nearest 1/100th ft or 1 mm
9—Unknown
Z—Other (describe)

6.2.4.4 *Status*—If applicable, document the status of the ground-water site as it relates to the water level. Suggested status components and representative codes are as follows:

A—Static water level (site is in equilibrium)
D—Site dry (no water level recorded)
E—Flowed recently
F—Site flowing (water level could not be measured) (no water level recorded)
G—Nearby site tapping same aquifer was flowing
H—Nearby site tapping same aquifer flowed recently
I—Injector site (recharge water being injected)
J—Injector site monitor (nearby site tapping same aquifer injecting recharge water)
N—Measurements discontinued
O—Obstruction encountered in well above water surface (no water level recorded)
P—Site being pumped
R—Site pumped recently
S—Nearby site tapping same aquifer being pumped
T—Nearby site tapping same aquifer pumped recently
V—Foreign substance present on surface of water
W—Site destroyed
X—Affected by stage in nearby surface-water body
Z—Other conditions affect water level (describe)

6.2.4.5 *Method of Measurement and Instrumentation*—If applicable, document the instruments and method designating the means by which the water level was measured. Suggested method of measurement components and representative codes are as follows:

A—Air line
B—Analog or graphic recorder
C—Calibrated air line
E—Estimated
F—Fiberglass tape
G—Pressure gage, mechanical
H—Calibrated pressure gage
I—Interface probe
L—Interpreted from geophysical logs
M—Manometer
N—Nonrecording gage (for example, staff gage)
P—Pressure transducer with data logger
R—Reported, method not known
Q—Sonar sounder
S—Steel tape
T—Electric tape
V—Calibrated electric tape
Z—Other (describe)

6.2.4.6 *Instrumentation*—If applicable, document the type of all permanently mounted data collection instruments at the ground-water site used for water-level monitoring. Suggested instrument components and representative codes are as follows:

B—Bubble gage
C—Crest-stage gage
D—Digital recorder (mechanical or electronic)
E—Continuous-record type recorder
G—Graphic or analog recorder
I—In situ, without readout or data logger
M—Data logger
P—Pressure transducer
R—Radio relay
S—Satellite relay
T—Telemetry
Z—Other (describe)

6.2.4.7 *Source of Data*—If applicable, document the source of the discharge data. Suggested source of data components and representative codes are as follows:

A—Government agency
C—Consultant
D—Driller's log or report
L—Personnel of source agency or company
M—Memory (owner, operator, driller)
O—Records by owner
R—Person other than owner, driller, or another government agency
S—Personnel of reporting agency or company
U—University associate
Z—Other source (describe)

6.2.4.8 *Statistics Method*—If applicable, document the method which describes how the measurement was selected from a continuous recorder (for example, analog, digital or micrologger) available for that day. Suggested statistics code components and representative codes are as follows:

M—Water level shown is a daily maximum (for example, deepest water level for the day)
N—Water level shown is a daily minimum (for example, shallowest water level for the day)
A—Water level is 12:00 noon reading
P—Water level is 12:00 midnight reading
N—Mean, daily, monthly, etc. (specify)
O—Other (describe)

6.2.5 *Discharge Record*—This record is used to document instantaneous discharge measurements for ground-water sites. Each discharge record requires most of the following data elements to thoroughly document the measurement. A single site may have many records **(3–6, 10, 12, 18, 19)**.

6.2.5.1 *Measurement Date and Time*—Document the date and time of day (standard time and 2400 clock) of the instantaneous discharge measurement. Many historical measurements do not have the time of day information and are accurate only to the nearest day. Some measurements, accurate to only the nearest year, are extremely valuable in documenting long-term water usage. Unless obvious, document the accuracy of the date of measurement.

6.2.5.2 *Discharge*—Document the discharge value for the ground-water site in gallons per minute, cubic feet (or metres) per second, litres per second or any other standard volume/time unit. This value must correspond to the remainder of the discharge event record for the site. Document the volume/time unit used. Document the accuracy or confidence classification for this data element.

6.2.5.3 *Type of Discharge*—If applicable, document the method of discharge. Suggested type of discharge components and representative codes are as follows:

P—Pumped
F—Flow
Z—Other (describe)

6.2.5.4 *Source of Data*—If applicable, document the source of the discharge and related (water level) data. Suggested source of the data components and representative codes are as follows:

A—Government agency
C—Consultant
D—Driller's log or report
L—Personnel of source agency or company
M—Memory (owner, operator, driller)
O—Records by owner
R—Person other than owner, driller, or another government agency
S—Personnel of reporting agency or company
U—University associate
Z—Other source (describe)

6.2.5.5 *Method of Discharge Measurement*—If applicable, document the method used to measure the discharge. Suggested method of discharge measurement components and representative codes are as follows:

A—Acoustic or sonic meter (transient-time meter)
B—Bailer
C—Current meter; either propeller-type meter in the discharge pipe, or propeller- or cup-type meter in the discharge channel
D—Doppler meter
E—Estimated
F—Flume
M—Totaling meter
O—Orifice
P—Pitot-tube meter, includes Cox meter, Collins meter, and the like
R—Reported, method not known
T—Trajectory method (free-fall method)
U—Venturi meter
V—Volumetric; bucket or barrel and stopwatch
W—Weir
Z—Other (describe)

6.2.5.6 *Instrumentation*—Ground-water sites used for discharge monitoring can have permanently mounted data collection instruments. Document the type and use of all instruments at the site as follows:

D—Digital recorder (mechanical and electronic)
E—Continuous-record type recorder
G—Graphic or analog recorder
M—Data logger
R—Radio relay
S—Satellite relay
T—Telemetry
Z—Other (describe)

6.2.5.7 *Production or Pumping Level*—If applicable, document the water level, in feet or metres in reference to the measuring point, measured while the ground-water site was discharging at the amount transcribed for this discharge record. If the water level is above the measuring point, for example a flowing artesian well, precede the value with a minus (−) sign. Document the accuracy or confidence classification for this data element.

6.2.5.8 *Static Level*—Document the water level, in feet or metres in reference to the measuring point, that relates to the remainder of the even record at the ground-water site. The measurement should be made before production begins or after the production has been stopped and the water level has reached equilibrium. For those water levels that are above the measuring point (normally flowing artesian wells), precede the value with a minus (−) sign to distinguish those water levels from one at or below the measuring point. Document the accuracy or confidence classification for this data element.

6.2.5.9 *Method of Water-Level Measurement*—If applicable, document the method indicating how the water level was measured. Suggested method of water-level measurement components and representative codes are as follows:

A—Airline
B—Analog or graphic recorder
C—Calibrated airline
E—Estimated
G—Pressure gage
H—Calibrated pressure gage
L—Interpreted from geophysical logs
M—Manometer
N—Nonrecording gage
R—Reported, method not known
S—Steel tape
T—Electric tape
V—Calibrated electric tape
Z—Other (describe)

6.2.5.10 *Pumping Period*—If applicable, document length of time, in hours and minutes, that the ground-water site was pumped or allowed to flow prior to the measurement of the production water level. Document the accuracy or confidence classification for this data element.

6.2.5.11 *Specific Capacity*—If applicable, document the specific capacity of the ground-water site. The value is computed by dividing the yield of the well in gallons per minute by the drawdown in feet ((yield)/(production water level−static water level)). Document the time of test if different than pumping period and the accuracy or confidence classification for this data element.

6.2.5.12 *Drawdown*—If applicable, document the drawdown, in feet or metres, of the water level of the pumping or flowing ground-water site. The drawdown is equal to the production level minus the static level. Document the accuracy or confidence classification for this data element.

6.2.5.13 *Source of Water-Level Data*—If applicable, document the source of the water-level data. Suggested source of data components and representative codes are as follows:

A—Government agency
C—Consultant
D—Driller's log or report
L—Personnel of source agency or company
M—Memory (owner, operator, driller)
O—Records by owner
R—Person other than owner, driller, or another government agency
S—Personnel of reporting agency or company
U—University associate
Z—Other source (describe)

6.2.6 *Water-Use Record*—The water-use record is used to document a history of long-term water withdrawals from a ground-water site. These withdrawals are usually reported daily, monthly, quarterly, or yearly **(10, 20)**.

6.2.6.1 *Date Range of Water-Use Record*—If applicable, document the date interval, in year, month, and day (YYYYMMDD), that the water-use data were gathered for the ground-water site.

6.2.6.2 *Data Collection Interval*—If applicable, document the data collection interval for the cumulative long-term water use for the ground-water site. Suggested time-increment components and representative codes are as follows:

D—Daily
M—Monthly
Q—Quarterly
R—Seasonal
S—Semiannually
Y—Yearly
Z—Other (describe)

6.2.6.3 *Long-Term Water Use*—If applicable, document the cumulative long-term water use, in gallons or litres (for large amounts, use units of thousands, millions, or acre-feet), for the ground-water site. This value can be stated in volume per day, month, quarter, year, etc. If other volume units are used, identify the unit. Document the accuracy or confidence classification for this data element.

6.2.6.4 *Method Used to Determine Long-Term Water Use*—If applicable, document the method used to determine the long-term water use. Suggested method used to determine long-term water use components and representative codes are as follows:

E—Estimated from Periodic Measurements
M—Totaling Meter
O—Owners Meter Records
P—Estimated from Power Records
R—Reported, Method Not Known
Z—Other (describe)

6.2.7 *Water-Quality Record*—The process of collecting and analyzing water-quality samples requires that complete information be gathered to confirm that the sample meets quality assurance procedures that may be required by law. The following data elements may be used to document the water samples **(3, 4, 10, 14, 15, 18)**.

6.2.7.1 *Sample Date and Time*—If applicable, document the date (year, month, and day) and time of day (standard time and 2400 clock) that the sample was collected at the ground-water site. Many historical samples do not have the time of day information and are accurate only to the nearest day. Some measurements, accurate to only the nearest year, are extremely valuable in documenting long-term water-quality trends. Unless obvious, document the accuracy of the date of measurement **(2, 13)**.

6.2.7.2 *Agency That Analyzes Samples*—If applicable, document the name and address of the agency that performed the quality analyses on the water sample collected for the monitoring site.

6.2.7.3 *Type of Analyses*—If applicable, document the type of quality analyses conducted on the water collected at the monitoring site. Suggested type of analyses components and representative codes are as follows:

A—Physical properties
B—Common ions (major cations and anions)
C—Trace elements
D—Pesticides
E—Nutrients
F—Sanitary analysis (organisms)
H—Herbicides
R—Radioactive
T—Biological taxa
V—Volatile organic compounds
Z—Other (describe)

NOTE 12—The Environmental Protection Agency (EPA) has compiled lists of components required to fulfill the needs of various federal regulations. These lists are available through EPA publications **(16)**.

6.2.7.4 *Parameters Requested for Analysis*—If applicable, document the EPA's five-digit STORET or chemical name or symbol for those parameters requested for analysis. Normally, the results of the analyses will be stored in a related and separate file, for example, EPA's STORET data base.[6]

6.2.7.5 *Sample Depth/Interval*—If applicable, document the maximum depth or preferably, interval, in feet or metres below the datum, to specify the zone of origin of the water sample at the ground-water site. Document the accuracy or confidence classification for this data element.

6.2.7.6 *Water-Quality File Containing Analysis*—If applicable, document the location of the file that contains the final water-quality analysis. If stored in a computer data base, identify the file (for example, STORET). If stored in file cabinets or published, identify the file type or publication.

6.2.7.7 *Laboratory Number*—If applicable, document the number assigned by the analyzing laboratory to the water sample.

6.2.7.8 *Laboratory Name*—If applicable, document the name and address of the laboratory that analyzed the water sample.

6.2.7.9 *Replicate Sequence Number*—If applicable, document the sequence number for the replicate sample.

6.2.7.10 *Collecting Agency or Company*—If applicable, document the name and address of the agency or company that collected the sample.

6.2.7.11 *Agency or Company Code*—If applicable, document the code of the agency or company that collected the sample. This code is assigned by EPA's STORET User Assistance staff and is required before entry of the ground-water site data into EPA's computerized Water Quality File (WQF).[6]

6.2.7.12 *Chain of Custody*—If applicable, document the names of the people and agency or company that signed the chain of custody form and the dates signed.

6.2.7.13 *Sampling Purpose*—If applicable, document the purpose or reason for collecting the sample. Suggested sampling purpose components and representative codes are as follows:

B—Research
C—CERCLA
R—RCRA
D—Drinking water regulations
E—Exploration (water)
L—Local ordinance
S—State regulations, other than CERCLA or RCRA
F—Federal regulations, other than CERCLA or RCRA
H—Hydrologic benchmark
I—Environmental issues

[6] STORET Users Assistance, USEPA, Mail Code PM-2180, 401 M Street, S.W., Washington, DC 20406.

J—Judicial/litigation
M—Mining regulations
N—Natural resources exploration
U—Unknown
Z—Other (describe)

6.2.7.14 *Site Condition*—If applicable, document the condition of the ground-water site at the time the sample was collected. Suggested sampling condition components and representative codes are as follows:

P—Pond—fenced, but open to atmosphere
U—Pond—unprotected
S—Spring or tunnel—protected
V—Spring or tunnel—unprotected
W—Well—sealed
Y—Well—open casing
Z—Other (describe)

6.2.7.15 *Sample Appearance*—If applicable, document the appearance of the sample at the time of collection as to color and turbidity. Suggested sample appearance components and representative codes are as follows:

A—Clear or colorless
C—Colored, not turbid (give color)
S—Turbid suspended matter, particles visible (give color)
T—Turbid, suspended matter, particles not visible (give color)
U—Unknown
Z—Other (describe)

6.2.7.16 *Sample Odor*—If applicable, document the odor given off by the ground water during the collection of the sample. Suggested sample odor components and representative codes are as follows:

C—Chemical, unknown
D—Chlorine
H—Hydrogen-sulfide
M—Methane
N—None
P—Petroleum
Z—Other (describe)

6.2.7.17 *Presence of Immiscible Stratum*—If applicable, document the presence of an immiscible stratum that may be at the top or bottom of the water column at the time of sample collection. Suggested presence of immiscible stratum components and representative codes are as follows:

T—Top presence of immiscible stratum
B—Bottom presence of immiscible stratum
I—Indeterminable
M—Mixed as globules in water
N—None
S—Sheen
U—Unknown
Z—Other (describe)

6.2.7.18 *Thickness of Immiscible Stratum*—If applicable, document the thickness in inches or millimetres of an immiscible stratum that may be at the top or bottom of the water column at the time of sample collection.

6.2.7.19 *Sensors*—If applicable, document the type of sensors, manufacturer, and model numbers of sensors at the ground-water site used for sensing of water-quality parameters. Suggested sensor components and representative codes are as follows:

E—Electrical
I—Ion specific
O—Sensors, using fiber optics
U—Unknown
Z—Other (describe)

6.2.7.20 *Sample Preservation Method*—If applicable, document the method used to preserve the sample. Several methods of preservation may be used that depend upon the parameter analyzed. Each method should be documented. Suggested sample preservation method components and representative codes are as follows:

C—Cooled or iced
H—Hydrochloric acid
P—Hydrogen peroxide
N—Nitric acid
R—Phosphoric acid
S—Sulfuric acid
X—None
Z—Other (describe)

6.2.7.21 *Sample Filtration Material*—If applicable, document the filter size and sample filtration material used in-line or in the field following the sample collection. If more than one filter is used, document each material and filter size. Suggested sample filtration material components and representative codes are as follows:

D—Cellulose acetate
C—Cellulose nitrate
G—Glass
E—Polycarbonate
T—Teflon
X—None
Z—Other (describe)

6.2.7.22 *Pumped Period*—If applicable, document length of time, in hours and minutes, that the ground-water site was pumped or allowed to flow prior to the collection of the water sample. Document the accuracy or confidence classification for this data element.

6.2.7.23 *Casing Volume*—If applicable, document the amount of water that needs to be pumped to purge the casing one time, in gallons, cubic feet (or metres), litres, or any other standard volume/time unit. Document the volume unit used. For example, the amount of water needed to purge the system is in the well casing above the screen or open area. Document the accuracy or confidence classification for this data element.

6.2.7.24 *Amount Water Purged*—If applicable, document the amount of water that was pumped or allowed to flow before the water sample was collected at the ground-water site, in gallons per minute, cubic feet (or metres) per second, litres per second, or any other standard volume/time unit. Document the volume/time unit used. Document the accuracy or confidence classification for this data element.

6.2.7.25 *Sampling Method or Sampler Type*—If applicable, document the sampling method or sampler type used to collect the water sample at the ground-water site. Numerous sampling devices have been and are being developed. The following list was patterned after a draft USEPA reference guide **(4)**. Also see Note 5. Suggested type of sampling method or sampler type components and representative codes are as follows:

Portable Grab/Depth Specific Samplers
O—Open bailer
P—Point-source bailer
B—Bucket
S—Syringe sampler

W—Westbay sampler
K—Kemmerer/Van Dorn sampler
C—Collwasa sampler
T—Stratified thief sampler
A—Swabbing
D—Packer pumps

Portable Positive Displacement (Submersible) Samplers
E—Bladder pump
G—Grundfos centrifugal pump
F—Other centrifual pump
H—Helical rotor pump
I—Gas-drive piston pump
J—Gear-drive pump
L—Submersible rod pump

Other Portable Samplers
M—Peristaltic suction lift
N—Centrifugal suction
Q—Gas-drive/displacement
R—Inertial pump
V—Gas-lift pump
Y—Jet pump

Portable In Situ Samplers
1—Hydropunch
2—BAT sampler
3—Other cone penetrometer samplers
4—Other in situ samplers

Other or Unknown Methods
X—Natural flowing, spring or well
U—Unknown
Z—Other (describe)

6.2.7.26 *Sampler Material*—If applicable, document the material used in construction of the sampler. Suggested sampler material components and representative codes are as follows:

A—Aluminum
B—ABS (Plastic)
C—Copper
G—Galvanized iron
L—Steel
N—Nylon
O—Polypropylene
P—Polyvinyl chloride (PVC)
Q—Polyalkene
R—Rubber
S—Stainless steel
T—Teflon
W—Wood
U—Unknown
Z—Other (describe)

6.2.7.27 *Aquifer Sampled*—If applicable, document the identification of the aquifer or aquifers from where the water was obtained at the ground-water site. A convenient and systematic method of coding aquifer and geologic units is used by the U.S. Geological Survey in a national file (Catalog of Aquifer Names and Geologic Unit Codes used by the Water Resources Division) (for example, Edwards Limestone of Texas is coded 218EDRD). Information needed to obtain an ordered list of aquifers and related codes is available **(10)**:[7]

6.2.7.28 *Regulating Agency*—If applicable, document the name and address of the agency that regulates water quality at the ground-water site, for example, State Health Department.

6.2.8 *Field Water-Quality Record*—The field water-quality data denote those constituents or characteristics where the final values are determined in other than the laboratory. Many of the values for these parameters, such as temperature and pH, only can be measured at the time of sample collection because of rapidly changing conditions in the water upon removal from the natural environment. Properly measured, these parameters are extremely valuable in determining the true hydrologic conditions in the aquifer at the ground-water site **(3, 7, 10, 14, 15)**.

NOTE 13—Many of the supporting data elements listed under the water-quality record also apply and can be used for documenting the field water-quality record. Normally, field water-quality characteristics (for example, temperature, pH) are determined during the same site visit that water-quality samples are collected for laboratory analysis. Therefore, when samples are collected for laboratory analysis, those supporting data elements (for example, sample depth/interval, site condition, aquifer sampled) need not be repeated for the field water-quality record.

6.2.8.1 *Field Sample Date and Time*—Document the date (year, month, and day) and time of day (standard time and 2400 clock) that the water-quality characteristic was determined at the ground-water site. Many historical determinations do not have the time of day information and are accurate only to the nearest day. Some measurements, accurate to only the nearest year, are extremely valuable in documenting long-term water-quality trends. Unless obvious, document the accuracy of the date of measurement **(2, 13)**.

6.2.8.2 *Parameter Code*—Document the EPA's five-digit STORET number, CAS number, or chemical name or symbol for those water-quality parameters analyzed or measured in the field. The following are some of the common field-determined water-quality characteristics and the corresponding STORET number code:

00010—Temperature, water (°C)
00095—Specific conductance (microsiemens/cm at 25°C)
00300—Oxygen, dissolved (DO), milligrams per litre (mg/L)
00400—pH (standard units)
00405—Carbon dioxide, dissolved (mg/L as CO_2)
00410—Alkalinity, water, whole, total, field, as $CaCO_3$, mg/L
00430—Alkalinity, carbonate (mg/L as $CaCO_3$)
00440—Bicarbonate, water, whole, total, field, as HCO_3, mg/L
00445—Carbonate, water, whole, total, field, as CO_3, mg/L
00900—Hardness, total (mg/L as $CaCO_3$)
00940—Chloride, dissolved (mg/L as Cl)
00945—Sulfate, dissolved (mg/L as SO_4
00950—Fluoride, dissolved (mg/L as F)
01045—Iron, total (micrograms per litre (µg/L) as Fe)
31501—Coliform, membrane filter, immediate M-endo medium (colonies/100 mL)
31625—Coliform, fecal, 0.7 µm membrane filter (UM-MF) (colonies/100 mL)
31673—Streptococci, fecal, membrane filter, KF agar (colonies/100 mL)
71820—Density—grams per millilitre (g/mL) at 20°C
71830—Hydroxide, water, whole, total, field, as OH, mg/L

6.2.8.3 *Value of Parameter*—Document the value for the field determined water-quality constituent or characteristic. Document the accuracy or confidence classification for this water-quality parameter.

6.2.8.4 *Unit for Parameter*—Include with each water-

[7] Geologic Names Unit, U.S. Geological Survey, 439 National Center, Reston, VA 22092.

quality constituent or characteristic the unit used for recording the value, for example, degrees celsius for temperature, micromhos (or microsiemens) per centimetre at 25° (25° is the laboratory standard) celsius for specific conductance or milligrams per litre (mg/L) for chloride.

6.2.8.5 *Instrumentation or Method of Determination*—Document the method or function, manufacturer, model, and accuracy of the instruments used to obtain field water-quality characteristics at the ground-water site. A number of instruments are currently available to measure these data. Some generic examples of the more common field instruments are as follows **(17, 19)**:

pH meter
Thermometer
Specific-conductance meter
Dissolved-oxygen metre
Kits for field measurement of constituent values
Kits for field determination of biological values

NOTE 14—New apparatus become available daily and many instruments of laboratory quality have become portable and are suitable for use in the field. Whether the field-determined chemical analyses can be classed as laboratory equivalent depends upon the quality control maintained on the instruments used for the measurement.

6.2.9 *Monitoring Site at Waste-Facility Record*—The monitoring site at waste-facility record includes the documentation of data elements of ground-water sites that were constructed for the primary purpose of monitoring the pollutants and hydraulics at hazardous and solid-waste facilities **(11, 19)**.

NOTE 15—Many of the other records listed under the monitoring record group and other standards describing ground-water data components also apply and can be used for documenting the ground-water sites at waste facilities.

6.2.9.1 *Date Interval in Service*—If applicable, document the date interval, in year, month, and day (YYYYMMDD), that the ground-water site was in service at the waste facility. If still in use, document the date that the site was put into service.

6.2.9.2 *State Regulatory Agency*—If applicable, document the name and address of the state regulatory agency that has jurisdiction over the site.

6.2.9.3 *State Registration Identification*—If applicable, document the number or identification assigned to the site by the state regulatory agency.

6.2.9.4 *EPA Registration Identification*—If applicable, document the number or identification assigned to the site by the U.S. Environmental Protection Agency.

6.2.9.5 *Responsible Company*—If applicable, document the name and address of the company or agency that owns or has control of the ground-water site.

6.2.9.6 *Company's Site Identification*—If applicable, document the company or agency number or identification of the ground-water site.

6.2.9.6 *Site Location in Relationship to Waste Facility*—If applicable, document the relative location of the ground-water site at the waste facility. Common site locations in relationship to the ground-water hydraulic gradient and waste facility and representative codes are as follows:

A—Upgradient
D—Downgradient
S—Side
W—Within boundaries
U—Unknown
Z—Other (describe)

6.2.9.8 *Status of Site*—If applicable, document the status of the ground-water site at the waste facility. Suggested status components and representative codes are as follows:

A—Active
D—Destroyed
E—Damaged
I—Inactive
U—Unknown
Z—Other (describe)

6.2.9.9 *Sampled Interval*—If applicable, document the monitored interval, in feet or metres. Document the accuracy or confidence classification for this data element.

NOTE 16—For example, the interval might be 15 to 18 ft (4.57 to 5.49 m) below the measuring point. The measuring point is defined under the measuring-point record and is related to the altitude, therefore, if the supporting records are utilized, the sampled interval can be converted to altitude values.

6.3 *Irrigation Record Group:*

6.3.1 *Introduction*—The irrigation record group includes the record for documenting the data elements pertaining to irrigation at a ground-water site.

6.3.2 *Irrigation Record*—The irrigation record includes data elements that relate to the withdrawal of water to be used for irrigation at the ground-water site **(10)**.

6.3.2.1 *Irrigated Land Area*—If applicable, document the land area, in acres or hectares, that is irrigated by water from the ground-water site.

6.3.2.2 *Allowance for Irrigating*—If applicable, document the maximum amount of water, in acre-feet per year or another unit, that is allocated for withdrawal from the ground-water site. Document the measurement unit and accuracy or confidence classification for this data element.

6.3.2.3 *Date Legal Irrigation Begins*—If applicable, document the date, in year, month, and day (YYYYMMDD), that irrigation can begin in the spring or at the start of the irrigation season (for example, 19870301).

6.3.2.4 *Date Legal Irrigation Ends*—If applicable, document the date, in year, month, and day (YYYYMMDD), that irrigation ends in the fall or at the end of the irrigation season (for example, 19871001).

6.4 *Decommissioning Record Group:*

6.4.1 *Introduction*—The decommissioning record group includes the record for documenting data elements that pertain to the permanent decommissioning or closure of a ground-water site.

6.4.2 *Decommissioning Record*—The decommissioning record includes data elements that pertain to permanent closure of a ground-water site. Although the decommissioning normally applies to sites at solid or hazardous waste facilities, the record can be used for any ground-water site, such as a public supply well[3] **(18)**.

6.4.2.1 *Date Decommissioned*—If applicable, document the date, in year, month, and day (YYYYMMDD), that the ground-water site was decommissioned or closed.

6.4.2.2 *Method Used for Decommissioning*—If applicable, document the method used to decommission or close the ground-water site. Suggested method used for decommissioning components and representative codes are as follows:

L—Casing left in place and plugged
R—Casing removed and hole plugged
U—Unknown
Z—Other (describe)

6.4.2.3 *Reason for Decommissioning*—If applicable, document the reason for decommissioning or closing the ground-water site. Suggested reason for decommissioning components and representative codes are as follows:

C—Remove chance of unauthorized use
P—Prevent migration of contaminants
M—Reduce chance of vertical or horizontal migration
N—No longer required
F—Failed or damaged
R—Regulatory requirement
U—Unknown
Z—Other (describe)

6.4.2.4 *Plugging Material*—If applicable, document the material used to fill the opening, for example the borehole opening. Suggested generic plugging material components and representative codes are as follows:

P—Portland cement (describe)
E—Expansive cement (describe)
A—API cement (describe)
B—Other cement (describe)
D—Gypsum cement (describe)
F—Epoxy resin cement (describe)
C—Clay (describe)
G—Coarse-grained material (describe)
U—Unknown
Z—Other (describe)

NOTE 17—Plugging materials, such as Portland cement, have a number of different types. In addition, each type can have a number of possible additives that are used for extenders, accelerators, retarders, density improver, fluid loss controllers, and friction reducers. The various materials and additives should be described.

6.4.2.5 *Name and Address of Decommissioner*—If applicable, document the company or agency name and address of the decommissioner of the ground-water site.

6.4.2.6 *Step-by-Step Procedures*—If applicable, document the step-by-step procedure used to decommission the ground-water site.

6.4.2.7 *Availability of Decommissioning Report*—If applicable, document the availability of the decommissioning report. Include the name and address of the source of the report.

6.4.2.8 *Regulations Followed, Federal, State, Local*—If applicable, document the regulations followed in decommissioning the ground-water site. Include the name and address of the regulator.

7. Keywords

7.1 confidence classification; decommissioning of ground-water site; field water-quality sample; ground water; ground-water site; irrigation; monitoring network; regulatory agency; site identification; waste facility; water-quality sample; water use

REFERENCES

(1) Bates, R. L., and Jackson, J. A., *Glossary of Geology*, Third Edition; American Geological Institute, Alexandria, Virginia, 1987.

(2) *American National Standard for Information Systems—Representations of Local Time of Day for Information Interchange*, American National Standards Institute, Inc. Publication ANSI X3.43-1986, 1430 Broadway, New York, NY 10018.

(3) Barcelona, M. J., Gibb, J. P., Helfrich, J. A., and Garske, E. E., *Practical Guide for Ground-Water Sampling*, Office of Ground-Water Protection, U.S. EPA, EPA/600/2-85/104, Robert S. Kerr Environmental Research Laboratory, Ada, OK, February 1985.

(4) Boulding, J. R., (in preparation), *Subsurface Field Characterization and Monitoring Techniques: A Desk Reference Guide, Vol I: Solids and Ground Water, Vol II: The Vadose Zone Chemical Field Screening and Analysis*, Prepared for U.S. EPA Center for Environmental Research Information, Cincinnati, OH.

(5) Bureau of Reclamation, *Ground-Water Manual*, A Water Resources Technical Publication, Revised Reprint, U.S. Department of Interior, Bureau of Reclamation, Washington, DC, 1981.

(6) Campbell, M. D., and Lehr, J. H., *Water Well Technology*, New York, NY, McGraw-Hill, 1973.

(7) Fishman, M. J., and Bradford, W. L., *Methods of Determination of Inorganic Substances in Water and Fluvial Substances*, 3rd Edition, Techniques of Water-Resources Investigation of the United States Geological Survey, Chapter A1, Book 5, 1989.

(8) Edwards, M. D., and Josefson, B. M., *Identification Codes for Organizations Listed in Computerized Data Systems of the U.S. Geological Survey*, U.S. Geological Survey, Open-File Report 82-921, 1982.

(9) Heath, Ralph C., *Basic Ground-Water Hydrology*, U.S. Geological Survey Water-Supply Paper 2220, 1983.

(10) Mathey, Sharon B. (editor), *National Water Information System User's Manual, Vol 2, Chapter 4, Ground-Water Site Inventory System*, U.S. Geological Survey, Open-File Report, 1990, pp. 89–587.

(11) Perry, R. A., and Williams, O. O., *Data Index Maintained by the National Water Data Exchange*, U.S. Geological Survey, Open-File Report 82-327, 1982.

(12) *Ground-Water Data INTERFACE, Users Reference Manual*, Texas Natural Resources Information System, Nov. 20, 1986.

(13) U.S. Department of Commerce, *Representation of Local Time of Day for Information Interchange, Federal Information Standards (FIPS) Publication 58-1*, National Institute for Standards and Technology, Washington, DC, Jan. 27, 1988.

(14) U.S. Environmental Protection Agency, *STORET Users Handbook*, Vols 1 and 2, U.S. EPA, Washington, D.C., February 1982.

(15) U.S. Environmental Protection Agency, *Ground-Water Data Management with STORET*, Office of Ground-Water Protection, U.S. EPA, Washington, DC, March 1986.

(16) U.S. Environmental Protection Agency, *Ground Water Monitoring in SW-846, Field Manual Physical/Chemical Methods, Test Methods for Evaluating Soil Wastes*, Vol II, Chapter Eleven, Third Edition; Office of Solid Wastes and Emergency Responses, U.S. EPA, Washington, DC, November 1986.

(17) U.S. Environmental Protection Agency, *Field Screening Methods for Hazardous Waste Site Investigations, Proceedings of the First International Symposium*, presented by U.S. Environmental Protection Agency, Environmental Monitoring Systems Laboratory-Las Vegas and U.S. Army Toxic and Hazardous Materials Agency at Las Vegas, NV, Oct. 11–13, 1988.

(18) U.S. Environmental Protection Agency, *Handbook of Suggested Practices for the Design and Installation of Ground-Water Monitoring Wells*, Office of Research and Development, U.S. EPA, Washington, DC, March 1991.

(19) U.S. Geological Survey, *National Handbook of Recommended Methods for Water-Data Acquisition, Chapter 2—Ground Water*, Office of Data Coordination, Reston, VA, 1980, pp. 2-1 to 2-149.

(20) U.S. Geological Survey, *Guide to Obtaining USGS Information*, U.S. Geological Survey Circular 900, 1989.

(21) van der Leedan, F., Troise, F. L., and Todd, D. K., *The Water Encyclopedia, Geraghty and Miller Ground-Water Series*, 2nd Edition, Third Printing, Lewis Publishers, Inc., Chelsea, MI, 1991.

Designation: F 480 – 93

Standard Specification for Thermoplastic Well Casing Pipe and Couplings Made in Standard Dimension Ratios (SDR), SCH 40 and SCH 80[1]

This standard is issued under the fixed designation F 480; the number immediately following the designation indicates the year of original adoption or, in the case of revision, the year of last revision. A number in parentheses indicates the year of last reapproval. A superscript epsilon (ϵ) indicates an editorial change since the last revision or reapproval.

1. Scope

1.1 This specification covers water well casing pipe and couplings made from thermoplastic materials in standard dimension ratios (SDR), SCH 40 and SCH 80.

1.2 Specifications are provided for the application of these materials to water well and ground water monitoring applications. Flush threaded joint systems are included for screen and casing used primarily in the construction of ground water monitoring wells (see Practice D 5092).

1.3 The values stated in inch-pound units are to be regarded as the standard. The values given in parentheses are provided for information only (see Practice E 380).

NOTE 1—Certain field conditions may require alternative materials to ensure safe long-term use. The user should consult federal, state, and local codes governing the use of thermoplastic materials for well casing or monitor pipe.

1.4 Although the pipe sizes and SDR values listed in this specification are generally available, numerous other plastic pipes in Schedule 40 and 80 wall, other SDR values and various outside diameters have been used for well casing. Such products are often selected because they fulfill certain needs and Annex A1 includes a list of these Plastic Pipe Well Casing Specials.

1.5 The following safety hazards caveat pertains only to the test method portion, Section 6, of this specification: *This standard does not purport to address all of the safety problems, if any, associated with its use. It is the responsibility of the user of this standard to establish appropriate safety and health practices and determine the applicability of regulatory limitations prior to use.*

2. Referenced Documents

2.1 *ASTM Standards:*

D 256 Test Methods for Impact Resistance of Plastics and Electrical Insulating Materials[2]
D 618 Practice for Conditioning Plastics and Electrical Insulating Materials for Testing[2]
D 638 Test Method for Tensile Properties of Plastics[2]
D 648 Test Method for Deflection Temperature of Plastics Under Flexural Load[2]
D 653 Terminology Relating to Soil, Rock, and Contained Fluids[3]
D 1527 Specification for Acrylonitrile-Butadiene-Styrene (ABS) Plastic Pipe, Schedules 40 and 80[4]
D 1600 Terminology for Abbreviated Terms Relating to Plastics[2,4]
D 1784 Specification for Rigid Poly(Vinyl Chloride) (PVC) Compounds and Chlorinated Poly(Vinyl Chloride) (CPVC) Compounds[4,5]
D 1785 Specification for Poly(Vinyl Chloride) (PVC) Plastic Pipe, Schedules 40, 80, and 120[4]
D 1892 Specification for Styrene-Butadiene Molding and Extrusion Materials[6]
D 1898 Practice for Sampling of Plastics[5]
D 2122 Test Method for Determining Dimensions of Thermoplastic Pipe and Fittings[4]
D 2235 Specification for Solvent Cement for Acrylonitrile-Butadiene-Styrene (ABS) Plastic Pipe and Fittings[4]
D 2241 Specification for Poly(Vinyl Chloride) (PVC) Pressure-Rated Pipe (SDR Series)[4]
D 2282 Specification for Acrylonitrile-Butadiene-Styrene (ABS) Plastic Pipe (SDR-PR)[4]
D 2412 Test Method for Determination of External Loading Characteristics of Plastic Pipe by Parallel-Plate Loading[4]
D 2444 Test Method for Impact Resistance of Thermoplastic Pipe and Fittings by Means of a Tup (Falling Weight)[4]
D 2564 Specification for Solvent Cements for Poly(Vinyl Chloride) (PVC) Plastic Piping Systems[4]
D 2672 Specification for Joints for IPS PVC Pipe Using Solvent Cement[4]
D 2855 Practice for Making Solvent-Cemented Joints with Poly(Vinyl Chloride) (PVC) Pipe and Fittings[4]
D 3122 Specification for Solvent Cements for Styrene-Rubber (SR) Plastic Pipe and Fittings[4]
D 3965 Specification for Rigid Acrylonitrile-Butadiene-Styrene (ABS) Compounds for Pipe and Fittings[4]
D 5092 Practice for Design and Installation of Ground Water Monitoring Wells in Aquifers[3]
E 380 Practice for Use of the International System of Units (SI) (The Modernized Metric System)[7]
F 402 Practice for Safe Handling of Solvent Cements, Primers, and Cleaners Used for Joining Thermoplastic Pipe and Fittings[4]
F 412 Terminology Relating to Plastic Piping Systems[4]

[1] This specification is under the jurisdiction of ASTM Committee F-17 on Plastic Piping Systems and is the direct responsibility of Subcommittee F17.61 on Water Systems.

Current edition approved March 15, 1993. Published August 1993. Originally published as F 480 – 76. Last previous edition F 480 – 91.

[2] *Annual Book of ASTM Standards*, Vol 08.01.

[3] *Annual Book of ASTM Standards*, Vol 04.08.

[4] *Annual Book of ASTM Standards*, Vol 08.04.

[5] *Annual Book of ASTM Standards*, Vol 08.02.

[6] *Discontinued*—Replaced by Specification D 4549.

[7] *Annual Book of ASTM Standards*, Vol 14.02.

2.2 *ANSI Standards:*
B1.5 ACME Screw Threads[8]
B1.8 Stub ACME Screw Threads[8]
B1.9 Buttress Inch Screw Threads[8]
2.3 *Federal Standard:*
Fed. Std. No. 123 Marking for Shipment (Civil Agencies)[9]
2.4 *Military Standard:*
MIL-STD-129 Marking for Shipment and Storage[11]
Z17.1 Preferred Numbers[9]
2.5 *Other Documents:*
NWWA/PPI, *Manual on the Selection and Installation of Thermoplastic Water Well Casing*[10]
Screw-Threads Standards for Federal Services 1957, Handbook H28, Part III[11]
NSF Standard No. 61 Drinking Water System Components[12]

3. Terminology

3.1 Definitions are in accordance with Terminology F 412 and abbreviations are in accordance with Abbreviations D 1600, unless otherwise specified. The abbreviation for acrylonitrile-butadiene-styrene plastic is ABS. The abbreviation for poly(vinyl chloride) is PVC. The abbreviation for styrene-rubber is SR.

3.2 Ground water investigation terms are in accordance with Terminology D 653.

4. Classification

4.1 Well casing is produced in either plain end, belled end, or threaded, and is used for water wells, ground water monitoring, leak detection, recovery systems, dewatering systems, and waste disposal.

5. Materials and Manufacture

5.1 *Specification*—The material described shall meet or exceed the requirements of (*1*) Specification D 3965 for ABS with a cell classification of 44322 or 33333, (*2*) Specification D 1784 for PVC with a cell classification of 12454C or 14333C or D, or (*3*) Specification D 1892 for SR with a cell classification of 4434A. The material so described shall be approved for potable water. (See NWWA/PPI manual.)

NOTE 2—Thermoplastic water well casing pipe, couplings, and components intended for use in the transport of potable water should be evaluated and certified as safe for this purpose by a testing agency acceptable to the local health authority. The evaluation should be in accordance with requirements for chemical extraction that are no less restrictive than those included in National Sanitation Foundation NSF Standard 61. The referenced standard and the seal or mark of the laboratory making the evaluation should be included on the thermoplastic well casing pipe, couplings, and components. For ground water monitoring applications the pipe should not be marked with an ink. A certificate of compliance, confirming evaluation for chemical extraction that is no less restrictive than NSF Standard No. 61 is suitable documentation to be provided by the pipe manufacturer.

NOTE 3—Caution should be exercised to control heat of hydration during grouting as thermoplastic materials are heat sensitive. Accelerators tend to increase the heat of hydration and are not recommended.

5.2 Acrylonitrile-butadiene-styrene (ABS) well casing pipe and couplings plastic shall be virgin plastic produced by the original compounder (see Specification D 1527). The minimum butadiene content is 6 %; the minimum acrylonitrile content is 15 %; the minimum styrene or substituted styrene content, or both, is 15 %; and the maximum content of other monomers is 5 % and lubricants, stabilizers, and colorants.

5.3 Poly(vinyl chloride) (PVC) well casing pipe and couplings plastic shall be made of virgin plastic produced by the original compounder. It shall contain poly(vinyl chloride) homopolymer, and such additives—stabilizers, lubricants, processing aids, impact improvers, and colorants—as needed to provide the required processing and toughness characteristics (see Test Method D 638).

5.4 The SR plastics compound shall contain at least 50 % styrene plastics, combined with rubbers to a minimum rubber content of 5 %, and compounding materials such as antioxidants and lubricants, and may contain up to 15 % acrylonitrile combined in the styrene plastics or rubbers, or both. The rubbers shall be of the poly-butadiene or butadiene-styrene type, or both, with a maximum styrene content of 25 % or nitrile type, or both. The combined styrene plastics and rubber content shall be not less than 90 %.

5.5 *Rework Material*—Clean rework material generated from the manufacturer's own well casing pipe and couplings production may be used by the same manufacturer, provided the well casing pipe and couplings produced meet all the requirements of this specification.

5.6 *Solvent Cement:*

5.6.1 *Specification*—The solvent cement shall meet the requirements of Specification D 2235 for ABS, Specification D 2564 for PVC, or Specification D 3122 for SR (Note 2).

6. Requirements

6.1 *Workmanship*—The pipe shall be homogeneous throughout and essentially uniform in color, opacity, density, and other properties. The inside and outside surfaces shall be semi-matte or glossy in appearance (depending on the type of plastic) and free of chalking, sticky, or tacky material. The surfaces shall be free of excessive bloom, that is, slight bloom is acceptable. The pipe walls shall be free of cracks, holes, blisters, voids, foreign inclusion, or other defects that are visible to the naked eye and that may affect the wall integrity. Machined slots or holes deliberately placed in pipe are acceptable. Bloom or chalking may develop in pipe exposed to direct rays of the sun (ultraviolet radiant energy) for extended periods, and consequently these requirements do not apply to pipe after extended exposure to direct rays of the sun.

6.1.1 *Ground Water Investigations*—Pipe manufactured into products used in ground water investigations should have surfaces that are visually free of oils, grease, dust, and marks imparted as a result of the manufacturing process.

6.2 *Well Casing Pipe:*

[8] Available from American National Standards Institute, 11 West 42nd St., 13th Floor, New York, NY 10036. Sponsored by the American Society of Mechanical Engineering, United Engineering Center, 345 East 47th St., New York, NY 10017.

[9] Available from Standardization Documents Order Desk, Bldg. 4 Section D, 700 Robbins Ave., Philadelphia, PA 19111-5094.

[10] Available from the Plastic Pipe Institute, 155 Rt. 46 W., Wayne, NJ 07470.

[11] Available from the Superintendent of Documents, Washington, DC.

[12] Available from The National Sanitation Foundation, 3475 Plymouth Rd., PO Box 1468, Ann Arbor, MI 48106.

TABLE 1 Outside Diameters and Tolerance for Thermoplastic Well Casing Pipe, in.

Nominal Pipe Size	Outside Diameter		Out-of-Roundness Maximum Diameter Minus Minimum Diameter		
	Average	Tolerance on Average	SDR41 SDR32.5 SDR26	SDR21 SDR17 SDR13.5	SCH40 and SCH80[A]
2	2.375	±0.006	0.060	0.024	0.024
2½	2.875	±0.007	0.060	0.030	0.030
3	3.500	±0.008	0.060	0.030	0.030
3½	4.000	±0.008	0.100	0.030	0.030
4	4.500	±0.009	0.100	0.030	0.030
5	5.563	±0.010	0.100	0.060	0.060
6	6.625	±0.011	0.100	0.070	0.070
8	8.625	±0.015	0.150	0.090	0.090
10	10.750	±0.015	0.150	0.100	0.100
12	12.750	±0.015	0.150	0.120	0.120
14	14.000	±0.020	0.150	0.150	0.150
15	15.300	±0.020	0.150	0.150	
16	16.000	±0.020	0.150	0.150	0.150

[A] Reference D 1527 for ABS and D 1785 for PVC.

6.2.1 *Dimensions*—The outside diameter and wall thickness of the well casing pipe shall meet the requirements given in Table 1 or Table 2 when measured in accordance with Method D 2122. (See Specification D 2282.)

6.2.2 *Wall Thickness Eccentricity*—The wall thickness eccentricity of the pipe shall be within 12 %.

6.2.3 *Length*—The well casing pipe shall be in either 10 or 20-ft (3.05 or 6.10-m) lengths, unless otherwise specified. The allowable tolerance on length shall be +½, −0 in. (+13, −0 mm) when measured in accordance with Test Method D 2122.

6.2.4 *Flush Joint Threaded Length*—If specified by the manufacturer or purchaser, the assembled length of flush threaded casing or screen shall be a nominal length such as 5, 10, or 20 ft. Any given laying length the purchaser specifies will constitute an assembled length. The allowable tolerance shall be +½, −⅛ in. (+13, −3 mm) on the components of the assembled laying length. The overall length of all flush threaded screen and casing shall be the nominal or specified laying length plus the length of the exposed male thread (pin).

NOTE 4—The purchaser should specify whether the length is to be the laying length or the overall length. The term "laying length" refers to the overall length less the length required to complete the assembly.

6.3 *Well Casing Pipe Couplings:*

6.3.1 *Socket Dimensions*—The socket dimensions of couplings shall conform to the requirements given in Table 3 when measured in accordance with Test Method D 2122.

6.3.2 *Bell Socket Dimensions*—The socket dimensions of well casing pipe bell couplings shall be as shown in Table 4 when measured in accordance with Test Method D 2122.

6.3.3 *Bell Socket Wall Thickness*—The wall thickness of an integral bell shall be considered satisfactory if formed from pipe that meets the requirements of this specification.

6.3.4 *Laying Length Dimensions*—The laying length dimensions of well casing pipe couplings shall conform to the requirements given in Table 3 and Table 4 when measured in accordance with Test Method D 2122.

6.3.5 *Socket Concentricity or Alignment*—The maximum misalignment of axis of couplings with the pipe measured in the plane of the coupling face shall not exceed ¾ in./20 ft (3 mm/1 m) of projected axis when measured in accordance with 7.4.

6.4 *Pipe Stiffness and Crush Resistance:*

6.4.1 *Well Casing Pipe*—The well casing pipe shall have a pipe stiffness at 5 % deflection equal to that shown in Table 5 and shall deflect 60 % of the original diameter (crush resistance) without cracking, rupture, or other visible evidence of failure when tested in accordance with Test Method D 2412. Three specimens shall be tested and all shall pass.

NOTE 5—This test is intended for use as a quality control test, not as a simulated service test.

6.4.2 *Couplings and Bells* shall meet all the designated dimensional requirements of Table 3 or Table 4. Molded couplings shall have a pipe stiffness at 5 % deflection equal to that shown in Table 5 and shall deflect 15 % without cracking, rupture, or other visible evidence of failure when tested in accordance with Test Method D 2412. Three specimens shall be tested and all shall pass.

6.5 *Impact Resistance Classification*—The impact resistance classification (IC) value for well casing pipe shall be selected from Table 6 by the manufacturer based on the measured average impact values determined in accordance with 7.5.

6.6 *Tup Puncture Resistance*—The well casing pipe and well casing couplings shall deflect 30 % (puncture resistance)

TABLE 2 Minimum Wall Thickness for Thermoplastic Well Casing Pipe, in.[A]

Nominal Pipe Size	SDR41	SDR32.5	SDR26	SDR21	SDR17	SDR13.5	SCH40[B]	SCH80[B]
	Min	Min	Min	Min	Min	Min	Min	Min
2	..	..	..	0.113	0.140	0.176	0.154	0.218
2½	..	..	..	0.137	0.169	0.213	0.203	0.276
3	..	..	..	0.167	0.206	0.259	0.216	0.300
3½	..	..	..	0.190	0.235	0.296	0.226	0.318
4	0.110	0.138	0.173	0.214	0.265	0.333	0.237	0.337
5	0.136	0.171	0.214	0.265	0.327	0.412	0.258	0.375
6	0.162	0.204	0.255	0.316	0.390	0.491	0.280	0.432
8	0.210	0.265	0.332	0.410	0.508	..	0.322	0.500
10	0.262	0.331	0.413	0.511	0.632	..	0.365	0.593
12	0.311	0.392	0.490	0.606	0.750	..	0.406	0.687
14	0.342	0.430	0.539	0.667	..	..	0.437	0.750
15	0.373	0.471	..	..	..	..	..	..
16	0.390	0.493	0.616	0.762	..	..	0.500	0.843

[A] The minimum is the lowest wall thickness of the well casing pipe at any cross section.
[B] Reference: D 1527 for SCH40 and 80 ABS
D 1785 for SCH40 and 80 PVC
D 2241 for SDR PVC

TABLE 3 Thermoplastic Water Well Casing Pipe Couplings Socket Dimensions and Laying Length Dimensions, in. (see Specification D 2672)

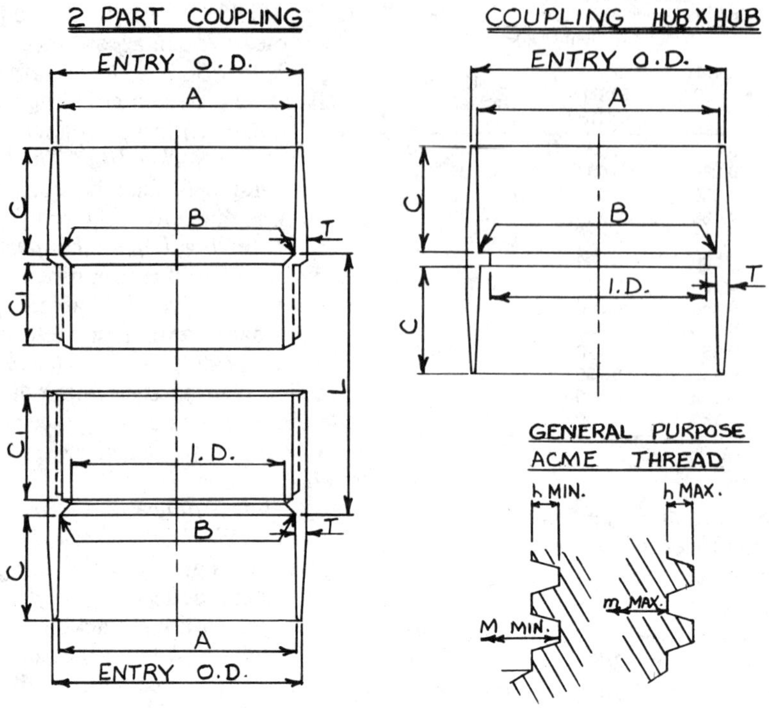

S	2	2½	3	3½	4	5	6	8	10	12
A	2.386	2.887	3.514	4.015	4.517	5.584	6.648	8.649	10.796	12.778
A_1	±0.006	±0.007	±0.008	±0.008	±0.009	±0.010	±0.011	±0.015	±0.015	±0.015
B	2.370	2.869	3.493	3.992	4.491	5.553	6.614	8.613	12.737	12.736
B_1	±0.006	±0.007	±0.008	±0.008	±0.009	±0.010	±0.011	±0.015	±0.015	±0.015
C	1.500	1.750	2.000	2.250	2.500	3.000	3.500	4.500	5.000	6.000
C_1	1.000	1.250	1.500	1.750	2.000	2.500	3.000	4.000	4.500	5.500
ID	2.149	2.601	3.166	3.620	4.072	5.033	5.993	7.805	9.728	11.538
T	0.113	0.137	0.167	0.190	0.214	0.265	0.316	0.410	0.511	0.606
T_1	+0.020	+0.020	+0.020	+0.023	+0.026	+0.032	+0.038	+0.049	+0.061	+0.073
h	0.083	0.100	0.100	0.125	0.125	0.166	0.200	0.250	0.333	0.375
M	2.465	2.982	3.606	4.131	4.630	5.735	6.830	8.878	11.086	13.127
m	2.289	2.772	3.396	3.871	4.370	5.393	6.420	8.368	10.410	12.367
L	1.500	1.750	2.000	2.250	2.500	3.000	3.500	4.500	5.000	6.000
OD	2.612	3.161	3.848	4.395	4.945	6.114	7.280	9.469	11.818	13.900

S = nominal pipe size
A = socket entrance diameter
A_1 = tolerance on diameter
B = socket bottom diameter
B_1 = tolerance on diameter
C = socket depth
C_1 = thread length
ID = bore diameter
T = minimum wall thickness
T_1 = tolerance on wall thickness
h = thread height—for ACME 2G screw thread (Note)
M = major diameter of internal thread—for ACME 2G screw thread (Note)
m = minor diameter of external thread—for ACME 2G screw thread (Note)
L = lay length
OD = outside diameter at entry of hub

NOTE—See 5.7 for thread dimensions other than ACME 2G screw threads.

without cracking, rupture, or other visible evidence of failure when tested in accordance with 7.6 (Note 5). Three specimens shall be tested and all shall pass.

6.7 *Threads*—Well casing, screens, and couplings having threads shall have either American Standard ACME 2G screw threads, American Standard Stub ACME 2G screw threads, or Buttress screw threads, Class 2, or square form flush joint threads, in accordance with ANSI B1.5 for ACME 2G screw threads, ANSI B1.8 for Stub ACME 2G screw threads, and ANSI B1.9 for Buttress screw threads. Examples of acceptable square form flush joint thread patterns for monitoring well construction are included in the annex.

6.7.1 All ACME, Stub ACME, and Butress screw threads shall be gaged in accordance with 7.7.

6.7.2 Machining flush joint square threads directly into the wall of the pipe may cause difficulty in measuring the

TABLE 3a Metric Equivalents for Table 3, mm

S	2	2½	3	3½	4	5	6	8	10	12
A	60.60	73.32	89.26	101.98	114.74	141.84	168.86	219.68	273.71	324.56
A_1	±0.16	±0.18	±0.20	±0.20	±0.22	±0.26	±0.28	±0.38	±0.38	±0.38
B	60.20	72.88	88.72	101.40	114.08	141.04	168.00	218.78	272.72	323.50
B_1	±0.16	±0.18	±0.20	±0.20	±0.22	±0.26	±0.28	±0.38	±0.38	±0.38
C	38	44	51	57	64	76	89	114	127	152
C_1	25	32	38	44	51	64	76	102	114	140
ID	54.58	66.06	80.42	91.94	103.42	127.84	152.22	198.24	247.10	293.06
T	2.88	3.48	4.24	4.82	5.44	6.74	8.02	10.42	12.98	15.40
T_1	+0.50	+0.50	+0.50	+0.58	+0.66	+0.82	+0.96	+1.24	+1.54	+1.86
h	2.10	2.54	2.54	3.18	3.18	4.22	5.08	6.34	8.46	9.52
M	62.62	75.74	91.60	104.92	117.60	145.66	173.48	225.50	281.58	333.42
m	58.14	70.40	86.26	98.32	111.00	136.98	163.06	212.54	264.40	314.12
L	38	44	51	57	64	76	89	114	127	152
OD	66.34	80.28	97.74	111.64	125.60	155.30	184.92	240.52	300.18	355.34

TABLE 4 Tapered Sockets for Bell-End Pipe, in.[A]

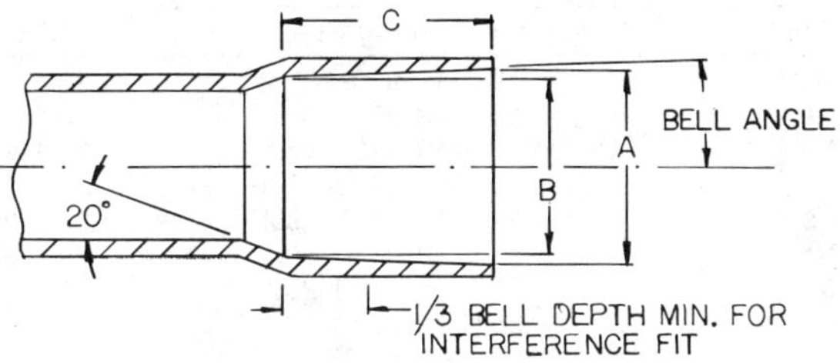

Nominal Pipe Size	A Socket Entrance Diameter			B Socket Bottom Diameter			C Socket Length, min
	Diameter	Tolerance on Diameter	Max Out-of-Round	Diameter	Tolerance on Diameter	Max Out-of-Round	
2	2.386	±0.006	±0.012	2.363	±0.006	±0.012	3.000
2½	2.887	±0.007	±0.015	2.861	±0.007	±0.015	3.500
3	3.514	±0.008	±0.015	3.484	±0.008	±0.015	4.000
3½	4.015	±0.008	±0.015	3.984	±0.008	±0.015	4.500
4	4.517	±0.009	±0.015	4.482	±0.009	±0.015	5.000
4½	4.968	±0.009	±0.030	4.932	±0.009	±0.030	5.500
5	5.584	±0.010	±0.030	5.543	±0.010	±0.030	6.000
6	6.648	±0.011	±0.030	6.603	±0.011	±0.030	6.500
6¼	6.922	±0.011	±0.030	6.878	±0.011	±0.030	7.000
8	8.649	±0.015	±0.045	8.598	±0.015	±0.045	7.000
10	10.776	±0.015	±0.050	10.722	±0.015	±0.050	7.500
12	12.778	±0.015	±0.060	12.721	±0.015	±0.060	8.000
14	14.035	±0.015	±0.060	13.970	±0.015	±0.060	8.000
16	16.045	±0.015	±0.060	15.970	±0.015	±0.060	8.000

[A] Minimum dimensions have zero negative tolerance. The sketches and designs of fittings are illustrative only.

TABLE 5 Minimum Pipe Stiffness at 5 % Deflection, lbf/(in. · in.) (kN/(m · m))

NOTE— The PS values are computed on the basis of minimum pipe wall thickness with the following material moduli: SR, 300 000 psi (2.07 GPa); ABS, 250 000 and 350 000 psi (1.72 to 2.41 GPa); and PVC, 400 000 psi (2.76 GPa).

	SDR 13.5	SDR 17	SDR 21	SDR 26	SDR 32.5	SDR 41
SR	684 (4720)	339 (2340)	168 (1160)	84 (580)	...	...
ABS 250 000	570 (3935)	273 (1942)	140 (964)	70 (485)	...	...
ABS 350 000	798 (5510)	395 (2720)	196 (1350)	98 (677)	...	...
PVC	912 (6290)	452 (3120)	224 (1550)	112 (774)	56	28

thread dimensions when the pipe is removed from the threading device. The inherent out-of-round condition of the pipe will cause the thread dimensions to conform to the irregularities of the pipe. Measurements must be taken at many points and averaged. Alternatively gages of metal or other rigid material may be used when gage dimensions or methods are available.

6.7.3 *Thread Out-of-Roundness*—Out of roundness for threaded dimensions on Schedule 40 and Schedule 80 threads, described in the annex must conform to the "Out-of-Roundness Maximum Diameter Minus Minimum Diameter" figures found in Table 1.

6.7.4 *Thread Eccentricity*—The wall thickness eccentricity as well as the flush joint threads found in the annex shall be within 12 %.

NOTE 6—Ground water monitoring wells are usually constructed with flush joints to prevent bridging of materials placed in the well annulus during well installation. Interior surfaces are a flush finish to prevent equipment from becoming lodged at points where the inside diameter might constrict. Externally coupled pipe may be used, but the special needs of wells constructed for ground water monitoring, for example, no gravel or grout bridging, hollow stem auger inside diameter,

TABLE 5a Minimum Pipe Stiffness for SCH 40 and SCH 80 Well Casing Pipe, lbf/(in.·in.) (kN/(m·m))[A]

Nominal Pipe Size, in.	SCH40	SCH80	SCH40	SCH80	SCH40	SCH80
	ABS 250 000	ABS 250 000	ABS 350 000	ABS 350 000	PVC	PVC
2	54 (373)	166 (1155)	79 (543)	244 (1680)	90 (620)	278 (1920)
2½	71 (490)	194 (1340)	91 (630)	284 (1960)	116 (800)	324 (2240)
3	46 (318)	133 (921)	67 (464)	193 (1330)	77 (530)	220 (1520)
3½	35 (240)	104 (720)	51 (350)	150 (1033)	58 (400)	171 (1180)
4	28 (192)	86 (593)	41 (284)	121 (823)	47 (324)	135 (940)
5	19 (128)	61 (422)	27 (182)	89 (613)	30 (208)	102 (702)
6	14 (96)	48 (329)	20 (137)	80 (553)	23 (156)	91 (632)
8	9 (65)	38 (260)	13 (91)	55 (378)	15 (104)	62 (432)
10	7 (48)	32 (222)	10 (68)	47 (326)	11 (78)	54 (372)
12	7 (49)	30 (206)	8 (58)	44 (301)	9 (63)	50 (344)
14	...	...	...	...	9 (59)	49 (336)
16	...	...	...	...	9 (59)	47 (324)

[A] Reference D 1527 for ABS and D 1785 for PVC.

TABLE 6 Impact Classification (IC) Cell Classification (see Test Methods D 256)

Nominal Pipe Size, in.	IC-0	IC-1, m·kg(ft·lb)	IC-2, m·kg(ft·lb)	IC-3, min, m·kg(ft·lb)
2	Ns[A]	9.7–12.4(70–90)	12.4–15.2(90–110)	>15.2(110)
2½	Ns	11.0–13.8(80–100)	13.8–18.0(100–130)	>18.0(130)
3	Ns	16.6–19.4(120–140)	19.4–24.9(140–180)	>24.9(180)
3½	Ns	18.0–22.1(130–160)	22.1–27.7(160–200)	>27.7(200)
4	Ns	22.1–27.7(160–200)	27.7–33.2(200–240)	>33.2(240)
5	Ns	24.9–30.4(180–220)	30.4–36.0(220–260)	>36.0(260)
6	Ns	27.7–36.0(200–260)	36.0–41.5(260–300)	>41.5(300)
8	Ns	36.0–41.5(260–300)	>41.5(300)	>41.5(300)
10	Ns	>41.5(300)	>41.5(300)	>41.5(300)
12, 14, 16	Ns	>41.5(300)	>41.5(300)	>41.5(300)

[A] Ns—Not specified.

solvent-free couplings, should be addressed.

6.8 *Threaded Joints*—Threaded joints including flush threaded joints made with well casing pipe and well casing couplings shall not leak. For ACME, Stub ACME, Buttress thread and Flush Thread joints the test shall be an internal pressure of 25 psi (170 kPa) for 1 h in accordance with 7.8.

6.9 *Joint Strength*—Joint tensile strength requirements will be added when test method and test results are available.

6.10 *Well Screens*—Screens manufactured from pipe or casing shall have uniform slots placed perpendicular to the long axis of the pipe. Slot width, length, inter-slot spacing, number of rows around the screen circumference and overall screen length shall comprise the product specification.

7. Test Methods

7.1 *Sampling*—A sample of the well casing pipe and coupling sufficient to determine conformance with this specification shall be taken at random from each lot in accordance with Section 9 of Practice D 1898.

7.2 *Conditioning*—Unless otherwise specified, condition the specimens prior to test at 72.4 ± 3.6°F (23 ± 2°C) and 50 ± 5 % relative humidity for not less than 40 h in accordance with Procedure A of Practice D 618. The manufacturer may use shorter conditioning time, but in case of disagreement Procedure A of Practice D 618 shall be used.

7.3 *Test Conditions*—Conduct tests in the standard laboratory atmosphere 73.4 ± 3.6°F (23 ± 2°C) and 50 ± 5 % relative humidity, unless otherwise specified in the test methods or in this specification.

7.4 *Socket Concentricity or Alignment (see Practice D 2855)*—Select three bell socket couplings with sufficient pipe so that concentricity and alignment tests for bell socket couplings may be made. Select three well casing pipe specimens, each specimen 12 ± ⅛ in. (300 ± 3 mm) in length. Cut the ends square, free of burrs and jagged edges. Solvent cement the bell socket coupling and well casing pipe together following recommended procedures. Center the coupling onto a test mandrel that has been accurately centered on a lathe or lathe-type spindle. While the assembly is being rotated, eccentricity in the plane of the bell socket coupling face may be determined by means of a dial gage bearing radially against the bell socket coupling. The extent of the angular misalignment may be determined by means of a dial gage bearing radially against the well casing pipe approximately 1 ft (0.30 m) from the bell socket coupling face or a dial gage parallel to the axis against the bell socket coupling face. Other test methods may be used when agreed upon between the purchaser and the seller.

7.5 *Impact Classification (see Test Methods D 256)*—Determine the impact classification in accordance with Test Method D 2444, using Tup B weighing 30 lb and Holder B. Select ten well casing pipe specimens of each size with each specimen 6 ± ⅛ in. (150 ± 3 mm) in length. Condition the test specimens in a low-temperature environmental chamber maintaining a test temperature of 32 to 35.6°F a minimum of 2 h or in a mixture of ice and water at 32 to 35.6°F for 1 h before testing and test immediately on removal. Test ten specimens in accordance with Test Method D 2444; nine of the ten specimens shall be above the lower limit of the IC cell. Examine the results for conformance with 5.5.

7.6 *Tup Puncture Resistance Test:*

7.6.1 *Procedure (see Test Method D 648)*—Select three pipe specimens and three couplings. Determine whether the specimens are resistant to tup puncture at 30 % deflection by using the apparatus required for Test Method D 2412 and

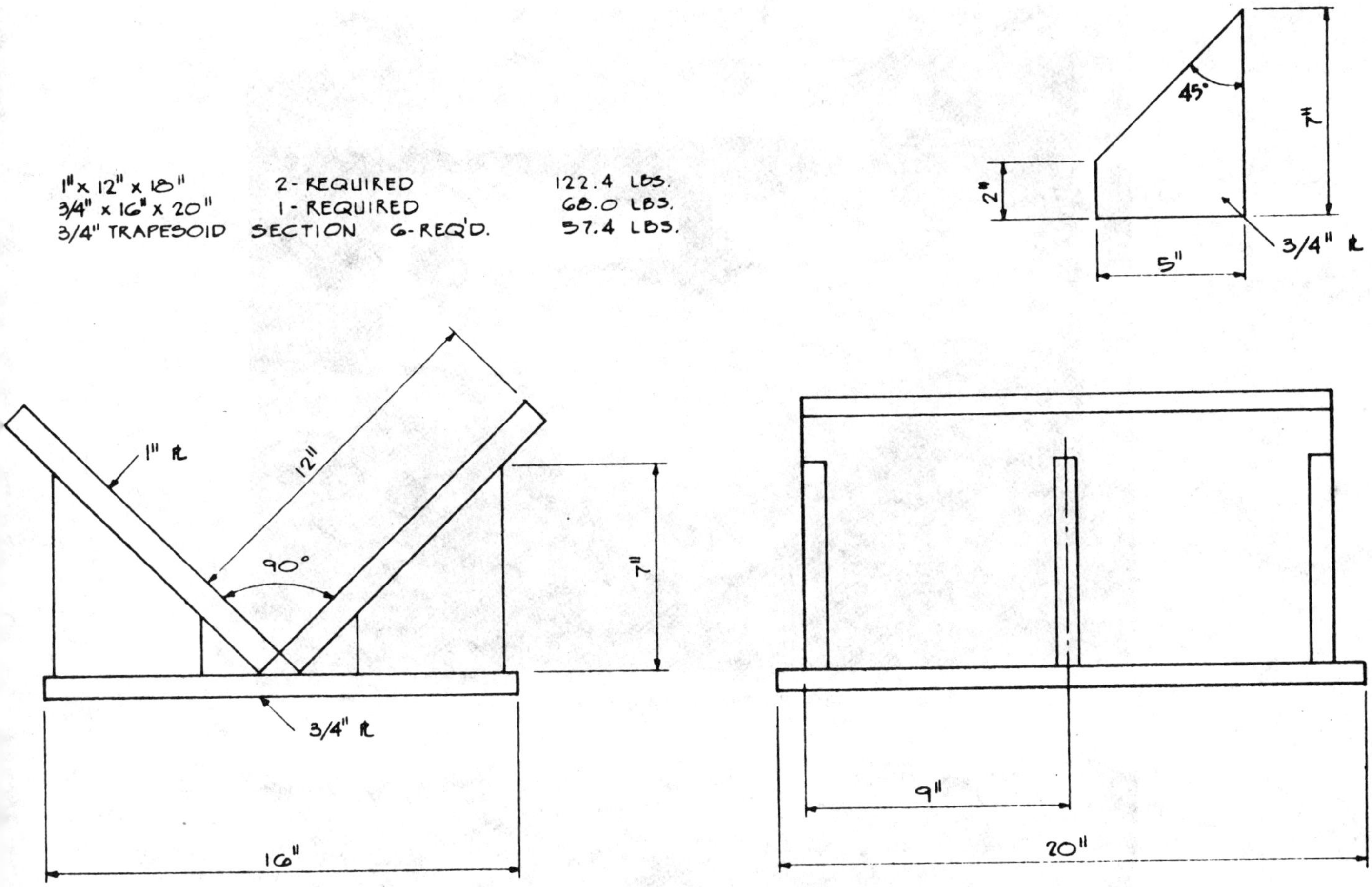

FIG. 1 V-Block Support (No Scale)

Configuration of Tup A Impact Resistance of Thermoplastic Pipe (Test Method D 2444)

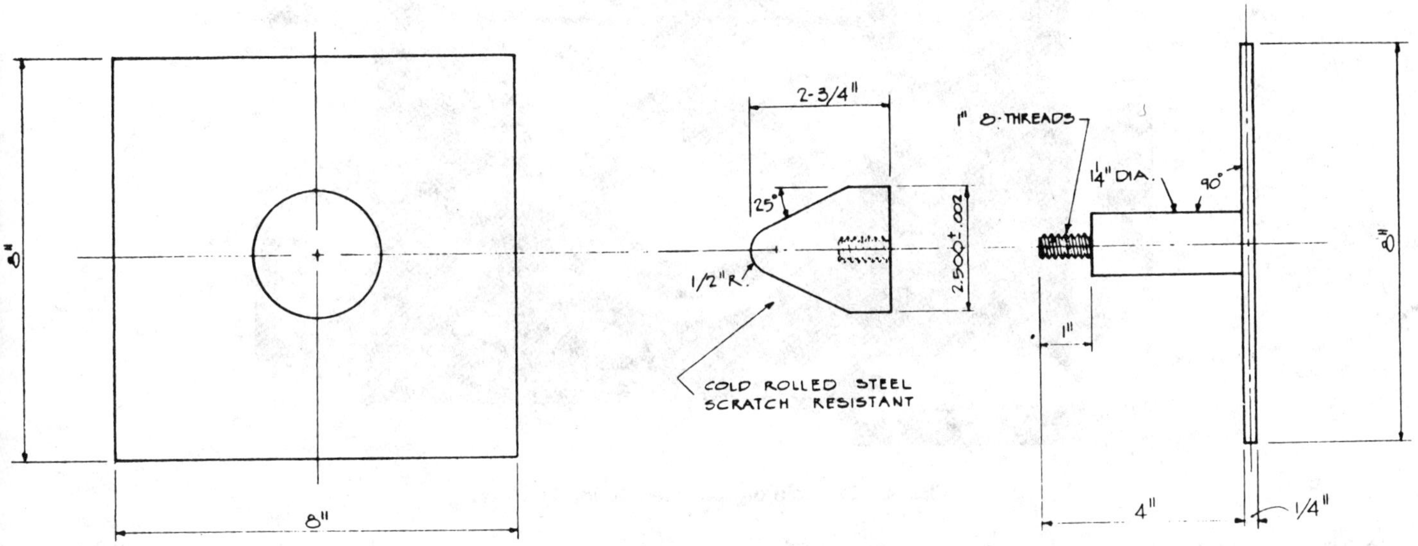

FIG. 2 Tup Test Apparatus (No Scale)

Tup A as defined in Test Method D 2444. The test method utilizes a constant-load rate instead of an impact load. The Vee-block base has been enlarged to accommodate larger specimens as seen in Fig. 1. Machine the base so that the two sides of the Vee form an angle of 90 ± 0.01°. Mount Tup A of Test Method D 2444 on a square steel plate as shown in Fig. 2. Attach two Federal D01S dial gages to the square steel plate 180° apart as seen in Figs. 3 and 4. Measure the specimens to determine the point of minimum wall thickness. Place the pipe or coupling specimen of 6 ± ⅛ in. (150

FIG. 3 Tup Test Apparatus

FIG. 4 Example of Specimen During Tup Test

± 3 mm) in length in the universal load machine with the minimum wall of the specimen positioned directly under the nose of the tup. Place spacers between the dial gage stems and the base so that the deflection of the tup versus load can be measured. The speed of testing shall be 0.5 ± 0.02 in. (12.5 ± 0.5 mm)/min. Continue the test until the diameter is deflected 30 % of its original diameter (puncture resistance). Examine the test results for each specimen of pipe and coupling for conformance to 5.6.

NOTE 7—The tup puncture test for point load is derived from Test Methods D 2412 and D 2444, combined to achieve a meaningful design parameter for well casing pipe used in water well construction. There are many possibilities for a point load to be exerted on the well casing pipe. Perhaps the most significant of these is the stringing of a well casing pipe through a boulder field. The sides of the well hole are seldom smooth surfaces, but rather pieces of rock are embedded in the surrounding soil layers. When these rock particles come in contact with the well casing pipe, a point loading situation can develop. Natural earth movements can impose high stresses over a small area of well casing pipe surface. This type of localized load is an entirely different situation from a uniform load.

7.7 *Threads*—Gage the threads using gages specified in accordance with recommended gages and gaging practice for external and internal threads as specified in the following standards:

ANSI B1.5 1973
ANSI B1.8 1973
ANSI B1.9 1973
(Not Flush Threaded)
Screw-Threads, Standards for Federal Service 1957 Part III, Handbook H28

7.7.1 Machining flush joint threads directly into the wall of the pipe may cause difficulty in measuring the thread dimensions when the pipe is removed from the threading device. The inherent out-of-round condition of the pipe will cause the thread dimensions to conform to the irregularities of the pipe. Measurements must be taken at many points and averaged. Alternatively gages of metal or other rigid material may be used when gage dimensions or methods are available. Gages should be polished and free of all lubricants when used. The male gage should be fitted with an "O" ring of the same material, durometer, and dimensions as that specified for the manufactured product.

7.8 *Threaded Joints*—Assemble two specimens of well casing pipe, 6 ± 1/8 in. (150 ± 3 mm), together with a male and female two-piece threaded well casing pipe coupling. Solvent cement the male and female well casing pipe coupling to each specimen of well casing pipe and allow to stand for 24 h at room temperature. Apply an approved thread lubricant, specifically intended for use with the designated plastic material, to the threads of the male and female two-piece well casing pipe coupling and assemble. Subject the specimen to an internal pressure of 25 psi (170 kPa) with water as a medium, for 1 h. Should the threads be an integral part of the casing, the assembling of the coupling to the pipe shall be omitted. Examine the well casing pipe coupling joint for leakage to determine conformance to the requirements of 6.8.

7.8.1 *Flush Threaded Joints*—Assemble the specimens of well casing, one threaded female and one threaded male with an "O" ring. No adhesives, solvents, or sealants may be used. Water only may be used, if necessary, to ensure a smooth closure. The joint should then be torqued to the maximum value to be specified for that joint pattern. Allow the joint to stand for 24 h at room temperature. Subject the specimen to an internal pressure of 25 psi (170 kPa) with water as a medium, for 1 h. Examine the well casing pipe joint for leakage to determine conformance to the requirements of 6.8.

8. Retest and Rejection

8.1 If the results of any test(s) do not meet the requirements of this specification, the test(s) may be conducted again in accordance with an agreement between the purchaser and the seller. In retesting, the product requirements of this specification shall be met, and the test methods designated in the specification shall be followed. If, upon retest, failure occurs, the quantity of product represented by the test(s) shall be rejected.

9. Product Marking

9.1 *Quality Assurance*—When the product is marked with this designation, ASTM F 480, or, in the case of pipe manufactured for ground water monitoring products, when the pipe manufacturer so certifies in writing to the purchaser, it affirms that the product was manufactured, inspected, sampled, and tested in accordance with this specification and has been found to meet the requirements of this specification.

9.2 *Quality of Product Marking*—The product marking shall be applied to the pipe in such a manner that it remains legible (easily read) after installation and inspection.

9.3 *Content of Product Marking:*

9.3.1 *Well Casing Pipe*—The well casing pipe shall be marked at least every 5 ft (1.5 m) in letters not less than 3/16 in. (5 mm) high in a contrasting color with the following information:

9.3.1.1 Nominal well casing pipe size (for example, 2 in.),

9.3.1.2 Well casing pipe standard dimension ratio, in accordance with designation code given in 3.2 (for example, SDR 17),

9.3.1.3 Type of plastic well casing pipe material (for example, ABS250, ABS350, PVC, or SR),

9.3.1.4 The wording—well casing—followed by the impact classification (for example, IC-3),

9.3.1.5 Designation ASTM F 480 including the year of issue, with which the well casing pipe complies,

9.3.1.6 Manufacturer's name (or trademark), and

9.3.1.7 Manufacturer's code for resin manufacture, lot number, and date of manufacture.

9.3.1.8 Well casing pipe intended for potable water shall also include the seal or mark of the laboratory making the evaluation for this purpose spaced at intervals specified by the laboratory (Note 2).

9.3.1.9 Well casing pipe intended for manufacture into screen or casing for ground water monitoring use shall not be marked with any foreign material, for example, ink, unless it can be independently proven that the marking material will not contaminate the ground water sample or affect the subsequent analysis of the water sample for pollutants.

9.4 *Well Casing Pipe Coupling*—Well casing pipe couplings shall be marked in letters not less than 3/16 in. (5 mm) high, with the following information:

9.4.1 Nominal well casing pipe coupling size (for example, 2 in.),

9.4.2 Type of plastic well casing pipe coupling material (for example, ABS250, ABS350, PVC, or SR),

9.4.3 ASTM Designation F 480, including year of issue with which the well casing pipe coupling complies, and

9.4.4 Manufacturer's name (or trademark).

9.4.5 Well casing pipe couplings intended for potable water shall also include the seal or mark of the laboratory making the evaluation for this purpose spaced at intervals specified by the laboratory (Note 2).

GOVERNMENT/MILITARY PROCUREMENT

These requirements apply *only* to federal/military procurement, not domestic sales or transfers.

S1. *Responsibility for Inspection*—Unless otherwise specified in the contract or purchase order, the producer is responsible for the performance of all inspection and test requirements specified herein. The producer may use his own or any other suitable facilities for the performance of the inspection and test requirements specified herein, unless the purchaser disapproves. The purchaser shall have the right to perform any of the inspections and tests set forth in this specification where such inspections are deemed necessary to ensure that material conforms to prescribed requirements.

NOTE S1.1—In U.S. federal contracts, the contractor is responsible for inspection.

S2. *Packaging and Marking for U.S. Government Procurement:*

S2.1 *Packaging*—Unless otherwise specified in the contract, the material shall be packaged in accordance with the supplier's standard practice in a manner ensuring arrival at destination in satisfactory condition and which will be acceptable to the carrier at lowest rates. Containers and packing shall comply with Uniform Freight Classification rules or National Motor Freight Classification rules.

S2.2 *Marking*—Marking for shipment shall be in accordance with Fed. Std. No. 123 for civil agencies and MIL-STD-129 for military agencies.

NOTE S2.1—The inclusion of U.S. Government procurement requirements shall not be construed as an indication that the U.S. Government uses or endorses the products described in this specification.

ANNEX

(Mandatory Information)

A1. PLASTIC PIPE WELL CASING SPECIALS

A1.1 To fill the needs of the water well industry, plastic pipe having a variety of sizes and wall thicknesses have been used as well casing over the past 20 years. Some of these casings were standard products, for example, Schedule 40 and 80 Iron Pipe Size (IPS) OD pipe and Plastic Irrigation Pipe (PIP) OD SDR pipe. Other products were made to special outside diameters and wall thicknesses expressly to fill the needs of various well drillers. Because the tables in this specification include only IPS-OD SDR pipe and because many of these other products have a long service history and fulfill special needs, this listing of Well Casing specials is offered here. These well casings can be tested to verify that they meet the quality requirements in this specification by using the minimum Pipe Stiffness (PS) values in these tables and the other quality control tests in the specification.

TABLE A1.1 PVC Well Casing Specials Referencing Specifications, in.

Reference Standard	Nominal Size	Outside Diameter	SDR 41		SDR 32.5	
			Minimum Wall	Minimum PS	Minimum Wall	Minimum PS
ASAE	6	6.140	0.150	28	0.189	56
S376	8	8.160	0.199	28	0.251	56
and	10	10.200	0.249	28	0.314	56
SCS	12	12.240	0.299	28	0.377	56
432-D	15	15.300	0.373	28	0.471	56
Miscellaneous					SDR	
	4½	4.950	0.190	112		
		4.950	0.236	224	21	
		4.950	0.248	268		
		4.950	0.291	452	17	
	5	5.563	0.190	78		
	6	6.625	0.190	60		
	6¼	6.900	0.250	95		
		6.900	0.329	224	21	
		6.900	0.406	452	17	
	7	7.000	0.300	164		
	15	15.300	0.410	38		
	16	16.000	0.429	38		

TABLE A1.2 SR Well Casing Specials, in.

Nominal Size	Outside Diameter	Minimum Wall	Minimum PS
4	4.500	0.175	87
	4.500	0.200	138
	4.500	0.250	285
4½	4.886	0.200	105
	4.886	0.230	165
5	5.300	0.175	53
	5.300	0.200	79
	5.300	0.250	162
	5.300	0.320	366
5½	5.563	0.225	99
6	6.275	0.175	32
	6.275	0.200	47
	6.275	0.250	96
	6.275	0.320	216
6⅝	6.625	0.250	80
	6.625	0.320	177
7	7.000	0.250	68
8⅝	8.625	0.250	36

A2. FLUSH THREAD DESIGN DATA FOR GROUND WATER MONITORING CASING AND SCREEN

A2.1 *Rationale*—The publication of the major revision of this specification in 1988 resulted in approaches to the responsible Subcommittee F17.61, from manufacturers of well screen and casing used primarily for ground water monitoring. The resulting task group has proposed several changes to the specification that address the needs of the ground water monitoring well construction industry. The needs of the ground water monitoring industry were surveyed through Subcommittee D18.21. This survey concluded that there was need for alternate thread patterns and more extensive testing than was originally included in F 480 – 88. The basic specification is modified to more closely reflect the needs of the industry defined by the task group and the industry survey as follows:

A2.1.1 Bias, by the provision of a single flush thread design, is removed.

A2.2.2 All flush thread design data is now included in the annex.

A2.2.3 A new standard is being written to specifically address the needs of the ground water monitoring industry for the application of thermoplastic well casing pipe. This new standard will define design and testing criteria in performance terms for the application of thermoplastic well casing pipe in ground water monitoring.

A2.2 The following tables provide basic dimensional information for flush thread forms, (single entry with "O" ring). The performance of these designs will equal or better that specified in Section 6 of this specification. Pipe sizes larger than 12 in. have been eliminated from the tables as they are not used for ground water monitoring wells. Tolerances are included and data in SI units are added.

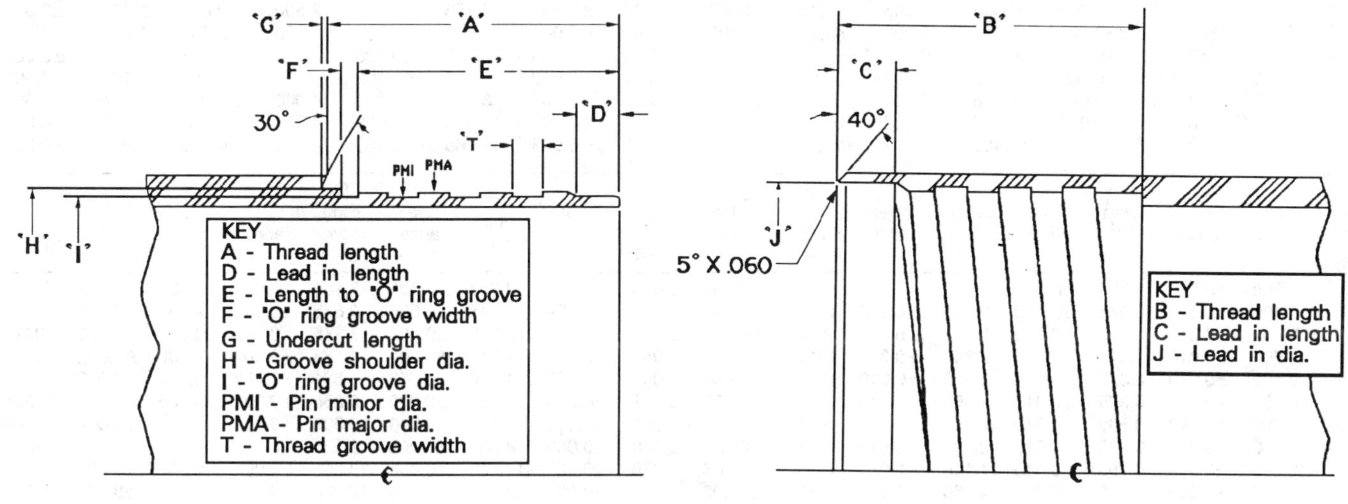

FIG. A2.1 Square Flush Thread Design

TABLE A2.1a Flush Thread Design Data (Nominal 2 Threads/in.)[A]

NOTE—For Tables A2.1a through d, designs incorporating alternate "O" rings and compensating dimensional changes are currently available as follows, tolerances as in the main tables:

ALTERNATE "O" RING FLUSH THREAD DESIGN DATA (Nominal 2 threads/in.)

Schedule	Size	"O" Ring "F"	"O" Ring I	"O" Number
40	4	0.075	4.207	045
40	4½	0.075	4.467	047
40	5	0.075	5.248	049
40	6	0.075	6.302	050
80	4	0.075	4.133	045
80	5	0.075	5.073	047
80	6	0.075	6.155	050

Size S40	TPI	Thread Groove Width "T"	Toler-ance	Pin Major Diameter	Toler-ance	Pin Minor Diameter	Toler-ance	Box Major Diameter	Toler-ance	Box Minor Diameter	Toler-ance	Thread "A"	Toler-ance	Thread "B"	Toler-ance
2	2	0.255	±0.003	2.239	±0.003	2.155	±0.003	2.249	±0.003	2.165	±0.003	1.500	±0.003	1.535	±0.003
2½	2	0.255	±0.003	2.729	±0.003	2.645	±0.003	2.739	±0.003	2.655	±0.003	2.375	±0.003	2.415	±0.003
3	2	0.255	±0.003	3.320	±0.003	3.236	±0.003	3.330	±0.003	3.246	±0.003	2.375	±0.003	2.420	±0.003
4	2	0.255	±0.003	4.300	±0.003	4.216	±0.003	4.310	±0.003	4.226	±0.003	2.500	±0.003	2.555	±0.003
4½	2	0.255	±0.003	4.739	±0.003	4.655	±0.003	4.749	±0.003	4.665	±0.003	2.500	±0.003	2.552	±0.003
5	2	0.255	±0.003	5.348	±0.003	5.248	±0.003	5.362	±0.003	5.262	±0.003	2.750	±0.003	2.802	±0.003
6	2	0.255	±0.003	6.393	±0.003	6.283	±0.003	6.407	±0.003	6.297	±0.002	3.000	±0.003	3.052	±0.003
8	2	0.255	±0.003	8.346	±0.003	8.226	±0.003	8.360	±0.003	8.240	±0.003	4.000	±0.003	4.070	±0.003
10	2	0.255	±0.003	10.447	±0.003	10.307	±0.003	10.463	±0.003	10.323	±0.003	4.375	±0.003	4.445	±0.003
12	2	0.255	±0.003	12.420	±0.003	12.250	±0.003	12.438	±0.003	12.268	±0.003	4.500	±0.003	4.580	±0.003

Size S40	Start "C"	Toler-ance	Start "D"	Toler-ance	Length "E"	Toler-ance	"O" Ring Width "F"	Toler-ance	Length "G"	Toler-ance	Diam-eter "H"	Toler-ance	"O" Ring Diam-eter "I"	Toler-ance	Diam-eter "J"	Toler-ance	"O" Ring Number
2	0.500	±0.010	0.375	±0.010	1.350	±0.010	0.075	±0.003	0.035	±0.005	2.252	±0.005	2.169	±0.002	2.285	±0.002	032
2½	0.500	±0.010	0.375	±0.010	2.225	±0.010	0.075	±0.003	0.040	±0.005	2.745	±0.005	2.645	±0.002	2.765	±0.002	036
3	0.500	±0.010	0.375	±0.010	2.225	±0.010	0.075	±0.003	0.045	±0.005	3.350	±0.005	3.245	±0.002	3.365	±0.002	041
4	0.500	±0.010	0.375	±0.010	2.350	±0.010	0.110	±0.003	0.055	±0.005	4.315	±0.005	4.184	±0.002	4.330	±0.002	151
4½	0.500	±0.010	0.375	±0.010	2.300	±0.010	0.110	±0.003	0.052	±0.005	4.770	±0.005	4.631	±0.002	4.795	±0.002	155
5	0.500	±0.010	0.375	±0.010	2.550	±0.010	0.110	±0.003	0.052	±0.005	5.380	±0.005	5.265	±0.002	5.435	±0.002	156
6	0.750	±0.010	0.500	±0.010	2.750	±0.010	0.110	±0.003	0.052	±0.005	6.445	±0.005	6.305	±0.002	6.465	±0.002	158
8	0.750	±0.010	0.500	±0.010	3.750	±0.010	0.110	±0.003	0.070	±0.005	8.385	±0.005	8.235	±0.002	8.405	±0.002	168
10	0.750	±0.010	0.750	±0.010	4.125	±0.010	0.150	±0.003	0.070	±0.005	10.510	±0.005	10.315	±0.002	10.535	±0.002	272
12	0.750	±0.010	0.750	±0.010	4.250	±0.010	0.150	±0.003	0.080	±0.005	12.470	±0.005	12.260	±0.002	12.485	±0.002	277

[A] Dimensions in inches. Use illustrations of thread patterns and keys as shown on Fig. A2.1.

TABLE A2.1b Flush Thread Design Data (Nominal 2 Threads/in.)[A]

Size S80	TPI	Thread Groove Width "T"	Toler-ance	Pin Major Diameter	Toler-ance	Pin Minor Diameter	Toler-ance	Box Major Diameter	Toler-ance	Box Minor Diameter	Toler-ance	Thread "A"	Toler-ance	Thread "B"	Toler-ance
½	4	0.136	±0.003	0.717	±0.003	0.663	±0.003	0.723	±0.003	0.669	±0.003	0.825	±0.003	0.825	±0.003
¾	4	0.136	±0.003	0.920	±0.003	0.866	±0.003	0.926	±0.003	0.812	±0.003	1.125	±0.003	1.125	±0.003
1	4	0.136	±0.003	1.159	±0.003	1.105	±0.003	1.167	±0.003	1.113	±0.003	1.125	±0.003	1.125	±0.003
1¼	2	0.255	±0.003	1.507	±0.003	1.423	±0.003	1.575	±0.003	1.431	±0.003	1.500	±0.003	1.500	±0.003
1½	2	0.255	±0.003	1.737	±0.003	1.653	±0.003	1.747	±0.003	1.663	±0.003	1.500	±0.003	1.500	±0.003
2	2	0.255	±0.003	2.164	±0.003	2.085	±0.003	2.179	±0.003	2.095	±0.003	2.000	±0.003	2.050	±0.003
2½	2	0.255	±0.003	2.644	±0.003	2.544	±0.003	2.654	±0.003	2.554	±0.003	2.375	±0.003	2.430	±0.003
3	2	0.255	±0.003	3.250	±0.003	3.140	±0.003	3.260	±0.003	3.150	±0.003	2.375	±0.003	2.444	±0.003
4	2	0.255	±0.003	4.213	±0.003	4.103	±0.003	4.223	±0.003	4.113	±0.003	2.500	±0.003	2.565	±0.003
4½								NOT APPLICABLE							
5	2	0.255	±0.003	5.183	±0.003	5.073	±0.003	5.193	±0.003	5.083	±0.003	2.750	±0.003	2.845	±0.003
6	2	0.255	±0.003	6.255	±0.003	6.120	±0.003	6.265	±0.003	6.130	±0.003	3.000	±0.003	3.095	±0.003
8	2	0.255	±0.003	8.193	±0.003	8.043	±0.003	8.207	±0.003	8.057	±0.003	4.000	±0.003	4.115	±0.003
10	2	0.255	±0.003	10.250	±0.003	10.050	±0.003	10.264	±0.003	10.064	±0.003	4.375	±0.003	4.500	±0.003
12	2	0.255	±0.003	12.175	±0.003	11.935	±0.003	12.191	±0.003	11.951	±0.003	4.500	±0.003	4.650	±0.003

Size S80	Start "C"	Toler-ance	Start "D"	Toler-ance	Length "E"	Toler-ance	"O" Ring Width "F"	Toler-ance	Length "G"	Toler-ance	Diam-eter "H"	Toler-ance	"O" Ring Diam-eter "I"	Toler-ance	Diam-eter "J"	Toler-ance	"O" Ring Number
½	0.200	±0.010	0.200	±0.010	...	...	...	...	0.020	...	...	...	0.660	±0.003	...	...	014
¾	0.200	±0.010	0.200	±0.010	...	...	...	...	0.020	...	...	...	0.863	±0.003	...	...	016
1	0.200	±0.010	0.200	±0.010	...	...	...	...	0.020	...	...	...	1.102	±0.003	...	...	020
1¼	0.250	±0.010	0.250	±0.010	...	...	...	...	0.020	...	...	...	1.420	±0.003	...	...	028
1½	0.250	±0.010	0.250	±0.010	...	...	...	...	0.020	...	...	...	1.650	±0.003	...	...	030
2	0.500	±0.010	0.375	±0.010	1.850	±0.010	0.075	±0.003	0.050	±0.005	2.194	±0.005	2.105	±0.002	2.225	±0.002	032
2½	0.500	±0.010	0.375	±0.010	2.225	±0.010	0.075	±0.003	0.055	±0.005	2.685	±0.005	2.570	±0.002	2.700	±0.002	036
3	0.500	±0.010	0.375	±0.010	2.225	±0.010	0.075	±0.003	0.065	±0.005	3.275	±0.005	3.150	±0.002	3.282	±0.002	041
4	0.500	±0.010	0.375	±0.010	2.350	±0.010	0.120	±0.003	0.065	±0.005	4.281	±0.005	4.113	±0.002	4.296	±0.002	155
4½									NOT APPLICABLE								
5	0.650	±0.010	0.500	±0.010	2.550	±0.010	0.150	±0.003	0.095	±0.005	5.225	±0.005	5.070	±0.002	5.298	±0.002	242
6	0.750	±0.010	0.750	±0.010	2.750	±0.010	0.150	±0.003	0.095	±0.005	6.295	±0.005	6.120	±0.002	6.316	±0.002	251
8	0.750	±0.010	0.750	±0.010	3.750	±0.010	0.150	±0.003	0.115	±0.005	8.229	±0.005	8.048	±0.002	8.244	±0.002	262
10	0.750	±0.010	0.750	±0.010	4.125	±0.010	0.150	±0.003	0.125	±0.005	10.295	±0.005	10.110	±0.002	10.310	±0.002	270
12	0.750	±0.010	0.750	±0.010	4.250	±0.010	0.150	±0.003	0.150	±0.005	12.220	±0.005	12.035	±0.002	12.235	±0.002	276

[A] Dimensions in inches. Use illustrations of thread patterns and keys as shown on Fig. A2.1.

TABLE A2.1c Flush Thread Design Data (Nominal 2 Threads/in.) SI units[A]

Size S40	TPI	Thread Groove Width "T"	Toler-ance	Pin Major Diameter	Toler-ance	Pin Minor Diameter	Toler-ance	Box Major Diameter	Toler-ance	Box Minor Diameter	Toler-ance	Thread "A"	Toler-ance	Thread "B"	Toler-ance	Start "C"
2	2	6.48	±0.08	56.87	±0.08	54.74	±0.08	57.12	±0.08	54.99	±0.08	38.1	±0.08	38.99	±0.08	12.70
2½	2	6.48	±0.08	69.32	±0.08	67.18	±0.08	69.57	±0.08	67.44	±0.08	60.33	±0.08	61.34	±0.08	12.70
3	2	6.48	±0.08	84.33	±0.08	82.19	±0.08	84.58	±0.08	82.47	±0.08	60.33	±0.08	61.47	±0.08	12.70
4	2	6.48	±0.08	109.22	±0.08	107.09	±0.08	109.47	±0.08	107.34	±0.08	63.50	±0.08	64.90	±0.08	12.70
4½	2	6.48	±0.08	120.37	±0.08	118.24	±0.08	120.62	±0.08	118.49	±0.08	63.50	±0.08	64.82	±0.08	12.70
5	2	6.48	±0.08	135.84	±0.08	133.30	±0.08	136.19	±0.08	133.65	±0.08	69.85	±0.08	71.17	±0.08	12.70
6	2	6.48	±0.08	162.38	±0.08	159.59	±0.08	162.74	±0.08	159.94	±0.08	76.20	±0.08	77.52	±0.08	19.05
8	2	6.48	±0.08	211.99	±0.08	208.94	±0.08	212.34	±0.08	209.30	±0.08	101.60	±0.08	103.38	±0.08	19.05
10	2	6.48	±0.08	265.35	±0.08	261.80	±0.08	265.76	±0.08	303.50	±0.08	111.13	±0.08	112.90	±0.08	19.05
12	2	6.48	±0.08	315.47	±0.08	311.15	±0.08	315.93	±0.08	311.61	±0.08	114.30	±0.08	116.33	±0.08	19.05

Size S40	Toler-ance	Start "D"	Toler-ance	Length "E"	Toler-ance	"O" Ring Width "F"	Toler-ance	Length "G"	Toler-ance	Diameter "H"	Toler-ance	"O" Ring Diameter "I"	Toler-ance	Diameter "J"	Toler-ance	"O" Number
2	±0.25	9.53	±0.25	34.29	±0.25	1.91	±0.08	0.89	±0.13	57.20	±0.13	55.09	±0.05	58.04	±0.05	032
2½	±0.25	9.53	±0.25	56.52	±0.25	1.91	±0.08	1.02	±0.13	69.72	±0.13	67.18	±0.05	70.23	±0.05	036
3	±0.25	9.53	±0.25	56.52	±0.25	1.91	±0.08	1.14	±0.13	85.09	±0.13	82.42	±0.05	85.47	±0.05	041
4	±0.25	9.53	±0.25	59.69	±0.25	2.79	±0.08	1.40	±0.13	109.60	±0.13	106.27	±0.05	109.98	±0.05	151
4½	±0.25	9.53	±0.25	58.42	±0.25	2.79	±0.08	1.32	±0.13	121.16	±0.13	117.63	±0.05	121.87	±0.05	155
5	±0.25	9.53	±0.25	64.77	±0.25	2.79	±0.08	1.32	±0.13	136.65	±0.13	133.73	±0.05	138.05	±0.05	156
6	±0.25	12.70	±0.25	69.85	±0.25	2.79	±0.08	1.32	±0.13	163.70	±0.13	160.15	±0.05	164.21	±0.05	158
8	±0.25	12.70	±0.25	95.25	±0.25	2.79	±0.08	1.78	±0.13	212.98	±0.13	209.17	±0.05	213.49	±0.05	168
10	±0.25	19.05	±0.25	104.78	±0.25	3.81	±0.08	1.78	±0.13	266.95	±0.13	262.00	±0.05	267.59	±0.05	272
12	±0.25	19.05	±0.25	107.95	±0.25	3.81	±0.08	2.03	±0.13	316.74	±0.13	311.40	±0.05	317.12	±0.05	277

[A] Dimensions in millimetres. Use illustrations of thread patterns and keys as shown on Fig. A2.1.

TABLE A2.1d Flush Thread Design Data (Nominal 2 Threads/in.) SI units[A]

Size S80	TPI	Thread Groove Width "T"	Toler-ance	Pin Major Diameter	Toler-ance	Pin Minor Diameter	Toler-ance	Box Major Diameter	Toler-ance	Box Minor Diameter	Toler-ance	Thread "A"	Toler-ance	Thread "B"	Toler-ance	Start "C"
½	4	3.45	±0.08	18.21	±0.08	16.84	±0.08	18.42	±0.08	16.99	±0.08	20.96	±0.08	20.96	±0.08	5.08
¾	4	3.45	±0.08	23.37	±0.08	22.00	±0.08	23.52	±0.08	20.62	±0.08	28.58	±0.08	28.58	±0.08	5.08
1	4	3.45	±0.08	29.44	±0.08	28.07	±0.08	29.64	±0.08	28.27	±0.08	28.58	±0.08	28.58	±0.08	5.08
1¼	2	6.48	±0.08	38.28	±0.08	36.14	±0.08	40.00	±0.08	36.35	±0.08	38.10	±0.08	38.10	±0.08	6.35
1½	2	6.48	±0.08	44.12	±0.08	41.99	±0.08	44.37	±0.08	42.24	±0.08	38.10	±0.08	38.10	±0.08	6.35
2	2	6.48	±0.08	54.97	±0.08	52.96	±0.08	55.35	±0.08	53.21	±0.08	50.80	±0.08	52.07	±0.08	12.70
2½	2	6.48	±0.08	67.16	±0.08	64.62	±0.08	67.41	±0.08	64.87	±0.08	60.33	±0.08	61.72	±0.08	12.70
3	2	6.48	±0.08	82.55	±0.08	79.76	±0.08	82.80	±0.08	80.01	±0.08	60.33	±0.08	62.08	±0.08	12.70
4	2	6.48	±0.08	107.01	±0.08	104.22	±0.08	107.26	±0.08	104.47	±0.08	63.50	±0.08	65.15	±0.08	12.70
4½								NOT APPLICABLE								
5	2	6.48	±0.08	131.65	±0.08	128.85	±0.08	131.90	±0.08	129.11	±0.08	69.85	±0.08	72.26	±0.08	16.51
6	2	6.48	±0.08	158.88	±0.08	155.45	±0.08	159.13	±0.08	155.70	±0.08	76.20	±0.08	78.61	±0.08	19.05
8	2	6.48	±0.08	208.10	±0.08	204.29	±0.08	208.46	±0.08	204.65	±0.08	101.60	±0.08	104.52	±0.08	19.05
10	2	6.48	±0.08	260.35	±0.08	255.27	±0.08	260.71	±0.08	255.63	±0.08	111.13	±0.08	114.30	±0.08	19.05
12	2	6.48	±0.08	309.25	±0.08	303.15	±0.08	309.65	±0.08	305.56	±0.08	114.30	±0.08	118.11	±0.08	19.05

Size S80	Toler-ance	Start "D"	Toler-ance	Length "E"	Toler-ance	"O" Ring Width "F"	Toler-ance	Length "G"	Toler-ance	Diameter "H"	Toler-ance	"O" Ring Diameter "I"	Toler-ance	Diameter "J"	Toler-ance	"O" Ring Number
½	±0.25	5.08	±0.25	...	...	...	...	0.51	...	...	...	16.76	±0.08	...	...	014
¾	±0.25	5.08	±0.25	...	...	...	...	0.51	...	...	...	21.92	±0.08	...	...	016
1	±0.25	5.08	±0.25	...	...	...	...	0.51	...	...	...	27.99	±0.08	...	...	020
1¼	±0.25	6.35	±0.25	...	...	...	...	0.51	...	...	...	36.07	±0.08	...	...	028
1½	±0.25	6.35	±0.25	...	...	...	...	0.51	...	...	...	41.91	±0.08	...	...	030
2	±0.25	12.70	±0.25	46.99	±0.25	1.91	±0.08	1.27	±0.13	55.73	±0.13	53.47	±0.05	56.52	±0.05	032
2½	±0.25	12.70	±0.25	56.52	±0.25	1.91	±0.08	1.40	±0.13	68.20	±0.13	65.28	±0.05	68.58	±0.05	036
3	±0.25	12.70	±0.25	56.52	±0.25	1.91	±0.08	1.65	±0.13	83.19	±0.13	80.01	±0.05	83.36	±0.05	041
4	±0.25	12.70	±0.25	59.69	±0.25	3.05	±0.08	1.65	±0.13	108.74	±0.13	104.47	±0.05	109.12	±0.05	155
4½								NOT APPLICABLE								
5	±0.25	16.51	±0.25	64.77	±0.25	3.81	±0.08	2.41	±0.13	132.72	±0.13	128.78	±0.05	134.57	±0.05	242
6	±0.25	19.05	±0.25	69.85	±0.25	3.81	±0.08	2.41	±0.13	159.89	±0.13	155.45	±0.05	160.43	±0.05	251
8	±0.25	19.05	±0.25	95.25	±0.25	3.81	±0.08	2.92	±0.13	209.02	±0.13	204.42	±0.05	209.40	±0.05	262
10	±0.25	19.05	±0.25	104.78	±0.25	3.81	±0.08	3.18	±0.13	261.49	±0.13	256.79	±0.05	261.87	±0.05	270
12	±0.25	19.05	±0.25	107.95	±0.25	3.81	±0.08	3.81	±0.13	310.39	±0.13	305.69	±0.05	310.77	±0.05	276

[A] Dimensions in millimetres. Use illustrations of thread patterns and keys as shown on Fig. A2.1.

TABLE A2.2a Flush Thread Design Data (Nominal 4 Threads/in.)[A]

Size S40	TPI	Thread Groove Width "T"	Toler-ance	Pin Major Diameter	Toler-ance	Pin Minor Diameter	Toler-ance	Box Major Diameter	Toler-ance	Box Minor Diameter	Toler-ance	Thread "A"	Toler-ance	Thread "B"	Toler-ance	Start "C"
1	8	0.070	+0.004 −0.002	1.200	±0.003	1.150	±0.003	1.210	±0.003	1.160	±0.003	1.000	+0.000 −0.010	1.000	+0.010 −0.000	0.200
1¼	8	0.070	+0.004 −0.002	1.530	±0.003	1.470	±0.003	1.545	±0.003	1.485	±0.003	1.250	+0.000 −0.010	1.250	+0.010 −0.000	0.250
1½	4	0.130	+0.000 −0.002	1.775	+0.000 −0.003	1.715	+0.000 −0.003	1.790	+0.003 −0.000	1.730	+0.003 −0.000	1.625	+0.000 −0.002	1.625	+0.002 −0.000	0.250
2	4	0.130	+0.000 −0.002	2.243	+0.000 −0.005	2.187	+0.000 −0.005	2.255	+0.005 −0.000	2.200	+0.005 −0.000	1.690	+0.000 −0.002	1.690	+0.002 −0.000	0.375
2½	4	0.130	+0.000 −0.002	2.722	+0.000 −0.005	2.622	+0.000 −0.005	2.737	+0.005 −0.000	2.637	+0.005 −0.000	2.375	+0.000 −0.005	2.375	+0.005 −0.000	0.580
3	4	0.130	+0.000 −0.002	3.310	+0.000 −0.005	3.220	+0.000 −0.005	3.330	+0.005 −0.000	3.240	+0.005 −0.000	2.375	+0.000 −0.005	2.375	+0.005 −0.000	0.588
4	4	0.130	+0.000 −0.002	4.312	+0.000 −0.005	4.181	+0.000 −0.005	4.332	+0.005 −0.000	4.208	+0.005 −0.000	2.500	+0.000 −0.002	2.500	+0.002 −0.000	0.625
4½	4	0.130	+0.000 −0.002	4.711	+0.000 −0.005	4.580	+0.000 −0.005	4.731	+0.005 −0.000	4.600	+0.005 −0.000	2.500	+0.000 −0.005	2.500	+0.005 −0.000	0.625
5	4	0.130	+0.000 −0.002	5.341	+0.000 −0.005	5.249	+0.000 −0.005	5.361	+0.005 −0.000	5.269	+0.005 −0.000	2.955	+0.000 −0.005	2.955	+0.005 −0.000	0.641
6	3	0.165	+0.000 −0.002	6.357	+0.000 −0.005	6.227	+0.000 −0.005	6.392	+0.005 −0.000	6.262	+0.005 −0.000	3.000	+0.000 −0.010	3.000	+0.010 −0.000	0.625

Size S40	Toler-ance	Start "D"	Toler-ance	Length "E"	Toler-ance	"O" Ring Width "F"	Toler-ance	Length "G"	Toler-ance	Diameter "H"	Toler-ance	"O" Ring Diameter "I"	Toler-ance	Diameter "J"	Toler-ance	"O" Ring
1	±0.005	0.187	±0.005	0.880	±0.005	0.070	±0.003	NA[B]	NA	NA	NA	1.140	±0.003	1.220	±0.001	018
1¼	±0.005	0.187	±0.005	1.130	±0.005	0.070	±0.003	NA	NA	NA	NA	1.465	±0.003	1.565	±0.001	024
1½	±0.005	0.250	±0.005	1.482	±0.005	0.070	±0.003	NA	NA	NA	NA	1.716	±0.003	1.815	+0.002 −0.000	026
2	±0.005	0.250	±0.005	1.505	±0.005	0.080	±0.003	0.030	+0.000 −0.005	NA	NA	2.172	±0.001	2.280	+0.002 −0.000	032
2½	±0.005	0.437	±0.005	2.125	±0.005	0.125	±0.003	0.044	+0.000 −0.005	NA	NA	2.590	±0.001	2.757	+0.002 −0.000	145
3	±0.005	0.437	±0.005	2.125	±0.005	0.125	±0.003	0.055	+0.000 −0.005	NA	NA	3.195	±0.001	3.350	+0.002 −0.000	140
4	±0.005	0.437	±0.005	2.250	±0.005	0.125	±0.003	0.050	+0.000 −0.005	NA	NA	4.199	±0.001	4.350	+0.002 −0.000	151
4½	±0.005	0.437	±0.005	2.250	±0.005	0.125	±0.003	0.069	+0.000 −0.005	NA	NA	4.590	±0.001	4.750	+0.002 −0.000	153
5	±0.005	0.466	±0.005	2.700	±0.005	0.125	±0.003	0.064	+0.000 −0.005	NA	NA	5.214	±0.001	5.380	+0.002 −0.000	158
6	±0.005	0.500	±0.005	2.750	±0.005	0.125	±0.003	NA	NA	NA	NA	6.230	±0.002	6.415	+0.004 −0.000	163

[A] Dimensions in inches. Use illustrations of thread patterns and keys as shown on Fig. A2.1.
[B] Not applicable.

TABLE A2.2b Flush Thread Design Data (Nominal 4 Threads/in.)[A]

Size S80	TPI	Thread Groove Width "T"	Tolerance	Pin Major Diameter	Tolerance	Pin Minor Diameter	Tolerance	Box Major Diameter	Tolerance	Box Minor Diameter	Tolerance	Thread "A"	Tolerance	Thread "B"	Tolerance	Start "C"
½	8	0.070	+0.004 −0.002	0.715	+0.000 −0.003	0.655	+0.000 −0.003	0.725	+0.003 −0.000	0.665	+0.003 −0.000	0.800	+0.000 −0.003	0.800	+0.002 −0.000	0.250
¾	8	0.070	+0.004 −0.002	0.918	±0.002	0.857	±0.002	0.933	±0.002	0.870	±0.002	1.000	+0.000 −0.010	1.000	+0.010 −0.000	0.250
1	8	0.070	+0.004 −0.002	1.200	±0.003	1.150	±0.003	1.210	±0.003	1.160	±0.003	1.000	+0.000 −0.010	1.000	+0.010 −0.000	0.200
1¼	8	0.070	+0.004 −0.002	1.530	±0.003	1.470	±0.003	1.545	±0.003	1.485	±0.003	1.250	+0.000 −0.010	1.250	+0.010 −0.000	0.250
1½	4	0.130	+0.000 −0.002	1.775	+0.000 −0.003	1.715	+0.000 −0.003	1.790	+0.003 −0.000	1.730	+0.003 −0.000	1.625	+0.000 −0.002	1.625	+0.002 −0.000	0.250
2	4	0.130	+0.000 −0.002	2.243	+0.000 −0.005	2.187	+0.000 −0.005	2.255	+0.005 −0.000	2.200	+0.005 −0.000	1.690	+0.000 −0.002	1.690	+0.002 −0.000	0.375
2½	4	0.130	+0.000 −0.002	2.722	+0.000 −0.005	2.622	+0.000 −0.005	2.737	+0.005 −0.000	2.637	+0.005 −0.000	2.375	+0.000 −0.005	2.375	+0.005 −0.000	0.580
3	4	0.130	+0.000 −0.002	3.310	+0.000 −0.005	3.220	+0.000 −0.005	3.330	+0.005 −0.000	3.240	+0.005 −0.000	2.375	+0.000 −0.005	2.375	+0.005 −0.000	0.588
4	4	0.130	+0.000 −0.002	4.312	+0.000 −0.005	4.181	+0.000 −0.005	4.332	+0.005 −0.000	4.208	+0.005 −0.000	2.500	+0.000 −0.002	2.500	+0.002 −0.000	0.625
4½								NOT APPLICABLE								
5	4	0.130	+0.000 −0.002	5.341	+0.000 −0.005	5.249	+0.000 −0.005	5.361	+0.005 −0.000	5.269	+0.005 −0.000	2.955	+0.000 −0.005	2.955	+0.005 −0.000	0.641
6	3	0.163	+0.010 −0.002	6.357	+0.000 −0.005	6.227	+0.000 −0.005	6.392	+0.005 −0.000	6.262	+0.005 −0.000	3.000	+0.000 −0.010	3.000	+0.010 −0.000	0.625

Size S80	Tolerance	Start "D"	Tolerance	Length "E"	Tolerance	"O" Ring Width "F"	Tolerance	Length "G"	Tolerance	Diameter "H"	Tolerance	"O" Ring Diameter "I"	Tolerance	Diameter "J"	Tolerance	"O" Ring
½	±0.005	0.187	±0.005	0.685	±0.005	0.070	±0.003	NA[B]	NA	NA	NA	0.647	±0.002	0.725	±0.002 −0.000	013
¾	±0.005	0.187	±0.005	0.895	±0.005	0.070	±0.003	NA	NA	NA	NA	0.852	±0.002	0.945	±0.001	016
1	±0.005	0.187	±0.005	0.880	±0.005	0.070	±0.003	NA	NA	NA	NA	1.140	±0.003	1.220	±0.001	018
1¼	±0.005	0.187	±0.005	1.130	±0.005	0.070	±0.003	NA	NA	NA	NA	1.465	±0.003	1.565	±0.001	024
1½	±0.005	0.250	±0.005	1.482	±0.005	0.070	±0.003	NA	NA	NA	NA	1.716	±0.003	1.815	+0.002 −0.000	026
2	±0.005	0.250	±0.005	1.505	±0.005	0.080	±0.003	0.030	+0.000 −0.005	NA	NA	2.172	±0.001	2.280	+0.002 −0.000	032
2½	±0.005	0.437	±0.005	2.125	±0.005	0.125	±0.003	0.044	+0.000 −0.005	NA	NA	2.590	±0.001	2.757	+0.002 −0.000	145
3	±0.005	0.437	±0.005	2.125	±0.005	0.125	±0.003	0.055	+0.000 −0.005	NA	NA	3.195	±0.001	3.350	+0.002 −0.000	140
4	±0.005	0.437	±0.005	2.250	±0.005	0.125	±0.003	0.050	+0.000 −0.005	NA	NA	4.199	±0.001	4.350	+0.002 −0.000	151
4½								NOT APPLICABLE								
5	±0.005	0.466	±0.005	2.700	±0.005	0.125	±0.003	0.064	+0.000 −0.005	NA	NA	5.214	±0.001	5.380	+0.002 −0.000	158
6	±0.005	0.500	±0.005	2.750	±0.005	0.125	±0.003	NA	NA	NA	NA	6.230	±0.002	6.415	+0.004 −0.000	163

[A] Dimensions in inches. Use illustrations of thread patterns and keys as shown on Fig. A2.1.
[B] Not applicable.

TABLE A2.2c Flush Thread Design Data (Nominal 4 Threads/in.) SI units[A]

Size S40	TPI	Thread Groove Width "T"	Tolerance	Pin Major Diameter	Tolerance	Pin Minor Diameter	Tolerance	Box Major Diameter	Tolerance	Box Minor Diameter	Tolerance	Thread "A"	Tolerance	Thread "B"	Tolerance	Start "C"
1	8	1.79	+0.10 −0.05	30.48	±0.08	29.21	±0.08	30.73	±0.08	29.46	±0.08	25.40	+0.00 −0.25	25.40	+0.25 −0.00	5.08
1¼	8	1.79	+0.10 −0.05	38.86	±0.08	37.34	±0.08	39.24	±0.08	37.72	±0.08	31.75	+0.00 −0.25	31.75	+0.25 −0.00	6.35
1½	4	3.30	+0.00 −0.05	45.09	+0.00 −0.08	43.56	+0.00 −0.08	45.47	+0.08 −0.00	43.94	+0.08 −0.00	41.28	+0.00 −0.05	41.28	+0.05 −0.00	6.35
2	4	3.30	+0.00 −0.05	56.97	+0.00 −0.13	55.55	+0.00 −0.13	57.28	+0.13 −0.00	55.88	+0.13 −0.00	41.91	+0.00 −0.05	41.91	+0.05 −0.00	9.53
2½	4	3.30	+0.00 −0.05	69.14	+0.00 −0.13	66.60	+0.00 −0.13	69.52	+0.13 −0.00	66.98	+0.13 −0.00	60.33	+0.00 −0.13	60.33	+0.13 −0.00	14.73
3	4	3.30	+0.00 −0.05	84.07	+0.00 −0.13	81.79	+0.00 −0.13	84.58	+0.13 −0.00	82.30	+0.13 −0.00	60.33	+0.00 −0.13	60.33	+0.13 −0.00	14.95
4	4	3.30	+0.00 −0.05	109.52	+0.00 −0.13	106.20	+0.00 −0.13	110.03	+0.13 −0.00	106.88	+0.13 −0.00	63.50	+0.00 −0.13	63.50	+0.13 −0.00	15.89
4½	4	3.30	+0.00 −0.05	119.66	+0.00 −0.13	116.33	+0.00 −0.13	120.17	+0.13 −0.00	116.84	+0.13 −0.00	63.50	+0.00 −0.13	63.50	+0.13 −0.00	15.89
5	4	3.30	+0.00 −0.05	136.66	+0.00 −0.13	133.33	+0.00 −0.13	136.17	+0.13 −0.00	133.83	+0.13 −0.00	75.06	+0.00 −0.13	75.06	+0.13 −0.00	16.28
6	3	4.19	+0.00 −0.05	161.47	+0.00 −0.13	158.17	+0.00 −0.13	162.36	+0.13 −0.00	159.06	+0.13 −0.00	76.20	+0.00 −0.25	76.20	+0.25 −0.00	15.89

Size S40	Tolerance	Start "D"	Tolerance	Length "E"	Tolerance	"O" Ring Width "F"	Tolerance	Length "G"	Tolerance	Diameter "H"	Tolerance	"O" Ring Diameter "I"	Tolerance	Diameter "J"	Tolerance	"O" Number
1	±0.13	4.75	±0.13	22.35	±0.13	1.78	±0.08	NA[B]	NA	NA	NA	28.96	±0.08	30.99	±0.03	018
1¼	±0.13	4.75	±0.13	28.70	±0.13	1.78	±0.08	NA	NA	NA	NA	37.21	±0.08	39.75	±0.03	024
1½	±0.13	6.35	±0.13	37.64	±0.13	1.78	±0.08	NA	NA	NA	NA	43.59	±0.08	46.10	+0.05 −0.00	026
2	±0.13	6.35	±0.13	38.23	±0.13	2.03	±0.08	0.76	+0.00 −0.13	NA	NA	55.17	±0.03	57.91	+0.05 −0.00	032
2½	±0.13	11.10	±0.13	53.98	±0.13	3.18	±0.08	1.12	+0.00 −0.13	NA	NA	65.79	±0.03	70.03	+0.05 −0.00	145
3	±0.13	11.10	±0.13	53.98	±0.13	3.18	±0.08	1.40	+0.00 −0.13	NA	NA	81.15	±0.03	85.09	+0.05 −0.00	140
4	±0.13	11.10	±0.13	57.15	±0.13	3.18	±0.08	1.27	+0.00 −0.13	NA	NA	106.65	±0.03	110.49	+0.05 −0.00	151
4½	±0.13	11.10	±0.13	57.15	±0.13	3.18	±0.08	1.75	+0.00 −0.13	NA	NA	116.59	±0.03	120.65	+0.05 −0.00	153
5	±0.13	11.84	±0.13	68.58	±0.13	3.18	±0.08	1.63	+0.00 −0.13	NA	NA	132.44	±0.03	136.65	+0.05 −0.00	158
6	±0.13	12.7	±0.13	69.85	±0.13	3.18	±0.08	NA	NA	NA	NA	158.24	±0.05	162.94	+0.10 −0.00	163

[A] Dimensions in millimetres. Use illustrations of thread patterns and keys as shown on Fig. A2.1.
[B] Not applicable.

TABLE A2.2d Flush Thread Design Data (Nominal 4 Threads/in.) SI units[A]

Size S80	TPI	Thread Groove Width "T"	Toler-ance	Pin Major Diameter	Toler-ance	Pin Minor Diameter	Toler-ance	Box Major Diameter	Toler-ance	Box Minor Diameter	Toler-ance	Thread "A"	Toler-ance	Thread "B"	Toler-ance	Start "C"
1/2	8	1.79	+0.10 −0.05	18.16	±0.00 −0.08	16.64	+0.00 −0.08	18.42	+0.08 −0.00	16.89	+0.08 −0.00	20.32	+0.00 −0.08	20.32	+0.05 −0.00	6.35
3/4	8	1.79	+0.10 −0.05	23.32	±0.05	21.77	±0.05	23.70	±0.05	22.10	±0.05	25.40	+0.00 −0.25	25.40	+0.25 −0.00	6.35
1	8	1.79	+0.10 −0.05	30.48	±0.08	29.21	±0.08	30.73	±0.08	29.46	±0.08	25.40	+0.00 −0.25	25.40	+0.25 −0.00	5.08
1 1/4	8	1.79	+0.10 −0.05	38.86	±0.08	37.34	±0.08	39.24	±0.08	37.72	±0.08	31.75	+0.00 −0.25	31.75	+0.25 −0.00	6.35
1 1/2	4	3.30	+0.00 −0.05	45.09	+0.00 −0.08	43.56	+0.00 −0.08	45.47	+0.08 −0.00	43.94	+0.08 −0.00	41.28	+0.00 −0.05	41.28	+0.05 −0.00	6.35
2	4	3.30	+0.00 −0.05	56.97	+0.00 −0.13	55.55	+0.00 −0.13	57.28	+0.13 −0.00	55.88	+0.13 −0.00	41.91	+0.00 −0.05	41.91	+0.05 −0.00	9.53
2 1/2	4	3.30	+0.00 −0.05	69.14	+0.00 −0.13	66.60	+0.00 −0.13	69.52	+0.13 −0.00	66.98	+0.13 −0.00	60.33	+0.00 −0.13	60.33	+0.13 −0.00	14.73
3	4	3.30	+0.00 −0.05	84.07	+0.00 −0.13	81.79	+0.00 −0.13	84.58	+0.13 −0.00	82.30	+0.13 −0.00	60.33	+0.00 −0.13	60.33	+0.13 −0.00	14.95
4	4	3.30	+0.00 −0.05	109.52	+0.00 −0.13	106.20	+0.00 −0.13	110.03	+0.13 −0.00	106.88	+0.13 −0.00	63.50	+0.00 −0.13	63.50	+0.13 −0.00	15.89
4 1/2	4	3.30	+0.00 −0.05	119.66	+0.00 −0.13	116.33	+0.00 −0.13	120.17	+0.13 −0.00	116.84	+0.13 −0.00	63.50	+0.00 −0.13	63.50	+0.13 −0.00	15.89
5	4	3.30	+0.00 −0.05	136.66	+0.00 −0.13	133.33	+0.00 −0.13	136.17	+0.13 −0.00	133.83	+0.13 −0.00	75.06	+0.00 −0.13	75.06	+0.13 −0.00	16.28
6	3	4.19	+0.00 −0.05	161.47	+0.00 −0.13	158.17	+0.00 −0.13	162.36	+0.13 −0.00	159.06	+0.13 −0.00	76.20	+0.00 −0.25	76.20	+0.25 −0.00	15.89

Size S80	Toler-ance	Start "D"	Toler-ance	Length "E"	Toler-ance	"O" Ring Width "F"	Toler-ance	Length "G"	Toler-ance	Diameter "H"	Toler-ance	"O" Ring Diameter "I"	Toler-ance	Diameter "J"	Toler-ance	"O" Number
1/2	±0.13	4.75	±0.13	17.40	±0.13	1.78	±0.08	NA[B]	NA	NA	NA	16.44	±0.05	18.42	+0.05 −0.00	013
3/4	±0.13	4.75	±0.13	22.73	±0.13	1.78	±0.08	NA	NA	NA	NA	21.64	±0.05	24.00	±0.03	016
1	±0.13	4.75	±0.13	22.35	±0.13	1.78	±0.08	NA	NA	NA	NA	28.96	±0.08	30.99	±0.03	018
1 1/4	±0.13	4.75	±0.13	28.70	±0.13	1.78	±0.08	NA	NA	NA	NA	37.21	±0.08	39.75	±0.03	024
1 1/2	±0.13	6.35	±0.13	37.64	±0.13	1.78	±0.08	NA	NA	NA	NA	43.59	±0.08	46.10	+0.05 −0.00	026
2	±0.13	6.35	±0.13	38.23	±0.13	2.03	±0.08	0.76	+0.00 −0.13	NA	NA	55.17	±0.03	57.91	+0.05 −0.00	032
2 1/2	±0.13	11.10	±0.13	53.98	±0.13	3.18	±0.08	1.12	+0.00 −0.13	NA	NA	65.79	±0.03	70.03	+0.05 −0.00	145
3	±0.13	11.10	±0.13	53.98	±0.13	3.18	±0.08	1.40	+0.00 −0.13	NA	NA	81.15	±0.03	85.09	+0.05 −0.00	140
4	±0.13	11.10	±0.13	57.15	±0.13	3.18	±0.08	1.27	+0.00 −0.13	NA	NA	106.65	±0.03	110.49	+0.05 −0.00	151
4 1/2								NOT APPLICABLE								
5	±0.13	11.84	±0.13	68.58	±0.13	3.18	±0.08	1.63	+0.00 −0.13	NA	NA	132.44	±0.03	136.65	+0.05 −0.00	158
6	±0.13	12.70	±0.13	69.85	±0.13	3.18	±0.08	NA	NA	NA	NA	158.24	±0.05	162.94	+0.10 −0.00	163

[A] Dimensions in millimetres. Use illustrations of thread patterns and keys as shown on Fig. A2.1.
[B] Not applicable.

TABLE A2.3a Flush Thread Design Data (Nominal 8 Threads/in.)[A]

Size S40	TPI	Thread Groove Width "T"	Tolerance	Pin Major Diameter	Tolerance	Pin Minor Diameter	Tolerance	Box Major Diameter	Tolerance	Box Minor Diameter	Tolerance	Thread "A"	Tolerance	Thread "B"	Tolerance
½	8	0.082	+0.005 −0.003	0.750	±0.002	0.690	±0.002	0.760	±0.002	0.700	±0.002	1.000	±0.002	1.050	±0.002
¾	8	0.082	+0.005 −0.003	0.950	±0.002	0.890	±0.002	0.965	±0.002	0.905	±0.002	1.000	±0.002	1.050	±0.002
1	8	0.082	+0.005 −0.003	1.200	±0.002	1.105	±0.002	1.215	±0.002	1.124	±0.002	1.000	±0.002	1.050	±0.002
1¼	8	0.082	+0.005 −0.003	1.545	±0.002	1.570	±0.002	1.570	±0.002	1.475	±0.002	1.000	±0.002	1.050	±0.002
1½	8	0.082	+0.005 −0.003	1.770	±0.002	1.670	±0.002	1.785	±0.002	1.685	±0.002	1.125	±0.002	1.175	±0.002
2	8	0.082	+0.005 −0.003	2.250	±0.002	2.160	±0.002	2.272	±0.002	2.182	±0.002	1.250	±0.002	1.300	±0.002
2½	8	0.082	+0.005 −0.003	2.722	±0.002	2.600	±0.002	2.730	±0.002	2.615	±0.002	1.250	±0.002	1.300	±0.002
3	8	0.082	+0.005 −0.003	3.306	±0.002	3.182	±0.002	3.325	±0.002	3.204	±0.002	1.500	±0.002	1.550	±0.002
4	8	0.082	+0.005 −0.003	4.306	±0.002	4.186	±0.002	4.325	±0.002	4.208	±0.002	1.875	±0.002	1.925	±0.002
4½	8	0.082	+0.005 −0.003	4.730	±0.002	4.605	±0.002	4.750	±0.002	4.628	±0.002	1.875	±0.002	1.925	±0.002
5	8	0.082	+0.005 −0.003	5.310	±0.002	5.190	±0.002	5.340	±0.002	5.220	±0.002	2.250	±0.002	2.300	±0.002
6	8	0.082	+0.005 −0.003	6.360	±0.002	6.235	±0.002	6.385	±0.002	6.265	±0.002	2.500	±0.002	2.550	±0.002
8	4	0.130	+0.005 −0.003	8.346	±0.002	8.224	±0.002	8.362	±0.002	8.242	±0.002	4.075	±0.002	4.150	±0.002
10	4	0.130	+0.005 −0.003	10.447	±0.002	10.307	±0.002	10.465	±0.002	10.325	±0.002	4.375	±0.002	4.450	±0.002
12	4	0.130	+0.005 −0.003	12.420	±0.002	12.250	±0.002	12.440	±0.002	12.268	±0.002	4.555	±0.002	4.630	±0.002

Size S40	Start "C"	Tolerance	Start "D"	Tolerance	Length "E"	Tolerance	"O" Ring Width "F"	Tolerance	Length "G"	Tolerance	Diameter "H"	Tolerance	"O" Ring Diameter "I"	Tolerance	Diameter "J"	Tolerance	"O" Number
½	0.250	±0.002	0.250	±0.002	0.705	±0.002	0.095	±0.002	NA[B]	NA	0.750	±0.002	NA	NA	0.760	±0.002	016
¾	0.250	±0.002	0.250	±0.002	0.705	±0.002	0.095	±0.002	NA	NA	0.950	±0.002	0.870	±0.002	0.965	±0.002	019
1	0.250	±0.002	0.250	±0.002	0.705	±0.002	0.095	±0.002	NA	NA	1.200	±0.002	1.110	±0.002	1.215	±0.002	024
1¼	0.250	±0.002	0.250	±0.002	0.705	±0.002	0.095	±0.002	NA	NA	1.545	±0.002	1.455	±0.002	1.570	±0.002	028
1½	0.250	±0.002	0.250	±0.002	0.830	±0.002	0.095	±0.002	NA	NA	1.770	±0.002	1.685	±0.002	1.795	±0.002	030
2	0.500	±0.002	0.250	±0.002	0.955	±0.002	0.095	±0.002	NA	NA	2.250	±0.002	2.165	±0.002	2.272	±0.002	033
2½	0.250	±0.002	0.250	±0.002	0.955	±0.002	0.095	±0.002	NA	NA	2.722	±0.002	2.630	±0.002	2.730	±0.002	037
3	0.250	±0.002	0.250	±0.002	1.205	±0.002	0.095	±0.002	NA	NA	3.306	±0.002	3.220	±0.002	3.325	±0.002	041
4	0.500	±0.002	0.250	±0.002	1.580	±0.002	0.095	±0.002	NA	NA	4.306	±0.002	4.228	±0.002	4.325	±0.002	045
4½	0.250	±0.002	0.250	±0.002	1.580	±0.002	0.095	±0.002	NA	NA	4.730	±0.002	4.640	±0.002	4.750	±0.002	045
5	0.250	±0.002	0.250	±0.002	1.955	±0.002	0.095	±0.002	NA	NA	5.310	±0.002	5.250	±0.002	5.340	±0.002	046
6	0.250	±0.002	0.250	±0.002	1.805	±0.002	0.095	±0.002	NA	NA	6.360	±0.002	6.290	±0.002	6.385	±0.002	050
8	0.750	±0.002	0.500	±0.002	3.900	±0.002	0.110	±0.002	NA	NA	8.346	±0.002	8.235	±0.002	8.362	±0.002	168
10	0.750	±0.002	0.500	±0.002	4.200	±0.002	0.130	±0.002	NA	NA	10.447	±0.002	10.315	±0.002	10.465	±0.002	272
12	0.750	±0.002	0.500	±0.002	4.380	±0.002	0.130	±0.002	NA	NA	12.420	±0.002	12.260	±0.002	12.440	±0.002	277

[A] Dimensions in inches. Use illustrations of thread patterns and keys as shown on Fig. A2.1.

[B] Not applicable.

TABLE A2.3b Square Flush Thread Design Data (Nominal 8 Threads/in.)[A]

Size S80	TPI	Thread Groove Width "T"	Toler-ance	Pin Major Diameter	Toler-ance	Pin Minor Diameter	Toler-ance	Box Major Diameter	Toler-ance	Box Minor Diameter	Toler-ance	Thread "A"	Toler-ance	Thread "B"	Toler-ance
½	8	0.082	+0.005 −0.003	0.750	±0.002	0.650	±0.002	0.775	±0.002	0.675	±0.002	1.000	±0.002	1.050	±0.002
¾	8	0.082	+0.005 −0.003	0.947	±0.002	0.827	±0.002	0.965	±0.002	0.845	±0.002	1.000	±0.002	1.050	±0.002
1	8	0.082	+0.005 −0.003	1.200	±0.002	1.075	±0.002	1.215	±0.002	1.092	±0.002	1.000	±0.002	1.050	±0.002
1¼	8	0.082	+0.005 −0.003	1.545	±0.002	1.425	±0.002	1.565	±0.002	1.445	±0.002	1.000	±0.002	1.050	±0.002
1½	8	0.082	+0.005 −0.003	1.745	±0.002	1.625	±0.002	1.765	±0.002	1.645	±0.002	1.125	±0.002	1.175	±0.002
2	8	0.082	+0.005 −0.003	2.188	±0.002	2.066	±0.002	2.200	±0.002	2.080	±0.002	1.250	±0.002	1.300	±0.002
2½	8	0.082	+0.005 −0.003	2.626	±0.002	2.502	±0.002	2.645	±0.002	2.522	±0.002	1.250	±0.002	1.300	±0.002
3	8	0.082	+0.005 −0.003	3.238	±0.002	3.115	±0.002	3.245	±0.002	3.122	±0.002	1.500	±0.002	1.550	±0.002
4	8	0.082	+0.005 −0.003	4.206	±0.002	4.084	±0.002	4.225	±0.002	4.100	±0.002	1.875	±0.002	1.925	±0.002
4½								NOT APPLICABLE							
5	8	0.082	+0.005 −0.003	5.230	±0.002	5.105	±0.002	5.255	±0.002	5.135	±0.002	2.250	±0.002	2.300	±0.002
6	8	0.082	+0.005 −0.003	6.184	±0.002	6.060	±0.002	6.215	±0.002	6.095	±0.002	2.500	±0.002	2.550	±0.002
8	4	0.130	+0.005 −0.003	8.193	±0.002	8.043	±0.002	8.057	±0.002	8.207	±0.002	4.075	±0.002	4.150	±0.002
10	4	0.130	+0.005 −0.003	10.250	±0.002	10.048	±0.002	10.265	±0.002	10.066	±0.002	4.475	±0.002	4.550	±0.002
12	4	0.130	+0.005 −0.003	12.175	±0.002	11.935	±0.002	12.190	±0.002	11.950	±0.002	4.575	±0.002	4.650	±0.002

Size S80	Start "C"	Toler-ance	Start "D"	Toler-ance	Length "E"	Toler-ance	"O" Ring Width "F"	Toler-ance	Length "G"	Toler-ance	Diameter "H"	Toler-ance	"O" Ring Diameter "I"	Toler-ance	Diameter "J"	Toler-ance	"O" Number
½	0.250	±0.002	0.250	±0.002	0.705	±0.002	0.095	±0.002	NA[B]	NA	0.750	±0.002	0.648	NA	0.775	±0.002	016
¾	0.250	±0.002	0.250	±0.002	0.705	±0.002	0.095	±0.002	NA	NA	0.947	±0.002	0.820	±0.002	0.965	±0.002	019
1	0.250	±0.002	0.250	±0.002	0.705	±0.002	0.095	±0.002	NA	NA	1.200	±0.002	1.070	±0.002	1.215	±0.002	024
1¼	0.250	±0.002	0.250	±0.002	0.705	±0.002	0.095	±0.002	NA	NA	1.545	±0.002	1.420	±0.002	1.565	±0.002	028
1½	0.250	±0.002	0.250	±0.002	0.830	±0.002	0.095	±0.002	NA	NA	1.745	±0.002	1.625	±0.002	1.765	±0.002	030
2	0.500	±0.002	0.250	±0.002	0.955	±0.002	0.095	±0.002	NA	NA	2.188	±0.002	2.060	±0.002	2.200	±0.002	033
2½	0.250	±0.002	0.250	±0.002	0.955	±0.002	0.095	±0.002	NA	NA	2.626	±0.002	2.500	±0.002	2.645	±0.002	037
3	0.250	±0.002	0.250	±0.002	1.205	±0.002	0.095	±0.002	NA	NA	3.238	±0.002	3.110	±0.002	3.245	±0.002	041
4	0.500	±0.002	0.250	±0.002	1.580	±0.002	0.095	±0.002	NA	NA	4.206	±0.002	4.080	±0.002	4.225	±0.002	045
4½								NOT APPLICABLE									
5	0.250	±0.002	0.250	±0.002	1.955	±0.002	0.095	±0.002	NA	NA	5.230	±0.002	5.100	±0.002	5.255	±0.002	046
6	0.250	±0.002	0.250	±0.002	2.205	±0.002	0.095	±0.002	NA	NA	6.184	±0.002	6.060	±0.002	6.215	±0.002	050
8	0.750	±0.002	0.500	±0.002	3.900	±0.002	0.130	±0.002	NA	NA	8.193	±0.002	8.050	±0.002	8.057	±0.002	262
10	0.750	±0.002	0.500	±0.002	4.300	±0.002	0.130	±0.002	NA	NA	10.250	±0.002	10.110	±0.002	10.260	±0.002	270
12	0.750	±0.002	0.500	±0.002	4.400	±0.002	0.130	±0.002	NA	NA	12.175	±0.002	12.035	±0.002	12.190	±0.002	276

[A] Dimensions in inches. Use illustrations of thread patterns and keys as shown on Fig. A2.1.

[B] Not applicable.

TABLE A2.3c Flush Thread Design Data (Nominal 8 Threads/in.) SI units[A]

Size S40	TPI	Thread Groove Width "T"	Toler-ance	Pin Major Diameter	Toler-ance	Pin Minor Diameter	Toler-ance	Box Major Diameter	Toler-ance	Box Minor Diameter	Toler-ance	Thread "A"	Toler-ance	Thread "B"	Toler-ance
½	8	2.08	+0.13 −0.08	19.05	±0.05	17.53	±0.05	19.30	±0.05	17.78	±0.05	25.40	±0.05	26.67	±0.05
¾	8	2.08	+0.13 −0.08	24.13	±0.05	22.61	±0.05	24.51	±0.05	22.99	±0.05	25.40	±0.05	26.67	±0.05
1	8	2.08	+0.13 −0.08	30.48	±0.05	28.07	±0.05	30.86	±0.05	28.55	±0.05	25.40	±0.05	26.67	±0.05
1¼	8	2.08	+0.13 −0.08	39.74	±0.05	36.71	±0.05	39.88	±0.05	37.47	±0.05	25.40	±0.05	26.67	±0.05
1½	8	2.08	+0.13 −0.08	44.96	±0.05	42.42	±0.05	45.34	±0.05	42.80	±0.05	28.58	±0.05	29.85	±0.05
2	8	2.08	+0.13 −0.08	57.28	±0.05	54.81	±0.05	57.71	±0.05	55.42	±0.05	31.75	±0.05	33.02	±0.05
2½	8	2.08	+0.13 −0.08	69.14	±0.05	66.05	±0.05	69.34	±0.05	66.42	±0.05	31.75	±0.05	33.02	±0.05
3	8	2.08	+0.13 −0.08	83.97	±0.05	80.83	±0.05	84.46	±0.05	81.38	±0.05	38.10	±0.05	39.37	±0.05
4	8	2.08	+0.13 −0.08	109.37	±0.05	106.34	±0.05	109.86	±0.05	106.88	±0.05	47.63	±0.05	48.90	±0.05
4½	8	2.08	+0.13 −0.08	120.14	±0.05	116.98	±0.05	120.65	±0.05	117.55	±0.05	47.63	±0.05	48.90	±0.05
5	8	2.08	+0.13 −0.08	134.87	±0.05	131.83	±0.05	135.64	±0.05	132.59	±0.05	57.15	±0.05	58.42	±0.05
6	8	2.08	+0.13 −0.08	161.54	±0.05	158.37	±0.05	162.18	±0.05	159.13	±0.05	63.50	±0.05	64.47	±0.05
8	4	3.30	+0.13 −0.08	211.99	±0.05	208.89	±0.05	212.41	±0.05	209.35	±0.05	103.51	±0.05	105.41	±0.05
10	4	3.30	+0.13 −0.08	265.35	±0.05	261.80	±0.05	265.81	±0.05	262.26	±0.05	111.13	±0.05	113.03	±0.05
12	4	3.30	+0.13 −0.08	315.47	±0.05	311.15	±0.05	315.98	±0.05	311.61	±0.05	115.72	±0.05	117.60	±0.05

Size S40	Start "C"	Toler-ance	Start "D"	Toler-ance	Length "E"	Toler-ance	"O" Ring Width "F"	Toler-ance	Length "G"	Toler-ance	Diameter "H"	Toler-ance	"O" Ring Diameter "I"	Toler-ance	Diameter "J"	Toler-ance	"O" Number
½	6.35	±0.05	6.35	±0.05	17.91	±0.05	2.41	±0.05	NA[B]	NA	19.05	±0.05	NA	NA	19.30	±0.05	016
¾	6.35	±0.05	6.35	±0.05	17.91	±0.05	2.41	±0.05	NA	NA	24.13	±0.05	22.10	±0.05	24.51	±0.05	019
1	6.35	±0.05	6.35	±0.05	17.91	±0.05	2.41	±0.05	NA	NA	30.48	±0.05	28.19	±0.05	30.86	±0.05	024
1¼	6.35	±0.05	6.35	±0.05	17.91	±0.05	2.41	±0.05	NA	NA	39.24	±0.05	36.96	±0.05	39.88	±0.05	028
1½	6.35	±0.05	6.35	±0.05	20.40	±0.05	2.41	±0.05	NA	NA	44.96	±0.05	42.80	±0.05	45.34	±0.05	030
2	6.35	±0.05	6.35	±0.05	24.26	±0.05	2.41	±0.05	NA	NA	57.15	±0.05	54.99	±0.05	57.71	±0.05	033
2½	6.35	±0.05	6.35	±0.05	24.26	±0.05	2.41	±0.05	NA	NA	69.27	±0.05	66.80	±0.05	69.34	±0.05	037
3	6.35	±0.05	6.35	±0.05	30.61	±0.05	2.41	±0.05	NA	NA	83.97	±0.05	81.79	±0.05	84.46	±0.05	041
4	6.35	±0.05	6.35	±0.05	40.13	±0.05	2.41	±0.05	NA	NA	109.37	±0.05	107.39	±0.05	109.86	±0.05	045
4½	6.35	±0.05	6.35	±0.05	40.13	±0.05	2.41	±0.05	NA	NA	120.14	±0.05	117.86	±0.05	120.65	±0.05	045
5	6.35	±0.05	6.35	±0.05	49.66	±0.05	2.41	±0.05	NA	NA	134.87	±0.05	133.35	±0.05	135.64	±0.05	046
6	6.35	±0.05	6.35	±0.05	56.01	±0.05	2.41	±0.05	NA	NA	161.54	±0.05	159.77	±0.05	162.18	±0.05	050
8	19.05	±0.05	12.70	±0.05	99.06	±0.05	2.79	±0.05	NA	NA	211.99	±0.05	209.17	±0.05	212.41	±0.05	168
10	19.05	±0.05	12.70	±0.05	106.68	±0.05	3.30	±0.05	NA	NA	265.35	±0.05	262.00	±0.05	265.81	±0.05	272
12	19.05	±0.05	12.70	±0.05	111.25	±0.05	3.30	±0.05	NA	NA	315.47	±0.05	311.40	±0.05	315.98	±0.05	277

[A] Dimensions in inches. Use illustrations of thread patterns and keys as shown on Fig. A2.1.

[B] Not applicable.

TABLE A2.3d Flush Thread Design Data (Nominal 8 Threads/in.) SI units[A]

Size S80	TPI	Thread Groove Width "T"	Toler-ance	Pin Major Diameter	Toler-ance	Pin Minor Diameter	Toler-ance	Box Major Diameter	Toler-ance	Box Minor Diameter	Toler-ance	Thread "A"	Toler-ance	Thread "B"	Toler-ance
½	8	2.08	+0.13 −0.08	19.05	±0.05	16.51	±0.05	19.65	±0.05	17.15	±0.05	25.40	±0.05	26.67	±0.05
¾	8	2.08	+0.13 −0.08	24.05	±0.05	21.01	±0.05	24.51	±0.05	21.46	±0.05	25.40	±0.05	26.67	±0.05
1	8	2.08	+0.13 −0.08	30.48	±0.05	27.31	±0.05	30.86	±0.05	27.74	±0.05	25.40	±0.05	26.67	±0.05
1¼	8	2.08	+0.13 −0.08	39.24	±0.05	36.20	±0.05	39.75	±0.05	36.70	±0.05	25.40	±0.05	26.67	±0.05
1½	8	2.08	+0.13 −0.08	44.32	±0.05	41.28	±0.05	44.83	±0.05	41.78	±0.05	28.58	±0.05	29.85	±0.05
2	8	2.08	+0.13 −0.08	55.58	±0.05	52.48	±0.05	55.88	±0.05	52.83	±0.05	31.75	±0.05	33.02	±0.05
2½	8	2.08	+0.13 −0.08	66.70	±0.05	63.55	±0.05	67.18	±0.05	64.06	±0.05	31.75	±0.05	33.02	±0.05
3	8	2.08	+0.13 −0.08	82.25	±0.05	79.12	±0.05	82.42	±0.05	79.30	±0.05	38.10	±0.05	39.37	±0.05
4	8	2.08	+0.13 −0.08	106.63	±0.05	102.82	±0.05	108.08	±0.05	104.14	±0.05	47.63	±0.05	48.90	±0.05
4½							NOT APPLICABLE								
5	8	2.08	+0.13 −0.08	132.84	±0.05	129.67	±0.05	133.48	±0.05	130.43	±0.05	57.15	±0.05	58.42	±0.05
6	8	2.08	+0.13 −0.08	157.07	±0.05	153.92	±0.05	157.86	±0.05	154.81	±0.05	63.50	±0.05	64.77	±0.05
8	4	3.30	+0.13 −0.08	208.10	±0.05	204.29	±0.05	204.65	±0.05	210.06	±0.05	103.51	±0.05	105.41	±0.05
10	4	3.30	+0.13 −0.08	260.35	±0.05	255.22	±0.05	260.73	±0.05	255.68	±0.05	113.67	±0.05	115.57	±0.05
12	4	3.30	+0.13 −0.08	309.25	±0.05	303.15	±0.05	309.63	±0.05	303.53	±0.05	116.22	±0.05	118.11	±0.05

Size S80	Start "C"	Toler-ance	Start "D"	Toler-ance	Length "E"	Toler-ance	"O" Ring Width "F"	Toler-ance	Length "G"	Toler-ance	Diameter "H"	Toler-ance	"O" Ring Diameter "I"	Toler-ance	Diameter "J"	Toler-ance	"O" Number
½	6.35	±0.05	6.35	±0.05	17.91	±0.05	2.41	±0.05	NA[B]	NA	19.05	±0.05	16.46	±0.05	19.69	±0.05	016
¾	6.35	±0.05	6.35	±0.05	17.91	±0.05	2.41	±0.05	NA	NA	24.05	±0.05	20.83	±0.05	24.51	±0.05	019
1	6.35	±0.05	6.35	±0.05	17.91	±0.05	2.41	±0.05	NA	NA	30.98	±0.05	27.18	±0.05	30.86	±0.05	024
1¼	6.35	±0.05	6.35	±0.05	17.91	±0.05	2.41	±0.05	NA	NA	39.24	±0.05	36.07	±0.05	39.75	±0.05	028
1½	6.35	±0.05	6.35	±0.05	21.08	±0.05	2.41	±0.05	NA	NA	44.32	±0.05	41.28	±0.05	44.83	±0.05	030
2	6.35	±0.05	6.35	±0.05	24.26	±0.05	2.41	±0.05	NA	NA	55.88	±0.05	52.32	±0.05	55.88	±0.05	033
2½	6.35	±0.05	6.35	±0.05	24.26	±0.05	2.41	±0.05	NA	NA	66.70	±0.05	63.50	±0.05	67.18	±0.05	037
3	6.35	±0.05	6.35	±0.05	30.61	±0.05	2.41	±0.05	NA	NA	82.55	±0.05	78.99	±0.05	82.42	±0.05	041
4	6.35	±0.05	6.35	±0.05	40.13	±0.05	2.41	±0.05	NA	NA	106.83	±0.05	103.63	±0.05	107.32	±0.05	045
4½								NOT APPLICABLE									
5	6.35	±0.05	6.35	±0.05	49.66	±0.05	2.41	±0.05	NA	NA	132.84	±0.05	129.54	±0.05	132.48	±0.05	046
6	6.35	±0.05	6.35	±0.05	56.01	±0.05	2.41	±0.05	NA	NA	207.87	±0.05	153.92	±0.05	157.86	±0.05	050
8	19.05	±0.05	12.70	±0.05	99.06	±0.05	3.30	±0.05	NA	NA	208.10	±0.05	204.47	±0.05	204.65	±0.05	262
10	19.05	±0.05	12.70	±0.05	109.22	±0.05	3.30	±0.05	NA	NA	260.35	±0.05	256.79	±0.05	260.73	±0.05	270
12	19.05	±0.05	12.70	±0.05	111.76	±0.05	3.30	±0.05	NA	NA	309.25	±0.05	305.69	±0.05	309.63	±0.05	276

[A] Dimensions in millimetres. Use illustrations of thread patterns and keys as shown on Fig. A2.1.
[B] Not applicable.

APPENDIXES

(Nonmandatory Information)

X1. STORAGE, HANDLING, AND ASSEMBLY PROCEDURES FOR THERMOPLASTIC WATER WELL CASING PIPE AND COUPLINGS MADE IN STANDARD DIMENSION RATIOS (SDR)

X1.1 *Storage*—Do not store well casing pipe and well casing pipe couplings in direct sunlight for long periods. Store well casing pipe in such a manner as to prevent sagging or bending.

X1.2 *Field Inspection*—Well casing pipe should be carefully inspected for cuts, gouges, deep scratches, damaged ends, and other major imperfections before use. Defective well casing pipe should be rejected.

X1.3 *Solvent Cement*—Use solvent cements meeting the requirements of the specification for the material used. The solvent cement should provide sufficient open time for making good joints and connections but joints should be completed immediately upon applying solvent cement.

X1.4 *Socket Fit*—Well casing pipe and well casing pipe

couplings are manufactured to close tolerances. Close tolerances are required to ensure satisfactory "interference" fit between well casing pipe and well casing pipe coupling during the solvent cement joining. Use only well casing pipe and well casing pipe coupling combinations that give interference fits. Well casing pipe loose in the socket may not properly bond. The allowable tolerances ensure a forced fit and when solvent is applied will readily mate, this ensuring a chemical fusion equal in strength to well casing pipe and well casing pipe coupling. In the dry joint, the pipe should enter the socket to between one half and two thirds of the socket depth by hand. Attempting to correct a loose fit after assembly by additional cement may result in an unsatisfactory joint.

X1.5 *Joining Technique:*

X1.5.1 *Cutting*—Well casing pipe can be easily cut with an ordinary hacksaw or carpenter's saw. Fine-tooth blades with little or no set should be used for best results. The well casing pipe should be cut square and all burrs removed with a sharp knife, a fine-tooth file, or other suitable device. A miter box is useful to ensure square cut ends. Standard steel pipe or tubing cutters are not recommended for cutting well casing pipe since they might cause excessive heat and pressures that can result in cracked or irregular ends. There are special cutters available with extra-wide rollers and thin cutting wheels which have been designated for cutting plastic pipe.

X1.5.2 *Cleaning*—Remove burrs and wipe off all dust, dirt, and moisture from surfaces to be cemented. Chemical or mechanical cleaners may be used, but the recommendation of the manufacturer should be followed.

X1.5.3 *Primers*—Primers should be used where necessary to soften and dissolve the surface of the well casing pipe and coupling to form a continuous bond between the mating surfaces and shall be used when recommended by the solvent cement manufacturer.

X1.5.4 *Application of Cement*—Using an ordinary brush with a width of one half the pipe diameter, or the brush supplied with the can of solvent cement, first apply a moderate even coating of cement in the well casing pipe coupling covering the well casing pipe joining surfaces only. Heavy or excess applications of cement may become an obstruction inside the well casing pipe. Quickly apply cement to the outside of the well casing pipe. Make sure that the coated distance on the well casing pipe is equal to the depth of the well casing pipe coupling socket.

X1.5.5 *Assembly*—Make the joint as quickly as possible after application of the cement and before the cement dries. Should the cement dry partially before the joint is made up, reapply cement before assembling. Insert the well casing pipe into the well casing pipe coupling socket, turning the well casing pipe to ensure even distribution of cement. Make sure that the well casing pipe is inserted to the full depth of the well casing pipe coupling socket. Remove excess solvent cement from the exterior of the joint with a clean, dry cloth. Do not attempt to disturb the well casing pipe coupling joint until after the cement has set; damage to the joint and loss of fit may result. Allow sufficient time for the joint to develop good handling strength. Larger diameter water well casing pipe should be assembled by using commercially available plastic pipe joiners to draw the water well casing pipe and coupling together ensuring the proper bottoming of the water well casing pipe with the coupling.

X1.5.6 Handle newly assembled joints carefully until initial set has taken place. Recommended setting time allowed before handling or moving is related to temperature. After initial set, the joints will withstand the stresses of a normal installation. A badly misaligned installation will cause excessive stresses in the joint, pipe, and fittings.

X1.5.7 *Cure Time Before Press Testing*—Joint strength development is very rapid during the first 48 h. Short cure periods are satisfactory for high-ambient temperatures with low humidity and interference-type fittings. Longer cure periods are necessary for low temperatures, large pipe sizes, loose fits, and relatively high humidity.

X1.5.8 *Safety Requirements*—Follow Practice F 402.

X1.6 *Well Casing Pipe Coupling:*

X1.6.1 *Threaded Water Well Casing Pipe Couplings*—Molded or formed threads are permitted. Use only approved thread lubricant specifically intended for use with the designated plastic material used.

X1.6.2 *Connection to Nonplastic Water Well Screens*—When connecting plastic water well casing pipe to nonplastic water well screens, use only approved types of well casing pipe couplings designated for the specific transition intended.

X1.6.3 *Thread Tightness*—When a threaded joint is made, obtain tightness with a strap wrench not to exceed one full turn.

X1.7 *Installation*—It is recommended that installation be in accordance with the Manual of Recommended Water Well Construction Standards prepared by the National Water Well Association for the United States Environmental Protection Agency.

X1.8 *Thermal Expansion*—Allow for thermal expansion and movement in all piping installation by the use of expansion joints when deemed necessary.

X2. CONSIDERATIONS FOR USE OF THERMOPLASTIC WATER WELL CASING PIPE IN SDR RATIOS

X2.1 *General*—Thermoplastic materials are being used successfully as casing pipe for water wells. They offer many advantages to the user such as non-corrosion, resistance to micro- and macro-biological attack, light weight, competitive price, and approval for potable water use.

X2.2 *Collapsing Pressure*[13]—Thermoplastic materials have the ability to withstand collapsing pressure under external load. The ability of the thermoplastic material to withstand an external load is influenced by the following:

(*1*) The physical properties of the casing material,

(*2*) The ratio of diameter to wall thickness referred to as Standard Dimension Ratio (SDR),

(*3*) The amount of ovality and eccentricity in the casing pipe, and

(*4*) The magnitude and nature of the residual stress present in the casing pipe.

Of these four factors the physical properties of the material and the SDR ratios are used in arriving at the collapsing pressure the material will withstand. The modulus of elasticity of the material is important design information: typical values are 300 000 psi (2.07 GPa) for SR; 250 000 to 350 000 psi (1.72 to 2.41 GPa) for ABS; and 400 000 psi (2.76 GPa) for PVC.

X2.2.1 Many wells are drilled by the use of drilling mud to hold the hole from collapsing while the hole is being drilled. While the casing is being set, either water or drilling mud is allowed to fill the inside of the casing to equalize pressures. After the screen is set, the well is usually brought in with compressed air. Compressed air fed into the well casing reduces the specific gravity of the fluid inside the casing causing the liquid to rise out of the well. The difference in specific gravity between the inside fluid and outside fluid causes a collapsing pressure to be placed upon the casing. The faster the rate at which air is fed into the casing, the greater the difference in specific gravity of the two fluids and the greater the collapsing pressure. Since the drilling mud on the outside of the casing is in a near liquid state, it is possible to calculate the short-term critical collapse pressure of an unsupported thermoplastic water well casing pipe by the following equation:

$$P_c = \frac{2E}{1 - \mu_2} \times \frac{1}{\frac{d}{t}\left(\frac{d}{t} - 1\right)^2}$$

or

$$P_c = \frac{2E}{1 - \mu^2} \times \frac{1}{\mathrm{SDR}(\mathrm{SDR} - 1)^2}$$

where:

P_c = critical collapse pressure, psi,
E = modulus of elasticity, psi,
μ = Poisson's ratio,
d = outside diameter of the well casing pipe, in.,
t = wall thickness of the well casing pipe, in., and
SDR = standard dimension ratio.

X2.3 *Sample Calculation*—Calculate the critical collapse of an unsupported nominal 6-in. SDR17 well casing pipe made from PVC with a modulus of elasticity of 400 000 psi (2.76 GPa) and assume a Poisson's ratio of 0.38.[14]

$$P_c = \frac{2E}{1 - \mu^2} \times \frac{1}{\mathrm{SDR}(\mathrm{SDR} - 1)^2}$$

$$P_c = \frac{2 \times 400\,000}{1 - (0.38)^2} \times \frac{1}{17(17 - 1)^2}$$

$$P_c = \frac{800\,000}{0.8556} \times \frac{1}{4352}$$

$$P_c = 935016.4 \times 0.0002298$$

$$P_c = 214.85 \text{ psi}$$

X2.4 Displacing the mud around the outside of the well casing pipe with gravel, concrete, or other solid filler will give support to the well casing pipe and resistance to collapse. The reason for this is that as soon as the diameter of the well casing pipe starts to decrease at any point initiating a collapse, the diameter at right angle to this diameter must increase and, since this increase is prevented by the support material on the outside of the well casing pipe, the well casing pipe does not collapse. The ultimate resistance to collapse is obtained when the filler around the well casing pipe firmly holds the well casing pipe from getting out of round. Therefore, it is important that the packing material be carefully selected and the method of packing be performed with care.

X2.5 *Depth of Use*—The depth at which thermoplastic water well casing can be placed is a design judgment. Thermoplastic water well casing pipe has been placed over 2000 ft (609.6 m) deep, but the water level inside the well casing pipe is less than 100 ft (30.5 m) from the surface.

X2.5.1 For the nominal 6-in. SDR17 well casing pipe used in the sample, calculation indicates a maximum difference of hydrostatic pressure of 215 psi (1483 kPa) or a head of 496 ft (151 m). Thus, a well casing with packing to support the casing and prevent deformation will allow greater depth as long as the pump down or head differential is within the design limits. The restraint offered by the soil to deflection of the well casing pipe makes use at greater depths practical.

X2.5.2 For the 6-in. SDR17 SR well casing pipe used in the sample calculation, a water or fluid level outside the casing that is more than 496 ft (151 m) higher than the level inside the casing (215-psi (1483-kPa)) maximum hydrostatic pressure) could result in collapse of the casing if there is no earth or gravel or concrete packing around it. On the other hand, casing with gravel or concrete packing can operate with almost unlimited head differential.

[13] Clinedinst, W. O., "A Rational Expression for the Critical Collapsing Pressure of Pipe Under External Pressure," *API Drill and Prod. Practice*, 1939, pp. 383–391.

[14] McLean, Jasper T., "Determination of Poisson's Ratio," *Proceedings of the 27th Annual Meeting of American Society of Testing and Materials*, Vol 24, 1924, Part II, p. 1012.

The American Society for Testing and Materials takes no position respecting the validity of any patent rights asserted in connection with any item mentioned in this standard. Users of this standard are expressly advised that determination of the validity of any such patent rights, and the risk of infringement of such rights, are entirely their own responsibility.

This standard is subject to revision at any time by the responsible technical committee and must be reviewed every five years and if not revised, either reapproved or withdrawn. Your comments are invited either for revision of this standard or for additional standards and should be addressed to ASTM Headquarters. Your comments will receive careful consideration at a meeting of the responsible technical committee, which you may attend. If you feel that your comments have not received a fair hearing you should make your views known to the ASTM Committee on Standards, 1916 Race St., Philadelphia, PA 19103.

Index

ASTM Standards on Ground Water and Vadose Zone Investigation

This index covers the standards and related material appearing in this volume only. The boldface references are to the ASTM designations; the standards appear in each volume in alphanumeric order. Proposed methods and specifications carry the index reference, "(Proposed)"; they appear, as does other ancillary material, in the Related Material (gray-edged) section in the back of the book. A triangle (Δ) preceding the ASTM designation denotes that Adjunct Material for the standard is published separately; the Adjunct No. is given in the standard. A Combined Index, covering the standards appearing in all volumes of the *1993 Annual Book of ASTM Standards*, is issued as Volume 00.01.

Alphabetization in the index is letter-for-letter, with no consideration to punctuation or word division. Initial prepositions of (indented) subentries are ignored for alphabetization.

In the preparation of indexes, every attempt has been made to index standards on three levels: (1) by main subject, using general and specific search terms; (2) by test methods or other significant sections of ASTM standards; and (3) by cross-references to locate main subject entry terms. (*See also* references are abbreviated as *Sa* and appear under main entry terms.) Specifications A 333/A333M, A 789/A789M, and Test Methods B 789 and E 426 illustrate ASTM's method of indexing.

INDEX TERMS FOR SPECIFICATIONS

Steel pipe—carbon steel
pipe—carbon/alloy steel, seamless/welded (for low-temperature service), spec., **A 333/333M (01.01)**

Steel tube—ferritic stainless steel
tube—ferritic/austenitic stainless steel, seamless/welded (for general corrosive service), spec., **A 789/789M (01.01)**

INDEX TERMS FOR TESTS

Gold—electrodeposited coating
gold/palladium coatings on metal substrates—porosity by gel-bulk electrography, test, **B 789 (03.04)**

Effective elastic parameter (E_{eff})
effective elastic parameter (E_{eff})—x-ray diffraction measurements of residual stress, test, **E 1426 (03.01)**

CROSS-REFERENCES

Water—drinking
See **Drinking water**

PDB (pressure design basis)
See **Pressure design basis (PDB)**
Sa **Pressure testing** (*headings*)

INDEX

A

B

C

D

E

F

G

H

I

P

Q

R

S

T

U

V

1916 Race Street
Philadelphia, PA 19103
Phone: (215) 299-2632
FAX: (215) 977-9679

European Office
27-29 Knowl Piece, Wilbury Way
Hitchin, Herts SG4 OSX, England
Phone: 0462-437933
FAX: 0462-433678

1994 MEMBERSHIP APPLICATION

THIS APPLICATION IS FOR SOCIETY MEMBERSHIP ONLY.

EACH TECHNICAL COMMITTEE REQUIRES A SEPARATE APPLICATION.

PLEASE SEND APPLICATIONS FOR THE FOLLOWING

COMMITTEE (S) ______________________

SEE BENEFITS AND FEES ON REVERSE SIDE

APPLICATION IS MADE FOR MEMBERSHIP IN ASTM:

☐ **ORGANIZATIONAL MEMBERSHIP** — An organizational member shall pay an annual fee and can designate one official representative. This membership entitles your organization to the benefits listed on the reverse side.

☐ **INDIVIDUAL MEMBERSHIP** — An individual member shall pay an annual fee and can join and participate on ASTM technical committees representing his or her employer unless unassociated with any organization. This membership entitles the individual to the benefits listed on the reverse side.

PLEASE PRINT ALL INFORMATION CLEARLY. DO NOT EXCEED THE CAPACITY OF EACH LINE.

LAST NAME ______ FIRST NAME ______ INITIAL ___

AFFILIATION ______

FACILITY ______ ☐ CHECK HERE IF HOME ADDRESS

STREET ______

P.O. BOX ______

CITY ______ STATE ___ ZIP _____ – ____

PHONE ( ) – EXTENSION ____

FAX ( ) –

COUNTRY ______

JOB TITLE ______

NOTE: IF YOUR AFFILIATION IS A SUBSIDIARY, PLEASE IDENTIFY THE PARENT ORGANIZATION.

____

OFFICIAL REPRESENTATIVE (ORGANIZATIONAL MEMBERSHIP ONLY)

LAST NAME ______ FIRST NAME ______ INITIAL ___

TITLE ______

☐ ENCLOSED IS PAYMENT. ALL CHECKS MADE PAYABLE TO ASTM IN U.S. FUNDS ON U.S. BANKS.

OPTIONAL METHODS OF PAYMENT

☐ AMERICAN EXPRESS ☐ MASTER CARD ☐ VISA

ACCOUNT NO. (ALL DIGITS) ______ EXPIRATION DATE ______ SIGNATURE ______

ONE FREE VOLUME — BOOK OF ASTM STANDARDS — YOUR CHOICE — VOLUME # ______

MAIL TO: ASTM ATTN: MEMBER SERVICES

BENEFITS AND FEES FOR 1994

	Organizational	Member
Annual Administrative Fee	$350.00*†	$50.00*†
Participation on ASTM Technical Committees	One Official Rep. Only	Member Only
Annual Book of ASTM Standards	One Free Volume	One Free Volume
Complete Set of Book of ASTM Standards (Vols. 1–70)	$3,700.00 (Prepaid)†	$4,100.00 (Prepaid)†
Individual Book of Standards at Member Price	No Limit	None
Complete Sections at Member Price (Prepaid)	No Limit	No Limit
Other Publications at Member prices	No Limit	No Limit
Annual Subscription to ASTM *Standardization News*	1 Free	1 Free

* Administrative fee payments made to ASTM are not tax deductible as charitable contributions for federal income tax purposes. They may be deductible as ordinary and necessary business expense.

† Canadian Prospective Members – Please add 7% GST – Registration Number R129162244.

MEMBERSHIP

- Membership year is January 1 to December 31. Fees are payable in advance and are not prorated. Membership will become effective upon payment of fees.
- Only the official representative of an organizational membership may participate on committees. All other employees of the organization must maintain individual memberships in order to participate on committees.
- An organizational membership may change its representative as needed by notifying ASTM Headquarters.
- An individual membership may be transferred without payment of additional fees, if the fees were paid by an employer, by notifying ASTM Headquarters. However, no additional free volume of the Book of ASTM Standards will be allowed.
- Technical committee participation is not transferable. A committee application will have to be submitted by the new individual member, or the new official representative of an organizational membership.

NOTE: Membership benefits are subject to change.

Printed in the USA.